ELECTRONIC DEVICES AND CIRCUITS

SANJAY SHARMA
Associate Professor
Department of Electronics & Communication Engg.
Krishna Institute of Engg. & Technology,
Ghaziabad

Edited by
DEEKSHA SHARMA

S.K. KATARIA & SONS®
Publisher of Engineering Books
4885/109, Prakash Mahal, Dr. Subhash Bhargav Lane,
Opposite Delhi Medical Association, Daryaganj, New Delhi–110002 (INDIA)
Phone: +91-11-43551243; Mobile: +91-9871775858, 7678425209
e-mail: katariabook@gmail.com, katariabooks@yahoo.com
Website: www.skkatariaandsons.com
(e-books, subscriptions and bulk buying also available on our website)

Published by:

S.K. Kataria & Sons®

4885/109, Prakash Mahal, Dr. Subhash Bhargav Lane,
Opposite Delhi Medical Association, Daryaganj, New Delhi-110002 (INDIA)
Phone: +91-11-43551243; Mobile: +91-9871775858, 7678425209
e-mail: katariabook@gmail.com, katariabooks@yahoo.com
Website: www.skkatariaandsons.com

First Edition: July, 2010, Reprint: 2026

Laser Typesetting by : Plus Computers, Delhi *Printed at :* New A.S. Printing Press, Delhi

Preface to the First Edition

Electronic Devices and Circuits is a compulsory subject for the second year engineering students from all the disciplines. This book is primarily intended to serve as a textbook in accordance with the syllabus of Electronic Devices and Circuits offered by various universities in India.

The objective of the book is to provide a clear explanation of the operation of all important basic electronic devices in general use today.

This book has been written in a very simple and lucid manner. Every effort has been made to make the treatments simple and comprehensive. Throughout the book, the stress has been given on fundamental concepts through illustrative examples. Neat and Clear diagrams have been used for explanation.

Summary of each chapter is provided to help the students to grasp the subject clearly. A large number of self explanatory accurate diagrams and tables have been used to supplement the text. Exercises and problems are also provided. These have been selected from examination papers of various Indian Universities. This will help the students to prepare accordingly.

This book is also intended as a textbook for the use in Electronic and Communication courses in polytechnics, IETE, AMIE and for various Engg. Electronics examinations. Thus, primarily designated as an undergraduate textbook for students of science and technology, this study will provide balanced coverage of devices, analog circuits. It also serves as a quick reference for practising professionals.

The whole text has been logically organized and spreaded over 10 chapters :

Chapter 1 provides the Introduction to semiconductor materials.

Chapter 2 introduces the concept of p-n junction diode.

Chapter 3 explains the Rectifiers and Filters.

Chapter 4 explains the Zener diode and voltage stabilization.

Chapter 5 introduces the BJT.

Chapter 6 is devoted to biasing and stabilization.

Chapter 7 treates the Transistor as an Amplifier.

Chapter 8 deals with FETs.

Chapter 9 discusses FET Amplifiers.

Chapter 10 discusses Special Purpose Electronic Devices.

The author hopes that with these unique features the book will fulfill the genuine requirements of the student community. Lastly, much originality cannot be claimed in a book of this kind. The author takes this opportunity to place on record his indebtedness to the large number of books and journals that he has freely consulted in the preparation of this book.

Sanjay Sharma
KIET, Ghaziabad

Acknowledgement

I take this opportunity to express my gratitude and thanks to the respected Director General of KIET, **REAR ADMIRAL Dr. AJAY SHARMA** for his valuable technical suggestions and constant encouragement, without which this book would not have come into existence.

I am specially grateful to **Dr. S. NASEEM AHMAD**, Professor and Head, Department of Electronics and Communication Engg., Jamia Millia Islamia, New Delhi for his time-to-time, much needed valuable guidance.

I would like to express my sincere thanks to **Mr. Vibhav Sachan** (Associate Professor, E&C, KIET), and **Mr. Manish Gupta** (Lecturer, E&C, MIT, Modinagar) for constant inspiration and encouragement.

Special thanks to **Dr. B.P. Singh** (Former Professor, Electronics & Comm. Engg., MMMEC, Gorakhpur), **Dr. R.K. Chauhan** (Professor, Electronics & Comm. Engg., MMMEC, Gorakhpur), **Dr. G.S. Tripathi** (Electronics & Comm. Engg., MMMEC, Gorakhpur), **Dr. Arun Kumar** (Former Professor, Electronics & Comm. Engg., MMMEC, Gorakhpur) and **Dr. C.B. Tripathi** (Professor Electronics & Comm. Engg., MMMEC Gorakhpur), **Dr. Avinash Kumar** (Professor & Head, EC, KIET) and **Dr. Vipin Kumar** (H.O.D. Physics, KIET), for their guidance and valuable suggestions.

I am grateful to my younger brothers **Rajeev** and **Rakesh,** daughter **Ritika**, for inspiring me for this project.

I deeply express my heartful thanks to the publishers **S.K. Kataria & Sons** for publishing this book in such a beautiful get-up and well in time.

Lastly, I am also grateful to **Mr. Manoj Tyagi** and his staff from **Plus Computer's** for their hard work.

Sanjay Sharma

Contents

UNIT SIX 283–319

8. Field Effect Transistor 285-319

UNIT SEVEN 321–369

9. FET Amplifiers 323-369

UNIT—I

p-n Junction Diode

Qualitative Theory of p-n Junction, p-n Junction as a Diode, Diode equation, Volt-Ampere Characteristics, Temperature dependence of V–I characteristic, Ideal versus Practical-Resistance levels (Static and Dynamic), Transition and Diffusion Capacitances, Diode Equivalent Circuits, Load Line Analysis, Breakdown Mechanisms in Semi conductor Diodes, Zener Diode Characteristics.

Above Syllabus has been covered in following Chapter:

CHAPTER 1 : Semiconductor Materials and Properties

CHAPTER 2 : P-N Junction Diode

Chapter 1

Semiconductor Materials and Properties

SYLLABUS

Qualitative Theory of p-n Junction.

Chapter Outline

1.1. INTRODUCTION

It has been proved by scientists that the operation of electronic devices depends upon the motion of charged particles within them. Therefore, one should have an understanding of the elements which control the motion of these particles. But as a fact, it is the physical structure of a solid, which exerts this control. Therefore, every electronic engineer should study the structure and arrangement of atoms in solids.

1.2. STRUCTURE OF SOLIDS

A detailed study of X-ray scattering has revealed the fact that most of the solids are crystal-line in structure. It means that a solid consists of atoms or molecules which are arranged in a periodic manner. There is always some basic arrangement of atoms, which is repeated throughout the entire solid material. Such an arrangement of atoms within a solid is called crystal lattice. Such solids are called crystalline solids. But there are some other solid materials which do not have crystalline structure. Such solid materials are called non-crystalline or amorphous solids. All metals and semiconductors like silicon and germanium are crystalline materials while wood, plastic, paper, glass are amorphous solid materials.

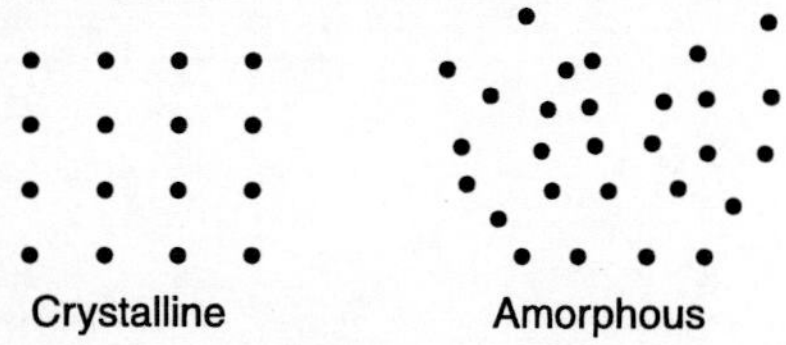

Fig. 1.1

- Conventional Current and Electron Flow
- Extrinsic Semiconductors
- N-type Semiconductor
- P-type Semiconductor
- Effect of Temperature on Extrinsic Semiconductors
- Intrinsic Concentration
- Electrical Properties of Semiconductors
- Comparison between Metals (Conductors) and Semiconductors
- Comparison of Intrinsic and Extrinsic Semicon-ductors
- Semiconductor Materials used for Opto Devices, Modern Semiconductor Devices and ICs

1.3. CLASSIFICATION OF SOLID MATERIALS

Basically, solid materials may be classified into three groups as under:

(*a*) Conductors

(*b*) Insulators

(*c*) Semiconductors

(*a*) Conductors: Conductors are those materials which are good conductors of electricity. This is due to the fact that in conductors there are a large number of mobile charge carriers or free electrons which carry electric current. When the temperature of a conductor is increased, its resistivity also increases. It means that conductors have positive temperature coefficient of resistance. Examples of good conductors are copper, silver, aluminium etc. When a constant electric field E is applied to a conductor, the electrons would be accelerated and the velocity would increase indefinitely with time. But due to collision of electrons, electrons loose energy and a steady-state condition is reached where a finite value of drift velocity v_d is attained.

The drift velocity v_d is in a direction opposite to that of the electric field and its magnitude is proportional to E.

Hence $\quad v_d = \mu E$

μ is mobility of the electron in m^2/volt-second. If the concentration of free electrons is n per cubic meter, the current density J is

$$J = nqv_d$$

But $\quad v_d = \mu E$

Therefore, $\quad J = nq\mu E$

or $\quad J = \sigma E \qquad ...(1.1)$

where $\quad \sigma = nq\mu$

σ is the conductivity of the conductor in (ohm-meter)$^{-1}$. For a good conductor n is very large, approximately, 10^{28} electrons/m^3.

Equation (1.1) is also known as Ohm's law which states that the conduction current density is proportional to the applied electric field.

(*b*) Insulators: Insulators are those materials which are bad conductors of electricity. In other words, they have very high resistivity. This is due to the fact that they have no charge carriers or free electrons to carry electric current. Examples of insulators are glass, quartz, rubber, bakelite etc.

(*c*) Semiconductors: Semiconductors are those materials whose conductivities lie between conductors and insulators. They have poor conductivity than conductors and higher than insulators. Hence these are neither good conductors nor good insulators. When the temperature of a semiconductor is increased, its resistivity decreases or conductivity increases. It means that at higher temperature, a semiconductor conducts better. Therefore, the semiconductors have negative temperature coefficient of resistance.

The semiconductors have another very interesting quality. The conductivity of a semiconductor can be increased by adding a very small amount of some particular materials. These specific materials are called impurities. Examples of semiconductors are silicon (Si) and germanium (Ge).

Table 1.1: Conductivity and Resistivity of Few Conductors, Semiconductors and Insulators at Room Temperature

Name of Materials	Conductivity (S/m)	Resistivity (Ωm)	Classification
Silver	6.25×10^{7}	1.6×10^{-8}	Conductors
Copper	5.88×10^{7}	1.7×10^{-8}	
Aluminium	3.85×10^{7}	2.6×10^{-8}	
Germanium	1.54	6.5×10^{-1}	Semiconductors
Silicon	5.0×10^{-4}	2.0×10^{3}	
Porcelain	3.33×10^{-10}	3.0×10^{9}	Insulators
Glass	5.88×10^{-12}	1.7×10^{11}	
Hard rubber	1.0×10^{-16}	1.0×10^{16}	

1.4. STRUCTURE OF AN ATOM

The most fundamental unit of matter is an atom. An atom consists of a central portion which is called the nucleus. A number of smaller particles called electrons move in approximately elliptical orbits around the nucelus. The nucleus of all elements except hydrogen, contains two types of particles: protons and neutrons. The mass of proton and neutron is approximately same. The difference between the two is that the proton is a positively charged particle but the neutron is an electrically neutral particle.

Almost total mass of an atom is concentrated in its nucleus. The electrons which revolve around the nucleus are very light in weight. An electron is almost 1850 times lighter than a proton or neutron. The amount of charge on proton is same as that on proton but the charge on electron is negative.

The matter in the normal state is electrically neutral. Since an atom is the most basic building block of matter. This should also be neutral. Therefore, to make an atom neutral, the number of orbiting electrons must be the same as the number of protons in the nucleus of the atom.

The three particles, electrons, protons and neutrons are called the fundamental particles of the universe. This means that there is no difference between an electron in an atom of silver and an electron in an atom of copper. Different elements behave differently because of the fact that there is a difference in the number and arrangement of the electrons, protons and neutrons in atoms of different elements.

PIONEERS IN ELECTRONICS

French natural philosopher *Charles Augustin Coulomb (1736-1806)* invented a method for measuring the force of attraction and repulsion between two electrically charged spheres. Coulomb established a law of inverse squares and defined the basic unit of charge quantity, the coulomb.

The number of protons or electrons in an atom is called its **atomic number**. For example, hydrogen (H) atom has one proton in its nucleus and one electron around its nucleus. Therefore, the atomic number of hydrogen atom is 1.

The atomic number of some important atoms are : $Z = 5$ for boron, $Z = 6$ for carbon, $Z = 13$ for aluminium, $Z = 14$ for silicon.

All the electrons of an atom do not move in the same orbit. All the electrons are arranged in different orbits or shells. The maximum number of electrons which can exist in the first orbit is two. This first orbit is the nearest to the nucleus. The second orbit can accommodate not more than eight electrons, the third not more than eighteen, the fourth orbit not more than thirty-two and so on. In general, an orbit or shell can contain a maximum of $2n^2$ electrons, where n is the number of the orbit. But, there is an exception. The outermost orbit in an atom cannot have more than eight electrons. The electrons present in the outermost shell are called **valence electrons.** These valence electrons take part in chemical bonding.

1.5. ATOMIC STRUCTURE OF FEW ELEMENTS

(*a*) Aluminium atom → atomic number → 13, Al (13)

Aluminium atom have 13 protons and 14 neutrons in its nucleus.

(*b*) Silicon atom → atomic number → 14, Si (14)

Silicon atom have 14 protons and 14 neutrons in its nucleus.

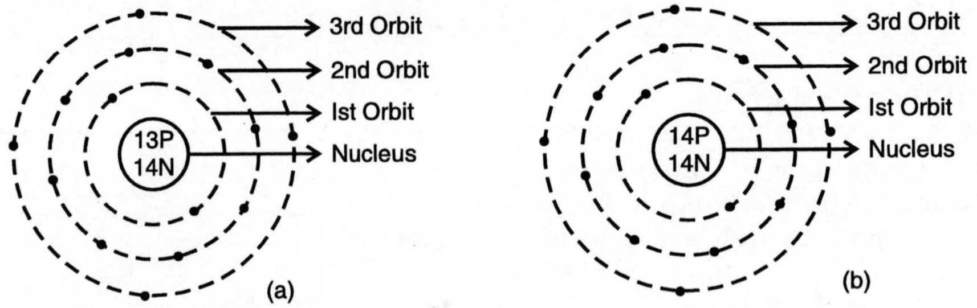

Fig. 1.2. (a) Atomic structure of Al(13) (b) Atomic structure of Si(14)

1.6. ENERGY LEVELS

According to modern physics, only discrete values of electron energies are possible. This means that each isolated atom can have only a certain number of orbits. These available orbits represent energy levels for the electrons. Electrons can revolve only in the permitted orbits and not in any intermediate orbit. Each orbit has fixed amount of energy associated with it and the electrons moving in a particular orbit possess the energy of that orbit. The larger the orbit, the greater is its energy. Hence an electron in the outer orbit have more energy than the electron in the inner orbit. Figure 1.3 shows the energy levels diagram for Aluminium atom. The permitted energy levels are numbered as $n = 1, 2...$in increasing order of energy.

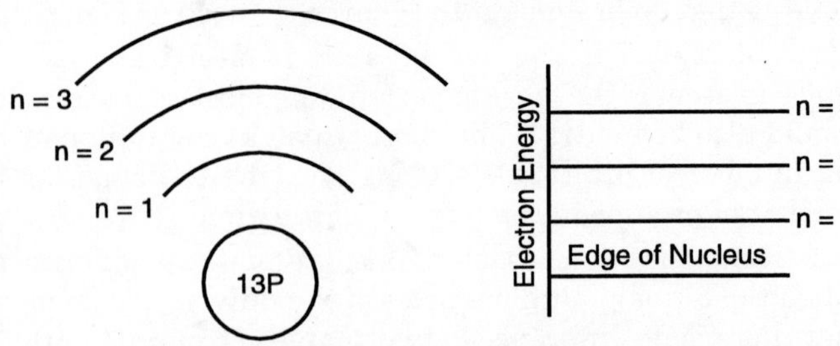

Fig. 1.3. Energy levels in an isolated Al atom

Nothing is more simple than greatness; indeed, to be simple is to be great. —R.W. Emerson

1.7. ATOMIC BONDS

Whether a material is a conductor, a semiconductor, or an insulator depends largely upon what happens to the outer shell electrons when the atoms bond themselves together to form a solid.

Atomic bonds are of two types: *i.e.* primary (or *chemical*) bonds and secondary (or molecular) bonds.

Primary bonds are strongest bonds between atoms and may be further subdivided into (*i*) metallic, (*ii*) covalent and (*iii*) ionic bonds. Secondary or molecular bonds are weaker than primary bonds. Attraction forces (also known as vander Waals forces) exist between atoms or molecules.

In metallic bonds, atoms give up their valence electrons to form clouds of electrons which is shared by all atoms. Such type of bond exist between elements having loosely held valence electrons so that they can be easily released to a common pool

As an example, in case of copper, the atoms give up easily detachable electrons and this produces a large mass of free electrons (or electron gas) drifting about through the spaces between the copper atoms. Now, each atom of copper becomes a positive ion because of loosing an electron. The electron gas is, of course, negatively charged, consequently, an electrostatic force of attraction exists between the positive ions and the electron gas. This is the *bonding force* that holds the material together in a solid. In the case of copper and other metals, the bonding force is termed as metallic bonding or sometimes electron gas bonding. Such a bonding is shown in figure 1.4(*a*).

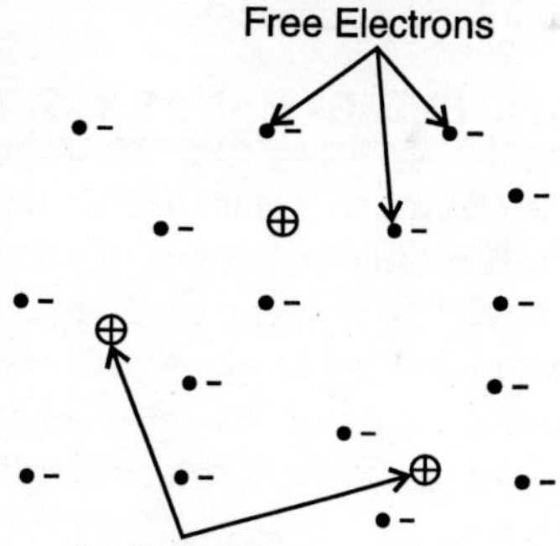

(a) Metallic bonding

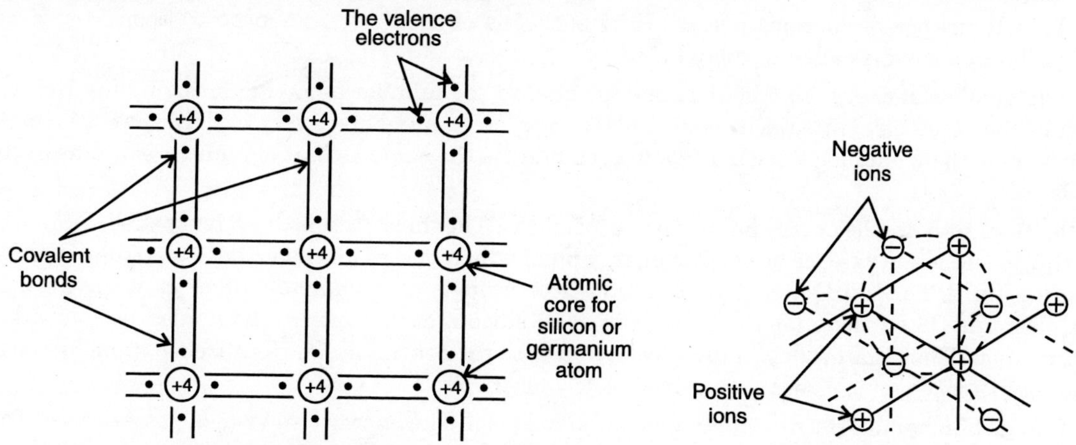

(b) Covalent Bonding

(c) Ionic Bonding

Fig. 1.4. Illustration of bonding between atoms in solids

Covalent Bond is formed by sharing of electrons by two or more atoms. It can be formed between atoms of same or different elements. This arrangement has been shown in figure 1.4(*b*). In covalent bonding every valence shell of every atom appears to be filled, and consequently there are no holes and no free electrons drifting about within the material. The same covalent bonding arrangement exists in silicon and germanium.

DO YOU KNOW?

The process of conversion from an electrically neutral atom to an ion is called as ionization.

The Si (or Ge) atom at the centre share one of its valence electrons with each of the valence electrons of the four Si (or Ge) atoms surrounding it as shown in figure 1.4(*b*). This sharing of valence electrons produces four pairs of electrons which are covalently bonded and hence completing its octlet.

In **ionic bonds,** atoms of one element transfer electrons to the other so that both have stable configurations.

In this arrangement, some atoms part with outer-shell electrons, but these are accepted into the orbit of other atoms. Hence, the atoms are ionized and those which give up electrons become *positive ions,* and those which accept the electrons become negative ions. This produces an electrostatic bonding force between the atoms, known as the *ionic bonding* as shown in figure 1.4(c). Such bonding occurs in some of the insulating materials.

DO YOU KNOW?

Conduction band electrons are actually responsible for the flow of current. More the number of electrons in the conduction band more will be the current flow.

1.8. ENERGY BANDS IN SOLID SILICON *(V. Important)*

As discussed in last article, in a single isolated atom, the electrons revolving in any shell possess a certain amount of energy. But when atoms form a solid, the orbit of an electron is affected not only by the charges of its own atom but by nucleus and electrons of each atom in a solid. Because of this, the electrons in the same orbits have a range of energies rather than a single energy. This is called as **energy band**.

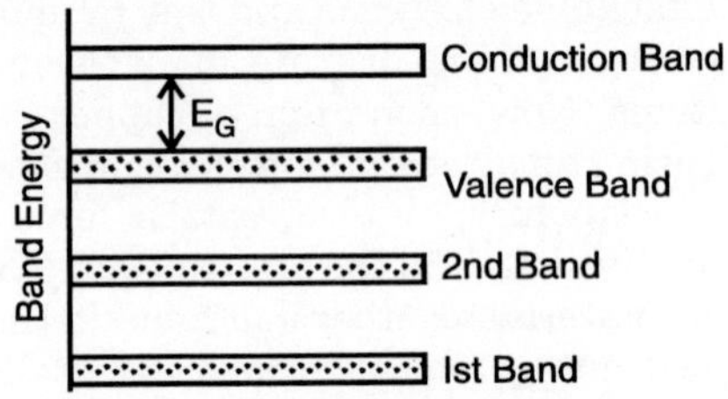

Fig. 1.5. Energy-band diagram of solid silicon

The closely placed energy levels which differ slightly in energy, therefore, form an energy band.

Thus the range of energies possessed by electrons of the same orbit in a solid is called **energy band.**

The number of energy bands in a solid is equal to the number of energy levels in any isolated atom of the same element. Each band consists of a large number of very closely spaced discrete energy levels. The number of such levels in each band is the same as that of number of atoms in the solids.

Figure 1.6 shows the energy band diagram of silicon. The third shell is the valence shell. Similarly, the third band is the valence band. The energy band which possesses the valence electrons is called **valence band**. The third shell of an isolated silicon atom is not completely filled. It is having only four electrons. The reason for this is that in solid silicon, each atom is surrounded by four other silicon atoms. These four atoms share an electron with the central atom. Hence, each atom has eight electrons in a solid and so valence band is completely filled.

Conduction band is the next higher energy band level. The energy band which possesses the free electrons is called **conduction band**. Electrons in this band take part in conduction. If a substance has empty conduction band, it means that current conduction is not possible in that substance.

1.9. FORBIDDEN ENERGY GAP

Between the valence band and the conduction band there is an energy gap E_G. This is called forbidden energy gap. An electron can be sent from the valence band to the conduction band if we provide some energy to the silicon. This energy should be kept larger than the energy gap E_G. If we provide less energy than E_G, the electron will not be lifted because there is no other energy level between valence band and conduction band. Due to above reason, this energy gap E_G is generally called **forbidden energy** gap.

In other words, forbidden energy gap is a region in which no electron can stay because there is no allowed energy state. To make the valence electrons free, some external energy through heat, light equal to the forbidden energy gap should be supplied. The forbidden energy gap E_G is **1.12 eV** for silicon and **0.72 eV** for germanium.

In considering the electrical properties, we are only interested in two upper sidebands, valence bands and conduction bands. The free electrons in the conduction band can move readily under the influence of an external field and thus take part in the conduction of current.

1.10. INSULATORS, METALS AND SEMICONDUCTORS ON THE BASIS OF BAND GAP

(Important)

For any given material, the forbidden energy-gap may be large, small or non-existent. Thus, the difference between insulators, metals and semiconductors is largely concerned with the relative widths of the forbidden energy gaps. In this article, we shall discuss the electrical properties of insulators, metals and semiconductors on the basis of band gap or forbidden energy gap.

1.10.1. Insulators

In case of insulators, there is generally no electron in the conduction band and the valence band is filled.* Figure 1.6(*a*) shows the energy band diagram of insulators. It may be observed from energy band diagram that there is a wide gap between valence and conduction bands (forbidden energy gap). It is generally 5 eV or more. Due to this wide gap, it is almost impossible for an electron to cross the gap to go from valence band to conduction band. At room temperature, the valence electrons of an insulator can not have so much energy that it is able to jump to the conduction band. Due to this fact, the insulators are not able to conduct electric current. This means that the insulators have very high resistivity or extremely low conductivity at room temperature. But an insulator may conduct if its temperature is very high or if a very high voltage is applied across it. This is also known as the **breakdown** of the insulators.

1.10.2. Metals or Conductors

Figure 1.6(*b*) shows the energy band diagram of metals. It may be observed that forbidden energy gap or energy gap between valence and conduction bands is zero. In fact, the valence and conduction bands overlap each other. In a metal, the valence band energies and conduction band energies are same. The orbits in the conduction band are very large. An electron in the conduction band experiences almost negligible nuclear attraction. Hence it is very easy for a valence band electron to become a conduction band electron. It means that a metal consists of a large number of free electrons without supplying any external energy. Because of this fact, a metal works as a very good conductor.

1.10.3. Semiconductors

Figure 1.6(*c*) shows the energy band diagram of a semiconductor. It may be observed that forbidden energy gap E_G is not very wide for semiconductors. It is 0.72 eV for germanium and 1.12 eV for silicon. At 0 K,** the semiconductors behave as insulators because there is no free electron to conduct in the conduction band. However, in semiconductor, an electron can be lifted from the valence band to the conduction band by imparting some amount of energy to it. This energy must be more than energy

> **DO YOU KNOW?**
>
> Another common semiconductor element is carbon (C), which is used mainly in the production of resistors.

* Valence band filled implies that all the permissible energy levels in this band are occupied by the electrons.

** At a temperature of absolute zero (–273°C).

gap E_G. If the energy imparted to the electron is less than E_G, it cannot be lifted from valence band to conduction band since no permissible energy levels exist between the two bands. Room temperature provides the sufficient amount of energy to lift electrons from the valence band to the conduction band. Some electrons jump to conduction band. Hence at room temperature, semiconductors are able to conduct some electric current. If temperature is further raised above room temperature, more and more valence electrons acquire energy and cross the energy to go to conduction band. Hence semiconductors have negative temperature coefficient of resistance.

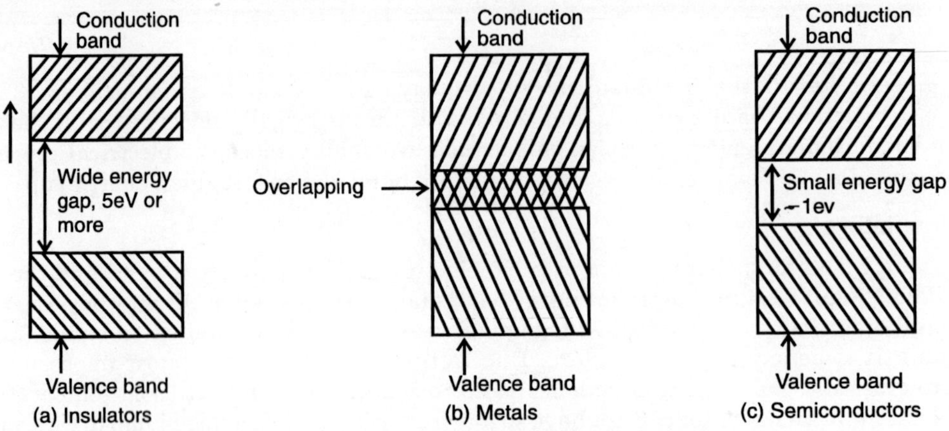

Fig. 1.6. Energy band diagram

1.10.3.1. Why Silicon is More Widely used Semiconductor Material?

As a matter of fact, the valence electrons in germanium are in the fourth shell while those in silicon are in the third shell, *i.e.*, closer to nucleus. This means that the germanium valence electrons are at higher energy levels than those in silicon. Hence, the germanium valence electrons will need smaller amount of additional energy to escape from the atom. Because of this reason, the germanium produces more number of electron hole pairs than silicon. Hence, the leakage current (reverse saturation current) is more in germanium than that in silicon. This property makes germanium more unstable at high temperatures. Therefore, silicon is more widely used material than germanium.

1.10.3.2. Elemental and Compound Semiconductors

Table 1.2 shows a portion of the periodic table in which the more common semiconductors are found. Silicon and germanium are in group IV and are the elemental semiconductors. But, the gallium arsenide (GaAs) is a group III-V compound semiconductor.

Table 1.2: A portion of Periodic Table

S.No.	Group III	Group IV	Group V
1.	B (Borium)	C (Carbon)	
2.	Al (Aluminium)	Si (Silicon)	P (Phosphorus)
3.	Ga (Gallium)	Ge (Germanium)	As (Arsenic)

A successful man is one who can lay a firm foundation with the bricks that others throw at him.
—**David Brinkley**

1.11. COMPARISON OF CONDUCTORS, INSULATORS AND SEMICONDUCTORS

(Important)

Table 1.3: Comparison of Metals (Conductors), Insulators and Semiconductors

S. No.	*Parameter of comparison*	*Conductors (Metals)*	*Insulators*	*(Semiconductors*
1.	Conductivity	Very high	Very low	Moderate
2.	Resistivity	Very low	Very high	Moderate
3.	Forbidden gap	No forbidden gap	Large gap ($E_G \cong 5$ eV)	Medium ($E_G = 1$ to 2 eV)
4.	Temperature coefficient of resistance	Positive	Negative	Negative
5.	Effect of temperature on resistance	R increases as T increases	R decreases as T increases	R decreases as T increases
6.	Number of electrons available for conduction	Very large	Very small	Moderate
7.	Conductivity at room temperature	Very good	Poor	Moderate
8.	Examples	Aluminium, copper	Paper, Mica, Glass	Silicon, Germanium
9.	Applications	As conductors, wires, bus bars.	Capacitors, insulation for wires	Semiconductor devices

1.12. CLASSIFICATION OF SEMICONDUCTORS: INTRINSIC AND EXTRINSIC*

A semiconductor material is one whose electrical properties lie in between those of insulators and good conductors. At room temperature, such materials have conductivities considerably lower than those of conductors and much higher than those of insulators. Infact, the conductivity of a semicondutor lies in the range of 10^5 to 10^{-4} siemens per meter. An important property of a semiconductor is that its resistance decreases with the increase in temperature. This means that a semiconductor has a negative temperature coefficient of reistance. An even more important property of a semiconductor is that their resistivity or conductivity changes considerably when even a very small amounts of certain other substances called the impurities, are added to them.

Basically Semiconductors are of Two Types: Elemental and compound. The examples of elemental semiconductors are germanium (Ge) and silicon (Si). Germanium was the mainly used material for the majority of the early solid state devices like transistors. Subsequently, germanium has been replaced by silicon in several applications because of the fact that silicon has a better thermal stability, is readily available and has an advanced technology. The example of compound semiconductor is GaAs.

Semiconductors may also be classified in following way:

1. Intrinsic semiconductor
2. Extrinsic semiconductor

1.13. INTRINSIC SEMICONDUCTOR**

A semiconductor in an extremely pure form is known as **intrinsic semiconductor.** A semiconductor

* Distinguish between intrinsic and extrinsic semiconductor. What is the effect of temperature on conductivity of a semiconductor? *(JNTU Hyderabad, Sem. Exam., 2004-05)*

** Define an intrinsic material, a negative temperature coefficient, and covalent bonding. List three materials that have negative temperature coefficient. *(JNTU Hyderabad, Sem. Exam., 2003-04)*

> **DO YOU KNOW?**
>
> Silicon as well as Germanium atoms contain 4 electrons in the valence shell. If there are 4-electrons in the outer-most orbit, the semiconductor material is called as pure or intrinsic semiconductor.

is not intrinsic unless its impurity content is less than one part impurity in 100 million parts of semiconductor.

The silicon and germanium are the two most widely used intrinsic semiconductors. Figure 1.7 shows the two dimensional crystal structure of silicon. There are four electrons in the outermost orbit or valence shell.

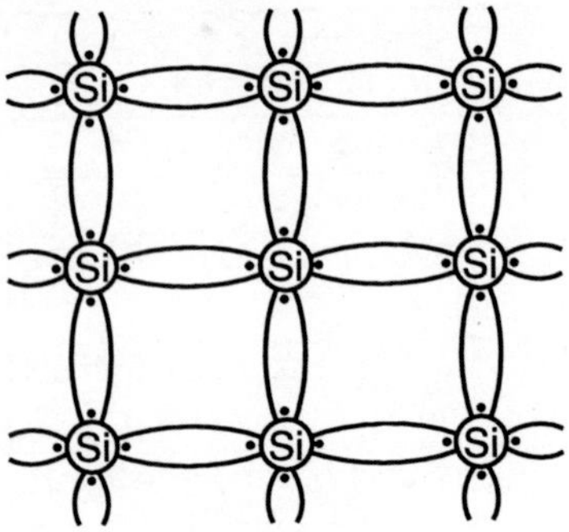

Fig. 1.7. Intrinsic semiconductor

Each of the four valence electrons takes part in forming covalent bonds with the four neighbouring atoms. A covalent bond consists of two electrons, one from each adjacent atoms. Atoms bond together to form molecules because each atom attempts to get eight electrons in the valence shell to become a stable structure.

At a low temperature such as absolute zero (0 K), all the valence electrons are tightly held by parent atoms and by covalent bonds with other atoms. Hence electrons cannot move through the crystal structure or in other words, no free electrons are available to conduct electricity. Thus, the intrinsic semiconductors, at absolute zero (0 K) behave as an **insulator**.

1.13.1. Crystalline Structure of Intrinsic Semiconductors

Some of the unique qualities of Silicon and Germanium are due to their special type of atomic structure. The atoms of both these materials form a very definite pattern which is periodic in nature *i.e.*, it continuously repeats itself. One such complete pattern is called as **crystal** and the periodic arrangement of atoms is called as lattice. For the Silicon and Germanium, the crystal has a three dimensional diamond shape as shown in figure 1.8.

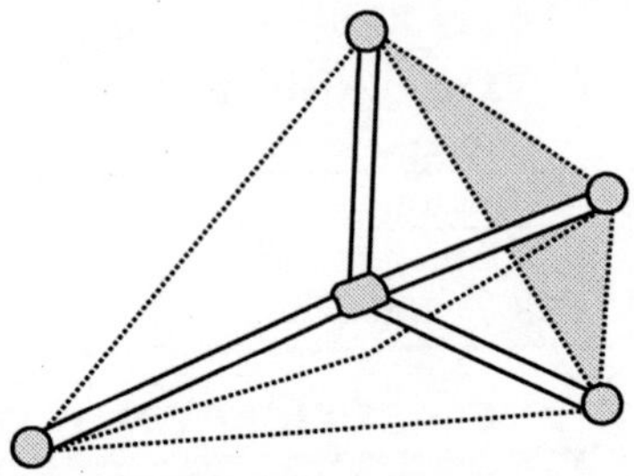

Fig. 1.8. Silicon and Germanium crystal structure

1.14. INTRINSIC SEMICONDUCTOR AT ROOM TEMPERATURE (300 K) OR THERMAL GENERATION OF ELECTRON-HOLE PAIRS

(Very Important)

At absolute zero, an intrinsic semiconductor behaves as a perfect insulator. But if the temperature is increased upto room temperature (300 K), some covalent bonds break. When a covalent bond breaks, the electron becomes free to move through the crystal. When an electron breaks a covalent bond and becomes free, a vacancy is also produced. This vacancy is known as a **hole**. Hence, when a free electron is produced, a hole is also produced simultaneously. This means that electrons and holes are produced in pairs.* In intrinsic semiconductors, the concentration of free electrons will always be equal to the concentration of holes. This type of simultaneous generation of electrons and holes due to temperature is also called **thermal generation**. Hence, at room temperature intrinsic semiconductor has some conductivity. Since, an electron is a negatively charged particle, the vacancy (or hole) created by this electron will be assumed as positively charged. These electrons and holes are also called **free charge carriers** because they are free to move throughout the crystal.

* The librated electron and the hole left behind is called an electron-hole pair.

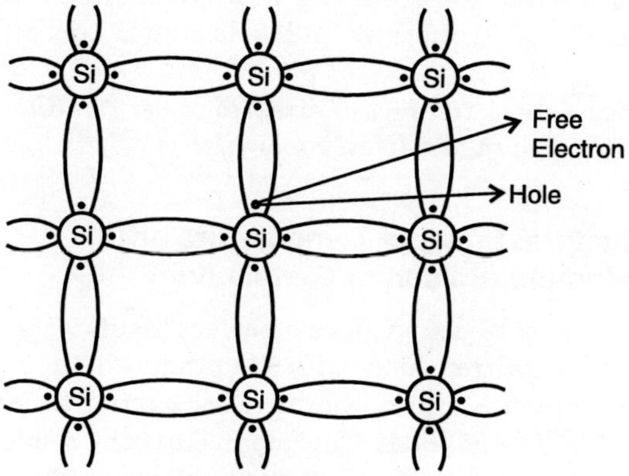

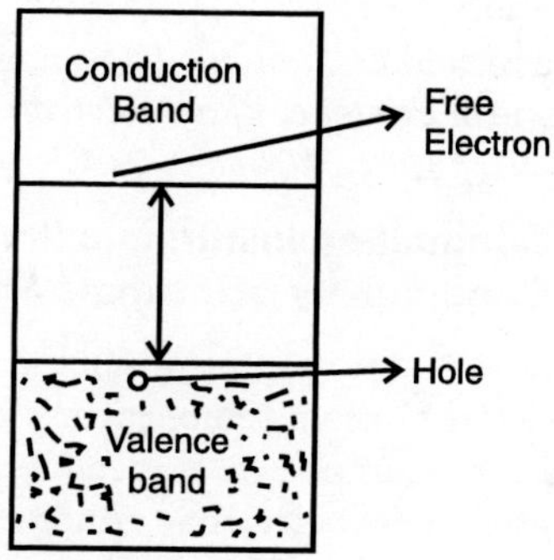

Fig. 1.9. Generation of electron-hole pair at room temperature in intrinsic semiconductor

Figure 1.9(*a*) shows the generation of electron-hole pair at room temperature in intrinsic semiconductor. To break a covalent bond in the crystal lattice, a certain amount of energy is required. This energy is 0.72 eV for germanium and 1.12 eV for silicon. Room temperature (300 K) is sufficient to provide this energy.

Figure 1.9(*b*) shows the energy band-diagram for electron-hole pair generation at room temperature in intrinsic semiconductor. Whenever an electron-hole pair generates, the hole remains in the valence band and electron lifts to the conduction band to take part in conduction of current. The energy to lift the electron from the valence band to conduction band is **0.72 eV for Ge** and **1.12 eV for Si.**

1.14.1. Why Intrinsic Semiconductors Behave Like Insulators at Low Temperatures?

(Important)

In each covalent bond, both the electrons involved are shared by both the parent atoms. Thus, with four covalent bonds associated with each atom, all the silicon atoms will effectively have eight electrons in their valence shells. Four of these electrons are their own and remaining four are due to the four covalent bonds. Due to this, the outermost shells of all the atoms are completely filled.

> **DO YOU KNOW?**
>
> A hole and an electron each possesses a charge of 0.16×10^{-18} C but of opposite polarity.

This makes the outermost shells of all atoms stable. Therefore, not a single free electron is available at the absolute zero temperature *i.e.*, at 0°K or – 273°C. Hence, the intrinsic semiconductor acts like an insulator at the absolute zero temperature.

1.14.2. Recombination

In case of an intrinsic semiconductor, electron-hole pairs are generated due to temperature or thermal agitation. These thermally generated electrons and holes move freely throughout the crystal. Due to this there is a possibility of collision between electrons and holes. Whenever, there is a collision, an electron takes the position of holes and both of them disappear. This process is called **recombination**. Thus, in the process of recombination, both the free electron and hole disappear and energy* is released as a quantum heat or light. This quantum is absorbed by

* Whenever an electron is freed from valence band, a certain amount of energy is absorbed. The same amount of energy is released when an electron falls into a hole.

another electron to breakaway from its valence band and creates a new election-hole pair. At a given temperature, the rate of generation of electrons and holes is equal to the rate of recombination.

Before recombination an electron or hole covers some distance or exists for a very small time. This average time of existence of an electron or hole is called **mean lifetime** *or lifetime. This lifetime falls in the range of 1μs to 10^3 μs.*

1.14.3. Conduction in Intrinsic Semiconductors: Effect of Temperature on Conductivity of Intrinsic Semiconductors (Silicon or Germanium)

We know that any semiconductor (germanium or silicon) acts as a perfect insulator at absolute zero. However, at room temperature (300 K) some electron-hole pairs are produced due to thermal energy. As an example, in a specimen (sample) of germanium, the intrinsic carrier concentration (concentration of free electrons or holes) is **$2.5 \times 10^{19}/m^3$** at 300 K. This means that the semiconductor has a small conductivity. But if the temperature is raised further, more electron-hole pairs are produced. Hence, the higher the temperature, the higher is the concentration of charge carriers.

Therefore, since more charge carriers are made available, the conductivity of an intrinsic semiconductor increases with temperature. This means that the resistivity decreases with the increase in temperature. Hence, the semiconductors have **negative temperature-coefficient** *of resistance.*

1.14.4. Electron and Hole Current

As we apply voltage across a piece of intrinsic semiconductor material, the thermally generated free electrons in the conduction band are attracted towards the positive end. The current constituted due to movement of free electrons is called as **electron current** as shown, in figure 1.10.

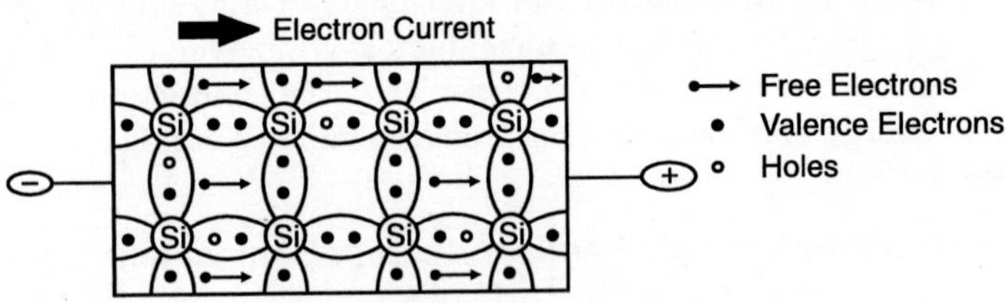

Fig. 1.10. Electron current

(i) Hole current

Figure 1.11 (*a*) shows three neigbouring silicon atoms A, B and C. The covalent bonds do exist but for convenience they have not been shown. A hole exists in a broken covalent bond with atom *A* as shown in figure 1.11 (*a*). A valence electron may jump from another atom B to fill up this hole as shown in figure 1.11 (*b*). When it jumps, the electron leaves a hole behind it and thus the hole has moved in the direction opposite to that of the electron. Similarly the hole passes from atom *B* to atom *C* as shown in figure 1.11 (*c*). The current constituted due to this hole movement is called as the **hole current**. Thus, holes move in the opposite direction to that of electrons.

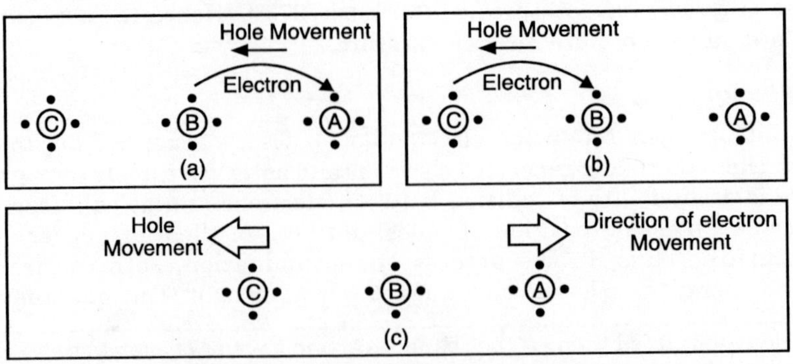

Fig. 1.11. Movement of electrons and holes.

Important Point : *Here, it may be noted that the hole current flows due to the valence electrons and not due to the free electrons. These electrons move due to the presence of holes. Therefore, holes are also considered as charge carriers like the electrons and the hole current always flows in opposite direction to that of the 'electron' current.*

(ii) Total current

The flow of an electric current is due to the movement of electrons in the conduction band and movement of holes in the valence band.

Thus, we can write

Total current = electron current + hole current

1.14.5. Semiconductors have a Negative Temperature Coefficient of Resistivity

Just now we have seen that as the temperature rises, more number of electrons will absorb the thermal energy to break the covalent bonds and contribute to the number of free carriers. The increased number of free electrons and holes will increase the conductivity of the semiconductor and results in a lower resistance level. Thus, resistance of the semiconductors decreases with increase in temperature. Therefore, semiconductors are said to have a *negative temperature coefficient of resistivity*.

1.14.6. Carrier Life Time

In a pure semiconductor, the number of free electrons is equal to the number of holes. The thermal energy keeps producing new electron-hole pairs whereas some of the hole electron pairs will disappear due to the process of recombination. Thus, thermal agitation gives rise to production of electron-hole pairs while the recombination will make them disappear. The time duration between these two processes is called as the **carrier lifetime**. On an average, a hole or an electron will exist for τ_p (for holes) or τ_n (electron) respectively. This time is called as **mean lifetime** of holes and electrons respectively. The value of carrier lifetime ranges from a few nanoseconds to few microseconds. The carrier lifetime is very important because it indicates the time taken by the electron or hole concentrations, to return to their equilibrium value. The carrier lifetime can be precisely adjusted to a desired value by introducing gold as recombination agent into the silicon under a controlled condition.

1.15. CONVENTIONAL CURRENT AND ELECTRON FLOW

(i) Conventional Current

The current flow from positive to negative is referred to as conventional current. Every graphical symbol used to represent an electronic device has an arrowhead which indicates conventional current direction. Therefore, the electronic circuits are most easily explained by using the conventional current flow.

(ii) Electron Flow

Under the influence of the external dc source or battery, the free electrons in the semiconductor slab, which are negatively charged, will get attracted towards the positive terminal and holes being positively charged will be attracted towards the negative terminal of the external battery. Thus, as shown in figure 1.12, the electrons flow

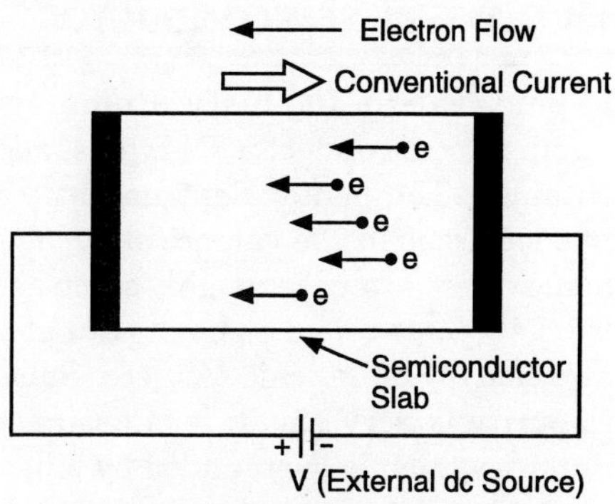

Fig. 1.12. Conventional current and electron flow in

exactly in the opposite direction to that of the conventional current. The study of electron movement is essential in understanding the operation of any electronic device.

1.16. EXTRINSIC SEMICONDUCTORS

As discussed earlier, intrinsic or pure semiconductors have small conductivity at room temperature. Therefore, they are not of much use. By adding some amount of impurity atoms to a pure or intrinsic semiconductor, we can change its conductivity or characteristics. This process of adding impurity to a pure semiconductor is called **doping.** Hence, the conductivity is increased by means of the doping process. Doping is done at a rate such that only one atom of impurity is added per 10^6 to 10^{10} semiconductor atoms.

> **DO YOU KNOW?**
>
> The extrinsic semiconductors have a better conductivity than the intrinsic semiconductors. Therefore, these are used in manufacturing of all the electronic components such as diodes, transistors etc.

We add impurities either to increase the number of electrons or holes in a semiconductor. The impurities are the materials used to dope the intrinsic semiconductor materials. These impurity materials can be of two types:

(*i*) Donor impurity

(*ii*) Acceptor impurity

(*i*) Donor impurity (Pentavalent impurity)

The material which is being used as impurity in the process of doping is called as *dopant.* When the dopant is a pentavalent atom *i.e.*, the atom containing five valence electrons then it is called as the *donor impurity* and the doping is called as *donor doping.* Donor doping is used to manufacture *N*-type extrinsic semiconductor. The examples of pentavalent or donor impurities are Arsenic, Phosphorous and Antimony.

(*ii*) Acceptor impurity (Trivalent impurity)

When the dopant is a trivalent atom *i.e.*, the atom consisting of only three valence electrons, then it is called as the *acceptor impurity* and the doping is called as *acceptor doping.* Acceptor doping is used to manufacture *P*-type extrinsic semiconductors. The examples of trivalent or acceptor impurities are Boron, Gallium, Aluminium and Indium.

A doped semiconductor is called **extrinsic semiconductor.** Depending upon the types of impurity added, there are two types of extrinsic semiconductors :

(*a*) N-type Semiconductor;

(*b*) P-type Semiconductor.

1.17. N-TYPE SEMICONDUCTOR

If pentavalent impurity atoms are added to intrinsic semiconductor, N-type semiconductor is obtained. The pentavalent impurity atoms have five electrons in the valence shell. The size of the pentavalent atoms is roughly same as that of Si or Ge. Few pentavelent elements are phosphorus (P) antimony (Sb), arsenic (As) etc. Since amount of impurity is very small, it is assumed that each impurity atom is surrounded by silicon atoms. In

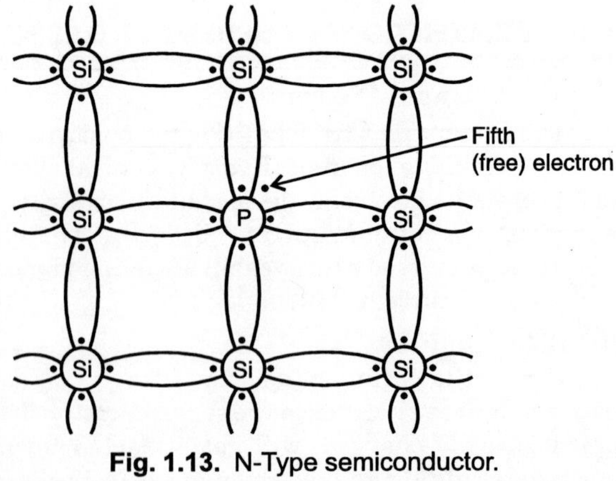

Fig. 1.13. N-Type semiconductor.

figure 1.13 an impurity phosphorus atom is surrounded by silicon atoms. The phosphorus atom has five valence electrons. Four, out of these five electrons form covalent bonds with four silicon atoms. The fifth electron has no chance of forming a covalent bond. This electron is loosely bound to the parent atom. A very little amount of energy is required to de-attach this electron from the nucleus of its parent atom. This energy is of the order of 0.05 eV for silicon and 0.01 eV for germanium. Practically, at room temperature, all such electrons become free.

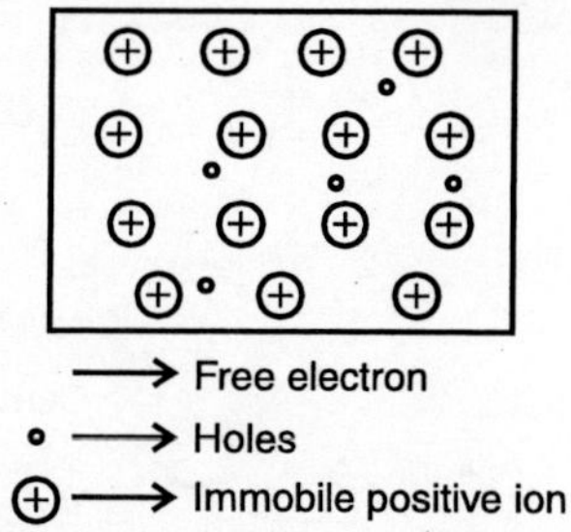

Fig. 1.14. Representation of N-type semiconductor

In this way, each impurity atom donates one electron to the conduction band. Due to this reason, pentavalent impurity is also called **donor type impurity.** Besides a large number of donated electrons, there are also some thermally-generated, electron-hole pairs. A large number of electrons increases the rate of recombination of electrons with holes. Because of this, the net concentration of holes is much less than intrinsic value. Hence, in a **N-type semiconductor, electrons are majority carriers and holes are minority carriers.**

The pentavalent impurity atom donates one electron. After donating this electron, the impurity atom becomes a positively charged ion. This is also called **donor ion.** This donor ion is held tightly by four covalent bonds in crystal. This is unable to move and hence is called positively charged **immobile ion.**

1.17.1. Representation of N-type Semiconductor

We know that in a N-type semiconductor, electrons are the majority carriers and holes are the minority carriers. Also, there are a large number of immobile positive ions. N-type semiconductor is not negatively charged but is electrically neutral. The total number of electrons is exactly equal to the total number of holes and immobile ions. Hence, negative charge is exactly balanced by the positive charge.

1.17.2. Changes in the Energy Band Diagram due to Doping

Addition of donor impurity to an intrinsic semiconductor is called *doping*. The process of doping will alter the energy band diagram to a great extent as shown in figure 1.15. The donor energy level (E_D) shown by the dotted lines is very close to the conduction band energy level (E_C). Therefore, the unbonded valence electrons of the impurity atoms can very easily jump into the conduction band and become free electrons. Thus, even at room temperature, almost all the extra electrons of pentavalent impurity will jump to the conduction band. Thus, a large number of free electrons are available for conduction in an N-type semiconductor.

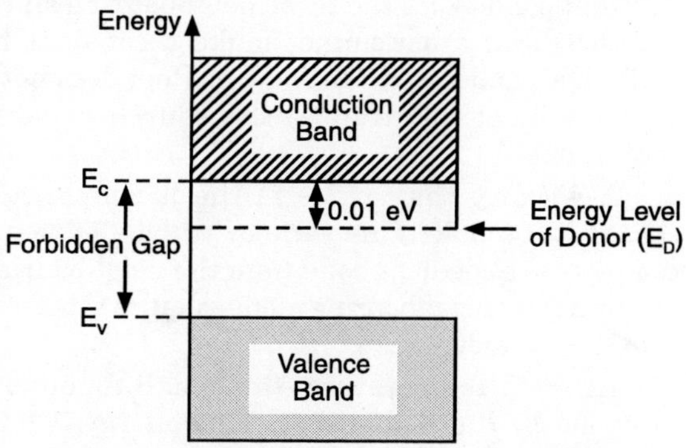

Fig. 1.15. Energy band diagram of N-type semiconductor

1.17.3. Conduction in N-type Material

When an external DC voltage is applied to the N-type semiconductor material, the free electrons move towards the positive terminal of the source and holes move towards the negative end as shown in figure 1.16.

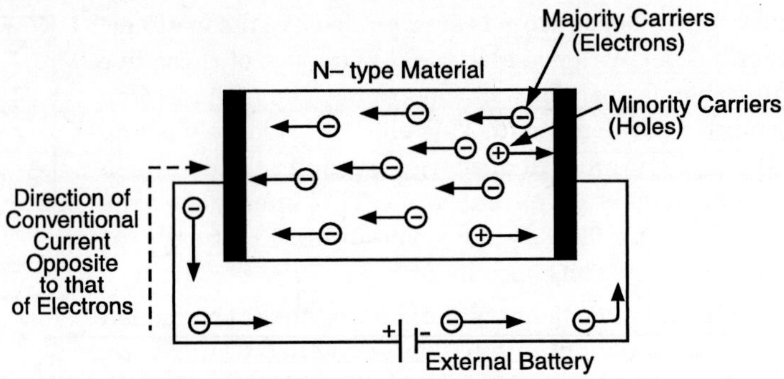

Fig. 1.16. Conduction in N-type material.

As electrons outnumber the holes, the conduction in *N*-type material is mainly due to the majority charge carriers *electrons.*

1.18. P-TYPE SEMICONDUCTOR

If trivalent impurity atoms are added to intrinsic semiconductor, P-type semiconductor is obtained. The trivalent impurity atoms have three electrons in the valence shell. The size of the trivalent atoms is roughly same as that of Si or Ge. The trivalent impurities are boron, aluminium, gallium and indium. As an example, consider a sample of intrinsic silicon to which a very small amount of trivalent impurity such as indium is added. Since amount of impurity is very small, it may be assumed that each impurity atom is surrounded by silicon atoms. In figure, an impurity atom (Indium In) is surrounded by silicon atoms. The Indium atom has three valence electrons. These three electrons form covalent bonds with the three neighbouring silicon atoms. The fourth silicon atom cannot make a covalent bond with the Indium atom because the Indium atom does not have fourth valence electron. Hence, the fourth covalent bond is incomplete.

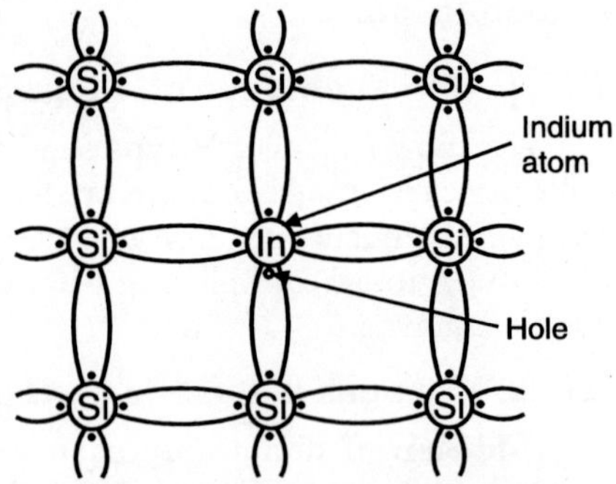

Fig. 1.17. P-type semiconductor.

> **DO YOU KNOW?**
>
> For both N- and P-type semiconductors, an increase in temperature produces an identical increase in the number of minority and majority current carriers.

A vacancy that exists in the incomple covalent bond constitutes a hole. This vacancy or hole have a tendency to complete the covalent bond from the neighbouring atoms. An electron from neighbouring atoms require some energy to jump into the vacancy.

At room temperature, this small amount of energy is provided by thermal energy. This energy is 0.05 eV for silicon and 0.01 eV for germanium. When an electron from the neighbouring atoms jump into the vacancy around the indium atom to complete the covalent bonds, the effect is two fold. Firstly another vacancy or hole is created in the covalent bond of surrounding atom from where the electron had jumped. Secondly, after acquiring an electron, the indium atom becomes a negative ion. Since indium atom accepts one electron to become negative ion, it is also called acceptor ion or **acceptor type impurity**. Negative ion is also immobile because it is held tightly in the crystal structure by covalent bonds.

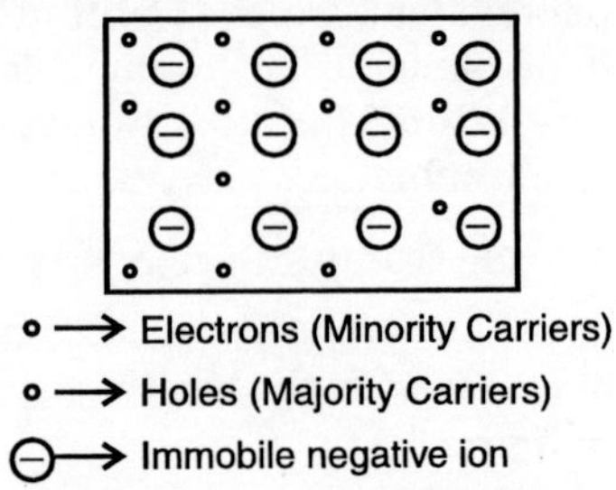

Fig. 1.18. Representation of P-type semiconductor

Hence, from above it is clear that in a P-type semiconductor, a large number of holes are created due to acceptor type impurities. This large number of holes increases the rate of recombination of holes with electrons so that number of electrons is further reduced than intrinsic level.

Therefore, in a P-type semiconductor, holes are the **majority carriers and electrons are the minority carriers.**

As a whole, a P-type material has holes as majority carriers, electrons as minority carriers and negative immobile ions.

1.18.1. Representation of P-type Semiconductor

We know that if a trivalent impurity is added to an intrinsic semiconductor, a P-type semiconductor is produced. In P-type semiconductor, holes are the majority carriers, electrons are the minority carriers. In addition to these two types of charge carriers, there are also immobile negative ions.

1.18.2. Changes in the Energy Band Diagram Due to Doping

The changes in the energy band diagram due to doping with acceptor impurities is as shown in figure 1.19. The energy level of acceptor impurity (E_A) shown by dotted lines is very close to the energy level of valence band. Therefore, the valence band electrons can very easily jump to the acceptor energy level E_A leaving behind holes in the valence band. This process of hole creation can take place even at the room temperature. Thus, even at room temperature, a large number of holes are created inside a P-type semiconductor. As holes are positively charged, the presence of a large number of holes will make the semiconductor a P-type semiconductor.

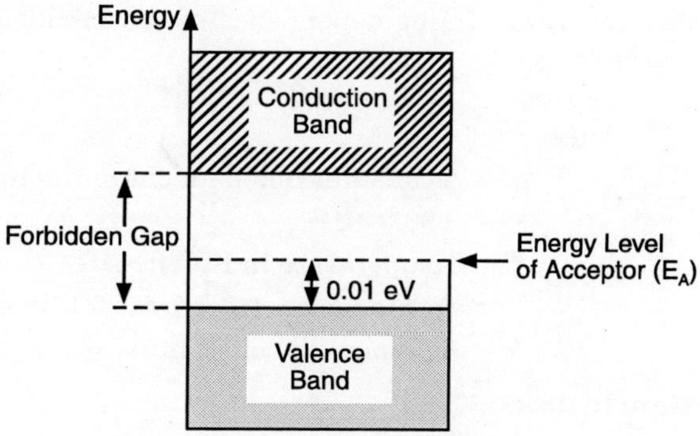

Fig. 1.19. Energy band diagram or P-type semiconductor.

1.18.3. Conduction in P-type Semiconductors

The conduction inside a *P*-type semiconductor is as shown in figure 1.20. When an external dc source is applied across a *P*-type material, the holes move in the valence band. The movement of holes is towards the negative terminal of the supply. The free electrons move towards the positive terminal of the battery, as shown in figure 1.20. Note that the *P* and *N*-type semiconductors are electrically neutral. *P* or *N* does not mean that they are positively or negatively charged.

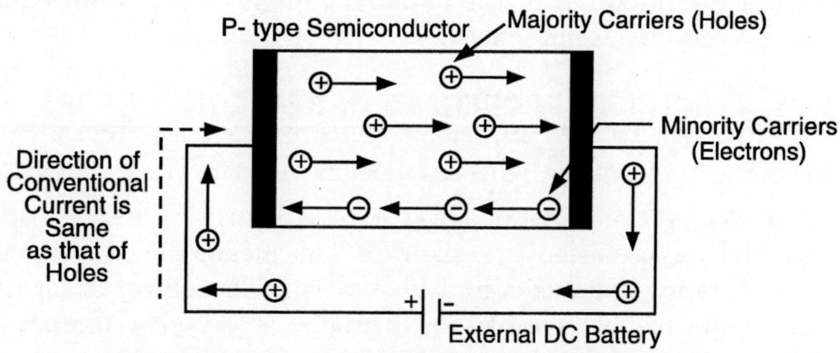

Fig. 1.20. Conduction in a P-type semiconductor.

1.19. EFFECT OF TEMPERATURE ON EXTRINSIC SEMICONDUCTORS

We know that addition of a small amount of donor or acceptor impurity generates a large number of charge-carriers in any extrinsic semiconductor. Actually, due to this large number of charge carriers, the conductivity of an extrinsic semiconductor is several times that of an intrinsic semiconductor at room temperature (300 K).

Now, let us observe what happens if we increase the temperature of extrinsic semiconductor (say N-type semiconductor). In this case because all the donor atoms have already donated their free electrons (at 300 K), the additional thermal energy due to increase in temperature only serves to increase the thermally produced carriers.

Due to this, the concentration of minority carriers increases. In this process a temperature is reached when the number of covalent bonds broken is quite large such that the number of holes is nearly equal to the number of electrons. Therefore, the extrinsic semiconductor now acts like an intrinsic semiconductor with higher conductivity. This critical temperature is 200°C for Si and 85°C for Ge.

1.20. INTRINSIC CONCENTRATION

It may be observed that in an intrinsic semiconductor, the concentration n_i increases with the increase in temperature. This dependence of the intrinsic concentration n_i on the temperature T may be expressed as

$$n_i = A_0 . T^{3/2} . e^{-E_G/2kT} \quad ...(1.2)$$

where

A_0 = A constant independent of the temperature

$= 3.83 \times 10^{21}$

T = Temperature in Kelvin

E_G = Forbidden energy gap at 0 K in eV

k = Boltzmann's constant

Conclusion

From equation (1.2), we can draw following conclusions:

(*i*) The intrinsic concentration n_i is independent of Fermi level position.

(*ii*) The intrinsic concentration n_i depends upon the band gap E_G which represents the energy required to break a bond.

(*iii*) The intrinsic concentration n_i depends strongly upon the temperature ($n_i \propto T^{3/2}$)

(*iv*) The factor 2 in the exponent means that when one covalent bond breaks two charge carriers are produced.

1.21. ELECTRICAL PROPERTIES OF SEMICONDUCTORS

In this section, we shall discuss various electrical properties of semiconductors:

1. Negative temperature co-efficient of resistance: Semiconductors have negative temperature co-efficient of resistance. This means that with the rise in temperature of semiconductor, its resistance decreases and vice versa. Therefore, according to this electrical property, the semiconductors behave like an insulator at very low temperature whereas at high temperature, semiconductors behave as a conductor.

2. Doping: We know that addition of an impurity to a semiconductor is called doping. When a suitable impurity like gallium, arsenic, antimony is added to a semiconductor, it changes the current conducting properties of semiconductor.

3. Resistivity: The resistivity of a semiconductor is more than a conductor but is less than an insulator.

4. Mobility: Mobility μ of a semiconductor varies as T^{-m} over a temperature range of 100°K to 400°K. For silicon Si, m = 2.5 for electrons and 2.7 for holes. For germanium Ge, m = 1.66 for electrons and 2.33 for holes. The mobility is also a function of electric field intensity E and remains constant only if $E < 10^3$ volt/cm in N-type silicon Si.

For $10^3 < E < 10^4$ volt/cm, μ varies approximately as $E^{-1/2}$.

5. Variation of conductivity with temperature: We know that, in an intrinsic semiconductor, both types of carriers, electrons and holes contribute to the conductivity. This total conductivity of intrinsic semiconductor is expressed as

$$\sigma = q.n_i\,(\mu_n + \mu_p)$$

But we know that

$$n_i \;\alpha\; e^{-E_G/2kT}$$

Therefore,

$$\sigma = K.e^{-E_G/2kT}$$

Here K is some constant.

This shows that if the temperature increases, the conductivity of the semiconductor increases exponentially.

Table 1.4 shows various electrical properties of Silicon (Si) and Germanium (Ge).

Table 1.4: Electrical Properties of Silicon (Si) and Germanium (Ge)

S.No.	Property	Silicon (Si)	Germanium (Ge)
1.	Atomic weight	28.1	72.6
2.	Atomic number	14	32
3.	Atoms/cm^3	5.0×10^{22}	3.4×10^{22}
4.	Relative dielectric constant	12	16
5.	E_G(eV) at 0°K	1.21	0.785
6.	E_G(eV) at 300°K	1.1	0.72
7.	Intrinsic concentration n_i at 300°K/cm^3	1.5×10^{10}	2.5×10^{13}
8.	Intrinsic resistivity at 300°K, Ω-cm	2,30,000	45
9.	Electron mobility μ_n at 300°K, cm^2/V-s	1,300	3,800
10.	Hole mobility μ_p at 300°K, cm^2/V-s	500	1,800

1.22. COMPARISON BETWEEN METALS (CONDUCTORS) AND SEMICONDUCTORS

Table 1.5: Comparison Between Metals and Semiconductors

S.No.	Metals (Conductors)	Semiconductors
1.	Charge carriers are only electrons	Charge carriers are electrons and holes
2.	Crystalline	Crystalline
3.	Has low resistivity or high conductivity	Has moderate conductivity
4.	Carrier density is high	Carrier density is moderate
5.	Resistivity increases with temperature	Resistivity decreases with temperature.
6.	Current flows only by drift process	Current flows due to both diffusion and drift

* A life, which does not go into action, is a failure.

— **Arnold J. Toynbee**

1.23. COMPARISON OF INTRINSIC AND EXTRINSIC SEMICONDUCTORS

Table 1.6 gives the comparison between intrinsic and extrinsic semiconductors.

Table. 1.6.

S.No.	Parameter	Intrinsic semiconductors	Extrinsic semiconductros
1.	Purity	Extremely pure, No impurity added.	Not pure as it contains impurity atoms.
2.	Doping	Not done	Is done
3.	Type	–	*P* or *N* type
4.	Number of electrons and holes	Number of electrons is always equal to number of holes	They are never equal to each other.
5.	Conductivity	Poor	Much higher
6.	Current conduction	Due to electrons and holes.	Mainly due to electrons or holes depending on type of semiconductors
7.	Position of Fermi level	At the centre of the forbidden gap	It is either near the conduction band or near valence band
8.	Applications	None	For manufacturing of electronic components.

1.24. SEMICONDUCTOR MATERIALS USED FOR OPTO DEVICES, MODERN SEMICONDUCTOR DEVICES AND ICs

Elements in the periodic table are grouped according to the number of valence electrons. In table 1.7 commonly used semiconductor materials have been shown. Silicon and Germanium are the popular semiconductors. They are called as elemental semiconductors and appear in group IV of the periodic table.

There is another important type of semiconductors. They are called as group III-V compound semiconductors. The GaAs (Gallium Arsenide) and Gallium Arsenide Phosphide (GaAsP) are some of the examples of group III-V compound semiconductors.

Table 1.7: A Portion of the Periodic Table

S.No.	Group III	Group IV	Group V
1.	B (Boron)	C (Carbon)	
2.	Al (Aluminium)	Si (Silicon)	P (Phosphorous)
3.	Ga (Gallium)	Ge (Germanium)	As (Arsenic)

These compound semiconductor materials are used for the following applications:

(*i*) For producing opto electronic devices such as LED, Photo transistors, Photodiodes etc.

(*ii*) For producing certain semiconductor devices such as MOSFET and integrated circuits (ICs).

SUMMARY

- A solid material can be classified in the form of conductors, insulators and semiconductors.
- Conductors are those materials which are good conductors of electricity.
- Insulators are those materials which are bad conductors of electricity.
- Semiconductors are those materials whose conductivites lie between conductors and insulators.

- In a solid, the closely placed energy levels, which differ slightly in energy, form an energy band.
- The range of energies possessed by electrons of the same orbit in a solid, is called energy band.
- The energy band which possesses the free electrons is called valence band.
- The energy band which possesses the free electrons is called conduction band.
- Electrons present in the conduction band take part in conduction.
- Between the valence band and the conduction band there is an energy gap. This energy gap is called forbidden energy gap and is denoted by E_G.
- Insulators, metals and semiconductors can also be defined on the basis of band gap.
- In insulators, there is generally no electron in the conduction band and the valence band is filled.
- In metals, the forbidden energy gap between the valence band and conduction band is zero. Infact, the valence and conduction bands overlap each other.
- In case of semiconductors, forbidden energy gap is not very wide.
- Semiconductors can be classified in two forms: intrinsic semiconductors and extrinsic semiconductors.
- A semiconductor in an extremely pure form is known as intrinsic semiconductor.
- At absolute zero (*i.e.* 0 K) the intrinsic semiconductor behaves as an insulator.
- At room temperature the intrinsic semiconductor has some conductivity.
- In intrinsic semiconductors, the concentration of free electrons will always be equal to the concentration of holes. This type of simultaneous generation of electrons and holes due to temperature is called thermal generation.
- At a given temperature, the rate of generation of electrons and holes is equal to the rate of recombination.
- Before recombination, an electron or hole covers some distance or exists for a very small time. This average time of existence of an electron or hole is called mean life time.
- The conductivity of an intrinsic semiconductor increases with temperature.
- The semiconductors have negative temperature-coefficient of resistance.
- By adding some amount of impurity atoms to a pure or intrinsic semiconductor, we can change its conductivity or characteristics.
- The process of adding impurity to a pure semiconductor is called doping.
- A doped semiconductor is called extrinsic semiconductor.
- Extrinsic semiconductors are of two types: N-type semiconductor and P-type semiconductor.
- The pentavalent impurities are Phosphorus (P), antimony (Sb), Arsenic (As) etc.
- The pentavalent impurity is also called donor type impurity.
- In a N-type semiconductor, the electrons are the majority carriers and holes are minority carriers.
- In a P-type semiconductor, holes are the majority carriers and electrons are the minority carriers.
- The conductivity of a material is proportional to the concentration of free electrons.
- The conductivity of metal, σ is expressed as
$$\sigma = n\,q\,\mu$$
- The conductivity of intrinsic semiconductor is
$$\sigma = q.n_i(\mu_n + \mu_p)$$
- The conductivity of N-type semiconductor is
$$\sigma = q.n.\mu_n$$
- The conductivity of P-type semiconductor is
$$\sigma = q.p.\mu_p$$

SHORT QUESTIONS WITH ANSWERS

Q. 1. What is forbidden energy gap? Explain.

Ans. Between the valence band and the conduction band there is an energy gap E_G. This is called forbidden energy gap. An electron can be sent from the valence band to the conduction band if we provide some energy to the silicon. This energy should be kept larger than the energy gap E_G. If we provide less energy than E_G, the electron will not be lifted because there is no other energy level between valence band and conduction band. Due to above reason, this energy gap E_G is generally called **forbidden energy** gap.

Q. 2. Classify insulators, metals and semiconductors on the basis of band gap.

Ans. In case of insulators, there is generally no

electron in the conduction band and the valence band is filled. A metal consists of a large number of free electrons without supplying any external energy. Because of this fact, a metal works as a very good conductor. That forbidden energy gap E_G is not very wide for semiconductors. It is 0.72 eV for germanium and 1.12 eV for silicon.

Q. 3. What are the types of semiconductors?

Ans. Basically semiconductors are of two types: Elemental and compound. The examples of elemental semiconductors are germanium (Ge) and silicon (Si). Germanium was the mainly used material for the majority of the early solid state devices like transistors. Subsequently, germanium has been replaced by silicon in several applications because of the fact that silicon has a better thermal stability, is readily available and has an advanced technology. The example of compound semiconductor is GaAs.

Semiconductors may also be classified in following way:

1. Intrinsic semiconductor
2. Extrinsic semiconductor

Q. 4. What is the effect of temperature on conductivity of intrinsic semiconductors?

Ans. We know that any semiconductor (germanium or silicon) acts as a perfect insulator at absolute zero. However, at room temperature (300 K) some electron-hole pairs are produced due to thermal energy. As an example, in a specimen (sample) of germanium, the intrinsic carrier concentration (concentration of free electrons or holes) is $\mathbf{2.5 \times 10^{19}/m^3}$ at 300 K. This means that the semiconductor has a small conductivity. But if the temperature is raised further, more electron-hole pairs are produced. Hence, the higher the temperature, the higher is the concentration of charge carriers.

Q. 5. What is doping?

Ans. The process of adding impurity to a pure semiconductor is called doping. Doping is done at a rate such that only one atom of impurity is added per 10^6 to 10^{10} semiconductor atoms.

Q. 6. What are the types of extrinsic semiconductor ?

Ans. Depending upon the types of impurity added, there are two types of extrinsic semiconductors:

(*a*) N-type semiconductor;
(*b*) P-type semiconductor

7. What is the energy gap?

Ans. The energy gap between the valence band and conduction band is known as forbidden energy gap. Forbidden energy gap is a region in which no electron can stay as there is no allowed energy state. The greater the forbidden energy gap more tightly the valence electrons are bound to the nucleus.

Q. 8. Define a hole in a semiconductor.

Ans. When an energy is supplied to a semiconductor, a valence electron is lifted to a higher energy level, the departing electron leaves a vacancy in the valence band. The vacancy is called a hole. Thus, a vacancy left in the valence band, because of lifting of an electron from valence band to conduction band, is known as hole.

Q. 9. For the same order of doping, why does N-type semiconductor exhibit larger conductivity than P-type semiconductor?

Ans. Since the mobility of electrons is greater than that of holes, for same level of doping, N-type semiconductor exhibits larger conductivity.

Q. 10. What is hole current?

Ans. The movement of the hole (positively charged vacancy in the valence band) from positive terminal of the supply to the negative terminal through semiconductor constitutes hole current.

Q. 11. In which bands do the movement of electrons and holes take place?

Ans. Free electrons move in conduction band while holes in valence bands.

Q. 12. For the same order of doping, why does N-type semiconductor exhibit larger conductivity than P-type semiconductor?

Ans. Since the mobility of electrons is greater than that of holes, for same level of doping, N-type semiconductor exhibits larger conductivity.

Q. 13. Why Silicon is more widely used Semiconductor Material ?

Ans. As a matter of fact, the valence electrons in germanium are in the fourth shell while those in silicon are in the third shell, *i.e.*, closer to nucleus. This means that the germanium valence electrons are at higher energy levels than those in silicon. Hence, the germanium valence electrons will need smaller amount of additional energy to escape from the atom. Because of this reason, the germanium produces more number of electron hole pairs than silicon. Hence, the leakage current (reverse saturation current) is more in germanium than that of silicon. This property makes the germanium more unstable at high temperatures. Therefore, silicon is more widely used material than germanium.

Q. 14. Why intrinsic semiconductors behave like insulators at low temperatures?

Ans. In each covalent bond, both the electrons involved are shared by both the parent atoms. Thus, with four covalent bonds associated with each atom, all the silicon atoms will effectively have eight electrons in their valence shells. Four of these electrons are their own and remaining four are due to the four covalent bonds. Due to this, the outermost shells of all the atoms are completely filled.

This makes the outermost shells of all atoms stable. Therefore, not a single free electron is available at the absolute zero temperature *i.e.*, at 0°K or – 273°C. Hence, the intrinsic semiconductor acts like an insulator at the absolute zero temperature.

REVIEW QUESTIONS

1. Define the term atomic number.
2. Explain the procedure for bonding in a solid.
3. Define valence band and conduction band for a solid.
4. Explain the concept of forbidden energy gap.
5. Differentiate between insulators, conductors and semiconductors using forbidden energy gap.
6. Explain the importance of valence electrons.
7. Define drift velocity.
8. Describe the electronic structure of elements.
9. Explain the concept of energy bands in solids.
10. Give the energy-band structure of insulators, semiconductors and conductors.
11. Explain with the help of a neat sketch why copper is a good conductor. Classify the following as conductor or semiconductor:

    (*i*) Gold (*ii*) Germanium
    (*iii*) Silver (*iv*) Silicon

    Ans. Copper is a metal and it has many free electrons. As the free electrons are responsible for flow of current, copper acts as a good conductor.

Material	Type
Gold	Conductor
Germanium	Semiconductor
Silver	Conductor
Silicon	Semiconductor

12. Describe the difference between P-type and N-type semiconductor materials. Sketch the atomic structure of silicon and insert the impurity of arsenic in silicon.
13. Explain clearly the difference between acceptor and donor impurities and state what type of carriers are contributed by them.
14. Define diffusion current in a semiconductor.
15. Compare insulators, conductors and semiconductors on the basis of energy band diagram.
16. Write note on P-type and N-type semiconductors.
17. Define mobility. Give its dimensions. A semiconductor is doped with both donors and acceptors of concentration N_D and N_A. Derive the equation to determine electron and hole concentration.
18. Distinguish between intrinsic and extrinsic semiconductor. What is the effect of temperature on conductivity of a semiconductor?
19. How are the carrier mobilities related with resistivity of a semiconductor. Does a hole in a semiconductor contribute to a flow of current? If yes, how and if no, how ?
20. Explain, why the energy levels of an atom become energy bands in a solid.
21. Define a hole in a semiconductor. Explain, why a semiconductor acts as an insulator at 0 K and why its conductivity increases with increasing temperature.
22. Differentiate semiconductors, conductors and insulators on the basis of band gap. What is the mechanism by which conduction takes place inside the semiconductor?
23. Which of the two semiconductor materials Si or Ge has larger conductivity at room temperature and why? What happens to the conductivity of the semiconductor with the rise in the temperature?
24. Is hole a fundamental particle in an atom? Between electron and hole which one has a greater mobility? Give a reason for the difference.

Ans. The hole is not a fundamental particle in an atom. Mobility is defined as,

$\mu = v/E$

where v is the velocity and E is electric field. Electrons are in the conduction band while holes are in the valence band. So the velocity of electrons is higher than that of holes if the applied field intensity is same.

Hence, mobility of electrons is higher than that of the holes.

25. What do you mean by the term semiconductor? Explain about intrinsic and extrinsic semiconductors.
26. Explain the effect of temperature on semiconductors.
27. Explain the mechanism of electrical conduction in a semiconductor such as silicon or germanium. How the conductivity or resistivity of a pure semiconductor is affected by adding pentavalent and trivalent impurities?
28. What is meant by N-type impurity and P-type impurity in a semiconductor?
29. Name three elements which are used as N-type and P-type impurities?

OBJECTIVE TYPE QUESTIONS

I. Fill up the Blanks

1. The electrons in the outer most orbit of an atom are called .... electrons.
2. Electrons in the conduction band have .... energy than those in the valence band.
3. The maximum number of electrons in the orbit preceding the valence orbit cannot exceed ....
4. An isolated germanium atom contains .... orbits.
5. A semiconductor has almost empty .... band.
6. N-type germanium is obtained on doping intrinsic germanium by ....
7. The energy gap in a semiconductor .... with temperature.
8. In an N-type semiconductor, the position of fermi level .... than the centre of energy gap.
9. Free electrons move in .... band.
10. In N-type semiconductor, the current conduction is by ....
11. In P-type semiconductor, the majority carriers are ....
12. Donor type semiconductor is obtained by adding impurity of valency ....
13. Semiconductors have .... bonds.

State whether the Statement is True or False

14. Germanium is a conductor.
15. Aluminium is an insulator.
16. Conduction band and valence band of Si overlap each other.
17. A 0°K, Ge acts like an insulator.
18. Electrons in the conduction band have higher energy than those in the valence band.

II. Multiple Choice Questions

1. In a P-type semiconductors, the majority carriers are
   (*a*) electrons (*b*) holes
   (*c*) none of the above
   (*d*) both
2. Which of the following atoms may be used as a *P*-type impurity
   (*a*) arsenic (*b*) boron
   (*c*) phosphorous (*d*) antimony
3. In an intrinsic semiconductor, Fermi level lies
   (*a*) in the middle of conduction band and valence band
   (*b*) near conduction band
   (*c*) near valence band
   (*d*) none of the above
4. The mobility of charge-carriers has the unit
   (*a*)m/volt-sec^2 (*b*) m volt-sec
   (*c*)m^3/volt-sec (*d*) m^2/volt-sec
5. A semiconductor is electrically neutral since it has
   (*a*) no minority carriers
   (*b*) no majority carriers
   (*c*) no free carriers
   (*d*) equal number of positive and negative carriers.
6. To obtain a P-type semiconductor, the impurity added to a pure semiconductor is
   (*a*) pentavalent (*b*) trivalent
   (*c*) tetravalent (*d*) none of the above.

7. The energy gap in a semiconductor
   (*a*) does not change with temperature
   (*b*) increases with temperature
   (*c*) decreases with temperature
   (*d*) is zero
8. With a rise in temperature around 300 K, the conductivity of an intrinsic semiconductor
   (*a*) remains constant
   (*b*) decreases
   (*c*) increases
   (*d*) becomes zero.
9. Impurities like boron and indium are added to intrinsic semiconductor to form
   (*a*) P-type doped semiconductor
   (*b*) N-type doped semiconductor
   (*c*) a junction diode
   (*d*) none of the above.

Answers to Self Test

I. Fill up the Blanks

1. Valence, 2. higher, 3. $2n^2$, 3. Four, 5. Conduction, 6. Pentavalent impurity, 7. decreases, 8. higher, 9. conduction, 10. Electrons, 11. Holes, 12. Five, 13. Covalent, 14. False, 15. False, 16. False, 17. True, 18. True.

II. Multiple Choice Questions

1. (*b*), 2. (*b*), 3. (*a*), 3. (*d*), 5. (*d*), 6. (*b*), 7. (*c*), 8. (*c*), 9. (*a*).

Chapter

2 P-N Junction Diode

SYLLABUS

P-N junction as a Diode, Diode equation, Volt-Ampere Characteristics, Temperature dependence of V–I characteristic, Ideal versus Practical–Resistance levels (Static and Dynamic), Transition and Diffusion Capacitances, Diode Equivalent Circuits, Load Line Analysis, Breakdown Mechanisms in Semiconductors Diodes, Zener Diode Characteristics.

Chapter Outline

- PN-Junction
- The Depletion Region Formation
- Barrier Potential or Junction Potential (V_j)
- Energy Diagrams of a P-N Junction
- An Open Circuited Step-Graded Junction
- Currents in a P-N Junction with no External Battery
- Biasing of a P-N Junction
- Important Terms often used in P-N Junction
- P-N Junction Diode
- Circuit Symbol of a Diode
- V-I Characteristics of P-N Junction Diode
- The Current Components in a P-N Junction Diode
- Diode Current Equation
- Complete V-I Characteristics of Silicon and Germanium Diodes
- Effect of Temperature on the V-I Characteristics
- Power and Current Ratings of a Diode
- Diode Specification Sheets
- Some Commonly Used Diodes
- Diode Resistance
- Diode Capacitance
- Diode Equivalent Circuit
- The Diode Ratings

2.1. P-N JUNCTION

We know about P-type semiconductors and N-type semiconductors. But P-type and N-type semiconductors, taken separately, are of a very little use. If a piece of P-type semiconductor is joined to a piece of N-type semiconductor in such a manner that the crystal structure remains continuous at the boundary, then a new structure called P-N junction is formed. Such a PN-junction makes a versatile device which is called a semiconductor diode, P-N junction diode or simply crystal diode.

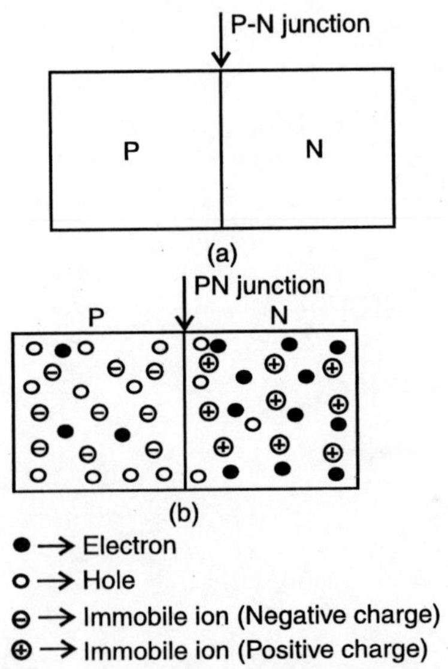

Fig. 2.1. PN-junction

The P-N junction cannot be made simply by joining two pieces. Special fabrication techniques are required to form a P-N junction. A semiconductor P-N junction is formed when a wafer of the semiconductor material such as silicon is doped so that one region is N-type and the other region is P-type.

The P-N junction is an important device because, all other semiconductor devices contain at least one P-N junction.

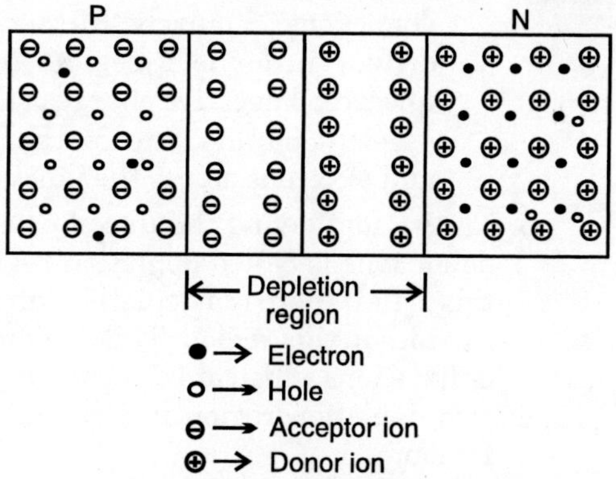

Fig. 2.2. Formation of depletion region in a P-N junction

2.2. THE DEPLETION REGION FORMATION

(Important)

Figure 2.2 shows a P-N junction with depletion region formed at the junction. The P-region contains holes as majority carriers, electrons as minority carriers and negative charged immobile ions called acceptor ions. The N-region contains electrons as majority carriers, holes as minority carriers and positive charged immobile ions called donor ions. The sample as a whole is the neutral so the P-region and N-region separately. In figure 2.2 no voltage is applied to the P-N junction. As soon as the PN-junction is formed, the following actions take place:

> **DO YOU KNOW?**
>
> The depletion region is also known as space charge region. In the unbiased P-N junction, the depletion region gets formed very quickly after the formation of the junction.

1. Holes from the P-region diffuse into the N-region. In the N-region, they combine with the electrons.
2. Electrons from the N-region diffuse into the P-region. In the P-region, they combine with the free holes.
3. The P-region has more number of holes and N-region has more number of electrons, therefore, there is a difference of concentrations in two regions. Due to this difference, the diffusion of holes and electrons takes place. A concentration gradient is produced due to the difference in concentration. Besides this, the electrons and holes move at random in all directions because of thermal energy. Some charge carriers cross the junction.
4. The process of diffusion continues only for a short period of time. After a few recombination of holes and electrons, a restraining force is set up automatically in the neighbourhood of the junction. This restraining force is called barrier. As a result of this force, further diffusion of holes and electrons from one region to other is checked. The formation of this barrier may be given in following steps:
   (*i*) As the PN-junction is formed, some of holes in the P-region and some of the electrons in the N-region diffuse in each other and recombine.
   (*ii*) Each recombination eliminates a hole and a free electron.
   (*iii*) In the above process, the negative acceptor ions in the P-region and positive donor ions in the N-region are left uncovered or uncompensated in the neighbourhood of the junction.
   (*iv*) Further holes trying to diffuse into N-region are repelled by the uncovered positive charge of the

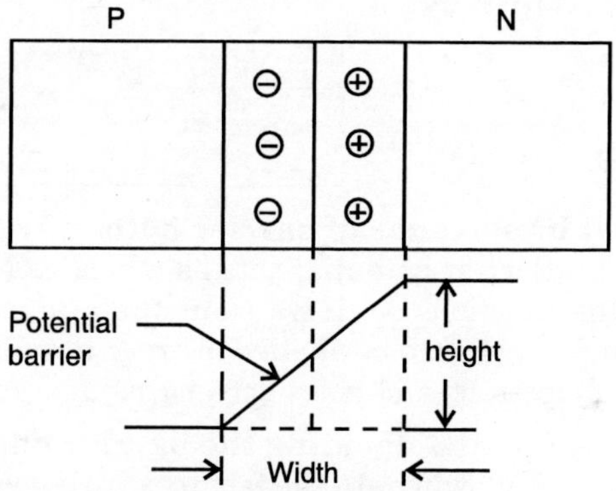

Fig. 2.3. Potential barrier.

donor ions. Similarly, further electrons trying to diffuse into the P-region are repelled by the uncovered negative charge of the acceptor ions. As a result of this, the further diffusion of holes and electrons across the junction is stopped.

5. The region having the uncompensated accepted and donor ions is called depletion region. The reason for this is that there is a depletion (dirth) of mobile charge carriers in this region. This region contains immobile or fixed ions which are electrically charged. Therefore, this depletion region is also called as space-charge region.
6. The width of depletion region depends upon the doping level of impurity in N-type or P-type semiconductor. The greater the doping level, the thinner will be the depletion region. This is due to the reason that a highly doped P-N junction contains a large number of electrons and holes.

PIONEERS IN ELECTRONICS

In 1796 Italian physicist *Alessandro Volta (1745-1827)* invented the first chemical battery, which provided the first practical source of electricity.

2.3. BARRIER POTENTIAL OR JUNCTION POTENTIAL (V_j)

Due to the presence of immobile positive and negative ions on opposite sides of the junction, an electric field is created across the junction. This electric field is known as the *barrier potential* or *junction potential* or *cut in voltage*. It has fixed polarities as shown in figure 2.3. The polarities of barrier potential are decided by the type of immobile ions present on the two sides of the junction. Thus the negative terminal of the barrier potential is on the *P* side and positive side is on the N-side as shown in figure 2.4. This is called as barrier potential because it acts as a barrier to oppose the flow of electrons and holes across the junction. The barrier potential represents the height of the barrier that is to be overcome for commencement of flow of electrons and holes. Barrier potential is measured in volts. The barrier potential for silicon is about 0.7 volt whereas its value for the Germanium is 0.3 volt at 25°C as shown in Table 2.1.

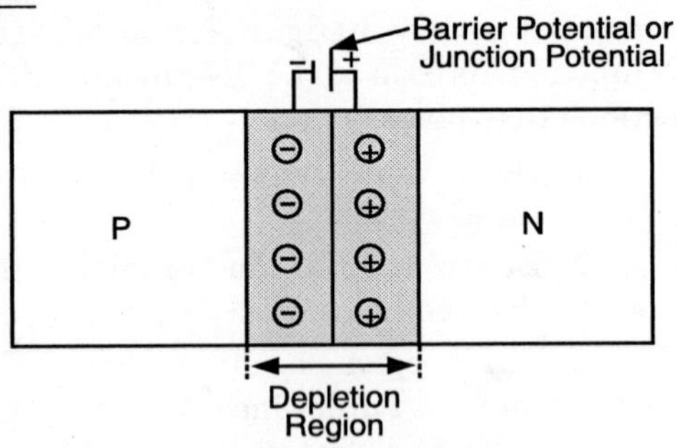

Fig. 2.4. Barrier potential

Table 2.1: Barrier Potential for Different Materials

S.No.	Material	Barrier potential
1.	Silicon	0.7 volt
2.	Germanium	0.3 volt

(*i*) Importance of barrier potential

Barrier potential acts as a barrier (wall) which does not allow the electrons and holes to cross the junction. So, if we want them to cross the junction, then an external voltage of appropriate polarity has to be applied in order to overcome the opposition of the barrier potential. Then, the flow of electrons and holes across the junction can restart again.

(*ii*) Factors deciding the barrier potential value

Following are the factors which decide the value of the barrier potential:

(*i*) Semiconductor material used (Silicon or Germanium).

(*ii*) The intrinsic concentration of Si or Ge before doping.

(*iii*) The level of doping on *P* and *N* sides

(*iv*) Temperaure.

The complete unbiased P-N junction, immobile ions and the variation of barrier potential along the length is shown in figure 2.5.

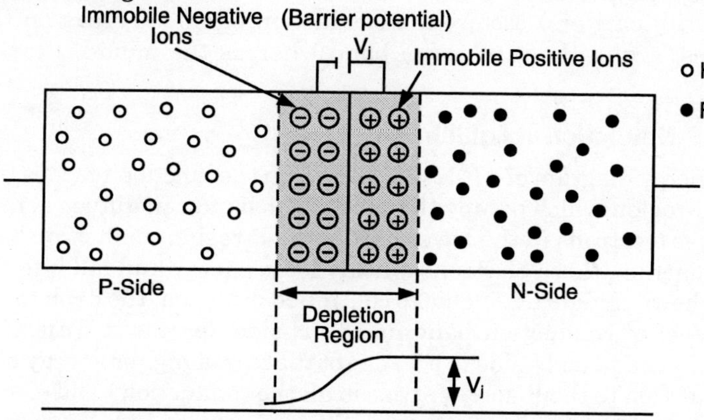

Fig. 2.5. Open circuited P-N junction and variation of junction potential

(*iii*) Penetration of depletion region

The penetration of the depletion region into *P* or *N* side depends on the doping levels of those sides. If both these sides are equally doped then the depletion region penetrates equally on both the sides as shown in figure 2.6 (*a*). But, if *P*-region is lightly doped as compared to the N-region then the penetration depletion region is more on the *P*-side as shown in figure 2.6 (*b*). Similarly, if *N*-side is lightly doped as compared to *P*-side then the depletion region extends more into the *N*-side as shown in figure 2.6 (c).

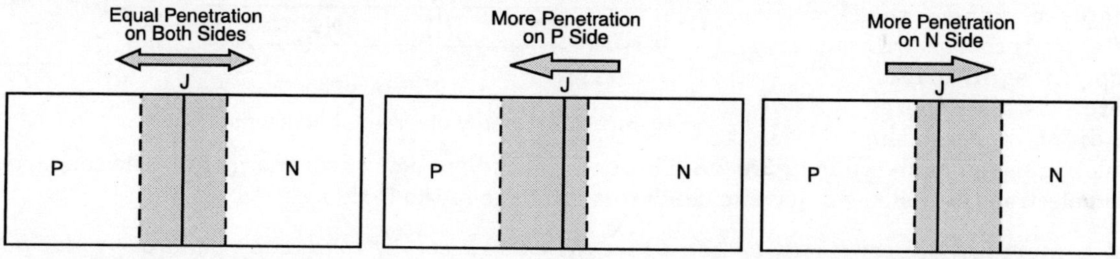

(a) Both sides equally doped (b) P side is lightly doped (c) N side is lightly doped

Fig. 2.6.

Thus, the depletion region always penetrates more on the side which is lightly doped as to the other.

2.4. ENERGY DIAGRAMS OF A P-N JUNCTION

(Important)

The valence and conduction bands in an *N* type material are at slightly lower energy levels than the valence and conduction bands in the *P*-type material. This is due to the difference in the atomic characteristics of the pentavalent (*N*-type) and trivalent (*P*-type) impurity atoms. The energy

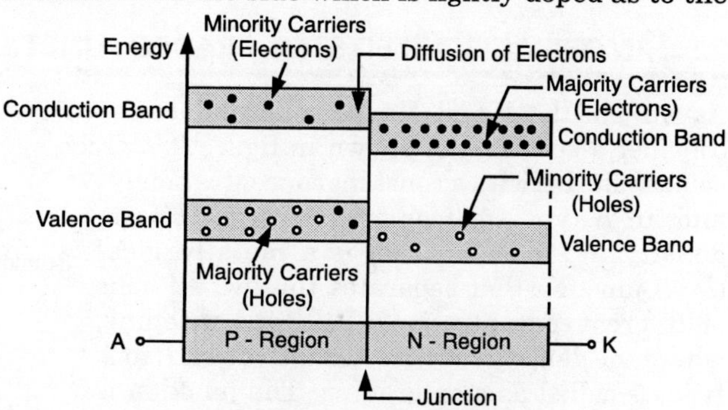

Fig. 2.7. Energy diagram of a P-N junction at the instant of junction formation.

diagram for a *P-N* junction at the instant of formation is shown in figure 2.7. The valence and conduction bands on the N-side are at lower energy levels than those on the *P*-side. However, the conduction bands on both sides and valence bands on both sides overlap to a considerable extent.

On P-side, the majority carriers *i.e.*, holes occupy the valence band whereas the thermally generated electrons (minority carriers) occupy the conduction band. Whereas on the N-side, the majority carriers (electrons) are in the conduction band whereas the minority carriers *i.e.*, holes occupy the valence band.

2.4.1. Energy Diagram of P-N Junction at Equilibrium

Figure 2.8 shows the energy diagram of a P-N junction when the junction reaches the equilibrium. The free electrons in the N-region which occupy the conduction band can diffuse across the junction and temporarily become free electrons in the lower part of the P-region conduction band. As soon as these electrons cross the junction, these electrons quickly loose energy and fall into the holes in the P-region valence band as shown in figure 2.7. Due to continued diffusion, the depletion region begins to form and the energy level of conduction band on the N-side decreases. This reduction in the energy level of the conduction band on N-side is because of the loss of higher energy electrons, which have diffused across the junction to the P-side. Very soon all the conduction band electrons on *N* side having enough energy to cross the junction will cross the junction. So, no such electrons will be left on N-side.

At this point, the junction is at equilibrium and the formation of depletion region is complete and the diffusion process will stop. At equilibrium, there exists an energy gradient across the depletion region as shown in figure 2.8. This energy gradient acts as an energy hill. The conduction band electrons of *N*-side will have to climb this hill to go to the P-side.

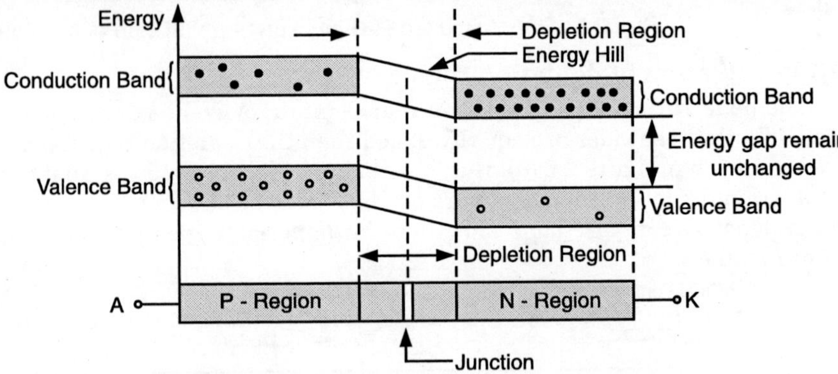

Fig. 2.8. Energy diagram at equilibrium.

Important Point : *It may be noted that the energy level of conduction band on N-side has shifted downwards as shown in figure 2.8. But the energy level of the valence band also has shifted down with the same amount. Hence, the energy gap between the conduction and valence bands will remain unchanged.*

2.5. AN OPEN CIRCUITED STEP-GRADED JUNCTION

As we join the *P* and *N*-type semiconductors, at the boundary a *P-N* junction is formed. Let us consider a special case shown in figure 2.9 which shows a *P-N* junction formed between a *P*-type semiconductor with a constant concentration N_A and an *N*-type semicondctor with a uniform density N_D. The junction is a metallurgical (*P-N*) junction that separates the two sections of different concentrations. This type of doping, where the density changes abruptly from *P* to *N* type, is called as step grading. The junction is located at the plane where the concentration is zero.

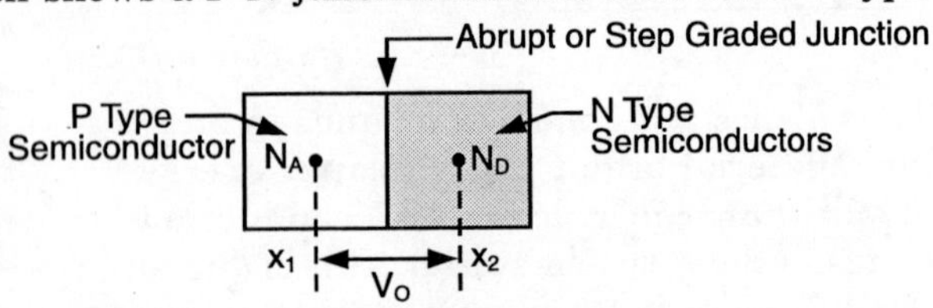

Fig. 2.9. An abrupt or step graded *P-N* junction.

(i) A step graded junction

When a junction is formed between a P-type and an N-type material, there is an abrupt change in the concentration of holes and electrons. Holes are the majority carriers on P-side and hence available in plenty on that side. But as we cross the junction and enter into the N-side, the holes will be minority carriers. The concentration of holes on the N-side will be low. Thus, the concentration of holes will suddenly decrease at the junction as we move from P-side to N-side. This is shown in figure 2.10. Similarly, the concentration of electrons is high on the N-side and reduces drastically at the junction as they become minority carriers on the P-side.

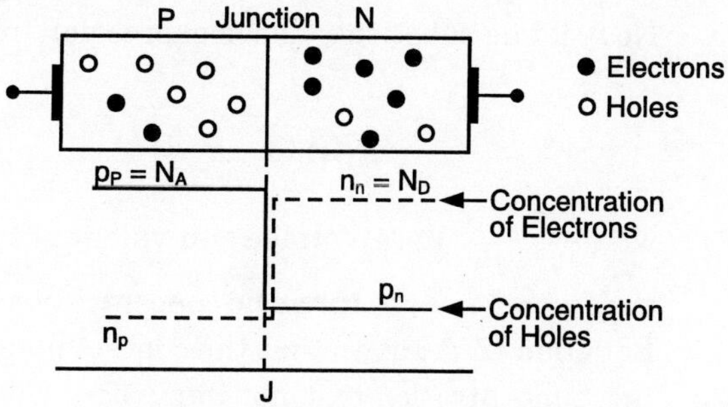

Fig. 2.10. A step graded P-N junction.

Because there is a sudden change or step change in the concentrations of electrons and holes at the junction, this P-N junction is known as the step graded or an abrupt P-N junction.

The concentration of holes on the P-side is equal to the concentration of the acceptor impurity atoms.

$$\therefore \qquad p_p = N_A \qquad \text{...on } p\text{-side} \qquad \text{...(2.1)}$$

The concentration of electrons on the P-side is denoted by n_p which is very low as compared to p_p.

$$\therefore \qquad p_p >> n_p \qquad \text{...(2.2)}$$

Similarly, the concentrations of electrons and holes on the N-side are denoted by n_n and p_n respectively and they are given by,

$$n_n = N_D \qquad \text{..on } N\text{-side} \qquad \text{...(2.3)}$$

$$\text{and} \qquad n_n >> p_n \qquad \text{..on } N\text{-side} \qquad \text{...(2.4)}$$

(ii) Contact difference potential or junction potential

In the previous section, we have proved that in a graded semiconductor, there exists an internal potential difference which is proportional to change in concentration. So, here due to the step or abrupt change in concentration at the junction, a potential gets developed which is called as contact difference potential or junction potential or barrier potential which is denoted by V_0 or V_j.

(iii) Expression for contact difference potential

Let us now obtain the expression for the barrier potential of the abrupt junction. We have

$$V_{21} = V_T \log_e \left[\frac{p_1}{p_2}\right] \qquad \text{...(2.5)}$$

At the junction, the hole concentration changes suddenly from p_p to p_n s we move from P to N side.

Substituting $p_1 = p_p = N_A$ and $p_2 = p_n$ in equation (2.5) we get,

$$V_j \text{ or } V_0 = V_T \log_e \left[\frac{N_A}{p_n}\right] \qquad \text{...(2.6)}$$

where V_j or V_0 is known as the potential barrier or junction potential or contact potential.

According to the law of mass action, we have

$$n_n \times p_n = n_i^2$$

Therefore,

$$p_n = \frac{n_i^2}{n_n}$$

But, $n_n = N_D$, so, we write

$$p_n = \frac{n_i^2}{N_D}$$

Now, let us substitute above expression into equation (2.6) to get,

$$V_j = V_T \log_e \left[\frac{N_A N_D}{n_i^2}\right] \text{ volts} \quad \text{...(2.7)}$$

where, V_T = Voltage equivalent of temperature = $\frac{T}{11.600}$

and n_i = Intrinsic concentration.

Equation (2.7) shows that the contact potential depends on the following four parameters

(*i*) Concentration of donor impurities N_D

(*ii*) Concentration of intrinsic impurities N_A

(*iii*) Intrinsic concentration n_i.

(*iv*) V_T

(*iv*) Conclusion

The important conclusions from equation (2.7) may be listed as under:

(*i*) The junction potential depends on the voltage equivalent of temperature V_T. This makes it dependent on temperature.

(*ii*) The junction potential is decided by the doping levels on the *P* and *N* sides (N_A and N_D).

(*iii*) The junction potential for Silicon diodes is 0.6 to 0.7 V while for Germanium diodes, it is 0.2 to 0.3 V.

EXAMPLE 2.1. Given a step graded Germanium *P-N* junction. It has $N_D = 10^3 N_A$ and N_A corresponds to 1 atom, per 10^8 Germanium atoms. Evaluate the junction potential. Assume: $n_i = 2.5 \times 10^{13}$ per cm³. Atom density of Ge = 4.4×10^{22} atoms per cm³.

Solution: Given that atom density of Ge = 4.4×10^{22} atoms/cm³

N_A corresponds to 1 atom per 10^8 Ge atoms.

Hence, corresponding to 4.4×10^{22} atoms of Ge, the number of atoms of acceptor impurity will be,

$$N_A = \frac{4.4 \times 10^{22}}{10^8} = 4.4 \times 10^{14}/\text{cm}^3$$

or $$N_A = 4.4 \times 10^{14} \times 10^6/\text{m}^3 = 4.4 \times 10^{20}/\text{m}^3 \quad \text{...(i)}$$

It is given that, $$N_D = 10^3 N_A$$

or $$N_D = 4.4 \times 10^{23}/\text{m}^3 \quad \text{...(ii)}$$

Intrinsic concentration, $$n_i = 2.5 \times 10^{13}/\text{cm}^3 = 2.5 \times 10^{19}/\text{m}^3 \quad \text{...(iii)}$$

The junction potential V_j is given by,

$$V_j = V_T \log_e \left[\frac{N_A N_D}{n_i^2}\right]$$

Assuming V_T = 26 mV, we have

$$V_j = 26 \times 10^{-3} \log_e \left[\frac{4.4 \times 10^{20} \times 4.4 \times 10^{23}}{(2.5 \times 10^{19})^2}\right] = 0.3287 \text{ Volts or } 328.7 \text{ mV}$$ **Ans.**

To accomplish great things, we must not only act, but also dream: not only plan, but also believe.
— **Anatole France**

2.6. CURRENTS IN A P-N JUNCTION WITH NO EXTERNAL BATTERY

When a PN-junction is formed, a depletion region is established across the junction. Because of this a barrier potential exists across the junction which stops the further diffusion of holes and electrons across the junction. Few majority carriers have sufficiently high kinetic energy to balance the barrier potential and cross the junction. Hence due to majority carriers small current flows which is called majority carrier current.

> **DO YOU KNOW?**
>
> When the P-N junction is formed, the depletion region gets created and the movement of electrons and holes stops. Thus, the current flowing through an unbiased P-N junction is zero.

There is another current which flows across the junction. This current is due to the minority carriers. Since, there are some minority carriers in the P-region and in the N-region, they are accelerated across the junction by the barrier potential. This current due to minority carriers is called minority carrier current.

But the majority carrier current flows in a direction opposite to the direction of minority carrier current. Hence flow of minority carriers across the junction is balanced by the flow of some majority carriers. Therefore, there is no net flow of current across the junction when no external battery is applied.

2.7. BIASING OF A P-N JUNCTION

When we apply a battery across the PN-junction, this process is called biasing of a PN-junction. By applying external voltage source across the PN-junction, the width of the depletion region can be controlled. By controlling the width of depletion region, P-N junction finds a large number of applications in the field of electronics.

The P-N junction can be biased in two ways:

1. Forward-Bias
2. Reverse-Bias

2.7.1. Forward-Biasing of P-N Junction

(Important)

In forward-biasing of a P-N junction, positive terminal of the battery is connected to the P-side and the negative terminal to the N-side as shown in the figure 2.11. In this case, holes are repelled from the positive terminal of the battery and forced towards the junction. Similarly electrons are repelled from the negative terminal of the battery and forced towards the junction. Because of this increased energy, some holes and electrons enter the depletion region. This reduces the potential barrier. Therefore, width of depletion region also reduces. As a result of this, more majority carriers diffuse across the junction. Hence, this causes a large current to flow across the junction.

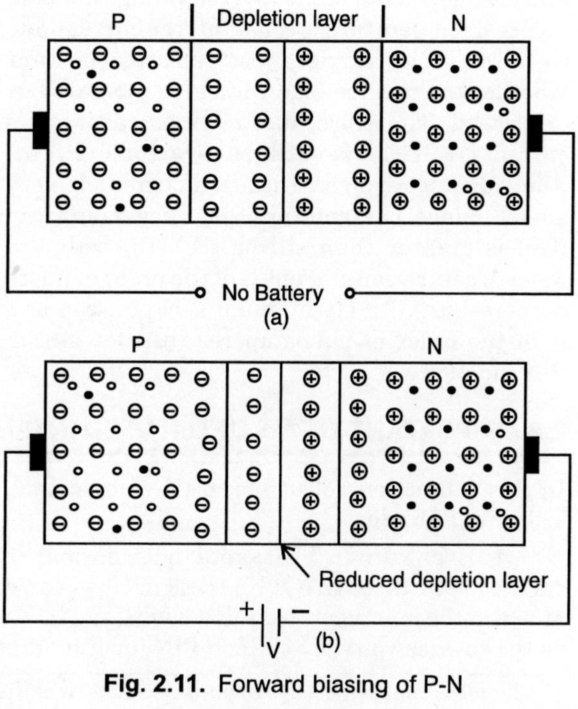

Fig. 2.11. Forward biasing of P-N

For each recombination of free electron and hole, an electron from the negative terminal of the battery enters the N-type region. This then moves towards the junction. Similarly, in the P-type region near the positive terminal of the battery, an electron breaks a bond in the crystal structure and enters the positive terminal of the battery. For each electron which breaks its bond, a hole is created. This hole moves towards the junction. Therefore, the current through the external circuit is due to the movement of electrons only. On the other hand, the current in the P-type region is due to the movement of holes and the current in the N-type region is due to the movement of electrons. Hence, there is a large current in the external circuit as long as the external battery is present in the circuit.

2.7.2. Reverse Biasing of P-N Junction

In reverse biasing of a PN-junction, a battery is connected across the P-N junction in such a way that its positive terminal is connected to the N-region and negative terminal is connected to the P-region.

In reverse bias the holes in the P-region are attracted towards the negative terminal of the battery whereas the electrons are attracted towards the positive terminal of the battery. In this way majority charge carriers are drawn away from the PN-junction. Due to this, the depletion region becomes wider. This increases the barrier potential.

Due to the increased barrier potential, the majority charge carriers are not able to cross the junction. Hence, in reverse bias, there is no current due to the majority charge carriers.

But there are few thermally generated minority carriers in both the regions. The increased barrier potential enhances the flow of minority carriers across the junction. As a minority carrier is generated, it is drifted towards the P-N junction due to the barrier potenial. Therefore, a very small amount of current flows through the P-N junction. The generation of minority carriers depends upon the temperature and is independent of the reverse voltage applied. Hence if temperature is constant, the current due to the minority carriers also remains constant whether the reverse bias voltage is increased or decreased. Therefore, due to this reason, this current is called as **reverse saturation current**. The value of reverse saturation current is very small. Since the conductivity of germanium (Ge) is greater than silicon (Si), the value of reverse saturation current is of the order of micro amperes (μA) for Germanium P-N junction and is of the order of nanoamperes (nA) for silicon P-N junction.

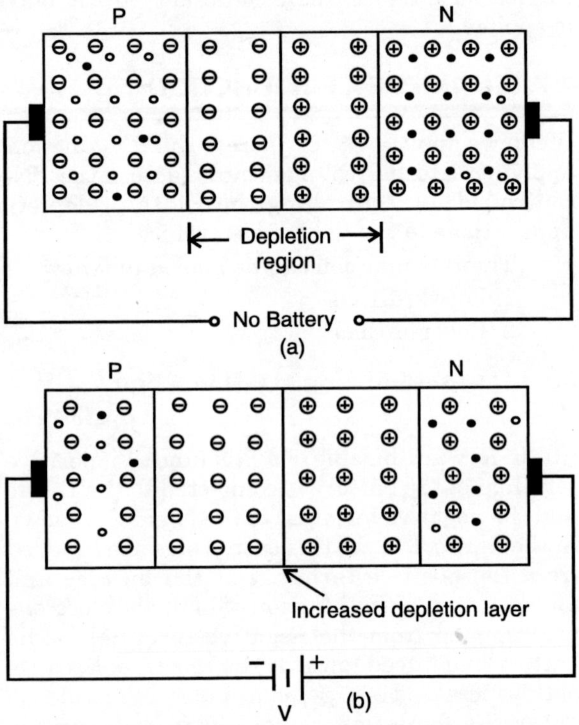

Fig. 2.12. Reverse biasing of PN-junction

2.8. IMPORTANT TERMS OFTEN USED IN P-N JUNCTION

In this subsection, let us discuss few important terms which are generally used for P-N junction. They are as under:

1. Breakdown Voltage. Under normal reverse bias voltage, a very little amount of reverse current flows through a P-N junction. However, on increasing the reverse voltage, a point may reach at which the junction breaks down with sudden rise in reverse current. Breakdown voltage is defined as the reverse voltage at which P-N junction breaks down with sudden rise in reverse current.

* The force that rules the world is conduct, whether it be a moral or immoral. —**Nichola Murray**

2. Knee Voltage. As explained earlier, with forward bias to P-N junction a very little current flows until the forward voltage exceeds the junction barrier potential (0.3 V in case of Ge and 0.7 V in case of Si). The forward voltage, at which the current through the junction starts increasing rapidly, is known as the ***knee voltage*** or the ***cut-in-voltage.***

3. Maximum Forward Current. It is the highest instantaneous forward current that a P-N junction can conduct without damage to the junction. In case the forward current exceeds this rating, the junction will get destroyed due to over-heating.

4. Peak Inverse Voltage. It is the maximum reverse voltage that can be applied to the P-N junction without damage to the junction. If the reverse voltage across the junction exceeds its peak inverse voltage (PIV), the junction may get destroyed owing to excessive heat.

5. Maximum Power Rating. It is the maximum power that can be dissipated at the junction without any damage to it.

It may also be noted that every P-N junction has limiting values of maximum forward current, peak inverse voltage and maximum power rating. The P-N junction gives satisfactory performance when operated within these limiting values. The P-N junction may get destroyed due to excessive heat if any of these values is exceeded.

2.9. P-N JUNCTION DIODE

A P-N junction is called as P-N junction diode or a semiconductor diode. This is also called as crystal diode because it is formed in a crystal (Silicon or germanium). A P-N junction diode is an unidirectional two terminals semiconductor device. It conducts only in one direction and offers high resistance in other direction.

2.10. CIRCUIT SYMBOL OF A DIODE

When diode is used as a circuit element, it is convenient to represent it by a symbol.

The two terminals of diode are called anode and cathode. The P-type region is called the anode and the N-type region is called the cathode. The arrowhead in the symbol points from the P-region towards the N-region. This shows that the direction of conventional current is from the P-region to the N-region which is also the direction of flow of holes.

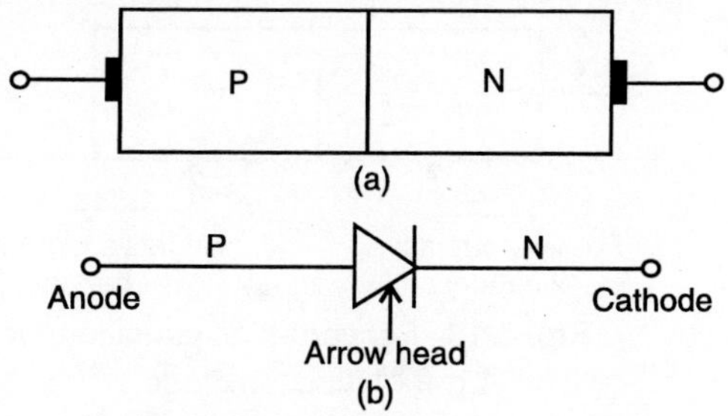

Fig. 2.13. Circuit symbol of diode.

2.11. VOLT AMPERE (V-I) CHARACTERISTICS OF P-N JUNCTION DIODE*

If diode is used as a circuit element in an electrical circuit, there may be a requirement to know how it responds or behaves in electric circuit. This type of information may be obtained by a curve known as Volt-Ampere (V-I) characteristics of a practical diode.

The V-I characteristics of a diode is simply a curve or graph between the voltage applied across its terminals and the current that flows through the diode due to this applied voltage. The entire V-I characteristics may be divided into two parts namely:

1. Forward characteristics
2. Reverse characteristics

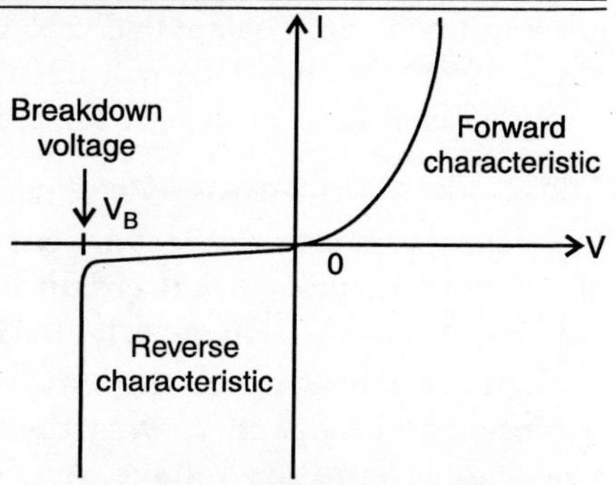

Fig. 2.14. Entire V-I characteristics of diode

* Explain the behaviour of p-n junction at no bias, reverse bias and forward bias. Sketch the V-I characteristics of p-n junction diode. ***(JNTU, Hyderabad, Sem. Exam., 2002-03)***

2.11.1. Forward Characteristic

The circuit arrangement shown in figure 2.15 is used in the laboratory to obtain forward characteristic of a diode. The diode in the circuit is in forward bias state because the battery V_{FF} is pushing the current in the direction of the arrowhead.

In the circuit, the battery V_{FF} is connected to the diode through potentiometer P. The potentiometer helps in varying the voltage across the diode. The load resistance R_L is used in the circuit to limit the current through the diode. A voltmeter is connected across the diode to measure the voltage. A milliammeter is also connected to measure the current in the circuit.

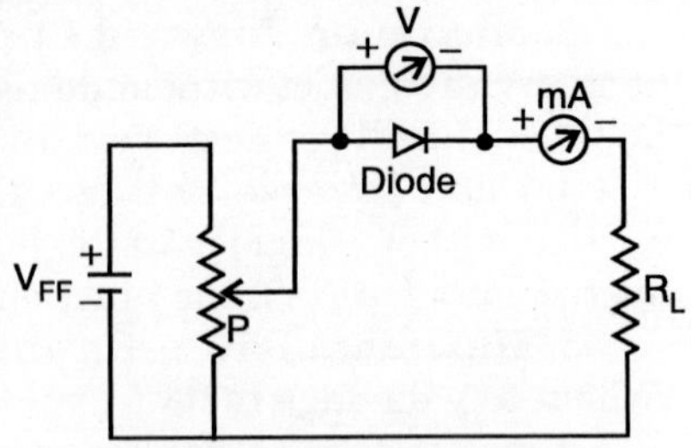

Fig. 2.15. Circuit arrangement for forward bias

To plot the characteristic, the voltage across the diode is increased gradually and the corresponding increase in the current is noted. The graph is plotted by taking voltage on the horizontal axis and current on the vertical axis.

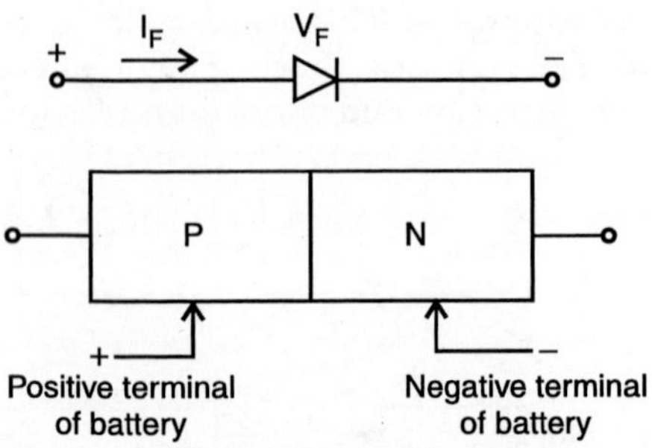

Fig. 2.16. Forward-bias condition for a P-N junction diode.

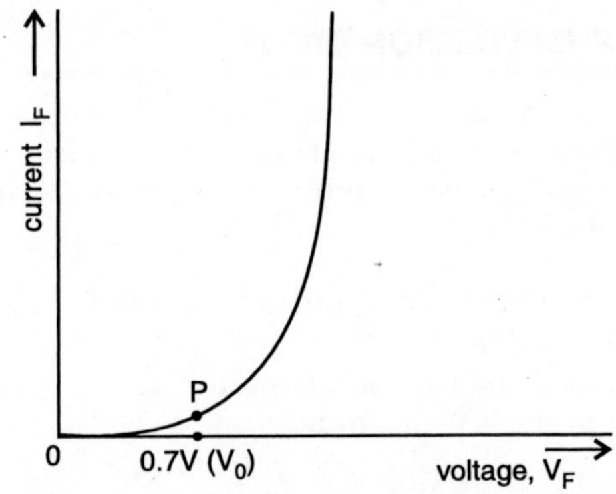

Fig. 2.17. Forward characteristics (Si diode)

A carefull look at the forward characteristic reveals the fact that upto point P, the diode current is very small. This is so because the applied voltage has to overcome the barrier potential and the diode conducts poorly. Once the applied voltage is slightly greater than the barrier potential, the diode current increases rapidly and diode conducts heavily. This voltage at which the current starts increasing is called the **knee voltage (V_0).** Its value is 0.7 V for a silicon diode and 0.3 V for a germanium diode. The applied voltage across the diode should not increase beyond a specified safe value otherwise the diode will burn out due to heat. This specified safe value is provided by the manufacturer for each type of the diode.

2.11.2. Reverse Characteristic

The circuit arrangement in figure 2.18(*a*) is used in the laboratory to obtain reverse characteristic of a diode. The diode in the circuit is in reverse bias state because the battery V_{RR} is pushing the current in the direction opposite to the arrowhead.

In the above circuit, the battery V_{RR} is connected to the diode through potentiometer P. The potentiometer helps in varying the voltage across the diode. A voltmeter is connected across the diode to measure the voltage. A microammeter is also connected to measure the current in the circuit.

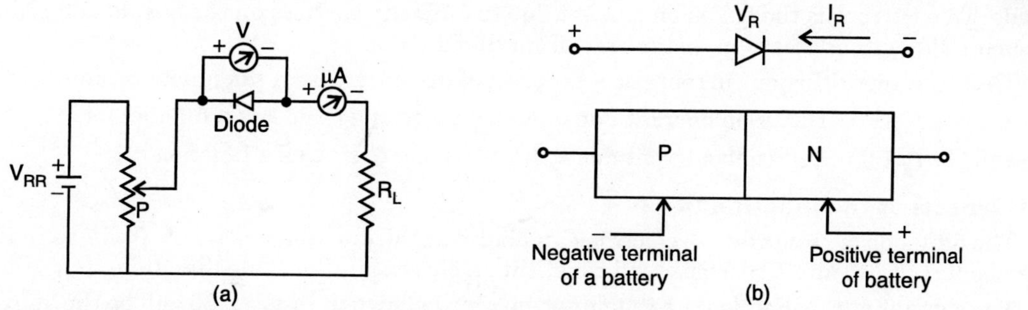

Fig. 2.18. (a) Circuit arrangement for Reverse bias (b) Reverse-bias condition for a P-N junction diode

A careful look at the reverse characteristic reveals the fact that below the breakdown voltage the diode current is very small and almost remains constant. This current is also called **reverse saturation current I_o.** It is of the order of μA for Germanium diode and nanoamperes for Silicon diode. It is due to the movement of minority carriers which are thermally generated. It means that reverse bias current is temperature dependent and does not depend upon the applied reverse bias voltage.

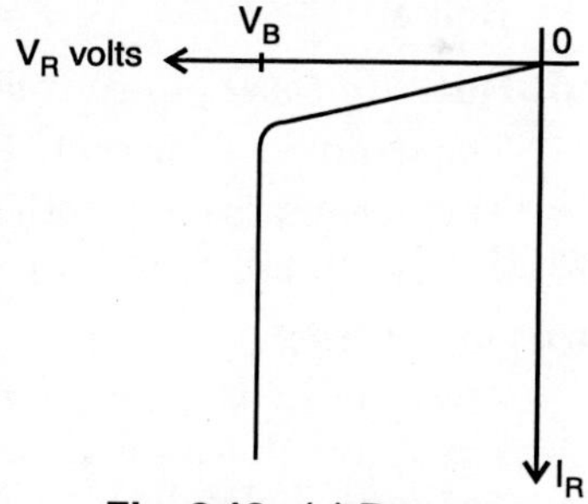

Fig. 2.18. (c) Reverse characteristics

But, if a large reverse bias voltage is applied, a process known as **junction breakdown** occurs. Due to this, the diode reverse current increases rapidly. The applied reverse bias voltage at which this occurs is called as **breakdown voltage (V_B)** of a diode.

2.12. THE CURRENT COMPONENTS IN A P-N JUNCTION DIODE*

(JNTU, Hyderabad, Sem. Exam. 2001-02, 2002-03) (05 marks)

Figure 2.19 shows a forward biased P-N junction diode. The current I circulating in the circuit is the forward current of the diode. This current flows due to the majority carriers.

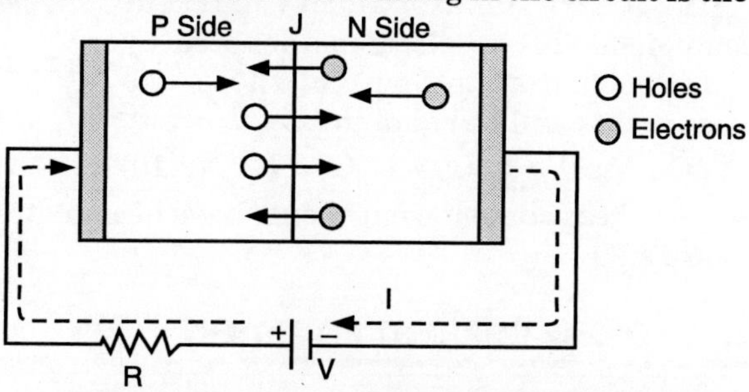

Fig. 2.19. A forward biased P-N junction.

Diffusion Currents on the *P*-and *N*-side

Under the forward biased condition, the depletion region becomes extremely narrow and holes from *P* side diffuse into the *N* side and the electrons from the *N* side diffuse into the *P* side as shown in figure 2.19. The current carried by these diffused electrons, which become minority carriers once they enter into the *P*-side, is denoted by I_{np} where *n* stands for electrons and *p* for the *P*-side.

This current is the diffusion current on *P*-side and it decreases exponentially with increase in distance (x) from the junction. Similarly the current carried by the diffused holes which become minority carriers once they enter into the *N*-side is denoted by I_{pn} where *p* stands for holes and *n* for the

DO YOU KNOW?

The forward current through a P-N junction diode flows due to the majority carriers and its direction of flow (conventional) is always from anode to cathode.

* Discuss the current flow mechanism in a p-n junction under:
(*i*) no bias (*ii*) forward bias (*ii*) reverse bias conditions ***(JNTU, Hyderabad, Sem. Exam., 2006-07)***

N-side. This current is the diffusion current due to minority carriers on the N-side and it decreases exponentially with increase in distance (x) from the junction.

Thus, the two diffusion currents as a function of distance x from the junction can be defined as:

$I_{np}(x)$ = Diffusion current due to electrons on the P-side as a function of x.

and $I_{pn}(x)$ = Diffusion current due to holes on the N-side as a function of x.

(*i*) Currents at the junction (x = 0)

The diffusion currents due to electrons and holes at the junction *i.e.*, for $x = 0$ will be in the same direction *i.e.*, $I_{np}(0)$ and $I_{pn}(0)$ both will constitute a current in the same direction.

The current at the junction, i.e., the total current I shown in figure 2.20 will be the sum of these two diffusion currents $I_{np}(0)$ and $I_{pn}(0)$.

Hence, $$I = I_{np}(0) + I_{pn}(0) \quad \text{... (2.8)}$$

(*ii*) But, the total current should remain constant

These diffusion currents $I_{np}(x)$ and $I_{pn}(x)$ decrease exponentially with increase in x.

This means that on both the sides, there must be some other components of current present which can actually maintain the current I constant.

(*iii*) On *P*-side

Along with the minority diffusion current $I_{np}(x)$ there must be a majority current due to holes denoted by $I_{pp}(x)$ such that the sum of these two components will be equal to I.

On P-side : $I_{np}(x) + I_{pp}(x) = I$ ... (2.9)

(*iv*)) On the *N*-side

Similarly, on the N-side along with the minority diffusion current due to holes *i.e.*, $I_{pn}(x)$, there is a majority current component due to electrons denoted by $I_{nn}(x)$ such that the sum of these two components will be equal to I.

On N-side : $I_{pn}(x) + I_{nn}(x) = I$ ... (2.10)

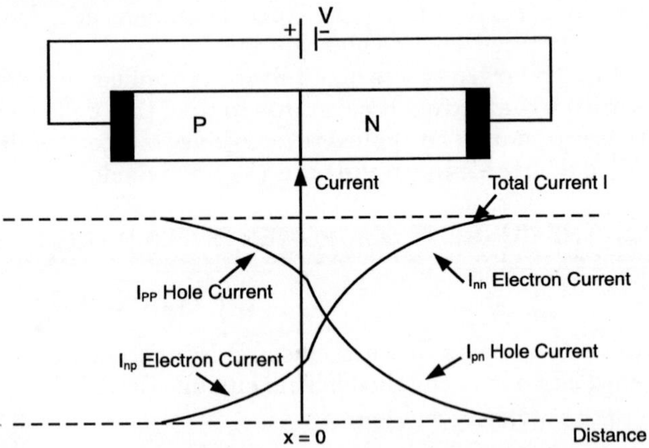

Fig. 2.20. Various current components.

All these current components have been plotted as a function of distance (x) from the junction in figure 2.20.

2.13. DIODE CURRENT EQUATION

With the help of solid state physics, the diode current equation, relating the voltage V and current I for the forward-and reverse-bias regions, can be given by

$$I = I_o\left[e^{V/\eta V_T} - 1\right] \quad \text{...(2.11)}$$

Here, I = diode current

I_o = diode reverse saturation current at room temperature

V = external voltage applied to the diode

η = a constant

It is 1 for germanium and 2 for silicon

and $$V_T = \frac{kT}{q} = \frac{T}{11,600} \text{ Volt equivalent of temperature.} \quad \text{...(2.12)}$$

where k = Boltzmann's constant = 1.38066×10^{-23} J/Kelvin

q = Electronic Charge = 1.6×10^{-19} Coulomb

T = Diode junction temperature in (°K)

Now, at room temperature, (T = 300°K), V_T = 26 mV

Putting these two values in current equation, we get

$$I = I_0[e^{40V/\eta} - 1] \qquad ...(2.13)$$

For germanium diode, $\eta = 1$

Therefore, $$I = I_0[e^{40V} - 1] \qquad ...(2.14)$$

and for silicon diode, $\eta = 2$

Therefore, $$I = I_0[e^{20V} - 1] \qquad ...(2.15)$$

If the value of applied voltage is greater than unity then the diode current equation for germanium will be

$$I = I_0[e^{40V}]$$

and for silicon, $$I = I_0[e^{40V}] \qquad ...(2.16)$$

If the diode is reverse biased, the current equation can be obtained by reversing the sign of the applied voltage V.

Hence, diode current with reverse bias is

$$I = I_0\,[e^{(-V/\eta V_T)} - 1] \qquad ...(2.17)$$

Now, if $V >> V_T$

then the term $e^{(-V/\eta V_T)} << 1$

Therefore, $$I \cong -I_0 \qquad ...(2.18)$$

This equation is valid as long as the external voltage is below the breakdown value.

EXAMPLE 2.2 The reverse saturation current at room temperature is 0.3 μA when a reverse bias is applied to a germanium diode. Find the value of current flowing in the diode when 0.15 V forward bias is applied at room temperature.

Solution: Given that $I_o = 0.3\ \mu A = 0.3 \times 10^{-6}$ A

and forward voltage, V_F = 0.15 Volt. The current flowing through the diode under forward bias is given by

$$I = I_o\left(e^{40V_F} - 1\right) = 0.3 \times 10^{-6}\,(e^{40 \times 0.15} - 1)$$

$$I = 120.73\ \mu A \qquad \textbf{Ans.}$$

2.14. COMPLETE V-I CHARACTERISTICS OF SILICON AND GERMANIUM DIODES

The complete V-I characteristics is the combination of forward and reverse characteristics. The typical V-I characteristics of Silicon and Germanium diodes are as shown in figure 2.21.

* The happiness of every country depends upon the character of its people rather than the form of its government.
— **Thomas C. Haliburton**

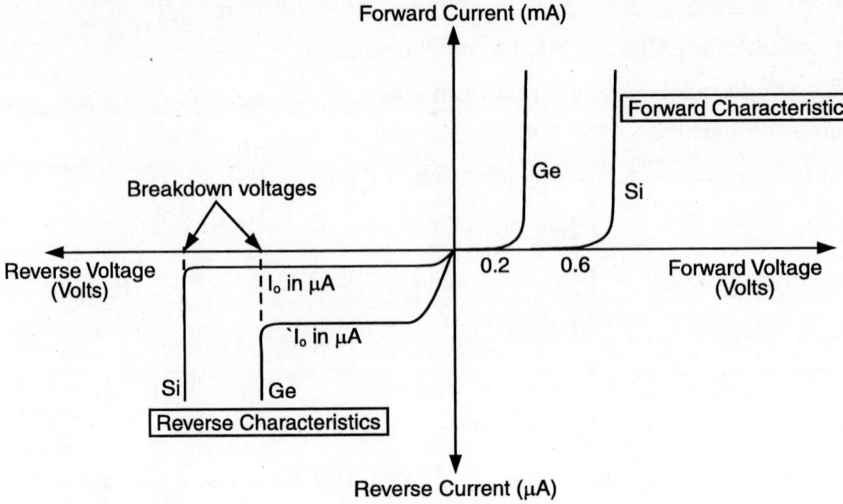

Fig. 2.21. V-I Characteristics of Silicon and Germanium diodes.

Conclusions

We can draw following conclusions from V-I characteristics of figure 2.21:

(*i*) Cut-in voltages for Silicon and Germanium diodes are 0.6 and 0.2 V respectively.

(*ii*) Breakdown voltage of Silicon diode is higher than that of the Germanium diode. Therefore, Silicon diodes can withstand to a higher reverse voltage.

(*iii*) The reverse saturation current I_0 for a Germanium diode is few μA whereas that for a Silicon diode, it is in nA at room temperature.

2.15. COMPARISON OF SILICON AND GERMANIUM DIODES

S.No.	Parameter of comparison	Silicon Diode	Ge Diode
1.	Material used	Silicon	Germanium
2.	Cut-in voltage	0.6 V	0.2 V
3.	Reverse saturation current	In nanoamp	In microamp
4.	Effect of temperature	Less	More
5.	Breakdown voltage	Higher	Lower
6.	Applications	Rectifiers, clippers clampers, etc.	Low voltage Low temperature applications

2.16. TEMPERATURE DEPENDENCE OF V-I CHARACTERISTICS : EFFECT OF TEMPERATURE ON THE V-I CHARACTERISTICS

We know that the expression for diode current is given by

$$I_D = I_0 \left[e^{V/\eta V_T} - 1\right]$$

where I_0 = Reverse saturation current

$$V_T = \frac{T}{11600}$$

η = 1 for G_e diode = 2 for S_i diode

V = diode voltage

The diode characteristics is mathematically expressed by the equation of I_D.

Two parameters I_0 and V_T are temperature dependent. Hence, the characteristics is dependent on the temperature. The effect of change in temperature on the V-I characteristics are as shown in figure 2.22.

DO YOU KNOW?

For any diode, the forward voltage, V_F, decreases as the temperature of the diode increases. As a rough approxima-tion, V_F decreases by 2 mV for each degree Celsius rise in temperature.

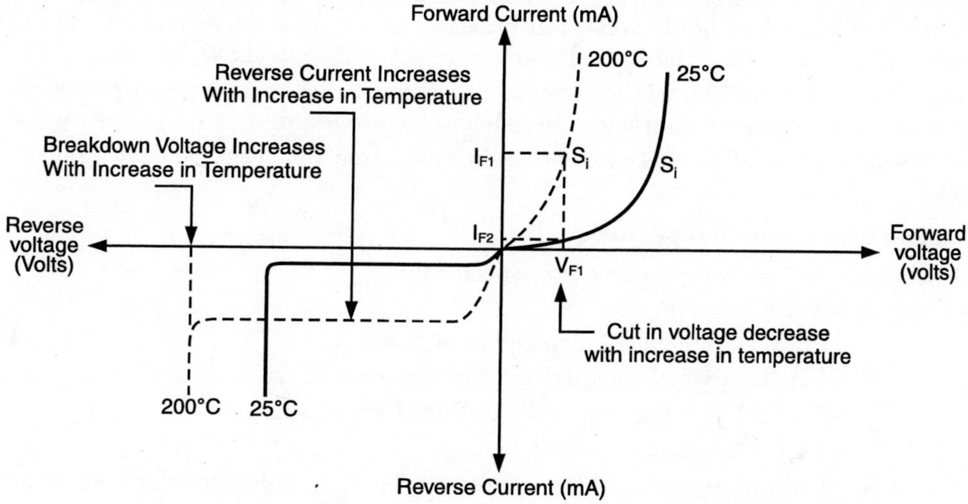

Fig. 2.22. Effects of temperature on V-I characteristics of a silicon diode.

Conclusions

As Observed from figure 2.22, the effects of temperature may be listed as under:

(*i*) Reduction in the cut-in voltage takes place with increase in temperature. Therefore, at the same forward voltage V_{F1}, a larger current I_{F1} flows through the diode at increased temperature.

(*ii*) The breakdown voltage increases with increase in temperature.

(*iii*) Reverse saturation current increases with increase in temperature.

Important Point: *In the expression for diode current, there are two factors which depend on the temperature. They are I_0 and V_T. Hence, the diode characteristics becomes the function of temperature.*

2.16.1. Germanium Diodes and Effect of Temperature

As a matter of fact, junction diodes fabricated from germanium are also available. The principle of operation is same as that of the silicon diodes and the expression for volt-ampere characteristic is also same as that of the silicon diodes. The value of η = 1 for germanium (G_e). The reverse saturation current of the G_e diodes is three to four times higher than the silicon diodes of same dimensions and doping densities. The cut-in voltage V_γ = 0.2 V for the G_e diodes.

2.16.2. G_e Diodes Produce Higher Reverse Saturation Current

We know that the four valence electrons of G_e are in the fourth shell whereas those of a Si atom are in the third shell. Hence, the force of attraction between the nucleus and valence electrons is weak in the generation atoms than that in the silicon atoms. The forbidden energy gap is smaller in G_e than Si atoms. Therefore, at the same temperature, more valence electrons will jump to the conduction band to produce higher reverse saturation current in case of G_e diode.

2.16.3. S_i Diode is More Popular than G_e Diode *(V. Important)*

Following are the reasons for the popularity of silicon (S_i) diode:

(*i*) The reverse saturation current for a Silicon diode is much lower than that of a G_e diode. Therefore, even with the two fold increase in I_0 after every 10°C rise in temperature, the reverse saturation current through Silicon diodes will still remain very low. But, at increased temperature, the reverse saturation current through a G_e diode is very high. In fact, it is of the order of 100 μA or so. However, in practice, this level of reverse saturation current is unacceptable. Therefore, the S_i diodes are more popular than the G_e diodes.

(*ii*) Ideally the diode characteristics should not change due to change in temperature and practically the change in characteristics should be minimum.

(*iii*) The characteristics of G_e diode are more dependant on temperature than that of a silicon diode.

2.16.4. Dependence of I_0 on Temperature

The dependence of I_0 on temperature is expressed by

$$I_0 = K\,T^m\, e^{-V_{GO}/\eta V_T} \qquad \text{... (2.19)}$$

where K = Constant which is independent of temperature.

m = Constant which is equal to 2 for Germanium and 1.5 for Silicon

V_{GO} = Forbidden energy gap = 0.785 V for Germanium

= 1.21 V for Silicon

Practically we need not always use the expression for I_0 to calculate the reverse saturation current at a new temperature. As we know that there is a 7% rise in I_0 for every 1°C rise in temperature, we can write this as under:

$$I_{01} = (1.07)^{\Delta T} I_{02} \qquad \text{... (2.20)}$$

where I_{01} = Reverse saturation current at temperature T_1

I_{02} = Reverse saturation current at temperature T_2

and $\Delta T = |T_2 - T_1|$

2.16.5. Conclusions

Some important conclusions related to the reverse saturation current may be listed as under:

(*i*) The reverse saturation current increases at a rate of 7% for every 1°C rise in temperature. In other words, I_0 doubles its value for every 10°C rise in temperature.

(*ii*) Reverse saturation current for silicon diode is lower than that for a germanium diode.

EXAMPLE 2.3 The forward current through a Silicon diode is 10 mA at room temperature (27°C). The corresponding forward voltage is 0.75 volts. Calculate the reverse saturation current I_0.

Solution: Given that

$$I_F = 10\text{ mA},\quad V_F = 0.75\text{ Volt},\quad T = 27°\text{C} = 300°\text{K},\quad \eta = 2 \text{ for a Silicon diode}$$

We know that

$$V_T = \frac{T}{11{,}600} = \frac{300}{11{,}600}$$

or

$$V_T = 26\text{ mV} \qquad \text{... (i)}$$

The forward current through a diode is given by

$$I_F = (I_0\, e^{V_F/\eta V_T} - 1) \qquad \text{... (ii)}$$

Substituting all the values, we get

$$10 \times 10^{-3} = I_0 \, |e^{0.75/2 \times 26 \times 10-3} - 1|$$

or $\qquad I_0 = 5.446 \text{ nA}.$ **Ans.**

The reverse saturation current for silicon diode = 5.446 nA.

EXAMPLE 2.4 Explain the effect of temperature on a P-N junction diode. Determine the forward current for a Germanium diode at room temperature when the voltage across it is 0.3 V. Assume room temperature is 27°C and reverse saturation current is 1 nA. Volt-equivalent of temperature is $V_T = \dfrac{T}{11{,}600}$.

Solution: Given that $\qquad I_0 = 1 \text{ nA} = 1 \times 10^{-9} \text{ A at } T = 27° \text{ C} = 300° \text{ K}$

We know that

$$V_T = \frac{T}{11{,}600} = \frac{300}{11{,}600} = 25.86 \text{ mV},$$

$$V_F = 0.3 \text{ V}$$

For G_e diode $\qquad \eta = 1$

The forward current is given by

$$I_F = I_0 \left[e^{(V_F/\eta V_T)} - 1\right]$$

or

$$I_F = 1 \times 10^{-9} \, [e^{300/25.86} - 1]$$

or

$$I_F = 1.091 \times 10^{-4} \text{ A or } 0.1091 \text{ mA}.$$ **Ans.**

EXAMPLE 2.5 A diode operating at 300°K has V (forward) of 0.4 V across it when the current through it is 10 mA and 0.42 V when the current is twice as large. What values of I_0 and η allow the diode to be modelled by the diode equation ?

Solution: Given that $\qquad T = 300° \text{ K}, \quad V_1 = 0.4 \text{ V}, \quad V_2 = 0.42 \text{ V}, \quad I_1 = 10 \text{ mA}, \quad I_2 = 20 \text{ mA}$

At $T = 300°\text{K}$, we have

$$V_T = \frac{T}{11{,}600} = \frac{300}{11{,}600} = 26 \text{ mV} \qquad \text{... }(i)$$

First, let us calculate the value of η.

From the given data we can write two equations of current for different values of V as under:

$$I_1 = I_0 \left[e^{V_1/\eta V_T} - 1\right]$$

and

$$I_2 = I_0 \left[e^{V_2/\eta V_T} - 1\right]$$

Substituting the given values, we get

$$10 \text{ mA} = I_0 \left[e^{0.4/\eta \times 26 \times 10^{-3}} - 1\right] \qquad \text{... }(ii)$$

$$20 \text{ mA} = I_0 \left[e^{0.42/\eta \times 26 \times 10^{-3}} - 1\right] \qquad \text{... }(iii)$$

Taking the ratio of equations (*iii*) and (*ii*), we obtain

$$\frac{20}{10} = \frac{I_0 \left[e^{0.42/\eta \times 26 \times 10^{-3}} - 1\right]}{I_0 \left[e^{0.4/\eta \times 26 \times 10^{-3}} - 1\right]}$$

Neglecting – 1 term, we have

$$2 = \frac{\left[e^{0.42/\eta \times 26 \times 10^{-3}}\right]}{\left[e^{0.4/\eta \times 26 \times 10^{-3}}\right]}$$

Taking log on both the sides, we obtain

$$0.693 = \frac{16.15}{\eta} - \frac{15.38}{\eta}$$

or $$0.693\,\eta = 0.77$$

or $$\eta = 1.11$$

Now, substituting this value of η into equation (*ii*), we get

$$10 \times 10^{-3} = I_0 \left[e^{0.4/1.11 \times 26 \times 10^{-3}} - 1\right]$$

Solving the equation, we get

$$I_0 = 9.69 \times 10^{-9} \text{ Amp} \quad \text{or} \quad 9.69 \text{ nA.}$$ **Ans.**

2.17. POWER AND CURRENT RATINGS OF A DIODE

Any diode having certain current, dissipates some amount of power, the value of which is expressed as a product of voltage and the current through it.

Mathematically, for a forward-bias diode, the power dissipation is

$$P_D = V_F \times I_F$$

Here V_F = Forward voltage drop

I_F = Forward current

For reverse-bias diode, the power dissipation is

$$P_D = V_R \times I_R$$

Here V_R = Reverse-bias voltage drop

I_R = Reverse-bias current

> **DO YOU KNOW?**
>
> As with conventional diodes, the manufacturer places a band on the cathode end of the zener diode for terminal identification.

The maximum value of power which a diode can withstand without failure, is known as the **power-rating** of diode. This means that the power dissipation must not exceed power rating in any condition. If this occurs then the diode will destroy. The manufacturers provide power rating of a diode on data sheets.

As an example, the power rating of diode IN 914 is 250 mW. This indicates that the diode IN 914 may dissipate a maximum power of 250 mW.

More often, the manufacturers provide the maximum current, which a diode can handle safely. It is called current rating rather than power rating. It is so since it is easy to measure the current rating than power rating.

As an example, for diode IN 4003, the data sheet does not show power rating, rather it shows a current rating of 1 ampere.

From above, it is evident that the data sheets provide two categories of diodes, small signal diodes and rectifiers or power diodes. Actually the small signal diodes are those diodes which have a power rating below 0.5 watt whereas the rectifier or power diodes have a power rating above 0.5 W.

* Change is not made without inconvenience even from worse to better. —**Richard Hooker**

Therefore, the diode IN 914 is a small signal diode since its power rating is 0.25 Watt whereas IN 4003 is a rectifier diode since its power rating is 1 Watt.

2.18. DIODE SPECIFICATION SHEETS

All manufacturers of diode provide a data sheet for a specific diode. This data sheet enables the users to make proper utilization of the device. The data sheet provides the information arranged in graphs, tables etc. This data sheet generally has data for the followings :

1. Forward voltage, V_F at any specific current and temperature.
2. Reverse saturation current I_R at any specific voltage and temperature.
3. Maximum forward current $I_{F\,max}$ at a specific temperature.
4. Reverse voltage rating PIV at a specific temperature.
5. Maximum power dissipation at a particular temperature.
6. Reverse recovery time, t_{rr}.
7. Capacitance values.
8. Operating temperature range.

DO YOU KNOW?

Internet search engines, such as Google, can quickly help locate semiconductor specifications.

In addition to this, depending upon the type of a particular diode more data like frequency range, switching time, noise level and thermal resistance may also be provided. Further, if maximum power rating is also provided, it is equal to

$$P_{D\,max} = V_D . I_D$$

Here V_D and I_D are the diode voltage and current at a particular operating point.

2.19. SOME COMMONLY USED DIODES

In electronic circuits, a large number of semiconductor diodes with different ratings are used according to the requirement. These diodes have different sizes, shapes and colours. Figure 2.23 shows such few important semiconductor diodes.

Figure 2.23(*a*) shows a BY 127 diode which can carry maximum 1 A forward current with a PIV of 1000 V. The colour of its body is green and the direction in which it can conduct is marked by a symbol printed on its body as shown in figure. Another diode in the same series is BY 118.

Figure 2.23(*b*) shows a IN 4007 diode. This diode can carry 1 A maximum forward current with a PIV of 1000 V. The colour of its body is black. On one side of the diode, a silver colour band is printed which shows the negattive end (*i.e.* cathode) of the diode. Other diodes in the same series are IN 4001, IN 4002, IN 4003, IN 4004 etc. There is one other diode of the same series IN 5406 which can carry 6 A current with a PIV of 200 V.

DO YOU KNOW?

The part number for a diode will always begin with the prefix-1N, where the number 1 indicates one junction and the lelter N indicates that the device is a semiconductor diode.

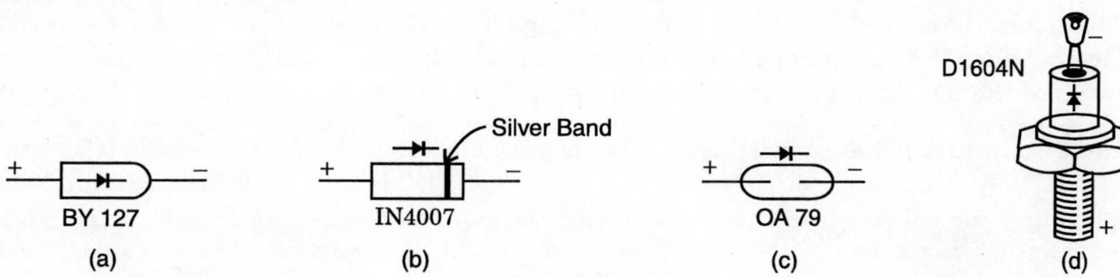

Fig. 2.23.

Figure 2.23(*c*) shows an OA 79 diode. The body of this is made of transparent glass. There is a red mark on its body which denotes the positive terminal. In the same series, the other diodes are OA 80, OA 85 etc.

Figure 2.23(*d*) shows a power diode D 1604 N. This diode has metallic body and can handle large amount of power. It can carry 16A maximum current with a PIV of 400 V. Its polarity is marked in figure. There is another power diode 10 KLR 12 which can carry 10 A maximum current with a PIV of 1200 V. Similarly, there are several other diodes of different ratings. The rating of a diode may be determined from the manufacturer's data book.

According to the ratings, a particular diode is placed in a given circuit.

2.20. CHECKING OF DIODE TERMINALS WITH AN OHMMETER

The polarity of diode terminals can be determined with the help of an ohm-meter (*i.e.*, multimeter). The internal battery in ohmmeters would forward bias or reverse-vias a diode. This allows a simple and quick check for proper functioning.

To check the diode in the forward direction, the positive meter lead is connected to the anode and the negative lead to the cathode of the diode as shown in figure. If a diode is forward biased, its internal resistance is low. Typically, this is around 100 ohms.

In other case, when the meter leads are reversed, the internal ohmmeter battery reverse biases the diode. The meter indicates a very high resistance (ideally infinite). If a low resistance is indicated in both forward and reverse bias conditions, the PN-junction is short-circuited. Also, if a high resistance is indicated in both forward and reverse bias conditions, the diode is open.

DO YOU KNOW?

When you troubleshoot a circuit that contains a silicon diode that is supposed to be forward biased, a diode voltage measurement much greater than 0.7 V means that the diode has failed and is in fact open.

2.21. DIODE RESISTANCE* : STATIC AND DYNAMIC

An ideal diode offers zero resistance when forward bias and infinite resistance when it is reverse biased. But practical or real diode does not behave ideally. It means that the forward resistance of a real diode is quite small and reverse resistance is very high.

2.21.1. Forward Resistance

The resistance offered by the diode in the forward bias is called forward resistance. This resistance is different for the flow of direct current (d.c.) and for alternating current (a.c.). Hence, this resistance is of two types:

(*a*) d.c. forward resistance

(*b*) a.c. forward resistance

2.21.2. D.C. Forward Resistance or Static Resistance**

When a dc voltage is applied to a diode, a dc current will flow through it and the operating point on the characteristic curve of the diode will not change its position with time. This is the resistance offered by the diode to dc in forward bias condition. It is also known as **static resistance.** It is given by the ratio of the d.c. forward voltage across the diode to the d.c. current flowing through it at a

* What is static and dynamic resistance in the forward biasing of the diode? Explain with suitable diagram. ***(JNTU, Hyderabad, Sem. Exam., 2003-04)***

** Explain how you will determine the static resistance and dynamic resistance of p-n junction diode after defining both of than. ***(JNTU, Hyderabad, Sem. Exam., 2003-04)***

particular point or operating point on the forward characteristic curve.

Mathematically, the static forward resistance from the curve at point P

$$R_F = \frac{V_F}{I_F} = \frac{OA}{OB} \quad \text{...(2.21)}$$

Because of the non-linear shape of the diode characteristic, the value of R_F depends upon the exact location of the operating point on the curve. That typical value of forward static resistance R_F is between 10 Ω to 50 Ω.

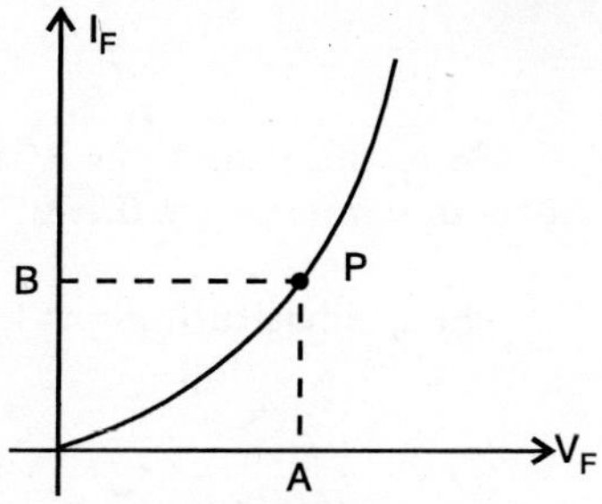

Fig. 2.24 (a). Static resistance of diode

2.21.3. A.C. Forward Resistance

It is the resistance offered by the diode to the a.c. in forward bias. It is also called **dynamic resistance.** It is given by the ratio of change in voltage across the diode to the resulting change in current through diode.

$$\text{a.c. forward resistance} = \frac{\text{Change in voltage across diode}}{\text{Resulting change in current through diode}}$$

From figure 2.24 (*b*), if P is operating point, then

change in forward voltage = CE

change in forward current = DF

$$\text{Therefore, a.c. forward resistance} = \frac{CE}{DF}$$

or
$$r_F = \frac{\Delta V_F}{\Delta I_F} \quad \text{...(2.22)}$$

Because of the non-linear shape of the forward characteristic, the value of a.c. resistance of a diode is in range of 1 to 25 Ω.

Using equation (2.11), the diode current equation is given as

$$I = I_0\left(e^{V/\eta V_T} - 1\right)$$

Let for forward-bias, the above equation be

$$I_F = I_0\left(e^{V_F/\eta V_T} - 1\right)$$

Differentiating, we get

$$\frac{dV_F}{dI_F} = \frac{\eta V_T}{I_0 . e^{V_F/\eta V_T}} = \frac{\eta V_T}{I + I_0}$$

Using equation (2.33), we get

$$\frac{dV_F}{dI_F} = r_F$$

Therefore,
$$r_F = \frac{\eta V_T}{I + I_0}$$

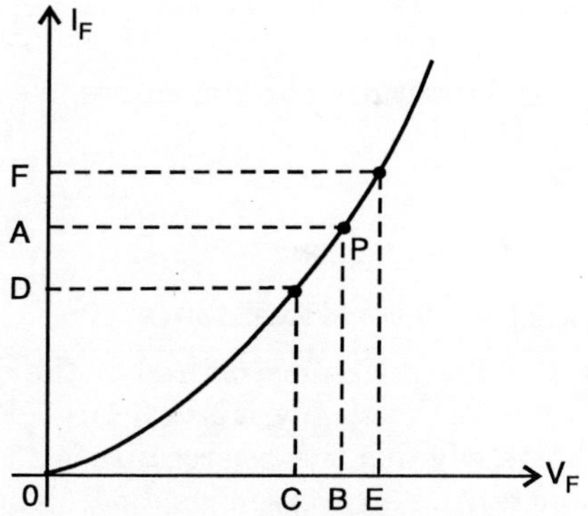

Fig. 2.24 (*b*). A.C. forward resistance

For a forward-bias greater than a few tenths of volt, $\left|\frac{V_F}{\eta V_T}\right|$ >> 1 and according to equation (2.19) $I_F >> I_0$, and thus, dynamic resistance r_F is given approximately by

Natural ability without education has more often raised a man to glory and virtue than education without natural ability.

— **M.T. Cicero**

$$r_F = \frac{\eta V_T}{I_F}$$

From this equation, it is clear that for forward-bias, r_F varies inversely with the current I_F. At room temperature (*i.e.* 27°C or 300°K).

$$V_T = 0.026 \text{ volt or } 26 \text{ m Volt}$$

Thus, substituting $\eta = 1$ and $V_T = 26$ m Volt in the last equation, we get

$$r_F = \frac{26 \text{ mV}}{I_F \text{ (mA)}}$$

EXAMPLE 2.6 Determine the dynamic resistance of a P-N junction diode at a forward current of 2 mA. Assume that *kT/e* = 25 mV.

Solution: Given that

Forward current, $I = 25 \text{ mA } = 0.025 \text{ A}$

and volt-equivalent of temperature,

$$V_T = \frac{kT}{e} = 2.5 \text{ mV} = 0.025 \text{ V}$$

We know that the dynamic resistance is given approximately as

$$r_F = \frac{\eta V_T}{I_F}$$

Substituting, $\eta = 1$, $V_T = 0.025$ V and $I = 0.002$ A in above equation, we get

Dynamic resistance, $r_F = \frac{0.025}{0.002} = 12.5$ ohms **Ans.**

EXAMPLE 2.7 Find ac resistance for a semiconductor diode having a forward bias of 200 mV and reverse saturation current of 1 μ_A at room temperature.

Solution: At room temperature (*i.e.* 300°K)

$$V_T = 26 \text{ mV} = 0.026 \text{ V}$$

Applied forward voltage $V_F = 200 \text{ mV} = 0.2 \text{ V}$

Reverse saturation current $I_0 = 1 \text{ μA } = 1 \times 10^{-6} \text{ A}$

We know that the a.c. resistance for the diode is given as

$$r_F = \frac{\eta V_T}{I_0 e^{V_F/\eta V_T}}$$

Substituting all the values, we get

$$r_F = \frac{0.026}{1 \times 10^{-6} \times e^{0.2/0.026}}$$

On solving, we get $r_F = 11.86$ ohms **Ans.**

2.21.4. Reverse Resistance

The resistance offered by the diode in reverse bias is known as reverse resistance. Ideally it is infinite. Practically, it is very large. It can be d.c. reverse resistance or a.c. reverse resistance depending upon whether the reverse bias is d.c. or a.c. This reverse resistance is due to the existence of leakage current in a reverse bias diode. The reverse bias resistance is very large compared to the forward resistance.

2.22. DIODE CAPACITANCE

Diode offers two types of capacitance, one in forward bias and other in reverse bias. They are

(*i*) Diffusion or Storage Capacitance

(*ii*) Depletion or Transition Capacitance

2.22.1. Diffusion or Storage Capacitance*

The capacitance which exists in a forward-biased junction is called a diffusion or storage capacitance. It is called diffusion capacitance to account for the time delay in moving charges across the junction by diffusion process.

If in a forward-biased junction the applied voltage is suddenly reversed, then forward current I_F ceases at once but a lot of majority charge carriers are left in the depletion region. This charge represents a stored charge in the reverse bias condition and should be removed from the space charge region. This removal of charge takes a finite time. This effect is similar to the discharge of a capacitor. Therefore, the amount of stored charge represents the magnitude of diffusion capacitance. It may be expressed as

$$C_D = \frac{\tau . I_F}{\eta . V_T} \qquad ...(2.23)$$

where C_D = Diffusion capacitance

η = Constant (η = 2 for Si, η = 1 for Ge)

V_T = Volt equivalent of temperature

τ = Mean life time of carriers

I_F = Forward current

Hence, the diffusion capacitance is directly proportional to the value of forward current I_F. The typical value of diffusion capacitance is about 0.02 μF. The effect of diffusion capacitance is negligible for a reverse-biased PN-junction.

2.22.2. Transition or Depletion Capacitance

When a PN-junction is formed, there exists a depletion region at the junction. This depletion region or layer consists of positive and negative immobile ions. This depletion layer is non-conductive and hence acts as a dielectric medium between P-region and N-region. The P-region and N-region act as the two plates of a capacitor because they have a low resistance. These two plates are separated by a dielectric (depletion layer).

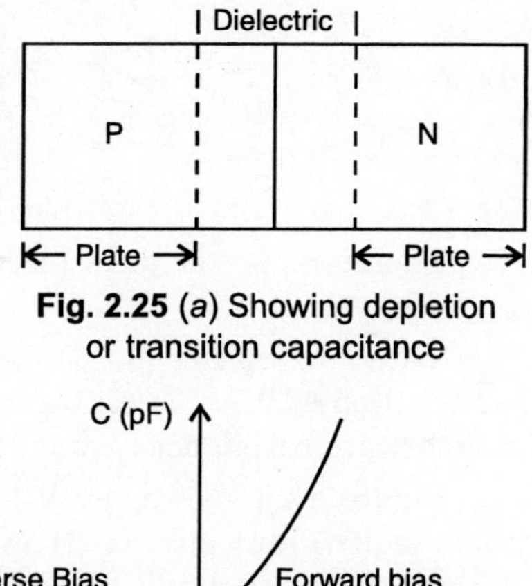

Fig. 2.25 (*a*) Showing depletion or transition capacitance

The capacitance of a parallel plate capacitor is

$$C = \frac{\in . A}{d} \qquad ...(2.24)$$

Here $\in$ is the permittivity of dielectric.

A is the area of plates and d is distance between plates.

Since the depletion layer width *(d)* increases with increase in reverse bias voltage, the depletion capacitance should decrease with the increase in reverse bias. In other words, the depletion capacitance dominates in reverse-bias.

Depletion capacitance is given as

$$C_T = \frac{K}{\sqrt{V}} \qquad ...(2.25)$$

where K is a constant.

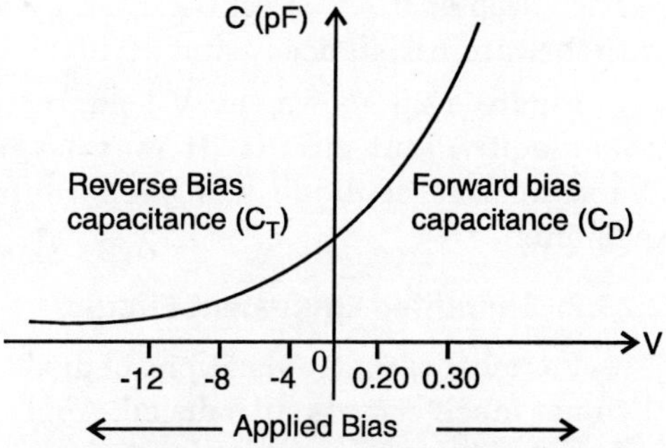

Fig. 2.25 (b). Diffusion capacitance and depletion layer capacitance versus applied bias voltage for a diode

* Explain the transition capacitance and diffusion capacitance of a P-N junction.

2.23. DIODE EQUIVALENT CIRCUITS

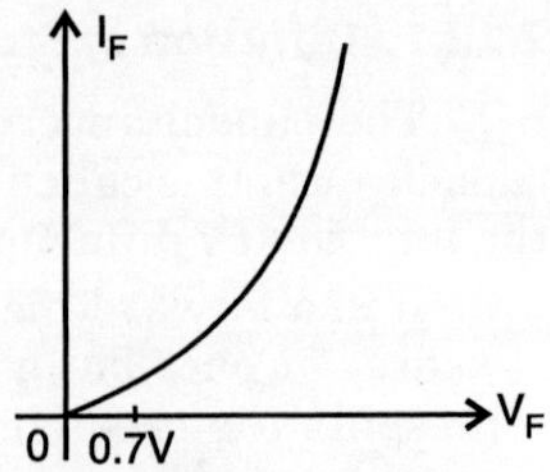

Fig. 2.26. Forward characteristics of diode

A system or device may be represented by its equivalent circuit. An equivalent circuit of a device consists of electric components or elements. The actual behaviour of the device is not affected by its equivalent circuit. With the help of equivalent circuit generally a type of network is obtained which may be solved for various parameters. We will now discuss the equivalent circuits of diode:

2.23.1. Piece-wise Linear Equivalent Circuit

Figure 2.26 shows the forward characteristics of a diode.

Here, the forward voltage drop V_F has to overcome two voltage drops as under:

(*i*) potential barrier V_T (0.7 V for S_i and 0.3 V for G_e)

(*ii*) forward resistance drop or internal drop $I_F r_F$*

Therefore, $V_F = V_T + I_F r_F$ ...(2.26)

where r_F = forward resistance which is varying with point of operation because the *V-I* characteristic curve is non-linear.

Now if the *V-I* characteristic is assumed as linear then forward resistance r_F is constant or it can be taken as average a.c. dynamic forward resistance. Hence due to this concept, we can replace the non-linear curve by straight-line segments. This type of equivalent circuit is known as **piece-wise-linear equivalent circuit**, as shown in figure 2.27.

Since, $V_F = V_T + I_F r_F$ ...(2.27)

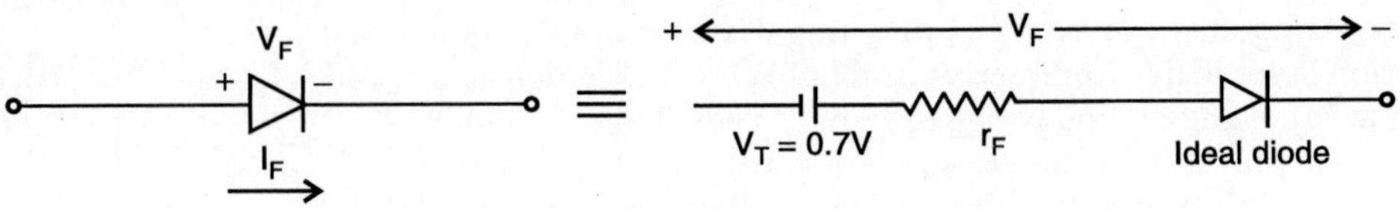

Fig. 2.27. Piece-wise linear equivalent circuit

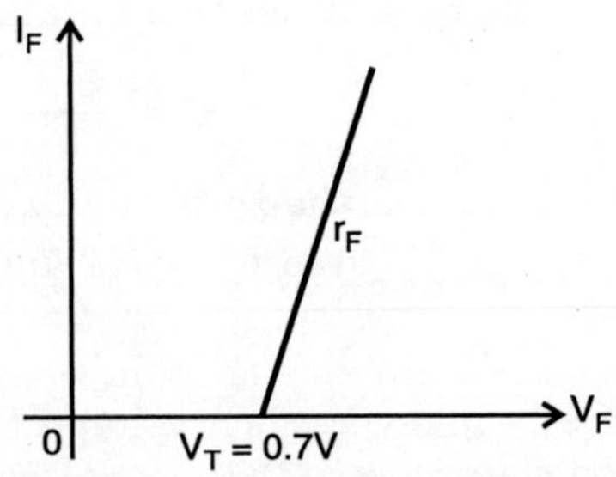

Fig. 2.28. V-I characteristics for piece-wise linear equivalent circuit

Therefore, we can draw piece-wise linear equivalent circuit as above.

Thus, in finding piece-wise linear equivalent circuit, a barrier drop of 0.7 V is represented by a battery put in series with forward resistance r_F and an ideal diode.

Figure 2.28 shows the V-I characteristics for piece-wise linear equivalent circuit. It may be noted that non-linear characteristic of diode has been replaced by straight-line segments.

2.23.2. Simplified Equivalent Circuit

For most of the applications of diode, the forward resistance r_F is very small. If this is neglected then not much error is introduced. Applying this concept, the piece-wise linear equivalent circuit is reduced to another equivalent circuit known as simplified equivalent circuit as shown in figure 2.29.

---

* Here r_F is the dynamic or a.c. resistance.

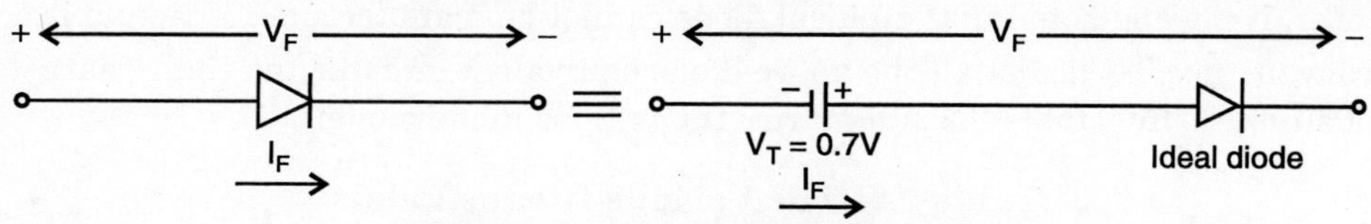

Fig. 2.29. Simplified equivalent circuit

Figure 2.30 shows the V-I characteristics of a diode for simplified equivalent circuit. From figure 2.30, it may be noted that if forward resistance r_F is 0, then at V_T (*i.e.* knee voltage) the curve becomes a vertical straight line.

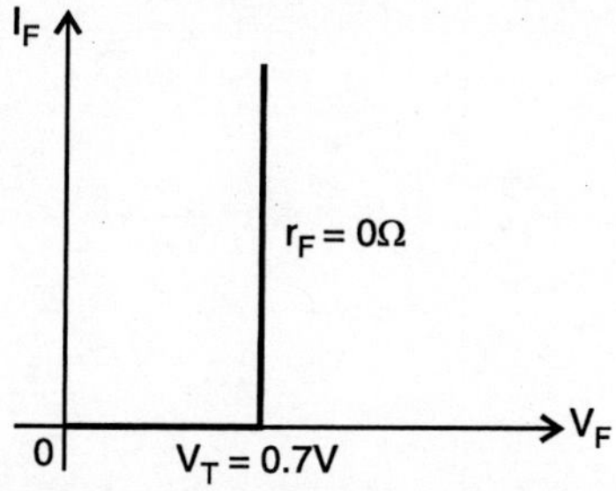

Fig. 2.30. V-I characteristics of a silicon diode for simplified equivalent circuit

2.23.3. Ideal Diode: Ideal Equivalent Circuit

(V. Important)

Taking one step further, since barrier potential V_T is very small, we can neglect this also. In this case, the equivalent circuit is called ideal equivalent circuit as shown in figure 2.31.

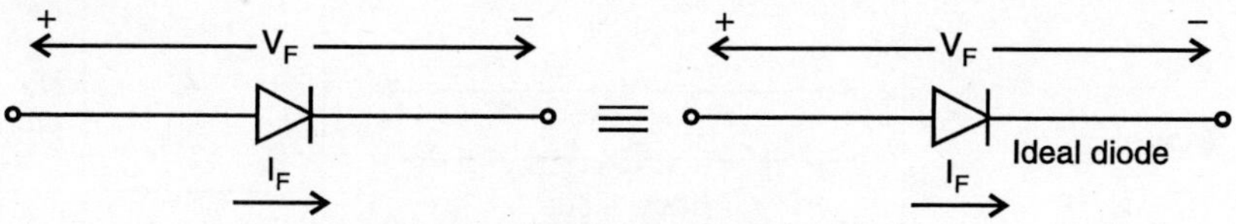

Fig. 2.31. Ideal equivalent circuit

From the characteristics of ideal diode, it may be observed that in forward bias there is zero voltage drop across the diode and the current flowing through the diode is maximum. It means that in forward bias, diode is working as a perfect conductor. In reverse bias, there is no current through the diode and it works as perfect insulator.

Ideal diode can also be used as a perfect switch. The switch is closed if the diode is forward bias and is open if the diode is reverse bias as shown in figure 2.33.

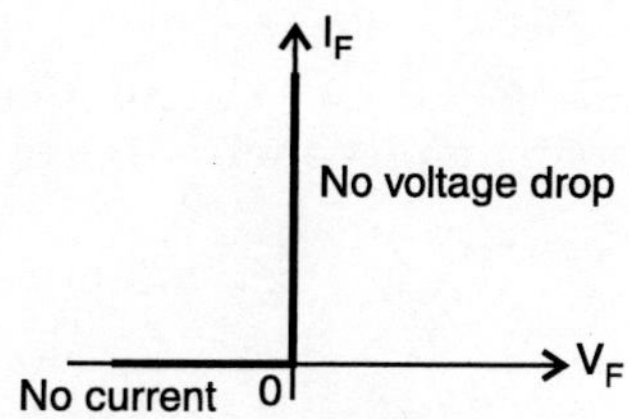

Fig. 2.32. Characteristic for ideal diode.

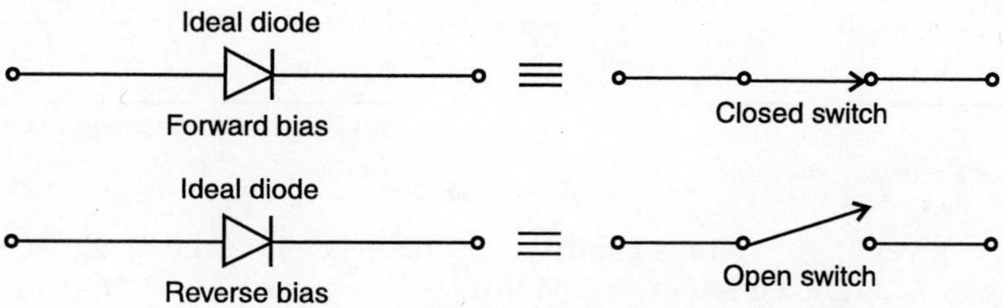

Fig. 2.33. Ideal diode as a switch

From above it is clear that an ideal diode offers zero resistance in forward bias* because there is zero voltage drop ($R = V/I$) and infinite resistance in reverse bias because there is zero current ($R = V/I$).**

* Zero resistance in forward-bias means that an ideal diode works as a short-circuit in forward-bias condition.

**Infine resistance in reverse-bias means that an ideal diode works as an open-circuit in reverse-bias condition.

However, it may be noted that an ideal diode cannot be manufactured in actual practice.

For convenience, let us list all the above diode equivalent circuits and their characteristics in the form of a table. Thus, table 2.2 summaries the various diode models.

Table 2.2: The Various Diode Models

S.No.	Type of Model	Equivalent Circuit	Characteristics
1.	Piece-wise linear model	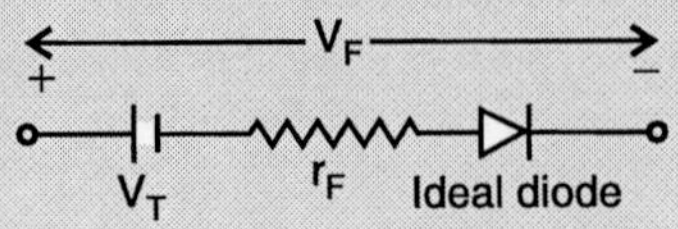	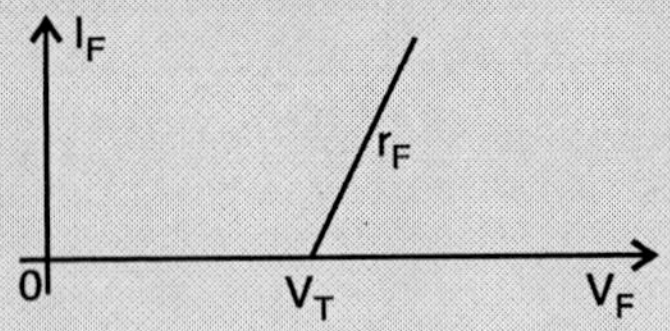
2.	Simplified Model	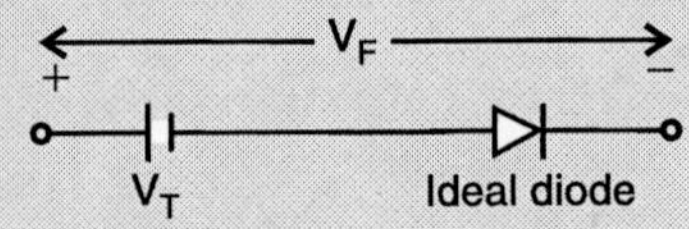	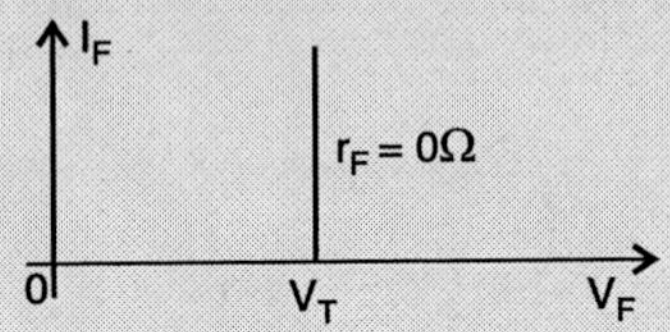
3.	Ideal Model	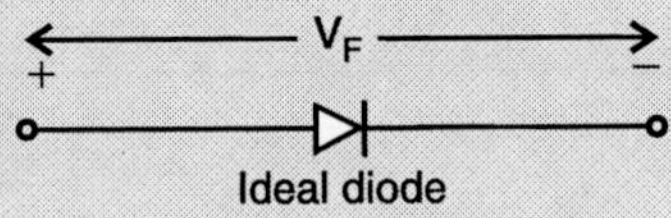	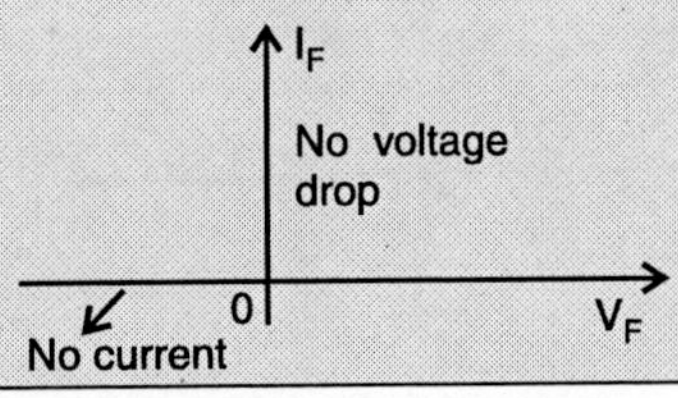

EXAMPLE 2.8 Calculate the diode voltage and current in the circuit shown in figure 2.34(*a*) using a piece-wise linear model. Assume the piece-wise linear diode parameters as

$$V_\gamma = 0.6 \text{ V} \quad \text{and} \quad r_F = 12\ \Omega.$$

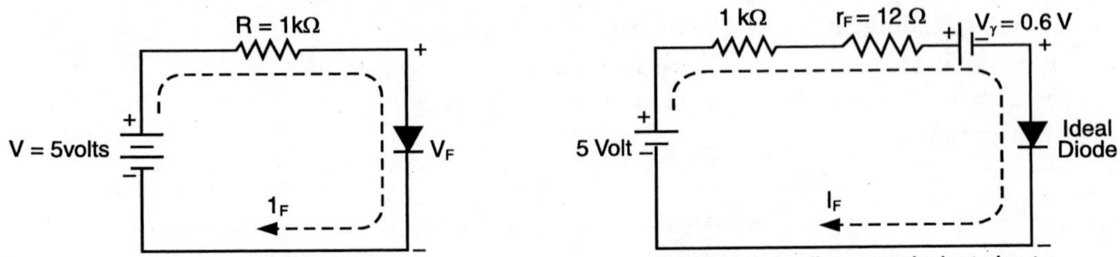

(a)

(b) Piece-wise linear equivalent circut

Fig. 2.34.

Solution: With the given input voltage polarity, the diode is forward biased. So, $I_F > 0$. Therefore, the equivalent circuit is as shown in figure 2.34 (*b*).

The forward diode current is, therefore, given by,

$$I_F = \frac{5 - V_\gamma}{1000 + 12} = \frac{5 - 0.6}{1012} = 4.3 \text{ mA} \qquad \textbf{Ans.}$$

And the voltage across the diode is given by

$$V_F = V_\gamma + I_F r_F = 0.6 + (4.3 \times 10^{-3} \times 12)$$

or $\quad V_F = 0.65$ Volts. **Ans.**

2.23.4. Ideal Versus Practical Diodes: Comparison of Ideal and Practical Diodes

S.No.	Parameter of comparison	Ideal Diode	Practical Diode
1.	Forward resistance	0 Ω	Few Ω
2.	Reverse resistance	∞	Few hundred kΩ
3.	Cut in voltage	0 V	0.6 V for Si and 0.2 V for Ge diode
4.	Reverse saturation current	Zero	Few nA for S_i diode Few mA for G_e diode
5.	Equivalent circuit in the forward biased state	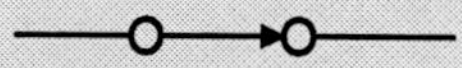	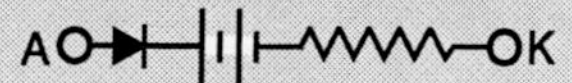
6.	Equivalent circuit in the reverse biased state		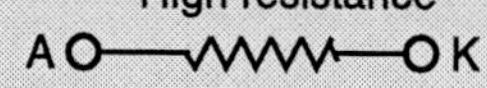

2.24. THE DIODE RATINGS*

(*i*) Average Current

It is defined as average value of a periodic function given by area under one cycle of the function divided by the base.

(*ii*) Maximum forward current

The maximum value of diode forward current, which a P-N junction (or diode) can carry without damaging itself, is known as its **maximum forward current.** *It is generally denoted by I_F.*

The rating of a P-N junction or diode is specified by the manufacturer in its data sheet. However, if the forward current in a diode is increased beyond this specified value, the junction will be destroyed.

(*iii*) Peak inverse voltage (PIV)

The maximum value of reverse voltage that a P-N junction (or diode) can withstand without damaging it is called its **Peak Inverse Voltage (PIV).**

This rating of a P-N junction or diode is also specified by the manufacturer in its data sheet. However, if the voltage coming across the junction at reverse bias condition increases beyond this specified value, the junction will be destroyed.

(*iv*) Maximum power rating

The maximum power that a P-N junction or diode can dissipate without damaging it is called as **maximum power rating.**

Generally, it is specified by the manufacturer in its data sheet. The power dissipated at the junction is equal to the product of junction current and the voltage across the junction. If the power developed across the junction is more than the maximum power dissipated by it, the junction will be overheated and may be destroyed. It may be noted here that more often, the data sheets list only the maximum current that a diode can handle since it is more convenient to measure and to design with.

(*v*) Peak repetitive forward current

This is the maximum instantaneous value of the repetitive forward current. Since it is at this level for a brief period of time, its level can be higher than the continuous level.

* What do you understand by average current, repetitive peak current, non repetitive current, peak inverse voltage and reverse saturation current? ***(JNTU, Hyderabad, Sem. Exam., 2001-02)***

(*vi*) Peak forward surge current or non-repetitive peak forward curren

During turn-on, the device malfunctions and so there will be very high currents through the device for very brief intervals of time (that are non-repetitive). The time interval during which such a current appears across the device is of the order of μs or *ns* or even less. This rating defines the maximum value and the time interval for such surge in the current level.

(vii) Reverse saturation current

Under reverse biased condition, small number of electrons in p-region are attracted by positive end of battery and small number of holes in N-region are attracted by negative terminal of supply. This phenomenon consisititute the reverse current which is called as *reverse saturation current.*

2.24.1 Applications of P-N Junction Diode

Some of the applications of a P-N junction diode may be listed as under:

(*i*) Rectifier circuits
(*ii*) Clipping and clamping circuits.
(*iii*) Voltage multipliers
(*iv*) A.M. detection
(*v*) Feedback diodes
(*vi*) Freewheeling diodes
(*vii*) Log and antilog amplifiers using Op-Amp.
(*viii*) Precision rectifiers using Op-Amp.

2.24.2. Diode Switching Times

Diodes are used in applications such as logic gates. In such applications diodes are used as switches and not for the purpose of rectification. Hence, the diode is expected to change its status from ON to OFF or OFF to ON instantaneously. But, this is possible only if the diode is ideal. However, the practical diodes will require some finite time to switch from one state to the other. The time required to turn OFF is longer than the time required to turn ON a diode. Therefore, the turn OFF time and turn OFF characteristics are more important. In the forward biased state, a large number of electrons from N-type material are progressing through the P-side and a large number of holes from P-side are progressing through the N-region. This is essential to ensure conduction. The electrons in P-side and holes in the N-side establish a large number of minority carriers in each side, if the applied voltage is reversed. Hence, even if the applied voltage is reversed to establish a reverse bias condition, the diode does not switch off instantaneously. However, due to the large number of minority carriers on either side the current will simply reverse as shown in figure 2.35.

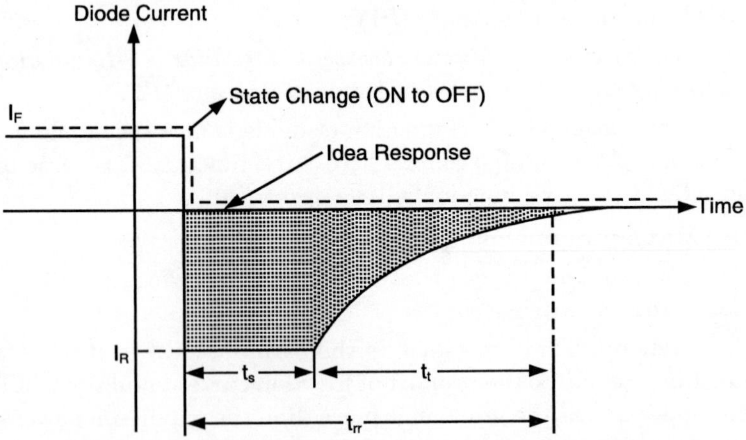

Fig. 2.35. Reverse recovery characteristics of a diode.

The current will be negative and at a measurable level for the time period t_s called storage time. This much time is required for the minority carriers to return to their majority carrier state in the opposite materials. After the storage time, this current will reduce gradually to almost zero as shown in figure 2.35. This period is called as the transition time (t_t)

The reverse recovery time is the sum of these two time intervals.

Thus, $$t_{rr} = t_s + t_t \quad \text{... (2.28)}$$

where t_{rr} = reverse recovery time

t_s = storage time

t_t = transition time.

Forward Recovery Time t_{rf}

The forward recovery time (t_{rf}) is defined as the time required for the diode current to change from 10 to 90 percent of its final value when the diode is switched from OFF to ON state. In practice, the forward recovery time is very short as compared to the reverse recovery time. Hence, we can neglect it. Hence, the turn ON time of a diode is always shorter than its turn OFF time. Manufacturers normally specify the reverse recovery time t_{rr} of a diode in a typical operating condition in terms of the current waveform shown in figure 2.35.

Switching Diodes

Commercial switching type diodes are available with time t_{rr} ranging from 1 nanosecond (ns) upto as high as 1 microsecond (μs). Higher values of t_{rr} correspond to the diodes intended to handle larger currents. IN 4153 is a switching diode. It has a reverse recovery time of a few nanosecond under the test conditions. IN 4153 is high speed switching diode for computer and general purpose applications. Some of the important features of this switching diode are as follows :

Important Characteristics of Switching Diode 1N 4153

(*i*) Reverse breakdown voltage : V_{br} = 75 Volts.

(*ii*) Static forward voltage : 0.5 to 0.8 volts depending on the forward current.

(*iii*) Total capacitance : 2 pF

(*iv*) Reverse recovery time : 4 ns.

Why can not a diode change its state instantaneously ?

From the discussion so far, it is clear that a diode needs some finite time to change its state from ON to OFF or from OFF to ON. The question is why can not a diode change its state instantly ? The answer is due to the internal capacitances associated with the diode. The internal diode capacitances will not allow the device to change the state instantly. Hence, the internal capacitances should be as possible to reduce the switching times and make the diode faster.

For switching diodes, the internal capacitance is few pF.

2.24.3. Importance of Diode Switching Times

The reverse recovery time decides the time required to switch the diode off. So it decides, how fast or at what maximum rate can we switch the diode on and off. Thus, reverse recovery time decides the highest switching frequency of a diode. The reverse recovery time depends on the transition capacitance of the diode. So, the transition capacitance of switching diodes is small. The maximum switching frequency is given by,

$$f_{max} = \frac{1}{10\, t_{rr}} \quad \text{... (2.29)}$$

So, as t_{rr} decreases f_{max} will increase.

EXAMPLE 2.9 Figure 2.36 shows the circuit of series diode configuration. Find the value of V_F, V_R and I_F.

Solution: From figure 2.36, it is clear that the battery V_S is pushing the current in the direction of diode arrow. Therefore, the diode is forward biased or in 'ON' state.

Because diode is a S_i diode, therefore, we have

$$V_F = 0.7 \text{ Volt}$$

Applying KVL across the circuit, we write

$$V_S = V_F + V_R$$
$$V_R = V_S - V_F$$
$$= 8 - 0.7 = 7.3 \text{ Volt}$$

Also
$$I_F = I_R = \frac{V_R}{R} = \frac{7.3}{2\times10^3}$$
$$I_F \cong 3.65 \times 10^{-3}$$
$$I_F \cong 3.65 \text{ mA} \quad \textbf{Ans.}$$

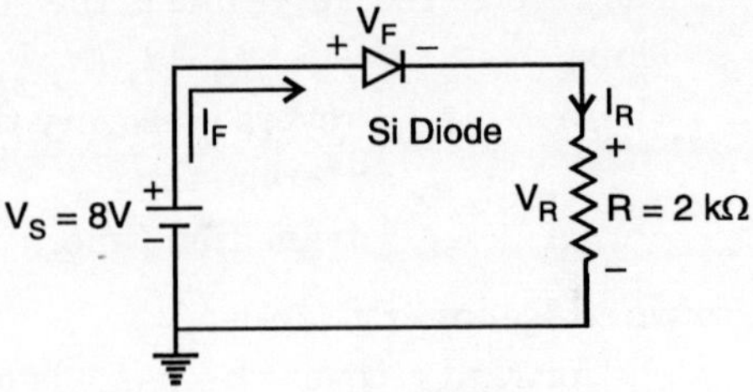

Fig. 2.36.

EXAMPLE 2.10 Figure 2.37 shows the circuit for series diode configuration. Find the value of V_R and I_F.

Solution: From figure 2.37 it is clear that the battery V_S is pushing the current in the direction of diode arrow. Hence, both the diodes are in forward bias or are in 'ON' state.

We know that $V_{F1} = 0.7$ V and $V_{F2} = 0.3$ V

Applying KVL across the circuit

$$V_S = V_{F1} + V_{F2} + V_R$$
$$12 = 0.7 + 0.3 + V_R$$

or
$$V_R = 12 - 1.0 = 11 \text{ Volt}$$

∴
$$I_F = I_R = \frac{V_R}{R} = \frac{11}{4\times10^3}$$
$$I_F = 2.75 \times 10^{-3} = 2.75 \text{ mA} \quad \textbf{Ans.}$$

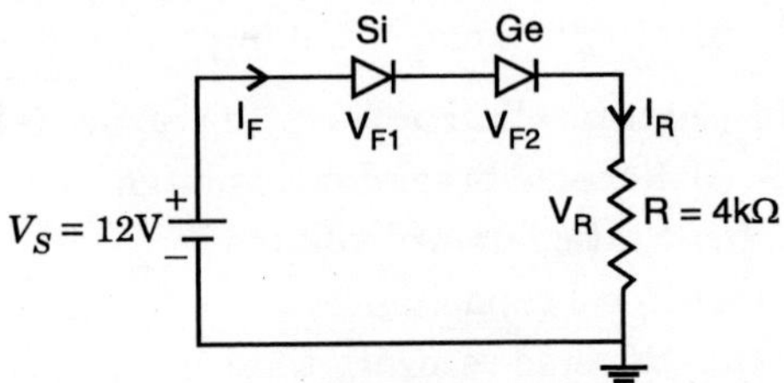

Fig. 2.37.

EXAMPLE 2.11 Figure 2.38 shows the circuit of series diode configuration. Find the value of I, V_{R1}, V_{R2} and V_0.

Solution: The polarities of batteries indicate that the diode is in forward bias. Since diode is a S_i diode, $V_F = 0.7$ Volt.

Applying KVL across the circuit

$$V_{S1} + V_{S2} = V_F + V_{R1} + V_{R2}$$

or
$$V_{R1} + V_{R2} = V_{S1} + V_{S2} - V_F$$

or
$$I_{R1} + I_{R2} = V_{S1} + V_{S2} - V_F$$

$$I = \frac{V_{S1} + V_{S2} - V_F}{R_1 + R_2} = \frac{10 + 5 - 0.7}{(4+2)\times10^3}$$

$$I = \frac{14.3}{6\times10^3} = \frac{14.3\times10^{-3}}{6}$$

$$I = 2.38 \times 10^{-3} \text{ A} = 2.38 \text{ mA}$$

$$V_{R1} = IR_1 = 2.38 \times 10^{-3} \times 4 \times 10^3 = 9.52 \text{ Volt}$$

and
$$V_{R2} = IR_2 = 2.38 \times 10^{-3} \times 2 \times 10^3 = 4.76 \text{ Volt}$$

Voltage V_0 is given by $V_0 = V_{R2} - V_{S2} = 4.76 - 5 = -0.24$ Volt **Ans.**

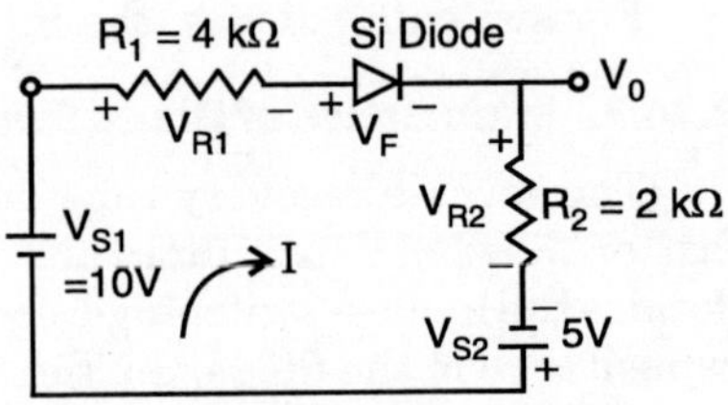

Fig. 2.38.

EXAMPLE 2.12 Figure 2.39 shows the circuit of parallel diode configuration. Find the values of V_0, I, I_{F1} and I_{F2}.

Solution: From figure, it is clear that the applied battery is pushing the currents in the directions of diode arrows. Hence, both diodes are in forward bias or are in 'ON' state.

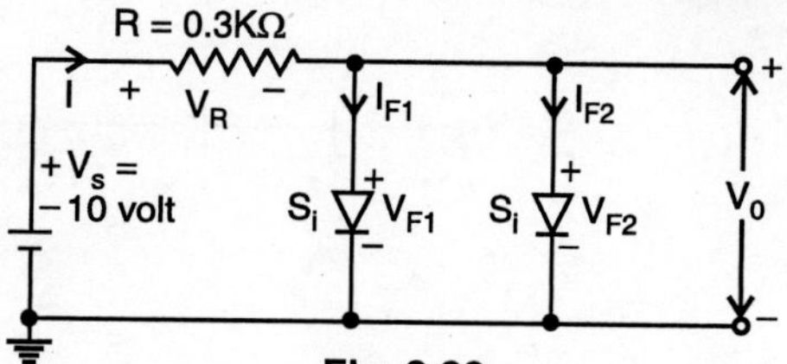

Fig. 2.39.

Because each diode is a silicon diode, therefore, voltage drop across each diode is 0.7 Volt.

Hence, $V_{F1} = V_{F2} = 0.7$ Volt

The current I is given by

$$I = \frac{V_R}{R} = \frac{V_S - V_{F1}}{R_1} = \frac{10-0.7}{0.3\times10^3} = \frac{9.30}{0.3\times10^3} = 31 \text{ mA}$$

Assuming both diodes are having similar characteristics, we have

$$I_{F1} = I_{F2} = \frac{I}{2} = \frac{33.1}{2} = 16.5 \text{ mA}$$

From figure, $V_0 = V_{F2} = 0.7$ Volt **Ans.**

EXAMPLE 2.13 Find the voltage V_A of the current given in figure 2.40.

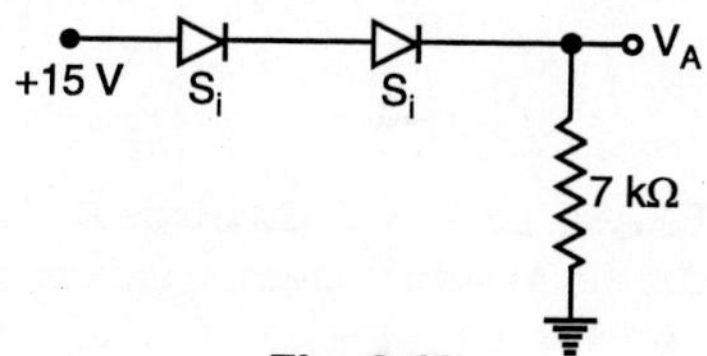

Fig. 2.40.

Solution: Redrawing the given circuit.

Since, diodes are silicon diodes, hence

$$V_{F1} = V_{F2} = 0.7 \text{ Volt}$$

From circuit, both diodes are forward biased. Now, applying KVL across the circuit,

$$V_S = V_{F1} + V_{F2} + V_A$$
$$V_A = V_S - V_{F1} - V_{F2}$$
$$= 15 - (0.7 + 0.7)$$
$$V_A = 15 - 1.4 = 13.6 \text{ Volt}$$

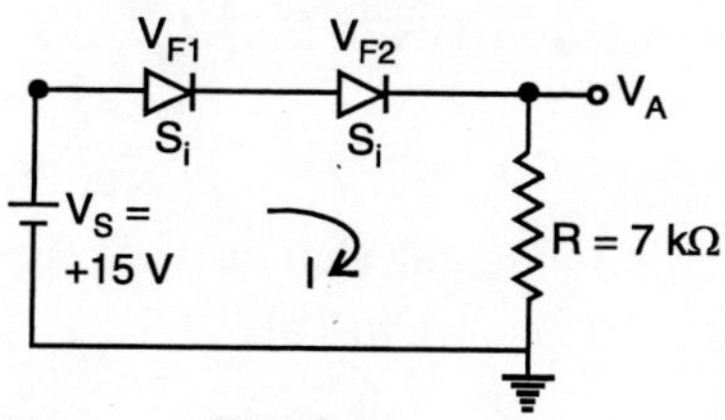

Fig. 2.41.

Also $I = \frac{V_A}{R}$

$$= \frac{13.6 \text{ V}}{7 \text{ k}\Omega} = 1.942 \text{ mA} \quad \textbf{Ans.}$$

EXAMPLE 2.14 What is the current in the following circuit shown in figure 2.42?

Solution: Since diode is a silicon diode, hence

$$V_F = 0.7 \text{ Volt}$$

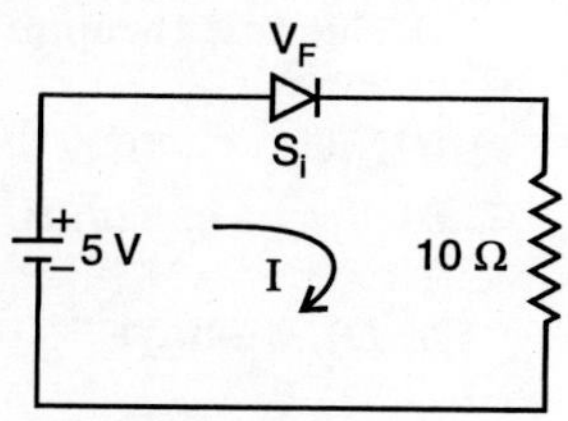

Fig. 2.42.

Applying KVL across the circuit

$$5 = V_F + I.10 = 0.7 + 10I$$
$$10\,I = 4.3$$
$$I = \frac{4.3}{10} = 0.43 \text{ A} \quad \textbf{Ans.}$$

EXAMPLE 2.15 Find the voltage V_A and current in the circuit of figure 2.43.

Solution: Redrawing the given circuit, we get

Since in the given circuit, both the diodes are silicon (S_i) diodes, therefore,

$$V_{F1} = V_{F2} = 0.7 \text{ Volt}$$

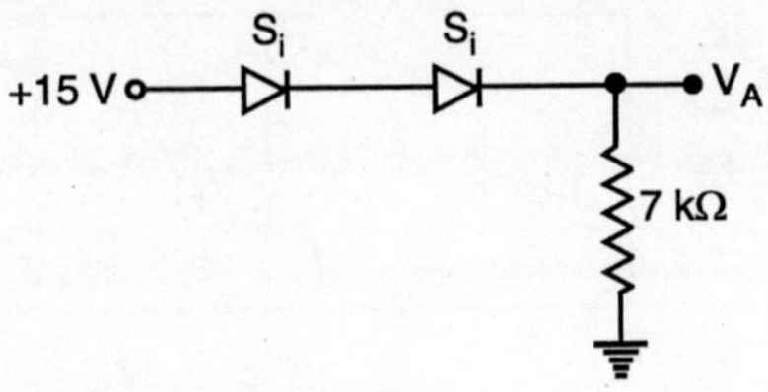

Fig. 2.43.

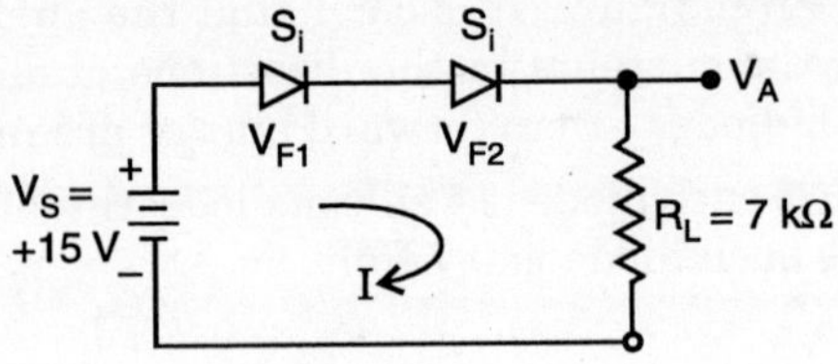

Fig. 2.44.

Since in the given circuit, the voltage source is pushing the current in the direction of diode arrows, therefore, both the diodes are foward biased.

Applying KVL across the circuit, we have

$$V_S = V_{F1} + V_{F2} + V_A$$

or

$$V_A = V_S - V_{F1} - V_{F2}$$

$$= 15 - 0.7 - 0.7 = 15 - 1.4 = 13.6 \text{ Volt}$$

Also, from circuit we have $V_A = I.R_L$

$$I = \frac{V_A}{R_L} = \frac{13.6\text{V}}{7\text{k}\Omega} = \frac{13.6}{7 \times 10^3} = 1.942 \text{ mA} \quad \textbf{Ans.}$$

EXAMPLE 2.16 The circuit shown in figure 2.45 uses identical diodes D_1, D_2 and D_3. For each diode, assume reverse saturation current of 10^{-13} A at $T = 300°$K and $\eta = 1$.

Calculate V_{in} to get 600 mV across D_3.

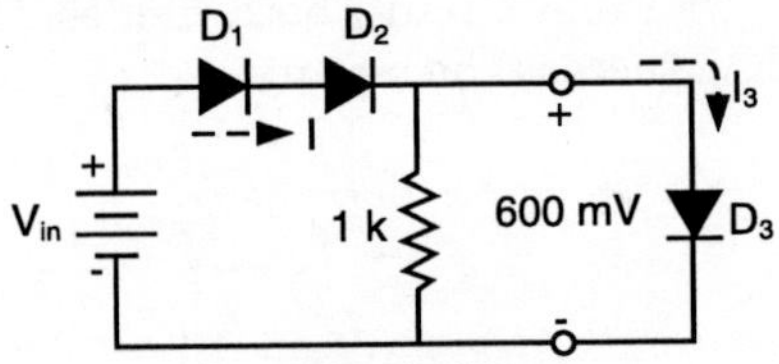

Fig. 2.45.

Solution : Given that $I_0 = 1 \times 10^{-13}$ A at $T = 300°$ K,

$\eta = 1$, $V = 0.6$ V,

V_T at 300° K is 26 mV.

(*i*) Assume that current through $D_3 = I_3$.

Hence, we have

$$I_3 = I_0 \left[e^{V/\eta V_T} - 1\right] = 10^{-3}\left[e^{0.6/26\times10^{-3}} - 1\right] = 1.052 \text{ mA} \quad \text{... (i)}$$

(*ii*) Current through 1 kΩ resistor $= \frac{600 \text{ mV}}{1 \text{ k}\Omega} = 0.6$ mA ... (*ii*)

(*iii*) Total current through D_1 and $D_2 = 1.052 + 0.6 = 1.652$ mA ... (*iii*)

(*iv*) Let us calculate voltage across diodes D_1 and D_2

For D_1, we have $I = I_0\left[e^{V/\eta V_T} - 1\right]$

or

$$1.652 \times 10^{-3} = 10^{-13}\left[e^{V/26\times10^{-3}} - 1\right]$$

or

$$\frac{V}{26\times10^{-3}} = \log_e [1.652 \times 10^{10}]$$

or

$$V = 0.6117 \text{ Volts}$$

As D_1 and D_2 are identical diodes, the voltage drop across D_2 will be the same i.e., 0.6117 Volt.

Therefore, $V_{in} = 0.6117 + 0.6117 + 0.6 = 1.823$ Volt. **Ans.**

EXAMPLE 2.17 Two Identical germanium diodes are connected in series opposition. A 10 V battery is impressed upon this series arrangement. Find the voltage across each diode at room temperature if :

(*i*) The zener breakdown voltage of each diode is 9.8 V.

(*ii*) The zener breakdown voltage of each diode is 10.2 V.

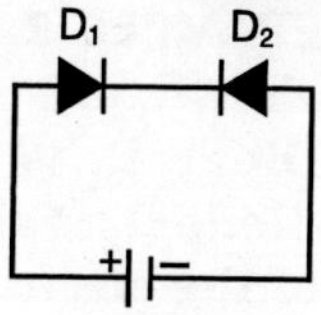

Fig. 2.46.

Solution :

Case I :

Diode D_1 is forward biased and D_2 is reverse biased. Hence, breakdown due to zener effect will take place in D_2. The voltage across it remains constant equal to $V_Z = 9.8$ V.

Hence, $\quad V_{D2} = 9.8$ Volt

and $\quad V_{D1} = 10 - V_{D2} = 0.2$ Volt.

Case II :

As the zener breakdown voltage $V_Z = 10.2$ volt, which is higher than the battery voltage, none of the diodes will breakdown.

D_1 is forward biased and D_2 is reverse biased. As they are connected in series, current through them remains same equal to reverse saturation current of D_2, which is I_0.

Thus, we have
$$I = I_0\left[e^{V_1/\eta V_T} - 1\right] = I_0 \quad \text{...}(i)$$

For Ge diodes $\eta = 1$ and $V_T = 26$ mV at room temperature.

Therefore, $\quad e^{V_1/V_T} - 1 = 1$

or $\quad e^{V_1/V_T} = 2$

or $$\frac{V_1}{V_T} = \log_e 2 = 0.693$$

or $\quad V_1 = 0.693 \times 26 \times 10^{-3} = 0.018$ Volts

and $\quad V_2 = 10 - V_1 = 10 - 0.018 = 9.982$ Volts

Answers:

Case I : $\quad V_{D1} = 0.2$ Volt

and $\quad V_{D2} = 9.8$ Volt

Case II : $\quad V_{D1} = 0.018$ Volt

and $\quad V_{D2} = 9.982$ Volt.

EXAMPLE 2.18 What is the magnitude of *r*. For silicon P-N junction diode at room temperature and for dc current of 1 mA.

Solution : We have
$$r = \frac{\eta V_T}{I + I_0} \quad \text{...}(i)$$

Let us assume $\quad V_T = 26$ mV at room temperature

$I + I_0 = I$ since I_0 is negligible

$\eta = 2$ for silicon

Substituting all the values, we get

$$r = \frac{2 \times 26 \times 10^{-3}}{1 \times 10^{-3}} = 52\ \Omega. \quad \textbf{Ans.}$$

EXAMPLE 2.19 Each diode is having linearised volt ampere characteristics with internal resistance r and offset voltage V_γ. Diode D_1 is germanium with $V_\gamma = 0.2$ V and $r = 20\ \Omega$ whereas diode D_2 is silicon with $V_\gamma = 0.6$ V and $r = 15\ \Omega$. Find diode current of the circuit shown in figure 2.47 (*a*) if $R = 10$ kΩ.

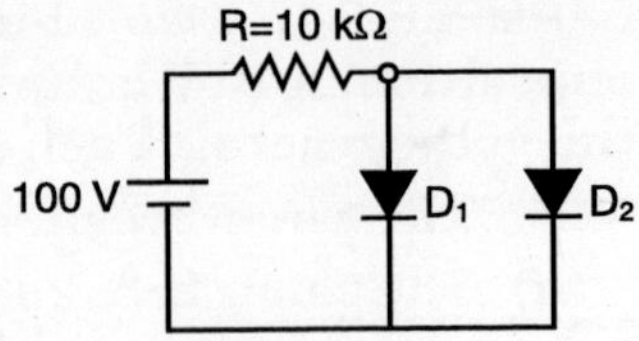

Fig. 2.47 (a)

Solution : First, let us replace the diodes by their equivalent circuits as shown in figure 2.47 (*b*)

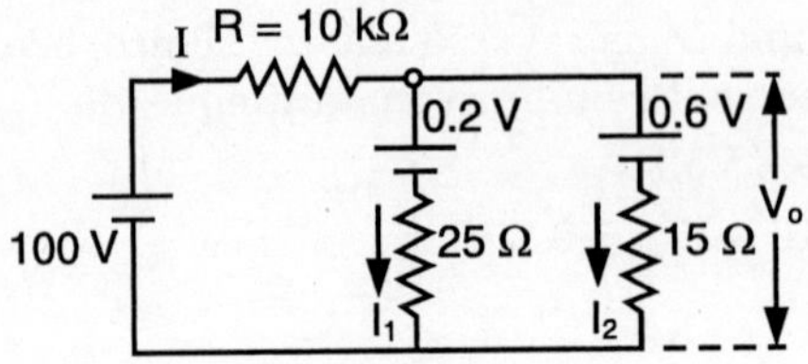

Fig. 2.47 (*b*)

Now, let us calculate V_D.

Here, we have $$V_D = 0.2 + 20\,I_1 = 0.6 + 15\,I_2 \qquad \text{... }(i)$$

Also $$I_1 + I_2 = I = \frac{100 - V_D}{100\text{ k}\Omega} \qquad \text{... }(ii)$$

or $$\frac{V_D - 0.2}{20} + \frac{V_D - 0.6}{15} = \frac{100 - V_D}{10 \times 10^3}$$

Simplifying, we get $10 \times 10^3 [3V_D - 0.6 + 4V_D - 2.4] = 60\ [100 - V_D]$

$$10 \times 10^3\ [7V_D - 3] = 6000 - 60V_D$$

or $$70 \times 10^3 V_D + 60\ V_D = (6000 + 30 \times 10^3)$$

or $$V_D = 0.5138 \text{ volts}$$

Lastly, we calculate diiode currents as under:

$$I_1 = \frac{V_D - 0.2}{20} = \frac{0.5138 - 0.2}{20} = 0.01569 \text{ A} = 15.69 \text{ mA}$$

$$I_2 = \frac{V_D - 0.6}{15} = \frac{0.5138 - 0.6}{15}$$

or $$I_2 = -\,5.7466 \text{ mA} \quad \textbf{Ans.}$$

EXAMPLE 2.20 A germanium diode carries a current of 1 mA of room temperature when a forward bias of 0.15 V is applied. Estimate the reverse saturation current at room temperature.

Solution : Given that Ge diode, so, we have

$$\eta = 1, \quad I = 1 \text{ mA}, \quad V_F = 0.15 \text{ Volts.}$$

The expression for forward diode current is given by

$$I_F = I_0\left[e^{V_F/\eta V_T} - 1\right]$$

Let $$V_T = 26 \text{ mV at room temperature}$$

Substituting all the values, we get

$$1 \times 10^{-30} = I_0\left[e^{0.15/1 \times 26 \times 10^{-3}} - 1\right]$$

or $$1 \times 10^{-3} = I_0\,[403.43 - 1]$$

or $$I_0 = 2.4849 \times 10^{-6}\ \text{A} \quad \text{or} \quad 2.4849\ \mu\text{A}. \qquad \textbf{Ans.}$$

EXAMPLE 2.21 Calculate V_{out} and I_D for the following figure 2.48.

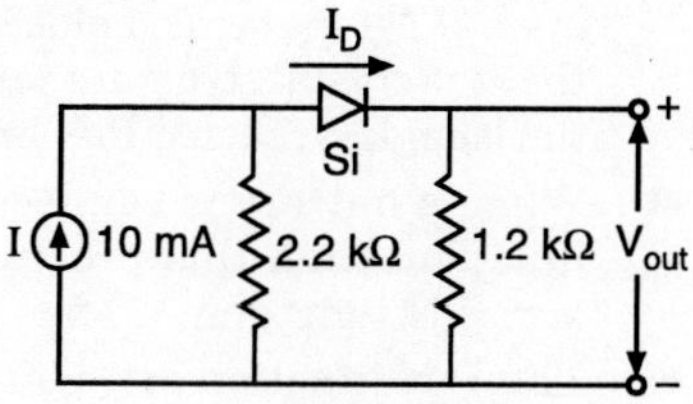

Fig. 2.48.

Solution: First, let us convert current source into voltage source in the given circuit. This has been shown in figure 2.49.

In figure 2.49, the value of V will be

$$V = I \times 2.2\ \text{k}\Omega = 10 \times 10^{-3} \times 2.2 \times 10^3$$

or $$V = 22\ \text{Volt} \quad \text{and}\ R_S = 2.2\ \text{k}\Omega$$

Applying KVL, we have

$$V - 2.2\,I_D - 0.7 - (1.2 \times I_D) = 0$$

or $$22 - 0.7 = 3.4\,I_D$$

or $$I_D = \frac{21.3}{3.4} = 6.26\ \text{mA} \quad \textbf{Ans.}$$

Also, $$V_{out} = I_D R_L = 6.26 \times 10^{-3} \times 1.2 \times 10^3$$

or $$V_{out} = 7.512\ \text{Volt} \quad \textbf{Ans.}$$

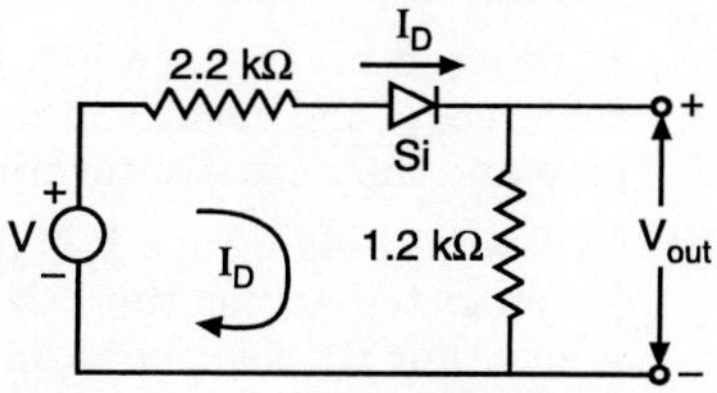

Fig. 2.49.

EXAMPLE 2.22 Explain the static and dynamic resistance in p-n junction diode. Determine the dc resistance levels for the diode of in figure 2.50 at

***(i)* I_D = 2 mA**

***(ii)* I_D = 20 mA**

***(iii)* V_D = – 10 V**

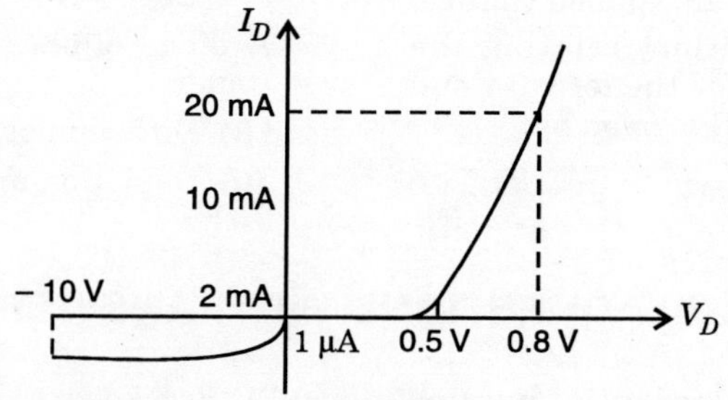

Fig. 2.50.

Solution: For first part, please go through article 2.21.

Now, we have

(i) DC resistance at $I_D = 2$ mA

$$r_{F1} = \frac{V_D}{I_D} = \frac{0.5\ \text{V}}{2\ \text{mA}} = 250\ \Omega \quad \textbf{Ans.}$$

(ii) DC resistance at $I_D = 20$ mA

$$r_{F2} = \frac{V_D}{I_D} = \frac{0.8\ \text{V}}{20\ \text{mV}} = 40\ \Omega \quad \textbf{Ans.}$$

(iii) DC resistance at $V_D = -10$ V

$$r_R = \frac{10\ \text{V}}{1\ \mu\text{A}} = 10\ \text{M}\Omega \quad \textbf{Ans.}$$

SUMMARY

- If a piece of *P*-type semiconductor is joined to a piece of *N*-type semiconductor in such a manner that the crystal structure remains continuous at the boundary, then a P-N junction is formed.
- When a battery is connected across the P-N junction, then this process is called biasing of a P-N junction.
- The P-N junction can be biased in two ways as under:

  (*i*) Forward-bias, (*ii*) Reverse-bias.
- In forward-biasing of a P-N junction, positive terminal of the battery is connected to the *P*-side and the negative terminal to the *N*-side.
- In reverse biasing of a P-N junction a battery is connected across the P-N junction in such a way that its positive terminal is connected to the *N*-region and negative terminal is connected to the *P*-region.
- A PN-junction is known as a P-N junction diode or a semiconductor diode.
- The V-I characteristics of a diode is simply a curve or graph between the voltage applied across its terminals and the current that flows through the diode due to this applied voltage.
- The diode current equation, relating the voltage *V* and current *I* for the forward and reverse-bias regions, can be given by

$$I = I_0\left(e^{V/\eta V_T} - 1\right)$$

where I = diode current

I_0 = diode reverse saturation current at room temperature

V = external voltage applied to the diode

η = a constant

It is 1 for G_e and 2 for S_i.

and $V_T = \frac{k_T}{q} = \frac{T}{11{,}600}$, volt-equivalent of temperature

where k = Boltzmann's constant = 1.38066 $\times 10^{-23}$ J/Kelvin

q = electronic charge = 1.6×10^{-19} Coulomb

T = Diode junction temperature in (°K)

- An ideal diode offers zero resistance when forward-biased and infinite resistance when it is reverse-biased.
- Practical or real diode does not behave ideally. Its forward resistance is quite small and reverse resistance is very high.
- A diode offers two types of capacitance as under:

  (*i*) Diffusion or storage capacitance.

  (*ii*) Depletion or transition capacitance.

SHORT QUESTIONS WITH ANSWERS

Q. 1. What do you mean by potential barrier for a P-N junction.

Ans. The electric field between the acceptor and the donor ions is called a barrier. The width of the barrier is the physical distance from one side of the barrier to the other side. The height of the barrier is defined as a difference of potential from one side of the barrier to the other side. When no external battery is connected across the junction, the height of the barrier is a few tenths of a volt.

The barrier potential for silicon (S_i) PN-junction is 0.7 V and that for Germanium (G_e) it is 0.3 V.

Q. 2. What are the important terms often used in P-N junction? Write names only.

Ans. The important terms which are generally used for P-N junction are as under:

1. Breakdown Voltage
2. Knee Voltage
3. Maximum Forward Current
4. Peak Inverse Voltage
5. Maximum Power Rating

Q. 3. Define diode resistance.

Ans. An ideal diode offers zero resistance when forward bias and infinite resistance when it is reverse biased. But practical or real diode does not behave ideally. It means that the forward resistance of a real diode is quite small and reverse resistance is very high.

Q. 4. What is forward resistance? Explain.

Ans. The resistance offered by the diode in the forward bias is called forward resistance. This resistance is different for the flow of direct current (d.c.) and for alternating current (a.c.). Hence, this resistance is of two types:

(*a*) d.c. forward resistance

(*b*) a.c. forward resistance.

Q. 5. What is reverse resistance? Explain.

Ans. The resistance offered by the diode in reverse bias is known as reverse resistance. Ideally it is infinite. Practically, it is very large. It can be d.c. reverse resistance or a.c. reverse resistance depending upon whether the reverse bias is d.c. or a.c. This reverse resistance is due to the existance of leakage current in a reverse bias diode. The reverse bias resistance is very large compared to the forward resistance.

Q. 6. What is diode capacitance? Explain.

Diode offers two types of capacitance, one in forward bias and other in reverse bias. They are

(*i*) Diffusion or Storage Capacitance

(*ii*) Depletion or Transition Capacitance.

Q7. What is a P-N junction?

Ans. The contact surface between the layers of P-type and N-type semiconductor pieces placed together so as to form a P-N junction is called the P-N junction.

Q.8. What is the effect of reverse bias on the width of a P-N junction?

Ans. When the P-N junction is reverse biased, the width of depletion region is increased.

Q.9. What types of carriers are present in space charge region.

Ans. No mobile carrier is present in the space charge region.

Q.10. What is cut-in voltage in semiconductors?

Ans. The forward voltage, at which the current through the P-N junction starts increasing rapidly, is called the cut-in voltage.

Q.11. What do you understand by reverse saturation current of a diode?

Ans. Reverse saturation current of a diode is due to minority carriers and is caused when the diode is reverse biased. Only a very small voltage is required to direct all minority carriers across the junction, and when all minority carriers are flowing across further increase in bias voltage will not cause increase in current. This current is referred to as a reverse saturation current.

Q.12. What is the effect of temperature on the reverse current of a P-N junction?

Ans. Reverse current of a P-N junction increases with the increase in junction temperature.

Q.13. Write the Importance of barrier potential.

Ans. Barrier potential acts as a barrier (wall) which does not allow the electrons and holes to cross the junction. So, if we want them to cross the junction, then an external voltage of appropriate polarity has to be applied in order to overcome the opposition of the barrier potential. Then, the flow of electrons and holes across the junction can restart again.

Q.14. Factors deciding the barrier potential value.

Ans. Following are the factors which decide the value of the barrier potential:

(*i*) Semiconductor material used (Silicon or Germanium).

(*ii*) The intrinsic concentration of Si or Ge before doping.

(*iii*) The level of doping on *P* and *N* sides

(*iv*) Temperature.

Q.15. What is a P-N junction Diode?

Ans. A P-N junction is called as P-N junction diode or a semiconductor diode. This is also called as crystal diode because it is formed in a crystal (Silicon or germanium). A P-N junction diode is an unidirectional two terminals semiconductor device. It conducts only in one direction and offers high resistance in other direction.

REVIEW QUESTIONS

1. What is a PN-junction ? How is it formed.
2. Explain the formation of depletion region in a PN-junction.
3. Explain how a barrier potential is developed at the PN-junction.
4. Explain the action of PN-junction diode under forward bias and reverse bias.
5. Draw and explain the V-I characteristics of a PN junction diode.
6. What is meant by biasing a P-N junction diode? Why is it necessary ?

7. What is meant by the term 'Barrier potential'? What are the values for Silicon and Germanium diodes ?
8. Define reverse saturation current of a diode. What are the factors that decide the magnitude of reverse saturation current ?
9. Why should the operation in the breakdown region be avoided ?
10. Why silicon is more widely used semiconductor material ?
11. Intrinsic semiconductors behave like insulators at low temperatures, explain it.
12. Explain the terms :

    (*i*) Diffusion (*ii*) Depletion region.
13. What is static resistance of the diode ? How will you find the dynamic resistance ?
14. Consider a step-graded junction under open-circuited condition. Upon what four parameters does the contact difference of potential depend ?
15. Plot the hole current, the electron current, and the total current as a function of distance on the both sides of a P-N junction. Indicate the transition region.
16. Draw the energy band diagram for a P-N junction under open-circuit condition. Clearly indicate various energy levels in P-region, space charge region and N-region. How will it be modified if *P-N*-junction is forward biased ?
17. Draw the characteristics of a junction diode at four different temperatures *e.g.*, 100°, 25°, 100°, 200° Celsius and comment on the curves.
18. For a semiconductor diode, define static and dynamic resistances.
19. Explain the formation of potential barrier in a P-N junction. Why is Silicon preferred to germanium in the manufacturing of semiconductor device ?
20. Plot and explain the hole current, the electron current and the total current as a function of distance on both sides of a P-N junction. Indicate the transition region.
21. Discuss 'Junction diode switching time' to justify diode reverse recovery time, storage time, and transition time, in brief. Elaborate these definitions by labelling on appropriate waveforms for diode.
22. Derive an expression for the dynamic resistance of a diode in forward bias condition and hence interpret graphically its variation with (*i*) Current and (*ii*) Temperature.
23. What are the various break down mechanisms in junction diode ? Give examples of, at least one type, for each of these mechanisms and elaborate their applications.
24. How the dynamic resistance *r* of a P-N junction diode vary with (*i*) current (*ii*) temperature ? What is the magnitude of *r* for Si P-N junction diode at room temperature and for a d.c. current of 1 mA.
25. Discuss the current flow mechanism in a P-N junction under :

    (*i*) no bias

    (*ii*) forward bias

    (*iii*) reverse bias conditions
26. A germanium diode carries a current of 1 mA at room temperature when a forward bias of 0.15 V is applied. Estimate the reverse saturation current at room temperature.
27. Which is more sensible to a charge in temperature, forward current or reverse current ? Explain why ?
28. Explain, the transition capacitance and diffusion capacitance of a P-N junction.
29. Define static and dynamic forward resistance for a PN-junction diode.
30. Write the volt-ampere equation for a PN-diode. Explain the meaning of each symbol.
31. Explain the term transition capacitance of a PN-diode.
32. Explain the term diffusion capacitance of a forward biased diode.
33. Distinguish between avalanche and zener mechanisms.
34. Write the applications of a PN-junction diode.
35. Differentiate between transition capacitance and diffusion capacitance of a PN-junction diode.
36. What is static resistance of diode? How will you find the dynamic resistance?
37. Explain the formation of potential barrier in *P-N* junction. Why is silicon preferred to germanium in the manufacturing of semiconductor device.

NUMERICAL PROBLEMS

1. Determine the forward bias voltage applied to a silicon diode to produce a forward current of 10 mA and reverse saturation current $I_0 = 25 \times 10^{-7}$ A at room temperature. **[Ans.** 0.4 Volt]
2. A germanium diode draws 50 mA with a forward bias of 0.27 V. The junction is at room temperature of 27° C. Determine the reverse saturation current of the diode. **[Ans.** 1.545 μA]
3. Determine ac resistance for a semiconductor diode with a forward bias of 0.25 V. Reverse saturation current at room temperature is of 1.2 μA. **[Ans.** 1.445 Ω]
4. Determine the current flowing in the circuit shown in figure 2.51. Also determine the potential of point A. **[Ans.** 5.5 mA, 11V]
5. Determine the current flowing in the curcuit shown in figure 2.51.

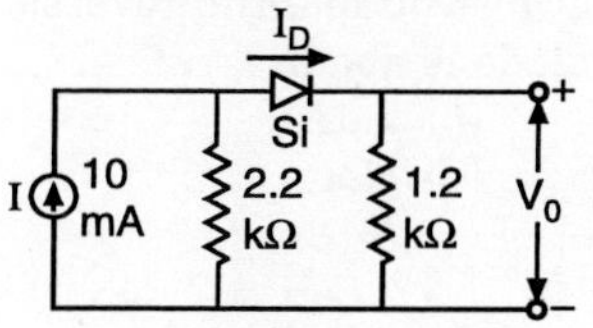

Fig. 2.51.

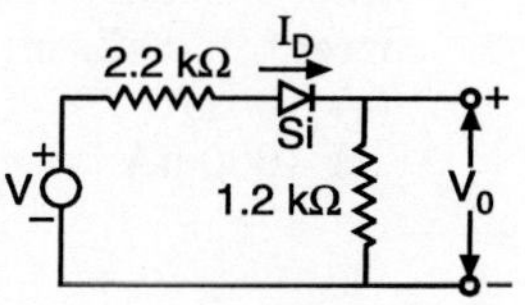

Fig. 2.52.

5. Determine the current through 2 kW resistor in the circuit shown in figure 2.52.

[Hint. $R_{Th} = \dfrac{6 \times 4}{6+4} = 2.4$ kΩ;

$V_{Th} = \dfrac{15}{6+4} \times 4 = 6$ V-s

$$I = \frac{V_{Th} - V_0}{R_{Th} + R} = \frac{6-0.7}{2.4+2} = 1.205 \text{ mA}]$$

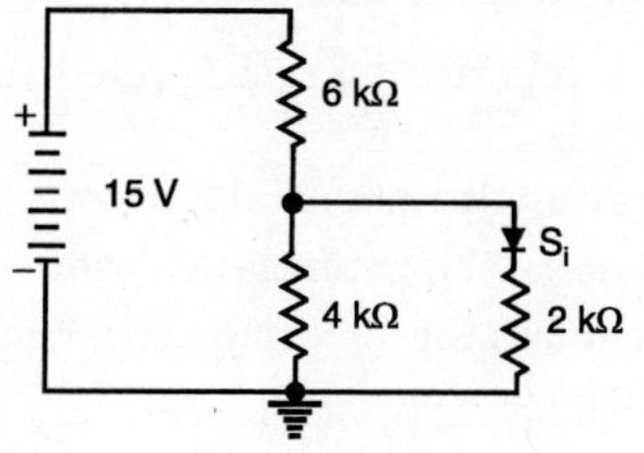

Fig. 2.53.

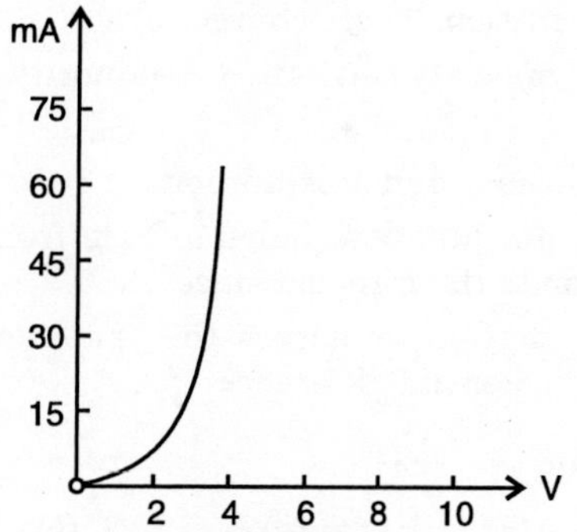

Fig. 2.54.

7. A diode is connected in series with a load resistance of 200 Ω across a dc supply voltage of 12 V. Draw dc load line and determine the quiescent point on the given characteristic of the diode (figure 2.53). **[Ans.** 3 V, 45 mA]

OBJECTIVE TYPE QUESTIONS

A : Choose one correct answer in the following:

1. In a semiconductor diode, the barrier potential offers oppositon to only
   (*a*) free electrons in the N-region
   (*b*) holes in the P-region
   (*c*) minority carriers in both regions
   (*d*) majority carriers in both regions
2. When reverse-bias is applied to a junction diode, it
   (*a*) raises the potential barrier
   (*b*) lowers the potential barrier
   (*c*) greatly increases the minority-carrier current
   (*d*) greatly increases the majority-carrier current
3. Avalanche breakdown in a semiconductor or diode occurs if
   (*a*) forward current exceeds a certain value
   (*b*) reverse bias exceeds a certain value.
   (*c*) forward bias exceeds a certain value
   (*d*) the potential barrier is reduced to zero.
4. The capacitance of a reverse-biased PN-junction
   (*a*) increases as the reverse bias is increased
   (*b*) increases as the reverse bias is decreased
   (*c*) depends mainly on the reverse saturation
   (*d*) makes the PN-junction more effective at current high frequencies
5. In reverse bias, the number of minority carriers crossing the junction of a diode depends primarily upon the

(*a*) concentration of doping impurities.
(*b*) rate of thermal generation of electron-hole pairs
(*c*) magnitude of the potential barrier
(*d*) None of the above

6. When a forward bias is applied to a crystal diode, it
   (*a*) raises the potential barrier
   (*b*) lowers the potential barrier
   (*c*) reduces the majority-carrier current ratio
   (*d*) none of the above.
7. In a p-n junction, potential barrier is established due to the charge on either side of the junction. These charges are
   (*a*) majority carriers (*b*) minority carriers
   (*c*) both *a* and *b*
   (*d*) donor and acceptor ions
8. In a p-n junction, holes diffuse from the p-region to the n-region since
   (*a*) they move across the junction by the potential difference
   (*b*) free electron available in the n-region attract them
   (*c*) the holes concentration in the p-region is greater as compared to n-region
   (*d*) all of the above.
9. At reverse bias, the number of minority carriers crossing the junction of a diode depends primarily on the
   (*a*) concentration of doping impurities
   (*b*) rate of thermal generation of electron-hole pairs
   (*c*) magnitude of the potential barriers
   (*d*) all of the above.
10. When a p-n junction is forward biased
    (*a*) electrons in the n-region are injected into the p-region
    (*b*) holes in the p-region are injected into the n-region
    (*c*) both (*a*) and (*b*) (*d*) none of above
11. Under normal operating voltage, the reverse current in a silicon diode is about
    (*a*) 10 mA (*b*) 1 μA
    (*c*) 1000 μA (*d*) None

Answers

1. (*d*)	2. (*a*)	3. (*b*)	4. (*b*)	5. (*b*)	6. (*b*)	7. (*d*)
8. (*c*)	9. (*b*)	10. (*c*)	11. (*b*)			

UNIT—II

Rectifiers and Filters

The p-n junction as a Rectifier, Half wave Rectifier, Full wave Rectifier, Bridge Rectifier, Harmonic components in a Rectifier circuit, Inductor Filters, Capacitor Filters, L-Section Filters, π-section Filters, Comparison of Filters, Voltage Regulation using Zener Diode.

Above Syllabus has been covered in following Chapters:

CHAPTER 3 : Rectifiers and Filters

CHAPTER 4 : Zener Diode and Voltage Regulation

Chapter 3

Rectifiers and Filters

SYLLABUS

The p-n junction as a Rectifier, Half wave Rectifier, Full wave Rectifier, Bridge Rectifier, Harmonic components in a Rectifier circuit, Inductor Filters, Capacitor Filters, L-Section Filters, π-section Filters, Comparison of Filters, Voltage Regulation using Zener Diode.

Chapter Outline

3.1. INTRODUCTION

A diode has unidirectional property that it conducts only in forward bias and in reverse bias it is cut-off. This property makes the diode a versatile device. A few applications may be listed as ahead :

1. Diode is used in rectifiers which are used in d.c. power supplies.
2. Diode is used as a switch in logic circuits in computers.
3. Diode is used in waveshaping circuits such as clippers and clampers.
4. Zener diodes are used in voltage regulator circuits.
5. Varactor diodes are used in radio and TV receivers.

3.2. POWER SUPPLY

Most of the devices in electronic equipments require a d.c. voltage. This d.c. voltage may be obtained with the help of dry cells or battery. We may also use a battery eliminator instead of dry cells. This battery eliminator changes the a.c. voltage from mains into d.c. voltage and in this way removes the need for dry cells.

Today, almost all electronic equipments contain a circuit called power supply to change the mains a.c. into d.c. The block diagram of power supply is shown in the figure 3.1.

The function of different blocks is as under:

(*i*) **Transformer:** The transformer steps up or steps down the a.c. mains voltage according to the requirement of d.c. to the circuit. Transformer also provides isolation.

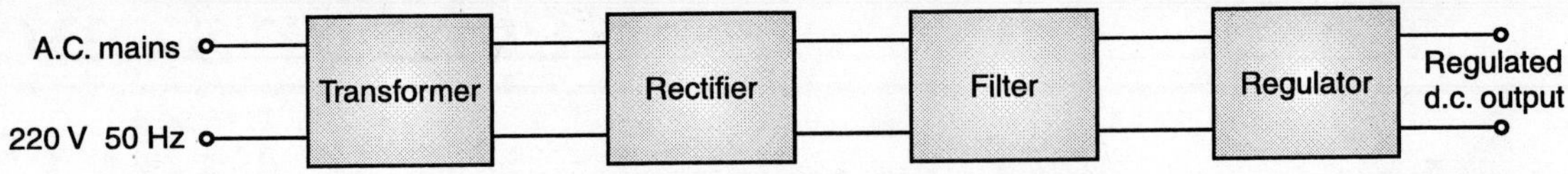

Fig. 3.1. Block diagram of a power supply

(*ii*) **Rectifier:** The function of the rectifier is to convert the a.c. voltage into d.c. voltage.

(*iii*) **Filter:** The d.c. output of rectifier may contain ripples. These ripples are removed with the help of filters.

(*iv*) **Regulator:** If there is no regulator in the circuit, the d.c. output will change if either input voltage or load changes. Therefore, the function of the regulator is to provide constant d.c. voltage or regulated d.c. voltage irrespective of the change in load or input voltage.

3.3. P-N JUNCTION AS A RECTIFIER

A rectifier is a circuit which is used to convert a.c. voltage into the pulsating d.c. voltage. A rectifier circuit uses one or more diodes. Broadly there are two types of rectifiers :

(*a*) Half-wave rectifier

(*b*) Full-wave rectifier.

3.4. HALF-WAVE RECTIFIER

Figure 3.2 shows the circuit of a half-wave rectifier. In half-wave rectifier circuit there is a transformer at the input of the circuit. The transformer couples a.c. input voltage to the rectifier circuit. The function of the transformer is two fold. Firstly, it allows us to step the source voltage up or down. Secondly, the transformer provides the isolation of a.c. power source from the rectifier circuit. It reduces the risk of electrical shock.

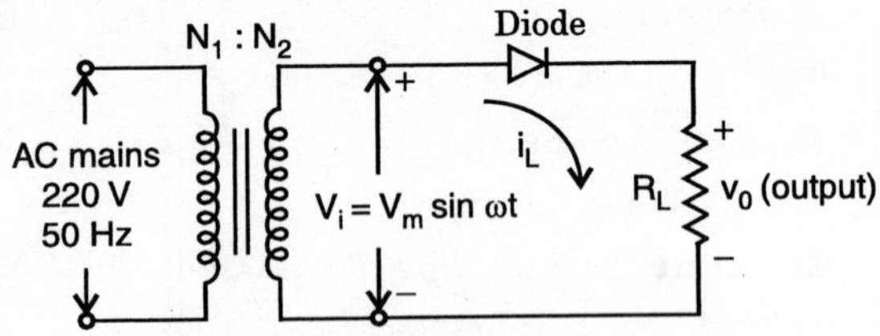

Fig. 3.2. Half wave rectifier

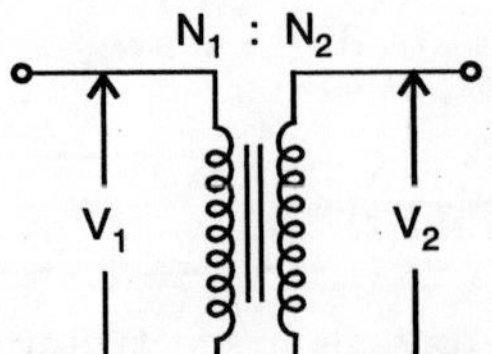

The primary and secondary voltages of transformer are related as

$$\frac{V_2}{V_1} = \frac{N_2}{N_1}$$

Fig. 3.3. Transformer

where V_1 = The r.m.s. primary voltage
N_1 = Number of turns in primary coil
V_2 = The r.m.s. secondary voltage
N_2 = Number of turns in secondary coil

The primary of the transformer is connected to the a.c. mains. An a.c. voltage is induced across the secondary coil of the transformer. This voltage may be stepped up or down depending upon the number of turns. The secondary a.c. voltage may be given as

$$v_i = V_m \sin \omega t \qquad ...(3.1)$$

During the positive half-cycle of the input voltage, the diode is forward-biased as shown in figure 3.4.

Since, diode is forward-biased, it conducts and a current i_L flows through the load resistor R_L to produce output voltage v_O. In forward-bias, the voltage drop across the diode is very small (0.7 V for S_i diode and 0.3 V for G_e diode). For practical purpose, we assume that diode is ideal so that the voltage drop across the diode is zero. Therefore, the output voltage v_o is nearly same as the input voltage v_i at every instant.

DO YOU KNOW?

An half wave rectifier is so called because it delivers power to the load during only one half cycle of the ac supply voltage.

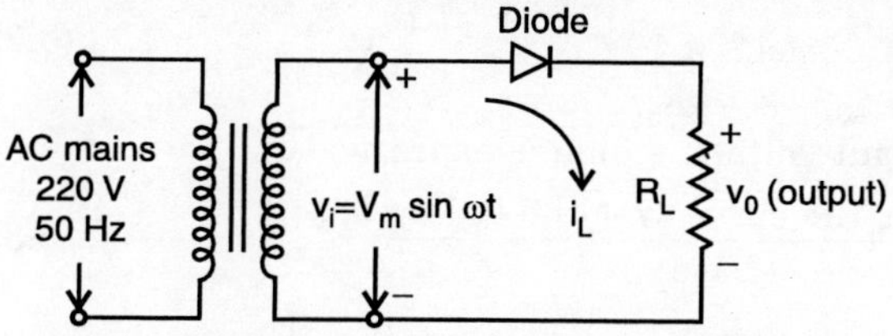

Fig. 3.4. Half-wave rectifier for positive half cycle

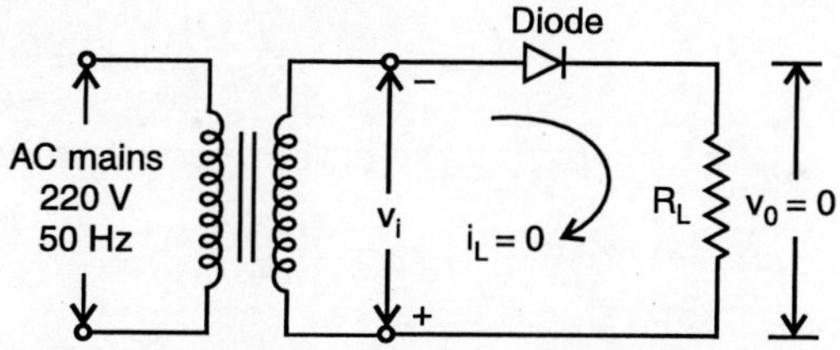

Fig. 3.5. Half-wave rectifier for negative half cycle

During the negative half-cycle of the input a.c. voltage, the diode is reverse-biased as shown in the figure 3.5.

Since, the diode is reverse-biased, it does not conduct. Therefore, no current flows in the circuit (i_L= 0) and the output voltage across load resistor is also zero (v_o = 0).

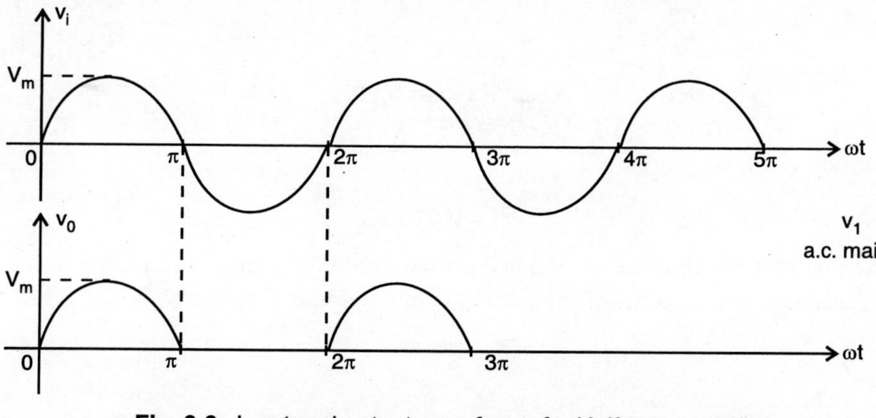

Fig. 3.6. Input and output waveforms for Half-wave rectifier

Thus, from above it is clear that only the positive half-cycle is utilized. During the positive cycle, output voltage available across the load is same as the input voltage. During negative cycle, no output voltage is available. From waveform, it is clear that output voltage is not a perfect d.c. It is unidirectional and pulsating voltage.

3.4.1. Average or D.C. Values of Output Voltage and Load Current for a Half-wave Rectifier

Let us consider the curves of output voltage v_0 and load current i_L for a half-wave rectifier.

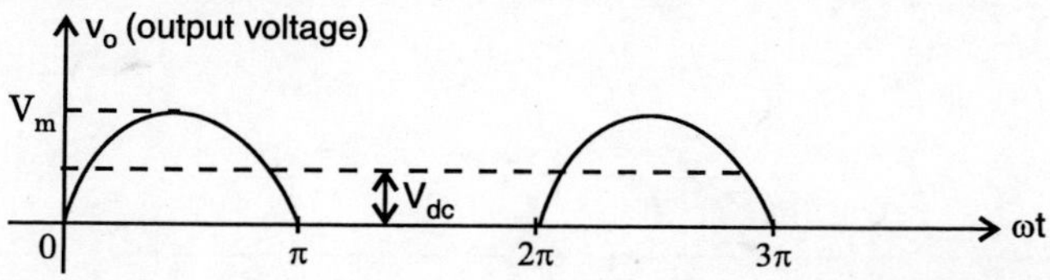

Fig. 3.7. Output voltage waveform for Half-wave rectifier

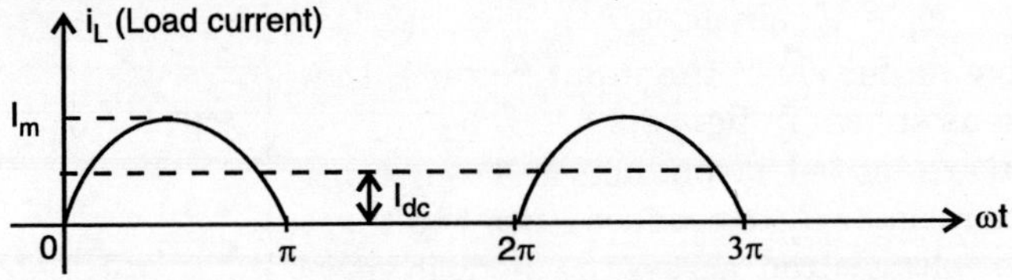

Fig. 3.8. Load current waveform for Half-wave rectifier

Figure 3.8 shows the output voltage of a Half-wave rectifier. Mathematically, it may be given as

$$v_o = \begin{cases} V_m \sin \omega t & \text{for } 0 \le \omega t \le \pi \\ 0 & \text{for } \pi \le \omega t \le 2\pi \end{cases} \quad \text{...(3.2)}$$

We have to find the d.c or average value of output voltage which is expressed as

we know that $$V_{ag} = V_{dc} = \frac{\text{Area under the curve over the full cycle}}{\text{Base}} \quad \text{....(3.3)}$$

But $$\text{Area} = \int_0^{2\pi} v_o.d(\omega t) = \int_0^{\pi} v_o.d(\omega t) + \int_{\pi}^{2\pi} v_o.d(\omega t)$$

Using the equation (3.2), we get

$$\text{Area} = \int_0^{\pi} V_m \sin \omega t.d(\omega t) + \int_{\pi}^{2\pi} 0.d(\omega t) \quad \text{...(3.4)}$$

$$\text{Area} = [-\cos \omega t]_0^{\pi} + 0$$

or $$\text{Area} = V_m[-\cos \pi - (-\cos 0)]$$

$$\text{Area} = V_m \cdot [1 + 1] = 2\, V_m$$

From equation (3.3), we get

$$V_{dc} = \frac{2V_m}{2\pi} = \frac{V_m}{\pi} = 0.318\, V_m \quad \text{...(3.5)}$$

Hence, the d.c. value of output voltage is 31.8 per cent of the maximum a.c. input voltage.

Similarly, average or d.c. value of load current may be found as follows:

$$I_{avg} = I_{dc} = \frac{V_{dc}}{R_L} \quad \text{...(3.6)}$$

But, V_{dc} is given as $$V_{dc} = \frac{V_m}{\pi}$$

Therefore, $$I_{dc} = \frac{V_m}{\pi R_L} = \frac{I_m}{\pi}^{*} \quad \text{..(3.7)}$$

$$I_{dc} = 0.318\, I_m$$

Hence, the d.c. value of load current is 31.8% of the maximum value of load current.

In all above calculations, we have assumed that the diode resistance in forward-bias is zero. If forward resistance r_F is taken into account, the expressions for V_{dc} and I_{dc} will be as

$$V_{dc} = \frac{V_m}{\pi} - I_{dc}.R_F \quad \text{...(3.8)}$$

$$I_{dc} = \frac{V_m}{\pi(R_L + R_F)} \quad \text{...(3.9)}$$

* Since $\frac{V_m}{R_L} = I_m$.

3.4.2. Peak Inverse Voltage for Half-Wave Rectifier

Figure 3.9 shows the half-wave rectifier during negative cycle of a.c. voltage. During negative cycle, the diode is reverse-biased and does not conduct. The current in the circuit is zero, and voltage across the load resistor is also zero. Hence, according to KVL, whatever is the applied input voltage, it will appear across the reverse-biased diode. The maximum value of this reverse voltage occurs at the peak of the negative cycle of the input voltage and is equal to V_m. This maximum voltage is called peak-inverse voltage (PIV). PIV represents the maximum voltage which a diode used in reverse-bias must withstand. Hence for a half-wave rectifier, we have

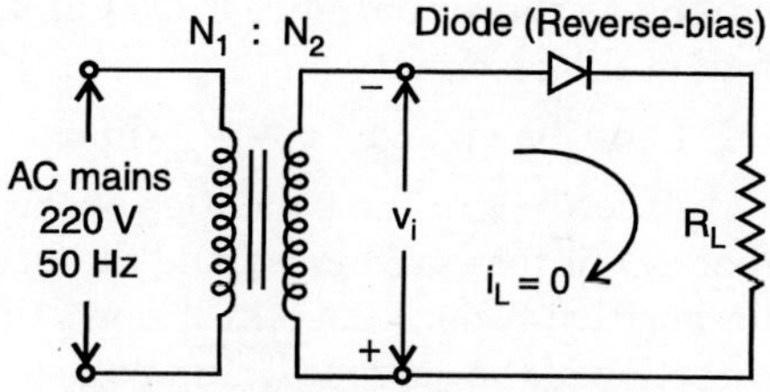

Fig. 3.9. Peak inverse voltage for half-wave rectifier

$$\textbf{Peak Inverse voltage (PIV)} = V_m \qquad \text{...(3.10)}$$

3.4.3. Drawbacks of Half-Wave Rectifier Circuit

Following are the drawbacks of a half wave rectifier circuit:

(*i*) The ripple factor of half wave rectifier circuit is 1.21, which is quite high. The output contains lot of varying components.

(*ii*) The maximum theoretical rectification efficiency is found to be 40%. The practical value will be less than this. This indicates that half wave rectifier circuit is quite inefficient.

(*iii*) The circuit has low transformer utilization factor, showing that the transformer is not fully utilized.

(*iv*) The d.c. current is flowing through the secondary winding of the transformer which may cause d.c. saturation of the core of the transformer. To minimize the saturation, the transformer size have to be increased accordingly. This increases the cost.

Because of all these drawbacks, the half-wave rectifier circuit is normally not used as a power rectifier circuit.

3.5. FULL-WAVE RECTIFIER

Full-wave rectifier is that type of rectifier which utilizes both the half cycle of a.c. input voltage. Hence, in a full-wave rectifier an unidirectional current flows through the load during the entire cycle of input voltage. There are two types of full wave rectifier circuits:

(*i*) Centre-tap full-wave rectifier

(*ii*) Full-wave Bridge Rectifier.

3.6. CENTRE-TAP FULL-WAVE RECTIFIER

The centre-tap full wave rectifier circuit makes use of two diodes D_1 and D_2 which are connected to the centre-tap secondary winding of the transformer. Generally, the centre-tap on the secondary is taken at zero voltage reference point. Input a.c. voltage is applied to the primary of the transformer. The voltage between one end of secondary and centre-tap is equal to half of the secondary voltage. This means that if the applied input voltage across the primary is v_1 and across the secondary is v_2, then the voltage between one end of secondary and centre-tap is equal to $v_2/2$. For centre-tap full-wave rectifier, this half of the secondary voltage (*i.e.* $v_2/2$) is taken as the instantaneous input voltage v_i.

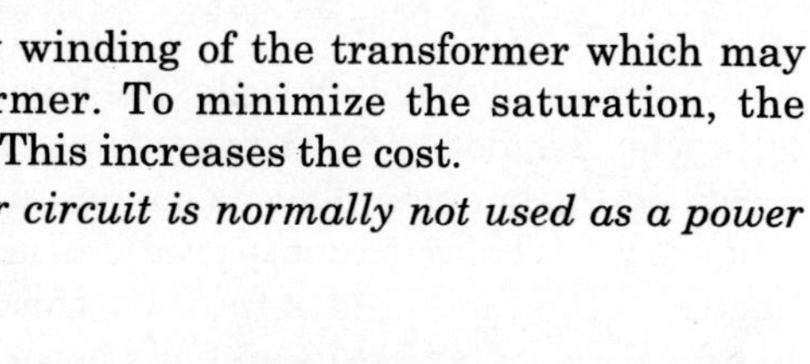

Fig. 3.10. Centre-tap full wave rectifier.

Working:

During positive half-cycle of the input voltage, the polarities of the voltages across the secondary will be as shown in figure 3.11. Let us consider that v_i is the instantaneous value of voltage across half of the secondary.

Let v_i be given as $v_i = V_m \sin \omega t$.

According to the polarities of the secondary winding for positive cycle of the input voltage, the diode D_1 is forward biased and the diode D_2 is reverse-biased. Hence, diode D_1 conducts and D_2 is cut-off for positive cycle. A current i_L flows through the load resistor R_L producing an output voltage v_0. The direction of the current i_L and polarity of the output voltage v_o, developed across the load resistor R_L will be as shown in figure 3.11.

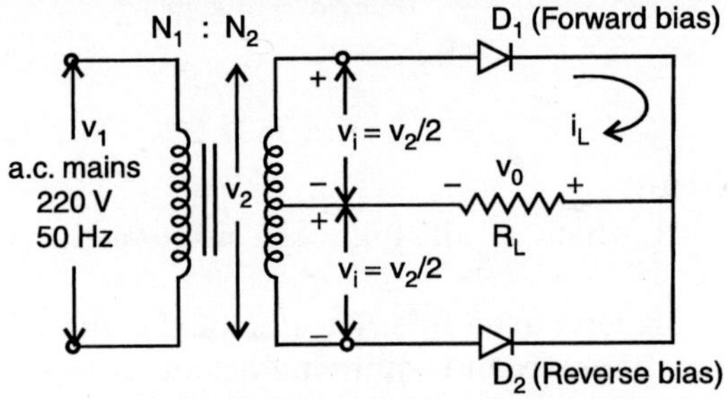

Fig. 3.11. Centre-tap full wave rectifier for positive cycle

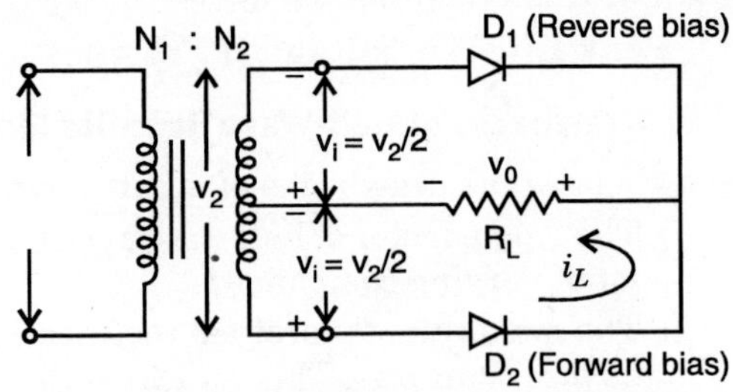

Fig. 3.12. Centre-tap full wave rectifier for negative half-cycle

During negative half cycle, the polarities will be as shown in the figure 3.12. During negative half-cycle, the diode D_1 is reverse-bias and the diode D_2 is forward-bias. Hence diode D_1 is cut-off and diode D_2 conducts. Thus, a current i_L flows through the load resistor R_L in the direction shown in figure 3.12. This current i_L produces an output voltage v_0 across the load resistor R_L. Figure 3.13 illustrates the input and output waveforms for centre-tap full-wave rectifier. It is clear from above that for positive half-cycle, one diode conducts and for negative half-cycle another diode conducts. Hence, an unidirectional current flows continuously in a full-wave rectifier.

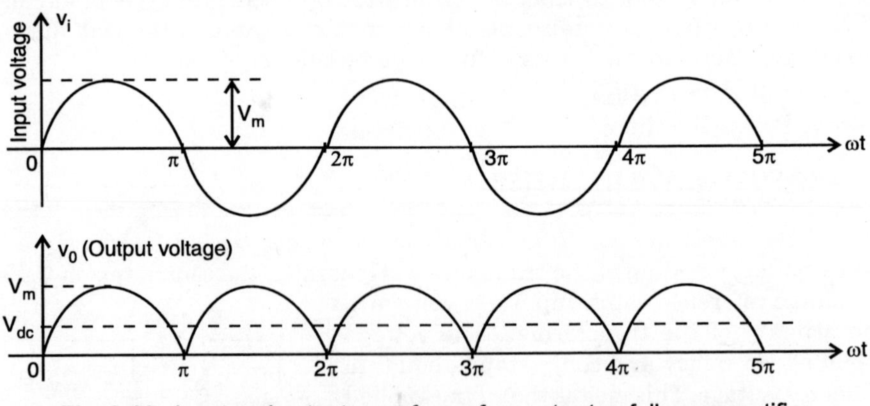

Fig. 3.13. Input and output waveforms for centre-tap full wave rectifier

3.6.1. Peak Inverse Voltage of a Diode in Centre-Tapped Full-Wave Rectifier

Let us consider the circuit of a centre-tap rectifier for negative half cycle as shown in figure 3.14.

In this case, diode D_1 conducts and diode D_2 is cut-off. Let v_i be the instantaneous value of voltage across half of the secondary given as

$$v_i = V_m . \sin \omega t$$

V_m is the maximum or peak value across half of the secondary winding. Diode D_1 is forward-bias and offers almost zero resistance. Therefore, the whole voltage V_m appears across the load resistor R_L. The diode D_2 is reverse-bias. The total voltage across diode D_2 is the sum of the voltage V_m across the lower half of the secondary and the voltage V_m across the load resistor R_L. Hence, the reverse voltage that appears across the non-conducting diode D_2 is $V_m + V_m = 2V_m$. This is peak inverse voltage for each diode in reverse-bias.

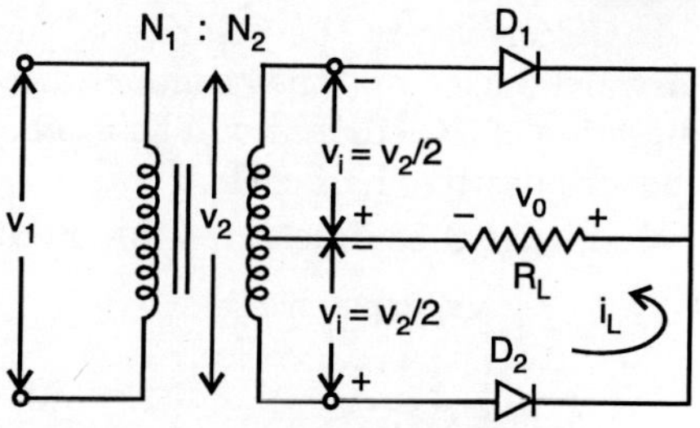

Fig. 3.14. Centre-tapped full wave rectifier for negative half cycle

$\therefore$ **PIV = $2V_m$ for each diode in centre-tapped full wave rectifier**

3.7. FULL-WAVE BRIDGE RECTIFIER*

Bridge rectifier uses four diodes which are connected across the secondary of the transformer. Bridge rectifier avoids the need of a centre-tapped transformer. Figure 3.15 shows the circuit of a full-wave rectifier.

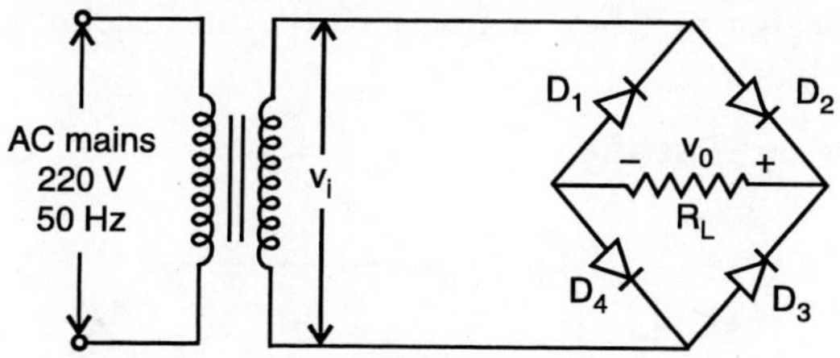

Fig. 3.15. Full-wave bridge rectifier

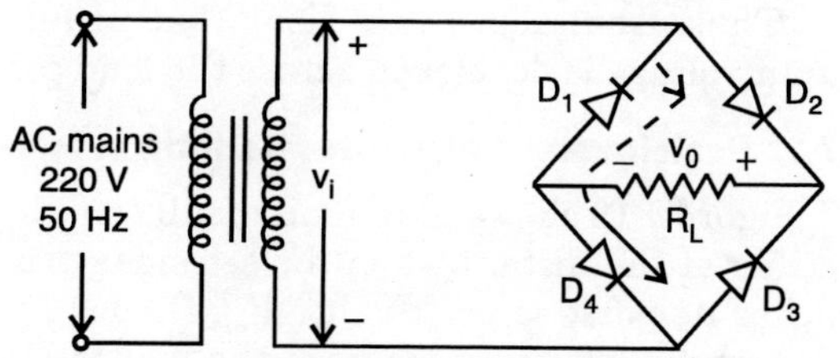

Fig. 3.16. Full-wave bridge rectifier for positive half cycle

During positive-half cycle, the polarity of the voltage across the secondary is as shown in the figure 3.16.

According to the polarity of the voltage v_i for positive half cycle, the diodes D_2 and D_4 are forward-biased and they conduct. Therefore, the direction of current will be as shown in figure. The output voltage v_O is developed across the resistor R_L.

For negative half-cycle, the polarity of the voltage across the secondary is as shown in the figure 3.17.

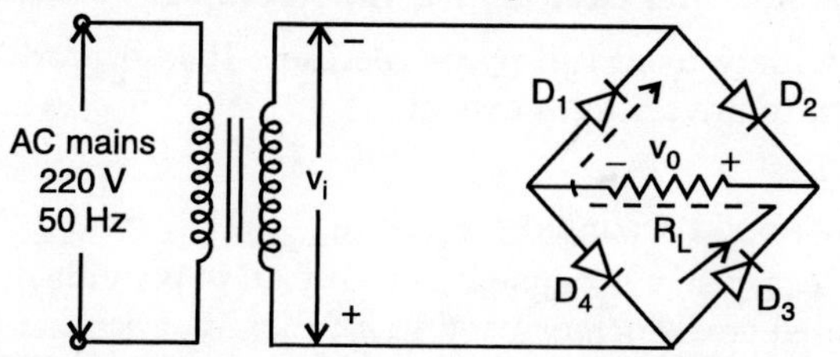

Fig. 3.17. Full-wave bridge rectifier for negative half cycle

DO YOU KNOW?

When a bridge rectifier, as opposed to a two-diode full-wave rectifier, is used, the same dc output voltage can be obtained with a transformer having a higher turns ratio N_1/N_0. This means that with a bridge rectifier, fewer turns of wire are needed in the transformer. Therefore, the transformer used with a bridge rectifier versus a two-diode full-wave rectifier will be small and lighter and will cost less. This benefit alone outweighs using four diodes instead of two in a conventional two-diode full-wave rectifier.

* Draw a neat diagram of a full wave bridge rectifier circuit. Explain its working in details, clearly, making the direction of flow of currents for positive and negative cycles.

According to the polarity of the voltage v_i for negative half cycle, the diodes D_1 and D_3 are forward-biased and they conduct. Diodes D_2 and D_4 are reverse-biased and they do not conduct. The direction of the current will be as shown in figure 3.17. The polarity of output voltage v_O is same as that for positive half cycle.

Figure 3.18 illustrates the input and output voltage waveform for a full-wave bridge rectifier.

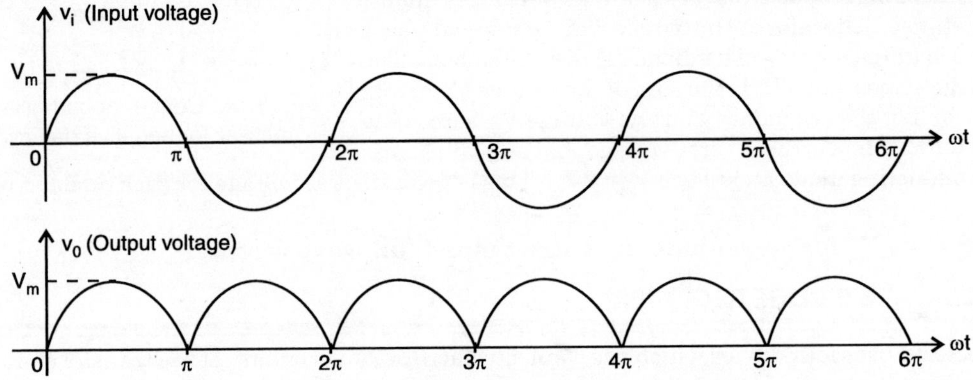

Fig. 3.18. Input and output waveforms for full wave bridge rectifier

Thus, from above it is clear that in full wave bridge rectifier, a fluctuating, unidirectional output voltage is developed across the load continuously.

3.7.1. Peak Inverse Voltage for a Diode in Full-wave Bridge Rectifier

Figure 3.19 shows the circuit of full wave bridge rectifier at the instant when the secondary voltage reaches its positive peak value V_m. Diodes D_1 and D_3 are not conducting whereas the diodes D_2 and D_4 are conducting. The voltage drops across D_2 and D_4 may be assumed zero as diodes D_2 and D_4 have almost zero resistance because they are in forward bias. Hence, points P and Q are at same potential and also, points R and S are at same potential. The total voltage V_m across the secondary winding appears across the load resistor R_L. The reverse voltage across the nonconducting diode D_1 or D_3 is also V_m. Therefore,

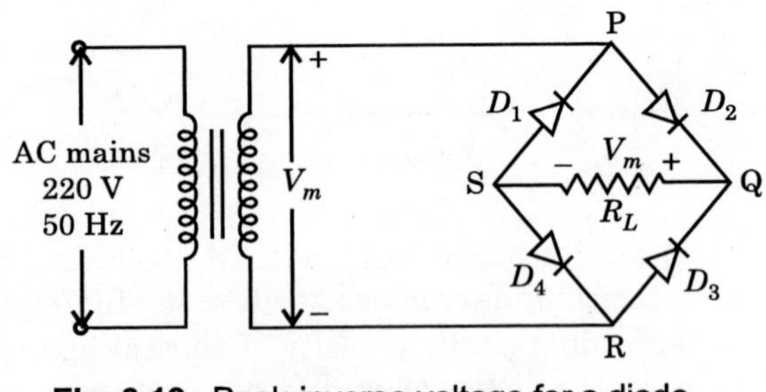

Fig. 3.19. Peak-inverse voltage for a diode in full-wave bridge rectifier

$$PIV = V_m \quad \text{...(3.11)}$$

3.7.2. Advantages and Disadvantages of Bridge Rectifier over Centre-Tap Rectifier

The full-wave bridge rectifier is the most widely used full-wave rectifier. It is preferred over centre-tap full-wave rectifier because it has certain advantages over it.

Advantages

(*i*) Bridge rectifier does not require centre-tapped secondary winding.

(*ii*) For bridge rectifier, the transformer is less costly because it is required to provide only half the voltage of an equivalent centre-tapped transformer used in a full wave rectifier circuit. It means that for a centre-tap rectifier, the transformer secondary must have double the number of turns.

(*iii*) The diode in a bridge rectifier is required to have PIV rating of only V_m. But each of two diodes in a centre-tap rectifier must have a PIV rating of $2V_m$. Therefore, the diodes used in a centre-tap rectifier are costlier than those used in bridge rectifier.

Disadvantages

The only disadvantage of bridge rectifier is that this needs four diodes. Two diodes out of four operate in alternate half cycles. When low d.c. voltages are required, there can be a problem that for low output dc voltage the secondary voltage should be low and the diode drops (1.4 V for S_i and 0.6 V for G_e diode) becomes important. Because of this reason, for low-voltage application, the centre-tap rectifiers are preferred.

3.7.3. Average or D.C. Values of Output Voltage and Load Current for a Full-wave Rectifier

Figure 3.20 shows the output voltage waveform of a full-wave rectifier,

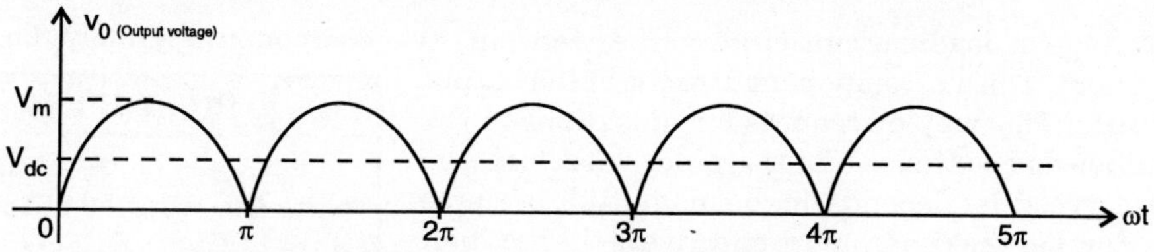

Fig. 3.20. Average or d.c. value of output voltage

Mathematically, the output voltage may be expressed as

$$v_o = \begin{cases} V_m \sin \omega t & \text{for} \quad 0 \le \omega t \le \pi \\ -V_m \sin \omega t & \text{for} \quad \pi \le \omega t \le 2\pi \end{cases} \qquad ...(3.12)$$

Here minus sign indicates that during the second-half cycle, the sine wave is inverted.

We know that average or d.c. value of voltage is given as

$$V_{ag} = V_{dc} = \frac{\text{Area under the curve over the full cycle}}{\text{Base}} \qquad ...(3.13)$$

But
$$\text{Area} = \int_0^{2\pi} v_o.d(\omega t) \qquad ...(3.14)$$

$$\text{Area} = \int_0^{\pi} v_o.d(\omega t) + \int_{\pi}^{2\pi} v_o.d(\omega t)$$

DO YOU KNOW?
The rms value of a full-wave signal is V_{rms} = 0.707 *Vp*, which is the same as V_{rms} for a full sine wave.

Using equation (3.12), we get

$$\text{Area} = \int_0^{\pi} V_m \sin \omega t.d(\omega t) + \int_{\pi}^{2\pi} -V_m \sin \omega t.d(\omega t)$$

$$\text{Area} = [-V_m \cos \omega t]_0^{\pi} + [V_m \cos \omega t]_{\pi}^{2\pi}$$

or
$$\text{Area} = V_m[-\cos \pi + \cos \theta] + [\cos 2\pi - \cos \theta]\, V_m = V_m\, [1+1+1+1] = 4\, V_m$$

Using equation (3.13), we get

$$V_{dc} = \frac{\text{Area}}{\text{Base}} = \frac{4.V_m}{2\pi} = \frac{2V_m}{\pi} = 0.636\, V_m \qquad ...(3.15)$$

Hence, the d.c. value of output voltage for full wave rectifier is 63.6 per cent of the maximum a.c. input voltage. ***This value is twice that of a half-wave rectifier.***

For every minute you are angry, you lose sixty seconds of happiness.

— **Ralph Waldo Emerson**

Similarly, average or d.c. value of load current is

$$I_{dc} = \frac{V_{dc}}{R_L} = \frac{2V_m}{\pi . R_L} = \frac{2I_m}{\pi} \qquad \left[\because \frac{V_m}{R_L} = I_m\right] \quad ...(3.16)$$

Therefore, $$I_{dc} = 0.636\, I_m \quad ...(3.17)$$

In all above calculations, the diodes have been **assumed ideals for simplicity.**

3.8. RIPPLE FACTOR

(V. Important)

(i) Definition

The output voltage or load current of a rectifier contains two components namely d.c. component and a.c. component. The a.c. component present in the output is called a **ripple**. This ripple accounts for pulsations in the output of rectifier and so this is undersirable. Therefore, the effectiveness of a rectifier depends on the magnitude of ripple in the output. Smaller the ripple, more effective is the rectifier. Ripple factor is a measure of purity of the output of a rectifier.

DO YOU KNOW?

The ripple factor indicates how close the rectified output is to the pure ideal dc voltage waveform.

(ii) Mathematical Expression

Mathematically, ripple factor may be expressed as

$$\text{Ripple factor, } r = \frac{\text{The r.m.s. value of a.c. component of output voltage}}{\text{The d.c. component of output voltage}} \quad ...(3.18)$$

$$r = \frac{V_r\,(rms)}{V_{dc}} = \frac{I_r\,(rms)}{I_{dc}} \quad ...(3.19)$$

Here $V_r\,(rms)$ = The r.m.s. value of the a.c. component of the output voltage

V_{dc} = The average or d.c. value of the output voltage

$I_r\,(rms)$ = The r.m.s. value of the a.c. component of current

I_{dc} = The average or d.c. value of the load current

But, the r.m.s. value of the rectified output load current is given by the expression

$$I_{rms} = \sqrt{I_{dc}^2 + I_r\,(rms)^2} \quad ...(3.20)$$

Dividing both sides by I_{dc}

$$\frac{I_{rms}}{I_{dc}} = \sqrt{\frac{I_{dc}^2 + I_r\,(rms)^2}{I_{dc}^2}} \quad ...(3.21)$$

$$\frac{I_{rms}}{I_{dc}} = \sqrt{1 + \left[\frac{I_r\,(rms)}{I_{dc}}\right]^2} \quad ...(3.22)$$

Simplifying the above equation, we get

$$\frac{I_r\,(rms)}{I_{dc}} = \sqrt{\left(\frac{I_{rms}}{I_{dc}}\right)^2 - 1} \quad ...(3.23)$$

or ripple factor, $$r = \sqrt{\left(\frac{I_{rms}}{I_{dc}}\right)^2 - 1} \quad ...(3.24)$$

with the help of above equation, ripple factor for half-wave and full-wave rectifier may be calculated.

3.8.1. Ripple Factor for a Half-wave Rectifier

We know that ripple factor is expressed as

$$r = \sqrt{\left(\frac{I_{rms}}{I_{dc}}\right)^2 - 1} \qquad ...(3.25)$$

For a half-wave rectifier, we know that

$$I_{dc} = \frac{I_m}{\pi} \qquad ...(3.26)$$

Also, I_{rms} = *rms* value of load current

But, the rms value of load current is expressed by the following expression :

$$I_{rms} = \sqrt{\frac{1}{2\pi}\int_0^{2\pi} i_L^2 . d(\omega t)} \qquad .(3.27)$$

Here $i_L = I_m \sin\omega t \; for \; 0 \le \omega t \le \pi = 0 \; for \; \pi \le \omega t \le 2\pi$ ...(3.28)

$$I_{rms} = \sqrt{\frac{1}{2\pi}\left[\int_0^{\pi} I_m^2 \sin^2 \omega t . d(\omega t) + \int_{\pi}^{2\pi} 0 . d(\omega t)\right]} \qquad ...(3.29)$$

$$I_{rms} = \sqrt{\frac{1}{2\pi}\int_0^{\pi} I_m^2 . \sin^2 \omega t . d(\omega t)} \qquad ...(3.30)$$

$$I_{rms} = \sqrt{\frac{I_m^2}{2\pi}\int_0^{\pi}\frac{(1-\cos 2\omega t)}{2}.d(\omega t)} = \sqrt{\frac{I_m^2}{2\pi \times 2}\left[\omega t - \frac{\sin 2\omega t}{2}\right]_0^{\pi}}$$

$$I_{rms} = \sqrt{\frac{I_m^2}{4\pi}[(\pi - 0) - (0 - 0)]} = \sqrt{\frac{I_m^2}{4}} = \frac{I_m}{2} \qquad ...(3.31)$$

Now, substituting the values of I_{dc} and I_{rms} in equation (3.31), we get

$$r = \sqrt{\left(\frac{I_{rms}}{I_{dc}}\right)^2 - 1} \qquad ...(3.32)$$

$$r = \sqrt{\left(\frac{I_m/2}{I_m/\pi}\right)^2 - 1} \qquad ...(3.33)$$

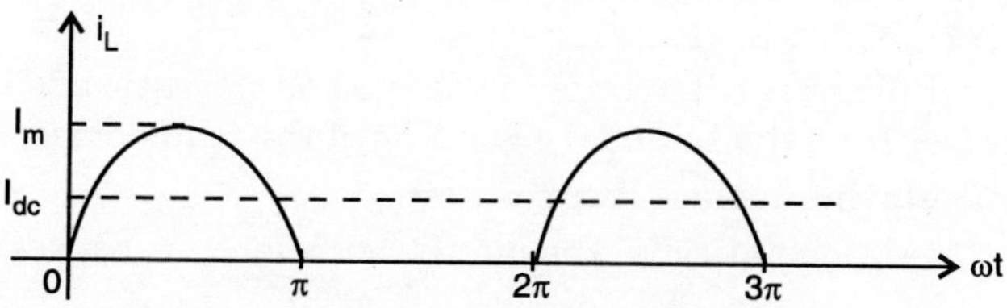

Fig. 3.21. Load current waveform for half rectifier.

Simplifying the above equation, we get

$r = 1.21$ ...(3.34)

Hence, ripple factor for half-wave rectifier is **1.21.**

3.8.2. Ripple Factor for a Full-Wave Rectifier

We know that ripple factor is given as

$$r = \sqrt{\left(\frac{I_{rms}}{I_{dc}}\right)^2 - 1} \qquad ...(3.35)$$

But for full-wave rectifier, we know that

$$I_{dc} = \frac{2I_m}{\pi} \quad ...(3.36)$$

Also, I_{rms} = rms value of load current

We know that the r.m.s. value is expressed as

$$I_{rms} = \sqrt{\frac{1}{\pi}\int_0^{\pi} i_L^2 \, d(\omega t)}$$

$\because$ $i_L = I_m \sin \omega t$ for $0 \le \omega t \le \pi$

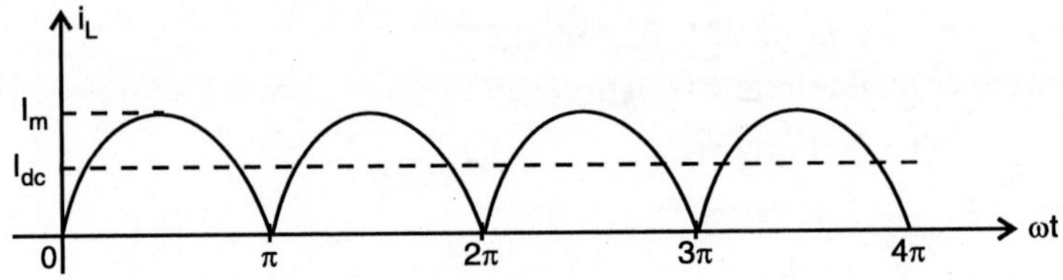

Fig. 3.22. Load current waveform for full-wave rectifier

Therefore, $$I_{rms} = \sqrt{\frac{1}{\pi}\int_0^{\pi} I_m^2 \sin^2 \omega t . d(\omega t)} = \sqrt{\frac{I_m^2}{\pi}\int_0^{\pi} \frac{1}{2}(1 - \cos 2\omega t) d(\omega t)}$$

$$I_{rms} = \sqrt{\frac{I_m^2}{\pi}\left[\frac{\omega t}{2} - \frac{\sin 2\omega t}{4}\right]_0^{\pi}} = \sqrt{\frac{I_m^2}{\pi} \times \frac{\pi}{2}} = \frac{I_m}{\sqrt{2}} \quad ...(3.37)$$

Now, substituting the values of I_{dc} and I_{rms} in equation (3.35), we get

$$r = \sqrt{\left(\frac{I_{rms}}{I_{dc}}\right)^2 - 1} = \sqrt{\left(\frac{I_m/\sqrt{2}}{2I_m/\pi}\right)^2 - 1} = \sqrt{\left(\frac{I_m . \pi}{\sqrt{2}.2I_m}\right)^2 - 1} = \sqrt{\frac{\pi^2}{8} - 1} = 0.482 \quad ...(3.38)$$

It is clear that the ripple factor of a full-wave rectifier is ***0.482*** *which is much smaller than that of a half-wave rectifier. Due to this reason, full wave rectifier is widely used.*

3.9. EFFICIENCY OF A RECTIFIER

(*i*) Definition

Efficiency of a rectifier is defined as the ratio of d.c. power delivered to the load to the a.c. input power from the secondary winding of the transformer.

(*ii*) Mathematical Expression

Mathematically, the rectifier efficiency is expressed as

Rectifier efficiency, $$\eta = \frac{\text{D.C. power delivered to the load}}{\text{A.C. input power from the transformer secondary}}$$

$$\eta = \frac{P_{dc}}{P_{ac}} \quad ...(3.39)$$

But $P_{dc} = I_{dc}^2 \times R_L$

and $P_{ac} = I_{rms}^2 (R_L + R_F)$ ...(3.40)

Here R_L = Load resistance

R_F = Forward resistance of diode

Putting the values of P_{dc} and P_{ac} in equation (3.39), we get

$$\eta = \frac{I_{dc}^2 \times R_L}{I_{rms}^2 (R_L + R_F)} \qquad ...(3.41)$$

With the help of this general expression, the efficiency of half-wave rectifier and full-wave rectifier may be calculated.

3.9.1. Efficiency of a Half-Wave Rectifier

General expression for rectifier efficiency is

$$\eta = \frac{I_{dc}^2 \times R_L}{I_{rms}^2 (R_L + R_F)} \qquad ...(3.42)$$

For half wave rectifier, we know that

$$I_{dc} = \frac{I_m}{\pi} \qquad ...(3.43)$$

and $$I_{rms} = \frac{I_m}{2} \qquad ...(3.44)$$

Putting the values of I_{dc} and I_{rms} in equation (3.42), we get

$$\eta = \frac{(I_m/\pi)^2 \times R_L}{(I_m/2)^2 \times (R_L + R_F)} = \frac{4}{\pi^2} \times \frac{R_L}{R_L + R_F} = \frac{0.406}{1 + \frac{R_F}{R_L}} \qquad ...(3.45)$$

This efficiency will be maximum when $R_L >> R_F$

$\therefore$ $$\eta_{max} = 0.406 = 40.6\% \qquad ...(3.46)$$

Hence, the efficiency of a half-wave rectifier is 40.6% provided that the value of load resistance is large enough as compared to the forward resistance of diode. But, in actual, the efficiency is always less than 40.6%.

3.9.2. Efficiency of a Full-Wave Rectifier

The general expression for rectifier efficiency is

$$\eta = \frac{I_{dc}^2 \times R_L}{I_{rms}^2 (R_L + R_F)} \qquad ...(3.47)$$

For full-wave rectifier, we know that

$$I_{dc} = \frac{2I_m}{\pi} \qquad ...(3.48)$$

and $$I_{rms} = \frac{I_m}{\sqrt{2}} \qquad ...(3.49)$$

Substituting the values of I_{dc} and I_{rms} in equation (3.47), we get

$$\eta = \frac{\left(\frac{2I_m}{\pi}\right)^2 \times R_L}{\left(\frac{I_m}{\sqrt{2}}\right)^2 \times (R_L + R_F)} = \frac{8}{\pi^2} \times \frac{R_L}{R_L + R_F} = \frac{0.812}{1 + \frac{R_F}{R_L}}$$

According to above equation, the efficiency will be maximum when $R_L >> R_F$

Hence $$\eta_{max} = 0.812$$

$$\eta_{max} = 81.2\% \quad ...(3.50)$$

Therefore, the maximum efficiency of a full wave rectifier is twice that of half-wave rectifier. Hence, a full wave rectifier is twice as effective as a half-wave rectifier.

3.9.3. Transformer Utilization Factor

(*i*) Definition

Transformer Utilization Factor may be defined as the ratio of d.c. power delivered to the load and the a.c. rating of the transformer secondary.

(*ii*) Mathematical Expression

$$\text{Mathematically, TUF} = \frac{\text{D.C. power delivered to the load}}{\text{A.C. power rating of the transformer}} = \frac{V_{dc}.I_{dc}}{V_{rms}.I_{rms}} \quad ...(3.51)$$

Assuming $R_L >> R_F$ for half-wave rectifier, we get

$$\text{TUF} = \frac{[V_m/\pi][I_m/\pi]}{\left[V_m/\sqrt{2}\right][I_m/2]}$$

$$\text{Simplifying, we get} \quad \text{TUF} = \frac{2\sqrt{2}}{\pi^2} \quad ...(3.52)$$

$$\text{Solving, we get} \quad \text{TUF} = 0.287 = 28.7\% \quad ...(3.53)$$

However, ideal value of TUF is 100% and practically it should be as high as possible.

If we take R_F into consideration, then we have

$$\text{TUF} = \frac{0.287\, R_L}{R_L + R_F}.$$

> **DO YOU KNOW?**
>
> The values of TUF is too low for the HWR. This value of TUF indicates that the transformer is being utilized to only 28.7% of its full capacity, which is gross under utilization.

3.9.4. Transformer Utilization Factor (TUF) For a Full-Wave Rectifier *(Important)*

TUF of a centre tapped transformer is calculated for the secondary and primary windings separately and then the average TUF is calculated.

$$\text{TUF of secondary windings} = \frac{\text{D.C. load power}}{\text{Power rating of secondary}} = \frac{I_{dc} \times V_{dc}}{I_{rms} \times V_{rms}}$$

Substituting the values, we get

$$\text{TUF of secondary winding} = \frac{(2I_m/\pi)(2V_m/\pi)}{\left(I_m/\sqrt{2}\right)\left(V_m/\sqrt{2}\right)}$$

$$\text{Therefore, Secondary TUF} = \frac{8}{\pi^2} = 0.812 = 81.2\% \quad ...(3.54)$$

As we know, a full-wave rectifier, consists of two half-wave rectifiers, working independently and feeding a common load.

Therefore,

TUF of primary winding = 2 × TUF of HWR = 2 × 0.287

Thus, primary TUF = 0.574

Hence,

$$\text{The average} \quad \text{TUF} = \frac{\text{TUF (primary)} + \text{TUF (secondary)}}{2} = \frac{0.574 + 0.812}{2}$$

$$\text{Thus, average} \quad \text{TUF} = 0.693 = 69.3\% \quad ...(3.55)$$

3.9.5. Voltage Regulation

The variation of d.c. output voltage as a function of d.c. load current is called voltage regulation. In other words, the degree to which a power supply varies in output voltage under conditions of load variations is measured by a term known as voltage regulation.

Voltage regulation is generally expressed as a percentage, *i.e.*,

$$\% \text{ Voltage Regulation } = \frac{V_{\text{no-load}} - V_{\text{full-load}}}{V_{\text{full-load}}} \times 100\% \qquad ...(3.56)$$

Now, for further illustration, let us draw a Thevenin circuit for the half-wave rectifier as shown in figure 3.23, where R_{INT} is the internal resistance of the power supply and, in this case, is equal to $R_S + R_F$.

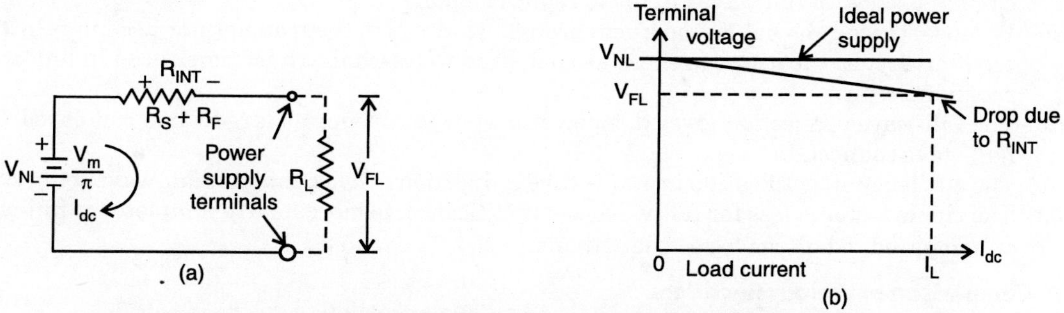

Fig. 3.23. (*a*) Thevenin equivalent circuit of a half-wave rectifier d.c. power supply. (*b*) Graph showing variation in terminal voltage with load current, for an ideal power supply and for a practical power supply.

It may be noted that a power supply which has good voltage regulation has a low percentage regulation. An ideal power supply has a full-load voltage equal to its no-load voltage, and hence zero percent regulation.

For example, if a power supply has a no-load voltage of 9 V and a full-load voltage of 8 V, then the percentage regulation will be:

$$\frac{9-8}{8} \times 100\% = 12.5\%$$

3.9.5.1. Experimental Determination of Internal Resistance R_{INT}

The value of the internal resistance can be determined experimentally as under :

First note the no-load (*i.e.*, open-circuit) voltage of the power supply with a high resistance voltmeter. Then load the power supply noting the full-load voltage and current. The difference between no-load and full-load voltage is the amount of voltage drop which occurred across the internal resistance of the power supply.

Thus, we have

$$V_{NL} - V_{FL} = I_L R_{INT} \qquad ...(3.57)$$

Therefore, R_{INT} will be

$$R_{INT} = \frac{V_{NL} - V_{FL}}{I_L} \text{ ohms} \qquad ...(3.58)$$

where I_L is specified in amperes.

Important Point : *By making use of suitable voltage regulators involving regulator tubes, Zener diodes and transistors, the internal resistance may be reduced to a fraction of an ohm.*

3.9.6. Comparison of Full-wave and Half-wave Rectifier Circuits

For comparison, we assume that the full-wave and half-wave rectifier circuits use identical diodes, identical load resistances and the voltage across half the secondary winding of transformer used in full-wave rectifier circuit is the same as the voltage across the secondary winding of the transformer used in the half-wave rectifier circuit.

(*i*) The d.c. load current in case of full-wave circuit is twice to that in half-wave circuit. Similarly, the d.c. load voltage in full wave circuit is twice that in half-wave circuit.

(*ii*) The lowest ripple frequency in full-wave circuit is twice that in half-wave circuit.
Now, to remove ripple, the additional circuits called filter circuits are used along with rectifier circuits. But, as the frequency is more in full-wave rectifier circuit, the capacitor values required in capacitance filter are much less. Hence, smaller elements are sufficient in filter circuits used with full wave circuit to reduce ripple.

(*iii*) Because, there is no net d.c. current through windings of the transformer used in full-wave rectifier circuit, the losses are less as compared to losses in transformer used in half-wave rectifier circuit.

(*iv*) The full-wave connection gives d.c. power output four times as large when compared with half-wave connection.

(*v*) The efficiency of rectification in a full-wave connection is twice that for half-wave connection.

(*vi*) The ripple factor is less for full-wave *i.e.*, rectification is more nearly complete for full-wave as compared to half-wave rectifier circuit.

3.9.7. Comparison of Various Rectifiers

We can compare various rectifiers with respect to some properties in the form of a table as follows :

Table 3.1: Comparison of Rectifiers

S.No.	Parameters of comparison	Half-wave rectifier	Full-wave rectifier	
			Centre-tap	Bridge
1.	No. of diodes	1	2	4
2.	Transformer necessary	no	yes	no
3.	Rectification efficiency	40.6%	81.2%	81.2%
4.	Ripple factor	1.21	0.482	0.482
5.	Peak inverse voltage	V_m	$2V_m$	V_m
6.	Output frequency	f_i	$2f_i$	$2f_i$
7.	R.M.S. current	$I_m/2$	$I_m/\sqrt{2}$	$I_m/\sqrt{2}$
8	D.C. current	I_m/π	$2I_m/\pi$	$2I_m/\pi$

3.10. FILTER CIRCUITS

As discussed earlier, output of a rectifier consists of a.c. as well as d.c. components. The function of rectifier is to change a.c. voltage into d.c. voltage. Therefore, the presence of a.c. components is most undesirable and hence this should be removed. Filter circuits are used for this purpose.

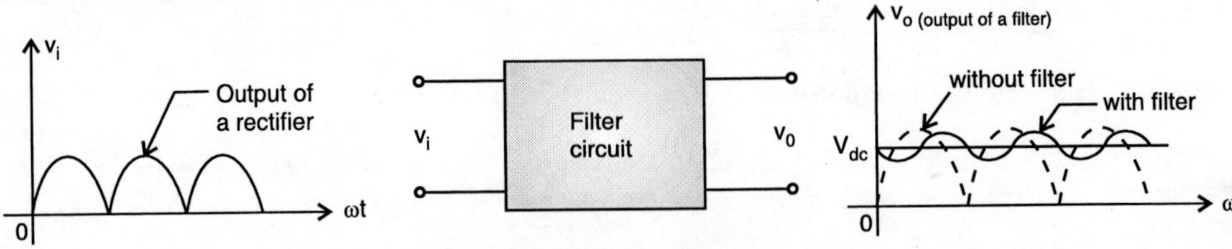

Fig. 3.24. Concept of filter circuit

Figure 3.24 explains the concept of a filter where the full-wave rectified voltage is applied at its input. The filter output although is not a perfect d.c. but it contains less a.c. components. Hence a filter may be defined as a circuit which removes or minimizes a.c. components from the output of a rectifier.

A filter circuit contains passive circuit elements such as inductors, capacitors and resistors. Students are aware of resistor. A brief idea about capacitor and inductor actions is necessary.

***(a)* Capacitor:** The capacitor offers opposition called capacitive reactance represented by X_C and is given as

$$X_C = \frac{1}{\omega C}$$

where ω = frequency,

and C = capacitance

or $$X_C = \frac{1}{2\pi f C} \quad \because \omega = 2\pi f$$

For d.c., we know that frequency $f = 0$

Therefore, $$X_C = \frac{1}{2\pi \times 0 \times C} = \infty$$

$X_C = \infty$ means that a capacitor for a d.c. acts as an **open circuit** and blocks the d.c.

On the other hand, for a.c., f is large, so X_C is small

Therefore, for a.c. a capacitor provides easy path or nearly acts as a **short circuit.***

***(b)* Inductor:** An inductor is also known as choke. The inductive reactance X_L may be expressed as

$$X_L = \omega_L = 2\pi f L$$

For d.c. $$f = 0, X_L = 0$$

Therefore, for d.c. an inductor acts as a **short circuit** or provides an easy path.

For a.c. f is large so X_L is also large

Therefore, for a.c. an inductor offers large opposition and acts almost as an **open circuit.**

3.11. THEORY AND ANALYSIS OF VARIOUS FILTERS

Before, we discuss various types of filters, let us first go through some basic aspects regarding filters. It will be interesting to know that the 121% ripple in the output of the half-wave rectifier and 48% in the full-wave rectifier is more than that can be normally tolerated (except in such applications as battery charging, motors, etc.). The filtering of the unwanted a.c. component to leave the desired d.c. output can be achieved in a number of different ways, according to the load demand. As a matter of fact, where low ripple output is desired it is obvious that full-wave rectification should be used. The following discussion considers only full-wave filtering, wherein the frequency of the ripple is 100 Hz for a 50 Hz a.c. line voltage. This is an advantage where either an inductor (choke) is used to prevent the passage of the ripple current (due to its high inductive reactance to a.c. but quite low resistance to d.c.), or a capacitor is used to short the ripple to ground but leave the d.c. to appear at the output. In addition to this, various combinations of L and C are also used.

DO YOU KNOW?

The filter circuits use resistor, capacitors and inductors. Therefore, they are called as passive filter.

Therefore, we shall discuss theory and analysis of the following filters in the sections to follow:

1. Inductor filter
2. Capacitor filter
3. L-C filter
4. CLC or π-filter

* Infact, for a perfect short-circuit, the opposition (*i.e.*, capacitive reactance) must be zero. However, even in case of small X_c, we also assume a short circuit (or nearly short-circuit).

3.12 INDUCTOR FILTER

The inductor filter has been shown in figure 3.25. An inductor (choke) L is connected in series with the load R. The property of an inductor is that it opposes any change in current through it, is being used here to reduce current ripple in the output of the rectifier.

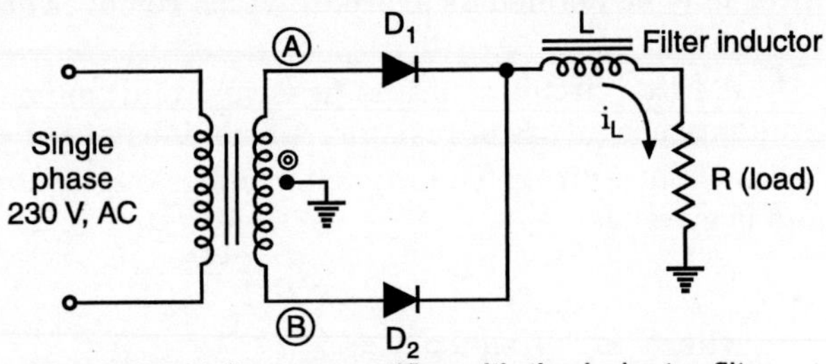

Fig. 3.25. Full wave rectifier with the inductor filter

3.12.1 Operation of the Circuit

The diodes D_1 and D_2 conduct alternately in the positive and negative half cycles to produce a unidirectional current through the load. Due to the inclusion of inductor L in series with the load, the current ripple reduces to a great extent and the load current is smooth as shown in figure 3.26. Higher the value of inductor is, lower will be the peak to peak ripple in the load current waveform.

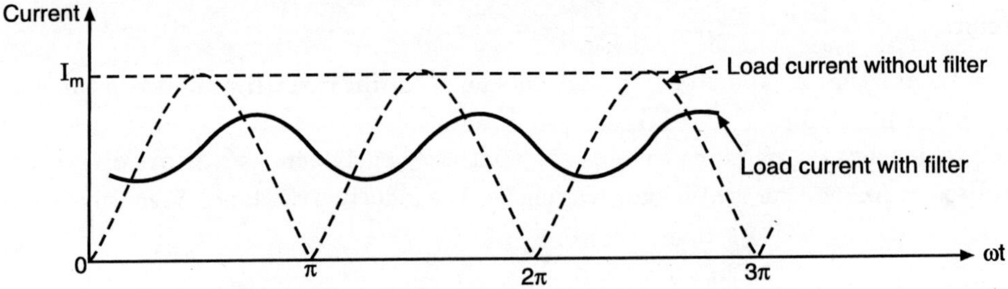

Fig. 3.26. Output current waveform for the inductor filter

3.12.3. Expression for the Ripple Factor

The output current of a full wave rectifier without filter may be expressed in the Fourier series form as under:

$$i_L = \frac{2I_m}{\pi} - \frac{4I_m}{3\pi}\cos 2\omega t - \frac{4I_m}{15\pi}\cos 4\omega t + \ldots \quad \ldots(3.59)$$

The first term in the above expression represents average value. Second term represents the second harmonic component, third term represents the fourth harmonic component of the load current. Neglecting the higher order harmonics, we can simplify equation (3.59) as :

$$i_L = \frac{2I_m}{\pi} - \frac{4I_m}{3\pi}\cos 2\omega t \quad \ldots(3.60)$$

Now, let us substitute the value of I_m in the above equation. Assuming that the forward resistance of diode and the resistance of the choke is quite small as compared to the load resistance, we can write:

$$I_m = \frac{V_m}{R}$$

Therefore, $$I_{L\,dc} = \frac{2I_m}{\pi} = \frac{2V_m}{\pi R} \quad \ldots(3.61)$$

For the second harmonic component, we can write:

$$I_m = \frac{V_m}{Z} \quad \ldots(3.62)$$

We cannot control the negative atmosphere of the world, but we can control the atmosphere of our minds.
— **John C. Maxwell**

Where, $Z = R + j2X_L = [R^2 + 4\,\omega^2\,L^2]^{1/2} \angle\theta$

Therefore, $$I_m = \frac{V_m}{[R^2 + 4\omega^2 L^2]^{1/2}} \qquad ...(3.63)$$

Therefore, substituting this value in equation (3.60), we get the peak value of second harmonic component as under :

$$\text{Peak value of second harmonic} = \frac{4I_m}{3\pi} = \frac{4V_m}{3\pi[R^2 + 4\omega^2 L^2]^{1/2}} \qquad ...(3.64)$$

Therefore, the rms value of the second harmonic component which represents ripple is given by

$$I_{rms} = \frac{\text{Peak value}}{\sqrt{2}} = \frac{4V_m}{3\sqrt{2}\,\pi\left[R^2 + 4\omega^2 L^2\right]^{1/2}} \qquad ...(3.65)$$

Now, the ripple factor is defined as,

Ripple factor $$r = \frac{I_{rms}}{I_{L\,dc}} \qquad ...(3.65)$$

Substituting the values of I_{rms} and $I_{L\,\text{dc}}$ from equation (3.65) and (3.61) respectively, we get:

$$r = \frac{[4V_m / 3\sqrt{2}\,\pi(R^2 + 4\omega^2 L^2)^{1/2}]}{[2\,V_m/\pi R]} = \frac{2}{3\sqrt{2}}\frac{1}{\left[1 + \dfrac{4\omega^2\,L^2}{R^2}\right]^{1/2}} \qquad ...(3.66)$$

Ripple factor on no load

On no load, load resistance $R = \infty$ and load current is zero.

Therefore, $$\frac{4\,\omega^2 L^2}{R^2} \rightarrow 0$$

Hence, $$r \text{ on no load} = \frac{2}{3\sqrt{2}} = 0.472 = 47.2\%$$

Ripple Factor on Load

As the laod resistance R is reduced, the load current increases. But now,

$$\frac{4\omega^2 L^2}{R^2} >> 1$$

Therefore, expression for ripple factor gets modified to

$$r = \frac{2}{3\sqrt{2}} \cdot \frac{1}{\left[\dfrac{4\omega^2 L^2}{R^2}\right]^{1/2}}$$

Therefore, $$r = \frac{1}{3\sqrt{2}}\frac{R}{\omega L} \qquad ...(3.67)$$

Conclusions

The important conclusions which can be drawn from equation (3.67) may be listed as under:

(*i*) The ripple factor is directly proportional to the laod resistance R.

(*ii*) The r is worst (maximum) at no load and improves with increase in the load current. This is exactly opposite behaviour as compared to the capacitor input filter.

(*iii*) This filter is suitable for high load current applications.

(*iv*) The ripple factor can be improved (reduced) by increasing the value of inductor L.

3.13. CAPACITOR INPUT FILTER

The capacitor input filter is used to reduce the ripple contents in the output of a rectifier to obtain a pure dc voltage. A full wave rectifier along with a capacitor input filter is as shown in figure 3.27. Here, C is the filter capacitor which is connected across the load resistance R. The value of C is very large i.e., a few hundred micro-farads in order to reduce the ripple successfully. The electrolytic capacitors are normally used as filter capacitors.

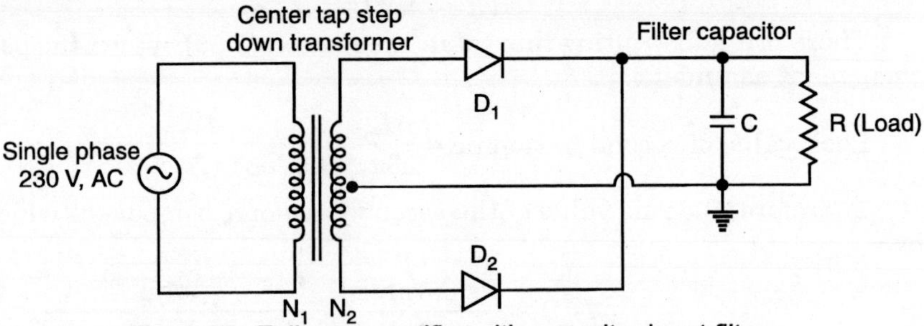

Fig. 3.27. Full wave rectifier with capacitor input filter.

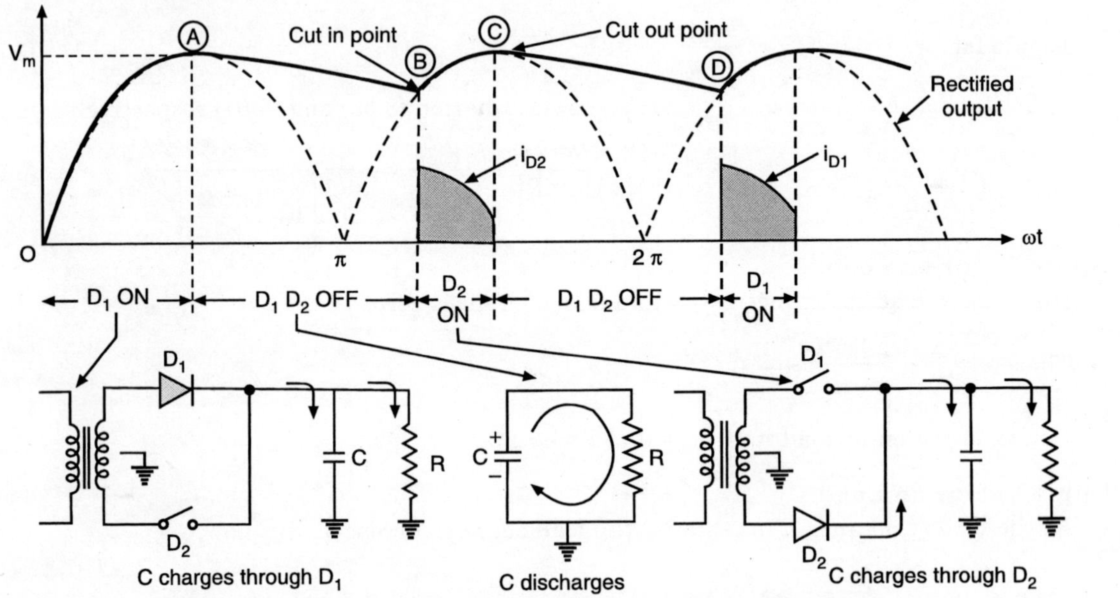

Fig. 3.28. Waveforms and equivalent circuits of capacitor input filter

3.13.1. Operation of the Circuit

Operation of the circuit can be explained in four different intervals. Let us consider the waveforms and equivalent circuits shown in figure 3.28 to understand the operation.

(*i*) Operation in the Interval 0 to A

The initial voltage on capacitor C is assumed to be zero. In the first half cycle of the supply, D_1 is forward biased and starts conducting, D_2 is reverse biased and acts as an open switch. The diode D_1 supplies for the charging current of the capacitor and the load current. Capacitor starts charging through D_1 and at the end of this interval *i.e.*, at A, it charges to the peak value of secondary voltage, *i.e.*, V_m.

Hence, At A, voltage on C *i.e.*, $V_C = V_m$

After point A, the instantaneous secondary voltage starts reducing as shown by the dotted waveform of rectifier output in figure, 3.28. This will reverse bias the diode D_1, hence, at A, the diode D_1 is turned off.

(ii)* Operation in the interval *A* to *B

During this interval, voltage on the capacitor is higher than rectifier output (shown by dotted lines). Hence, D_1 and D_2 both remain off. The capacitor discharges exponentially through the load resistance R. As the value of R is much higher than R_F, the capacitor discharge slowly. Discharging time constant is RC. Value of C is very large in order to make the discharging time constant as large as possible. This will reduce the ripple content in the output voltage.

(iii)* Operation in the Interval *B* to *C

At B, the instantaneous rectified voltage is equal to the voltage on capacitor and after B, it is greater than V_C. Therefore, diode D_2 starts conducting at instant B. The capacitor charges through D_2 and at the end of this interval *i.e.*, at point C, the voltage on capacitor is again equal to $+ V_m$. Due to this, D_2 is reverse biased and stops conducting at point C as shown in figure 3.28.

***(iv)* Operation in the interval C to D**

The operation in this interval is identical to that in the interval A to B.

Reduction in Ripple

Lety us look at the load voltage waveform of figure 3.28. The change in the instantaneous load voltage with time is much less than that for a full wave rectifier output without a filter. Thus, ripple has reduced and the output voltage is more close to a pure dc waveform.

Diode Currents

As shown in figure 3.28, the diodes D_1 and D_2 do not conduct for the period of complete half cycle. They conduct for a very short duration of time and the diode currents are in the form of pulses.

Cut in and cut out points

Cut in point is the point at which a diode starts conducting. As shown in figure 3.28, points B and D are the cut in points. Cut out points on the other hand are the points at which the conducting diodes are turned off due to reverse biasing. Point C in figure 3.28 is a cut out point.

3.13.2. Expression for the Ripple Factor

To obtain the expression for ripple factor of the output voltage using a capacitor filter, let us consider the two waveforms drawn in figure 3.29 (*a*) and (*b*). To simplify the mathematics, we will assume that ripple has a triangular shape.

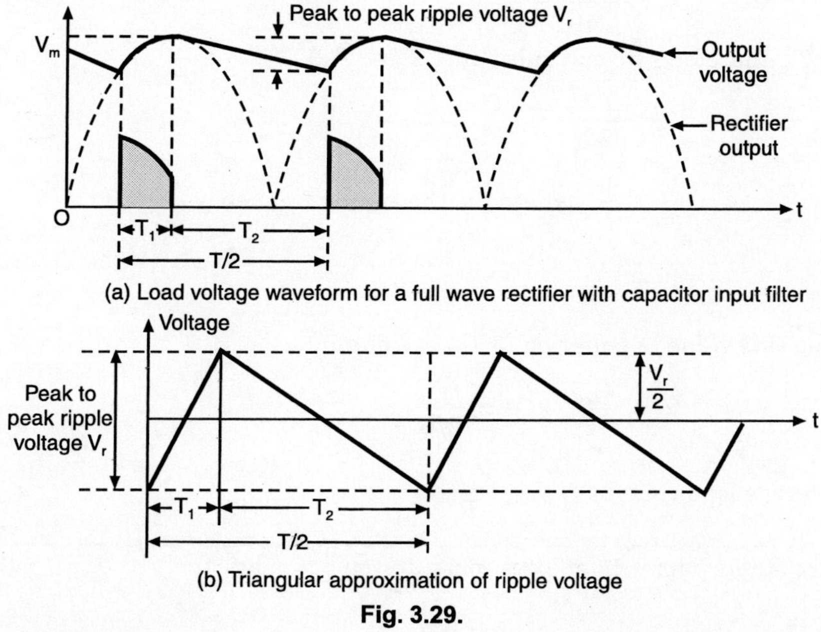

(a) Load voltage waveform for a full wave rectifier with capacitor input filter

(b) Triangular approximation of ripple voltage

Fig. 3.29.

Assumptions Made

Let us make the following assumptions before beginning the derivation for the ripple factor. The assumptions are:

(*i*) The capacitor charges linearly during period T_1 and discharges linearly during the period T_2 as shown in figure 3.29 (*b*).

(*ii*) The half cycle period is $T/2$ which is equal to the sum of T_1 and T_2.

(*iii*) $T_1 <<< T_2$, therefore, $(T/2) = (T_1 + T_2) \cong T_2$.

(*iv*) Peak to peak ripple voltage is V_r volts.

Derivation

The ripple factor is defined as, $r = \dfrac{V_{rms}}{V_{L\,dc}}$ ...(3.68)

Where V_{rms} = *rms* value of the ripple voltage

and $V_{L\,dc}$ = Average value of the load voltage

Let us now find out the *rms* value of the ripple voltage. To do so, we have to first obtain the peak to peak ripple voltage V_r.

Step 1 : To obtain peak to peak ripple voltage V_r

Let us consider figure 3.29 (*a*). Capacitor discharges through load R during the time interval T_2. The charge lost by the capacitor during this period is given by

$$Q = CV_r \quad ...(3.69)$$

The discharge current $i = dQ/dt$ ...(3.70)

Therefore, $$Q = \int_0^{T_2} i\,dt = I_{L\,dc}\, T_2 \quad ...(3.71)$$

This is because the integration gives us the average *i.e.*, dc value. Substituting this value of Q in equation (3.69) we get:

$$I_{L\,dc}\, T_2 = CV_r \quad ...(3.72)$$

Hence, $$V_r = \frac{I_{L\,dc}\, T_2}{C} \quad ...(3.73)$$

But, $T_1 << T_2$, therefore, we can substitute $T_2 = (T/2)$.

Hence, $$T_2 = \frac{T}{2}$$

But, $$T = \frac{1}{f},$$ where f is the supply frequency.

Therefore, $$T_2 = \frac{1}{2f} \quad ...(3.74)$$

Substituting this value in equation (3.73), we obtain

$$V_r = \frac{I_{L\,dc}}{C} \times \frac{T}{2} = \frac{I_{Ldc}}{2fC}$$

But, the average load current $I_{L\,dc} = \dfrac{V_{Ldc}}{R}$

Therefore, $$V_r = \frac{V_{Ldc}}{2fCR} \quad ...(3.75)$$

This is the required expression for peak to peak ripple voltage V_r

Step 2 : To obtain the *rms* value of ripple voltage (V_{rms}).

For a triangular ripple voltage shown in figure 3.29 (*b*), let us use a standard formula which states the *rms* value of ripple voltage. It is given by,

$$V_{rms} = \frac{V_r}{2\sqrt{3}} \quad ...(3.76)$$

Now, substituting the expression for V_r from equation (3.75),we get

$$V_{rms} = \frac{V_{Ldc}}{4\sqrt{3}\ fCR} \quad ...(3.77)$$

Step 3 : To obtain the ripple factor.

$$r = \frac{V_{rms}}{V_{Ldc}} = \frac{V_{L\,dc}}{4\sqrt{3}fCR\ V_{Ldc}}$$

This is the ripple factor for a full wave or bridge rectifier circuit. For the half wave rectifier the expression for r is as under:

$$r \text{ for HWR} = \frac{1}{2\sqrt{3}\ fCR} \quad ...(3.78)$$

How to reduce the ripple factor ?

The expressions for ripple factor stated in equations (3.77) and (3.78) show that the value of ripple factor can be reduced by :

(*i*) Increasing the value of filter capacitor C or

(*ii*) Increasing the load resistance R.

However, C cannot be too large because that will increase the diode surge current. The ripple factor is better for light loads i.e., when load current is small.

Average output voltage with capacitor Input Filter

The average load voltage at the output of capacitor input filter along with a full wave or bridge rectifier is given by:

$$V_{L\,dc} = V_m - \frac{I_{Ldc}}{4fC} \quad ...(3.79)$$

Important Conclusions

The conclusions from this equation are as follows:

(*i*) The average load voltage at zero load current *i.e.*, at no load is equal to the peak secondary voltage

Therefore, $V_{Ldc(N.L)} = V_m$ ...(3.80)

(*ii*) If C is constant, then, V_{Ldc} decrease with increase in load current. The second term in equation (3.79) represents drop in voltage. The variation in V_{Ldc} with change in load current is as shown in figure 3.30.

(*iii*) The load regulation can be calculated from the characteristics of figure 3.30. The regulation will be better with increase in the value of C or with reduction in the value of full load current.

Advantages of Capacitor Input Filter

The advantages of a capacitor input filter are as follows:

(i) Reduction in the ripple content of the output voltage waveform

(ii) Increase in the average load voltage.

Disadvantages

The disadvantages of a capacitor input filter are :

(i) Ripple factor is dependent on the load.

(ii) Regulation is relatively poor

(iii) Diodes have to handle large peak currents.

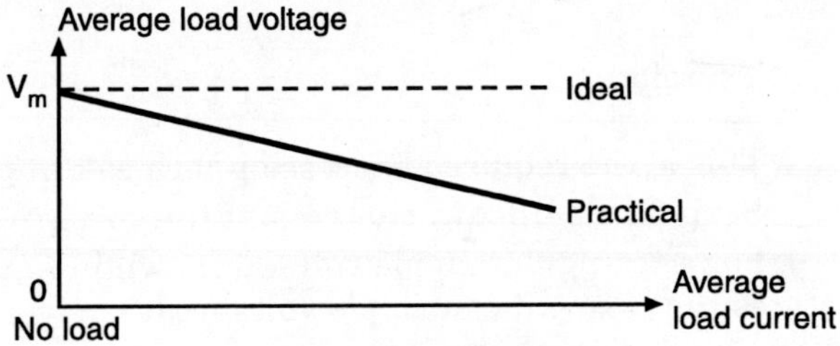

Fig. 3.30. Load regulation characteristics for capacitor input filter

Comparison of capacitor input and choke input filters

S. No.	*Parameter*	*Capacitor input filter*	*Choke input filter*
1.	Place of filter	Across the load	In series with the load.
2.	Useful in	Reducing ripple in load voltage	Reducing the ripple in the load current
3.	Lowest ripple in the load voltage at	No load or light loads	Heavy loads.
4.	Suitable for	Light load applications	Heavy load applications.
5.	Surge current through diodes	Very high and must be controlled	Low and need not be controlled.
6.	Expression for ripple factor	$RF = \dfrac{1}{4\sqrt{3}\, fCR}$	$RF = \dfrac{R}{3\sqrt{2}\, \omega L}$
7.	Size of filter	Small and compact	Bulky

3.14. L SECTION FILTER (LC FILTER)

The circuit diagram for an LC filter used with a full wave rectifier is as shown in figure 3.31. This is a combination of inductor filter and a capacitor input filter. We know that inductor filter is preferred for low values of R and the capacitor input filter is preferred for high values of load resistance. The LC filter can give low ripple factor irrespective of the load as it is a combination of the two filters.

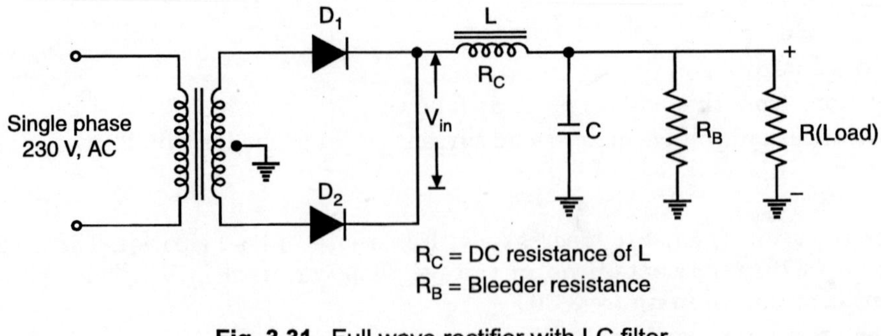

Fig. 3.31. Full wave rectifier with LC filter

The series connected inductor offers a high reactance to the harmonic components (ripple) in the output and attenuates them and the parallel capacitor provides a low reactance bypass path for them. This will reduce the ripple further.

Role of the Bleeder Resistance (R_B)

The resistance R_B connected across the capacitor is called as **bleeder resistance**. It is used to maintain a continuous current through the filter inductance L. If the current through L is not continuous *i.e.*, if it is interrupted, then a large back emf ($L\ di/dt$) will be developed across the inductor. This voltage may exceed the PIV rating of the rectifier diodes and damage them. This voltage may exceed the maximum rated voltage of the capacitor as well. Hence the induced back emf is dangerous for diodes as well as the capacitor. The back emf will not appear if the current through L is continuous. R_B will maintain a continuous current through L.

3.14.1 Expression for the Ripple Factor

The derivation of expression for the ripple of L section filter is based on the following assumptions:

Assumptions made

(*i*) While obtaining the expression for ripple factor we will consider only the second harmonic component and neglect all the other higher order harmonics.

(*ii*) As inductor L is reasonably large in value, its reactance at the second harmonic frequency is much higher than its dc resistance R_C.

Therefore, $$X_L = 2\,\omega L >>> R_C \qquad ...(3.81)$$

(*iii*) Let the parallel combination of the load resistance R and the bleeder resistance R_B be denoted by R_P.

Therefore, $$R_P = R\,||\,R_B = \frac{R \times R_B}{(R + R_B)} \qquad ...(3.82)$$

(*iv*) As the capacitor C is of large value, its reactance X_C at the second harmonic frequency is much less than the parallel combination R_P.

Therefore, $$X_C = \frac{1}{2\,\omega C} << R_P \qquad ...(3.83)$$

(*v*) The equivalent impedance of the parallel combination of X_C and R_P will be nearly equal to X_C.

Derivation

Step 1 : To express the load voltage in Fourier series form.

The output voltage of a full wave rectifier which acts as an input voltage to the filter can be expressed using the Fourier series as under :

$$V_{in} = \frac{2V_m}{\pi} - \frac{4V_m}{3\pi}\cos 2\,\omega t - \frac{4V_m}{15\pi}\cos 4\,\omega t... \qquad ...(3.84)$$

Where V_m = maximum value of half secondary voltage. The first term in equation (3.84) represents the dc or average value of the output, second term represents the second harmonic, third one represents the fourth harmonic and so on. Neglecting all the higher order frequency components except the second harmonic, we can write,

$$V_{in} = \frac{2V_m}{\pi} - \frac{4V_m}{3\pi}\cos 2\,\omega t \qquad ...(3.85)$$

Step 2 : To obtain the dc voltage across the load.

The dc current in the circuit is given by:

$$I_{Ldc} = \frac{2V_m/\pi}{(R_C + R_P)} \quad ...(3.86)$$

Where $R_P = R || R_B$

Therefore, the dc voltage developed across the load is given by,

$$V_{L\,dc} = I_{L\,dc} \times R_P = \frac{2V_m}{\pi(R_C + R_P)} \times R_P$$

Hence,

$$V_{L\,dc} = \frac{2\,V_m}{\pi\left[1 + \frac{R_C}{R_P}\right]} \quad ...(3.87)$$

But, $R_C << R_P$, therefore, equation (3.87) gets modified to,

$$V_{Ldc} = \frac{2V_m}{\pi} \quad ...(3.88)$$

Step 3: To obtain the *rms* value of ripple.

From equation (3.85), the peak value of second harmonic component of V_{in} is given by

$$\text{Peak value of second harmonic, } V_{m2} = \frac{4\,V_m}{3\pi} \quad ...(3.89)$$

Therefore, the peak value of second harmonic current is given by

$$I_{m2} = \frac{V_{m2}}{Z_2} \quad ...(3.90)$$

Where Z_2 = impedance offered by the filter circuit at second harmonic frequency.

$$Z_2 = R_C + j2\omega L + \left[\frac{1}{j2\omega C} \| R_P\right] \quad ...(3.91)$$

But, $\frac{1}{j2\omega C} << R_P$ and $R_C << j2\omega L$. Therefore, the above equation can be written as:

$$Z_2 = j2\omega L \quad ...(3.92)$$

Substituting this value in equation (3.90), we get

$$I_{m2} = \frac{4V_m}{3\pi \times 2\omega L} \quad ...(3.92A)$$

Therefore, the second harmonic voltage across the load is,

$$\text{Second harmonic voltage across load} = I_{m2} \times \left[\frac{1}{j2\omega C} \| R_P\right]$$

But, $\frac{1}{j2\omega C} << R_P$. Therefore, their parallel combination is approximately equal to $\frac{1}{j2\omega C}$

$$\text{Hence, Second harmonic voltage across load} = I_{m2} \times \frac{1}{j2\omega C}$$

Substituting the value of I_{m2}, we get

Second harmonic voltage across the load $= \dfrac{2V_m}{3\pi\omega L} \times \dfrac{1}{2\omega C} = \dfrac{V_m}{3\pi\omega^2 LC}$ ...(3.93)

But this is the peak value. To calculate the ripple factor, we need rms value.

Therefore, *rms* second harmonic voltage across laod $= \dfrac{V_m}{3\sqrt{2}\,\pi\,\omega^2 LC}$ ...(3.94)

Step 4 : To obtain expression for the ripple factor.

$$r = \frac{rms \text{ second harmonic voltage across load}}{V_{Ldc}} = \frac{V_m}{3\sqrt{2}\,\pi\,\omega^2 LC} \times \frac{\pi}{2V_m}$$

Hence, $$r = \frac{1}{6\sqrt{2}\,\omega^2 LC}$$

Note that the ripple factor does not depend on the laod at all. It is constant irrespective of the laod.

Load regulation of LC filter

As stated in equation (3.88), the average load voltage is approximately given as:

$$V_{L\,dc} = \frac{2V_m}{\pi}$$

Thus, it is not dependent on the load current. Therefore, the load regulation of LC filter is much better as compared to the load regulation of capacitor input filter.

Advantages of LC filter

(*i*) Very good load regulation.
(*ii*) Ripple factor is low and does not depend on the load.
(*iii*) This filter is suitable for light as well as heavy loads.

Disadvantages

(*i*) Due to the use of large value of L and C components, the circuit becomes costly.
(*ii*) Due to the use of inductor, the filter is bulky.
(*iii*) Bleeder resistance increases the rating of rectifier circuit.
(*iv*) Power loss takes place in the series inductor L due to its dc resistance.

3.15. π Type Filter

The π type filter which is a combination of capacitor input filter and LC filter is as shown in figure 3.32. It consists of two capacitors C_1 and C_2 along with the inductance L. Generally, both the capacitors are of same value.

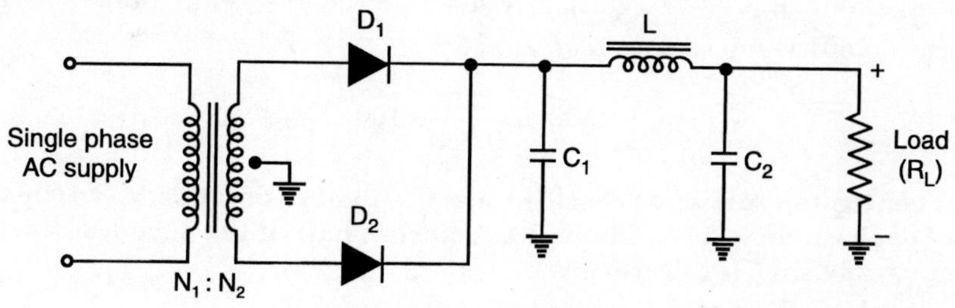

Fig. 3.32. π-type filter

(i) A full wave rectifier drives the π-type filter as shown in figure 3.32.

(ii) As C_1 comes first, looking from the rectifier side, the π-type filter behaves in a very similar way as the capacitor input filter. All the advantages and disadvantages of capacitor input filter are all applicable to the π-type filter as well.

(iii) Due the use of three filtering elements, the ripple factor of the π-type filter is very low as compared to the other filters.

The ripple factor for a π type filter is given by

$$r = \frac{\sqrt{2}}{8\omega^3 C_1 C_2 L R_L} \qquad \text{...(3.95)}$$

IMPORTANT SOLVED EXAMPLES

EXAMPLE 3.3 The forward resistance R_F of a diode is 20 Ω. This diode is used in a half-wave rectifier circuit. The applied input voltage is $v = 50 \sin \omega t$ and load resistance R_L is 800 Ω.

Determine: (i) I_m, I_{dc}, I_{rms}

(ii) a.c. power input and d.c. power output

(iii) Output d.c. voltage

(iv) Rectification efficiency

Solution: Given that $v = 50 \sin \omega t$

Hence $V_m = 50$ V

We know that $I_m = \dfrac{V_m}{R_L + R_F}$

Substituting all the given values, we get

(i) $$I_m = \frac{50}{800+20} = 0.06 \text{ A} \qquad \textbf{Ans.}$$

$$I_{dc} = \frac{I_m}{\pi} = \frac{0.06}{3.14} = 0.0194 \text{ A} \qquad \textbf{Ans.}$$

$$I_{rms} = \frac{I_m}{2} = \frac{0.06}{2} = 0.03 \text{ A} \qquad \textbf{Ans.}$$

(ii) A.C. power input, $P_{ac} = (I_{rms})^2 \times (R_L + R_F)$

Putting all the values, we have

$$P_{ac} = (30.5 \times 10^{-3})^2 \times (820) = 0.763 \text{ Watts} \qquad \textbf{Ans.}$$

D.C. power output $P_{dc} = I_{dc}^2 \times R_L = (19.4 \times 10^{-3})^2 \times 800 = 0.301$ Watt **Ans.**

(iii) Output d.c. voltage $= I_{dc} \times R_L = 19.4 \times 10^{-3} \times 800 = 15.52$ Volts **Ans.**

(iv) Rectification efficiency is expressed as

$$\eta = \frac{P_{dc}}{P_{ac}} \times 100 = \frac{0.301}{0.763} \times 100 = 39.5\% \qquad \textbf{Ans.}$$

EXAMPLE 3.4. A centre-tap full wave rectifier has the load resistance $R_L = 1000$ Ω. The forward resistance R_F of each diode is 20 Ω. The voltage across half of the secondary winding is given by the equation $v = 200 \sin 314t$. Determine :

(i) The peak value of current

(ii) The average or dc value of current

***(iii)* The rms value of current**

***(iv)* The ripple factor**

***(v)* The rectification efficiency**

Solution: Given that

Voltage across half of the secondary

$$v = 200 \sin 314 f \qquad ...(i)$$

$$R_L = 1000\ \Omega$$

$$R_F = 20\ \Omega$$

(i) The maximum or peak value of voltage using *(i)*

$$V_m = 200 \text{ Volts}$$

Hence, peak value of current will be

$$I_m = \frac{V_m}{R_L + R_F}$$

Putting all the given values, we get

$$I_m = \frac{200}{1000+20} = \frac{10}{51} \approx \frac{1}{5} \approx 0.2 \text{ A} \qquad \textbf{Ans.}$$

(ii) The average or d.c. value of current is given as

$$I_{dc} = \frac{2I_m}{\pi} = \frac{2\times 0.2}{3.14} = \frac{0.4}{3.14} = 0.127 \text{ A} \qquad \textbf{Ans.}$$

(iii) The r.m.s. value of current is expressed as

$$I_{rms} = \frac{I_m}{\sqrt{2}} = \frac{0.2}{\sqrt{2}} = \frac{0.2}{1.41} = 0.141 \text{ A} \qquad \textbf{Ans.}$$

(iv) The ripple factor is expressed as

$$r = \sqrt{\left(\frac{I_{rms}}{I_{dc}}\right)^2 - 1} = \sqrt{\left(\frac{0.141}{0.127}\right)^2 - 1} = \sqrt{(1.11)^2 - 1}$$

$$= \sqrt{1.24 - 1} = \sqrt{0.24} = \sqrt{1.116} \qquad \textbf{Ans.}$$

EXAMPLE 3.5. An a.c. supply of 230 V is connected to a half-wave rectifier circuit through a transformer having turns ratio 10:1 as shown in the figure 3.33.

Determine:

***(i)* The d.c. output voltage**

***(ii)* The peak inverse voltage of diode**

Assume the diode to be ideal one.

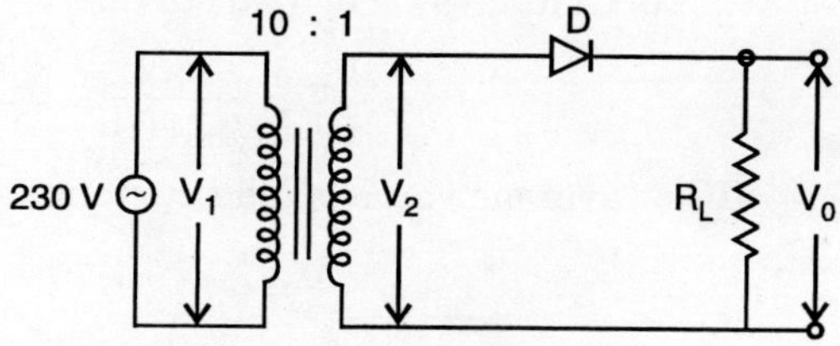

Fig. 3.33.

Solution:

Given that $\frac{N_1}{N_2} = 10$

Primary r.m.s. voltage $V_1 = 230$ V

∵ We know that $\frac{N_1}{N_2} = \frac{V_1}{V_2}$

∴ Secondary r.m.s. voltage

$$V_2 = \frac{N_2}{N_1} \times V_1 = \frac{1}{10} \times 230 = 23 \text{ V}$$

The maximum or peak value of secondary voltage

$$V_m = \sqrt{2} \times V_2 = \sqrt{2} \times 23 = 32.5 \text{ Volt}$$

(*i*) DC output voltage $V_{dc} = \frac{V_m}{\pi} = 0.318\ V_m = 0.318 \times 32.5$

$$V_{dc} = 10.3 \text{ V}$$

(*ii*) Peak inverse voltage of diode

$$PIV = V_m = 32.5 \text{ Volt}$$

EXAMPLE 3.6 **A half-wave rectifier circuit with a transformer coupled input is shown in the figure 3.34. Determine the value of d.c. load current.**

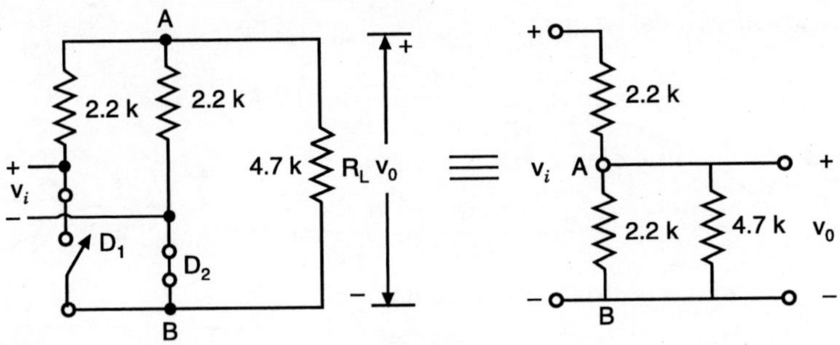

Fig. 3.34.

Solution: Given that

$$R_L = 20 \text{ k}\Omega = 20{,}000\ \Omega$$

r.m.s. value of secondary voltage

$$V_2 = 20 \text{ Volt}$$

∴ Maximum or peak value of secondary voltage

$$V_m = \sqrt{2} \times V_2 = \sqrt{2} \times 20$$

$$= 1.41 \times 20 = 28.2 \text{ Volt}$$

∴ Maximum value of load current

$$I_m = \frac{V_m}{R_L} = \frac{28.2}{20{,}000} = \frac{2.82 \times 10^{-4}}{2} = 14.1 \times 10^{-4} = 1.41 \text{ mA}$$

DC or average value of load current

$$I_{dc} = \frac{I_m}{\pi} = 0.318\ I_m \qquad \textbf{Ans.}$$

EXAMPLE 3.7 **A half-wave rectifier circuit with a transformer coupled input is shown in the figure 3.35. Determine the maximum and average values of power delivered to the load.**

(A.M.I.E. Examination 2000)

Progress is impossible without change, and those who cannot change their minds cannot change anything.

— **G.B. Shaw**

Solution:

Given that $\frac{N_1}{N_2} = 2$, $R_L = 200\ \Omega$

and *rms* value of primary voltage

$$V_1 = 230 \text{ V}$$

Since, we know that $\frac{N_1}{N_2} = \frac{V_1}{V_2}$

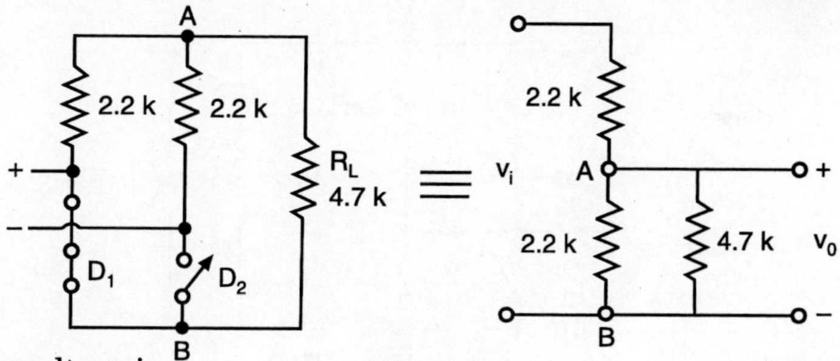

Fig. 3.35.

Therefore, r.m.s. value of secondary voltage is

$$V_2 = \frac{N_2}{N_1} \times V_1$$

$$= \frac{1}{2} \times 230 = 115 \text{ V}$$

Thus, maximum or peak value of secondary voltage,

$$V_m = \sqrt{2} \times V_2 = \sqrt{2} \times 115 = 162.6 \text{ V}$$

∵ Maximum value of load current is given by

$$I_m = \frac{V_m}{R_L} = \frac{162.6}{200} = 0.813 \text{ A}$$

Hence, maximum value of load power is

$$P_m = I_m^2 \times R_L = (0.813)^2 \times 200 = 132.3 \text{ Watts} \quad \textbf{Ans.}$$

Average value of load power is

$$P_{dc} = I_{dc}^2 \times R_L \quad ...(i)$$

But

$$I_{dc} = \frac{V_{dc}}{R_L} \quad ...(ii)$$

where average value of output voltage will be given by

$$V_{dc} = 0.318\ V_m = 0.318 \times 162.6 = 51.7 \text{ Volt}$$

Thus, average value of load current is [using equation (*ii*)]

$$I_{dc} = \frac{V_{dc}}{R_L} = \frac{51.7}{200} = 0.26 \text{ A}$$

Putting the value of I_{dc} in equation (*i*), we get the average value of load power

$$P_{dc} = (0.26)^2 \times 200 = 13.4 \text{ Watts} \quad \textbf{Ans.}$$

EXAMPLE 3.8. For a circuit shown in figure 3.36 sketch v_0 and determine the required PIV rating of each diode.

Solution: Let us assume that D_1 and D_2 are ideal diodes.

Operation in the positive half cycle:

Figure 3.37 shows the equivalent circuit in the positive half cycle.

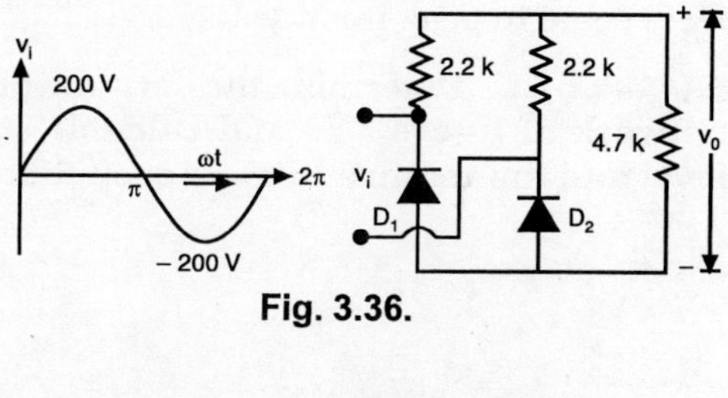

Fig. 3.36.

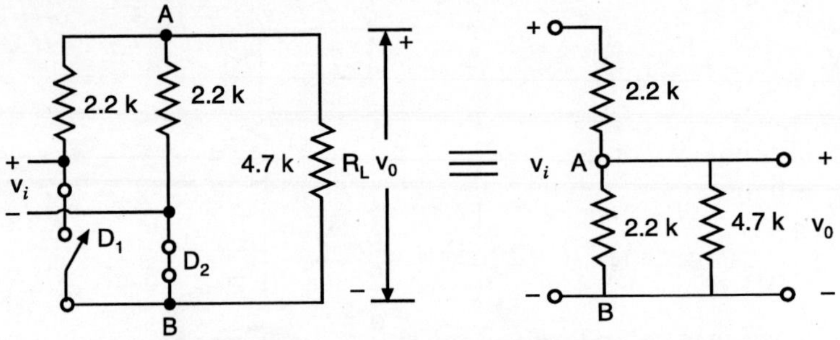

Fig. 3.37. Equivalent circuit in the positive half cycle.

$$\therefore \quad v_0 = v_{AB} = \frac{(2.2 \| 4.7)}{2.2 + (2.2 \| 4.7)} \times v_i = 0.405\ v_i$$

$$v_i = 200 \text{ V peak} \quad \text{or} \quad v_0 = 0.405 \times 200 = 81 \text{ V peak}$$

Operation in the negative half cycle:

Figure 3.38 shows the equivalent circuit in the negative half cycle.

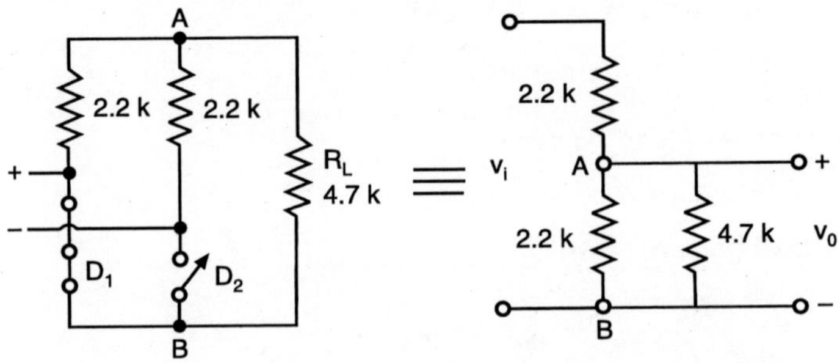

Fig. 3.38. Equivalent circuit in the negative half cycle

This circuit is same as that shown in figure 3.37.

$$\therefore \quad v_0 = 0.405\ V_i$$

And the output is positive.

$$\therefore \quad v_0 = 81 \text{ V peak}$$

Figure 3.39 shows the input and output voltage waveforms.

PIV of diodes: From figures 3.37 and 3.38, we can see that the maximum negative voltage across D_1 and D_2 is equal to peak input voltage.

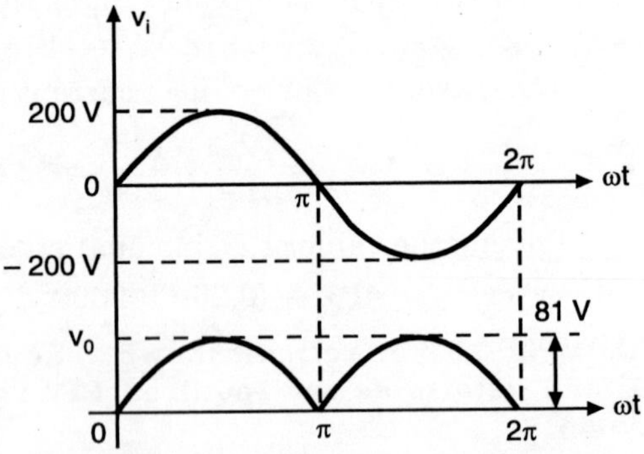

Fig. 3.39. Input and Output Waveforms

EXAMPLE 3.9 Determine the output waveform for the network of figure 3.40 and calculate the output dc level and the required PIV of each diode.

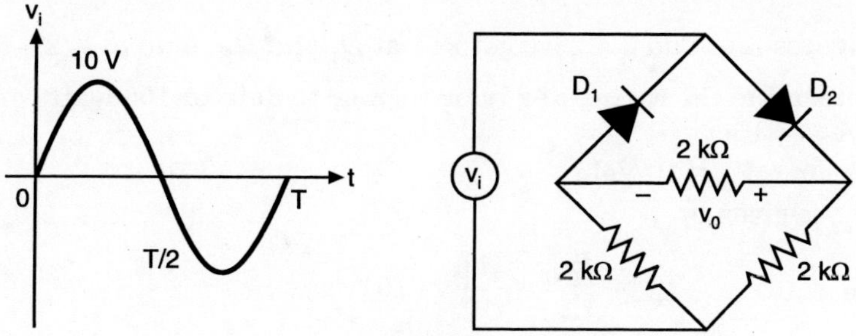

Fig. 3.40.

Solution: Let us assume that diodes D_1 and D_2 are ideal diodes.

Case (*i*) Positive half cycle:

Figure 3.41 shows the equivalent circuit in the positive half cycle. Here, the output voltage is given by

$$v_0 = \frac{2\text{ k}\Omega}{2\text{ k}\Omega + 2\text{ k}\Omega} \times v_i$$

$$\therefore \qquad v_0 = v_i/2$$

Hence, the output is positive

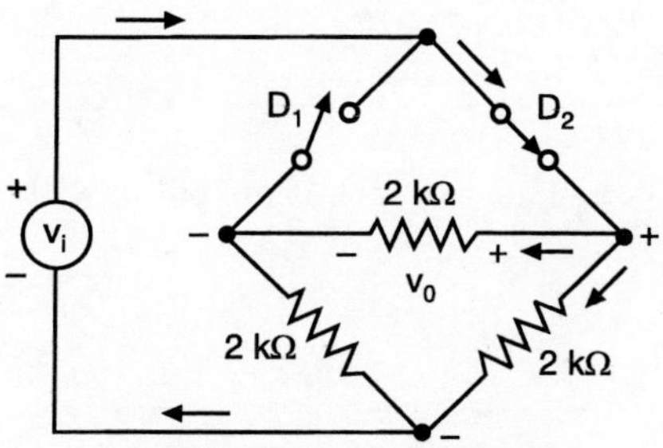

Fig. 3.41. Equivalent circuit in the positive half cycle.

Case (*ii*) Negative half cycle:

Figure 3.42 shows the equivalent circuit in the negative half cycle. Here, the output voltage is given by

$$\therefore \qquad v_0 = \frac{2\text{ k}\Omega}{2\text{ k}\Omega + 2\text{ k}\Omega} \times v_i = v_i/2$$

Hence, and the output is positive.

Figure 3.43 shows the input and output voltage waveforms.

Now, the average output voltage is given by

$$v_{0(\text{dc})} = \frac{1}{T/2} \int_0^{T/2} 5 \sin \omega t \, dt$$

$$v_{0(\text{dc})} = \frac{-2}{T} \times 5 \times \left[\frac{\cos \omega t}{\omega}\right]_0^{T/2}$$

But $\qquad \omega T = 2\pi fT = 2\pi$

Therefore, we have

$$v_{0(\text{dc})} = \frac{-10}{2\pi} [\cos(\omega T/2) - \cos 0]$$

$$v_{0(\text{dc})} = \frac{-5}{\pi} [\cos \pi - \cos 0] = \frac{-5 \times -2}{\pi}$$

$$v_{0(\text{dc})} = \frac{10}{\pi} = 3.183 \text{ Volts} \quad \textbf{Ans.}$$

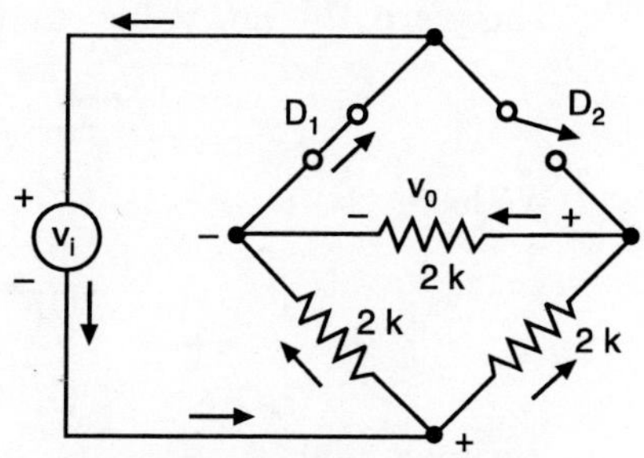

Fig. 3.42. Equivalent circuit in the negative half cycle.

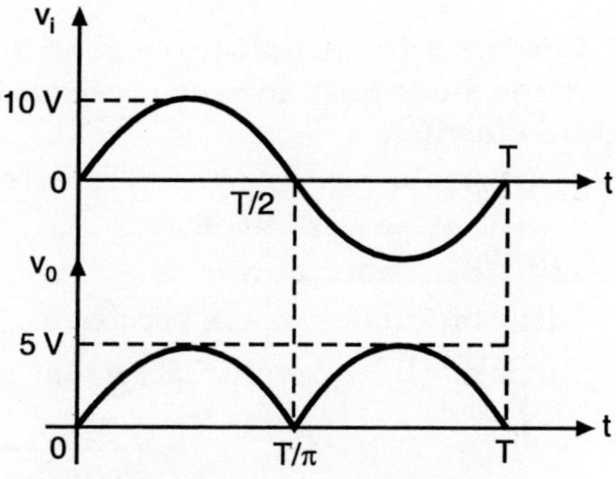

Fig. 3.43. Input and output Waveforms

PIV of diodes

Referring to figures 3.41 and 3.42, we get PIV at D_1 and D_2 equal to $v_i/2 = 5$ Volts.

EXAMPLE 3.10 Determine the rating of a transformer to deliver 100 watts of d.c. power to the load for a full wave rectifier.

Solution: Let as assume $V_{dc} = 10$ Volts

The value of I_{dc} is given by

$$I_{dc} = \frac{P_{dc}}{V_{dc}} = \frac{100}{10} = 10 \text{ A}$$

Also, $$V_{dc} = \frac{2V_m}{\pi}$$

$$\therefore \quad V_m = \frac{\pi}{2}V_{dc} = \frac{\pi}{2}\times 10 = 15.7 \text{ V}$$

Therefore, the rms value will be

$$V_{rms} = \frac{V_m}{\sqrt{2}} = 11.1 \text{ Volts}$$

$$I_{dc} = \frac{2I_m}{\pi}$$

$$\therefore \quad I_m = \frac{10\times\pi}{2} = 15.7 \text{ Amp.}$$

Therefore, the rms value of current is given by

$$I_{rms} = \frac{I_m}{\sqrt{2}} = \frac{15.7}{\sqrt{2}} = 11.1 \text{ Amp.}$$

We have

$$\text{TUF} = \frac{\text{DC power}}{\text{Secondary power rating}} = \frac{100}{11.1\times 11.1}\times 100$$

$$\therefore \quad \text{TUF} = 81.6\%$$

We know that

$$\frac{N_1}{N_2} = \frac{230}{11.1} = 20.72.$$

EXAMPLE 3.11 A half-wave rectifier rectifies an alternating voltage of 325 Volt peak value and the diode has a forward resistance of 100 Ω. The value of load resistance is 1000 Ω. Determine the following:

(i) peak, average and rms value of current

(ii) a.c. power output

(iii) a.c. input power

(iv) efficiency of the rectifier.

Solution: (i) Peak value of current is given as

$$I_m = \frac{V_m}{R_L + R_F} = \frac{325}{1000+100} = \frac{325}{1100} = 295.45 \text{ mA} \qquad \textbf{Ans.}$$

Average value of current or d.c. value of current

$$I_{avg} = I_{d.c.} = \frac{I_m}{\pi} = \frac{295.45}{3.14} = 94.046 \text{ mA} \qquad \textbf{Ans.}$$

rms value of current,

$$I_{rms} = \frac{I_m}{2} = \frac{295.45}{2} = 147.725 \text{ mA} \qquad \textbf{Ans.}$$

(*ii*) d.c. power output

$$P_{dc} = I_{dc}^2 \times R_L$$
$$= (94.046 \times 10^{-3})^2 \times 1000 = 8.845 \text{ Watts.} \qquad \textbf{Ans.}$$

(*iii*) A.C. input power

$$P_{a.c.} = (I_{rms})^2 \times (R_L + R_F)$$
$$= (147.725 \times 10^{-3})^2 \times (1000 + 100) = 24 \text{ Watts.} \qquad \textbf{Ans.}$$

(*iv*) Efficiency of rectification

$$\eta = \frac{P_{d.c.}}{P_{a.c.}} = \frac{8.845}{24} = 36.85\% \qquad \textbf{Ans.}$$

EXAMPLE 3.12 A.C. voltage of 200 Volt is applied to a half-wave rectifier circuit through transformer of turns ratio 4:1. The load resistance value is 300 Ω. Assuming the diode as ideal one, determine the following:

(*i*) d.c. output voltage

(*ii*) PIV

(*iii*) maximum value of power delivered to the load

(*iv*) average value of power delivered to the load.

Solution: Given that

$$\frac{N_1}{N_2} = \frac{4}{1}$$

Primary r.m.s. voltage $V_1 = 200$ Volt

We know that for a transformer

$$\frac{N_1}{N_2} = \frac{V_1}{V_2}$$

Therefore, secondary r.m.s. voltage $V_2 = \frac{N_2}{N_1} \times V_1$

$$V_2 = \frac{1}{4} \times 200 = 50 \text{ Volt}$$

The maximum or peak value of Secondary Voltage

$$V_m = \sqrt{2} \times V_2 = \sqrt{2} \times 50$$
$$V_m = 1.41 \times 50 = 70.7 \text{ Volt}$$

Therefore, d.c. output voltage

$$V_{d.c} = \frac{V_m}{\pi} = 0.318\, V_m$$
$$V_{d.c.} = 0.318 \times 70.7 = 22.4 \text{ Volt} \qquad \textbf{Ans.}$$

(*ii*) PIV of diode $= V_m = 70.7$ Volt **Ans.**

(*iii*) ∵ Maximum value of load current

$$I_m = \frac{V_m}{R_L} = \frac{70.7}{300} = 0.235 \text{ A}$$

Thus, maximum value of power delivered to the load

$$P_m = I_m^2 \times R_L = (0.235)^2 \times 300$$

or $$P_m = 0.055 \times 300 = 16.6 \text{ Watts} \qquad \textbf{Ans.}$$

Maximum $$I^2_m \times R_L = (0.235)^2 \times 300$$

$$P_m = 0.055 \times 300 = 16.6 \text{ Watts}$$

(iv) The average or d.c. value of load current

$$I_{d.c.} = \frac{V_{d.c.}}{R_L} = \frac{22.4}{300} = 0.0746 \text{ A}$$

Therefore, average value of power delivered to the load

$$P_{d.c.} = I^2_{dc} \times R_L = (0.0746)^2 \times 300 = 5.57 \times 300$$

$$P_{d.c.} = 1.67 \text{ Watts} \qquad \textbf{Ans.}$$

EXAMPLE 3.13 A half-wave rectifier is utilized to supply 20 volt d.c. to a resistive load of 400 Ω. The diode used in half-wave rectifier has a forward resistance of 40 Ω. Determine the maximum value of the a.c. voltage required at the input.

Solution : Given that $V_{dc} = 20$ Volt

$R_L = 400\ \Omega$

Forward resistance $R_F = 40\ \Omega$

We know that the average value of load current

$$I_{d.c.} = \frac{V_{d.c.}}{R_L} = \frac{20}{400} = \frac{2}{40} = 0.05 \text{ A} = 50 \text{ mA.}$$

Maximum value of load current

$$I_m = \pi \times I_{d.c.} = 3.14 \times 50 = 157 \text{ mA}$$

Therefore, maximum a.c. voltage required at the input

$$V_m = I_m \times (R_L + R_F) = 157 \times 10^{-3}\,(400 + 40)$$

$$V_m = 157 \times 10^{-3} \times 440 = 69080 \times 10^{-3} \text{ Volt}$$

$$V_m = 69.08 \text{ Volt} \qquad \textbf{Ans.}$$

EXAMPLE 3.14 A 220 V, 50 a.c. voltage is applied at the primary of a 4:1 step-down, center-tap transformer used in a full wave rectifier having a load resistance of 800 Ω. If the diode resistance has a value of 200 Ω, determine:

(*i*) d.c. voltage across the load

(*ii*) d.c. current flowing through the load

(*iii*) d.c. power delivered to the load

(*iv*) PIV across each diode

(*v*) ripple voltage and its frequency.

Solution: Given that

Primary r.m.s. voltage $V_1 = 220$ Volt

Secondary r.m.s. voltage $$V_2 = \frac{N_2}{N_1} \times V_1$$

$$V_2 = \frac{1}{4} \times 200 = 55 \text{ Volt}$$

r.m.s. voltage from center tapping to one end

$$V_{rms} = \frac{V_2}{2} = \frac{55}{2} = 27.5 \text{ Volt}$$

Peak value $V_m = V_{rms}\sqrt{2} = 27.5 \times 1.41 = 38.775$ Volt

(*i*) d.c. voltage across the load

$$V_{d.c.} = \frac{2V_m}{\pi} = \frac{2 \times 38.775}{3.14} = \frac{77.5}{3.14} = 24.6 \text{ Volt} \quad \textbf{Ans.}$$

(*ii*) d.c. current flowing through the load

$$I_{d.c.} = \frac{V_{dc}}{R_L + R_F} = \frac{24.6}{800 + 200}$$

$$I_{d.c.} = \frac{24.6}{1000} = 24.6 \text{ mA} \quad \textbf{Ans.}$$

(*iii*) d.c. power delivered to the load

$$P_{d.c.} = (I_{dc})^2 \times R_L = (24.6 \times 10^{-3})^2 \times 800$$

$$= 805.16 \times 10^{-6} \times 800 = 644128 \times 10^{6}$$

$$= 0.644 \text{ Watts} \quad \textbf{Ans.}$$

(*iv*) PIV across each diode $= 2V_m = 2 \times 38.775 = 77.5$ Volt **Ans.**

(*v*) Ripple voltage

$$V_r(rms) = \sqrt{(V_{rms})^2 - (V_{a.c.})^2} = \sqrt{(27.5)^2 - (24.6)^2}$$

$$V_r(rms) = \sqrt{(27.5 - 24.6)(27.5 + 24.6)} = \sqrt{(2.9)(52.1)}$$

$$V_r(rms) = \sqrt{51.09} = 7.147 \text{ Volt} \quad \textbf{Ans.}$$

Frequency of ripple voltage $= 2 \times 50 = 100$ Hz. **Ans.**

EXAMPLE 3.15 A full-wave rectifier circuit delivers 40 W to a load resistance of 150 W. If ripple factor is given 1% then determine the a.c. ripple voltage across the load.

Solution : We know that d.c. power delivered to the load,

$$P_{d.c} = \frac{V_{d.c.}^2}{R_L}$$

Hence, $V^2_{dc} = P_{d.c.} \times R_L$

or $V_{d.c} = \sqrt{P_{d.c.} \times R_L} = \sqrt{40 \times 150} = \sqrt{6000} = 77.4$ Volt

The ripple factor is given by

$$r = \frac{V_{a.c.}}{V_{d.c.}}$$

or $V_{a.c.} = r.V_{d.c.} = 0.01 \times 77.4 = 0.774$ Volt

Therefore, a.c. ripple voltage across the load,

$$V_{a.c.} = 0.774\ V \quad \textbf{Ans.}$$

EXAMPLE 3.16 A full-wave rectifier has a center-tap transformer of 50–0–50 Volt and each one of diodes is rated at I_{max} = 200 mA and I_{avg} = 75 mA. Neglecting the Voltage drop across the diodes, find the following :

(*i*) The value of load resistor that gives the largest d.c. power output

(*ii*) d.c. load voltage and current

(*iii*) PIV of each diode.

Solution: (*i*) We know that for normal operation, the maximum value of current flowing through the diode must not exceed 80% of its rated current.

But rated value is, $I_{max} = 200$ mA

Therefore, peak value of current,

$$I_m = 0.8 \times I_{max} = 0.8 \times 200$$

$$I_m = 160 \text{ mA}$$

Since, the rating of transformer is given as 50–0–50 V, therefore, the peak or maximum value of Secondary voltage is

$$V_m = \sqrt{2} \times 50 = 1.41 \times 50$$

or $$V_m = 70.50 = 70.5 \text{ Volt}$$

Therefore, the required value of load resistance that gives the largest d.c. power output

$$R_L = \frac{V_m}{I_m} = \frac{70.5}{160 \times 10^{-3}} = 221\ \Omega \quad \textbf{Ans.}$$

(*ii*) d.c. load voltage, $V_{d.c.} = \dfrac{2V_m}{\pi}$ **Ans.**

$$V_{d.c.} = \frac{2 \times 70.5}{3.14} = 45 \text{ Volt}$$

d.c. load current, $I_{d.c.} = \dfrac{V_{d.c.}}{R_L} = \dfrac{45}{221} = 0.204$ A **Ans.**

(*iii*) PIV of each diode $= 2V_m = 2 \times 70.5 = 143.0 = 143$ Volt **Ans.**

EXAMPLE 3.17 A 230 V, 50 Hz. a.c. voltage is applied to the primary of a 5:1 step-down transformer which is used in a bridge rectifier having a load resistor of value 500 Ω. Assuming the diodes to be ideal, determine the following:

(*i*) d.c. output voltage **(*ii*) d.c. power delivered to the load**

(*iii*) PIV of each diode **(*iv*) output frequency.** ***(M.D.U. Rohtak, 2005)***

Solution: (*i*) Given that, the rms value of primary voltage

$$V_1 = 230 \text{ Volt}$$

$$\frac{N_1}{N_2} = \frac{5}{1}$$

We know that $$\frac{V_1}{V_2} = \frac{N_1}{N_2}$$

Thus, the r.m.s. value of the transformer secondary voltage

$$V_2 = V_1 \times \frac{N_2}{N_1} = 230 \times \frac{1}{5} = 46 \text{ Volt}$$

the peak or maximum value of secondary voltage

$$V_m = \sqrt{2} \times V_2 = 1.41 \times 46 = 65 \text{ Volt}$$

Hence, d.c. output voltage,

$$V_{d.c.} = \frac{2V_m}{\pi} = \frac{2 \times 65}{3.14} = 41.4 \text{ Volt} \quad \textbf{Ans.}$$

(*ii*) d.c. power delivered to the load

$$P_{d.c.} = \frac{V_{d.c.}^2}{R_L} = \frac{(41.4)^2}{500} = \frac{1713.96}{500} = 3.42 \text{ Watts}$$

(*iii*) PIV across each diode

$$V_m = 65 \text{ Volt}$$

(*iv*) Output frequency $= 2 \times 50 = 100$ Hz. **Ans.**

EXAMPLE 3.18 A half-wave rectifier uses a diode with an equivalent forward resistance of 0.3 Ω. If the input ac voltage is 10 V (rms) and the load is a resistance of 2.0 Ω, calculate I_{dc} and I_{rms} in the load.

Solution: Supply voltage, $V_{rms} = 10$ V

Peak value of supply voltage, $V_m = 10\sqrt{2}$ V

Forward resistance, $R_F = 0.3\ \Omega$

Load resistance, $R_L = 2\ \Omega$

Peak value of current in load, $$I_m = \frac{V_m}{R_L + R_F} = \frac{10\sqrt{2}}{2+0.3} = 6.15 \text{ A}$$

DC output current, $$I_{dc} = \frac{I_m}{\pi} = \frac{6.15}{\pi} = 1.958 \text{ A} \quad \textbf{Ans.}$$

RMS value of output current, $$I_{rms} = \frac{I_m}{2} = \frac{6.15}{2} = 3.075 \text{ A} \quad \textbf{Ans.}$$

EXAMPLE 3.19 A half-wave rectifier uses a diode with a forward resistance of 100 Ω. If the input ac voltage is 220 V (rms) and the load resistance is of 2 kΩ, determine (*i*) I_{max}, I_{dc} and I_{rms} (*ii*) peak inverse voltage when the diode is ideal (*iii*) load output voltage (*iv*) dc output power and ac input power (*v*) ripple factor, (*vi*) transformer utilisation factor and (*vii*) rectification efficiency.

Solution: RMS value of supply voltage,

$$V_{rms} = 220 \text{ V}$$

Maximum value of supply voltage,

$$V_m = 220\sqrt{2} \text{ V}$$

Forward resistance, $R_F = 100\ \Omega$ or 0.1 kΩ

Load resistance, $R_L = 2$ KΩ

(*i*) Maximum value of current, $$I_m = \frac{V_m}{R_L + R_F} = \frac{220\sqrt{2}}{(2+0.1)} = 148.156 \text{ mA} \quad \textbf{Ans.}$$

Average value of output current, $$I_{dc} = \frac{I_m}{\pi} = \frac{148.156}{\pi} = 47.16 \text{ mA} \quad \textbf{Ans.}$$

RMS value of output current; $$I_{rms} = \frac{I_m}{\pi} = \frac{148.156}{\pi} = 74.078 \text{ mA} \quad \textbf{Ans.}$$

(*ii*) Peak inverse voltage, $\text{PIV} = V_m = 220\sqrt{2} = 311.127$ V **Ans.**

(*iii*) Load output voltage, $V_{dc} = I_{dc} R_L = 47.16 \times 10^{-3} \times 2 \times 10^3 = 94.32$ V **Ans.**

(*iv*) D.C. output power, $P_{dc} = I_{dc}^2 R_L = (47.16 \times 10^{-3})2 \times 2 \times 10^3 = 4.448$ W **Ans.**

A.C. input power, $$P_{ac} = \frac{I_m^2}{4}(R_F + R_L)$$

$$= \frac{(148.156 \times 10^{-3})^2}{4} \times (0.1+2) \times 10^3 = 11.524 \text{ W} \quad \textbf{Ans.}$$

(*v*) Ripple factor, $r = 1.21$ **Ans.**

(*vi*) Transformer utilisation factor,

$$\text{TUF} = \frac{0.286}{1+\frac{R_F}{R_L}} = \frac{0.286}{1+\frac{0.1}{2}} = 0.2724 \quad \textbf{Ans.}$$

(*vii*) Rectification efficiency, $\eta = \frac{P_{dc}}{P_{ac}} \times 100 = \frac{4.448}{11.524} \times 100 = 38.6\%$ **Ans.**

EXAMPLE 3.20 In the circuit shown in figure 3.44, calculate and sketch the waveform of current *i* over one period of the input voltage. Assume the diodes to be ideal.

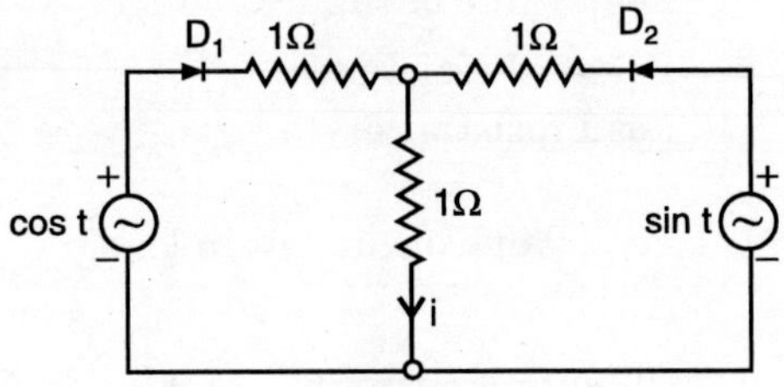

Fig. 3.44.

Solution: The two input voltages are shown in figure 3.45

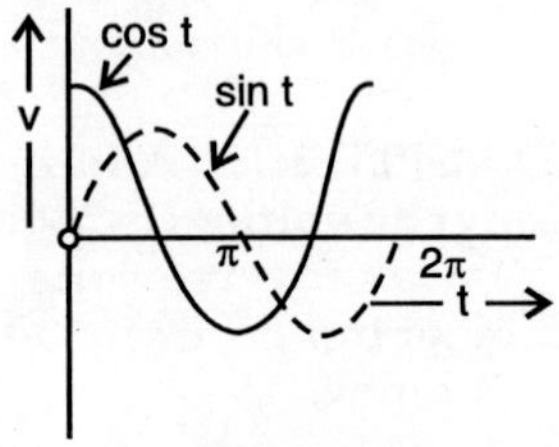

Fig. 3.45.

For $0 \le t \le \frac{\pi}{2}$, both diodes D_1 and D_2 are forward biased and are, therefore, conducting because both sin t and cos t are positive. The circuit is given in figure 3.46 (*a*). Let the potential of point *A* be V_A volts.

Applying Kirchhoff's current law at point *A*, we have

$$\frac{V_A - \cos t}{1} + \frac{V_A - \sin t}{1} + \frac{V_A}{1} = 0$$

or
$$3\,V_A = \cos t + \sin t$$

or
$$V_A = \frac{\cos t + \sin t}{3}$$

or
$$i = \frac{\cos t + \sin t}{3} \quad \text{...(i)}$$

For $\frac{\pi}{2} \le t \le \pi$, during this period, only sint is positive so only diode D_2 is forward biased and conducts. The circuit is given in figure 3.46 (*b*).

The current *i* is given as
$$i = \frac{\sin t}{2} \quad \text{...(ii)}$$

For
$$\pi \le t \le \frac{3\pi}{2}$$

Since during this period, both cost and sint are negative so both of the diodes are reverse-biased and do not conduct.

So,
$$i = 0 \quad \text{...(iii)}$$

For $\frac{3\pi}{2} \le t \le 2\,\pi$, during this period, only cost is positive so only diode D_1 is forward biased and conducts. The circuit is given in figure 3.46(*c*). The current *i* is given as

$$i = \frac{\cos t}{2} \quad \text{...(iv)}$$

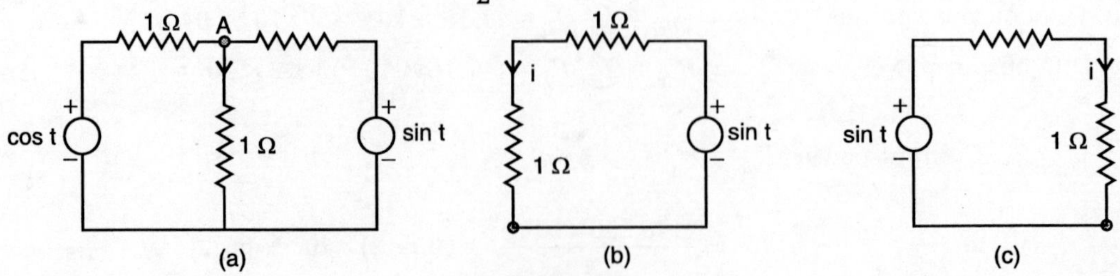

Fig. 3.46.

Waveform of output current i, as calculated above is drawn, as shown in figure 3.47.

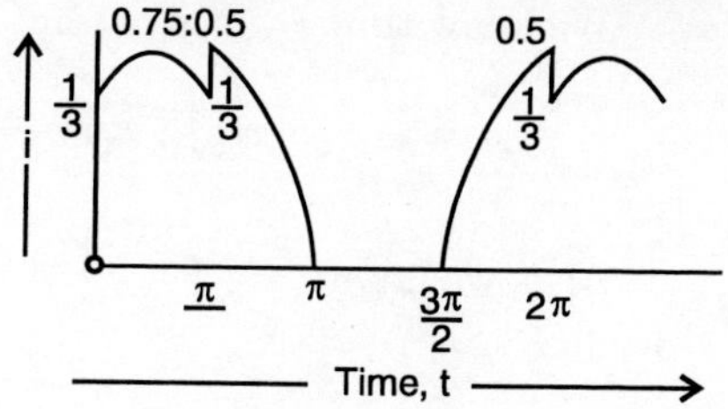

Fig. 3.47.

EXAMPLE 3.21 What is the ripple 2V on average of 50 V?

Solution: R.M.S. value of ac component,

$$V_{rms} = 2\text{ V}$$

Average value of output voltage,

$$V_{dc} = 50\text{ V}$$

Ripple factor,
$$\gamma = \frac{V_{rms}}{V_{dc}} = \frac{2}{50} = 0.04 \quad \textbf{Ans.}$$

EXAMPLE 3.22 In a power supply the dc output voltage drops from 44 V with no-load to 42 V at full-load. Calculate the percentage of voltage regulation.

Solution: No-load voltage, $V_{NL} = 44$ V

Full- load voltage, $V_{FL} = 42$ V

Percentage voltage regulation

$$= \frac{V_{NL} - V_{FL}}{V_{FL}} \times 100 = \frac{44 - 42}{42} \times 100 = 4.76\% \textbf{ Ans.}$$

EXAMPLE 3.23 A half-wave rectifier employs a diode having a forward resistance for 10 Ω. If the input voltage to the rectifier circuit is 12 V rms, find the d.c. output voltage at a load of 100 mA and PIV.

Solution: We have

$$\text{Peak inverse voltage (PIV)} = 12\sqrt{2} = v_0 = 16.9 \text{ Volt}$$

We know that the d.c. voltage is given by

$$V_{dc} = \frac{V_0}{\pi} - I_{dc} r_F = \frac{16.9}{\pi} - 100 \times 10^{-3} \times 10$$

or
$$V_{dc} = \frac{16.9}{\pi} - 1 = 4.4 \text{ Volt} \qquad \textbf{Ans.}$$

EXAMPLE 3.24 What is transformer utilisation factor? Determine the rating of a transformer to deliver a 100 Watts of d.c. power to a load under full wave rectifier.

Solution: Transformer utilisation factor

The transformer utilization factor (TUF) indicates how well the input transformer is being utilized. It may be defined as the ratio of dc output power to the ac power ratings of the transformer.

TUF may be defined as under:

$$\text{TUF} = \frac{\text{DC output power } (P_{dc})}{\text{AC power rating of the transformer}} = \frac{V_{dc}\, I_{dc}}{V_{rms}\, I_{rms}}$$

Assuming $R_L >> (R_S + R_F)$, we have

$$\text{TUF} = \frac{[V_m/\pi]\,[I_m/\pi]}{\left[\frac{V_m}{\sqrt{2}}\right]\left[\frac{I_m}{2}\right]} = \frac{2\sqrt{2}}{\pi^2}$$

Solving, we get TUF = 0.287 or 28.7%

Infact, ideal value of TUF is 100% and in practice, it must be as high as possible.

Important Point:

The value of TUF is too low for the HWR. This value of TUF indicates that the transformer is being utilized to only 28.7% of its full capacity which is gross underutilization.

Let us assume $V_{dc} = 10$ Volts.

We know that I_{dc} is given by

$$I_{dc} = \frac{P_{dc}}{V_{dc}} = \frac{100}{10} = 10 \text{ A}$$

Also, we have $$V_{dc} = \frac{2V_m}{\pi}$$

$$\therefore \quad V_m = \frac{\pi}{2}V_{dc} = \frac{\pi}{2}\times 10 = 15.7 \text{ V}$$

Therefore, secondary rms voltage will be

$$V_{\text{rms}} = \frac{V_m}{\sqrt{2}} = 11.1 \text{ Volts}$$

We know that $$I_{dc} = \frac{2I_m}{\pi}$$

$$\therefore \quad I_m = \frac{10\times\pi}{2} = 15.7 \text{ Amp.}$$

Hence, secondary rms current is given by

$$I_{\text{rms}} = \frac{I_m}{\sqrt{2}} = \frac{15.7}{\sqrt{2}} = 11.1 \text{ Amp.}$$

Also, TUF is given by

$$\text{TUF} = \frac{\text{DC power}}{\text{Secondary power rating}} = \frac{100}{11.1\times 11.1}\times 100$$

Thus, TUF = 81.6%

Also, truns ratio is given by

$$\frac{N_1}{N_2} = \frac{230}{11.1} = 20.72 \quad \textbf{Ans.}$$

EXAMPLE 3.25 Determine v_0 and required PIV rating of each of the diodes of the following circuit shown in figure 3.48.

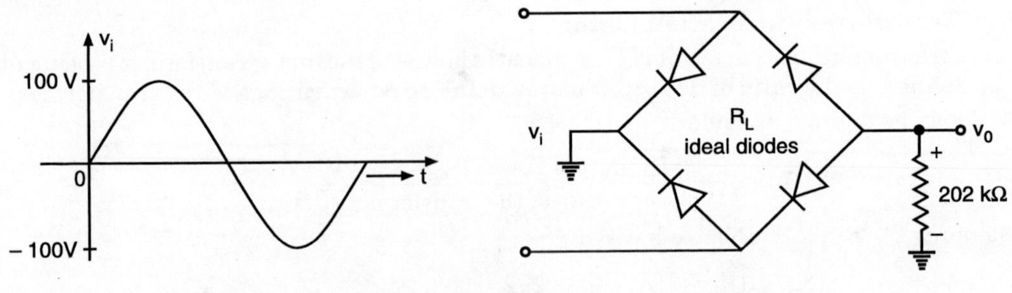

Fig. 3.48.

Solution: Given that V_{rms} = 100 V, R_L = 2.2 kΩ,

We know that the peak value of load current is given by

$$I_m = \frac{V_m}{R_s + 2R_f + R_L} \qquad (R_s = 0, R_f = 0, \because \text{ diodes are ideal})$$

Therefore, $$I_m = \frac{\sqrt{2}\,V_{rms}}{R_L} = \frac{\sqrt{2}\times 100}{2.2\times 10^3} = 64.28 \text{ mA}$$

DC load current $I_{dc} = \frac{2I_m}{\pi}$

or $$V_{dc} = \frac{2 \times 64.28 \times 10^{-3}}{\pi} = 40.92 \text{ mA.}$$

DC load voltage is given by

$$V_{dc} = I_{dc} \times R_L = 40.92 \times 10^{-3} \times 2.2 \times 10^3 = 90.02 \text{ V}$$

$\therefore$ $$v_0 = V_{dc} = 90.02 \text{ V} \quad \textbf{Ans.}$$

For bridge rectifier circuit, PIV for each diode is given by

$$\text{PIV} = V_m = \sqrt{2} \times V_{rms} = \sqrt{2} \times 100 = 141.42 \text{ V} \quad \textbf{Ans.}$$

SUMMARY

- A rectifier is a circuit which is used to convert a.c. voltage into the pulsating d.c. voltage.
- There are two types of rectifier : Half-wave rectifier and full wave rectifier.
- A Half-wave rectifier utilizes only positive half cycle.
- A full-wave rectifier utilizes both positive and negative half cycles.
- Peak-inverse voltage (PIV) represents the maximum voltage which a diode used in reverse-bias must withstand.
- For a half-wave rectifier: PIV = V_m.
- Full-wave rectifiers are of two types as under:
  (*i*) Centre-tap full wave rectifier,
  (*ii*) Full-wave bridge rectifier.
- For a full-wave bridge rectifier: PIV = V_m.
- The a.c. component present in the output of a rectifier is called a ripple.
- Ripple factor $r = \sqrt{\left(\frac{I_{rms}}{I_{dc}}\right)^2 - 1}$.
- Rectifier efficiency η is expressed as
  $$\eta = \frac{\text{D.C. power delivered to the load}}{\text{A.C. input power from the transformer secondary}} = \frac{P_{dc}}{P_{ac}}.$$
- A filter is a circuit which removes or minimizes a.c. components from the output of a rectifier.
- A clipper is a circuit with which the waveform is shaped by removing or clipping a portion of the applied input signal waveform without distorting the remaining part.
- Clamper circuits shift or clamp a signal to a different d.c. level.
- A voltage multiplier is a circuit which produces an output d.c. voltage whose value is multiple of peak a.c. input voltage.

SHORT QUESTIONS WITH ANSWERS

Q. 1. What is a rectifier circuit? Explain.

A rectifier circuit is a circuit which is used to convert a.c. voltage into the pulsating d.c. voltage. A rectifier circuit uses one or more diodes. Broadly there are two types of rectifiers:

(*a*) Half-wave rectifier

(*b*) Full-wave rectifier.

Q. 2. What are the disadvantages of Half wave rectifier circuit? Explain.

Ans. (*i*) The ripple factor of half wave rectifier circuit is 1.21, which is quite high. The output contains lot of varying components.

(*ii*) The maximum theoretical rectification efficiency is found to be 40%. The practical value will be less than this. This indicates that half wave rectifier circuit is quite inefficient.

(*iii*) The circuit has low transformer utilization factor, showing that the transformer is not fully utilized.

(*iv*) The dc current is flowing through the secondary winding of the transformer which may cause dc saturation of the core of the transformer. To minimize the saturation, the transformer size have to be increased accordingly. This increases the cost.

Q. 3. What is a full-wave rectifier? Explain.

Ans. Full-wave rectifier is that type of rectifier which utilizes both the half cycle of a.c. input voltage. Hence, in a full wave rectifier an unidirectional current flows through the load during the entire cycle of input voltage. There are two types of full wave rectifier circuits :

1. Centre-tap full-wave rectifier
2. Full-wave Bridge Rectifier.

Q. 4. What are the advantages and disadvantages of bridge rectifier over centre-tap rectifier.

Ans. The full-wave bridge rectifier is the most widely used full wave rectifier. It is preferred over centre-tap full-wave rectifier because it has certain advantages over it.

Advantages

1. Bridge rectifier does not require centre-tapped secondary winding.
2. For bridge rectifier, the transformer is less costly because it is required to provide only half the voltage of an equivalent centre-tapped transformer used in a full wave rectifier circuit. It means that for a centre-tap rectifier, the transformer secondary must have double the number of turns.
3. The diode in a bridge rectifier is required to have PIV rating of only V_m. But each of two diodes in a centre-tap rectifier must have a PIV rating of $2V_m$. Therefore, the diodes used in a centre-tap rectifier are costlier than those used in bridge rectifier.

Disadvantages

The only disadvantage of bridge rectifier is that this needs four diodes. Two diodes out of four operate in alternate half cycles. When low dc voltages are required, there can be a problem that for low output dc voltage the secondary voltage should be low and the diode drops (1.4 V for Si and 0.6 V for Ge diode) becomes important. Because of this reason, for low-voltage application, the centre-tap rectifiers are preferred.

Q. 5. What is a ripple factor? Explain.

(V. Important)

The output voltage or load current of a rectifier contains two components namely d.c. component and a.c. component. The a.c. component present in the output is called a **ripple**. This ripple accounts for pulsations in the output of rectifier and so this is undesirable. Therefore, the effectiveness of a rectifier depends on the magnitude of ripple in the output. Smaller the ripple, more effective is the rectifier. Ripple factor is a measure of purity of the output of a rectifier.

Mathematically, ripple factor may be expressed as

Ripple factor,

$$r = \frac{\text{The r.m.s. value of a.c. component of output voltage}}{\text{The d.c. component of output voltage}}$$

Q. 6. What do you mean by efficiency of a rectifier? Explain.

Ans. Efficiency of a rectifier is defined as the ratio of d.c. power delivered to the load to the a.c. input power from the secondary winding of the transformer.

Mathematically, the rectifier efficiency is expressed as rectifier efficiency,

$$\eta = \frac{\text{D.C. power delivered to the load}}{\text{A.C. input power from the transformer secondary}}$$

Q. 7. What is meant by Voltage Regulation ?

Ans. The variation of dc output voltage as a function of dc load current is called voltage regulation. In other words, the degree to which a power supply varies in output voltage under conditions of load variations is measured by a term known as voltage regulation.

Voltage regulation is generally expressed as a percentage, *i.e.*,

% Voltage Regulation =

$$\frac{V_{\text{no-load}} - V_{\text{full-load}}}{V_{\text{full-load}}} \times 100\%$$

Q. 8. What are filter circuits? Explain.

Ans. Output of a rectifier consists of a.c. as well as d.c. components. The function of rectifier is to change a.c. voltage into d.c. voltage. Therefore, the presence of a.c. components is most undesirable and hence this should be removed. Filter circuits are used for this purpose.

Q. 9. Define clipper circuits?

Ans. A clipper circuit is a circuit with which the waveform is shaped by removing or clipping a portion of the applied input signal waveform without distorting the remaining part. Clipper can remove signal voltages above or below a specified level.

They are of the following types :

(*i*) Positive clipper
(*ii*) Negative clipper
(*iii*) Biased clipper
(*iv*) Combination clipper.

Q. 10. What is positive clipper? Explain.

Ans. A positive clipper removes the positive half-cycles of the input voltage waveform. There are two types of positive clipper (*a*) series and (*b*) shunt.

Q. 11. What are clampers? Explain.

Clamping circuit shifts or clamps a signal to a different d.c. level. In other words, clamping circuit introduces a d.c. level to an a.c. signal. Therefore, the clamping circuit is also known as d.c. restorer. These circuits find application in television receivers to restore the d.c. reference signal to the video signal. The clamping circuit has various circuit components like a diode, a capacitor and a resistor. Basically clamping circuits are of three types :

(*i*) Positive clamper

(*ii*) Negative clamper

(*iii*) Biased clamper.

Q. 12. What are voltage multipliers? Explain.

Ans. A voltage multiplier is that of circuit which produces an output d.c. voltage whose value is multiple of peak a.c. input voltage (*i.e.*, 2 V_m, 3 V_m, 4 V_m and so on). Basically, a voltage multiplier circuit is a combination of two or more peak rectifier circuits. Each peak rectifier contains a diode and a capacitor. With the help of voltage multiplier circuits the voltage level may be raised into hundreds or thousands of volts.

Q. 13. What is a voltage doubler? Explain.

Ans. A voltage multiplier circuit whose output d.c. voltage is double of the peak a.c. input voltage is known a voltage doubler. A voltage doubler circuit may be classified as under:

(*i*) Half-wave voltage doubler

(*ii*) Full-wave voltage doubler.

Q. 14. Compare full-wave and half-wave rectifier circuits.

Ans. For comparison, we assume that the full-wave and half-wave rectifier circuits use identical diodes, identical load resistances and the voltage across half the secondary winding of transformer used in full-wave rectifier circuit is the same as the voltage across the secondary winding of the transformer used in the half-wave rectifier circuit.

(*i*) The d.c. load current in case of full-wave circuit is twice to that in half-wave circuit. Similarly, the d.c. load voltage in full wave circuit is twice that in half-wave circuit.

(*ii*) The lowest ripple frequency in full-wave circuit is twice that in half-wave circuit. Now, to remove ripple, the additional circuits called filter circuits are used along with rectifier circuits. But, as the frequency is more in full-wave rectifier circuit, the capacitor values required in capacitance filter are much less. Hence, smaller elements are sufficient in filter circuits used with full wave circuit to reduce ripple.

(*iii*) Because, there is no net d.c. current through windings of the transformer used in full-wave rectifier circuit, the losses are less as compared to losses in transformer used in half-wave rectifier circuit.

(*iv*) The full-wave connection gives d.c. power output four times as large when compared with half-wave connection.

(*v*) The efficiency of rectification in a full-wave connection is twice that for half-wave connection.

(*vi*) The ripple factor is less for full-wave *i.e.*, rectification is more nearly complete for full-wave as compared to half-wave rectifier circuit.

REVIEW QUESTIONS

1. What is a clipper? Describe the operation of positive and negative clippers with the help of circuit diagrams.
2. Explain the operation of biased clipper and combination clipper.
3. What is a clamper? Explain the operation of a clamping circuit with the help of circuit diagram and waveforms.
4. Differentiate between positive clamper and negative clamper.
5. Explain the following terms : (*i*) ripple factor, (*ii*) rectification efficiency, (*iii*) peak inverse voltage.
6. Draw the circuit diagram of an half wave rectifier and explain its operation.
7. Derive expression for rectification efficiency, ripple factor for half-wave rectifier.
8. Derive an expression for the efficiency of a full wave bridge rectifier.
9. What do you understand by reverse saturation current of diode?
10. What is the importance of peak inverse voltage ?
11. Explain the working of a full wave-bridge rectifier. Find the values of PIV of each diode.

12. What is a comparator circuit? How does such a circuit differ from a clipping circuit? Draw typical circuits for both.

13. Derive an expression for ripple factor of a rectifier in general.

14. For a semiconductor diode, define static and dynamic resistance.

NUMERICAL PROBLEMS

1. The turns ratio of the transformer used in a half wave rectifier is 2:1 and the primary is connected to 230 V, 50 Hz power mains. Assuming the diodes to be ideal, determine the following:
(*i*) d.c. voltage across the load, (*ii*) PIV of each diode, (*iii*) efficiency of the half-wave rectifier, (*iv*) output ripple frequency.
(Ans. 51.7 V, 162.2 V, 10.21%, 50 Hz)

2. In a full-wave rectifier, the voltage applied to each diode is 240 Sin 377t, the load resistance is R_L = 2000 W and each diode has a forward resistance of 400 W. Determine the following:
(*i*) peak value of current, (*ii*) d.c. value of current, (*iii*) rms value of current, (*iv*) rectifier efficiency, (*v*) ripple factor, (*vi*) output ripple frequency.
(Ans. 100 mA, 63.8 mA, 70.7 mA, 67.6%, 0.482, 120 Hz)

3. Draw the waveform for the output voltage for the clipping circuit given below and for the input waveform shown. Assume the diodes to be ideal.

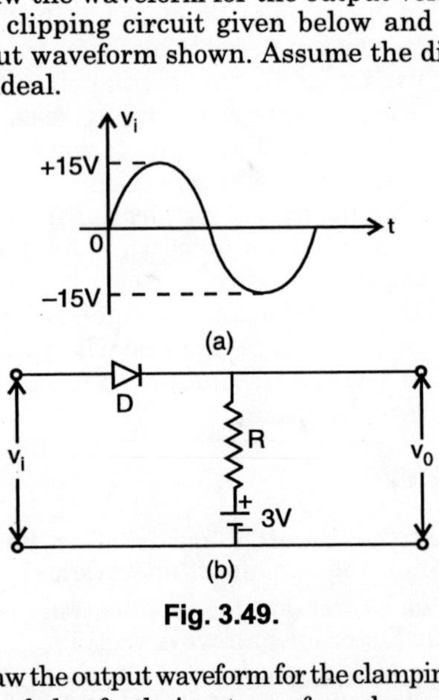

Fig. 3.49.

4. Draw the output waveform for the clamping circuit given below for the input waveform shown. Assume the diodes to be ideal.

5. In a bridge rectifier, the transformer is connected to 220 V, 60 Hz mains and the turns ratio of the strep down transformer is 11:1. Assuming the diodes to be ideal, determine the following :
(*i*) the voltage across the load, (*ii*) $I_{d.c.}$ (*iii*) PIV of each diode.
(Ans. 18 V, 18 mA, 28.28 Volt)

6. In a half-wave rectifier, an a.c. voltage of peak value 24 volt is connected in series with a silicon diode and load resistance of 480 W. If the forward resistance of the diode is 20 W. Determine the peak current flowing through the diode. **(Ans. 46.6 mA)**

7. What is comparator circuit? How does such a circuit differ from a clipping circuit? Draw typical circuits for both.

8. Explain the working of the following circuit. Write clearly the name of the circuit. List all the assumptions made.

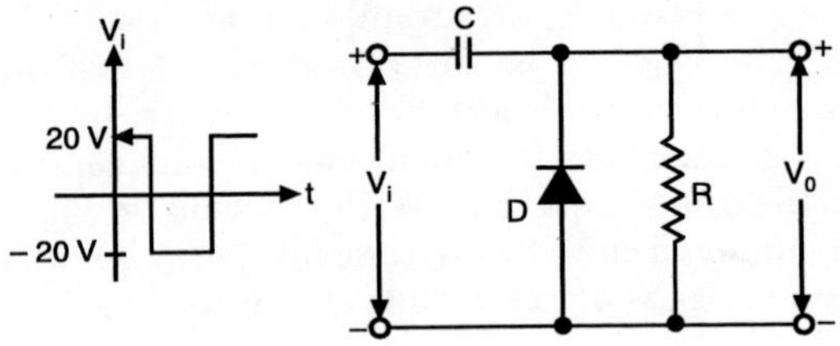

Fig. 3.50.

9. What is the difference between Clipping circuits and Clamping circuits? Determine V_0 for the netwrok of figure 3.51.

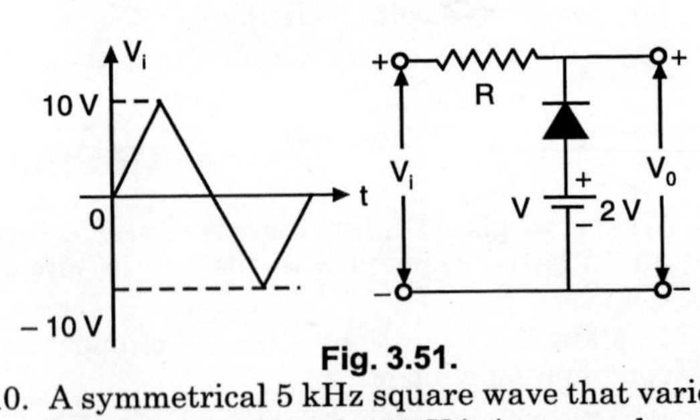

Fig. 3.51.

10. A symmetrical 5 kHz square wave that varies between + 10 V and – 10 V is impressed upon the clipping circuit shown in figure 3.52. Assume $R_f = 0$, $R_r = 2$ M and $V_1 = 0$. Sketch the steady state output waveform, indicating values of the maximum, minimum and constant portions.

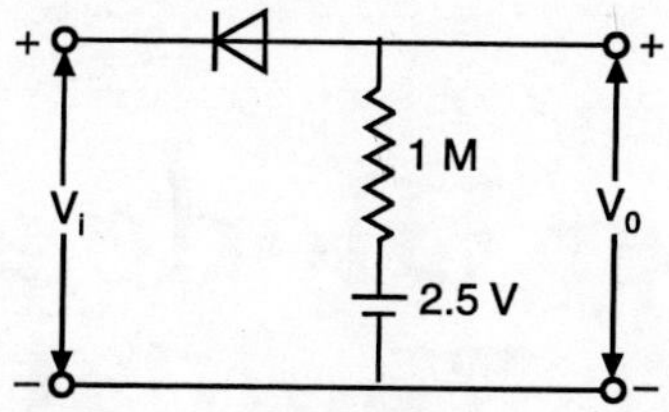

Fig. 3.52.

11. For the following circuit using ideal diode, sketch the output for the input shown. Label the most positive and most negative output levels. Assume $RC >> T$.

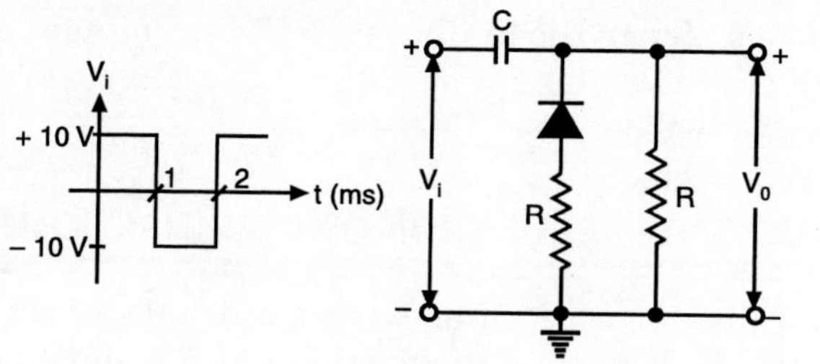

Fig. 3.53.

12. Draw the circuit and explain its working to obtain the output waveform for a sinusoidal input voltage shown in figure 3.53.

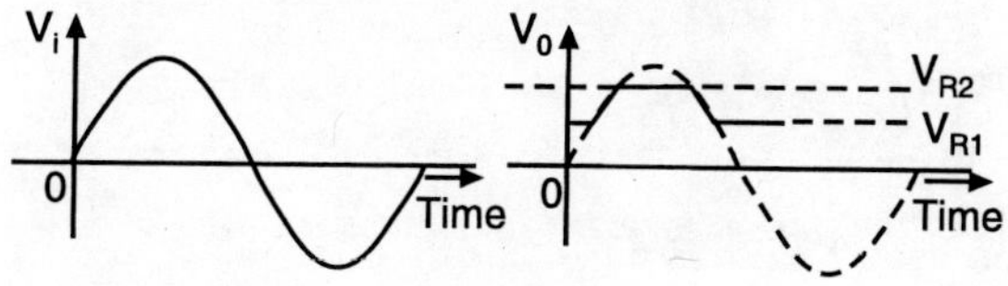

Fig. 3.54.

OBJECTIVE TYPE QUESTIONS

Choose one correct answer in the following:

1. In a half-wave rectifier, the load current flows
   (*a*) only for the positive half-cycle of the input signal
   (*b*) for less than half-cycle of the input signal
   (*c*) for more than half-cycle of the input signal
   (*d*) for whole cycle of the input signal

2. In a full-wave rectifier, the current in each diode flows for
   (*a*) whole cycle of the input signal
   (*b*) half-cycle of the input signal
   (*c*) more than half-cycle of the input signal
   (*d*) none of the above

3. For a full-wave rectifier, if the input frequency is 50 Hz, then the output frequency will be
   (*a*) 50 Hz (*b*) 75 Hz
   (*c*) 100 Hz (*d*) 200 Hz

4. In a centre-tap full-wave rectifier, if V_m is the peak voltage between the centre-tap and one end of the secondary, the maximum voltage coming across the reverse-biased diode is
   (*a*) V_m (*b*) $2\ V_m$
   (*c*) $\frac{1}{2} V_m$ (*d*) $\frac{V_m}{\sqrt{2}}$

5. The maximum efficiency of full wave rectification is
   (*a*) 40.6% (*b*) 100%
   (*c*) 81.2% (*d*) 85.6%

6. In a full-wave bridge rectifier, if V_m is the peak voltage across the secondary of the transformer, the maximum voltage coming across each reverse-biased diode is
   (*a*) V_m (b) $2\ V_m$
   (*c*) $\frac{1}{2} V_m$ (*d*) $\frac{V_m}{\sqrt{2}}$

Answer

1. (*a*) 2. (*b*) 3. (*c*) 4. (*b*) 5. (*c*) 6. (*a*)

Chapter 4

Zener Diode and Voltage Regulation

SYLLABUS

Breakdown Mechanisms in Semiconductor Diodes, Zener Diode Characteristics, Voltage Regulation using Zener Diode.

Chapter Outoine

- Voltage Regulation or Stabilisation
- Breakdown Mechanism
- Zener Breakdown
- Avalanche Breakdown in Zener Diodes
- Zener Diode
- Circuit Symbol and Biasing of a Zener Diode
- V-I Characteristics of a Zener Diode
- Applications of Zener Diodes
- Equivalent Circuit of a Zener Diode
- Analysis of Zener Diode Circuit
- Zener Diode as a Voltage Regulator
- Zener Diode Specifications
- Effect of Supply Voltage Variations
- Effect of Temperature on Zener Diodes
- Zener Diode Applications : Detailed Aspect
- Comparison of Zener Diode and P-N Junction Diode

4.1. VOLTAGE REGULATION OR STABILISATION

If we consider a rectifier with a suitable filter, then it will become a good source of d.c. output voltage. However, the main drawback of such a power supply is that the d.c. output voltage varies with the variation in load resistance or input voltage. Hence, if the applied input voltage increases, the d.c. output voltage also increases. In the same way, if the load-current increases, the output voltage falls because of the voltage drop in filter choke, rectifier element etc. However, in most of the electronic applications, it is desired that the d.c. output voltage of rectifier must remain constant irrespective of the variation in input a.c. voltage or load. Hence, to ensure this, a voltage stabilisation or regulation device known as voltage stabiliser is used. Although there are several voltage stabilising circuits which may be designed, but in this chapter, we shall discuss only zener diode as a voltage stabilizer or regulator.

4.2. BREAKDOWN MECHANISM

If the reverse-bias applied to a P-N-junction is increased, a point will reach when the junction breaks down and reverse current rises sharply to a value limited only by the external resistance connected in series. This specific value of the reverse bias voltage is called breakdown voltage (V_Z). After breakdown, a very small further increase the reverse current. The breakdown voltage depends upon the width of depletion layer. This width of depletion layer depends upon the doping level.

The following two processes cause junction breakdown due to the increase in reverse bias voltage:

(*i*) Zener Breakdown

(*ii*) Avalanche Breakdown

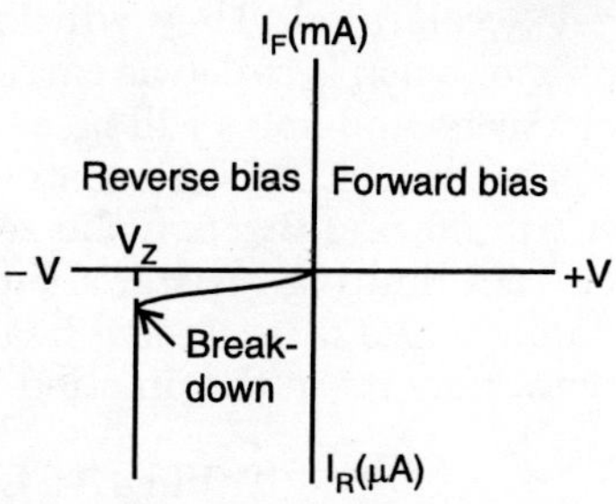

Fig. 4.1. Breakdown mechanism

4.3. ZENER BREAKDOWN*

The zener breakdown is observed in the zener diodes having V_z less than 5 V or between 5 to 8 Volts. Let us understand the zener breakdown mechanism for V_z less than 5 V. When a reverse voltage (5 V or less) is applied to a zener diode, it causes a very intense electric field to appear across a narrow depletion region. The field intensity is typically of the order of 3×10^5 V/cm. Such an intense electric field is strong enough to pull some of the valence electrons into the conduction band by breaking their covalent bonds. These electrons then become free electrons which are available for conduction. A large number of such free electrons will constitute a large reverse current through the zener diode and breakdown is said to have occurred due to the zener effect. A current limiting resistance should be connected in series with the zener diode to protect it against the damage due to excessive heating. In zener breakdown, the breakdown voltage depends on the temperature of P-N junction. The breakdown voltage decreases with increase in the junction temperature. The V-I characteristics in the reverse biased region is shown in figure 4.2 (*a*) which shows that the characteristics after breakdown is very sharp, almost vertical.

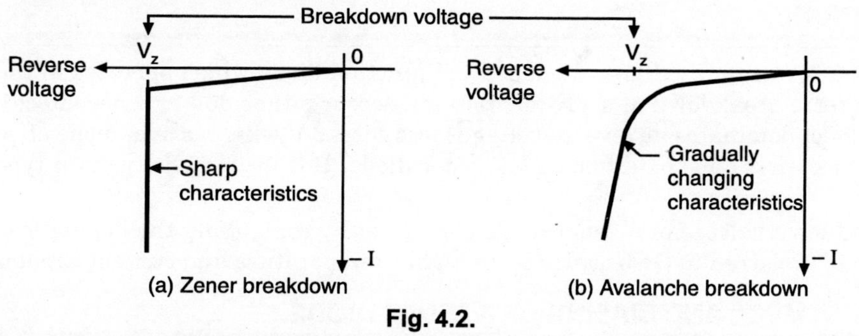

Fig. 4.2.

4.4. AVALANCHE BREAKDOWN IN ZENER DIODES**

The avalanche breakdown is observed in the zener diodes having V_z higher than 8 V. Even though the mechanism of breakdown has changed, the device is still called as zener diode. Let us now understand the mechanism. In the reverse biased condition, the conduction will take place only due to the minority carriers. As we increase the reverse voltage applied to the zener diode, these minority carriers tend to accelerate. Therefore, the kinetic energy associated with them increases. While travelling, these accelerated minority carriers will collide with the stationary atoms and impart some of the kinetic energy to the valence electrons present in the covalent bonds. Due to this additionally acquired energy, these valence electrons will break their covalent bonds and jump into the conduction bond to become free for conduction. Now these newly generated free electrons will

* Explain zener breakdown and Avalanche multiplication phenomena in *pn*-junction diode. "The breakdown phenomena for voltage less than 6 V is zener breakdown", explain your opinion on this statement.

** Explain 'Avalanche' breakdown.

get accelerated. They will knock out some more valence electrons by means of collision. This phenomenon is called as carrier multiplication. In a very short time, a large number of free minority electrons and holes will be available for conduction, and the carrier multiplication process becomes self sustained. This self sustained multiplication is called "Avalanche Effect". A large reverse current starts flowing through the zener diode and the avalanche breakdown is said to have occurred. A current limiting resistor should be connected in series with the zener diode to protect it against the damage due to excessive heating. The breakdown voltage in the avalanche breakdown increases with increase in the junction temperature.

Table 4.1: Comparison of Zener and Avalanche Breakdown

S. No.	Zener breakdown	Avalanche breakdown
1.	This is observed in zener diodes having V_z between 5 to 8 volts.	This is observed in zener diodes having V_z greater than 8 volts.
2.	The valence electrons are pulled into conduction band due to very intense electric field appearing across the narrow depletion region.	The valence electrons are pushed into conduction band due to the energy imparted by colliding accelerated minority carriers.
3.	V-I characteristics with the zener breakdown is very sharp.	The V-I characteristics with the avalanche breakdown increases gradually. It is not as sharp as that with the zener breakdown.
4.	The breakdown voltage decreases with increase in temperature.	The breakdown voltage increases with increase in temperature.

4.5. ZENER DIODE

Zener diode is a reverse-biased heavily-doped PN-junction diode which operates in the breakdown region. The reverse breakdown of a PN-junction may occur either due to zener effect or avalanche effect. Zener effect dominates at reverse voltages less than 6 V whereas avalanche effect dominates above 6 V. Hence, first one should be called zener diode. But for simplicity, both types are called zener diodes.

The breakdown voltage of a zener diode can be set by controlling the doping level. For zener diodes, silicon is preferred to Ge because of its higher temperature and current capability.

4.6. CIRCUIT SYMBOL AND BIASING OF A ZENER DIODE

The circuit symbol of a zener diode is as shown in figure 4.3 (*a*). It is a two terminal device and the terminals are anode and cathode. The arrowhead in the symbol points towards the conventional direction of current through the zener diode, when it is forward biased.

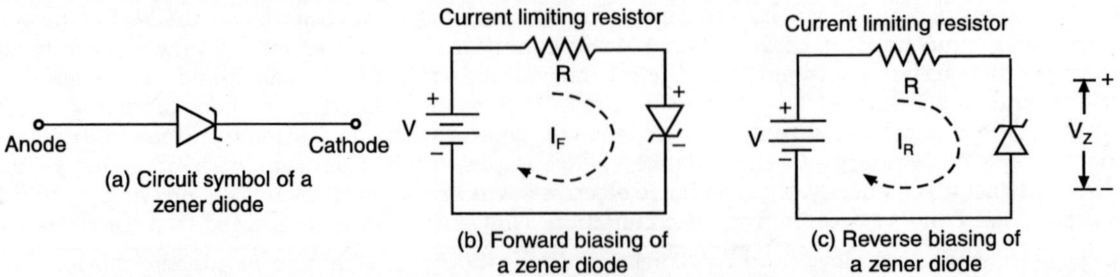

Fig. 4.3. Circuit symbol and biasing of a zener diode

(i) Forward biasing of zener diode

> **DO YOU KNOW?**
>
> Zener diodes with breakdown voltages near 7 V have the smallest zener impedance.

When the anode of the zener diode is connected to the positive terminal of the dc source and the cathode is connected to the negative terminal, the zener diode is said to be forward biased. The forward biased zener diode behaves identical to a forward biased diode. The forward biasing of a zener diode is shown in figure 4.3 (*b*). The zener diode is generally not used in the forward biased condition.

(ii) Reverse biasing of zener diode

When the cathode is connected to the positive terminal and anode is connected to the negative terminal of the dc source, the zener diode is said to be reverse biased. The operation of zener diode in the reverse biased condition is substantially different from that of a diode. The reverse biasing of a zener diode is shown in figure 4.3 (*c*). Zener diode in the reverse biased condition is used as a voltage regulator.

4.7. V-I CHARACTERISTICS OF A ZENER DIODE*

The V-I characteristics of a zener diode can be divided into two parts:

(*i*) Forward characteristics

(*ii*) Reverse characteristics.

(i) Forward characteristics

The forward characteristics of a zener diode is shown in figure 4.4 (*a*). It is almost identical to the forward characteristics of a *P-N* junction diode.

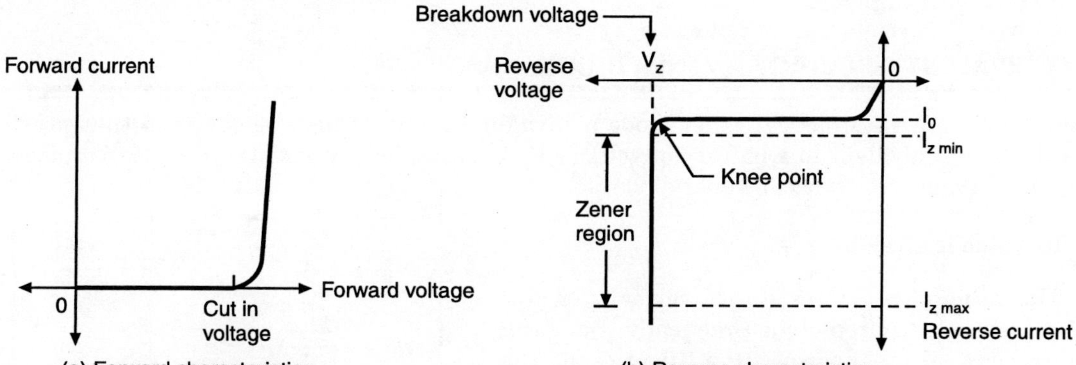

Fig. 4.4. V-I Characteristics of a zener diode

(ii) Reverse characteristics

As we increase the reverse voltage, initially a small reverse saturation current I_0, which is in μA, will flow. This current flows due to the thermally generated minority carriers. At a certain value of reverse voltage, the reverse current will increase suddenly and sharply. This is an indication that the breakdown has occurred. This breakdown voltage is called as **zener breakdown voltage or zener voltage and it is denoted by V_z**. The value of V_z can be precisely controlled by controlling the doping levels of *P* and *N* regions at the time of manufacturing a zener diode. After breakdown has occurred, the voltage across zener diode remains constant equal to V_z. Any increase in the source voltage will result in the increase in reverse zener current. The zener current after the reverse

* Draw the V-I characteristics of zener diode and explain how does a zener regulate a voltage ?

breakdown must be controlled by connecting a resistor R as shown in figure 4.3(*b*). This is essential to avoid any damage to the device due to excessive heating.

Zener Region and its importance

After reverse breakdown the zener diode operates in a region called zener region, as shown in figure 4.4 (*b*). In this region the voltage across zener diode remains constant but current changes depending on the supply voltage. Zener diode is operated in this region when it is being used as a voltage regulator. The complete V-I characteristics is as shown in figure 4.4 (*c*).

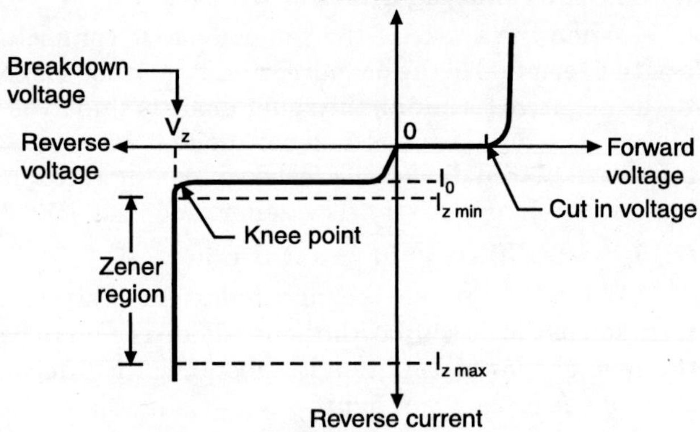

Fig. 4.4. (c) Complete V-I characteristics of a zener diode.

4.8. APPLICATIONS OF ZENER DIODES*

Zener diodes have a large number of applications. Few of them are :

(*i*) Zener diode is used as a voltage regulator

(*ii*) Zener diode is used as a peak clipper in waveshaping circuits

(*iii*) Zener diode is used as a fixed reference voltage in transistor biasing circuits

(*iv*) Zener diode is used for meter protection against damage from accidental application of excessive voltage.

4.9. EQUIVALENT CIRCUIT OF A ZENER DIODE

Figure 4.5(*a*) shows a practical zener diode equivalent circuit. It may be observed that practical zener diode is equivalent to a battery of voltage V_Z in series with a resistance r_z. Resistance r_z is known as dynamic resistance of zener diode.

Its value is given by $r_z = \dfrac{\Delta V_Z}{\Delta I_Z}$

The value of zener resistance varies from few ohms to several hundred ohms practically. But for a perfectly vertical breakdown curve (Ideal case), the value of zener resistance is zero.

Generally, the zener resistance value is quite small than external resistance of the circuit in which zener diode is connected. Therefore, zener resistance may be neglected generally. Figure 4.5(*b*) shows the ideal equivalent circuit of zener diode.

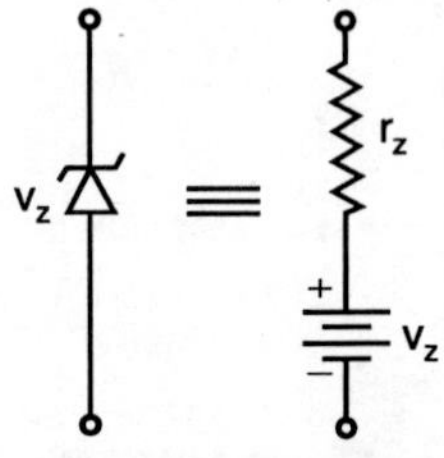

Fig. 4.5 (*a*). Practical equivalent circuit of zener diode

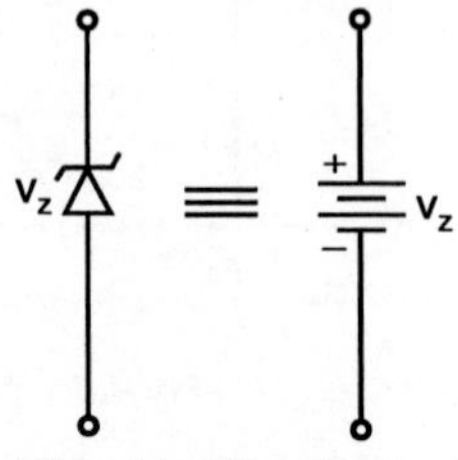

Fig. 4.5 (*b*). Ideal zener equivalent diode circuit

4.10. ANALYSIS OF ZENER DIODE CIRCUIT

We can analyze the circuits using zener diodes easily by replacing the zener diode with its equivalent circuit. We shall consider two states of zener diode.

---

* Write the applications of zener diodes.

(a) Zener diode circuit for $V > V_Z$ (b) Equivalent zener diode circuit in 'ON' state

Fig. 4.6.

***(i)* ON State:** If the reverse-bias voltage across a zener diode is equal to or more than breakdown voltage V_Z, the current will increase sharply. Hence, in this region the curve will be almost vertical. This indicates that the voltage across zener diode is constant at V_Z even if the current through it changes. Hence, in the breakdown region an ideal zener diode may be represented by a battery of voltage V_Z as shown figure 4.6. In this condition the zener diode is said to be in the 'ON' state.

***(ii)* OFF State :** If the reverse voltage across the zener diode is less than V_Z but greater than 0V. The zener diode will be in the 'OFF' state. In this condition, the zener diode may be represented by an open-circuit as shown in figure 4.7.

(a) Zener diode circuit for $V_Z > V > 0$ (b) Equivalent zener diode circuit in 'OFF' state

Fig. 4.7.

4.11. ZENER DIODE AS A VOLTAGE REGULATOR*

Voltage regulator is a circuit whose function is to maintain a constant output d.c. voltage inspite of the a.c. input voltage fluctuations or changes in load resistance values.

4.11.1. Zener Diode Shunt Regulator

A zener diode shunt regulator is shown in figure 4.8. This is called shunt regulator because the zener diode is connected in parallel or shunt with the load. The resistance R_S is connected to limit current in the circuit. The output or regulated voltage is obtained across the load resistor R_L. For the operation of the circuit, the input voltage V_S should be greater than zener diode voltage V_Z. Only then zener diode operates in breakdown region.

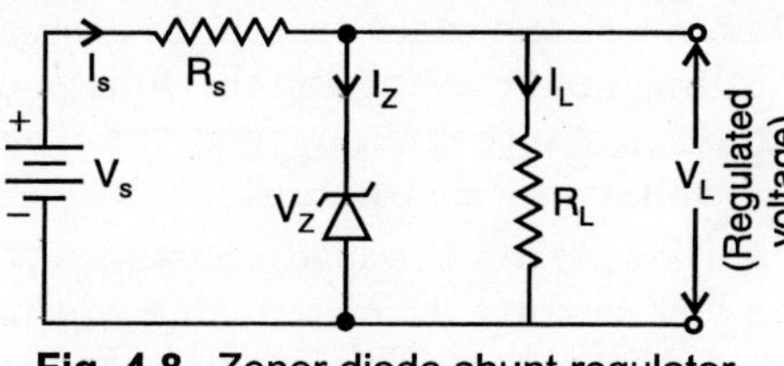

Fig. 4.8. Zener diode shunt regulator

The input current is calculated by

$$V_S = I_S R_S + V_Z$$

or

$$I_S = \frac{V_S - V_Z}{R_S}$$

Here V_S = Input d.c. unregulated voltage

V_Z = Voltage across zener diode

The ideal zener diode may be assumed as a constant voltage source of voltage V_Z. But in case of a practical zener diode, zener resistance is taken into account. Due to this zener resistance, there

* Draw a circuit diagram which uses a breakdown diode to regulate the output voltage across a load and explain the working.

is a voltage drop across it ($I_Z . r_Z$). The voltage across the terminals of zener diode which is also equal to the load voltage may be given as

$$V_L = V_Z + I_Z . r_Z$$

But, the zener resistance may be neglected.

Therefore, the load voltage is

$$V_L = V_Z$$

The current through the load resistance is

$$I_L = \frac{V_L}{R_L}$$

According to *KCL*, the input current will be given by

$$I_S = I_Z + I_L \quad \text{or} \quad I_Z = I_S - I_L$$

DO YOU KNOW?

Zener diodes can also be used to regulate the voltage from a power supply whose output is negative. All you need to do is reverse the connection of the zener so that the cathode is grounded and the anode is connected to the negative side of the unregulated supply.

4.11.2. Working of Zener Diode Shunt Regulator

To explain the working, we will consider two cases:

Case I: Regulation when input voltage is varried

In this case, the load resistance R_L is kept fixed and the input voltage V_S is varied. When the input voltage increases, the input current I_S is also increased.

Since $I_S = I_Z + I_L$

This increased current I_S increases the current through zener diode without affecting the load current I_L. Due to increase in input current I_S, the voltage drop across series resistance R_S also increases and hence keeping the load voltage V_L as constant.

Now, when the input voltage decreases, the input current also decreases. Due to this, the current through zener diode also decreases. Also the voltage drop across series resistance is reduced. Hence the load voltage V_L and load current I_L remain constant.

DO YOU KNOW?

The voltage across a zener diode does not remain perfectly constant as the zener current varies. The reason is that all zeners have some value of zener impedance, designated R_Z. R_Z acts like a small resistance in series with the zener. The effect of R_Z is that small changes in V_Z will occur when I_Z varies.

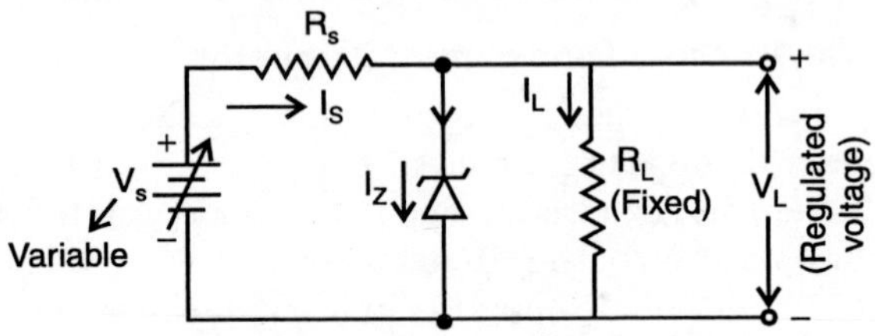

Fig. 4.9. Input voltage is varied

Case II: Regulation when load resistance is varied

In this case, the input voltage V_S is kept fixed and the load resistance R_L is varied. When the load resistance R_L decreases, the load current increases. This causes the zener current to decrease. Due to this, the input current and the voltage drop across series resistance remains constant. Therefore, the load voltage V_L is also constant.

Now, when the load resistance R_L increases, the load current decreases. Due to this, the zener current increases. This again keeps the values of input current and voltage drop across series resistance as a constant. Therefore, the load voltage remains constant.

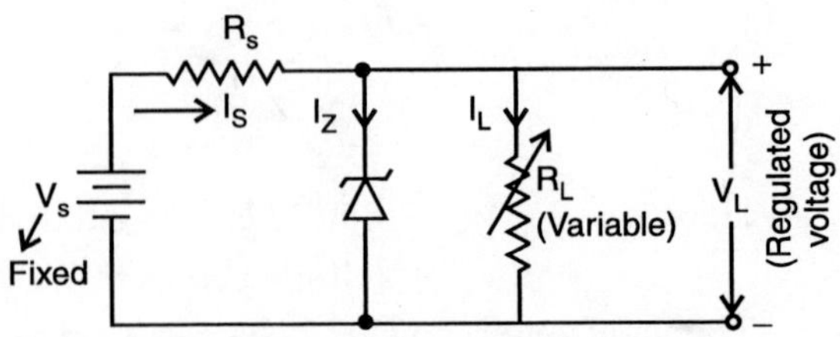

Fig. 4.10. Load resistance is varied

4.11.3. Limitations of Zener Regulators

Even though the zener diode provides a very simple means of voltage regulation, these regulators have following disadvantages:

(*i*) The output voltage of zener regulator is equal to V_Z. This is a constant voltage. Therefore, these voltage regulators cannot be made adjustable.

(*ii*) Large power gets dissipated in the series resistor R.

(*iii*) Corresponding to large changes in the load current, there will be large changes in the zener current. This will result in a large power wastage.

4.12. ZENER DIODE SPECIFICATIONS

Let us consider zener diode specifications in this section.

4.12.1. Zener Voltages

The manufacturers specify the value of breakdown voltage which is known as the Zener voltage, V_z at some value of test current, I_{ZT}. This is on the linear portion of the reverse characteristic and corresponds to approximately one quarter of the maximum power dissipation capability of the diode. Values of V_z are available at various values from 2.4 to 200 V with accuracies between 5 and 20%, depending upon cost.

> **DO YOU KNOW?**
>
> Exceeding the break-down voltage of a diode does not necessarily mean that you will destroy the diode. As long as the product of reverse voltage and reverse current does not exceed the diode's power rating, the diode will recover fully.

4.12.2. Power Dissipation

Power dissipation in the diode is the product of V_Z and reverse current I_Z, with maximum power ratings ranging from 150 mW to 50 W.

4.12.3. Breakover Current

Because there is some curvature of the reverse characteristic at low values of I_Z, there may be specified some value of current I_{ZK}, in the neighbourhood of the breakover knee, where the voltages across the diode starts to differ greatly from V_Z.

4.12.4. Dynamic Resistance

This is another specification provided by the manufacturer, which helps to describe how "vertical" is the reverse characteristic. This is known as the zener dynamic impedance, r_z, and is defined as:

$$r_Z = \frac{\Delta V_Z}{\Delta I_Z} \qquad \text{...(4.1)}$$

This impedance is generally evaluated in the region of the specified test current, I_{ZT}, and is the reciprocal of the slope of reverse characteristic. Ideal r_Z is zero for a perfectly vertical breakdown curve. However, in practice, it can vary from several ohms to several hundred ohms, depending upon the particular zener diode voltage and the operating current, as depicted in figure 4.11. In fact, the equation (4.1) is more useful in the following form:

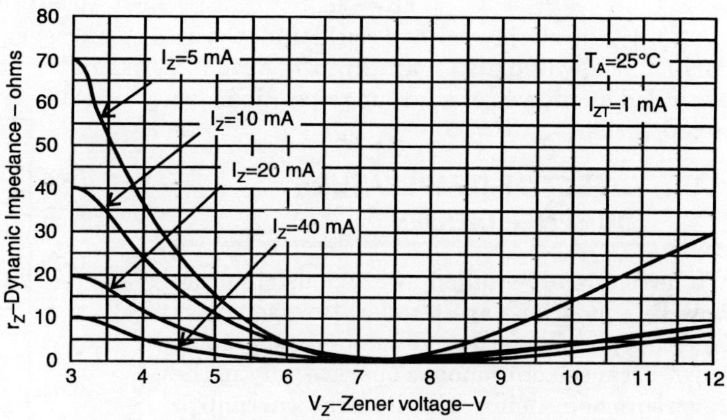

Fig. 4.11. Typical dynamic impedances versus zener voltage as a function of zener current

$$\Delta V_Z = r_Z \, \Delta I_Z \qquad \text{...(4.2)}$$

because it gives the change in voltage that will occur across the diode due to a change in current through the diode.

As an example, a 12 V zener diode operating at 5 mA has a r_Z of approximately 25 ohms (see figure 4.11) and will have a change of 25 mV across it if the current changes by 1 mA, whereas at 40 mA operations, the same diode will have only 5 mV change for the same change in current. Evidently a zener diode operated at a higher value of reverse current will have a more constant voltage across it in the face of any changes in diode current. This is an important property in the design of a voltage regulator circuit.

4.13. EFFECT OF SUPPLY VOLTAGE VARIATIONS

4.13.1. Increase in Supply Voltage

The circuit in figure 4.12 will also regulate the load voltage if the input voltage changes. Assuming that a 400 mW zener diode has been selected for our example (after determining that maximum dissipation would be 225 mW for the conditions specified), what maximum value of V_{in} could be tolerated without damage to the diode? Since $P = V_Z I_Z = 0.4$ W with $V_Z = 15$ V, then the maximum value of $I_Z = \frac{0.4A}{15} = 26$ mA. We must be sure that this zener current value is not exceeded, so that it may only occur when the load current is at its minimum of 10 mA. Thus, the maximum permissible value of $I_{in} = 26 + 10 = 36$ mA. Then the maximum voltage drop across R would be

$$V_R = I_{in} R = 36 \times 10^{-3} \times 200 = 7.2 \text{ V}$$

and the maximum input voltage would be

$$V_{in} = I_{in} R + V_{out} = 7.2 + 15 = 22.2 \text{ V}$$

The resistor power rating could be 0.5 W.

4.13.2. Decrease in Supply Voltage

Assuming the selected diode has an I_{ZK} of 2 mA (the zener current near the breakover knee), what would be the minimum value of V_{in} at which the zener diode could still maintain the output at a nominal 15 V under a 10 mA load current variation? We must now consider the Zener current at its minimum value of 20 mA. (Why?) Under this condition, $I_{in} = 22$ mA, and the voltage drop across R is

$$V_{in} = I_{in} R = 22 \times 10^{-3} \times 200 = 4.4 \text{ V}$$

and the minimum permissible input voltage is

$$V_{in} = I_{in} R + V_{out} = 4.4 + 15 = 19.4 \text{ V}$$

Thus, the supply voltage could undergo an 11% increase or a 3% decrease (from 20 V) and the zener diode would still maintain the output voltage at a nominal 15 V. This variation could be extended by the use of a larger zener diode, as noted earlier.

4.14. EFFECT OF TEMPERATURE ON ZENER DIODES

We have already studied in previous chapters how P-N junctions are temperature-sensitive in either the forward or reverse-biased mode. Now, because zener diodes operate only in the reverse mode, using the leakage current or minority carriers that are due to thermally generated electron hole pairs, we would expect temperature to have some profound effect on the zener voltage V_Z. This has been illustrated in figure 4.12. In this figure, the temperature coefficient α_Z in %/°C is shown as a function of zener voltage for three values of zener current.

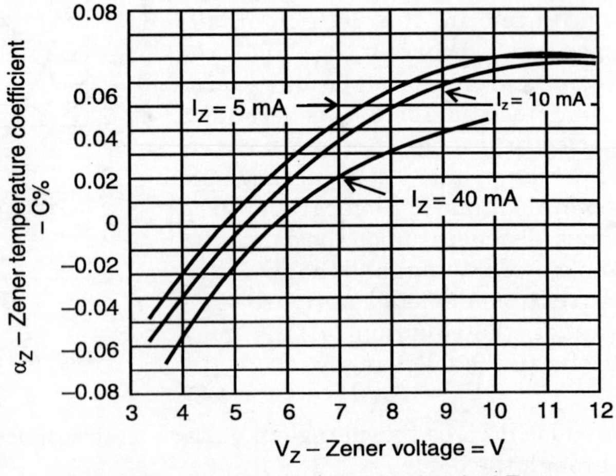

Fig. 4.12. Illustration of Graph of typical Zener temperature coefficients versus Zener voltage for silicon voltage regulator diodes.

This curve, typical for silicon voltage regulator diodes, may be interpretted as follows.

A 4 V Zener diode, for example, regardless of the level of current operation, will always be affected by temperature in a negative manner, decreasing zener voltage with an increase in temperature. On the other hand, zener diodes with V_Z in excess of 6 V will have an increase in breakdown voltage with an increase in temperature. If a diode is selected with a V_Z between 5 and 6 Volts and operated at the correct value of current, it will have very little change with temperature. The difference in sign of the temperature coefficient lies in the fact that reverse current is largely due to the true zener effect below 6 V, but is due to the avalanche effect above 6 V. Each of these mechanism is affected differently by temperature.

4.15. ZENER DIODE APPLICATIONS : DETAILED ASPECT *(Important)*

As a matter of fact, Zener diodes have numerous applications in transistor circuitry. Some of their common applications have been explained below.

1. Voltage Regulation

Voltage regulation is a measure of a circuit's ability to maintain a constant output voltage even when either input voltage or load current varies.

Figure 4.13 shows how a Zener diode can be used as a voltage regulator to provide a constant voltage from a source whose voltage may vary appreciably. A resistor R_S is necessary to limit the reverse current through the diode to safer value. The voltage source V_S and the resistor R_S are so selected that the diode operates in the breakdown region. The diode voltage in this region, which is also the voltage across the load R_L, is called *Zener voltage* V_Z and the diode current is called the Zener current I_Z. The series resistor R_S absorbs the output voltage fluctuations so as to maintain voltage across the load constant. Zener diode is reverse-connected across the input voltage whose variations are to be regulated.

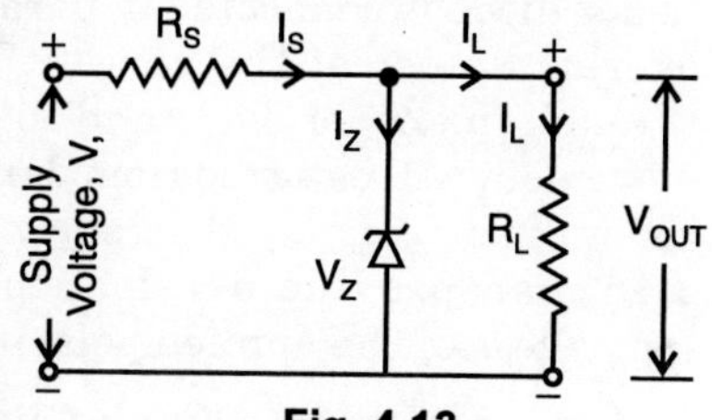

Fig. 4.13.

As long as voltage across the load resistor R_L is less than the breakdown voltage V_Z, the Zener diode does not conduct, the resistors R_S and R_L constitute a potential divider across V_S. At an increased supply voltage V_S, the voltage drop across load resistor (or Zener diode) becomes greater than the Zener breakdown voltage. It then operates in its breakdown region. The series resistor R_S limits the Zener current I_Z from exceeding its rated maximum $I_{Z\,max}$ because Zener current is given as

$$I_Z = \frac{V_S - V_Z}{R_S} \qquad ...(4.3)$$

The current from the power supply splits at the junction of Zener diode and the load resistor R_L.

So $$I_S = I_Z + I_L$$

When the Zener diode operates in its breakdown region, the voltage across it, V_Z remains fairly constant even though the current I_Z flowing through it may vary considerably.

When the supply voltage V_S increases, the current through both the Zener diode and load resistance R_L increases. At the same time, however, the zener diode resistance decreases and the current through the diode increases more than proportionately. As a result, a greater voltage drop will occur across the series resistor R_S and the output voltage V_{OUT} (voltage across the diode or load resistance R_L) will become very close to the original value. The reverse is also true. Thus, a Zener diode can maintain the output voltage V_{OUT} within a fraction of a volt when the supply or input voltage V_S may vary over a range of several volts. The Zener diode will maintain a constant voltage across the load as long as the supply voltage is more than the Zener voltage.

Let us examine the other cause of the output voltage variation. When the load resistance R_L decreases for constant input voltage V_S, load current I_L increases. This additional current is not supplied from the source of supply but the demand of additional load current is fulfilled by decrease in Zener current I_Z. This keeps the voltage drop across series resistance R_S constant and so the output voltage V_{OUT}. It may be noted that the diode current I_Z attains its maximum $I_{Z\,max}$ when load current I_L becomes zero. Thus the value of series resistor R_S can be determined from the following equation:

$$R_S = \frac{V_S - V_{OUT}}{I_{Z\,max}} \qquad ...(4.4)$$

The advantages of Zener diode voltage regulators over other voltage regulators are that they are smaller, lighter, more rugged and have a longer life. They are also simpler and inherently cheaper.

Such regulators have the following drawbacks :

1. Their efficiency is low for heavy load currents because there is a considerable power loss in their resistor R_S and the diode.
2. The output voltage varies slightly due to Zener impedance r_Z.
3. The output voltage cannot be choosen independently but depends upon the breakdown voltage of the Zener diode.

This regulator is employed when there are small variations in load current and supply voltage.

2. Meter Protection

Figure 4.14 shows how does a Zener diode protect a meter (AVO meter or multimeter) movement or circuit against damage from an accidental application of excessive voltage. In this arrangement, Zener diode is connected in parallel with the meter, as shown in figure 4.15. In case of accidental overload most of the current will pass through the Zener diode and so the meter movement will be saved from damage. Two zener diodes connected, as shown in figure 4.15, can provide overload protection regardless of the applied polarity.

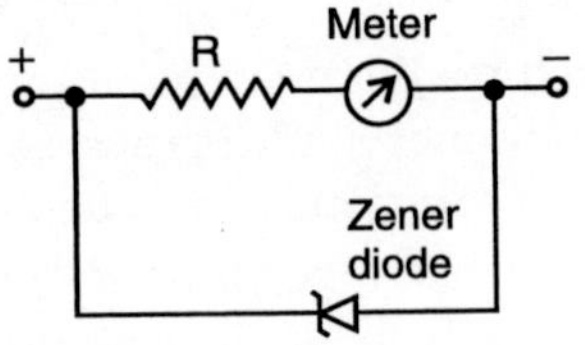

Fig. 4.14.

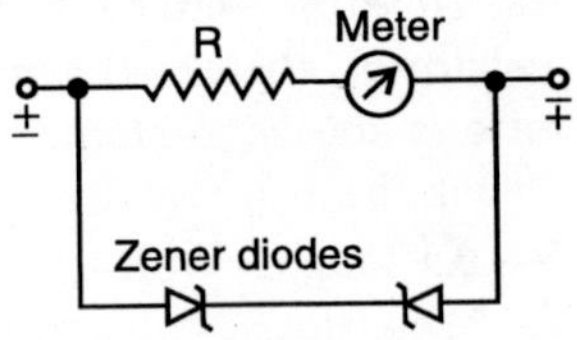

Fig. 4.15.

3. Zener Diode as a Peak Clipper

Zener diodes can be used for clipping off the input waveform in wave shaping circuits, as illustrated in figure 4.16.

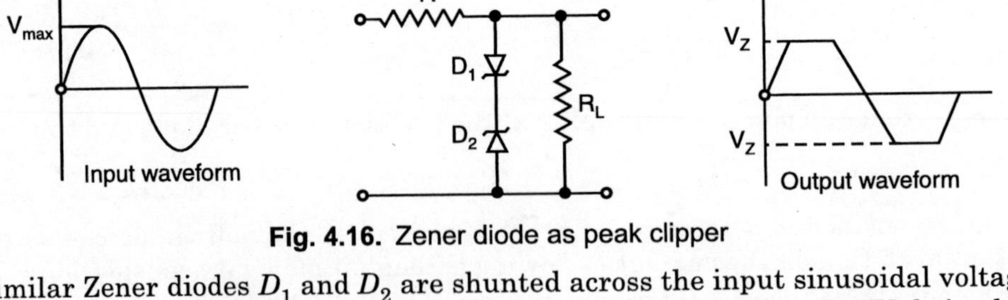

Fig. 4.16. Zener diode as peak clipper

Two similar Zener diodes D_1 and D_2 are shunted across the input sinusoidal voltage of peak value of V_{max} volts. Let the Zener voltage of each of Zener diodes be V_Z volts (V_Z being lesser than V_{max}).

We have seen that a Zener diode acts like a 'short' (or very low resistance) when forward biased and 'open' (or very high resistance) when reverse biased till it goes into breakdown at V_Z. The action of the circuit is explained below.

For positive input half cycle, diode D_1 is forward biased and acts like a 'short' while diode D_2 acts like an open (being reverse biased) upto V_Z. Thereafter diode D_2 goes into breakdown and holds the output voltage constant at V_Z till input voltage falls below V_Z in the later part of the positive half cycle. This is because during this duration of time both of the diodes act as 'short'. When the supply voltage falls below V_Z, diode D_2 comes out of the breakdown and again acts as an open across the load circuit. During the negative input half cycle, the actions of diodes D_1 and D_2 are reversed. The result is that the output voltage waveform is clipped off on both peaks, as illustrated in figure 4.16.

4. Zener Diode as a Reference Element

There are many applications in which it is desirable that a constant voltage is maintained between two points in a circuit, and use this voltage as reference for comparing other voltages against it. The difference between the two voltages (the voltage under comparison and reference voltages) is usually amplified and then used to perform some control function. Such an arrangement is employed in power supply voltage regulator circuits, measurement circuits and servomechanism circuits. The constant voltage characteristics of a zener diode in its breakdown region makes it desirable for such an application.

A circuit employing Zener diode as a reference element is given in figure 4.17. The value of R is so chosen that the diode operates well within its breakdown region. V_{REF} equals the Zener breakdown voltage V_Z. The difference of input voltage V_{IN} and reference voltage V_{REF} (*i.e.*, $V_{IN} - V_{REF}$) gives the control output. Obviously V_{IN} should be more than V_{REF}.

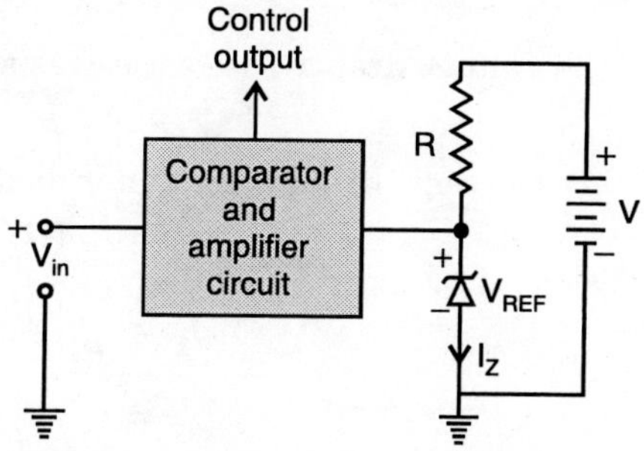

Fig. 4.17. Zener diode as reference element

5. Switching Operation

Zener diode can produce a sudden change from low current to high current, so it is useful for switching operation. The use of Zener diode in switching operation has made possible an extremely fast performance in computer applications.

4.16. COMPARISON OF ZENER DIODE AND P-N JUNCTION DIODE

Zener diode	P-N-junction diode
1. Operates in reverse-bias condition	1. Operates in forward-bias condition
2. The characteristics lies in third quadrant 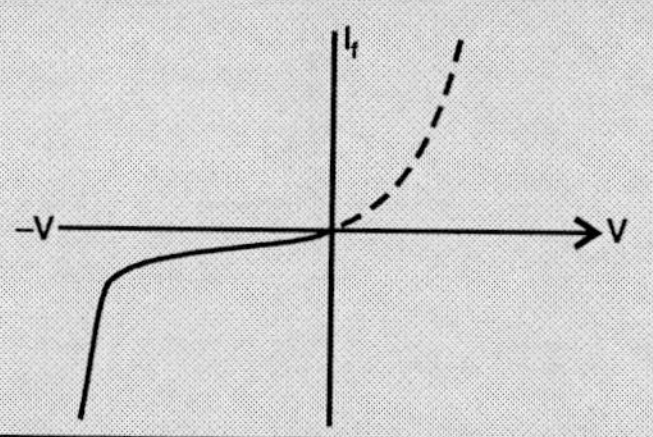	2. The characteristics lies in first quadrant 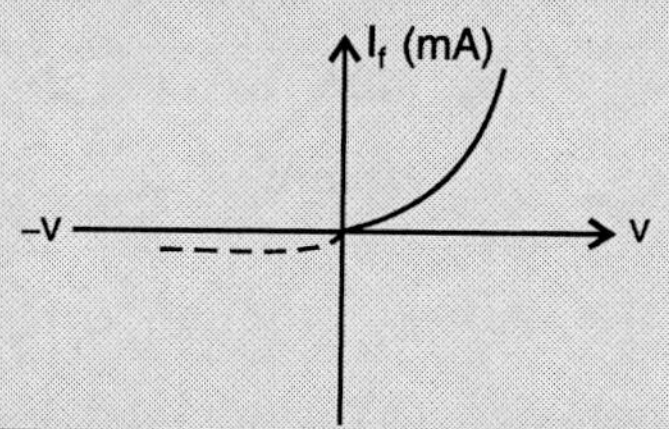
3. Dynamic Zener resistance is very small in reverse breakdown condition.	3. The diode resistance in reverse biased condition is very high
4. Zener diode symbol is 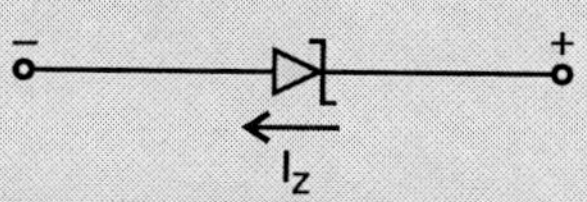	4. The P-N-junction diode symbol is 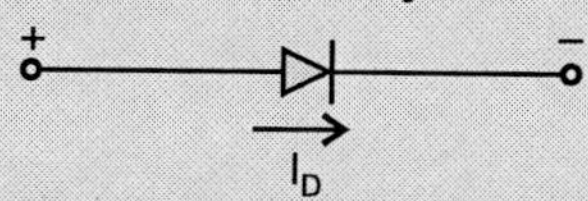
5. The conduction in zener is opposite to that of arrow in the symbol	5. The conduction when forward biased is in same direction as that of arrow in the symbol.
6. Applications of Zener diode are voltage regulator, protection circuits, voltage limiters etc.	6. Applications of P-N-junction diode are rectifiers, voltage multipliers, clippers, clampers and many electronic devices.

EXAMPLE 4.1 **A Zener diode shunt regulator is shown in figure 4.18. Determine: (*i*) The load voltage, (*ii*) Voltage drop across series resistance, (*iii*) Current through the diode.**

Solution: Given that

$V_S = 30$ V, $R_S = 240\ \Omega$, $V_Z = 12$ V, $R_L = 500\ \Omega$

(*i*) Load voltage $V_L = V_Z = 12$ V

(*ii*) $\because\ V_S = I_S R_S + V_Z$

or $V_S - V_Z = I_S R_S$

$$I_S = \frac{V_S - V_Z}{R_S} = \frac{30-12}{240} = \frac{18}{240}$$

$$= \frac{3}{40} = 0.075 \text{ A}$$

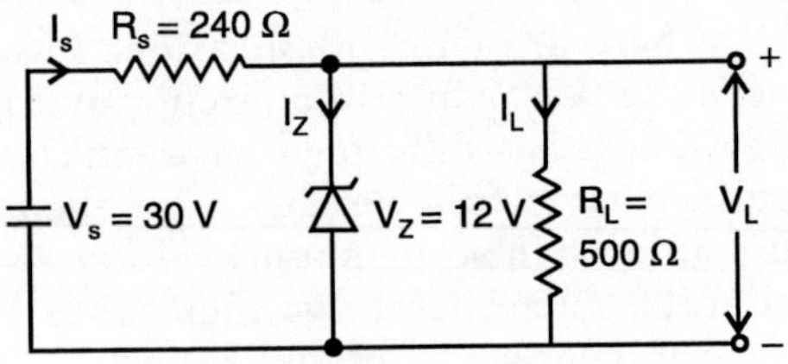

Fig. 4.18. Circuit for example 4.1

Voltage drop across series resistance R_S is

$$V_S = I_S . R_S = 0.075 \times 240 = 18 \text{ V}$$

Current through the Zener diode is given by

$$I_S = I_Z + I_L \quad (i)$$

$$I_L = \frac{V_L}{R_L} = \frac{12}{500} = 0.024 \text{ A}$$

$\therefore\ I_Z = I_S - I_L = 0.075 - 0.024 = 0.051\ A$ **Ans.**

EXAMPLE 4.2 **Figure 4.19 shows the circuit of zener diode shunt regulator. Determine the zener current if the load resistance R_L is**

(*i*) 30 kΩ, **(*ii*) 10 kΩ**

(*iii*) 3 kΩ.

Solution: $R_L = 30$ kΩ

$V_Z = 30$ Volt

From circuit, we have $V_S = I_S R_S + V_Z$

$$I_S = \frac{V_S - V_Z}{R_S} = \frac{50-30}{2\times 10^3} = \frac{20}{2\times 10^3}$$

$I_S = 10$ mA

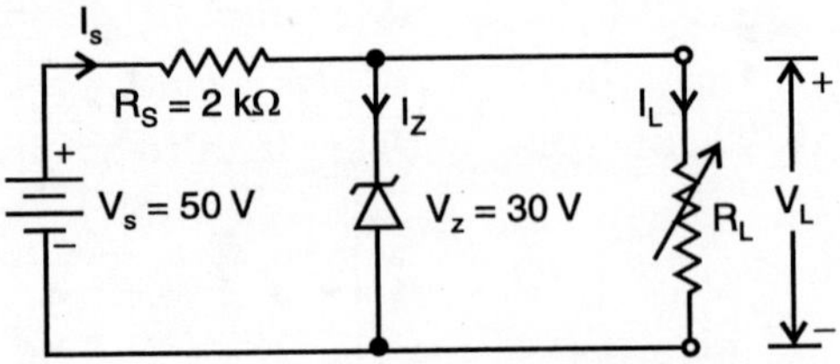

Fig. 4.19.

But we know that

$V_L = V_2 = 30 \text{ V} = V_Z = 30$ Volt

Also, $V_L = I_L . R_L$

Hence, $I_L = \dfrac{30}{30 \text{ k}\Omega} = 1$ mA

Since, $I_Z = I_S - I_L = 10 - 1 = 9$ mA **Ans.**

(*ii*) $R_L = 10$ kΩ ; $I_S = 10$ mA

$$I_L = \frac{V_L}{R_L} = \frac{30}{10} = 3 \text{ mA}$$

$\because\ I_S = I_Z + I_L$

or $$I_Z = I_S - I_L = 10 - 3 = 7 \text{ mA} \quad \textbf{Ans.}$$

(*iii*)
$$R_L = 3 \text{ k}\Omega$$
$$I_S = 10 \text{ mA}$$
$$I_L = \frac{V_L}{R_L} = \frac{30}{3} = 10 \text{ mA}$$

$\because$ $$I_S = I_Z + I_L$$

or $$I_Z = I_S - I_L = 10 - 10 = 0 \text{ mA} \quad \textbf{Ans.}$$

Since, $I_Z = 0$, this states that for this value of load resistance, the zener diode will come out of the break down region and so the Zener diode will no longer work as a voltage regulator.

EXAMPLE 4.3 For the circuit shown in figure 4.20. Find

(*a*) the output voltage

(*b*) the voltage drop across the series resistance

(*c*) the current through zener diode.

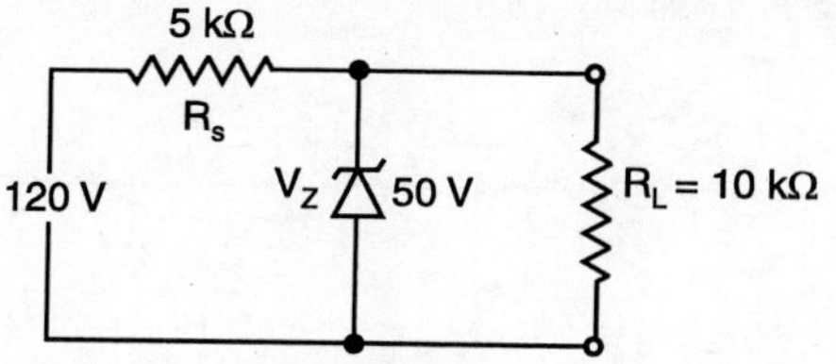

Fig. 4.20.

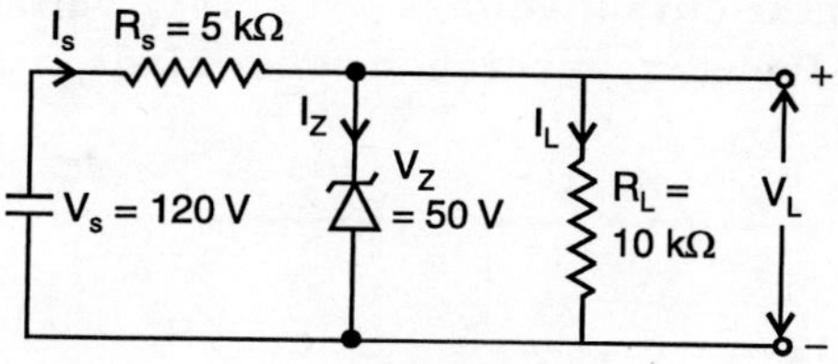

Fig. 4.21.

Solution: Let us redraw the given figure. This has been shown in figure 4.21.

(*a*) The output voltage $V_L = V_Z = 50$ Volt

(*b*) The voltage drop across the series resistance, R_S is

$$= V_S - V_Z = 120 \text{ V} - 50 \text{ V} = 70 \text{ Volts}$$

(c) Load current $$I_L = \frac{V_L}{R_L} = \frac{50\text{V}}{10\text{k}\Omega} = 5 \text{ mA}$$

The current through R_S is $I_S = \dfrac{70\text{V}}{5\text{k}\Omega} = 14 \text{ mA}$

Hence, the current through zener diode is given by

$$I_S = I_Z + I_L$$

$\therefore$ $$I_Z = I_S - I_L = 14 \text{ mA} - 5 \text{ mA} = 9 \text{ mA} \quad \textbf{Ans.}$$

EXAMPLE 4.4 For the circuit shown in figure 4.22, find the maximum and minimum values of zener diode current.

Solution: Let us redraw the given circuit.

From the circuit, we have

Minimum voltage drop across the series resistance R_S is

$$= V_S - V_Z = 80 - 50 = 30 \text{ Volts}$$

Maximum voltage drop across the series resistance R_S is

$$= V_S - V_Z = 120 - 50 = 70 \text{ Volts}$$

$\because$ Load current $I_L = \dfrac{V_L}{R_L} = \dfrac{V_Z}{R_L} = \dfrac{50 \text{ V}}{10 \text{ k}\Omega} = 5 \text{ mA}$

Minimum current through R_S

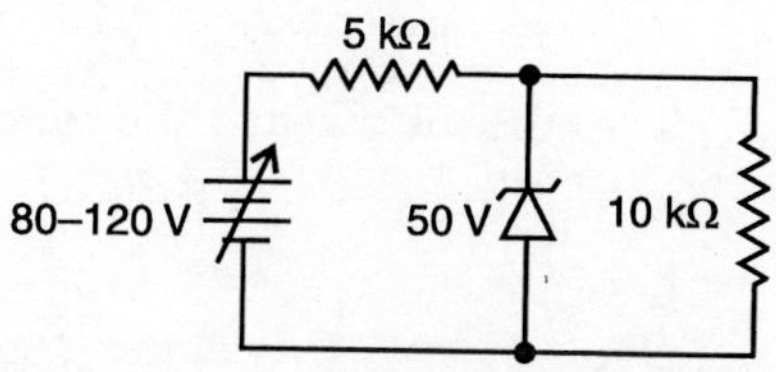

Fig. 4.22.

$$I_{S(min)} = \frac{30\text{ V}}{5\text{ k}\Omega} = 6\text{ mA}$$

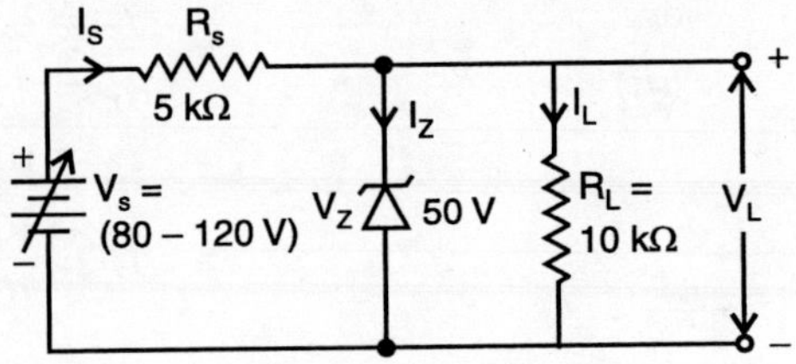

Fig. 4.23.

Therefore, minimum value of current though the zener diode

$$I_{Z(min)} = I_{S(min)} - I_L = 6\text{ mA} - 5\text{ mA} = 1\text{ mA}$$

Again, Maximum current through R_S

$$I_{S(max)} = \frac{70\text{V}}{5\text{ k}\Omega} = 14\text{ mA}$$

Therefore, maximum value of current through the zener diode

$$I_{Z(max)} = I_{S(max)} - I_L = 14\text{ mA} - 5\text{ mA}$$

$$\therefore \quad I_{Z(max)} = 9\text{ mA} \qquad \textbf{Ans.}$$

EXAMPLE 4.5 **For the circuit of the following figure 4.24, find**
(*a*) the output voltage **(*b*) voltage drop across R_S and**
(c) the current through zener diode.

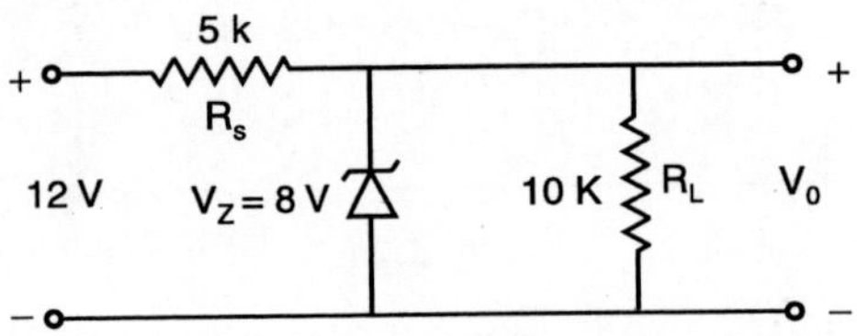

Fig. 4.24.

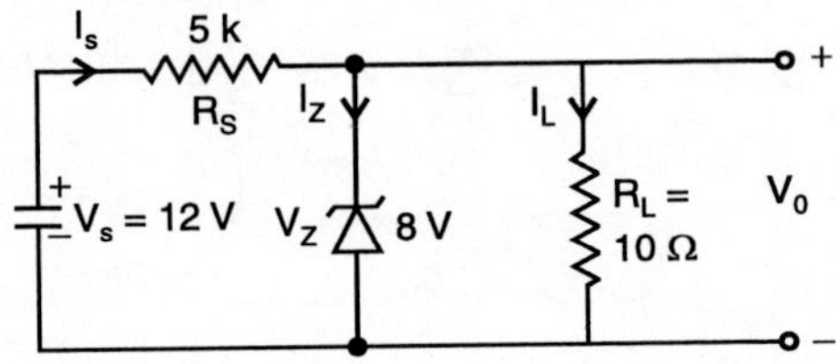

Fig. 4.25.

Solution: Let us redraw the given figure. This has been shown in figure 4.25.

Given that

$V_S = 12$ Volt $\qquad R_S = 5\text{ k}\Omega$,

$V_Z = 8\text{V} \qquad R_L = 10\text{ k}\Omega$

(*i*) The output voltage or load voltage

$$V_O = V_Z = 8\text{ Volt}$$

(*ii*) $\therefore V_S = I_S R_S + V_Z$

or $$I_S = \frac{V_S - V_Z}{R_S} = \frac{12-8}{5\text{ k}\Omega} = \frac{4}{5} \times 10^{-3} = 0.8\text{ mA}$$

Hence, voltage drop across series resistance R_S

$$= I_S . R_S = 8 \times 10^{-3} \times 5 \times 10^3 = 4\text{ Volt}$$

(*iii*) $\because \quad V_O = I_L . R_L$

$$\therefore \text{Load Current, } I_L = \frac{V_O}{R_L} = \frac{8}{10 \times 10^3} = 8 \times 10^{-4} = 0.8\text{ mA}$$

The current through the zener diode is given by

$$I_S = I_Z + I_L$$

$$I_Z = I_S - I_L = 0.8\text{ mA} - 0.8\text{ mA} = 0 \qquad \textbf{Ans.}$$

Since $I_Z = 0$, this states that for this given value of load resistance, the zener diode will come out of the breakdown region and so the zener diode will no longer work as a voltage regulator.

EXAMPLE 4.6 For the circuit shown in figure 4.26, determine the maximum and minimum value of zener diode current.

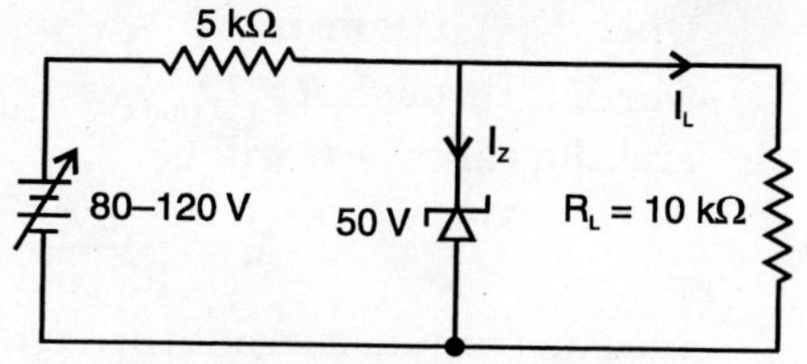

Fig. 4.26. Circuit for example 4.6

Solution: In the given circuit, the current through 5 kΩ resistance

$$I = \frac{120-50}{5\text{ k}\Omega} = 14 \text{ mA}$$

As the zener will conduct maximum current when input voltage is maximum *i.e.*, 120 V.

We have

$$\text{Load current} \quad I_L = \frac{50}{10\text{ k}\Omega} = 5 \text{ mA}$$

Therefore, maximum zener current = 14 – 5 = 9 mA **Ans.**

Further, for minimum input voltage, the zener current would be minimum.

Hence, we have

$$I = \frac{80-50}{5\text{ k}\Omega} = 6 \text{ mA}$$

Therefore, minimum zener current = 6 – 5 = 1 mA **Ans.**

EXAMPLE 4.7 The zener diode shown in figure 4.27 has a fixed voltage drop of 18 V across it so long as the zener current is maintained between 200 mA and 2 amp. Now,

(*i*) Determine the value of resistor *R* so that the load voltage remains 18 V as input voltage is free to vary from 22 Volts to 28 Volts.

(*ii*) Determine the maximum power dissipated by the zener diode.

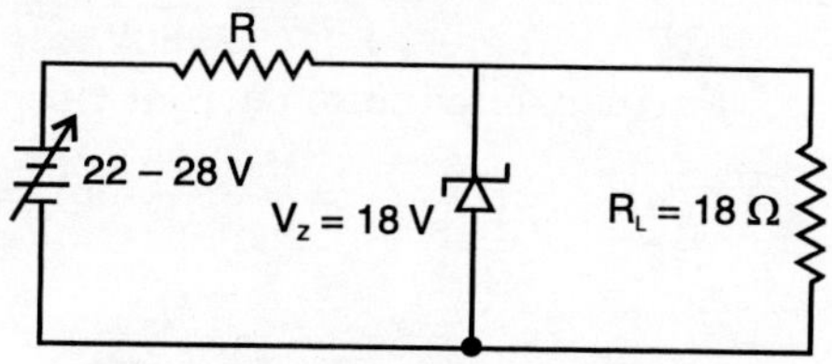

Fig. 4.27. Circuit for example 4.7

Solution: Minimum voltage across R = 22 – 18 = 4 V

Current through R, when $I_{L(\max)}$ flows in R_L, will be

$$I_{L(\max)} + I_{z(\min)} = \frac{18V}{18\Omega} + 200 \text{ mA} = 1200 \text{ mA}$$

Therefore, we have

(*i*) $$R = \frac{4V}{1200 \text{ mA}} = 3.33 \ \Omega$$

Also, maximum current through resistor R

$$= \frac{28-18}{3.33} = 3000 \text{ mA}$$

Also, $$I_{Z(\max)} = 3000 - 1000 = 2000 \text{ mA}$$

which is within the limit of $I_{Z(\max)}$ provided.

(*ii*) Maximum power dissipated = $V_z \ I_{z(\max)}$ = 18 V × 2000 mA = 36 W **Ans.**

EXAMPLE 4.8 A 10 volt regulated d.c. supply of 10 mA is required from a d.c. source of 12 – 15 V by using a pair of zener diodes. Assuming $I_{Z(\min)}$ = 0.2 mA, design the circuit.

Solution: Let us use zener diodes (5 V) in series as shown in figure 4.28.

Minimum voltage across

$$R = 12 - 10 = 2 \text{ V}$$

when $I_{L(max)}$ flows in R_L

Current through $R = I_{L(max)} + I_{z(min)} = 10 + 0.2 = 10.2$ mA

and the value of R will be

$$R = \frac{2V}{10.2 \text{ mA}} = 196\ \Omega$$

Also, maximum current through

$$R = \frac{5 \text{ V}}{196\ \Omega} = 26.5 \text{ mA}$$

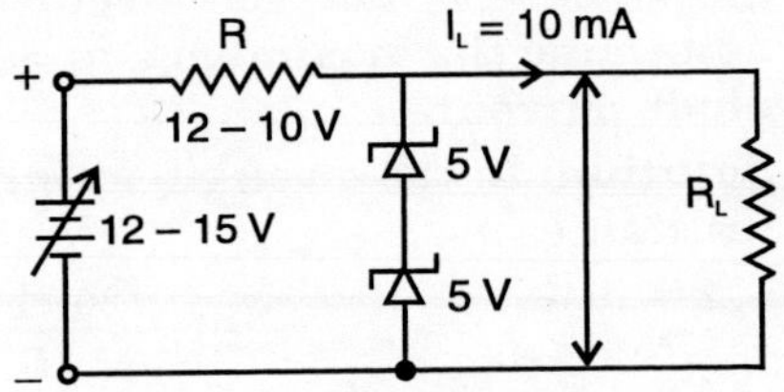

Fig. 4.28. Circuit for example 4.8.

Power rating of zener = 10 V × 26.5 mA = 255 mW **Ans.**

This is the required design of the circuit.

EXAMPLE 4.9 Design a zener voltage regulator circuit that will maintain an output voltage of 20 V across 1 kΩ load resistor when the input voltage is 30 – 50 V. Assume the zener knee current to be negligible compared to the load current. Indicate the maximum power of the diode.

Solution: The voltage regulator circuit has been shown in figure 4.29.

Therefore, we have

$$I_L = \frac{20}{1 \text{ k}\Omega} = 20 \text{ mA}$$

and $$R_S = \frac{(30-20)\text{V}}{20\text{mA}} = 0.5 \text{ k}\Omega = 500\ \Omega$$

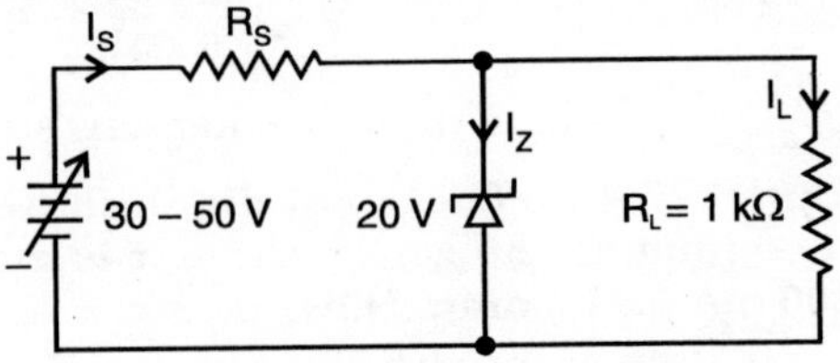

Fig. 4.29. Circuit for example 4.9

Further, maximum current through R_S will be

$$I_{S_{max}} = \frac{50-20}{0.5} \text{ mA} = 60 \text{ mA}$$

Again, $I_S = I_z + I_L$

So, $I_{Z_{max}} = I_{S_{max}} - I_L = (60 - 20) \text{ mA} = 40 \text{ mA}$

Thus, maximum power rating of zener diode will be

40 mA × 20 V = 800 mW = 0.8 watt **Ans.**

EXAMPLE 4.10 Consider the circuit shown in figure 4.30. The operating currents of the two zener diodes, Z_1 and Z_2, are 30 mA (V_Z = 10 V) and 15 mA (V_Z = 5 V), respectively. Compute the values of the limiting resistance R_1 and R_2. If the maximum operating current (or knee current) of the Zener diode Z_1 (V_Z = 10 V) is 5 mA, find the lowest power supply voltage V_B for which the current through the load R_L will remain nearly constant. ***(Important)***

Solution: From the given problem, we have

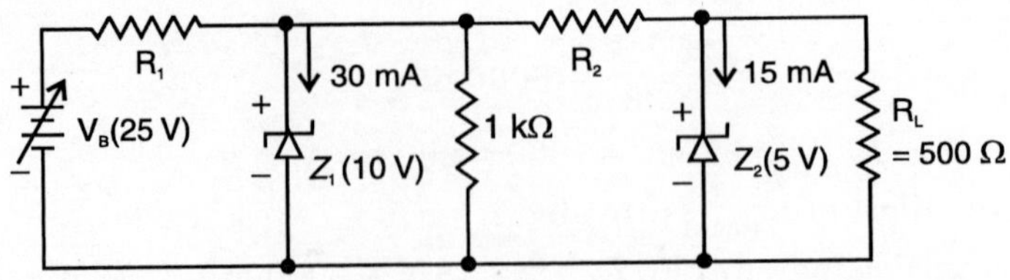

Fig. 4.30. Circuit for example 4.10.

Current through $R_L = \frac{5\text{V}}{500\ \Omega}$

Current through $R_2 = 15 \text{ mA} + \frac{5 \text{ V}}{500\ \Omega} = 25 \text{ mA}$

Voltage across $R_2 = 10 - 5 = 5$ V

Therefore, resistance $R = \dfrac{5 \text{ V}}{25 \text{ mA}} = 200\ \Omega$

Now, current through $R_1 = 30 \text{ mA} + \dfrac{10 \text{ V}}{1 \text{ k}\Omega} + 25 \text{ mA} = 65 \text{ mA}$

Voltage across $R_1 = 25 - 10 = 15$ V

Thus, Resistance $R_1 = \dfrac{15\text{V}}{65 \text{ mA}} = 230.77\ \Omega$

Knee current of zener $Z_1 = 5$ mA

Therefore, we have

Current through $R_1 = 5 \text{ mA} + \dfrac{10 \text{ V}}{1 \text{ k}\Omega} + 25 \text{ mA} = 40 \text{ mA}$

and voltage drop across

$$R_1 = 40 \text{ mA} \times 230.77 = 9.23 \text{ Volt}$$

Also, we have $V_{B(\min)} = V_{Z_1}$ + Voltage drop across $R_1 = 10 + 9.23 = 19.23$ Volt **Ans.**

EXAMPLE 4.11 Determine the temperature coefficient of a 5 V Zener diode at 25°C, if the nominal voltage drops to 4.8 V at a temperature of 100°C.

Solution: Change in Zener voltage, $\Delta V_Z = 4.8 - 5 = -0.2$ V

Change in temperature., $(\Delta T) = 100 - 25 = 75$°C.

Temperature coefficient of Zener diode

$$\alpha_Z = \frac{\Delta V_Z}{V_Z(\Delta T)} \times 100 = \frac{-0.2}{5 \times 75} \times 100 = -0.533\% \textbf{ Ans.}$$

EXAMPLE 4.12 Determine V_L, I_Z and P_Z for the circuit shown in figure 4.31.

Solution: Assuming that the Zener diode is not conducting the given circuit can be redrawn as shown in figure 4.32.

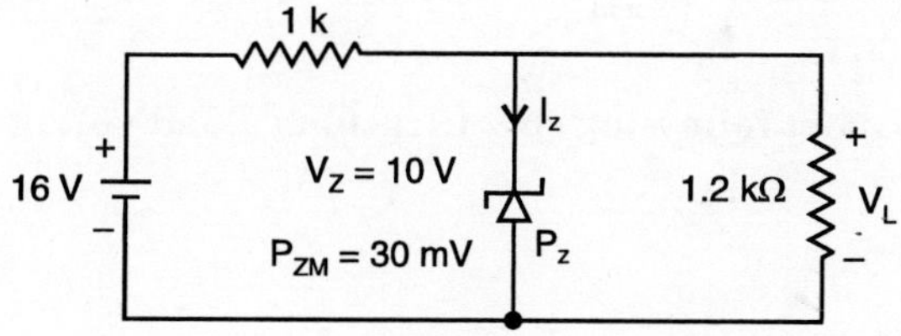

Fig. 4.31

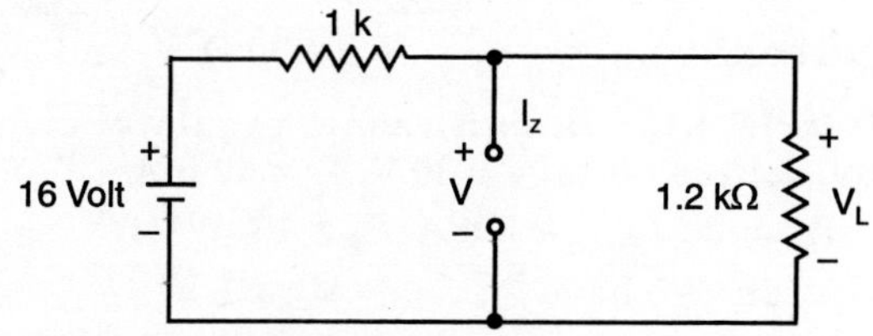

Fig. 4.32.

Now, let us apply voltage divider rule in circuit of figure 4.32.

Thus, we have

$$V_L = \frac{16 \times 1.2}{2.2} = 8.73 \text{ Volt}$$

Since the voltage across load, *i.e.*, V_L is less than the breakdown voltage of zener diode, *i.e.*, $V_L < V_Z$, therefore, the zener diode will not conduct.

Hence, the current I_Z will be given by

$$I_Z = 0$$

Also, $P_Z = I_Z \times V_Z = 0 \times 8.73 = 0$ Watt **Ans.**

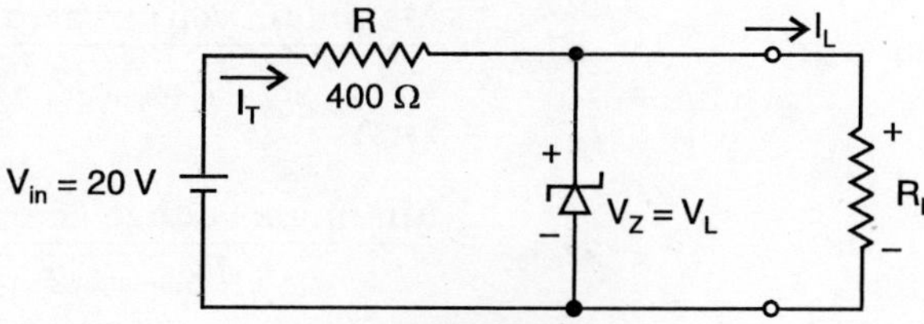

Fig. 4.33.

EXAMPLE 4.13 Determine the minimum and the maximum load currents for which the zener diode in figure 4.33 will maintain regulation. What is the minimum R_L that can be used? Given that $V_Z = 10$ V, $I_{Z\,min} = 5$ mA, and $I_{Z\,max} = 50$ mA. Assume $r_z = 0\ \Omega$.

Solution: When $I_L = 0$, I_Z is maximum and equal to the total circuit current I_T

$$I_Z = \frac{V_{in} - V_Z}{R} = \frac{20-10}{400} = 25 \text{ mA}$$

Since, it is less than $I_{Z\,max}$, assumption $I_L = 0$ is acceptable.

$$\therefore \qquad I_{L\,min} = 0\text{A}$$

The maximum value of I_L occurs when I_Z is minimum.

$$I_{L\,max} = I_T - I_{Z\,min} = 25 \text{ mA} - 5 \text{ mA} = 20 \text{ mA}$$

Therefore, the minimum value of R_L is

$$R_{L\,min} = \frac{V_Z}{I_{L\,max}} = \frac{10}{20 \text{ mA}} = 500\ \Omega \qquad \textbf{Ans.}$$

EXAMPLE 4.14 Design a zener regulator for the following specifications:

Output voltage, $V_0 = 5$ **V**
Input voltage, $V_i = 12 \pm 3$ **V**
Load current, $I_L = 20$ **mA**
Zener wattage, $P_Z = 500$ **mW.**

Solution: We have

$$I_{Z\,max} = \frac{P_Z}{V_0} = \frac{500 \text{ mW}}{5\text{V}} = 100 \text{ mA}$$

$$R_{min} = \frac{\text{Maximum voltage across } R}{I_{Zmax}} = \frac{V_{i\,max} - V_0}{I_{Z\,max}} = \frac{15-5}{100 \text{ mA}} = 100\ \Omega$$

Assuming $I_{Zmin} = 20$ mA, we have

$$R_{max} = \frac{\text{Minimum voltage across } R}{I_{Z\,min}} = \frac{V_{i\,min} - V_0}{I_{Z\,min}} = \frac{9-5}{20 \text{ mA}} = 200 \text{ W}$$

Therefore, $R \leq 200\ \Omega$ and $R \geq 100\ \Omega$.

EXAMPLE 4.15 Design a zener regulator circuit to meet following specifications. Load voltage = 8V, source voltage = 30 V, $I_L = 50$ mA.

Assume $I_{Zmin} = 5$ mA, $P_Z = 1.0$ Watt.

Solution: We have

$$V_0 = 8 \text{ V}$$
$$V_i = 30 \text{ V}$$
$$I_L = 50 \text{ mA},$$
$$P_Z = 1 \text{ W}$$

Now, $$I_{Zmax} = \frac{P_Z}{V_0} = \frac{1}{8} = 125 \text{ mA}$$

Also, $$R_{min} = \frac{\text{Maximum voltage across } R}{I_{Z\,max}} = \frac{V_i - V_0}{I_{Z\,max}} = \frac{30-8}{125 \times 10^{-3}} = 176\ \Omega$$

$$I_{Z\,min} = 5 \text{ mA}$$

$$\therefore \qquad R_{max} = \frac{\text{Minimum voltage across } R}{I_{Zmin}} = \frac{V_{i\,min} - V_0}{I_{Z\,min}} = \frac{30-8}{5 \times 10^{-3}} = 4400\ \Omega$$

Therefore, $R \geq 176\ \Omega$ and $R \leq 4400\ \Omega$

EXAMPLE 4.16 Describe the physical mechanism for Zener breakdown. For the circuit shown in figure 4.34, find the voltage drop across the resistance 5 kΩ.

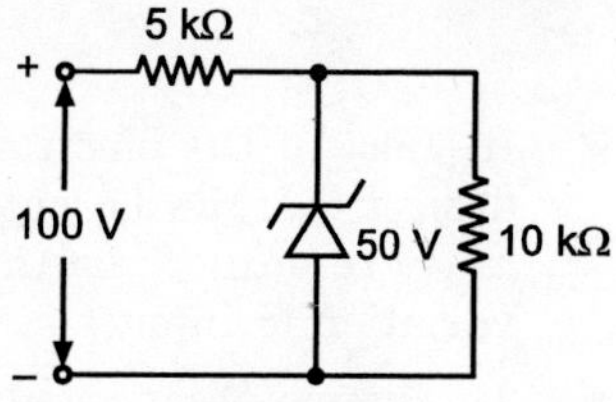

Fig. 4.34.

Solution: (*i*) For zener breakdown, go through article 4.3.

(*ii*) Voltage drop across 5 kΩ resistance = $V_S - V_Z$
$= 100 - 50 = 50$ Volts.

EXAMPLE 4.17 Calculate the following in the circuit given in figure 4.35.

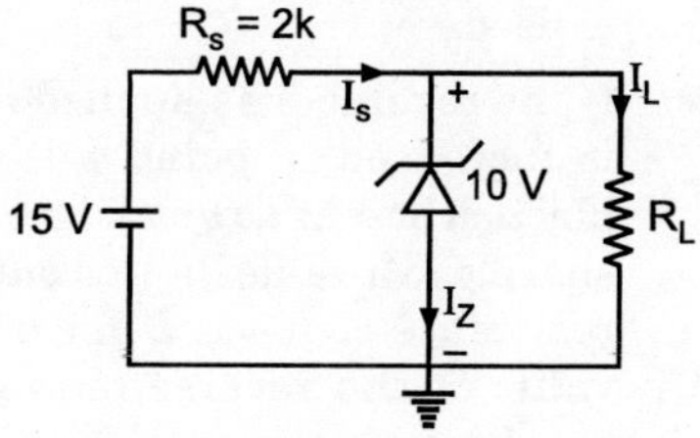

Fig. 4.35.

I_S, I_L, I_Z and P_Z, when
(*i*) $R_L = 2$ kΩ
(*ii*) $R_L = 10$ kΩ.

Solution: We have $I_S = \frac{V_S - V_Z}{R_S} = \frac{15-10}{2\text{ k}\Omega} = 2.5$ mA **Ans.**

Thus, I_S does not depend of the value of R_L.

Now, load current $I_L = \frac{V_Z}{R_L} = \frac{10\text{ V}}{2\text{ k}\Omega} = 5$ mA for $R_L = 2$ kΩ

$= \frac{10\text{ V}}{2\text{ k}\Omega} = 1$ mA for $R_L = 10$ kΩ

Also, we know that $I_Z = I_S - I_L = 2.5\text{ mA} - 1\text{ mA} = 1.5$ mA for $R_L = 10$ kΩ

Again, $P_Z = V_Z \times I_Z = 10 \times 1.5 \times 10^{-3} = 15$ mW for $R_L = 10$ kΩ **Ans.**

EXAMPLE 4.18 Explain 'Avalanche' breakdown. Draw the v-i characteristics of zener diode and explain how does a zener regulate a voltage ? The input voltage for the following figure 4.36 varies from 35 V to 45 V, $V_z = 20$ V, $r_z = 50$ Ω, I_L (min) = 0 mA, I_L(max) = 100 mA, I_z(min) = 10 mA, I_z(max) = 400 mA. Find the values of R and P_z(max).

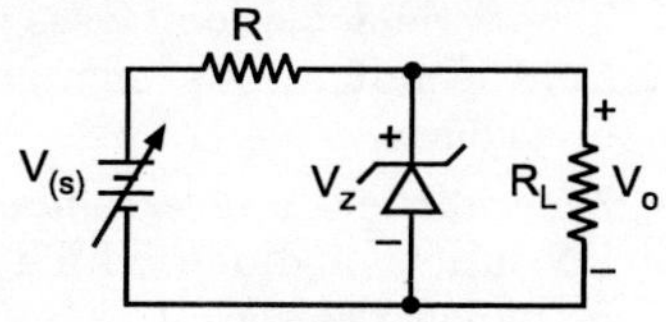

Fig. 4.36.

Solution : For theory part, please go through the text.

Given that $V_{in(min)} = 35$ V, $V_{in(max)} = 45$ V,
$V_z = 20$ V
$r_z = 50\ \Omega$, $I_{L(min)} = 0$ mA,
$I_{L(max)} = 100$ mA
$I_{z(min)} = 10$ mA, $I_{z(max)} = 400$ mA

We know that $R_{min} = \frac{V_{in(max)} - V_z}{I_{L(min)} + I_{z(max)}} = \frac{45-20}{0+(400\times10^{-3})} = 62.5\ \Omega$

We know that $R_{max} = \frac{V_{in(min)} - V_z}{I_{L(max)} + I_{z(min)}} = \frac{35-20}{100\times10^{-3}} = 150\ \Omega$

or $62.5 \le R \le 150$

Select $R = 100\ \Omega$ **Ans.**

Further, we have $P_{z(max)} = V_z \times I_{z(max)}$
$= 20 \times 400 \times 10^{-3} = 8$ Watts. **Ans.**

An intellectual is a man who takes more words than necessary to tell more than be knows.

— **Dwight D. Eisenhower**

SUMMARY

- In most of the electronic applications, it is desired that the d.c. output voltage of rectifier must remain constant irrespective of the variation in input a.c. voltage or load.
- To ensure this, a voltage stabilisation or regulation device known as voltage stabiliser is used.
- If the reverse-bias applied to a P-N junction is increased, a point will reach when the junction breaks down and reverse current rises sharply to a value limited only by the external resistance connected in series. This specific value of the reverse bias voltage is called breakdown voltage (V_B).
- The breakdown voltage depends upon the width of depletion layer.
- This width of depletion layer depends upon the doping level.
- The following two processes cause junction breakdown due to the increase in reverse bias voltage :

  (*i*) Zener Breakdown, (*ii*) Avalanche Breakdown.
- The depletion layer is narrow for heavily doped junctions.
- This form of breakdown occurs in junctions which are lightly doped. Lighly-doped junctions have wide depletion layers where the electric field is not strong enough to cause Zener breakdown.
- Zener diode is a reverse-biased heavily-doped PN-junction diode which operates in the breakdown region.
- The reverse breakdown of a PN-junction may occur either due to zener effect or avalanche effect.
- Zener effect dominates at reverse voltages less than 6 V whereas avalanche effect dominates above 6 V.
- The breakdown voltage of a Zener diode can be set by controlling the doping level.
- For zener diodes, silicon is preferred to G_e because of its higher temperature and current capability.
- Zener diodes have a large number of applications. Few of them are:

  (*i*) Zener diode is used as a voltage regulator.

  (*ii*) Zener diode is used as a peak clipper in wave shaping circuits.

  (*iii*) Zener diode is used as a fixed reference voltage in transistor biasing circuits.

  (*iv*) Zener diode is used for meter protection against damage from accidental application of excessive voltage.
- Generally, the Zener resistance value is quite small than external resistance of the circuit in which zener diode is connected.
- We can analyze the circuits using zener diodes easily by replacing the zener diode with its equivalent circuit.
- Voltage regulator is a circuit whose function is to maintain a constant output d.c. voltage inspite of the a.c. input voltage fluctuations or changes in load resistance values.

SHORT QUESTIONS WITH ANSWERS

Q. 1. What do you mean by breakdown mechanism? Explain.

Ans. If the reverse-bias applied to a P-N-junction is increased, a point will reach when the junction breaks down and reverse current rises sharply to a value limited only by the external resistance connected in series. This specific value of the reverse bias voltage is called breakdown voltage (V_B). The breakdown voltage depends upon the width of depletion layer. This width of depletion layer depends upon the doping level.

The following two processes cause junction breakdown due to the increase in reverse bias voltage:

(*i*) Zener Breakdown (*ii*) Avalanche Breakdown.

Q. 2. What is Zener breakdown? Explain.

Ans. This form of breakdown occurs in junctions that are heavily doped. The depletion layer is narrow for heavily doped junctions. When the reverse bias voltage is increased, electric field also increases across the narrow depletion layer. This strong electric field (about 10^8 V/m) causes covalent bonds to break in the crystal structure. Due to this reason a large number of carriers are generated. This causes a large current to flow.

Q. 3. What are the applications of zener diodes? Explain.

Ans. Zener diodes have a large number of applications. Few of them are:

(*i*) Zener diode is used as a voltage regulator.

(*ii*) Zener diode is used as a peak clipper in waveshaping circuits.

(*iii*) Zener diode is used as a fixed reference voltage in transistor biasing circuits.

(*iv*) Zener diode is used for meter protection against damage from accidental application of excessive voltage.

Q. 4. What is zener voltage ?

Ans. The voltage at which the zener diode breaks down is called the zener voltage.

Q. 5. What happens to the series current, load current and zener current when the dc input voltage of a zener regulator increases ?

Ans. Zener current and series current increase while the load current remains unchanged.

Q. 6. What is a zener diode ?

Ans. Zener diode is a P-N junction diode specially designed for operation in its breakdown region.

REVIEW QUESTIONS

1. Distinguish between Avalanche and Zener mechanisms.
2. Explain how zener diode maintains constant voltage across the load.
3. Draw the equivalent circuit of an ideal Zener diode in the breakdown region.
4. What is Zener effect? Explain the function of a zener diode and draw its characteristics.
5. Why is Zener diode used as a voltage regulator?
6. State and explain the characteristics of a zener diode. How it can be used as a voltage regulator?
7. Describe the physical mechanism for Zener breakdown.

NUMERICAL PROBLEMS

1. A Zener diode shunt regulator is shown in figure 4.37. Determine
(*i*) load voltage; (*ii*) voltage drop across series resistance; (*iii*) current through the diode.
[Ans. (*i*) 12 V (*ii*) 18 V (*iii*) 51 mA]

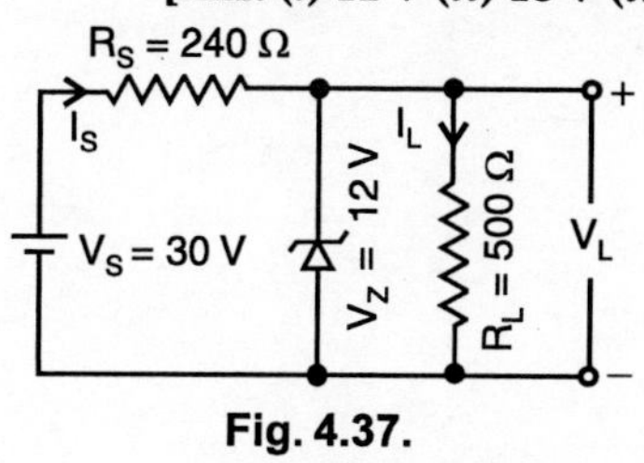

Fig. 4.37.

2. For the circuit shown in figure 4.38. Find
(*a*) the output voltage,

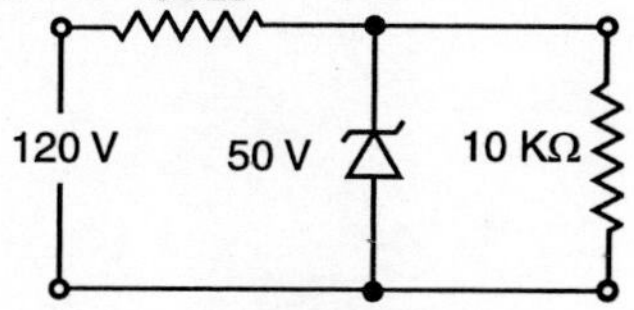

Fig. 4.38.

(*b*) the voltage drop across series resistance,
(*c*) the current through Zener diode.
[Ans. (*a*) 50 V (*b*) 70 V (*c*) 9 mA]

3. Determine the maximum and minimum values of Zener current if in the circuit shown in fig. 6.25 the load resistance R_L = 5 kΩ, series resistance is 2 kΩ, input voltage varies between 80 V and 100 V and the output voltage is 50 V.
[Ans. 15 mA and 5 mA]

4. In a zener diode shunt regulator the series resistance is 2 kΩ, load resistance is 5 kΩ, input voltage is 60 V and output voltage is 35 V. Determine (*i*) voltage drop across series resistance (*ii*) load current (*iii*) current through series resistance and (*iv*) Zener diode current.
[Ans. (*i*) 25 V (*ii*) 7.0 mA (*iii*) 12.5 mA (*iv*) 5.5 mA]

5. Calculate the following in the circuit given below:
I_S, I_L, I_Z and P_Z, when; (*i*) R_L = 2 kΩ ; (*ii*) R_L = 10 kΩ

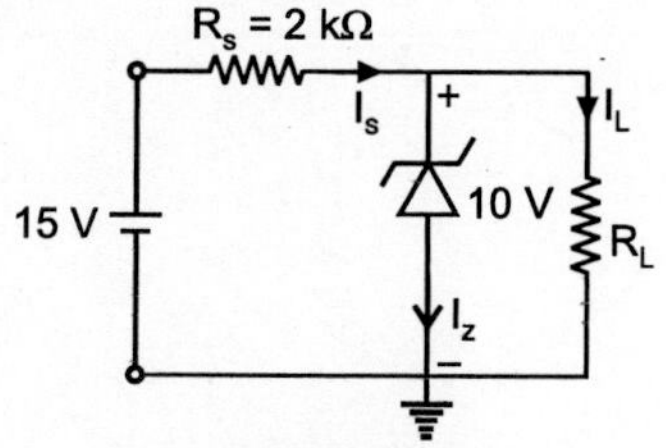

Fig. 4.39.

6. Explain Zener breakdown and Avalanche multiplication phenomena in PN-junction diode. The breakdown phenomena for voltages

less than 6 V is Zener breakdown, explain your opinion on this statement.

7. Explain 'Avalanche' breakdown. Draw the V-I characteristics of zener diode and explain how does a zener regulate a voltage? The input voltage for the following figure 4.40 varies from 35 V to 45 V, $V_Z = 20$ V, $r_z = 5\ \Omega$, $I_{L\,(min)} = 0$ mA, $I_{L\,(max)} = 100$ mA, $I_{z\,(min)} = 10$ mA, $I_{z\,(max)} = 400$ mA, Find the values of R and $P_{z\,(max)}$.

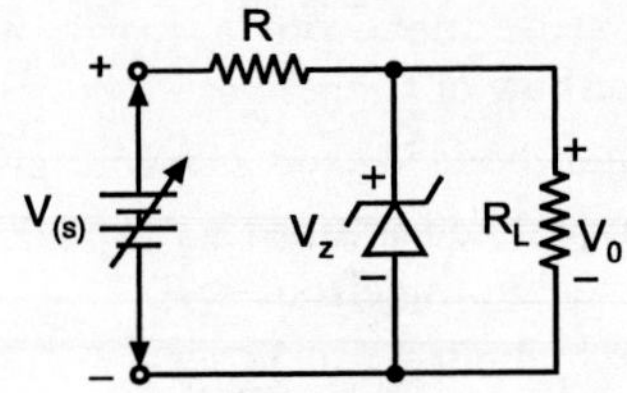

Fig. 4.40.

[Ans. $R = 100^{-2}$ $P_{z(max)}$ **= 8 watts**

■ ■ ■

UNIT—III

Bipolar Junction Transistor (BJT)

The Junction Transistor, Transistor Current Components, Transistors as an Amplifier, Transistor Construction,BJT Operation, BJT Symbol, Common Base,Common Emitter and Common Collector Configurations, Limits of Operation, BJT Specifications.

Above Syllabus has been covered in following Chapters:

CHAPTER 5 : Bipolar Junction Transistor (BJT)

Chapter 5

Bipolar Junction Transistor (BJT)

SYLLABUS

The Junction Transistor, Transistor Current Components, Transistors as an Amplifier, Transistor Construction,BJT Operation, BJT Symbol, Common Base,Common Emitter and Common Collector Configurations, Limits of Operation, BJT Specifications.

Chapter Outline

5.1 INTRODUCTION

The transistor was invented By Dr. William Shockley and Dr. John Bardeen at Bell Laboratory in America in 1951. First time, in 1952 transistor was used in telephone switching circuits. Since then, it has revolutionized the field of Electronics.

The transistor has replaced the bulky vaccum tubes in most of the electronic circuits. The transistor is a basic building block of all modern electronic systems. It is a three terminal device. The output voltage, current or power are controlled by the input current in a transistor. Therefore, it is also called a current-controlled device.

In short, transistor is also called as BJT. BJT stands for Bipolar junction transistor because the transistor operation is carried out by two types of charge carriers : majority carriers and minority carriers.

A transistor has a very important property that it can raise the strength of an input weak signal. This property is called amplification. Due to this quality, the transistor is one of the most widely used semiconductor devices. At present, transistor is used in almost every field of electronics.

Few applications may be listed as under :

(*i*) Transistor is used in control systems.

(*ii*) Transistor is used in satellites and mobile phones.

(*iii*) In digital computer electronics, the transistor is used as a high-speed electronic switch.

(*iv*) In communication systems, it is widely used as the primary component in the amplifier.

- Complete Static Transistor Characteristics
- Transistor Characteristics in Common-Base Configuration
- Transistor Characteristics in Common-Emitter Configuration
- Basic Common Emitter Amplifier and D.C. Load Line
- Numbering System used for Naming the Semiconductor Devices
- Transistor as Swtich
- Transistor Biasing (*i.e.*, Biasing the BJT For Discrete-Circuit Design)
- Selection of Operating Point
- Bias Stabilization
- Need of Stabilisation
- Requirement of Biasing Circuits
- The Stability Factor
- Methods of Transistor Biasing
- Fixed Bias Circuit
- Emitter Resistor Biasing
- Stability Factor for Emitter Resistor Biasing
- Collector to Base Bias Circuit
- Self Bias or Voltage Divider Bias
- Emitter Bias Circuit
- Stability of Emitter Bias Circuit
- Stability Factor of Emitter Bias Circuit

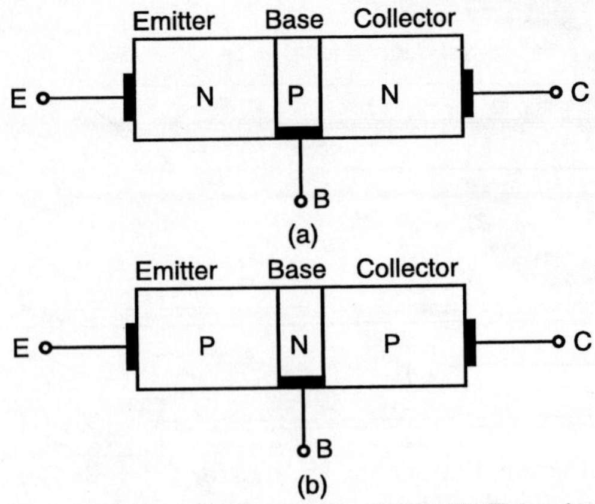

FIG. 5.1. *(a) NPN Transistor (b) PNP Transistor.*

5.1.1. Salient Features of a Transistor

The transistors are more desirable than the vacuum tubes because of their following salient features:

(*i*) Small size and ruggedness.

(*ii*) Do not require any filament power.

(*iii*) Operate at a low voltage.

(*iv*) Higher efficiency.

5.1.2. Why is it called as a Transistor ?

The term *transistor* was derived from the words **TRANSFER** and **RESISTOR**. This term was adopted because it best describes the operation of a transistor, which is the transfer of an input signal current from a low reistance circuit to a high resistance circuit.

5.2 JUNCTION TRANSISTOR CONSTRUCTION

A transistor is a single crystal of silicon (S_i) or Germanium (G_e). A transistor can be of two types : NPN or PNP. A transistor consists of two PN junctions. The junctions are formed by sandwitching either P-type or N-type semiconductor layers between a pair of opposite types as shown in figure 5.1(*a*) and (*b*).

A transistor has three regions. They are emitter, base and collector. A brief introduction and functioning of these three junctions may be discussed as ahead:

DO YOU KNOW?

During the afternoon of December 23, 1947, Walter H. Brattain and John Bardeen demonstrated the amplifying action of the first transistor at the Bell Telephone Laboratories. The first transistor was called a point-contact transistor, which was the predecessor to the junction transistor invented by Schockley

PIONEERS IN ELECTRONICS :
GEORGE SIMON OHM (1787-1854)

The unit of measure for resistance, the ohm, was named for German physicist *George Simon Ohm (1787-1854)*. Ohm is also known for his development of Ohm's law: $I = \frac{V}{R}$.

1. Emitter : It is an outer region situated in one side of transistor. The function of emitter is to inject charge carriers (electrons in case of NPN transistor and holes in case of PNP transistors) into the base. Since emitter has to supply a large number of charge carriers, so it is heavily doped.

"More is the doping, more will be the available charge carriers."

2. Base : It is the middle region of the transistor. The base is a very thin and lightly doped region. The function of this region is to pass all the charge carriers (electrons or holes) onto the collector.

3. Collector : It is the other outer region situated in the other side of transistor. The doping of collector is between the heavy doping of emitter and light doping of Base. The function of collector is to collect charge carriers (electrons or holes). The collector region is physically larger than the emitter region. The reason for this is that the collector has to dissipate more heat. Hence, it is clear that although a BJT has two same type of outer regions, their function cannot be interchanged.

DO YOU KNOW?

In a transistor, the emitter base depletion layer is narrower than the collector-base depletion layer. The reason can be attributed to the different doping levels of the emitter and collector regions. With much heavier doping in the emitter region, the penetration into the *n* material is minimal because of the availability of many more free electrons. However, on the collector side, fewer free electrons are available and the depletion layer must penetrate more deeply in order to set up the barrier potential.

A transistor has two PN-junctions. One junction is formed between the emitter and the base. This is known as emitter-base junction. Another junction is formed between the base and the collector. This is known as collector-base junction.

5.3 TRANSISTOR SYMBOLS

There are two types of transistors known as NPN or PNP. When the transistor is used as a circuit element in any electronic circuit, it is always represented by its circuit symbol as shown in figure 5.2 (*a*) and (*b*).

Figure 5.2(*a*) represents the symbol for NPN transistor. The emitter has an arrowhead. The arrowhead indicates the direction of a conventional current flow in a transistor. The arrowhead points from the P-region towards the N-region. Therefore, in a NPN transistor the conventional current will flow from the base to the emitter.

Figure 5.2(*b*) shows the symbol for PNP transistor. The arrowhead in the emitter points from the P-region towards the N-region. Therefore in a PNP transistor the conventional current will flow from the emitter to the base.

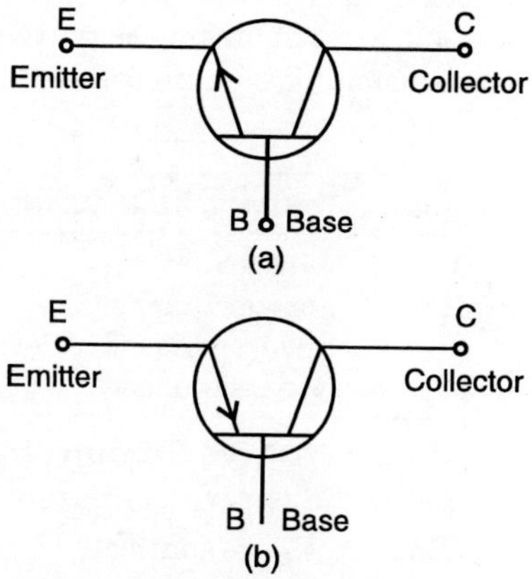

Fig. 3.2. (a) Symbol for NPN Transistor (b) Symbol for PNP Transistor

The PNP transistor is a complement of the NPN transistor. Hence, in NPN transistor, the majority carriers are free electrons but in PNP transistor majority carriers are the holes. The currents and voltages involved in the action of PNP transistor are opposite to those of NPN transistor.

Out of these two types of transistor, the NPN transistor is mostly used. The reason for this is that in NPN transistors, the current conduction is mainly by electrons whereas in PNP transistors the current conduction is mainly by holes. Since the electrons are more mobile than holes, the conduction is higher in NPN transistors than PNP transistors.

DO YOU KNOW?

A transistor has three doped regions: an emitter, a base, and a collector. A *pn*-junction exists between the base and the emitter; this part of the transistor is called the emitter diode. Another *pn* junction exists between the base and the collector; this part of the transistor is called the collector diode.

5.4 UNBIASED TRANSISTOR

When no battery is connected between the different terminals of a transistor then the transistor is said to be in an unbiased state or in an open-circuit state. In this condition, the diffusion of charge carriers across the junction produces two depletion regions. The barrier potential for each of these two depletion layers is **0.7 V for S_i** and **0.3 V for G_e**. But the three regions have different doping levels, so the depletion widths are not the same for two junctions. Since emitter is heavily doped than base and collector, the emitter-base depletion region width is smaller than that of collector-base depletion region width.

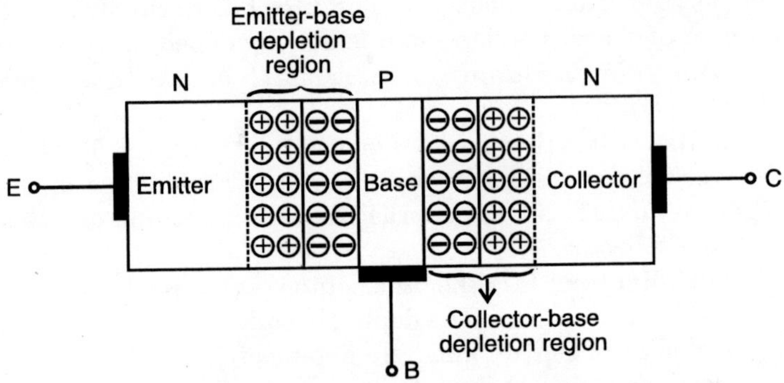

FIG. 5.3. *Unbiased Transistor*

5.5 BIASING OF A TRANSISTOR

When dc voltages are applied across the different terminals of a transistor, then this process is called biasing. Since, there are two junctions in a transistor namely emitter-base junction and collector-base junction, each of these two junctions may be forward-biased or reverse-biased. Therefore, there are four possible ways of biasing these two junctions. These possible ways are also called **modes** of operation of a transistor. These different modes are listed in Table 5.1 given below:

Table 5.1.

Cases	Emitter-base junction	Collector-base junction	Region of operation
I	Forward-biased	Reverse-biased	Active
II	Forward-biased	Forward-biased	Saturation
III	Reverse-biased	Reverse-biased	Cut-off
IV	Reverse-biased	Forward-biased	Inverted

Case I - Active Region : In this mode the emitter-base junction is forward-biased and collector-base junction is reverse-biased.

Battery V_{EB} is connected between emitter and base so as to make emitter-base junction forward-biased. Similarly battery V_{CB} makes collector-base junction reverse-biased. In this region, transistor

The very first step towards success in any occupation is to become interested in it.

—**Sir William Osler**

is used for amplification. In this region, the collector current depends upon the base current (input current).

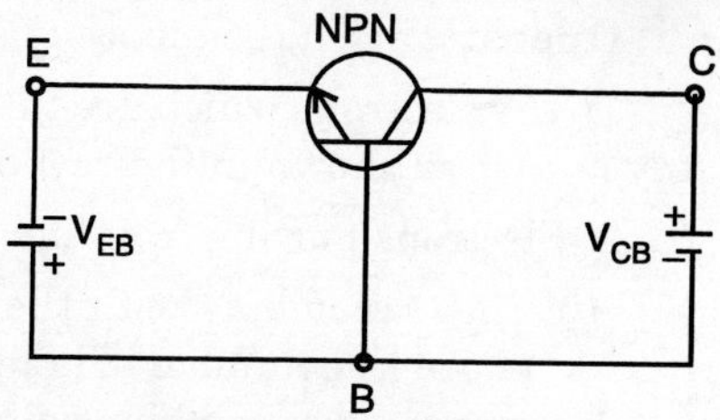

Fig. 5.4 (a). Active region

Case II - Saturation Region : In this region, both the junctions are forward-biased. The transistor works in saturation. In this region, the collector current becomes independent of the base current. Therefore, the transistor acts like a closed switch.

Case III - Cut-off Region : In this mode, both the junctions are reverse-biased. In this region, the transistor has practically zero current because the emitter does not emit charge carriers into the base and no charge carriers are collected by the collector except a few thermally generated minority carriers. The transistor is open in this mode when it is used as an open switch.

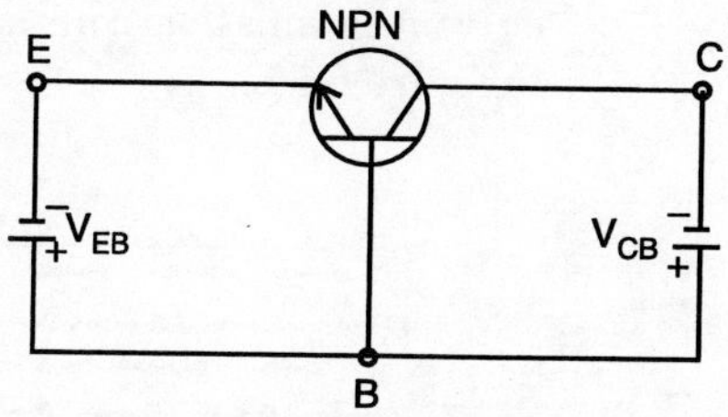

FIG. 5.4 *(b). Saturation region*

Case IV - Inverted Region : In this condition, the emitter-base junction is reverse-biased and the collector-base junction is forward-biased. Since, the doping level of emitter and collector is not the same, therefore, the collector cannot inject the majority charge carriers into the base. In this region, the action of transistor is poor. This region is of little importance.

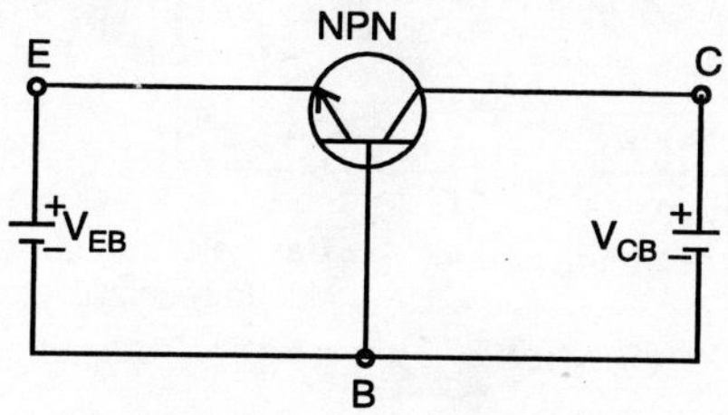

Fig. 5.4 (c). Cut-off region

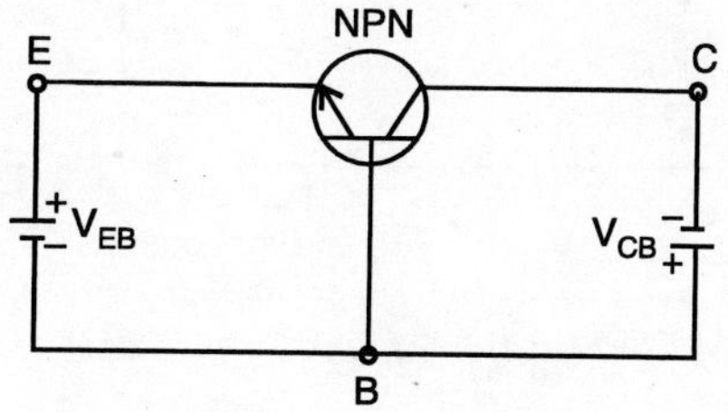

Fig. 5.4. (d). Inverted region

5.6 TRANSISTOR OPERATION IN ACTIVE REGION

5.6.1 Operation of n-p-n-Transistor

The positive supply V_{EE} forward biases the base-emitter junction and the voltage V_{CC} reverse biases the collector to base junction as shown in figure 5.5 (*a*). Hence, the width of depletion region for base-emitter junction is very small, but that at the collector-base junction is large.

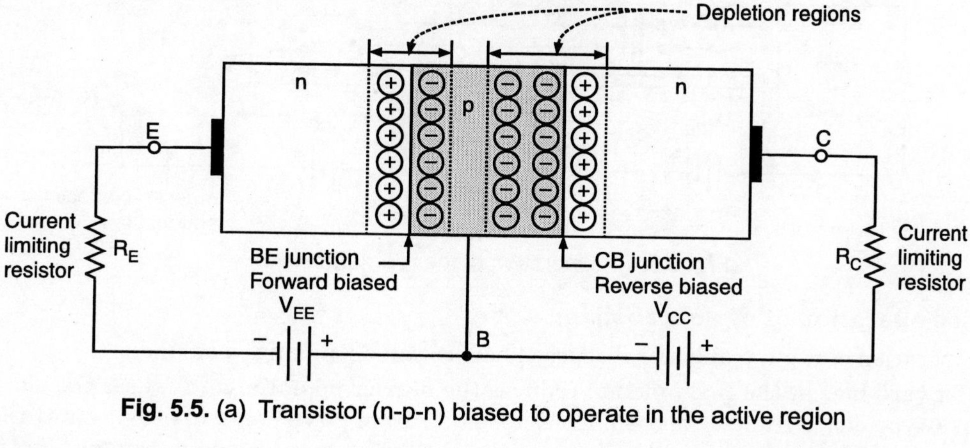

Fig. 5.5. (a) Transistor (n-p-n) biased to operate in the active region

Operation: The sequence of operation for an n-p-n transistor is as follows :

1. The electrons which are the majority carriers in the n-type emitter start flowing towards the p-type base as shown in figure 5.5 (*b*). This constitutes the emitter current I_E.

2. Electrons moving from the emitter to base have three options as under :

(*i*) They recombine with the holes present in the base. As the base region is thin and lightly doped, the number of holes is few. Hence, out of the total injected electrons from the emitter a very few recombine with the holes in the base region. This constitutes the base current I_B. Thus, base current flows due to recombination of electrons and holes. Therefore, the base current is, small as compared to the emitter current (typically 2% of total I_E). This is shown in figure 5.5(c).

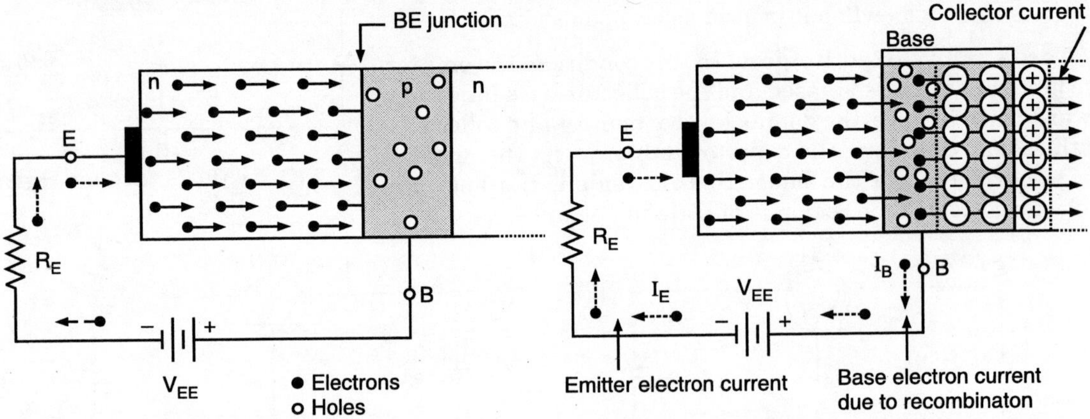

(*b*) *Constitution of emitter current* (*c*) *Constitution of base and collector currents*

Fig. 5.5. (b) and (c).

(*ii*) Some of the electrons diffuse through the base and out of the base connection.

(*iii*) The remaining large number of electrons pass through the depletion region of CB junction and pass through the collector region to the positive end of the external power supply V_{CC} as shown in figure 5.5 (*d*). The collector current I_C is much larger than the base current I_B.

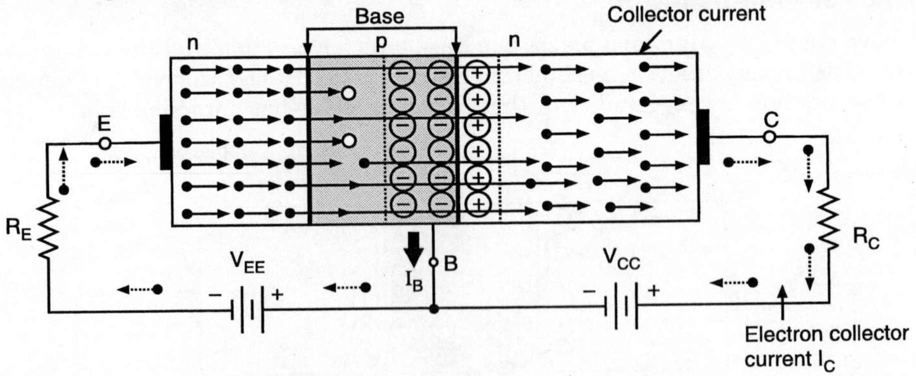

Fig. 5.5. (d) Operation of the n-p-n transistor

Simplified operation of n-p-n transistor

The operation of n-p-n transistor discussed earlier can be simplified as under :

The forward bias at the B-E junction reduces the barrier potential and causes the electrons to flow from n-type emitter to p-type base. Holes also flow from p-type base to n-type emitter. But, as the base is more lightly doped than the emitter, almost all the current flowing across the B-E junction consists of electrons entering the base from the emitter. Hence, electrons are the majority carriers in an n-p-n transistor. Some of the electrons entering into the base region do not reach the collector region. Instead they flow out of the base terminal via the base connection as shown in figure 5.6, due to recombination. As the base region is very thin and lightly doped, there are very few holes available in the base region for recombination. Hence, about 2% electrons flow out of base due to recombination. The remaining 98% electrons cross the reverse biased collector junction to constitute the collector current. They cross the collector region and are collected by the supply V_{CC}.

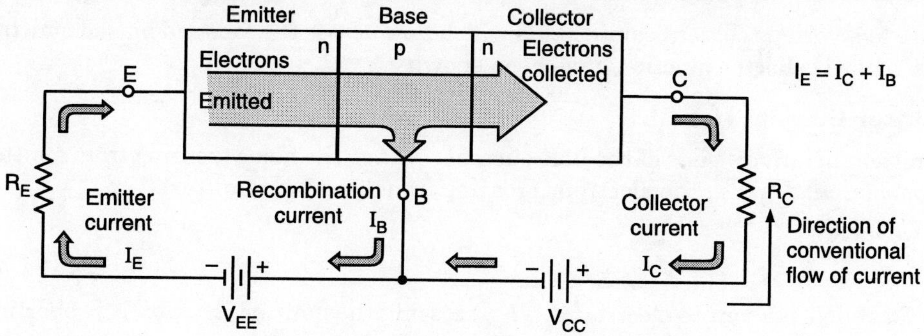

Fig. 5.6. Operation of an n-p-n transistor

The emitter current is equal to the sum of collector and base currents.

Therefore, $I_E = I_C + I_B$

5.6.2. Operation of p-n-p Transistor

The p-n-p transistor behaves exactly in the same way as the n-p-n device. The only difference is the majority charge carriers are holes instead of electrons. As shown in figure 5.7, holes are emitted from the p-type emitter across the forward biased EB junction, into the base.

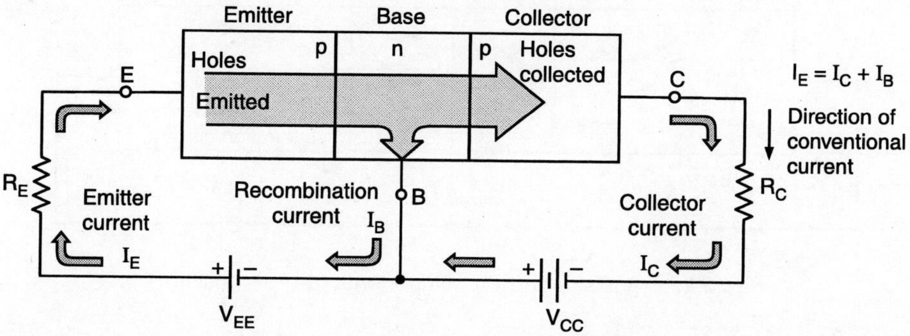

Fig. 5.7. Operation of a p-n-p transistor

In the lightly doped base there are very few number of electrons available for recombination. Therefore, about 2% of total emitted holes flow out via the base terminal and the remaining are drawn across the collector by the electric field at the reverse biased collector junction. As in case of n-p-n transistor, the forward bias at the *EB* junction controls the collector and emitter currents.

5.7 RELATION BETWEEN TRANSISTOR CURRENTS

As discussed earlier, the electrons injected from emitter into the base constitute the emitter current (I_E). Out of these electrons, very few combine with the holes in the thin base region to constitute the base current (I_B). The remaining electrons pass through to the collector region and then to the positive end of V_{CC} to constitute the collector current (I_C). Therefore we can write that,

$$I_E = I_C + I_B \quad \text{...(5.1)}$$

This means that emitter current is always equal to the sum of collector current and base current. As I_B is very small compared to I_E, we can assume the collector current to be nearly equal to the emitter current.

Hence, $$I_C \cong I_E \quad \text{...(5.2)}$$

5.7.1. Transistor Current Components

In figure 5.8, various current components which flow across the forward biased emitter junction and reverse biased collector junction have been shown.

1. The Emitter Current (I_E)

The emitter current consists of the hole current I_{pE} (*i.e.*, the holes crossing from emitter to base) and electron current I_{nE} (*i.e.*, the electrons crossing from base into emitter).

Thus, we have

$$I_E = I_{pE} + I_{nE} \quad \text{...(5.3)}$$

The ratio of hole current to electron, I_{pE}/I_{nE} crossing the emitter junction is proportional to the ratio of conductivity of the p material to that of *n* material. As the doping level of emitter (*p*) side is much higher than that of the base (*n*) layer, $I_{pE} >> I_{nE}$ for a p-n-p transistor. Therefore, the emitter current is approximately equal to the hole current I_{pE}.

Therefore, $$I_E \cong I_{pE} \quad \text{...(5.4)}$$

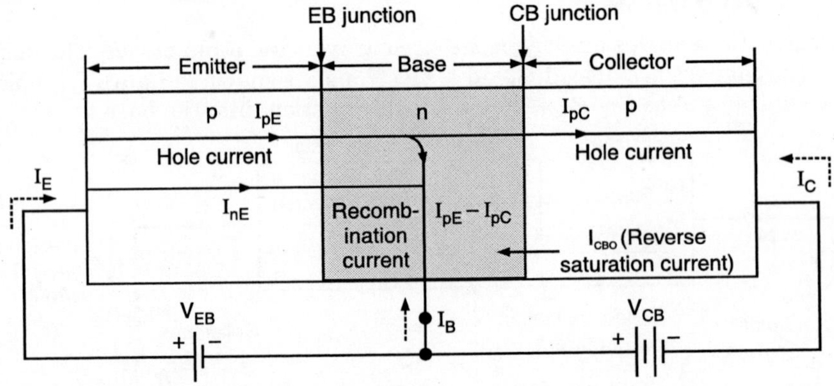

Fig. 5.8. Transistor current components (p-n-p transistor)

2. The base current (I_B)

Out of the holes crossing the emitter junction, some of the holes recombine with the electrons in the n-type base region, to constitute the base current. This is shown as recombination current ($I_{pE} - I_{pC}$) in figure 5.8.

3. The collector current (I_C)

The holes which do not recombine in the base region pass through the collector junction to constitute the collector current. It has been indicated as I_{pC} in figure 5.8. There is one more current component flowing inside the transistor which is called as the *Reverse saturation current* (I_{CBO}). I_{CBO} is the reverse saturation current that flows across reverse biased collector junction when emitter is open circuited as shown in figure 5.8.

Therefore, $$I_C = I_{CBO} \text{ when emitter is open circuited} \quad \text{...(5.5)}$$

The reverse saturation current I_{CBO} flows due to the minority carriers and it increases with increase in temperature. I_{CBO} is also called as collector to base leakage curtent. It is made up of minority carriers. The minority carriers in case of p-n-p transistors are electrons moving from the n-type base to p-type collector. The collector current (I_C) when emitter junction is forward biased will be the sum of I_{pC} and I_{CBO}.

Therefore, $$I_C = I_{pC} + I_{CBO} \quad \text{...(5.6)}$$

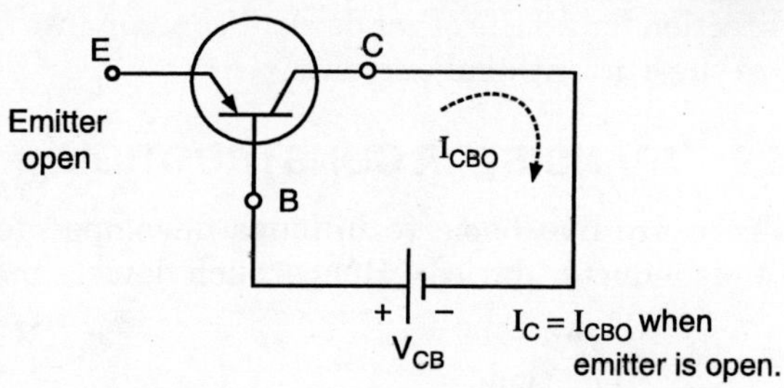

Fig. 5.9. $I_C = I_{CBO}$ when emitter is open

5.8 CIRCUIT SYMBOLS AND CONVENTIONS

The block diagram and conventional circuit symbols of n-p-n and p-n-p bipolar transistors have been shown in figure 5.10.

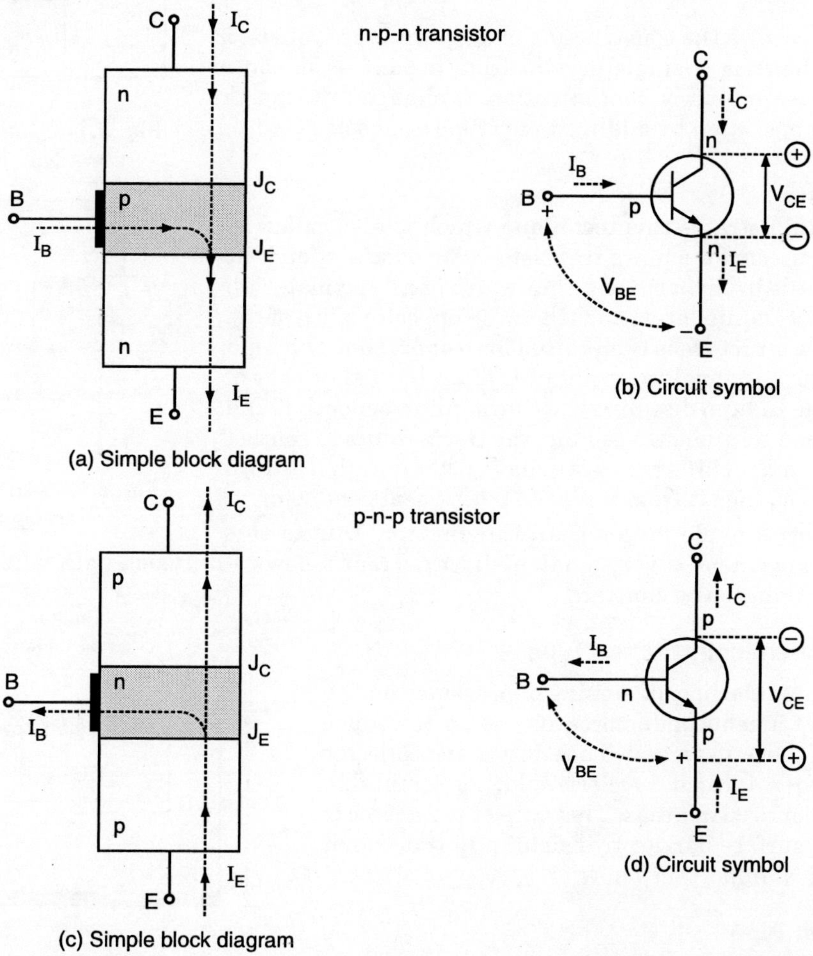

Fig. 5.10. Circuit symbols and conventions of a bipolar junction transistor (BJT)

The arrowhead is always placed on the emitter terminal and it indicates the direction of flow of the conventional emitter current. For the n-p-n transistor, the emitter current flows out of the emitter terminal whereas for the P-N-P transistor, the emitter flows into the emitter terminal. The current direction for all the other current components have been shown. Here, it may be noted that all these are the conventional currents.

5.9 TRANSISTOR CONSTRUCTION

There are five basic techniques developed for the manufactured diodes, transistors and other semiconductor devices. Hence, such devices may be classified into one of the following types:

1. Grown type
2. Alloy type
3. Electrochemically etched type
4. Diffusion type
5. Epitaxial type

Let us discuss them one by one

5.9.1. Grown Type

Figure 5.11 shows the construction of a grown type transistor. It is made by drawing a single crystal from a melt of silicon or germanium whose impurity concentration is changed during the crystal drawing operation by adding *n* or *p* type atoms as required.

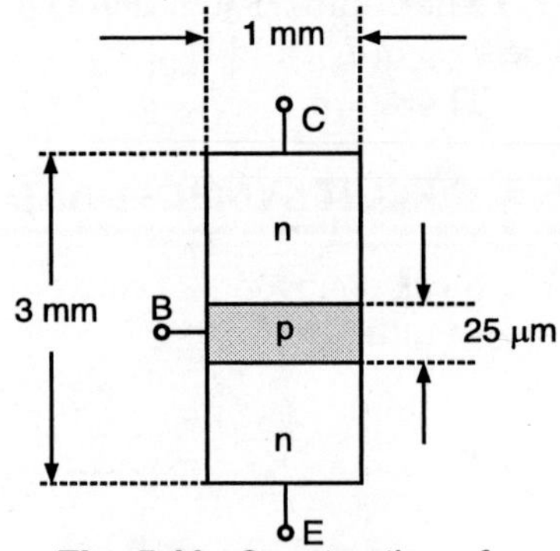

Fig. 5.11. Construction of grown type transistor

5.9.2. Alloy Type

Figure 5.12 illustrates this technique which is also called as the fused construction for a p-n-p transistor. The base section is at the center and it is in the form of a thin wafer of n-type material. Two small dots of Indium are attached to the opposite sides of the wafer. The whole structure is then raised for a short time to a high temperature above the melting point of Indium but below that of Germanium. The Indium dissolves the Germanium beneath it and forms a saturation solution. On cooling, the Germanium in contact with the base material recrystalizes with enough Indium concentration to change it from *n* type to *p* type semiconductor.

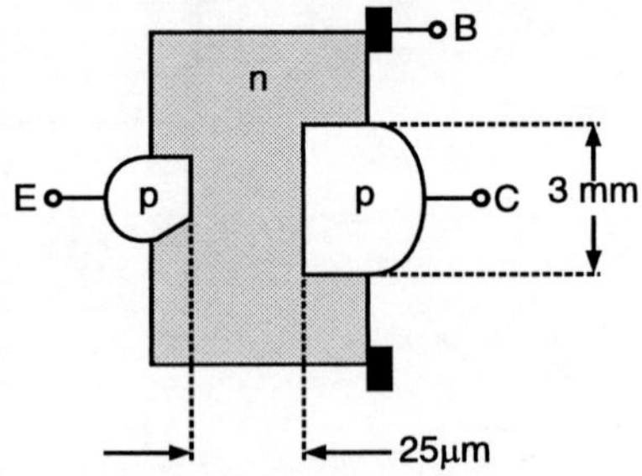

Fig. 5.12. Alloy type p-n-p transistor

The collector is made larger than the emitter. Due to this geometrical arrangement, a very small emitter current follows a diffusion path which carries it to the base rather than to the emitter.

5.9.3. Electrochemically Etched Type

This technique consists of etching depressions on the opposite sides of a semiconductor wafer so as to reduce the thickness of the base section. The emitter and collector junctions are then formed by electroplating a suitable metal into the depression areas. This type of transistor is also called as a surface barrier transistor and it does not have any practical importance now.

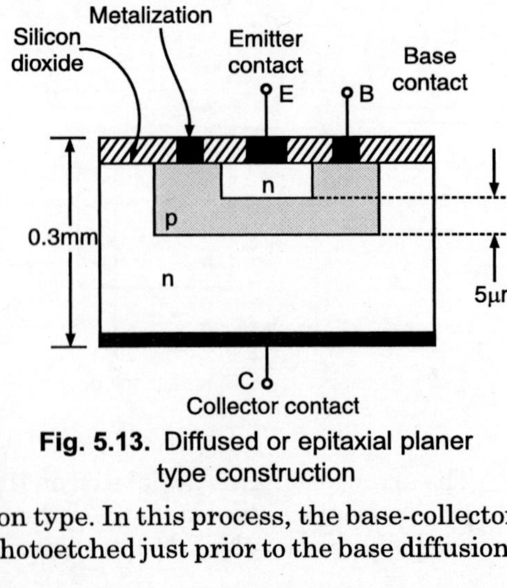

Fig. 5.13. Diffused or epitaxial planer type construction

5.9.4. Diffusion Type

In this technique, a semiconductor wafer is subjected to gaseous diffusions on both *n* and *p* type impurities to from the emitter and collector junctions. Figure 5.13 shows the construction of a planer silicon transistor of the diffusion type. In this process, the base-collector junction, area is determined by a diffusion mask, which is photoetched just prior to the base diffusion.

The emitter is then diffused on the base, and a final layer of silicon oxide is thermally grown over the entire surface. Due to this layer, most surface problems are avoided and a very low leakage current results. There is also an improvement in the current gain at low currents and in the noise figure.

5.9.5. Epitaxial Type

In this technique, a very thin, high purity, single crystal layer of silicon or germanium is grown on a heavily doped substrate of the same material. This augmented crystal forms the collector. The base and emitter may be diffused on it.

5.9.6. Combinations of Tecniques

The techniques discussed earlier can be combined to form a large number of methods for constructing the transistors. The examples of such combinations are as under :

(*i*) Diffused alloy type.

(*ii*) Grown diffused devices.

(*iii*) Alloy emitter epitaxial base transistor etc.

The V-I characteristics at low frequencies of all types of junction transistors are the same.

5.10 TRANSISTOR CONFIGURATIONS (Important)

We know that a transistor may be represented as a two port network. A transistor has three terminals. Any of these three terminals can be made common to input and output. This common terminal is generally grounded or connected to the chasis. Hence, depending upon the common terminal, there are three transistor circuit configurations namely common-base configuration, common-emitter configuration and common collector configuration. One thing should be noted that in all three configurations the transistor works in active region. This means that the emitter-base junction is always forward-biased and the collector-base junction is always reverse biased.

1. Common-base Configuration : In common-base configuration, the base of the transistor is made common. The input is applied between the emitter and base and the output is taken between the collector and base.

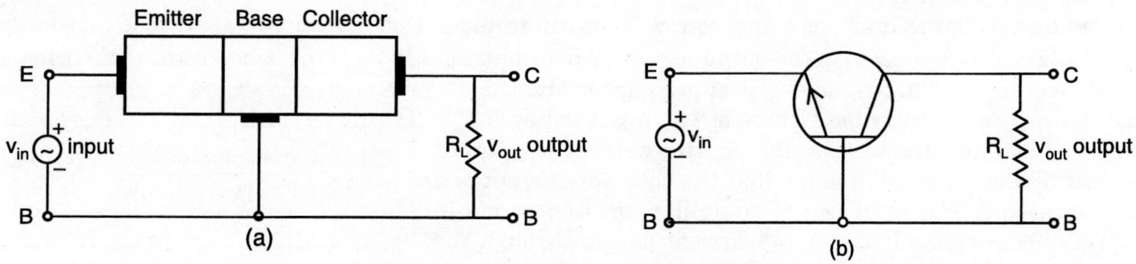

Fig. 5.14. Common Base Configuration.

2. Common-Emitter Configuration : In common-emitter configuration, the emitter of the transistor is made common. The input is applied between base and emitter and the output is taken between collector and emitter. This configuration is most widely used.

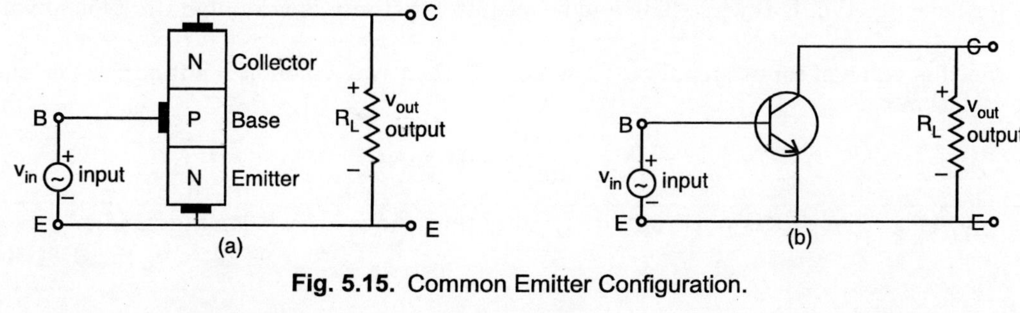

Fig. 5.15. Common Emitter Configuration.

3. Common Collector Configuration : In common collector configuration, the collector is made common. The input is applied between base and collector and the output is taken between emitter and collector.

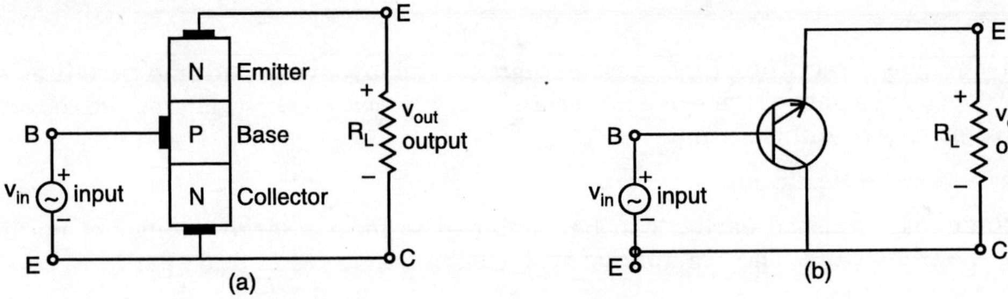

Fig. 5.16. Common Collector Configuration.

5.11 TRANSISTOR AS AN AMPLIFIER

A transistor may be used to perform a number of functions. But mainly transistor is used in amplifying electrical signals. Figure 5.17 shows the basic circuit of a transistor amplifier in which an NPN transistor is connected in common base configuration. The emitter is the input terminal and the collector is the output terminal. The weak input signal is applied between emitter-base junction and output is taken across the load R_L which is connected in the collector circuit. In order to get faithful amplification, the transistor is biased to operate in the active region. This means that the battery V_{EE} forward-biases the emitter-base junction and the battery V_{CC} reverse-biases the collector-base junction.

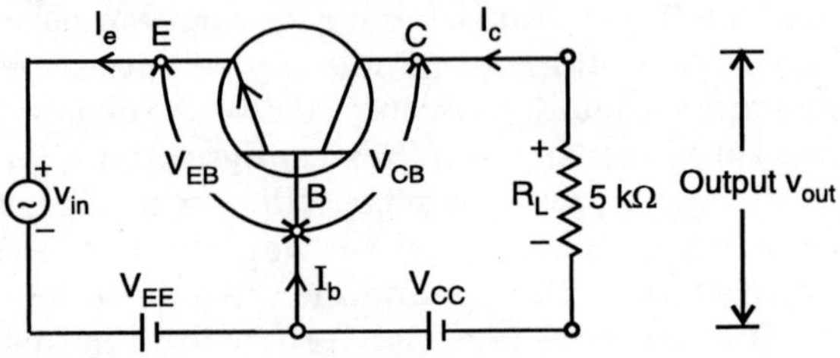

Fig. 5.17.

Let us consider that an a.c. signal source V_{in} is connected in the input circuit and a load resistance $R_L = 5\ \text{k}\Omega$ is also connected in the output circuit. An output voltage V_{out} is obtained across this resistor R_L. When the input a.c. signal V_{in} is super imposed on the d.c. voltage V_{EE}, the emitter-base voltage (*i.e.*, input voltage) V_{EB} varies with time. Due to this, the emitter current (*i.e.*, input current) I_e also varies. We know that the collector current is a function of emitter current, therefore similar variations occur in the collector current. This varying current passes through the load resistor R_L and so a varying voltage is developed across the load resistor R_L. This varying voltage is the output voltage V_{out}. The output voltage V_{out} will be several times greater than the input signal V_{in}.

> **DO YOU KNOW?**
>
> A bipolar transistor is frequently used as a constant current source.

The emitter-base junction is forward-bias, so it offers very low resistance to the input signal V_{in}. In *CB* configuration, typical average value of input resistance is 40 ohms. The output junction or collector-base junction is reverse-bias and so output resistance is very high. Its typical value is 500 kΩ.

Let the rms value of input signal be V_{in} = 20 mV, then rms value of input current or emitter current variations.

$$I_E = \frac{V_{in}}{r_i} = \frac{20\text{mV}}{40\Omega} = 0.5 \text{ mA}$$

A little knowledge that acts is worth infinitely more than much knowledge that is idle.

— **Khalil Gibran**

We know that collector current is equal to the emitter current. Hence, the rms value of input or collector current variations is given by

$$I_c \cong I_e = 0.5\text{mA}$$

Since, the output resistance of transistor is very high than load resistance, therefore, almost all the current I_C passes through the load resistance R_L.

Hence, the rms value of output voltage will be

$$V_{out} = I_c.R_L = (0.5 \times 10^{-3}) \times (5 \times 10^3) = 2.5 \text{ volt}$$

Now, voltage gain or voltage amplification A of amplifier is given as ratio of output voltage V_0 to the input voltage V_{in}.

Therefore, $$A = \frac{V_{out}}{V_{in}} = \frac{2.5}{20 \times 10^{-3}} = 125$$

Hence, amplitude of input voltages increases by 125 times.

5.12 CURRENT GAIN OF A TRANSISTOR IN COMMON-BASE CONFIGURATION

1. Definition

Figure 5.18 shows the circuit of common-base configuration using NPN transistor and the actual directions of different currents in the circuit. In this configuration, emitter current is the input current and collector current is the output current. The current gain of a transistor is defined as the ratio of the transistor output current to the input current. Current gain may be of two types namely d.c. current gain and a.c. current gain depending upon whether the input and output currents are d.c. currents or a.c. currents.

2. d.c. current gain in common-base configuration

d.c. current gain may be defined as the ratio of collector current I_C to emitter current I_E. It is denoted by α or α_{dc} or h_{FB}.

$$\alpha = \frac{I_C}{I_E} = h_{FB} \qquad ...(5.7)$$

But we know that the value of collector current is always less than the value of emitter current. Hence d.c. current gain in *CB* configuration is always less than unity.

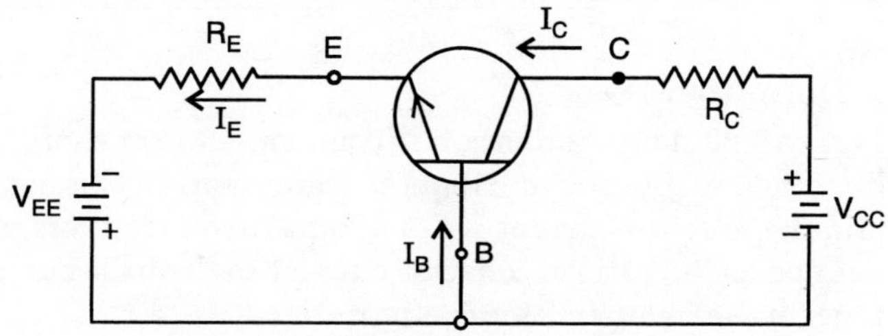

Fig. 5.18. Common-base configurations.

3. a.c. current gain in common-base configuration

A.C. current gain may be defined as the ratio of change in collector current (ΔI_C) to the change in emitter current (ΔI_E) for a constant collector-to-base voltage (V_{CB}).

It is represented by α_0, α_{ac} or h_{fb}

$$\alpha_0 = \alpha_{ac} = \frac{\Delta I_C}{\Delta I_E}\bigg|_{h_{fb} \text{ constant}} \qquad ...(5.8)$$

α_0 is also called **common-base short-circuit current gain.** The value of α_0 is also less than unity. α_0 is approximately same as α.

Therefore, for all practical purposes, we can write

$$\alpha = \alpha_0$$

d.c. current gain = a.c. current gain ...(5.9)

EXAMPLE 5.1. In a common base configuration, the value of collector current is 3.10 mA and the value of emitter current is 3.18 mA. Calculate the common-base d.c. current gain.

Solution : Common-base d.c. current gain is given as

$$\alpha_{dc} = \alpha = \frac{I_C}{I_E} = \frac{5.10}{5.18} \quad \textbf{Ans.}$$

EXAMPLE 5.2. Find the value of base current if the common-base d.c. current gain of a transistor is 0.987 and emitter current is 10 mA.

Solution : Common-base d.c. current gain is given as

$$\alpha = \frac{I_C}{I_E} \Rightarrow I_C = \alpha I_E$$

$$I_C = 0.987 \times 10 = 9.87 \text{ mA}$$

Emitter current is expressed as

$$I_E = I_B + I_C$$

$$I_B = I_E - I_C = 10 - 9.87 = 0.13 \text{ mA} \quad \textbf{Ans.}$$

EXAMPLE 5.3. Figure 5.19 shows a NPN transistor for which I_E = 10 mA, d.c. current gain α = 0.967. Find the values of I_C and I_B.

Solution : Common-base d.c. current gain is

$$\alpha = \frac{I_C}{I_E}$$

$$I_C = \alpha I_E = 0.967 \times 10$$

or $$I_C = 9.67 \text{ mA}$$

But $$I_E = I_B + I_C$$

$$I_B = I_E - I_C = 10 - 9.67 = 0.33 \text{ mA} \quad \textbf{Ans.}$$

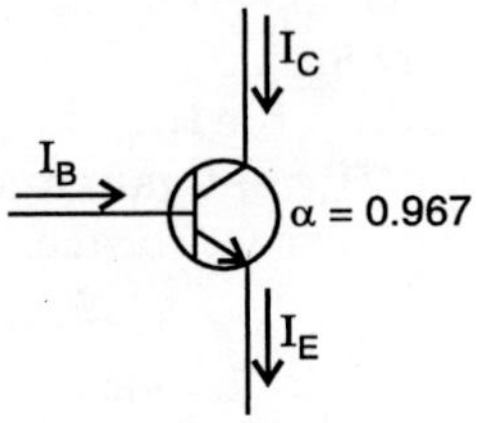

Fig. 5.19.

5.13 CURRENT GAIN OF A TRANSISTOR IN COMMON-EMITTER CONFIGURATION

(BHU, 2000)

1. Definition

Figure 5.20 shows the circuit of common-emitter configuration using NPN transistor and the actual directions of different currents in the circuit. In this configuration, base current is the input current and the collector current is the output current. In common-emitter configuration the current gain is defined as the ratio of collector current to the base current. Like *CB* configuration, current is of two types in common-emitter configuration.

2. Common-Emitter d.c. current gain.

d.c. current gain in common-emitter configuration maybe defined as the ratio of collector current I_C to base current I_B. It is denoted by β or β_{dc} or h_{FE}.

$$\beta_{dc} = \beta = \frac{I_C}{I_B} \quad ...(5.10)$$

Since, the collector current is much larger than the base current, the value of β is much greater than unity. Typical range of β may be from 20 to 300.

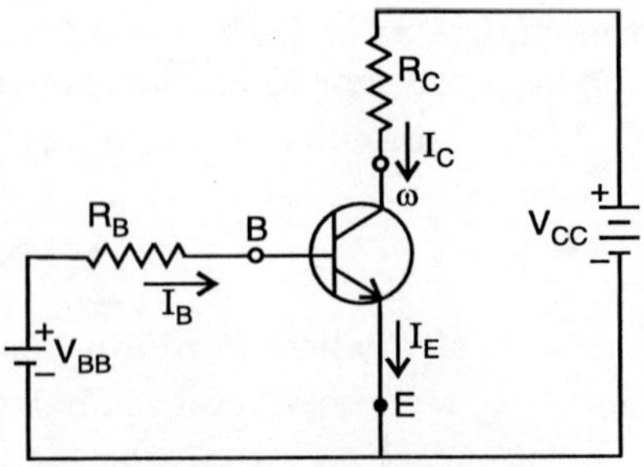

Fig. 5.20. Common emitter configuration

3. a.c. current gain in common emitter configuration

a.c. current gain in common emitter configuration may be

defined as the ratio of change in collector current ΔI_C to the change in base current ΔI_B keeping collector to emitter voltage V_{CE} constant. It is denoted by β_0, β_{ac} or h_{fe}.

$$\beta_0 = \frac{\Delta I_C}{\Delta I_B} \quad ...(5.11)$$

β_o is also known as **common-emitter short circuit current gain.** The values of β_0 are approximately same as that of β.

For all practical purposes, we can write

$$\beta = \beta_0 \quad ...(5.12)$$

d.c. current gain = a.c. current gain

> **DO YOU KNOW?**
>
> The ratio of the collector current to the base current is called the current gain, symbolized as β_{dc} or h_{FE}. For low-power transistors, this is typically 100 to 300. The emitter current is the largest of the three currents, the collector current is almost as large, and the base current is much smaller.

5.14 RELATION BETWEEN CURRENT GAIN α AND β

We know that the emitter current I_E of a transistor is given as

$$I_E = I_B + I_C \quad ...(5.13)$$

Here I_B = Base Current

I_C = Collector Current

Dividing above equation by I_C on both sides, we have

$$\frac{I_E}{I_C} = \frac{I_B}{I_C} + 1 \quad ...(5.14)$$

But $$\frac{I_C}{I_E} = \alpha$$

and $$\frac{I_C}{I_B} = \beta$$

Putting these values in the last expression, we get

$$\frac{1}{\alpha} = \frac{1}{\beta} + 1 = \frac{1+\beta}{\beta}$$

or $$\alpha = \frac{\beta}{1+\beta}$$

or $$\alpha(1+\beta) = \beta$$

$$\alpha + \alpha\beta = \beta$$

$$\alpha = \beta - \alpha\beta = \beta(1-\alpha)$$

or $$\beta = \frac{\alpha}{1-\alpha} \quad ...(5.15)$$

This is desired expression between current gain α and β.

EXAMPLE 5.4. The value of β for a transistor is 100. If the value of emitter current is 10 mA, then determine the values of collector and base current.

Solution : Given that $\beta = 100$ and $I_E = 10$ mA

value of β is given as $$\beta = \frac{\alpha}{1-\alpha}$$

so that $$\alpha = \frac{\beta}{\beta+1} = \frac{100}{100+1}$$

The good life is one inspired by love and guidead by knowledge. — **Bertrand Russell**

$$\alpha = \frac{100}{101} = 0.98 \quad \textbf{Ans.}$$

common-base current gain α is given as

$$\alpha = \frac{I_C}{I_E}$$

$$I_C = \alpha I_E = 0.98 \times 10 \times 10^{-3}$$

$$I_C = 0.98 \times 10^{-4} = 9.8 \text{ mA} \quad \textbf{Ans.}$$

value of emitter current is given as

$$I_E = I_B + I_C$$

$$I_B = I_E - I_C = 10 - 9.8 = 0.2 \text{ mA} \quad \textbf{Ans.}$$

EXAMPLE 5.5. The value of α for a transistor is 0.950. Find the value of β. Also find the value of α if β changes to 100.

Solution : The relation between α and β is given as

$$\beta = \frac{\alpha}{1-\alpha} = \frac{0.950}{1-0.950} = \frac{0.950}{0.050} = 19$$

For second case

$$\because \quad \beta = \frac{\alpha}{1-\alpha} \quad \text{or} \quad \alpha = \frac{\beta}{\beta+1}$$

$$\alpha = \frac{100}{100+1} = \frac{100}{101} = 0.99 \quad \textbf{Ans.}$$

EXAMPLE 5.6. Figure 5.21 shows a NPN transistor for which I_E = 10 mA and β = 100. Determine the values of I_C and I_B. ***(C.C.S. University, 2003)***

Solution :

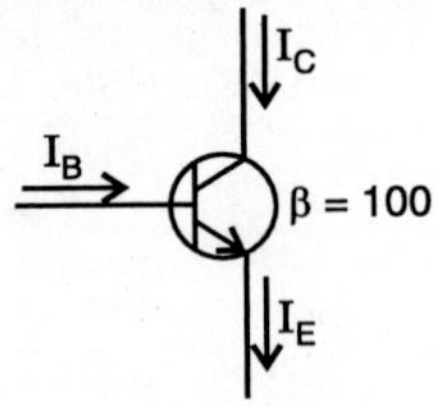

Fig. 5.21.

We know that $\alpha = \dfrac{\beta}{1+\beta}$

But $\alpha = \dfrac{I_C}{I_E}$

$$\beta = \frac{I_C}{I_B}$$

Putting the value of α and β, we get

$$\frac{I_C}{I_E} = \frac{I_C}{I_B} \cdot \left(\frac{1}{1+\beta}\right)$$

or $I_B = \dfrac{I_E}{1+\beta}$

Given that β = 100, I_E = 10 mA

$$I_B = \frac{10}{1+100} = \frac{10}{101} = 0.098 mA$$

Emitter current is given as

$$I_E = I_C + I_B$$

$$I_C = I_E - I_B = 10 - 0.098 = 9.902 \text{ mA} \quad \textbf{Ans.}$$

5.15 LEAKAGE CURRENT IN A COMMON-BASE TRANSISTOR

Figure 5.22 shows the circuit of a common base NPN transistor. The polarity of battery V_{EE} is such that it makes the emitter-base junction forward biased. The battery V_{CC} sets the collector-base junction in reverse biased. Now, we consider two cases

***(i)* Switch *S* is in closed position:** In this case, electrons are injected into base region due to forward-bias. The most of these electrons are attracted by collector due to reverse-bias.

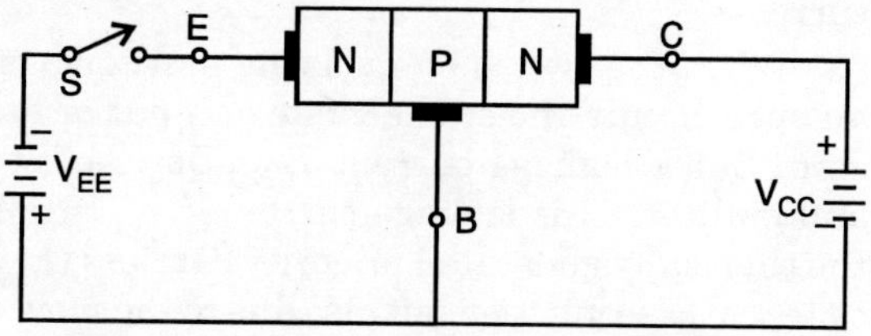

Fig. 5.22. *Common-base transistor circuit*

This accounts for collector current. Few electrons are combined with holes in base region and constitute base current. Therefore, total emitter current breaksup into two parts namely the collector current I_C and base current I_B.

$$I_E = I_B + I_C \qquad ...(5.16)$$

> **DO YOU KNOW?**
>
> The graph of base current versus base-emitter voltage looks like the graph of an ordinary diode. Because of this, we can use any of the three diode approximations to calculate the base current. Most of the time, the ideal and the second approximation are all that is necessary.

***(ii)* Switch S is opened:** Let us assume that the switch is opened.

In this case, emitter-base junction of transistor is open-circuited. Therefore there is no emitter current which means that there is no base current and collector current. But since the collector-base junction is reverse-biased, minority carriers diffuse across the collector base junction and this produce a small current. This small current is called leakage current and is denoted by I_{CBO} or I_{CO}. I_{CBO} means the leakage current between collector and base with emitter open. It is also known as reverse saturation current and it is like reverse saturation current in a PN-junction.

Hence, now it is clear that total collector current for a transistor contains following parts:

(a) The injected current due to majority carriers or a fraction of emitter current is equal to $\alpha \cdot I_E$.

(b) The leakage current I_{CBO} or I_{CO} produced by thermally generated minority carriers.

Therefore,
$$I_C = \alpha I_E + I_{CBO} \qquad ...(5.17)$$

or
$$\alpha I_E = I_C - I_{CBO}$$

$$\alpha = \frac{I_C - I_{CBO}}{I_E} = \frac{I_C - I_{CBO}}{I_B + I_C} \qquad \because I_E = I_B + I_C$$

or
$$\alpha I_B + \alpha . I_C + I_{CBO} = I_C$$

$$\text{or } I_C(1-\alpha) = \alpha I_B + I_{CBO}$$

$$I_C = \frac{\alpha . I_B}{1-\alpha} + \frac{I_{CBO}}{1-\alpha} \qquad ...(5.18)$$

For silicon transistor, at room temperature, the value of I_{CBO} is of the order of nanoamperes and for germanium transistor, it is of the order of microamperes. For both S_i and G_e, the value of I_{CBO} strongly depends on temperature. It becomes doubled for every 10° *C* rise in temperature. Since the value of I_{CBO} is less for S_i than G_e, the silicon transistors can be used upto 200° C whereas G_e transistors can only be used upto 100°C. Therefore silicon transistors are preferred over germanium transistors.

5.16 LEAKAGE CURRENT IN COMMON-EMITTER CONFIGURATION

Figure 5.23 shows a circuit of common-emitter configuration with NPN transistor. If the switch *S* is closed, then, the usual currents I_E, I_C and I_B flow. It means that emitter current is the sum of collector and base currents.

When the switch is open, the emitter-base junction is an open circuit and so the value of input or base current is zero. But a leakage current I_{CEO} flows between collector and emitter. This leakage current I_{CEO} is not only due to the thermally generated minority carriers (I_{CBO}) across the collector-base junction but also due to the movement of holes which flow across the base-emitter junction. Due to flow of holes, the emitter base junction is slightly forward biased. Due to this reason, it can be assumed that a small base current I_{CBO} is supplied to the transistor. This small base current I_{CBO} is equal to additional collector current $\beta.I_{CBO}$.

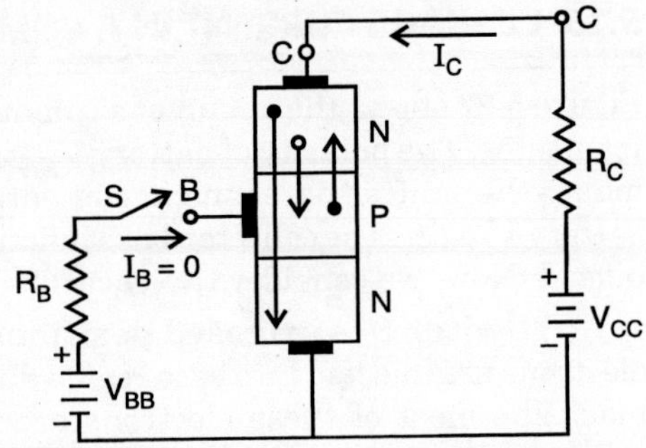

Fig. 5.23. Common emitter transistor ckt

Therefore, with emitter-base junction open, the total leakage current flowing through the transistor

$$I_{CEO} = I_{CBO} + \beta.I_{CBO} \quad ...(5.19)$$

$$I_{CEO} = I_{CBO}(1+\beta) \quad ...(5.20)$$

Hence, the total collector current in transistor is the sum of injected current $\beta.I_B$ and I_{CEO}

$$I_C = \beta.I_B + I_{CBO}(1+\beta)$$

$$I_C = \beta.I_B + I_{CBO}(1+\beta) \quad ...(5.21)$$

The above relation can also be proved in the following way :

We know that $I_E = I_C + I_B$ ...(5.22)

For *CB* configuration, we have

$$I_C = \alpha I_E + I_{CBO} \quad ...(5.23)$$

In *CE* configuration, I_B is the input current and I_C is the output current. Therefore, we have to find the relation between I_B and I_C.

Putting the value of I_E from equation (5.22) in equation (5.23), we get

$$I_C = \alpha(I_C + I_B) + I_{CBO} = \alpha I_C + \alpha I_B + I_{CBO}$$

$$I_C(1-\alpha) = \alpha I_B + I_{CBO}$$

or

$$I_C = \frac{\alpha}{1-\alpha}.I_B + \frac{1}{1-\alpha}.I_{CBO}$$

To simplify the above equation, defining

$$\frac{\alpha}{1-\alpha} = \beta \quad ...(5.24)$$

and

$$\frac{I_{CBO}}{1-\alpha} = I_{CEO} \quad ...(5.25)$$

$$I_C = \beta I_B + I_{CEO} \quad ...(5.26)$$

where β = common-emitter d.c. current gain

I_{CEO} = Leakage current in common emitter configuration.

Using equation (5.24), we have

$$\alpha = \frac{\beta}{1+\beta}$$

From equation (5.25), we have

$$I_{CEO} = \frac{I_{CBO}}{1-\alpha} = \frac{I_{CBO}}{1-\left(\frac{\beta}{1+\beta}\right)} = \frac{I_{CBO}(1+\beta)}{1+\beta-\beta}$$

$$I_{CEO} = (1 + \beta)\, I_{CBO} \quad ...(5.27)$$

Putting the value of I_{CEO} from (5.27) in (5.26) we get

$$I_C = \beta I_B + (1+\beta) I_{CBO} \quad ...(5.28)$$

5.17 COMMON COLLECTOR (CC) CONFIGURATION

Figure 5.24 shows the common collector (CC) configuration for NPN transistor. In this configuration, the input is applied between base and collector whereas the output signal is taken across the emitter and collector. This means that the collector of the transistor is common to both input and output circuits and hence the name common-collector configuration.

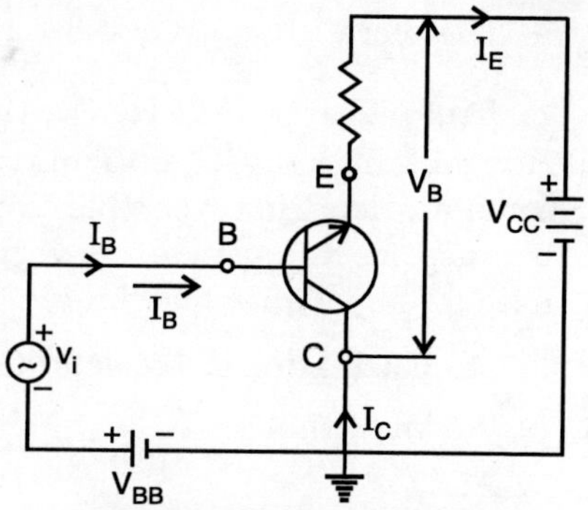

Fig. 5.24. Common-collector configuration for NPN transistor

5.17.1. Current Gain or Current Amplification Factor in CC Configuration

In CC configuration, the ratio of change in emitter current to the change in base current is called as current-gain or current amplification factor. It is denoted by letter γ (Gamma).

> **DO YOU KNOW?**
>
> The four distinct operating regions of a transistor are the active region, the sturation region, the cutoff region, and the breakdown region. When it is used as an amplifier, the transistor operates in the active region. When it is used in digital circuits, the transistor usually operates in the saturation and cutoff regions. The breakdown region is avoided because the risk of transistor destruction is too high.

Mathematically, $\gamma = \dfrac{\Delta I_E}{\Delta I_B}$

Relation between current gain γ and α

We know that

$$\alpha = \frac{\Delta I_C}{\Delta I_E} \quad \text{and} \quad \gamma = \frac{\Delta I_E}{\Delta I_B}$$

$$I_E = I_C + I_B$$

Therefore $\quad \Delta I_E = \Delta I_C + \Delta I_B = \Delta I_E - \Delta I_C$

Putting this value of ΔI_B in equation, $\gamma = \dfrac{DI_E}{DI_B}$, we get

$$\gamma = \frac{\Delta I_E}{\Delta I_E - \Delta I_C}$$

$$= \frac{\Delta I_E / \Delta I_E}{\Delta I_E / \Delta I_E - \Delta I_C / \Delta I_E} = \frac{1}{1 - \dfrac{\Delta I_C}{\Delta I_E}}$$

But $\quad \dfrac{\Delta I_C}{\Delta I_E} = \alpha$

Therefore, $\quad \gamma = \dfrac{1}{1-\alpha}$

Also, we know that $\beta = \dfrac{\alpha}{1-\alpha}$

or $\quad \beta + 1 = \dfrac{\alpha}{1-\alpha} + 1 = \dfrac{1}{1-\alpha}$

* Common collector (CC) configuration is also popularly known as emitter follower. It is so called because here output follows the input (gain is almost one). Although, it is not used in amplification, yet it is used in so many other applications. We shall discuss emitter follower in detail in chapter 6.

Therefore, $\gamma = \beta + 1$

Important Point: Hence, the value of γ is almost equal to β. This means that the value of current gain γ in CC configuration is very high and is comparable to the current gain β in CE configuration. However, this circuit in CC configuration is rarely used for amplification purpose because the voltage gain is very low (less than 1)*. Common-Collector configuration is mainly used for impedance-matching.

5.17.2. Calculation of Collector Current

We know that $I_C = \alpha I_E + I_{CBO}$

Also, $I_E = I_C + I_B$

or $I_E = (\alpha I_E + I_{CBO}) + I_B$

or $I_E - \alpha I_E = I_B + I_{CBO}$

$I_E(1-\alpha) = I_B + I_{CBO}$

or $$I_E = \frac{I_B}{1-\alpha} + \frac{I_{CBO}}{1-\alpha}$$

But $$\frac{1}{1-\alpha} = \beta + 1 = \gamma$$

Therefore, $I_E = I_B(\beta + 1) + I_{CBO}\,(\beta + 1)$

5.18 COMPARISON BETWEEN THREE TRANSISTOR CONFIGURATIONS *(Very Important)*

It may be observed that a transistor may be connected in any one of the three configurations namely *CB*, *CE* and *CC* configuration. The transistor behaves differently in different configuration. The configuration in which a transistor has to be connected depends upon the requirement or application. A configuration may be suitable for a particular application, but it may not be suitable for the other.

Some important parameters which govern the suitability of a configuration are input dynamic resistance, output dynamic resistance, dc current gain, ac current gain, ac voltage gain and leakage current of the transistor.

For example, out of the three configuration, the common collector configuration has very high input resistance but a small value of voltage gain. Hence it is best suited for impedance matching. Therefore, for the application of a particular configuration, we must know its characteristics.

Table 5.3. shows the comparison of the characteristics of three configurations.

Table 5.3. Comparison between the three configurations

S.No.	Characteristics	*CB* configuration	*CE* configuration	*CC* configuration
1.	Input resistance	Very low (40Ω)	Low (50 kΩ)	Very high (750 kΩ)
2.	Output resistance	Very high (1MΩ)	High (10 kΩ)	Low (50 Ω)
3.	Current Gain	Less than unity (0.98)	High (100)	High (100)
4.	Voltage Gain	Small (150)	High (500)	Less than unity
5.	Leakage Current	Very small (5 μA for G_e and 1μA for S_i)	Very large (500 μA for G_e and 20 μA for S_i)	Very large (500 μA for G_e and 20 μA for S_i)
6.	Applications	For high-frequency applications	For audio freq. applications	For impedance matching

5.19 CE CONFIGURATION IS MOST WIDELY USED IN AMPLIFIER CIRCUITS

For an ideal amplifier circuit, the input impedance should be as high as possible and the output impedance should be as low as possible. Although the input impedance for *CC* configuration is highest but gain for *CC* configuration is less than unity. So generally this is not preferred for amplifier circuits.

The input impedance of *CE* configuration is much higher than *CB* configuration since

$$r_i \text{ for } CB = \left.\frac{\Delta v_{EB}}{\Delta i_E}\right|_{V_{CB}=Const.} \qquad ...(5.29)$$

$$r_i \text{ for } CE = \left.\frac{\Delta v_{BE}}{\Delta i_B}\right|_{V_{CE}=Const.} \qquad ...(5.30)$$

The numerators are same in above two equations. But the denominator of the first is of the order of few μA, whereas that of the second is of the order of a few μA. Hence r_i is much higher for *CE* than *CB*.

Similarly output impedance is much lower for *CE* than *CB* since the output characteristic curves are more horizontal for *CB* than *CE*. Except this, the current gain of *CB* configuration is less than unity. It is generally 0.98. The current gain of *CE* configuration is quite high. It is generally 100 and may be as high as 250. Such high current gain in the *CE* configuration makes it possible to have quite high voltage gain as well as high power gain.

Hence, for the above reasons, CE configuration is most popular in amplifier circuits.

5.20 COMPLETE STATIC TRANSISTOR CHARACTERISTICS *(Expected)*

We know about some parameters of transistor such as α and β and leakage current. They give a small idea of transistor behaviour. The complete behaviour of transistor may be observed with the help of some types of curves known as characteristics curves. These characteristics curves relate the transistor currents and voltages. There are two types of characteristics curve:

(*a*) **Input characteristics :** Input characteristics curves relate the input current and input voltage for a given value of output voltage.

(*b*) **Output characteristics :** Output characteristics curves relate the output current and output voltage for a given value of input current.

With the help of these two types of characteristics curves the complete static operation of transistor can be explained. Hence, these curves are also known as static characteristics curves.

5.21 TRANSISTOR CHARACTERISTICS IN COMMON-BASE CONFIGURATION

In common-base configuration, two types of characteristics are :

(*a*) **Input characteristics :** Input characteristics curves relate the input or emitter current I_E and input or emitter-to-base voltage V_{EB} keeping output or collector-to-base voltage V_{CB} constant.

(*b*) **Output Characteristics :** Output characteristics curves relate the output or collector current I_C and output or collector-to-base voltage V_{CB} keeping the input or emitter current I_E constant.

Both these characteristics curves can be drawn with the same circuit arrangement shown in figure 5.25.

* Man is the highest animal. Indeed, it is intelligence that separates man from the other animals.
—**J. L. Nehru**

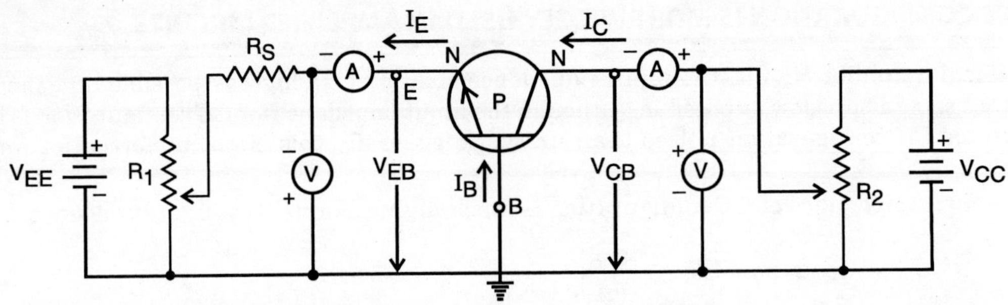

Fig. 5.25. Circuit arrangement for input and output characteristics curve in CB configuration

5.21.1. Transistor Input Characteristics in CB Configuration

Input characteristics curves are obtained with the help of circuit arrangement shown in figure 5.25. In this circuit NPN transistor is used. The series resistor R_s is connected to limit the emitter current. The value of voltage V_{EB} may be varied with the help of a potentiometer R_1. The collector voltage V_{CB} is varied by adjusting the potentiometer R_2.

The d.c. millimetres and d.c. voltmeters have been connected in the emitter and collector circuits to measure the voltages and currents.

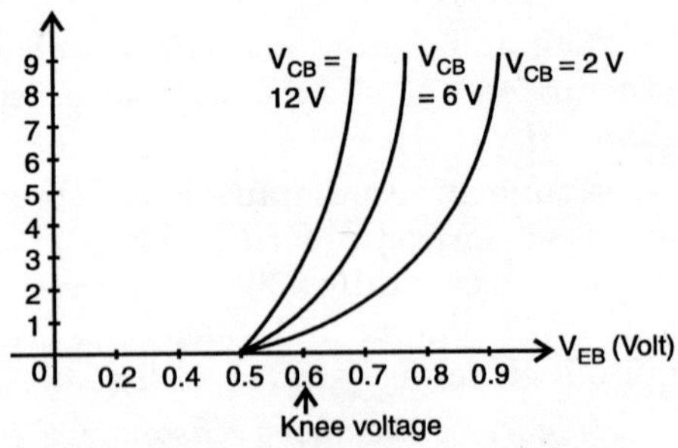

Fig. 5.26. Input characteristics in CB configurations.

The input characteristics in *CB* configuration are plotted between emitter current I_E and the emitter-base voltage V_{EB}, for different values of collector-base voltage V_{CB}. Input characteristics are shown in figure 5.26.

Few Points

(*i*) Below the cut-in voltage or knee voltage the emitter current is negligibly small. This value of cut-in-voltage is approximately 0.7 V for silicon and 0.3 V for germanium transistors.

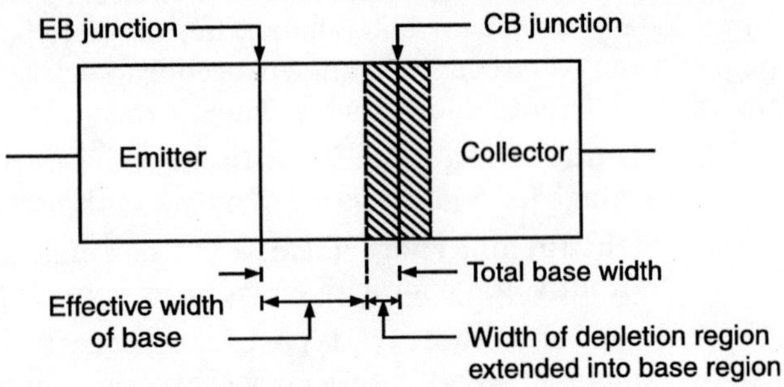

Fig. 5.27. Width of the base region.

(*ii*) After the knee voltage, the emitter current I_E increases rapidly with a small increase in emitter-to-base voltage V_{EB}. This indicates that input resistance in CB configuration is very small.

(*iii*) With the increase in V_{CB}, the curves shift upwards or becomes vertical. Hence r_i decreases.

(*iv*) At any point on the curve, the value of a.c. or dynamic input resistance is given as

$$r_i = \left.\frac{\Delta V_{EB}}{\Delta I_E}\right| V_{CB} \text{ constant} \qquad ...(5.31)$$

where, $\Delta V_{EB} \rightarrow$ Small change in emitter-to-base voltage V_{EB}.

$\Delta I_E \rightarrow$ Small change in emitter current I_E.

The dynamic input resistance r_i is very small nearly 25 to 100 Ω.

5.21.2. The Early Effect or Base Width Modulation

Referring to figure 5.27 and the width of the region, the total base width is equal to the sum of the widths of the depletion regions extended into the base region from the collector and emitter side and the width of the region occupied by the free charge particles. But as the *EB* junction is forward biased, the width of its depletion region is very narrow. As the *CB* junction is reverse biased, the width of its depletion region is much larger. Thus neglecting the width of depletion region at the *EB* junction we can write :

Total base width = Width of depletion region at the *CB* junction inside the base region + width of the region containing free charge particles

(*i*) Now as V_{CB} is increased, the reverse voltage applied to the *CB* junction increases. This widens the depletion region at the collector junction. Due to this, effective width of the base region decreases as shown in figure 5.27.

(*ii*) This will increase the charge concentration gradient in the base region. Due to increased charge carrier concentration, more number of electrons diffuse from the emitter to base *i.e.*, emitter current increases.

(*iii*) Thus with increase in V_{CB}, the input current I_E increases slightly.

The reduction in effective base width due to increase in V_{CB} is called as **"Early" effect** or **"Base Width Modulation"**.

Other effects of base width modulation :

There are two more effects of the base width modulation as under:

(*i*) As the base width decreases due to early effect, the possibility of recombination in the base region will reduce. This will increase the values of both, the amplification factor β and the common base current amplification factor α_{dc}, with increase in the voltage V_{CB}.

(*ii*) Due to increase in the charge gradient within the base, more number of "minority" charge carriers are injected across the junction. This will increase the current of minority carriers.

5.21.3. Output Characteristics in CB Configuration

The output characteristics curves in *CB* configuration can be obtained with the help of same circuit arrangement shown in figure 5.25. First, a suitable value of input current (emitter current) is fixed. Now keeping emitter current I_E constant. Collector-to-base voltage V_{CB} is increased and corresponding changes in collector current are noted and a graph is plotted. Figure 5.28 shows the output characteristics curve for CB configuration.

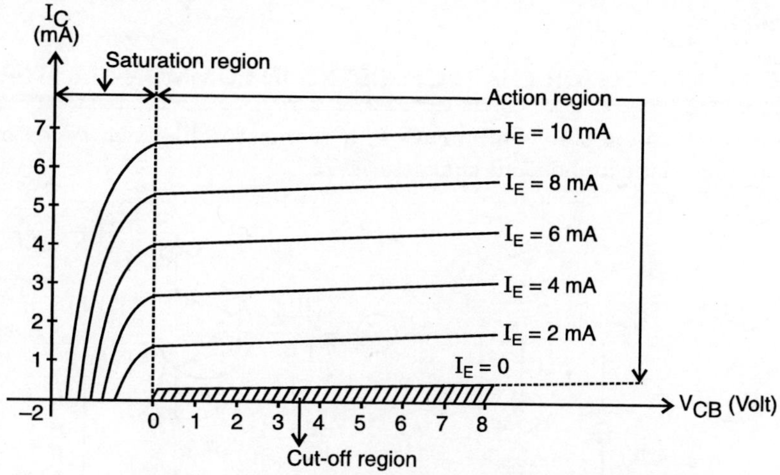

Fig. 5.28. Output characteristics in a CB configuration.

5.21.4. Few Points

(*i*) The characteristics curves show the three regions namely active region, cut-off region and saturation region.

(*ii*) In active region, the collector current is approximately equal to the emitter current I_E.

(*iii*) Emitter current increases slowly for increase in V_{CB} in active region. Hence, in active region, the curves are nearly flat. This also means that the output resistance in this configuration is very large.

(*iv*) If V_{CB} becomes negative, the collector-base junction is set in forward bias which causes collector current to decrease rapidly. This region is known as saturation region. In saturation region the collector current does not depend much on emitter current.

(*v*) In saturation region, since collector-base junction is also forward biased, a small increase in V_{CB} results in a large increase in collector current.

(*vi*) If emitter current I_E is made zero, then the collector current is not zero. It has a small value. This small value is leakage current I_{CBO}. The curve for $I_E = 0$ corresponds to cut-off region. In this region both the junctions are reverse-biased.

(*vii*) With the help of output characteristics, the dynamic output resistance r_0 can be found by the expression

$$r_0 = \left.\frac{\Delta V_{CB}}{\Delta I_C}\right|_{I_E=\text{Constant}} \quad \text{...(5.32)}$$

Here ΔV_{CB} and ΔI_C are small changes in collector-to-emitter voltage and collector current for a given point.

(*viii*) With the help of output characteristics curve d.c. and a.c. current gains in *CB* configuration can also be determined.

d.c. current gain

$$\alpha_{dc} = \alpha = \frac{I_C}{I_E} \quad \text{...(5.33)}$$

where I_C and I_E are the values of collector and emitter currents at any point on the curve. a.c. current gain is

$$\alpha_0 = \left.\frac{\Delta I_C}{\Delta I_E}\right|_{V_{CB}=\text{Constant}} \quad \text{...(5.34)}$$

where ΔI_C and ΔI_E are small changes in collector and emitter currents around a given point on the curve.

5.22 TRANSISTOR CHARACTERISTICS IN COMMON-EMITTER CONFIGURATION*

In common-emitter configuration, a transistor has two types of characteristics namely input characteristic and output characteristic.

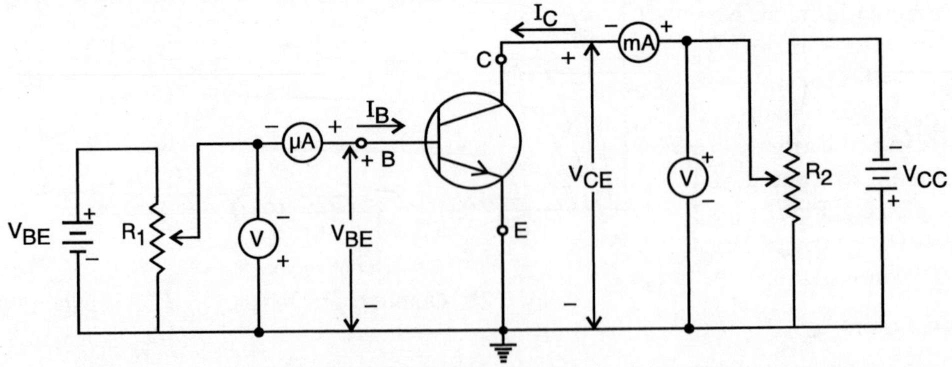

Fig. 5.29. Circuit arrangement for input and output characteristics.

* Draw the output characteristics of a common emitter amplifier. Also draw the load line.

(*a*) **Input characteristics :** Input characteristics curves relate the base current I_B and the base-to-emitter voltage V_{BE} keeping collector-to-emitter voltage V_{CE} constant.

(*b*) **Output characteristics :** Output characteristics curves relate the collector current I_C and collector-to-emitter voltage V_{CE} keeping base current I_B constant.

Both the characteristics may be obtained with the help of single circuit arrangement shown in above figure. Potentiometer R_2 is used to vary the collector-to-emitter voltage V_{CE} and potentiometer R_1 is used to vary base-to-emitter voltage V_{BE}. To measure different voltages and currents, d.c. millimetres and d.c. voltmeters are connected in base and collector circuit of transistor.

5.22.1. Input Characteristics in Common-Emitter (CE) Configuration

Input characteristics curves may be obtained with the help of circuit arrangement shown in figure 5.29 by increasing the V_{BE} in steps and recording the corresponding values of I_B while keeping V_{CE} constant. In figure 5.30, the typical input characteristics curves which relate I_B and V_{BE} for various values of V_{CE}.

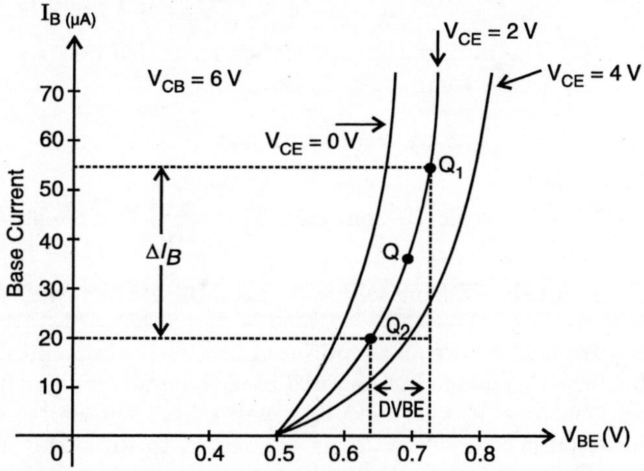

Fig. 5.30. Input characteristics in CE configuration

With the help of input characteristics, we can find the value expressed dynamic input resistance at a given point. At a given point, it is expressed as the ratio of small change in V_{BE} to the resulting small change in I_B keeping V_{CE} constant.

$$r_i = \frac{\Delta V_{BE}}{\Delta I_B}\bigg|_{V_{CE}\text{ constant}} \qquad ...(5.35)$$

5.22.2. Ouput Characteristics in Common-Emitter Configuration

Output characteristics in *CE* configuration can be plotted with the help of circuit arrangement shown in figure 5.29. The related output current I_C to the collector-to-emitter voltage for different values of input current I_B.

Figure 5.31 shows the typical output characteristics.

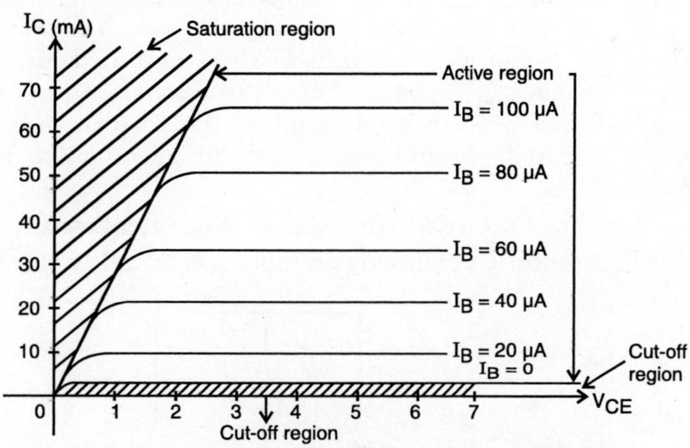

Fig. 5.31. Output characteristics in CE configuration.

From the output characteristics, we can draw few points :

(*i*) In active region, the collector current increases slowly with the increase in V_{CE}. Output characteristics in *CE* configuration are less horizontal than those in *CB* configuration. This indicates that output resistance is small in *CE* configuration than *CB* configuration.

(*ii*) As the value of V_{CE} decreases below zero, I_C decreases rapidly. This occurs due to reason that when V_{CE} decreases below zero, collector-base junction becomes forward-biased. This means that now both the junctions are forward-biased. This is called saturation region. In saturation region, the collector current I_C does not depend upon the input current I_B.

(*iii*) If base current is made zero, then the collector current is not zero. It has a value known as reverse leakage current I_{CEO}. This condition is called the cut-off region of transistor.

(*iv*) The output characteristics may be used to find the dynamic output resistance at any given point. Output resistance is given

$$r_0 = \left.\frac{\Delta V_{CE}}{\Delta I_C}\right| I_B \text{ constant} \quad \text{...(5.36)}$$

(*v*) The output characteristics may also be used to calculate the d.c. current gain β and a.c. current gain β_0 as follows :

$$\text{d.c. current gain } \beta = \frac{I_C}{I_B} \quad \text{...(5.37)}$$

$$\text{a.c. current gain } \beta_0 = \left.\frac{\Delta I_C}{\Delta I_B}\right| V_{CE} \text{ constant} \quad \text{...(5.38)}$$

5.23 BASIC COMMON EMITTER AMPLIFIER AND D.C. LOAD LINE

Figure 5.32 shows an amplifier circuit in common emitter configuration using NPN transistor. The battery V_{BB} is used to forward bias the emitter-base junction and battery V_{CC} is used to reverse bias the collector-base junction. The resistor R_B is connected to limit the input or base current. The resistor R_C acts as a load resistance. The amplified a.c. output voltage appears across load resistor R_C.

The input a.c. signal to be amplified is V_{in}. This is applied to the base through capacitor C_1. The capacitors C_1 and C_2 are called coupling capacitors. The coupling capacitor C_1 blocks d.c. Hence d.c. current flows only through R_B and is prevented from reaching the input signal source. The coupling capacitor C_2 also blocks d.c. from reaching the output terminals. Therefore, at the output terminals. Only amplified output a.c. signal voltage V_{out} appears, D.C. load line. With the help of d.c. load line, we can easily analyse the performance of an amplifier circuit. In this section, we will learn how to draw a d.c. load line.

DO YOU KNOW?

The emitter is grounded or common in a CE circuit. The base-emitter part of a transistor acts approximately like an ordinary diode. The base-collector part acts like a current source that is equal to β_{dc} times the base current. The transistor has an active region, a cutoff region, and a breakdown region. The active region is used in linear amplifiers. Saturation and cutoff are used in digital circuits.

For this, first we will find the d.c. equivalent circuit of the amplifier circuit shown in figure 5.32. This condition of having no input a.c. signal is also known as quiescent condition.

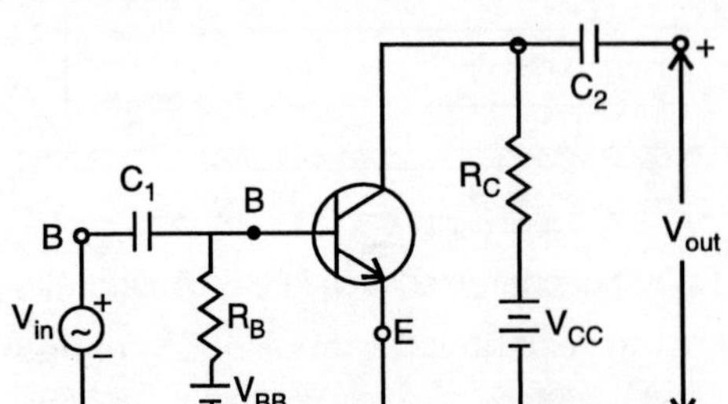

Fig. 5.32. Basic common emitter amplifier for NPN transistor.

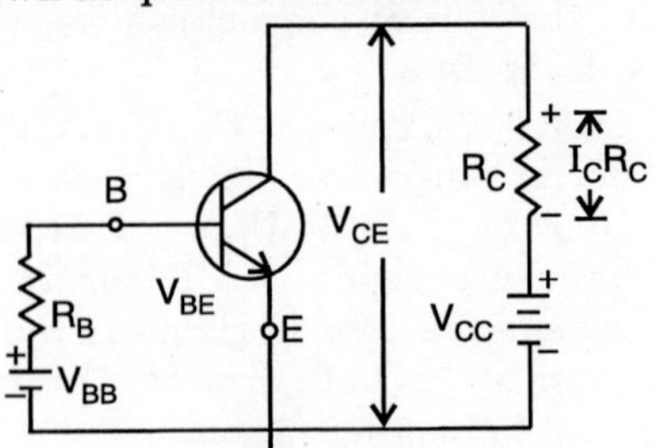

Fig. 5.33. Basic common emitter (d.c. equivalent circuit).

Figure 5.33 shows the d.c. equivalent circuit of one shown in figure 5.32. This circuit is obtained by opening the two capacitors and make the value of signal source equal to zero. For d.c. the capacitor works as an open circuit.

Applying KVL to the collector portion of the circuit, we get

$$I_C R_C + V_{CE} = V_{CC}$$

or
$$I_C = \left(-\frac{1}{R_C}\right)V_{CE} + \frac{V_{CC}}{R_C} \quad \text{...(5.39)}$$

The above equation is in the form of

$$Y = mx + c \quad \text{...(5.40)}$$

The equation (5.40) is an equation of straight line.

Comparing equations (5.39) and (5.40), the slope of straight line will be

$$m = -\frac{1}{R_C}$$

Therefore, if equation (5.39) is plotted on the output characteristics curves, we will get a straight line having slope equal to $\left(-\frac{1}{R_C}\right)$ and intercept on vertical axis equal to $\left(\frac{V_{CC}}{R_C}\right)$.

DO YOU KNOW?

Centering the Q-point on a transistor load line is important because it allows for the maximum ac output voltage from the amplifier. Centering the Q-point on the dc load line is sometimes referred to as mid-point bias.

This line plotted on the output characteristics is called d.c. load line. To plot the d.c. load line, we use the concept that on V_{CE} axis, the I_C is zero and on I_C axis, the V_{CE} is zero.

Substituting these conditions in the equation (5.39), *i.e.*

$$V_{CC} = V_{CE} + I_C R_C, \text{ we get}$$

when $I_C = 0 \Rightarrow V_{CE} = V_{CC}$, Point $A = (V_{CC}, 0)$

when $V_{CE} = 0 \Rightarrow V_{CC} = I_C R_C$

$$\text{or } I_C = \frac{V_{CC}}{R_C}, \textit{ Point } B \equiv \left(0, \frac{V_{CC}}{R_C}\right)$$

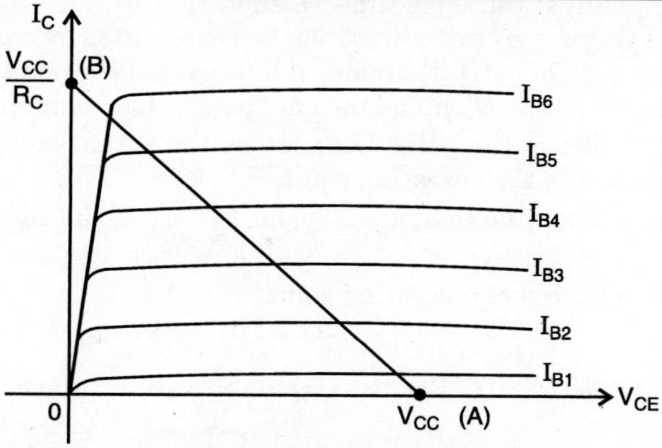

Fig. 5.34. The dc load line

If two points are located on the output characteristics and joined by a straight line, we get a d.c. load line. This line is called d.c. load line because its slope depends upon the value of resistor R_C which is the d.c. load of the circuit.

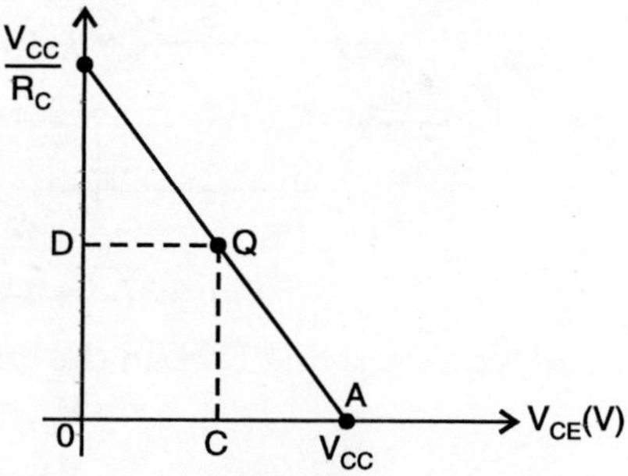

Fig. 5.35.

5.23.1. Importance of d.c. Load Line

The collector or output current I_C and collector to emitter voltage V_{CE} conditions in the transistor circuit are represented by a point on the output characteristics curve. This same information may be obtained from the d.c. load line.

Hence, when I_C is maximum $\left(I_C = \frac{V_{CC}}{R_C}\right)$ then V_{CE} will be zero ($V_{CE} = 0$) and when $I_C = 0$, then V_{CE} is maximum ($V_{CE} = V_{CC}$) as shown in figure 5.35.

For some other value of output or collector current I_C (say *OD*), the collector-to-emitter voltage

V_{CE} will be equal to OD. It means that d.c. load line provides a more convenient and direct solution to the problem.

From above, it also follows that if we plot the d.c. load line on the output characteristics curve of the transistor, we may investigate the behaviour of transistor amplifier.

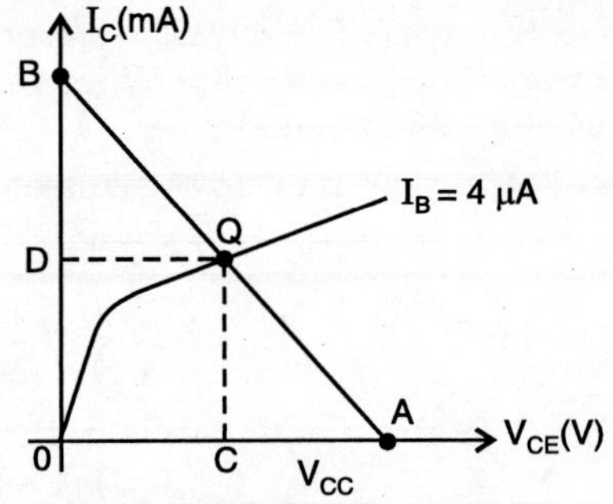

Fig. 5.36.

5.23.2. Operating Point

The zero signal values (no a.c. signal is applied) of collector current I_C and collector-to-emitter voltage V_{CE} are called as the **operating point.**

Since the variations of I_C and V_{CE} take place about this point, therefore this is known as the operating point.

This point is also called as quiescent (stable) point or simply Q-point since it is a point on output characteristics curve when the transistor is in a silent condition (*i.e.*, the absence of a.c. input signal).

Let us consider that in the absence of a.c. input signal, the base current is 4 μA. Then, in transistor circuit, the I_C and V_{CE} conditions must be represented by a point on I_B curve *i.e.*, $I_B = 4\mu A$ characteristic curve.

But at the same-time we know that I_C and V_{CE} conditions in transistor circuit must also be represented by a point on the d.c. load line AB. Therefore, it follows that the point Q where the d.c. load line and the characteristics curve intersect will be the only point which satisfies both of these conditions.

> **DO YOU KNOW?**
>
> The operating point of the transistor is on the dc load line. The exact location of this point is determined by the collector current and the collector-emitter voltage. With base bias, the Q point moves whenever any of the circuit values change.

Hence, the point Q describes the actual state of circuit in the zero signal conditions and is known as the operating point.

According to figure 5.36, for $I_B = 4\ \mu A$, the zero signal values are

$$V_{CE} = OC \text{ volts} = 10 \text{ Volts} \quad \text{and} \quad I_C = OD \text{ mA} = 6 \text{ mA}$$

Therefore, operating point

$$Q \equiv (V_{CE}, I_C) \equiv (10\text{ V}, 6\text{ mA})$$

EXAMPLE 5.7. For the circuit shown in figure 5.37, draw the d.c. load line.

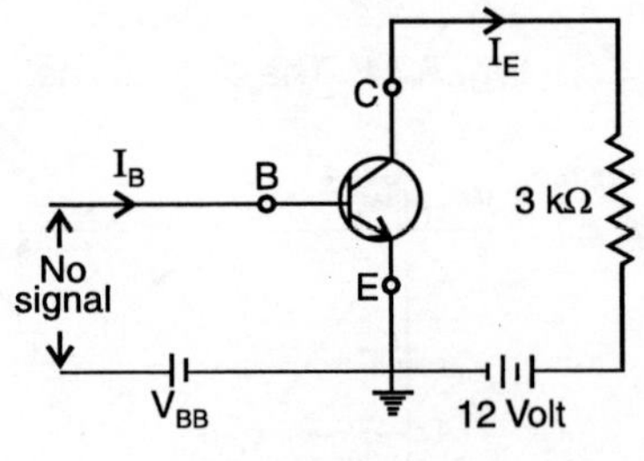

Fig. 5.37.

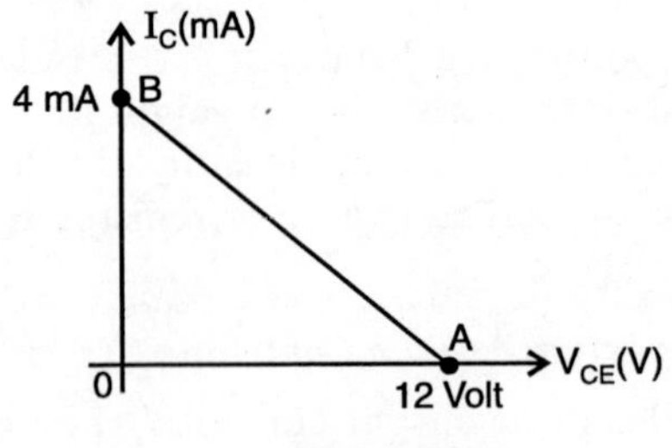

Fig. 5.38.

Solution : Applying KVL in the output portion, we get

$$V_{CC} = I_C R_C + V_{CE}$$

or $$V_{CE} = V_{CC} - I_C.R_C \quad ...(i)$$

Now, when $I_C = 0$

from equation (*i*) $$V_{CE} = V_{CC} - 0 = V_{CC} = 12 \text{ Volt}$$

This locates the point A of the d.c. load line on the collector-emitter voltage (V_{CE}) axis.

Similarly, when $V_{CE} = 0$

from equation (*i*) $\quad 0 = V_{CC} - I_C.R_C$

or $\quad I_C R_C = V_{CC}$

or $$I_C = \frac{V_{CC}}{R_C} = \frac{12\text{ V}}{3\text{ k}\Omega} = 4\text{ mA}$$

This located the point *B* of the d.c. load line on the collector current (I_C) axis.

Now, by joining these two points *A* and *B*, we get the d.c. load line *AB* as shown in figure 5.38.

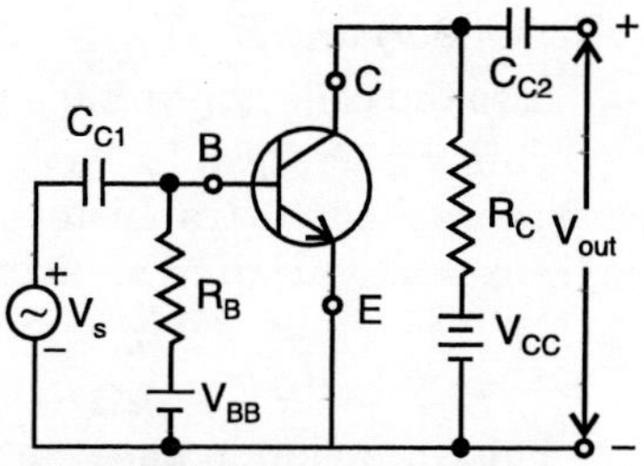

Fig. 5.39. A basic amplifier circuit

5.23.3. Analysis of Amplifier Using D.C. Load Line

As discussed earlier, a transistor may be used to amplify a.c. signals after fixing its operating point suitably.

In last article, we learnt how to fix the *Q*-point on the output characteristics of a transistor. In fact, the *Q*-point must lie in the middle portion on the output characteristics of a transistor. This helps the transistor to amplify a.c signals faithfully *i.e.*, without producing distortion in waveshape of input signal.

In quiescent condition, the base current is fixed at a constant d.c value. It is found from the *Q*-point. After this, when we apply an a.c signal at the input of the amplifier circuit, the base voltage varies as per the signal voltage V_s. Due to this, the base current will also vary. Now, as the base current varies, the instantaneous operating point of the transistor moves along the d.c. load line. This means that the instantaneous values of the collector current and voltage also vary according to the input signal. This variation in collector voltage is several times larger than the variation of the input signal.

DO YOU KNOW?

The dc load line contains all the possible dc operating points of a transistor circuit. The upper end of the load line is called saturation, and the lower end is called cutoff. The key step in finding the saturation current is to visualize a short between the collector and the emitter. The key step to finding the cutoff voltage is to visualize an open between the collector and emitter.

Also, this collector-voltage variation reaches the output terminals through coupling capacitor. Thus, the output signal is several times larger than the input signal. Let us consider an example. Figure 5.39 shows an amplifier circuit. Let us assume that

$V_{CC} = 12$ V, $R_C = 1$ kΩ,

$V_{BB} = 10.7$ V

and $R_B = 200$ kΩ.

Here the d.c. base current is found to be 50μA and the collector d.c current and collector d.c voltage are 6 *mA* and 6*V* respectively.

Now, let us apply a small a.c. signal voltage of 7mV at the input.

To show the variations in base current I_B on the output characteristic of the transistor, a line perpendicular

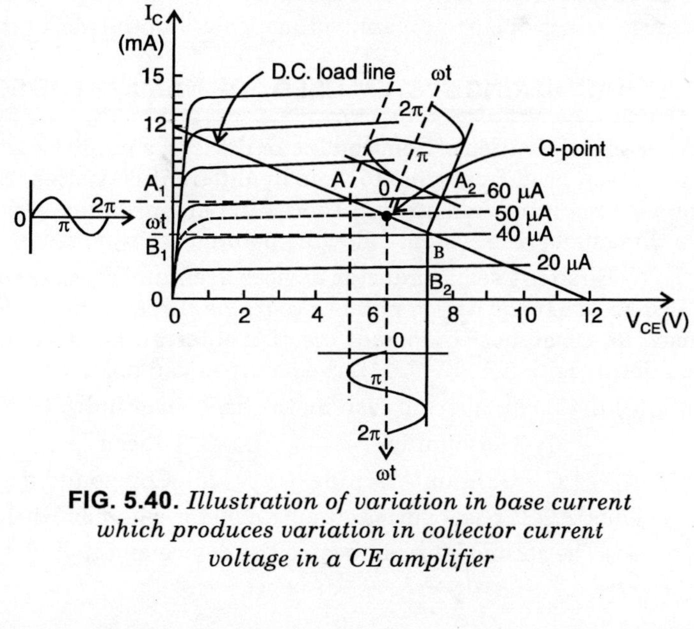

FIG. 5.40. *Illustration of variation in base current which produces variation in collector current voltage in a CE amplifier*

to the d.c. load line and passing through the Q-point is drawn as shown in figure 5.40. We take this line as ωt-axis and then variations in I_B are plotted.

As the instantaneous operating point moves along the d.c. load line between the points A and B, both the collector-current and collector voltage would vary.

The current I_C varies between the points A_1 ($I_C = 7.3$ mA) and B_1 ($I_C = 4.8$ mA). These variations are shown on the left side of the output characteristics. The voltage V_{CE} varies between points A_2 ($V_{CE} = 4.9$ V) and B_2 ($V_{CE} = 7.1$ V).

Also the collector-voltage variations are shown at the bottom of the output characteristics.

This voltage would have about 20 mV peak-to-peak variations. Now, if the input dynamic resistance r_i of the transistor is taken to be 1 kΩ, then the input voltage variation would produce a peak-to-peak variation of 20 μA in base current, since

$$\Delta I_B = \frac{20 mV}{1 k\Omega} = 20\ \mu A$$

This variation in base current exists around its quiescent value (*i.e.*, operating point) of 50 μA. Now as the base current varies, the instantaneous operating point moves along the d.c. load line between the points $A(I_B = 60\ \mu A)$ and $B(I_B = 40\ \mu A)$.

Now, we can compute the current gain, voltage gain and the power gain of the amplifier on the basis of peak-to-peak variation as under :

***(i)* Current gain**

$$A_i = \frac{\Delta I_C}{\Delta I_B}$$

Substituting all the values, we get

or $$A_i = \frac{(7.3 - 4.8) \times 10^{-3}}{(60 - 40) \times 10^{-6}} = 125$$

***(ii)* Voltage gain** $$A_v = \frac{\Delta V_{CE}}{\Delta V_{BE}} = \frac{7.1 - 4.9}{20 \times 10^{-3}} = 110$$

***(iii)* Power gain** $$= \frac{\text{Output ac power}}{\text{Input ac power}} = \frac{i_c . v_{ce}}{i_b . v_{be}} = A_i \times A_V = 125 \times 110 = 13750$$

Important Point: From above, it is obvious that a common emitter amplifier has sufficiently large values of current gain, voltage gain and the power gain.

5.24 NUMBERING SYSTEM USED FOR NAMING THE SEMICONDUCTOR DEVICES.

After the fabrication of Semiconductor devices, a numbering is required. Several numbering systems have been used from time to time by different countries. But, the numbering system invented by proelectron standardization Authority in Belgium, is accepted and adopted internationally. According to this numbering system following points may be noted :

(*i*) Different semiconductor devices are numbered by fine alpha-numeric symbols like BF 194, BFX 65 etc. If the device number contains three numeral digits (*i.e.*, BF 194), the device is usually used for consumer equipment (*i.e.*, for entertainment etc.) But if the device number contains two numeral digits (*i.e.*, BF X 15), the device is usually used for industrial or professional equipment.

(*ii*) In this numbering system, the first letter indicates the nature of Semiconductor material as

A – Germanium	*B*– Silicon
C – Gallium Arsenide	*R* – Compound (*e.g.*, cadmium sulphide)

Thus, AC 125 is a germanium transistor whereas BF 194 is a silicon transistor.

(*iii*) The second letter indicates the device and its function in the circuit as :

A – Diode
B – Varactor
C – A.F. low power transistor
D – A.F. Power transistor
E – Tunnel, diode
F – H.F. low power transistor
G – Multiple device
H – Magnetic Sensitive diode
K – Hall-effect modulator
L – H.F. power transistor
M – Hall-effect modulator
P – Radiation Sensitive diode
Q – Radiation generating diode
R – Thyristor (SCR or TRIAC)
S – Low power switching transistor
T – Thyristor, high power
U – switching transistor, power
X – diode, multiplier
Y – Power multiplier
Z – Zener diode

However, it may be noted that only this numbering system is not applied in electronics. Some other numbering systems are also used. As an example, IN 4007 is a silicon diode, 2N 3055 is a power transistor, BHT F 25 is a silicon controlled rectifier etc.

Thus, in this numbering system IN represents one junction device *i.e.*, diode, 2N represents 2-junctions device like transistor and 3N represents two gates device like FET.

5.25 LIMITS OF OPERATION

For any transistor, it is necessary to ensure that the maximum ratings are not being exceeded. A region of operation where transistor can operate safely is shown in figure 5.41.

The collector current I_C should be limited to $I_{C(max)}$ and the collector to emitter voltage should be limited to $V_{CE(max)}$ which is also called as V_{CEO} or $V_{(BR)\,CEO}$ in the specification sheet. The maximum power dissipation curve is obtained by taking the product.

$$P_{C\,max} = V_{CE} \times I_C$$

Various specifications of a transistor are as shown in table 5.4.

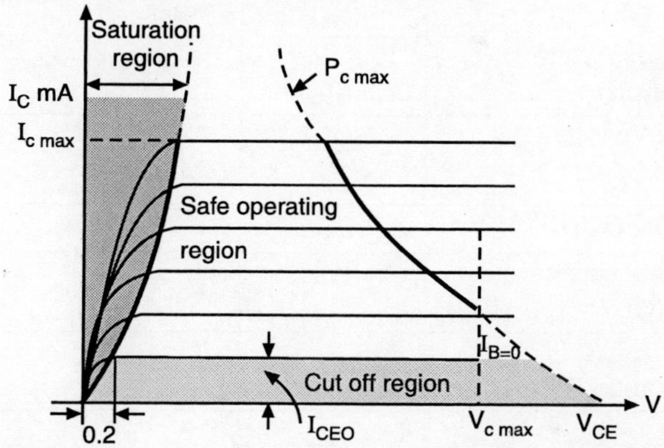

Fig. 5.41. Limits of operation of a transistor in CE configuration.

5.26 BJT SPECIFICATIONS

Table 5.4: Specifications or data sheet for 2N 4123 genral purpose transistor.

Maximum rating

Rating	Symbol	2N4123	Unit
Collector-emitter voltage	V_{CEO}	30	*Vdc*
Collector-Base Voltage	V_{EBO}	5.0	*Vdc*
Collector Current-Continuous	I_C	200	mAdc
Total device Dissipation @ T_A = 25°C Derate above 25°C	P_D	625 5.0	mW mW°C
Operating and storage junction temperature Range	T_j, T_{stg}	– 55 to + 150	°C

Thermal Charactristics

Characteristic	Symbol	Max	Unit
Thermal resistance, junction to case	$R_{\theta JC}$	83.3	°CW
Thermal Resistance, Junction to Ambient	$R_{\theta JA}$	200	°CW

ELECTRICAL CHARACTERISTICS (T_A = 25°C unless otherwise noted)

Characteristic	Symbol	Min	Maximum	Unit
OFF CHARACTERISTIC				
Collector-Emitter Breakdown Voltage (*i*) (I_C = 1.0 mAdC, I_E = 0)	$V_{(BR)CBO}$	40	...	*Vdc*
Collector-Base Breakdown Voltage (I_C = 1.0 μAdC, I_E = 0)	$V_{(BR)CBO}$	40	...	*Vdc*
Emitter-Base Breakdown Voltage (I_F = 1.0 μAdC, I_C = 0)	$V_{(BR)EBO}$	5.0	...	*Vdc*
Collector Cutoff Current (V_{CB} = 2.0 Vdc, I_E = 0)	I_{CBO}	...	50	*nAdc*
Emitter Cutoff Current (V_{BE} = 3.0 Vdc, I_C = 0)	I_{EBO}	...	50	*nAD*
ON CHARACTERISTIC				
DC Current Gain (1) (I_C = 2.0 mAdc, V_{CE} = 1.0 Vdc) (I_C = 50 mAdc, V_{CE} = 1.0 Vdc)	h_{FE}	50 25s	150	...
Collector-Emitter Saturation Voltage (1) (I_C = 50 mAdc, I_B = 5.0 mAdc)	$V_{CE(sat)}$	...	0.3	*Vdc*
Base-Emitter Saturation Voltage (1) (I_C = 50 mAdc, I_B = 5.0 mAdc)	$V_{BE(sat)}$	...	0.95	*Vd*
SMALL-SIGNAL CHARACTERISTICS				
Current-Gain-Bandwidth Product (I_C = 10 mAdc, V_{CE} = 20 Vd, *f* = 100 MHz)	f_T	250	...	MHz
Output Capacitance (V_{CB} = 5.0 Vdc, I_C = 0.*f* = 100 kHz)	C_{obo}	...	4.0	pF
Input Capacitance (V_{BE} = 0.5 Vdc, I_C = 0.*f* = 100 kHz)	C_{ibo}	...	8.0	pF

Collector-Base Capacitance (I_E = 2.0 V_{CB}, = 5.0, f = 100 kHz)	C_{cb}	...	4.0	pF
Small-Signal Current Gain (I_C = 2.0 mAdc V_{CE}, = 10 Vdc, f = 1.0 kHz)	h_{fe}	50	200	...
Current Gain-High frequency (I_C = 10 mAdc, V_{CE} = 20 Vdc, f = 100 kHz) (I_C = 2.0 V_{CE}, = 10 V, f = 1.0 kHz)	h_{fe}	2.5 5.0	200	...
Noise Figure (I_C = 100 μAdc, V_{CE}, = 5.0 Vdc, R_S = 1.0 k ohm, f = 1.0 kHz)	NF	...	6.0	dF

5.27. IMPORTANT FORMULAE

Table 5.5 Important formulae

S. No.	Formula/Expression
1.	$I_E = I_C + I_B$
2.	$I_C = \alpha_{dc} I_E + I_{CBO}$
3.	$\alpha_{dc} = \frac{I_C}{I_E}$
4.	$I_B \cong (1 - \alpha_{dc}) I_E$
5.	$\beta_{dc} = \left[\frac{\alpha_{dc}}{1 - \alpha_{dc}}\right]$
6.	$I_C = \beta_{dc} I_B + (1 + \beta_{dc}) I_{CBO}$
7.	$I_{CEO} = (1 + \beta_{dc}) I_{CBO}$
8.	$\alpha_{dc} = \frac{\beta_{dc}}{1 + \beta_{dc}}$
9.	$\beta_{dc} \cong \frac{I_C}{I_B}$
10.	$\frac{I_E}{I_B} = (1 + \beta_{dc})$

FEW IMPORTANT SOLVED EXAMPLES

EXAMPLE 5.8 A transistor has an emitter current of 10 mA and a collector current of 9.95 mA. Calculate its base current.

Solution: Given that

$$I_E = 10 \text{ mA} \quad \text{and} \quad I_C = 9.95 \text{ mA}$$

Emitter current is given by

$$I_E = I_C + I_B$$

or

$$I_B = I_E - I_C = 10 - 9.95 = 0.05 \text{ mA} \quad \textbf{Ans.}$$

EXAMPLE 5.9 Differentiate between I_{CO} and I_{CBO}. What is the effect of temperature on I_{CBO}?

Solution: I_{CO} is a reverse saturation current in a P-N-junction when it is reverse biased. I_{CBO} is a leakage current in a common emitter transistor which flows from collector to base due to minority charge carriers, when emitter is common to both input and output circuit.

Effect of Temperature on I_{CBO}

When the temperature of the junction increases, the number of minority carriers increases and hence, the current I_{CBO} increases. Therefore, it can be concluded that on increasing the temperature, the current I_{CBO} increases.

EXAMPLE 5.10 **Draw N-P-N and P-N-P transistors. Label all the currents and show the direction of flow. How are all the currents of a transistor related?**

Solution: NPN Transistor

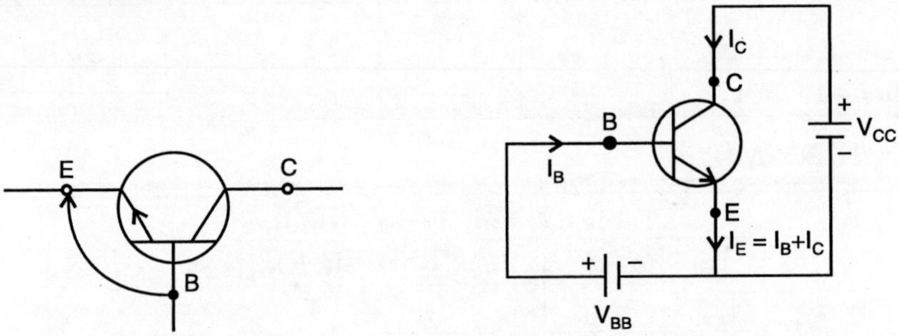

(a) Symbol (b) Common emitter NPN transistor

Fig. 5.42.

PNP Transistor

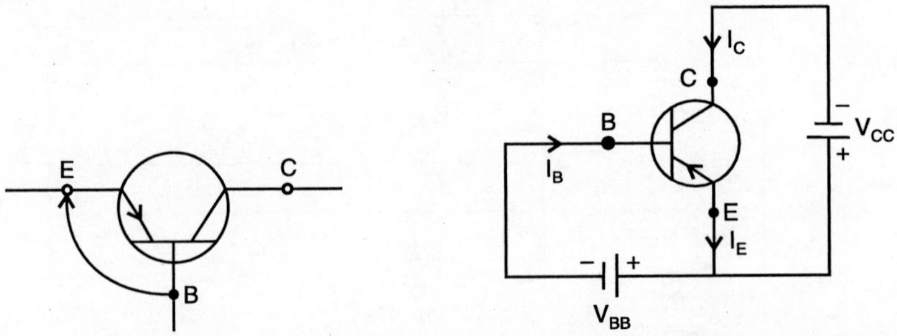

(a) Symbol (b) Common emitter PNP transistor

Fig. 5.43.

The currents of a transistor are related by the following relation:

$$I_E = I_B + I_C$$

where I_E = emitter current

I_C = collector current

I_B = base current.

EXAMPLE 5.11 **Find the value of β, V_{CC} and R_B in the given circuit shown in figure 5.44.**

Solution: Given that

$$I_B = 20\ \mu A = 0.02\ mA$$

$$R_C = 2.7\ k\Omega$$

$$V_{CE} = 7.3\ V,\ V_E = 2.1\ \text{volt}$$

$$R_E = 0.68\ k\Omega,\ V_{BE} = 0.7\ V\ \text{(assumed)}$$

From figure 5.44, we have

$$I_E R_E = V_E$$

or

$$I_E = \frac{V_E}{R_E}$$

Substituting values, we get

$$I_E = \frac{2.1}{0.68} = 3.09 \text{ mA}$$

Also, since $I_E = I_C + I_B$

Therefore, $I_C = I_E - I_B = 3.09 - 0.02$

or $I_C = 3.07 \text{ mA}$

Further, $I_C = \beta I_B$

So that $\beta = \frac{I_C}{I_B} = \frac{3.07}{0.02} = 154$

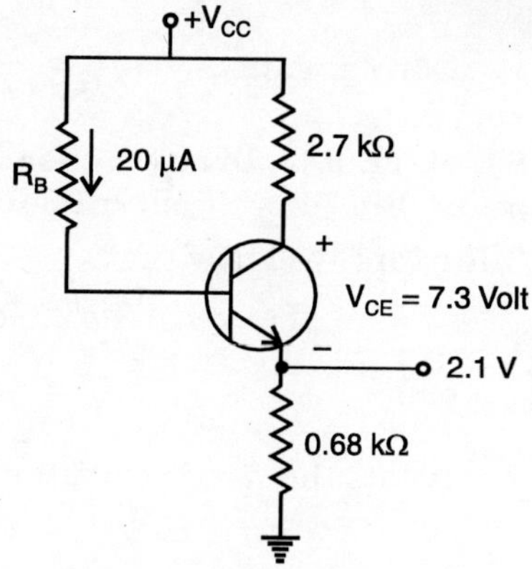

Fig. 5.44.

Applying KVL to the output side of the given circuit, we get

$$V_{CC} = I_C R_C + V_{CE} + V_E$$

Substituting values, we get

$$V_{CC} = 3.07 \times 2.7 + 7.3 + 2.1 = 17.7 \text{ Volts}$$

Also, applying KVL to the input side, we obtain

$$V_{CC} = I_B R_B + V_{BE} + V_E$$

or $$R_B = \frac{V_{CC} - V_{BE} - V_E}{I_B}$$

Putting values, we get

$$R_B \doteq \frac{17.7 - 0.7 - 2.1}{0.02 \times 10^{-3}} = 745 \text{ k}\Omega \quad \textbf{Ans.}$$

EXAMPLE 5.12 A transistor has an emitter current of 10 mA and a collector current of 9.95 mA. Calculate its base current.

Solution: Given, emitter current $I_E = 10$ mA

and collector current $I_C = 9.95$ mA

We know that $I_E = I_C + I_B$

$\therefore$ $I_B = I_E - I_C$

Then, base current = 10 – 9.95 = 0.05 mA **Ans.**

EXAMPLE 5.13 What is base current in the following figure 5.45 and if $I_C = 5$ mA, $I_B = 0.02$ mA, what is current gain?

Solution: Applying KVL to the base emitter section, we get

$$V_{BB} - I_B R_B - V_{BE} = 0$$

$\because$ $V_{BB} = 10 \text{ V}, R_B = 330 \text{ K}$

and $V_{BE} = 0.7 \text{ V (for } S_i)$

Therefore, $$I_B = \frac{V_{BB} - V_{BE}}{R_B} = \frac{10 - 0.7}{330} = \frac{9.3}{330}$$

or $I_B = 0.028 \text{ mA}$

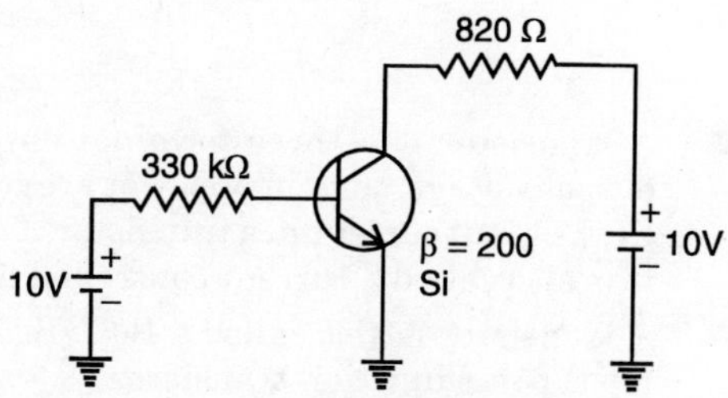

Fig. 5.45.

Now, current gain is given by

$$\beta = \frac{I_C}{I_B} = \frac{5}{0.02} = 250 \quad \textbf{Ans.}$$

EXAMPLE 5.14 If $I_C = 5$ mA and $I_B = 0.02$ mA, what is current gain?

Solution: Given $I_C = 5$ mA

and $I_B = 0.02$ mA

Current gain $\beta = \frac{I_C}{I_B} = \frac{5 \times 10^{-3}}{0.02 \times 10^{-3}} = 250$ **Ans.**

EXAMPLE 5.15 Draw the load line for the following figure 5.46. What is I_C at saturation point. Find V_{CE} at cut-off point.

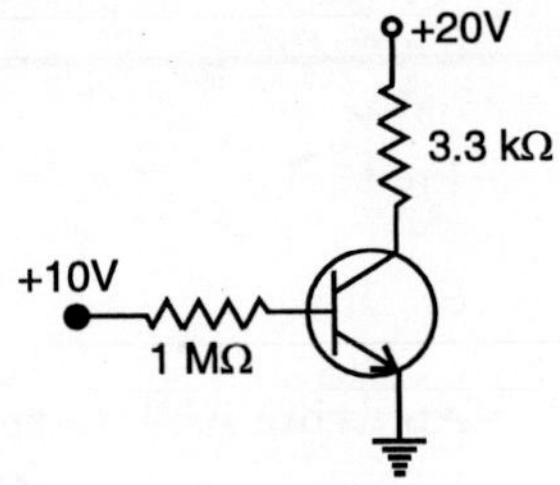

Fig. 5.46.

Solution: We know that

$$V_{CC} = I_C R_C + V_{CE}$$

or
$$I_C = -\frac{1}{R_C} V_{CE} + \frac{V_{CC}}{R_C}$$

It cuts the X-axis on output characteristic at

$$V_{CE} = V_{CC} \{V_{CE} \text{ at cut off}\}$$

$$V_{CE} = 20 \text{ V}$$

It cuts the Y-axis on output characteristic

$$I_C = \frac{V_{CC}}{R_C} = \frac{20}{3.3\ K}$$

$$I_C = 6.06 \text{ mA}$$

Load line can be drawn by joining two points having coordinates $\left(0, \frac{V_{CC}}{R_C}\right)$ and (V_{CC}, 0) *i.e.*, (0, 6.06 mA) and (20 V, 0).

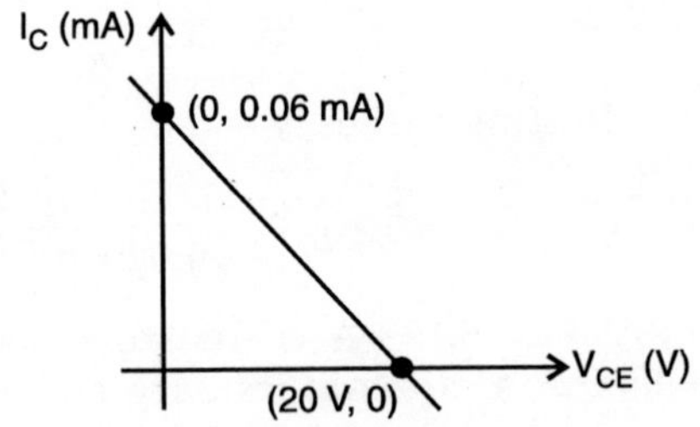

Fig. 5.47.

Here, we can plot load line as shown in figure 5.47 and slope of load line is

$$= -\frac{1}{R_C}$$

$$= -\frac{1}{3.3} = -0.3$$

Here, the co-ordinates at Y-axis and X-axis, represent the saturation of I_C and cut-off of V_C respectively.

$$V_{BE(sat)} = 0.8 \text{ V and } V_{CE(sat)} = 0.2$$

SUMMARY

- A transistor is a three terminal device. The output voltage, current or power are controlled by the input current in a transistor. Therefore, it is also called a current-controlled device.
- A transistor is also called a BJT. BJT stands for Bipolar junction transistor.
- A transistor has a very important property that it can raise the strength of an input weak signal.
- A transistor has three regions namely emitter, base and collector.
- When no battery is connected across the different terminals of a transistor then the transistor is said to be in an unbiased state or in an open-circuit state.
- When dc voltages are applied across the different terminals of a transistor, then this process is called biasing.
- There are two junctions in a transistor namely emitter-base junction and collector-base junction.
- There are four possible ways of biasing a transistor. They are as under:
  (*i*) active region,
  (*ii*) saturation region

(*iii*) cut-off region,

(*iv*) inverted region.

- A transistor can be treated as a two-port network.
- There are three transistor configurations namely Common-Base (*CB*) configuration, Common-Emitter (*CE*) configuration and Common-Collector (*CC*) configuration.
- A transistor can be used as an amplifier.
- In *CB* configuration the d.c. current gain is given as

$$\alpha = \frac{I_C}{I_E}$$

- In *CB* configuration, the a.c. current gain is given as

$$\alpha_0 = \left.\frac{\Delta I_C}{\Delta I_E}\right|_{hfb = \text{constant}}$$

SHORT QUESTIONS WITH ANSWERS

Q. 1. Write few applications of BJT.

Ans. 1. Transistor is used in control systems.

2. Transistor is used in satellites and mobile phones.

3. In digital computer electronics, the transistor is used as a high-speed electronic switch.

4. In communication systems, it is widely used as the primary component in the amplifier.

Q. 2. What do you mean by biasing of a transistor? Explain.

Ans. When dc voltages are applied across the different terminals of a transistor, then this process is called biasing. Since, there are two junctions in a transistor namely emitter-base junction and collector-base junction, each of these two junctions may be forward-biased or reverse-biased. Therefore, there are four possible ways of biasing these two junctions. These possible ways are also called modes of operation of a transistor.

Q. 3. What is common collector (CC) configuration? Explain.

Ans. In this configuration, the input is applied between base and collector whereas the output signal is taken across the emitter and collector.

This means that the collector of the transistor is common to both input and output circuits and hence the name common-collector configuration.

Q. 4. What do you mean by operating point? Explain.

Ans. The zero signal values (no a.c. signal is applied) of collector current I_C and collector-to-emitter voltage V_{CE} are known as the **operating point.**

Since the variations of I_C and V_{CE} take place about this point, therefore, this is known as the operating point.

This point is also called quiescent (stable) point or simply *Q*-point since it is a point on output characteristics curve when the transistor is in a silent condition (*i.e.* the absence of a.c. input signal).

Q. 5. Why biasing is required? Explain.

Ans. The transistor is used in a wide number of applications. The most basic application of a transistor is amplification. Amplification means to raise the strength or amplitude of a weak signal without any change in its original shape. However, for faithful amplification two things are necessary. Firstly, the transistor must operate in active region. It means that the emitter-base junction is forward-bias and collector-base junction is reverse bias. Active region for amplification is must because amplification is a linear process and transistor operates linearly only in active region. Secondly, we should get a fixed dc collector current at a fixed dc collector voltage. These fixed values of dc collector current and dc collector voltage are expressed by a term called operating point or quiescent point or *Q*-point.

Therefore, the primary aim of biasing is to obtain a fixed collector current at a fixed collector voltage. The circuits used for transistor biasing are simply called biasing circuits.

Q. 6. What is bias stabilization? Explain.

Ans. We know that for proper operation of transistor, the operating point should be properly fixed. However, only the fixing of operating point is not sufficient. The operating point must remain stable throughout the operation where it was fixed originally. This type of shift of operating point may drive the transistor into an unwanted region. Therefore, bias stabilization is of utmost importance. Following are the factors which are responsible for the shift of operating point.

Q. 7. What is stability factor? Explain.

The stability factor is expressed as the rate of change of collector current with respect to the reverse saturation current while keeping the common-emitter current gain β and base current I_B as constant.

Mathematically, the stability factor may be stated as

$$S = \frac{dI_C}{dI_{CO}}$$

The stability factor S is a measure of bias stability of a transistor circuit. It may be observed that higher the value of stability factor, poor is the stability whereas lower is the value of stability factor, good is the stability.

Q. 8. What are the requirements of biasing circuits? Explain.

Ans. Following are the requirements of a biasing circuit:

(*a*) A biasing circuit must set the operating point in the middle of the active region of the transistor characteristics.

(*b*) A biasing circuit must stabilize the collector current against temperature variations.

(*c*) A biasing circuit must ensure that the operating point is independent of the transistor parameters such as β so that operating point is not shifted if the transistor is replaced by another transistor of same type.

Q. 9. What are the methods of transistor biasing? Explain.

Ans. There are following four methods of transistor biasing which are most widely used:

1. Fixed bias
2. Collector to base bias
3. Self bias or voltage divider bias
4. Emitter bias

Q. 10. Define self bias or voltage divider bias?

Ans. For fixed bias circuit and collector-to-base bias circuit, the operating point or the values of d.c. collector current and collector voltage depend upon the value of current gain β of transistor. But the value of β is temperature dependent particularly for silicon transistors. Also, the value of β changes from transistor to transistor. This means that for same type of two transistors the values of β are different.

Q. 11. Explain why an ordinary transistor is called bipolar?

Ans. Because the transistor operation is carried out by two types of charge carriers (majority and minority carriers), an ordinary transistor is called bipolar.

Q. 12. What is done to the base region of a transistor to improve its operation ?

Ans. Base is made thin and very lightly doped in comparison to either emitter or collector so that it may pass most of the injected charge carriers to the collector.

Q. 13. Why transistor is called current controlled device ?

Ans. The output voltage, current or power is controlled by the input current in a transistor so it is called the current controlled device.

Q. 14. Why silicon type transistor are more often used than germanium type ?

Ans. Because, silicon has smaller cut-off current I_{CBO}, small variations in I_{CBO} due to variations in temperature as compared to those in case of germanium.

Q. 15. Why the width of the base region of a transistor is kept very small compared to other regions ? *(A.M.I.E. Examination, 1994)*

Ans. Base region of a transistor is kept very small and very lightly doped so as to pass most of the injected charge carriers to the collector.

REVIEW QUESTIONS

1. What is a Bipolar junction transistor?
2. Explain the different configurations of BJT.
3. Explain the input and output characteristics of a transistor in *CB* configuration.
4. Define α and β for a transistor.
5. Derive the relationship between α and β.
6. What is meant by *Q*-point?
7. What is the need for biasing a transistor?
8. Explain the requirements of a biasing circuit.
9. What are the factors responsible for the instability of operating point.
10. Draw a fixed bias circuit and obtain the value of d.c. Voltage and currents in the circuit.
11. Define, stability factor. Derive general expression for stability factor of a biasing circuit.
12. Draw a self-bias circuit and derive an expression for stability factor.
13. State various methods of improving stability.
14. Define Stability factor with respect to transistor biasing.
15. Explain the function of emitter in the operation of a junction transistor.

NUMERICAL PROBLEMS

1. For a fixed bias circuit using an NPN transistor, the value of β is 80. If $R_B = 390$ kΩ, R_C = 1.5 kΩ and V_{CC} = 30 V, find the co-ordinates of the Q-point. **(Ans. 21 V, 6 mA)**
2. Find the values of β, V_{CC} and R_B of the following circuit.

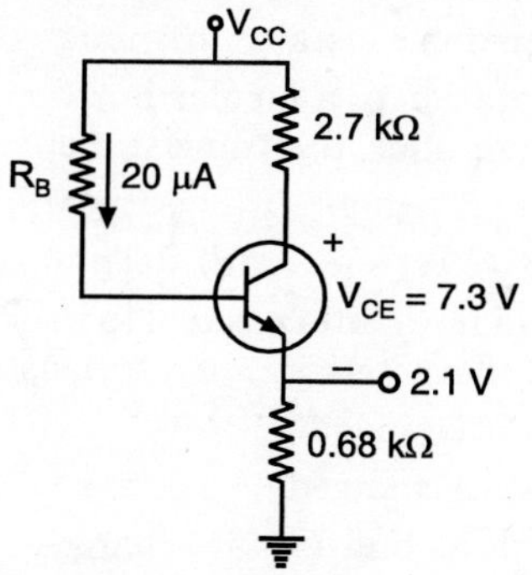

Fig. 5.48.

3. For a fixed bias circuit using a germanium NPN transistor, the value of β is 100. If V_{BE} = 0.2 V, V_{CC} = 16 V, R_C = 5 kΩ and R_B = 790 kΩ then determine its operating point.
4. Find I_C and V_{CE} for the following circuit. What happens to V_{CE} if β increases due to temperature?

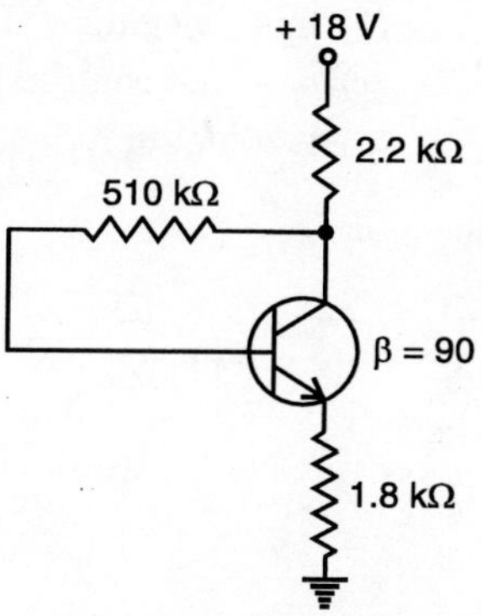

Fig. 5.49.

5. Find the value of operating point for the circuit of a potential divider bias arrangement with $R_2 = R_C = 5$ kΩ, $R_E = 1$ kΩ and $R_1 = 40$ kΩ. **(Ans. V_{CE} = 6 V, I_C = 1 mA)**
6. Consider a DC bias circuit with voltage feedback, given below. Determine the quiescient levels of I_{CQ} and V_{CEQ}. The β of the transistor is 90 and cut-in voltage is 0.7 V. What will be the effect on I_{CQ} and V_{CEQ} if β is increased by 50%.

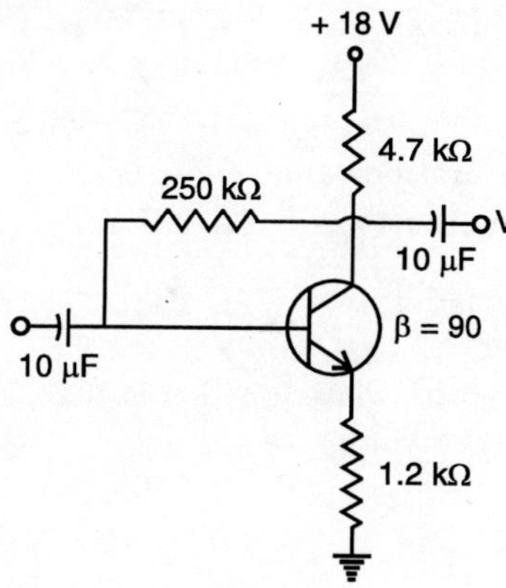

Fig. 5.50.

OBJECTIVE TYPE QUESTIONS

Choose one correct answer in the following:

1. The three terminals of a bipolar junction transistor are called
   (*a*) p,n,p (*b*) n,p,n
   (*c*) anode, cathode and gate
   (*d*) emitter, base and collector
2. For an n-p-n transistor, the n-regions are
   (*a*) emitter and base (*b*) base and collector
   (*c*) emitter and collector
   (*d*) none
3. For operation of p-n-p amplifier, the base of the amplifier must be
   (*a*) 0 V
   (*b*) positive with respect to collector
   (*c*) negative with respect to collector
   (*d*) greater than the collector current
4. The bias condition of a transistor for being used as an amplifier is
   (*a*) forward-reverse (*b*) reverse-reverse
   (*c*) reverse-forward (*d*) forward-forward

5. If β_{dc} is 50, the value of α_{dc} is
   (*a*) 51 (*b*) 50
   (*c*) $\frac{50}{51}$ (*d*) $\frac{1}{50}$
6. The relationship between α_{dc} and β_{dc} is given by
   (*a*) $\beta_{dc} = \frac{\alpha_{dc}}{1+\alpha_{dc}}$ (*b*) $\beta_{dc} = \frac{\alpha_{dc}}{1-\alpha_{dc}}$
   (*c*) $\alpha_{dc} = \frac{1+\beta_{dc}}{\beta_{dc}}$ (*d*) $\alpha_{dc} = \frac{\beta_{dc}}{\beta_{dc}-1}$
7. In cut-off, the value of V_{CE} is
   (*a*) 0 V (*b*) V_{CC}
   (*c*) minimum (*d*) $V_{BE} = 0.6$ V
8. The saturation value of V_{CE} is
   (*a*) equal to V_{CC} (*b*) 0 V
   (*c*) 0.7 V (*d*) $\frac{V_{CC}}{2}$
9. If the emitter-base junction is open, the value of collector voltage is
   (*a*) 0 V (*b*) 0.2 V
   (*c*) Floating (*d*) V_{CC}
10. If in a transistor $(I_C)_{sat} < I_B \beta_{dc}$, the transistor is in
    (*a*) cut off (*b*) saturation
    (*c*) amplification state
    (*d*) none
11. As temperature is increased, the *h*-parameters h_{ie}, h_{re}, h_{fe} and h_{oe} of a transistor
    (*a*) decreases (*b*) increases
    (*c*) remains constant (*d*) none
12. For operation of a bipolar junction transistor as an amplifier, the transistor operates in
    (*a*) cut off region (*b*) saturation region
    (*c*) active region (*d*) deep in saturation
13. The load line shifts parallel to itself on the CE output characteristics of a transistor when
    (*a*) R_L is changed
    (*b*) V_{CC} is changed
    (*c*) both R_L and V_{CC} are changed
    (*d*) base current is changed
14. The dc load line of a transistor
    (*a*) has a positive slope
    (*b*) has a negative slope
    (*c*) does not contain the Q-point
    (*d*) is a plot of I_C versus I_B

Answers

1. (*d*)	2. (*c*)	3. (*b*)	4. (*a*)	5. (*c*)	6. (*b*)
7. (*b*)	8. (*a*)	9. (*d*)	10. (*b*)	11. (*b*)	12. (*c*)
13. (*b*)	14. (*b*)				

❑❑❑

UNIT—IV

Transistor Biasing and Stabilization

Operating point, The DC and AC Load lines, Need for Biasing, Fixed Bias, Collector Feedback Bias, Emitter Feedback Bias, Collector-Emitter Feedback Bias, Voltage Divider Bias, Bias Stability, Stabilization Factors, Stabilization against variations in V_{BE} and β, Bias Compensation using Diode and Transistors, Thermal Runaway, Thermal Stability.

Above Syllabus has been covered in following Chapter:

CHAPTER 6 : Transistor Biasing and Stabilization

Chapter 6

Transistor Biasing and Stabilization

SYLLABUS

Operating point, The DC and AC Load lines, Need for Biasing, Fixed Bias, Collector Feedback Bias, Emitter Feedback Bias, Collector-Emitter Feedback Bias, Voltage Divider Bias, Bias Stability, Stabilization Factors, Stabilization against variations in V_{BE} and β, Bias Compensation using Diode and Transistors, Thermal Runaway, Thermal Stability.

Chapter Outline

6.1. BASIC COMMON EMITTER AMPLIFIER AND D.C. LOAD LINE

Figure 6.25 shows an amplifier circuit in common emitter configuration using NPN transistor. The battery V_{BB} is used to forward bias the emitter-base junction and battery V_{CC} is used to reverse bias the collector-base junction. The resistor R_B is connected to limit the input or base current. The resistor R_C acts as a load resistance. The amplified a.c. output voltage appears across load resistor R_C.

The input a.c. signal to be amplified is v_i. This is applied to the base through capacitor C_1. The capacitors C_1 and C_2 are called coupling capacitors. The coupling capacitor C_1 blocks d.c. Hence, d.c. current flows only through R_B and is prevented from reaching the input signal source. The coupling capacitor C_2 also blocks d.c. from reaching the output terminals. Therefore, at the output terminals only amplified output a.c. signal voltage V_o appears.

DO YOU KNOW?

Centering the Q-point on a transistor load line is important because it allows for the maximum ac output voltage from the amplifier. Centering the Q-point on the dc load line is sometimes referred to as mid point bias.

* Discussion is an exchange of knowledge; argument an excange of ignorance. — **Robert Quillen**

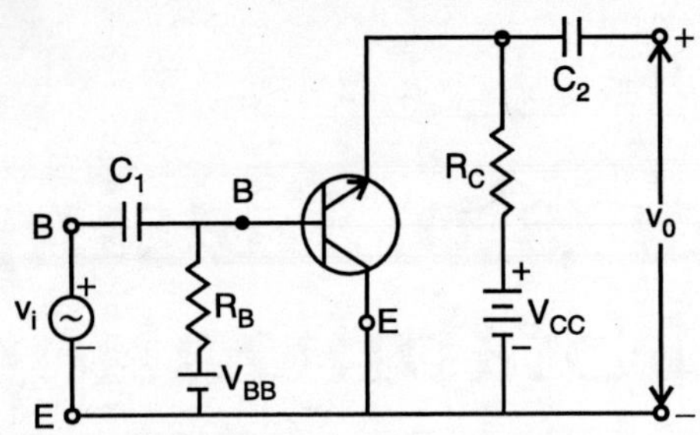

Fig. 6.1. Basic common emitter amplifier

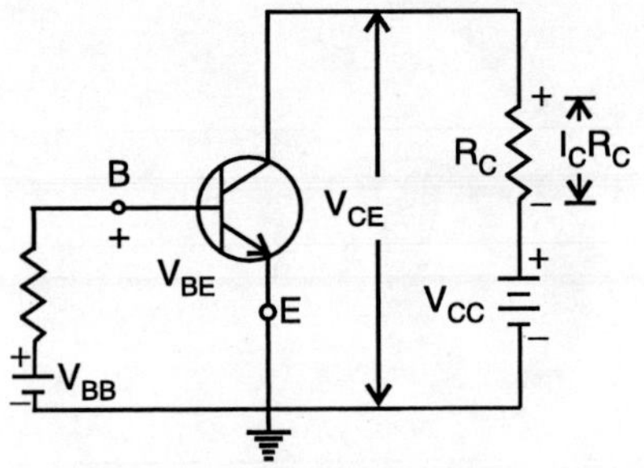

Fig. 6.2. Basic common emitter (D.C. equivalent circuit)

6.2. D.C. LOAD LINE

With the help of d.c. load line, we can easily analyse the performance of an amplifier circuit. In this section, we will learn how to draw a d.c. load line.

For this, first we will find the d.c. equivalent circuit of the amplifier circuit shown in figure 6.3.

This condition of having no input a.c. signal is also known as quiescent condition.

Figure 6.2 shows the d.c. equivalent circuit of one shown in figure 6.1. This circuit is obtained by opening the two capacitors and make the value of signal source equal to zero. For d.c. the capacitor works as an open circuit.

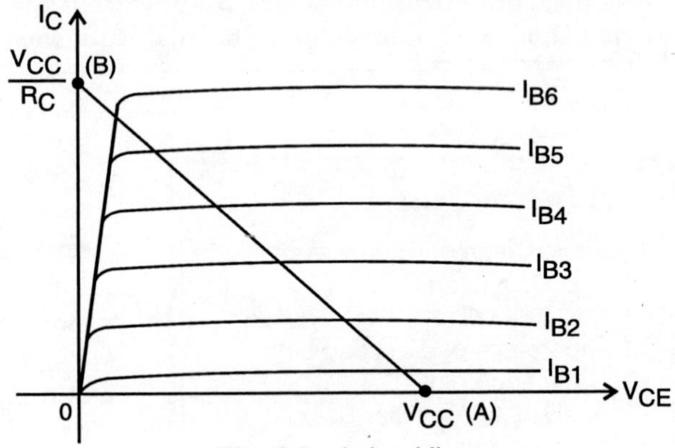

Fig. 6.3. dc load line

Applying KVL to the collector portion of the circuit, we have

$$I_C R_C + V_{CE} = V_{CC}$$

or
$$I_C R_C = -V_{CE} + V_{CC}$$

or
$$I_C = \left(-\frac{1}{R_C}\right) V_{CE} + \frac{V_{CC}}{R_C} \quad \text{...(6.1)}$$

The above equation is in the form of

$$Y = mx + c \quad \text{...(6.2)}$$

The equation (6.2) is an equation of straight line.

Comparing equation (6.1) and (6.2) slope of straight line, we have

$$m = -\frac{1}{R_C}$$

Therefore, if equation (6.1) is plotted on the output characteristics curves, we will get a straight line having slope equal to $\left(-\frac{1}{R_C}\right)$ and intercept on vertical axis equal to $\left(\frac{V_{CC}}{R_C}\right)$.

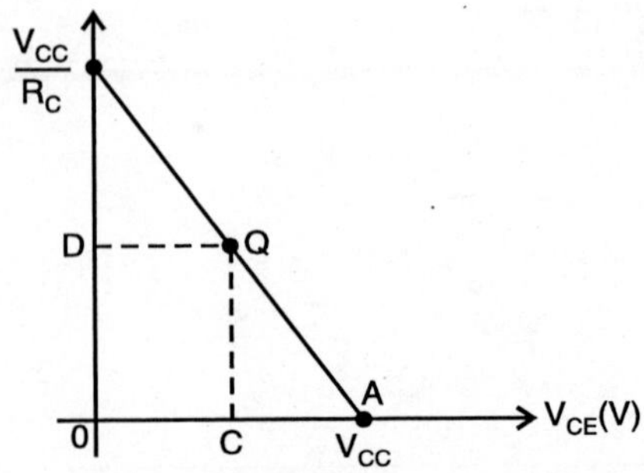

Fig. 6.4.

This line plotted on the output characteristics is called d.c. load line.

To plot the d.c. load line, we use the concept that on V_{CE} axis, the I_C is zero and on I_C axis, the V_{CE} is zero.

Substituting these conditions in the equation (6.32), we get

$$V_{CC} = V_{CE} + I_C R_C$$

when $I_C = 0 \Rightarrow V_{CE} = V_{CC}$

Point $A \equiv (V_{CC}, 0)$

when $V_{CE} = 0 \Rightarrow V_{CC} = I_C R_C$

or $$I_C = \frac{V_{CC}}{R_C}, \text{ Point } B \equiv \left(0, \frac{V_{CC}}{R_C}\right)$$

If two points are located on the output characteristics and joined by a straight line, we get a d.c. load line. This line is called d.c. load line because its slope depends upon the value of resistor R_C which is the d.c. load of the circuit.

6.3. IMPORTANCE OF D.C. LOAD LINE

The collector or output current I_C and collector to emitter voltage V_{CE} conditions in the transistor circuit are represented by a point on the output characteristics curve. This same information may be obtained from the d.c. load line.

Hence, when I_C is maximum $\left(I_C = \frac{V_{CC}}{R_C}\right)$ then V_{CE} will be zero ($V_{CE} = 0$) and when $I_C = 0$, then V_{CE} is maximum ($V_{CE} = V_{CC}$) as shown in figure 6.4.

For some other value of output or collector current I_C (say *OD*), the collector-to-emitter voltage V_{CE} will be equal to *OD*. It means that d.c. load line provides a more convenient and direct solution to the problem.

From above, it also follows that if we plot the d.c. load line on the output characteristics curve of the transistor, we may investigate the behaviour of transistor amplifier.

6.3.1. AC Load Line

Now, let us restrict our discussion to dc biasing in the absence of ac signal. However, later on, we shall understand how to use the transistor as an amplifier. Figure 6.5 shows an amplifeir circuit partially. The dc load resistance is R_C but the AC load resistance is ($R_C \,||\, R_L$). If a load line is drawn, the slope of which is $- 1/(R_C \,||\, R_L)$, then it is called as an AC load line and it is to be used when the transistor is operating as an amplifier. The AC load line, thus, represents the AC operating conditions of a circuit.

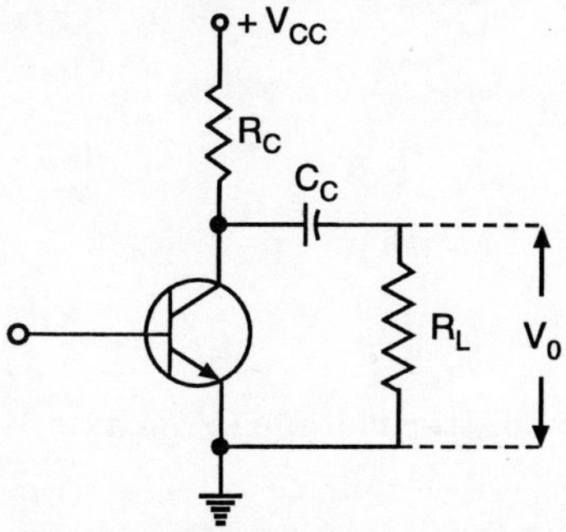

Fig. 6.5 Amplifier circuit drawn partially

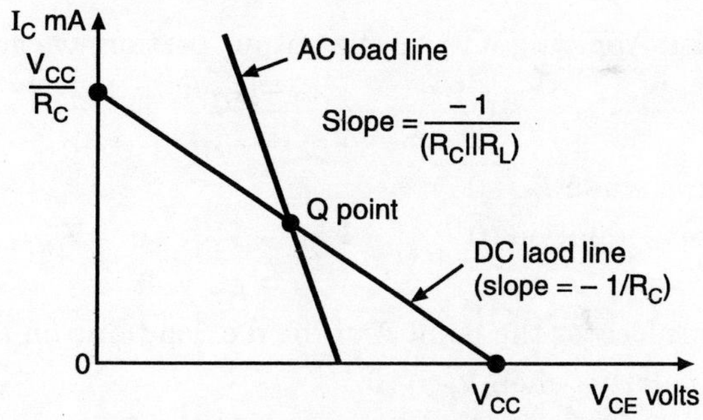

Fig. 4.6 AC and DC load lines.

The parallel combination ($R_C \,||\, R_L$) is always less than R_C. Therefore, the slop of AC laod line is higher than that of the DC load line as shown in figure 4.6.

6.4. OPERATING POINT

The zero signal values (no a.c. signal is applied) of collector current I_C and collector-to-emitter voltage V_{CE} are known as the **operating point.**

Since the variations of I_C and V_{CE} take place about this point, therefore, this is known as the operating point.

This point is also called quiescent (stable) point or simply Q-point since it is a point on output characteristics curve when the transistor is in a silent condition (*i.e.* the absence of a.c. input signal).

Let us consider that in the absence of a.c. input signal, the base current is 4 μA. Then, in transistor circuit, the I_C and V_{CE} conditions must be represented by a point on I_B curve *i.e.* $I_B = 4$ μA characteristics curve.

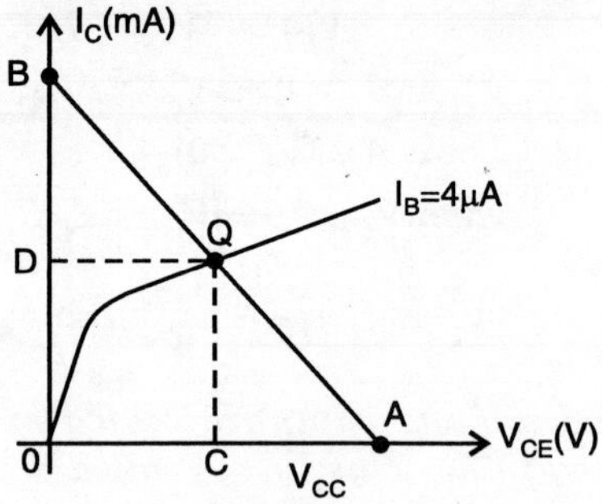

Fig. 6.7.

But at the sametime we know that I_C and V_{CE} conditions in transistor circuit must also be represented by a point on the d.c. load line *AB*. Therefore, it follows that the point *Q* where the d.c. load line and the characteristics curve intersect will be the only point which satisfies both of these conditions.

Hence, the point *Q* describes the actual state of circuit in the zero signal conditions and is known as the operating point.

According to figure 6.7, for $I_B = 4$ μA, the zero signal values are

$$V_{CE} = OC \text{ volts} = 10 \text{ Volts}$$

and

$$I_C = OD \text{ mA} = 6 \text{ mA}$$

Therefore, operating point will be

$$Q \equiv (V_{CE}, I_C) \equiv (10 \text{ V}, 6 \text{ mA})$$

EXAMPLE 6.7 **For the circuit shown in figure 6.8, draw the d.c. load line.**

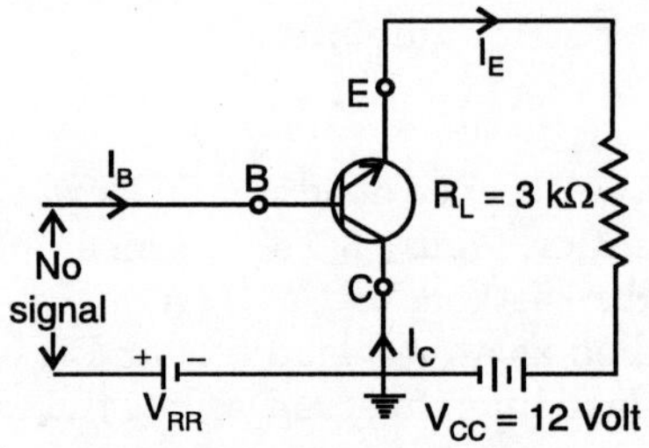

Fig. 6.8.

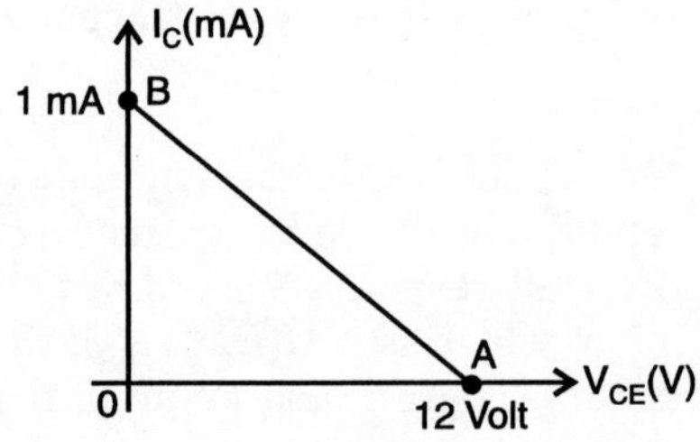

Fig. 6.9.

Solution: Applying KVL in the output portion, we get

$$V_{CC} = I_C R_C + V_{CE}$$

or

$$V_{CE} = V_{CC} - I_C.R_C \qquad ...(i)$$

Now, when $I_C = 0$

From equation (*i*)

$$V_{CE} = V_{CC} - 0 = V_{CC}$$

$$V_{CE} = 12 \text{ Volt}$$

This locates the point A of the d.c. load line on the collector-emitter voltage (V_{CE}) axis.

Similarly, when $V_{CE} = 0$.

From equation (*i*)

$$0 = V_{CC} - I_C.R_C$$

or

$$I_C R_C = V_{CC}$$

or $$I_C = \frac{V_{CC}}{R_C} = \frac{12\text{ V}}{3\text{ k}\Omega} = 4\text{ mA}$$

This locates the point B of the d.c. load line on the collector current (I_C) axis.

Now, by joining these two points A and B, we get the d.c. load line AB as shown in figure 6.9.

6.5. ANALYSIS OF AMPLIFIER USING D.C. LOAD LINE

As discussed earlier, a transistor may be used to amplify ac signals after fixing its operating point suitably.

In last article, we learnt how to fix the Q-point on the output characteristics of a transistor. In fact, the Q-point must lie in the middle portion on the output characteristics of a transistor. This helps the transistor to amplify a.c signals faithfully, *i.e.* without producing distortion in waveshape of input signal.

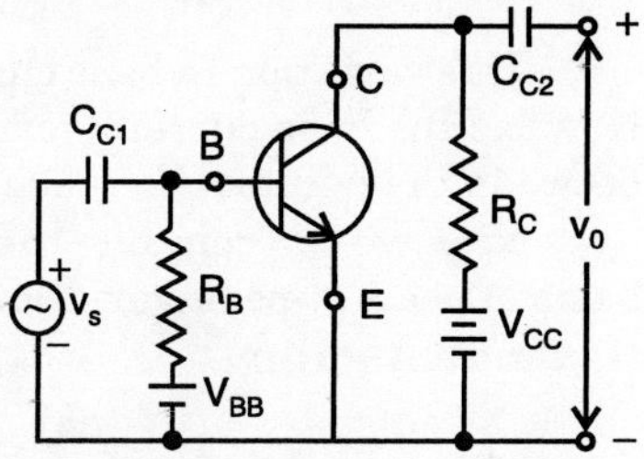

Fig. 6.10. A Basic amplifier circuit

In quiescent condition, the base current is fixed at a constant d.c. value. It is found from the Q-point. After this, when we apply an a.c signal at the input of the amplifier circuit, the base voltage varies as per the signal voltage v_s. Due to this, the base current will also vary. Now, as the base current varies, the instantaneous operating point of the transistor moves along the d.c. load line. This means that the instantaneous values of the collector current and voltage also vary according to the input signal. This variation in collector voltage is several times larger than the variation of the input signal.

Also, this collector-voltage variation reaches the output terminals through coupling capacitor. Thus, the output signal is several times larger than the input signal.

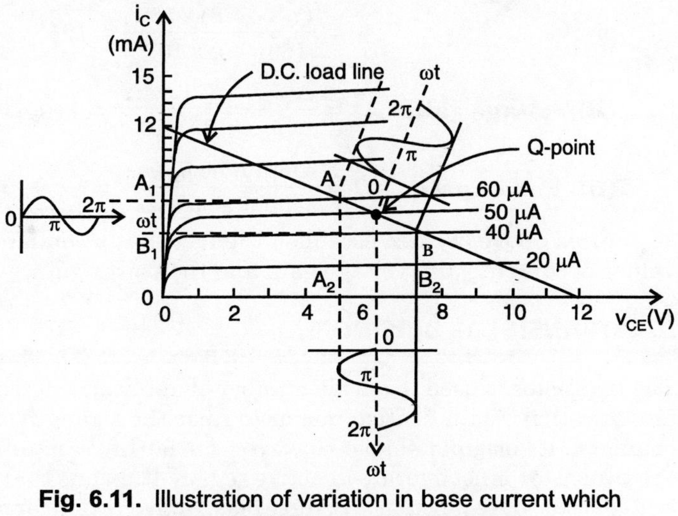

Fig. 6.11. Illustration of variation in base current which produces variation in collector current and voltage in a CE amplifier.

Let us consider an example. Figure 6.10 shows the circuit of a basic amplifier.

For the analysis purpose, let us assume that

$$V_{CC} = 12\text{ V}, R_C = 1\text{ k}\Omega,$$
$$V_{BB} = 10.7\text{ V and } R_B = 200\text{ k}\Omega$$

Here the d.c. base current is found to be 50 μA and the collector d.c current and collector d.c. voltage are 6 mA and 6 V respectively.

Now, let us apply a small a.c. signal voltage of 7 mV at the input.

To show the variations in base current I_B on the output characteristics of the transistor, a line perpendicular to the d.c. load line and passing through the Q-point is drawn. We take this line as ωt-axis and then variations in I_B are plotted.

As the instantaneous operating point moves along the d.c. load line between the points A and B, both the collector-current and collector voltage would vary.

The current i_C varies between the points A_1 (I_C = 7.3 mA) and B_1 (I_C = 4.8 mA). These variations are shown on the left side of the output characteristics. The voltage v_{CE} varies between points A_2 (v_{CE} = 4.9 V) and B_2 (v_{CE} = 7.1 V).

Also the collector-voltage variations are shown at the bottom of the output characteristics.

This voltage would have about 20 mV peak-to-peak variations. Now, if the input dynamic resistance r_i of the transistor is taken to be 1 kΩ, then the input voltage variation would produce a peak-to-peak variation of 20 μA in base current, since

$$\Delta i_B = \frac{20 \text{ mV}}{1 \text{ k}\Omega} = 20 \ \mu A$$

This variation in base current exists around its quiescent value (*i.e.*, operating point) of 50 μA. Now, as the base current varies, the instantaneous operating point moves along the d.c. load line between the points $A(I_B = 60 \ \mu\text{A})$ and $B(I_B = 40 \ \mu\text{A})$.

Now, we can compute the current gain, voltage gain and the power gain of the amplifier on the basis of peak-to-peak variation as under:

(*i*) Current gain:

$$A_i = \frac{\Delta i_C}{\Delta_{iB}}$$

Substituting all the values, we get

or
$$A_i = \frac{(7.3 - 4.8) \times 10^{-3}}{(60 - 40) \times 10^{-6}} = 125$$

(*ii*) Voltage gain: $A_v = \dfrac{\Delta v_{CE}}{\Delta v_{BE}} = \dfrac{7.1 - 4.9}{20 \times 10^{-3}} = 110$

(*iii*) Power gain: $Q = \dfrac{\text{Output ac power}}{\text{Input ac power}} = \dfrac{i_c . v_{ce}}{i_b . v_{be}} = A_i \times A_V = 125 \times 110 = 13750$

From above, we can establish the fact that a common emitter amplifier has sufficiently large values of current gain, voltage gain and the power gain.

6.6. TRANSISTOR BIASING

The transistor is used in a wide number of applications. The most basic application of a transistor is amplification. Amplification means to raise the strength or amplitude of a weak signal without any change in its original shape. However, for faithful amplification two things are necessary. Firstly, the transistor must operate in active region. It means that the emitter-base junction is forward-bias and collector-base junction is reverse bias. Active region for amplification is must because amplification is a linear process and transistor operates linearly only in active region. Secondly, we should get a fixed dc collector current at a fixed dc collector voltage. These fixed values of dc collector current and dc collector voltage are expressed by a term called operating point or quiescent point or Q-point.

Therefore, the primary aim of biasing is to obtain a fixed collector current at a fixed collector voltage. The circuits used for transistor biasing are simply called **biasing circuits.**

6.6.1. Need of Biasing

We have already learnt the basics of a bipolar junction transistor (BJT). In this chapter, we shall go one step ahead to learn the techniques to connect the dc power supplies to the transistor. We shall discuss various such techniques and analyze each one of them. Then the compensation techniqes for the BJT shall be discussed.

What is meant by dc biasing of a transistor?

We know that transistor can operate in any of the three regions of operations namely cutoff, active region and saturation. To operate the transistor in these regions, the two junctions of a transistor should be forward or reverse biased as shown in table 6.1.

Table 6.1 Regions of operation and applications

Region of operation	Base Emitter junction	Collector Base junction	Application
Cutoff	Reverse biasd	Reverse biased	As a switch
Active	Forward biased	Reverse biased	Amplifeir
Saturation	Forward biased	Forward biased	As a switch

In order to do so, we need to connect external dc power supplies with correct polarities and magnitudes. This process is called as **dc biasing of a transistor**.

6.7. SELECTION OF OPERATING POINT*

For faithful amplification of an input signal, a proper or judicious selection of operating point is of utmost importance.

Figure 6.12(*a*) shows the circuit of common emitter configuration and corresponding output characteristics of transistor.

To study the effects of biasing conditions on the performance of a transistor, it is necessary to draw a d.c. load line on the output characteristics of a transistor. Figure 6.12(*b*) shows that a d.c. load line is drawn on the characteristics. The operating point will essentially lie somewhere on this load line. Figure 6.12(*b*) shows three operating points. These three operating points can be set depending upon the values of base current. We will consider three cases to find which one of these three operating points is most suitable.

Case (*i*) If point Q_1 is the operating point, then Q_1 will lie near the saturation region. Therefore, the positive cycle of the signal will be clipped off. Hence, distortion will occur. Thus, the point Q_1 is not the suitable operating point.

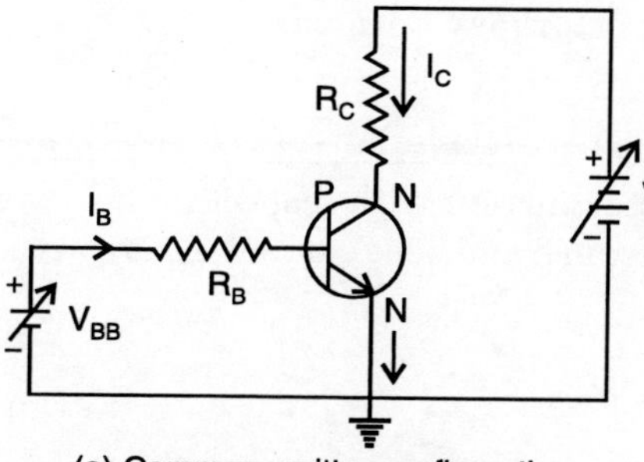

(a) Common emitter configuration

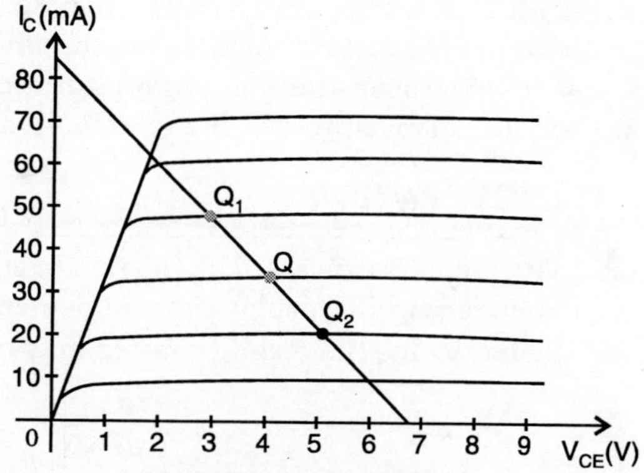

(b) Output characteristics

Fig. 6.12.

Case (*ii*) If Q_2 is considered the operating point, then this point will lie near the cut-off region. Therefore, the negative cycle will be clipped off. Hence, again distortion will occur. So, the point Q_2 is also not the suitable operating point.

Case (*iii*) If Q is chosen the operating point, then this will lie in the active region. Also this point is in the middle of active region. Therefore, the signal will not be clipped off. Hence, there will be no distortion at all. Therefore, point Q is the suitable operating point.

Hence, we can conclude that for faithful amplification, the operating point must lie in the middle of the active region.

* Explain how operating point is selected for amplification in CE mode using graphical method only.

6.8. BIAS STABILIZATION

We know that for proper operation of transistor, the operating point should be properly fixed. However, only the fixing of operating point is not sufficient. The operating point must remain stable throughout the operation where it was fixed originally. This type of shift of operating point may drive the transistor into an unwanted region. Therefore, bias stabilization is of utmost importance. Following are the factors which are responsible for the shift of operating point:

(*i*) Temperature dependence of transistor parameters

We know that the collector current is given as

$$I_C = \beta I_B + (1 + \beta)I_{CBO} \quad ...(6.3)$$

Here
$\beta \rightarrow$ Common emitter current gain
$I_B \rightarrow$ Base current
$I_{CBO} \rightarrow$ leakage current

The above three variables are strongly temperature dependent. Hence, with the increase in temperature, if any of these parameters changes, then collector current changes and so operating is shifted to any other region.

(*ii*) Variation in parameters between transistors of same type: Even after the tremendous advances in semiconductor technology, the transistor parameters such as β vary widely between transistors even of same type. Change in β will shift the operating point in any undesired region.

(*iii*) Thermal runway: The collector current is given as

$$I_C = \beta I_B + (1 + \beta)I_{CBO} \quad ...(6.4)$$

If temperature changes then I_{CBO} changes which in turn changes I_C and operating point. Flow of collector current in the collector circuit produces heat at the collector junction. This raises the temperature. Hence I_{CBO} increases which in turn increases collector current I_C. This again increases the temperature of collector junction and the whole process repeats again. Such a successive increase in I_C, will drive the operating point into saturation region. This process is also called thermal runway. This is very dangerous process due to excessive heat the transistor may burn out.

6.9. STABILITY FACTOR

The stability factor is expressed as the rate of change of collector current with respect to the reverse saturation current while keeping the common-emitter current gain β and base current I_B as constant.

Mathematically, the stability factor may be stated as

$$S = \left.\frac{dI_C}{dI_{CO}}\right|_{\beta \text{ and } I_B = \text{constant}} \quad ...(6.5)$$

The stability factor S is a measure of bias stability of a transistor circuit. It may be observed that higher the value of stability factor, poor is the stability whereas, lower is the value of stability factor, good is the stability.

Mathematical Expression for Stability Factor

Since, the value of collector current in a transistor is given as

$$I_C = \beta.I_B + (1 + \beta)I_{CO} \quad ...(6.6)$$

Here, I_C = Collector current
I_B = Base current

Every man has three characters-that which he exhibits, that which he has and that which he thinks he has.
— **Alpbonse Karr**

I_{CO} = Reverse saturation current

β = Common-emitter current gain

Differentiating equation (6.6) with respect to I_C

$$1 = \frac{d}{dI_C}(\beta . I_B) + \frac{d}{dI_C}[(1+\beta)I_{CO}]$$

Assuming β as constant, we have

$$1 = \beta \times \frac{dI_B}{dI_C} + (1+\beta).\frac{dI_{CO}}{dI_C} = \beta.\frac{dI_B}{dI_C} + (1+\beta).\frac{1}{S} \qquad \left[\because S = \frac{dI_C}{dI_{CO}}\right]$$

or
$$S = S.B\left(\frac{dI_B}{dI_C}\right) + (1+\beta)$$

or
$$S\left[1 - \beta\left(\frac{dI_B}{dI_C}\right)\right] = 1 + \beta$$

or
$$\boxed{S = \frac{1+\beta}{1-\beta\left(\frac{dI_B}{dI_C}\right)}} \qquad ...(6.7)$$

The above expression for stability factor is a general expression for any biasing circuit.

Stability Factor for Common Base Configuration

For common base configuration, the expression for collector current is

$$I_C = \alpha.\ I_E + I_{CO} \qquad ...(6.8)$$

where I_E = Emitter current

I_{CO} = Reverse saturation current

Differentiating equation (*i*) with respect to I_{CO}

$$\frac{dI_C}{dI_{CO}} = 0 + 1$$

$$\frac{dI_C}{dI_{CO}} = 1$$

or
$$\boxed{S = 1} \qquad ...(6.9)$$

The value of stability factor S is unity for common base circuits which means that these circuits are highly stable. Hence, there is no need of bias stabilization in these types of circuits.

Stability factor for common emitter configuration: For common emitter configuration, the expression for collector current is

$$I_C = \beta I_B + (1 + \beta)I_{CO} \qquad ...(6.10)$$

Differentiating the equation (*i*) with respect to I_{CO}

$$\frac{dI_C}{dI_{CO}} = 0 + (1 + \beta) = 1 + \beta$$

or
$$S = 1 + \beta \qquad ...(6.11)$$

Hence, according to above equation if $\beta = 50$ then value of $S = 1 + 50 = 51$

Which indicates that the collector current changes 51 times the change in reverse saturation current. Hence, in a *CE* configuration the collector current highly depends on I_{CO} which in turn depends on the temperature. Therefore, for *CE* configuration bias stabilization is of utmost importance to improve the stability factor.

6.10. REQUIREMENTS OF TRANSISTOR BIASING

6.10.1. Requirements

A transistor biasing circuit is expected to satisfy the following requirements.

(*i*) The transistor should be biased in the active region if it is to be used for amplification and in saturation and cut off if it is to be used as a switch.

(*ii*) The Q point should be adjusted approximately at the center of the load line for voltage amplifier application.

(*iii*) The value of stability factor (S) should be as small as possible.

(*iv*) Q-point should be stabilized by introducing a negative feedback in the biasing circuit.

(*v*) The Q-point shold not be affected due to temperature changes or device to device variation.

(*vi*) Bypass capacitor should be included to avoid reduction in voltage gain due to negative feedback.

(*vii*) Transistor should be biased in the linear region of the transfer characteristics.

6.10.2. Factors to be Considered to Design a Biasing Circuit

Following are some of the important factors to be considered while designing a biasing circuit.

(*i*) Position of a Q-point

(*ii*) Value of I_C (collector current) at quiescent point.

(*iii*) Value of every stabilty factor should be as low as possible.

(*iv*) Transistor should be biased in the linear portion of its transfer characteristics.

(*v*) Forward bias the BE junction and reverse bias the CB junction to bias the transistor in active region.

(*vi*) Maximum output swing without producing any distortion.

6.11. METHODS OF TRANSISTOR BIASING*

There are following four methods of transistor biasing which are most widely used:

1. Fixed bias
2. Collector to base bias
3. Self bias or voltage divider bias
4. Emitter bias

6.12. FIXED BIAS CIRCUIT

Figure 6.13 shows a fixed bias circuit. In this circuit two d.c. batteries V_{BB} and V_{CC} are used. The battery V_{BB} is a low value battery and V_{CC} is a high value battery.

The fixed bias circuit may be replaced with more practical circuit in which a single supply V_{CC} is used. In this circuit, both the base and collector resistors are connected to the positive side of the battery V_{CC} (Figure 6.14).

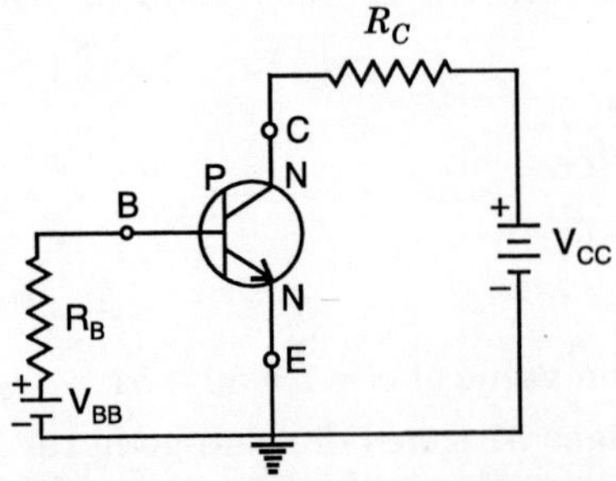

Fig. 6.13. Fixed bias circuit with two batteries

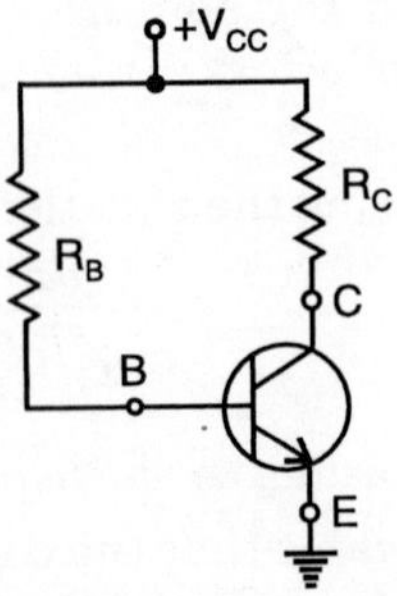

Fig. 6.14. Fixed bias circuit with single battery

* How many types of biasing are done on a BJT to work properly as an amplifier ? Which one is the best and why ?

Now, we will find the value of d.c. bias voltages and currents in the base and collector portions of transistor.

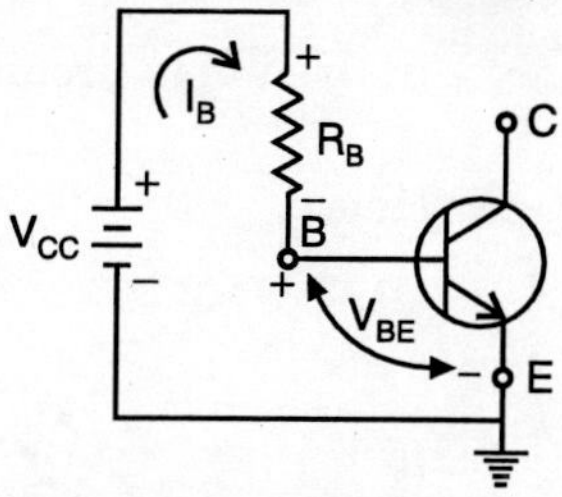

Fig. 6.15. Input portion of fixed bias circuit

Case (*i*): Considering the input portion of the fixed bias circuit (Figure 6.15)

Applying KVL for the input side, we get

$$V_{CC} = I_B R_B + V_{BE} \qquad ...(6.12)$$

or
$$I_B R_B = V_{CC} - V_{BE}$$

or
$$\text{Base Current } I_B = \frac{V_{CC} - V_{BE}}{R_B} \qquad ...(6.13)$$

The values of voltages V_{CC} and V_{BE} are fixed, therefore, the value of base current depends on the base resistor R_B. For simplification, we may neglect the value of V_{BE} because it is small.

Hence,

base current
$$I_B \simeq \frac{V_{CC}}{R_B} \qquad ...(6.14)$$

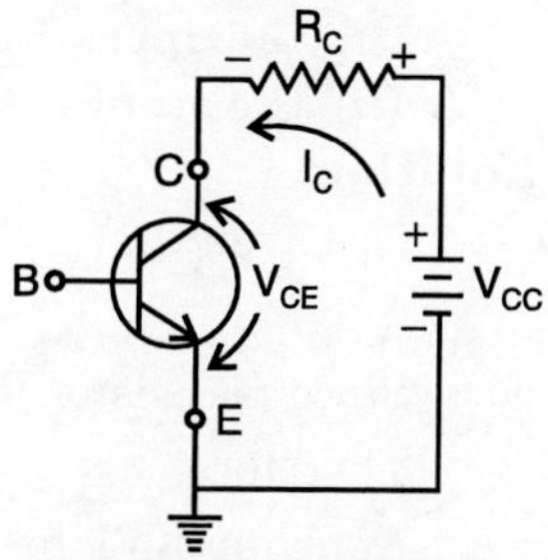

Fig. 6.16. Output portion of fixed bias circuit

Case (*ii*): Considering the output portion of the fixed bias circuit (Figure 6.16)

Applying KVL for output portion

$$I_C R_C + V_{CE} = V_{CC} \qquad ...(6.15)$$

$$\text{or } V_{CE} = V_{CC} - I_C.R_C$$

The collector current I_C is given by the relation

$$I_C = \beta I_B + I_{CO} \qquad ...(6.16)$$

As an approximation, neglecting I_{CO}

$$I_C = \beta I_B + I_{CO}$$

Putting the value of I_B,
$$I_C = \beta.\frac{V_{CC}}{R_B}$$

$$I_C = \frac{V_{CC}}{R_B/\beta} \qquad ...(6.17)$$

The collector current is β times greater than the base current and does not depend upon the collector resistor.

Hence, *Q*-point for fixed-bias circuit is

$$Q \equiv (V_{CE}, I_C)$$

where
$$V_{CE} = V_{CC} - I_C.R_C$$

and
$$I_C = \beta I_B = \frac{V_{CC}}{R_B/\beta}$$

From *Q*-point, it may be observed that the values of collector current I_C and collector-emitter voltage depend upon the value of β which in turn depends strongly on temperature. It means that *Q*-point will change with change in β due to variation in temperature.

This shows that a stable *Q*-point can not be achieved in a fixed-bias circuit. Hence, it is seldom used.

Stability Factor for Fixed-Bias Circuit

$$\text{Stability factor } S = \frac{\beta + 1}{1 - \beta\left(\frac{dI_B}{dI_C}\right)} \qquad ...(6.18)$$

In fixed bias circuit, $\frac{dI_B}{dI_C} = 0$ because I_B is independent of I_C.

Hence,
$$S = \frac{\beta + 1}{1 - \beta.0}$$

$$\boxed{S = \beta + 1} \qquad ...(6.19)$$

If $\beta = 100$, then $S = 101$

So, I_C changes 101 times for any change in I_{CO}. Due to large value of S, stability is poor.

EXAMPLE 6.8 Figure 6.17 shows a fixed bias circuit. Determine with β = 100

(*i*) the operating point

(*ii*) the stability factor

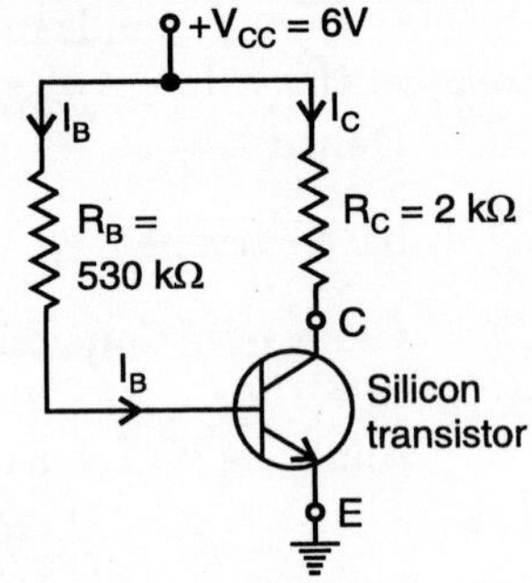

Fig. 6.17.

Solution:

(*i*) Given that $V_{CC} = 6$ V

$R_B = 530$ kΩ, $R_C = 2$ kΩ

Since transistor is silicon transistor

Therefore, $V_{BE} = 0.7$ V

Applying KVL for input side, we have

$$V_{CC} = I_B R_B + V_{BE}$$

or
$$I_B = \frac{V_{CC} - V_{BE}}{R_B} = \frac{6 - 0.7}{530 \text{ k}\Omega}$$

$$= \frac{5.3}{530 \times 10^3} = \frac{53 \times 10^{-5}}{53} = 10 \ \mu\text{A}$$

Collector current is given by

$$I_C = \beta I_B = \beta I_B = 100 \times 10 \ \mu\text{A} = 1000 \ \mu\text{A} = 1 \ \mu\text{A}$$

Applying KVL to the output side, we get

$$V_{CE} = V_{CC} - I_C R_C = 6 - 1 \times 10^{-3} \times 2 \times 10^3$$

$$V_{CE} = 6 - 2 = 4 \text{ V}$$

Thus, operationg point $Q \equiv (V_{CE}, I_C) \equiv (4 \text{ V}, 1 \ \mu\text{A})$

(ii)For fixed bias circuit, stability factor is given by

$$S = \beta + 1 = 100 + 1 = 101 \qquad \textbf{Ans.}$$

EXAMPLE 6.9 Figure 6.18 shows a fixed bias circuit with β = 75. Determine the following:

(*i*) Value of base current

(*ii*) Value of collector current

(*iii*) Value of collector-to-emitter voltage

(*iv*) Stability factor (neglect V_{BE})

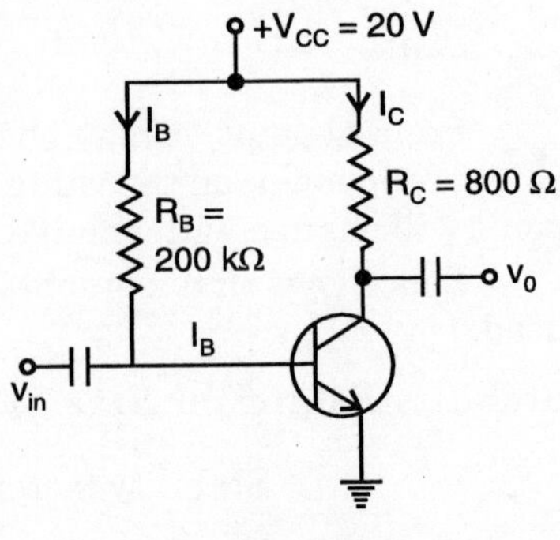

Fig. 6.18.

Solution: (*i*) Base current

From input section, we write

$$V_{CC} = I_B R_B + V_{BE}$$

But $V_{BE} = 0$

Thus, $V_{CC} = I_B R_B$

or
$$I_B = \frac{V_{CC}}{R_B} = \frac{20}{200 \times 10^3}$$

$$I_B = \frac{1}{10\times10^3} = 10^{-4} = 100\ \mu A \qquad \textbf{Ans.}$$

(*ii*) Value of collector current is given as

$\because \qquad I_C = \beta I_B$

$$I_C = 75 \times 100\ \mu A = 7500\ \mu A = 7.5\ mA \qquad \textbf{Ans.}$$

(*iii*) Collector to emitter voltage

From output section, we have

$$V_{CE} = V_{CC} - I_C R_C = 20 - 7.5 \times 10^{-3} \times 800$$

$$V_{CE} = 20 - 60.0 \times 10^{-1} = 20 - 6 = 14\ V \qquad \textbf{Ans.}$$

(*iv*) Stability factor for fixed bias circuit is expressed as

$$S = \beta + 1 = 75 + 1 = 76 \qquad \textbf{Ans.}$$

EXAMPLE 6.10 Figure 6.19 shows the circuit of fixed bias circuit with $\beta = 100$. Assume $V_{BE} \simeq 0$
Determine the following
(*i*) Value of bias resistor R_B
(*ii*) Value of voltage between the collector and ground
(*iii*) The stability factor

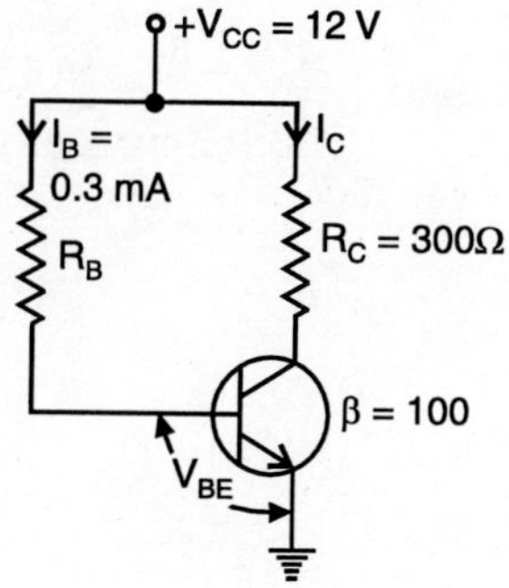

Fig. 6.19.

Solution: (*i*) From input section, we have

$$V_{BB} = I_B R_B + V_{BE}$$

But given that $\qquad V_{BE} \cong 0$

Therefore, $\qquad V_{BB} = I_B R_B$

or $\qquad I_B = \frac{V_{BB}}{R_B}$

or $\qquad R_B = \frac{V_{BB}}{I_B}$

Given $\qquad I_B = 0.3\ mA$

Thus, $\qquad R_B = \frac{12}{0.3 mA} = \frac{12}{0.3\times10^{-3}}$

$$R_B = 40 \times 10^3 = 40\ k\Omega \qquad \textbf{Ans.}$$

(*ii*) $\because \qquad I_C = \beta I_B = 100 \times 0.3 \times 10^{-3}$

$$I_C = 3 \times 10^{-2}\ A = 30\ mA$$

Because there is no emitter resistor, the voltage between collector and ground

= Voltage between collector and emitter

$= V_{CE} = V_{CC} - I_C R_C = 12 - 30 \times 10^{-3} \times 300$

$= 12 - 9 = 3\ V \qquad$ **Ans.**

(*iii*) Stability factor $\qquad S = \beta + 1 = 100 + 1 = 101 \qquad$ **Ans.**

6.13. COLLECTOR TO BASE BIAS CIRCUIT

Figure 6.20 shows the collector to base bias circuit. In this circuit, the base resistor is connected to the collector terminal of the transistor. This resistor R_B works as a feedback resistor. Figure 6.20 also shows the directions of all currents.

> **DO YOU KNOW?**
>
> Because the values of I_C and V_{CE} are dependent on the values of beta in a base-biased circuit, the circuit is said to be beta-dependent.

Now, we will find the values of d.c. currents and voltages in the base and collector portions of the transistor.

First, let us consider the input or base-emitter loop.

Applying KVL to this loop, we get

$$-V_{CC} + (I_B + I_C)\,R_C + I_B\,R_B + V_{BE} = 0 \text{ s} \quad \text{...(6.20)}$$

or $$(I_B + I_C)\,R_C + I_B = V_{CC} - V_{BE}$$

or $$I_B(R_B + R_C) + I_C R_C = V_{CC} - V_{BE}$$

or $$I_B(R_B + R_C) + \beta I_B\,R_C = V_{CC} - V_{BE}$$

or $$I_B\,[R_B + R_C\,(1 + \beta)] = V_{CC} - V_{BE}$$

or $$I_B = \frac{V_{CC} - V_{BE}}{R_B + R_C(1+\beta)} \quad \text{...(6.21)}$$

Since, β is quite large, therefore, $1 + \beta \cong \beta$

$$I_B = \frac{V_{CC} - V_{BE}}{R_B + R_C\,(\beta)} = \frac{V_{CC} - V_{BE}}{R_B + \beta R_C} \quad \text{...(6.22)}$$

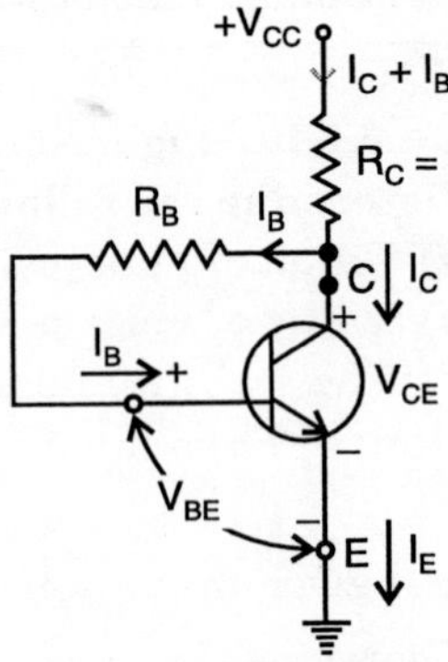

Fig. 6.20. Collector-to-base bias circuit

Collector current is given as

$$I_C = \beta I_B$$

$$I_C = \beta\left(\frac{V_{CC} - V_{BE}}{R_B + \beta R_C}\right)$$

$$I_C = \frac{V_{CC} - V_{BE}}{R_C + R_B/\beta} \quad \text{...(6.23)}$$

$\because$ $$V_{CC} >> V_{BE}$$

Therefore, $$I_C = \frac{V_{CC}}{R_C + R_B/\beta} \quad \text{...(6.24)}$$

Consider the output or collector-emitter loop. Applying, KVL

$$V_{CC} = V_{CE} + (I_C + I_B)\,R_C \quad \text{...(6.25)}$$

Since, $I_C >> I_B$, therefore, $I_C + I_B \cong I_C$

$$V_{CE} = V_{CC} - I_C R_C \quad \text{...(6.26)}$$

Stability factor for collector to base bias circuit

Stability factor is given by the expression

$$S = \frac{1+\beta}{1-\beta\left(\dfrac{dI_B}{dI_C}\right)} \quad \text{...(6.27)}$$

Applying KVL to the emitter-base portion of collector-to-base bias circuit of figure (6.20), we get

$$V_{CC} = V_{BE} + (I_B + I_C)R_C + I_B R_B \quad \text{...(6.28)}$$

or $$I_B(R_B + R_C) + I_C R_C = V_{CC} - V_{BE}$$

or $$I_B(R_B + R_C) = V_{CC} - V_{BE} - I_C R_C \quad \text{...(6.29)}$$

or $$I_B = \frac{V_{CC} - V_{BE} - I_C R_C}{R_B + R_C}$$

Differentiating the above equation with respect to I_C, we have

$$\frac{dI_B}{dI_C} = \frac{0 - 0 - 1.\,R_C}{R_B + R_C} = -\frac{R_C}{R_B + R_C} \quad \text{...(6.30)}$$

Putting the value of $\frac{dI_B}{dI_C}$ from equation (6.30) to (6.27), we get

$$S = \frac{1+\beta}{1+\beta\left(\frac{R_C}{R_B+R_C}\right)} \quad ...(6.31)$$

Important Point. *From the above expression, it is clear that stability factor of collector-base bias circuit is smaller than that of fixed bias circuit. Therefore, this biasing circuit has greater stability than fixed bias circuit.*

EXAMPLE 6.11 **Figure 6.21 shows the circuit of collector-to-base bias using N-P-N transistor. Assuming V_{BE} = 0.7 V, calculate the following**

(i) The collector current

(ii) The collector-to-emitter voltage

(iii) Stability factor

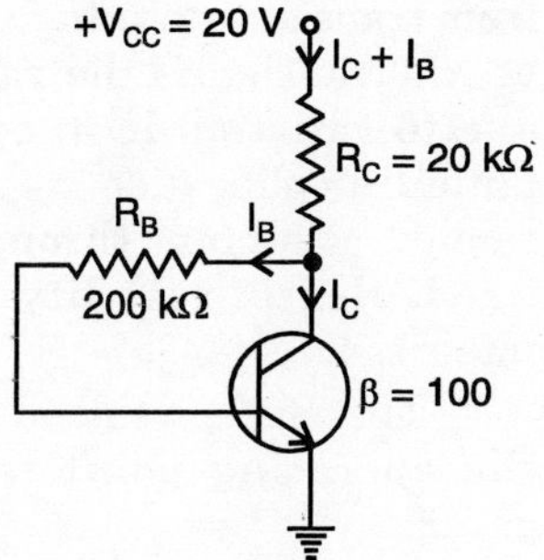

Fig. 6.21.

Solution: We know that, collector current is given as

(i) $$I_C = \frac{V_{CC}-V_{BE}}{R_C+\frac{R_B}{\beta}} = \frac{20-0.7}{\left(20+\frac{200}{100}\right)\times 10^3} = \frac{19.3}{(20+2)10^3} = \frac{19.3}{(22)10^3}$$

or $I_C = 0.877 \times 10^{-3} = 0.877$ mA **Ans.**

(ii) The collector-to-emitter voltage is expressed as

$$V_{CE} = V_{CC} - I_C.R_C = 20 - 0.877 \times 10^{-3} \times 20 \times 10^3$$

$$V_{CE} = 20 - 17.540 = 20 - 17.54 = 2.46 \text{ Volt}$$ **Ans.**

(iii) Stability factor, is expressed as

$$S = \frac{1+\beta}{1+\beta\left(\frac{R_C}{R_C+R_B}\right)} = \frac{1+100}{1+100\left(\frac{20}{20+200}\right)}$$

$$S = \frac{101}{1+\frac{100\times 20}{220}} = \frac{101\times 22}{22+200} = \frac{2222}{222} = 10.009$$ **Ans.**

EXAMPLE 6.12 **Figure 6.22 shows the collector-to-base bias circuit with β = 100. Assuming $V_{BE} \simeq 0$, determine the following:**

(i) the value of I_B

(ii) the value of I_C

(iii) the value of V_{CE}

(iv) the stability factor

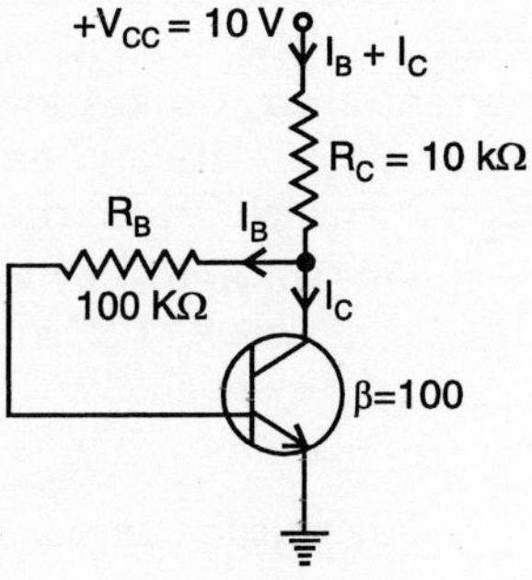

Fig. 6.22.

Solution: (i) Base current I_B is given as

$$I_B = \frac{V_{CC}-V_{BE}}{R_B+\beta.R_C} = \frac{10-0}{(100+100\times 10)10^3}$$

$$I_B = \frac{10\times 10^{-3}}{100+1000} = \frac{10^{-2}}{1100} = 0.09 \times 10^{-4} = 9\ \mu A$$

(ii) $I_C = \beta I_B = 100 \times 9\ \mu A$

$I_C = 900 \times 10^{-6} A = 0.9$ mA **Ans.**

(iii) $V_{CE} = V_{CC} - I_C.R_C = 10 - 0.9 - 10^{-3} \times 10 \times 10^3$

$V_{CE} = 10 - 9 = 1$ Volt **Ans.**

(*iv*) Stability factor
$$S = \frac{1+\beta}{1+\beta\left(\frac{R_C}{R_C+R_B}\right)} = \frac{1+100}{1+100\left(\frac{10}{10+100}\right)}$$

$$= \frac{101}{1+0.09} = \frac{101}{1.09} = 92.6 \qquad \textbf{Ans.}$$

6.14. SELF-BIAS OR VOLTAGE DIVIDER BIAS

For fixed bias circuit and collector-to-base bias circuit, the operating point or the values of d.c. collector current and collector voltage depend upon the value of current gain β of transistor. But the value of β is temperature dependent particularly for silicon transistors. Also, the value of β changes from transistor to transistor. This means that for same type of two transistors the values of β are different. Therefore, due to variation in β, operating point may vary. But for faithful amplification, we need that the operating point should remain at a point throughout the operation where it was fixed. Hence, stability of operating point is of utmost importance.

DO YOU KNOW?

When transistors are biased using well-designed voltage divider bias or emitter bias configurations, they are classified as beta-independent circuits. This is because the values of I_C and V_{CE} are unaffected by changes in the transistor beta.

Therefore, we need a particular type of d.c. bias circuit in which the operating point is independent of the transistor current gain β.

The voltage divider bias circuit or self bias circuit shown in figure 6.23 meets this requirement. Hence, this is most widely used d.c. bias circuit.

The name voltage divider comes from the fact that resistors R_1 and R_2 form a potential divider across the battery V_{CC}. The voltage drop across the resistor R_2 forward biases the emitter-base junction. The d.c. stability is provided by emitter resistor R_E.

Consider the emitter-base loop or input loop of voltage divider bias circuit.

Here, we take help of basic assumption that the resistance looking into the base or input resistance is much larger than the resistance R_2.

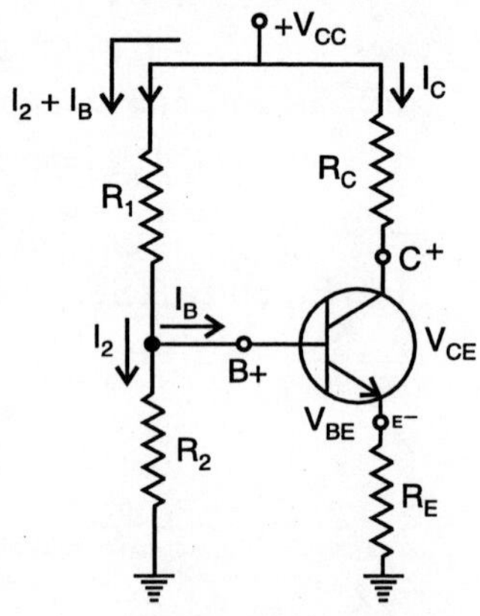

Fig. 6.23. Voltage divider or self bias circuit

According to above assumption, the base current flowing into the input resistance is negligibly small and the current through resistor R_1 flows almost completely into resistor R_2 because the current always takes low resistance path. Therefore, two resistors R_1 and R_2 effectively come in series. Let the voltage across resistor R_2 is V_2 which is also the base voltage V_B.

This voltage V_2 or V_B may be calculated with the help of voltage divider network of R_1 and R_2 and the supply voltage.

$$\therefore \qquad V_B = V_{CC} \times \frac{R_2}{R_1+R_2} \qquad \text{...(6.32)}$$

The voltage across the emitter resistor R_E is

$$V_E = V_B - V_{BE} \qquad \text{...(6.33)}$$

The emitter current is then

$$I_E = \frac{V_E}{R_E} = \frac{V_B - V_{BE}}{R_E} \qquad \text{...(6.34)}$$

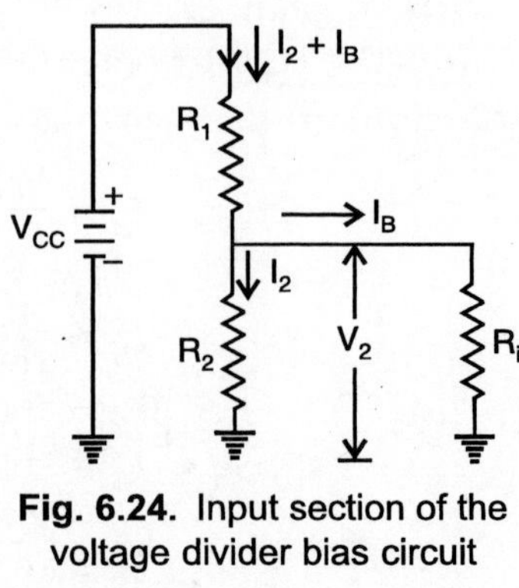

Fig. 6.24. Input section of the voltage divider bias circuit

Hence, the value of collector current is given by

$$I_C \cong I_E = \frac{V_B - V_{BE}}{R_E} \quad ...(6.35)$$

The voltage at the collector will be

$$V_C = V_{CC} - I_C R_C \quad ...(6.36)$$

The collector-to-emitter voltage is given by

$$V_{CE} = V_C - V_E = V_{CC} - I_C R_C - I_E R_E$$

Since $I_E = I_C$

Therefore, $$V_{CE} = V_{CC} - I_C R_C - I_C R_E = V_{CC} - I_C(R_C + R_E) \quad ...(6.37)$$

From the above it is clear that in no expression β appears. Hence, operating point is independent of β. Due to this reason, this voltage divider bias circuit or self-bias circuit is most widely used.

Further stabilization is provided by emitter resistor R_E.

Since $$V_B = V_E + V_{BE} \quad ...(6.38)$$

$$V_B = I_E R_E + V_{BE} = I_C R_E + V_{BE} \quad ...(6.39)$$

Important Note: *If collector current I_C increases due to rise in temperature, the voltage drop across emitter resistor R_E increases. But the voltage V_B (or V_2) is independent of I_C* $\left(\text{Since } V_B = \frac{V_{CC}.R_2}{R_1 + R_2}\right)$*, therefore, V_{BE} decreases. This will decrease the value of I_B which in turn reduce the value of I_C to its original level.*

Hence, due to above two reasons, the voltage divider bias circuit or self-bias circuit is widely used.

Stability factor of voltage divider bias or self bias: To obtain stability factor, first we will find Thevenin's equivalent voltage and Thevenin's equivalent resistance for voltage divider bias circuit.

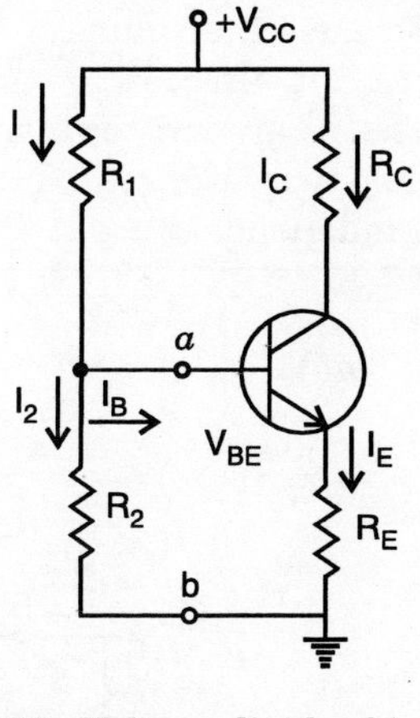

Fig. 6.25. Voltage divider bias or bias circuit.

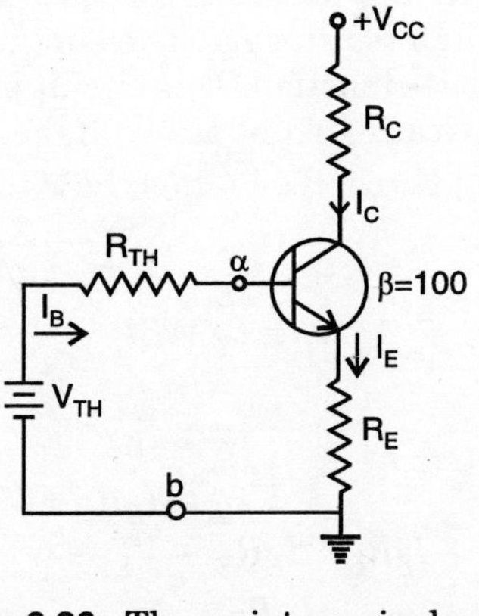

Fig. 6.26. Thevenin's equivalent.

Assuming two terminals *a* and *b* in the network. Thevenizing the circuit on the left of terminals *a* and *b*.

Thevenin's voltage is

$$V_{TH} = \frac{R_2}{R_1 + R_2} . V_{CC} \quad ...(6.40)$$

and Thevenin's equivalent resistance

$$R_{TH} = \frac{R_1 R_2}{R_1 + R_2} \quad \text{...(6.41)}$$

Now, applying KVL to the emitter base loop of the Thevenin's equivalent circuit, we get

$$V_{TH} = I_B.R_{TH} + V_{BE} + I_E.R_E \quad \text{...(6.42)}$$

$$I_B.R_{TH} + (I_B + I_C)R_E = V_{TH} - V_{BE} \qquad [\because I_E = I_B + I_C]$$

or $$I_B + (R_{TH} + R_E) = V_{TH} - V_{BE} - I_C R_E$$

or $$I_B = \frac{V_{TH} - V_{BE} - I_C R_E}{R_{TH} + R_E} \quad \text{...(6.43)}$$

Differentiating the above equation with respect to I_C.

$$\frac{dI_B}{dI_C} = \frac{0 - 0 - 1.R_E}{R_{TH} + R_E} = -\frac{R_E}{R_{TH} + R_E}$$

Since, stability factor is given by the expression

$$S = \frac{1+\beta}{1-\beta\left(\frac{dI_B}{dI_C}\right)} = \frac{1+\beta}{1-\beta\left(-\frac{R_E}{R_{TH}+R_E}\right)} = \frac{1+\beta}{1+\beta\left(\frac{R_E}{R_{TH}+R_E}\right)} \quad \text{...(6.44)}$$

or $$S = \frac{1+\beta(R_{TH}+R_E)}{R_{TH}+R_E+\beta R_E} = \frac{1+\beta\left(1+\frac{R_{TH}}{R_E}\right)}{1+\beta+\frac{R_{TH}}{R_E}}$$

6.15. EMITTER BIAS CIRCUIT

Figure 6.27 shows the emitter bias circuit. In this circuit both the positive ($+V_{CC}$) and negative ($-V_{EE}$) supply voltages are used. The name emitter bias comes from the fact that the negative supply V_{EE} is used to forward-bias the emitter-base junction through resistor R_E. The supply V_{CC} reverse-biases the collector-base junction. This circuit provides almost same stability of operating point as a voltage divider circuit.

DO YOU KNOW?

Because the values of I_C and V_{CE} are not affected by the value of beta in an emitter-biased circuit, this type of circuit is said to be β-independent.

Applying Kirchoff's Voltage Law to the emitter-base or input loop

$$V_{EE} = I_B R_B + V_{BE} + I_E R_E \quad \text{...(6.45)}$$

Putting $$I_C \cong I_E = \beta I_B$$

or $$I_B = \frac{I_C}{\beta} \simeq \frac{I_E}{\beta}$$

$$V_{EE} = I_B R_B + V_{BE} + I_E R_E$$

or $$I_B R_B + I_E R_E = V_{EE} - V_{BE}$$

$$I_E R_E + \frac{I_E R_B}{\beta} = V_{EE} - V_{BE}$$

or $$I_E = \frac{V_{ER} - V_{BE}}{R_E + R_{B/\beta}} \quad \text{...(6.46)}$$

To make the operating point independent of β, taking $R_E >> \frac{R_B}{\beta}$

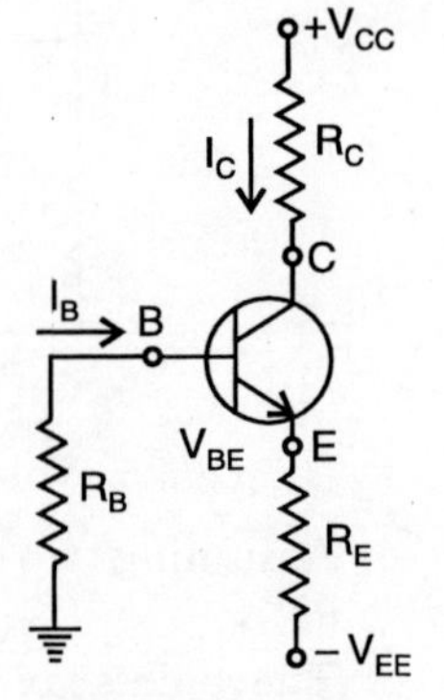

Fig. 6.27. Emitter Bias Circuit.

Also $V_{EE} >> V_{BE}$

Hence, $$I_E = \frac{V_{EE}}{R_E}$$

or $$I_C \cong I_E = \frac{V_{EE}}{R_E} \quad ...(6.47)$$

From last equation, $V_{EE} = I_E R_E$. This means that all the supply voltage V_{EE} appears across R_E. Hence, emitter is at ground potential effectively. Therefore, with emitter at ground potential, the voltage

$$V_{CE} = V_{CC} - I_C.R_C \quad ...(6.48)$$

6.16 BIAS COMPENSATION

The collector to base bias circuit and the voltage divider circuit which we have discussed earlier, are used to limit the variation in the Q point collector current I_{CQ} caused by variations in I_{CO}, V_{BE} and β_{dc}. These circuits are examples of feedback amplifiers. Due to the negative feedback present in these circuits, the amplification of the AC signals is reduced drastically. If this loss of signal cannot be tolerated, then the compensation techniques are used to reduce the drift in the operating point. Very often both stabilization and compensation techniques are used to provide maximum bias and thermal stabilization.

Types of compensation techniques

The compensation techniques that we are going to study may be listed as under :

(*i*) Compensation for changes in V_{BE} and

(*ii*) Compensation for changes in I_{CO}.

6.16.1 Diode Compensation for VBE

The voltage divider bias circuit with diode compensation is as shown in figure 6.28. The additional power supply V_{DD} is connected in order to forward bias the diode *D*. The diode is of same material and type as the transistor. The voltage V_F across the diode has the same temperature coefficient (– 25 mV/°C) as that of the base to emitter voltage V_{BE}.

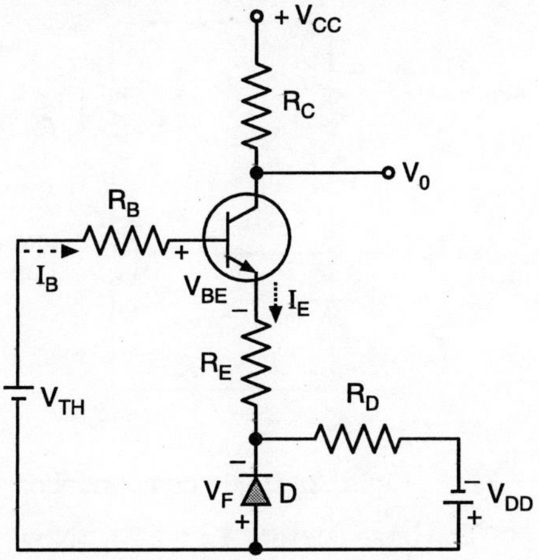

Fig. 6.28 Diode compensation for V_{BE}

How does the compensation take place ?

From figure 6.28, the KVL for the base circuit yields :

$$V_{TH} = I_B R_B + V_{BE} + (I_B + I_C)\, R_E - V_F \quad ...(6.49)$$

But, $$I_C = \beta_{dc}\, I_B + (1 + \beta_{dc})\, I_{CBO} \quad ...(6.50)$$

Therefore, $$V_{TH} = I_B\,(R_B + R_E) + V_{BE} + I_C R_E - V_F \quad ...(6.51)$$

From equation (6.50), $I_B = \dfrac{I_C}{\beta_{dc}} - \dfrac{(1+\beta_{dc})\, I_{CBO}}{\beta_{dc}}$

Substituting this value in equation (6.51), we get,

$$V_{TH} = \frac{IC}{\beta_{dc}}(R_B + R_E) - \frac{(1+\beta_{dc})(R_B + R_E)}{\beta_{dc}} I_{CBO} + V_{BE} + I_C R_E - V_F$$

Therefore, $$V_{TH} = \underbrace{V_{BE} - V_F}_{\downarrow} + \frac{R_B + R_E(1+\beta_{dc})}{\beta_{dc}} I_C - \frac{(R_B + R_E)}{\beta_{dc}}(1+\beta_{dc})\, I_{CBO} \quad ...(6.52)$$

These terms are equal and opposite, so they cancel each other.

The material used for the diode is same as that for the transistor and has same temperature coefficient. Hence, the change in voltage across the diode (V_F) will be equal to change in V_{BE} due to temperature changes. Hence, the first two terms in equation (6.52) cancel each other. Thus, the change in V_{BE} due to temperature is compensated by change in V_F and the collector current becomes insensitive to variation in V_{BE}.

6.16.2 Diode Compensation for I_{CO}

For the germanium transistors, changes in I_{CO} due to change in temperature is more prominent than changes in V_{BE} due to temperature. The diode compensation circuit shown in figure 6.29 offers the stabilization against variation in I_{CO}.

How is the compensation provided ?

The diode and the transistor are of same type and material. Therefore, the reverse saturation current of the diode *i.e.*, I_o will be increase with temperature at the same rate as the current I_{CO}. Referring to figure 6.29, we can write,

$$I = \frac{V_{CC} - V_{BE}}{R_1} \cong \frac{V_{CC}}{R_1} \equiv \text{constant}$$

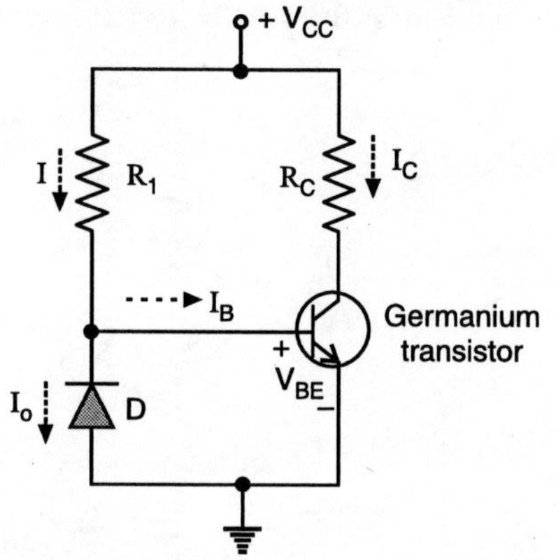

Fig. 6.29 Diode compensation for I_{CO}

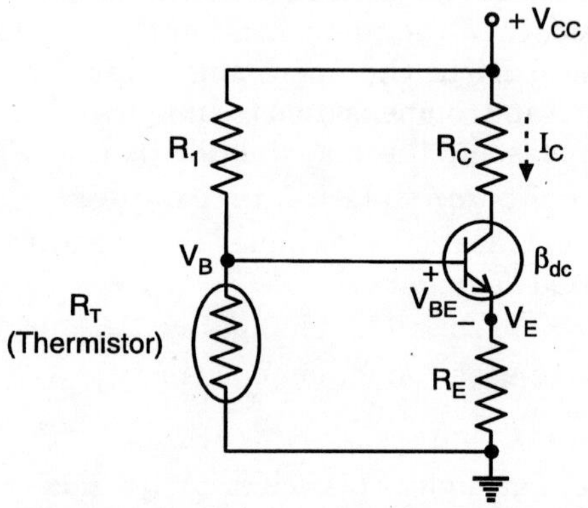

Fig. 6.30 Bias compensation using a thermistor

The base current $I_B = I - I_0$. Now, substituting this value of I_B in the following equation, we get

$$I_C = \beta_{dc} I_B + (1 + \beta_{dc}) I_{CO} = \beta_{dc} (I - I_o) + (1 + \beta_{dc}) I_{CO}$$

Therefore, $$I_C = \beta_{dc} I - \beta_{dc} I_o + (1 + \beta_{dc}) I_{CO} \quad ...(6.53)$$

If the saturation current of the diode (I_o) is equal to the leakage current I_{CO} of the transistor, then the last two terms in equation (6.53) will get cancelled.

Therefore, $$I_C = \beta_{dc} I \qquad [\text{Since } \beta_{dc} I_o \cong (1 + \beta_{dc}) I_{CO}]$$

As I is constant, I_C will also remain constant and compensation is provided. Looking at the last two terms of equation (6.53), we conclude that if $\beta_{dc} >> 1$ and if I_o of D and I_{CO} of the transistor track each other over the desired temperature range, then I_C will remain constant. Thus, compensation is provided.

6.16.3 Bias Compensation using Thermistor

In the two previous circuits, we have seen how to use a diode for compensation. Now, let us see how a thermistor can be used for compensation. Thermistor is a resistor the resistance of which is

Progress is impossible without change; and who cannot change their minds cannot change anything.

— G.B. Shaw

temperature dependent. It's resistance decreases exponentially with increase in temperature. Hence, it is said to have a negative temperature coefficient of resistivity. The variation of the thermistor resistance with change in temperature is shown in figure 6.31. The use of thermistor for bias compensation is demonstrated in figure 6.30.

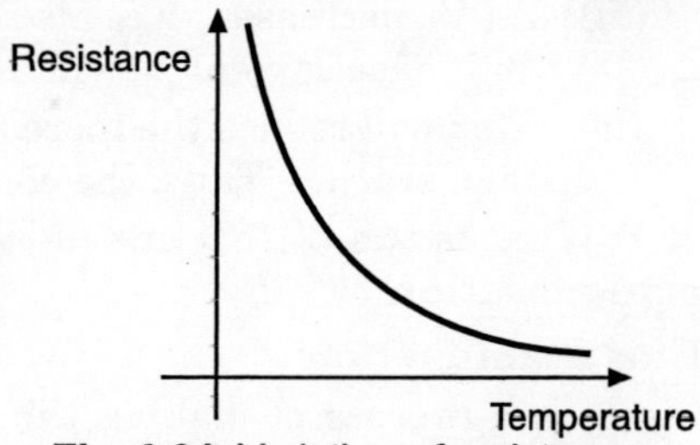

Fig. 6.31 Variation of resistance of a thermistor with temperature

The thermistor R_T alongwith R_1 forms a resistance divider across the supply voltage V_{CC}. Hence, the base voltage is given by,

$$V_B = \frac{R_T}{(R_1 + R_T)} V_{CC} \quad ...(6.54)$$

The emitter voltage V_E is given by,

$$V_E = I_E R_E \quad ...(6.55)$$

Also, the base to emitter voltage is given by,

$$V_{BE} = V_B - V_E \quad ...(6.56)$$

Note that I_B will increase if V_{BE} increases and it will decrease with reduction in V_{BE}.

Principle of Operation of the circuit

As we have seen earlier, with increase in temperature, the collector current I_C tends to increase. Now, let us see how this increased I_C is reduced.

(*i*) As temperature increases, I_C tends to increase.

↓

(*ii*) Due to increase in temperature, resistance of thermistor (R_T) decreases.

↓

(*iii*) This reduces the base voltage V_B as $V_B = \frac{R_T}{R_1 + R_T} V_{CC}$

↓

(*iv*) Hence, V_{BE} decreases as $V_{BE} = V_B - V_E$

↓

(*v*) Due to reduced V_{BE}, base current I_B decreases

↓

(*vi*) This will reduce I_C. Thus, compensation is achieved.

6.16.4 Bias Compensation using Sensistor

The bias compensation circuit using a sensistor is shown in figure 6.32. Sensistor is a temperature dependent resistance, the resistance of which increases with increase in temperature. Thus, it has a positive temperature coefficient of resistivity. The base voltage V_B is given by,

$$V_B = \frac{R_2}{R_2 + R_T} V_{CC} \quad ...(6.57)$$

- The emitter voltage $V_E = I_E R_E$ and the base to emitter voltage is given by,

$$V_{BE} = V_B - V_E \quad ...(6.58)$$

Operation of the circuit

(*i*) As the temperature increases, R_T increases.

(*ii*) As R_T increases, voltage drop across R_2 *i.e.*, V_B will decrease.

(iii) As V_B decreases, V_{BE} also will decrease. Due to reduction in V_{BE}, base current I_B will decrease.

(iv) This will reduce the increased I_C due to increase in I_{CBO} with temperature. Thus, the compensation is achieved.

What is the difference between bias stabilization and bias compensation ?

Bias Stabilization

It is a process of making the operating point Q independent of changes in temperature or variations of transistor parameters. Self bias, fixed bias and collector to base bias circuits are the bias stabilization circuits.

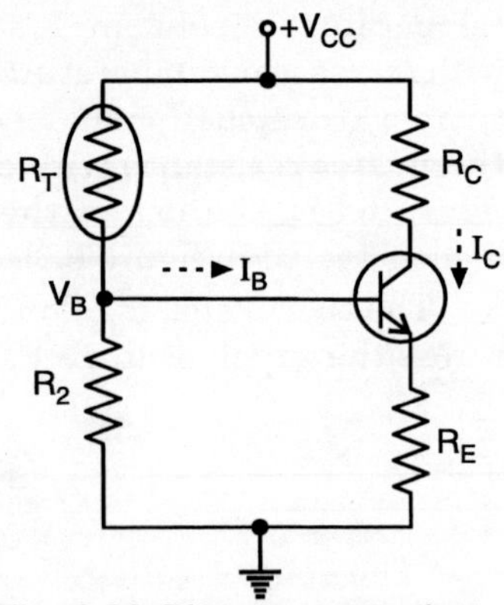

Fig. 6.32 Bias compensation using a sensistor

Bias Compensation

The bias stabilizing circuits use a negative feedback. This will reduce the amplification of AC signals drastically. If this loss of signal is not to be tolerated then bias compensation techniques are used to reduce the drift in the operating point. Bias compensation circuits are :

(i) Diode compensation, (ii) Thermistor compensation

6.17 THERMAL RUNAWAY : DETAILED ASPECTS

The maximum power that a transistor can dissipate without getting damaged depends largely on the maximum temperature that a collector-base junction can withstand. The rise in the collector-base junction takes place due to two reasons :

(i) Due to increase in the ambient temperature and

(ii) Due to the internal heating.

Out of them, the internal heating process is cumulative as explained below :

(i) An increase in collector current I_C increases the power dissipated in the collector-base junction of the transistor.

(ii) This will increase the temperature of C-B junction.

(iii) As the transistor has a negative temperature coefficient of resistivity, increased junction temperature reduces the resistance.

(iv) The reduced resistance will increase the collector current further.

This becomes a cumulative process which will finally damage the transistor due to excessive internal heating. This process is known as *Thermal Runaway*.

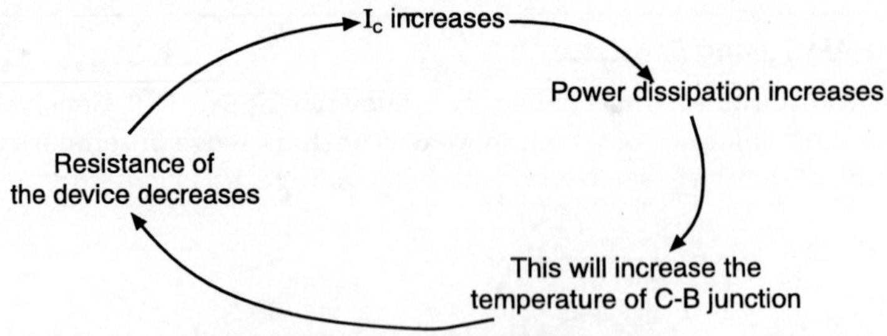

Fig. 6.33 Thermal runaway process

How to avoid the thermal runaway?

(i) Never exceed the collector current beyond a certain maximum value specified by the manufacturer.

(ii) Never exceed the internal dissipation above the maximum permissible value.

(iii) Use heat sink.

6.17.1 Thermal Resistance

It is observed that the steady state temperature rise at the junction of a transistor is proportional to the power dissipation at the junction. *i.e.*,

$$\Delta T \propto P_D$$

Substituting $\Delta T = T_j - T_A$ in the above equation, we get,

$$(T_j - T_A) \propto P_D$$

Therefore, $(T_j - T_A) = \theta \, P_D$ ...(6.59)

where, T_j = Junction temperature in °C

T_A = Ambient temperature in °C

P_D = Power in watts, dissipated at the collector junction

θ = Proportionality constant called as thermal resistance.

For smooth flow of heat from the junction of power transistor to the ambient, the value of thermal resistance should be as small as possible. The thermal resistance θ is expressed in °C/W. The thermal resistance of a power transistor is always less than that of a logic level transistor. The value of thermal resistance depends on :

(*i*) Size of the transistor

(*ii*) Type of heat transfer mechanism (convection, radiation etc).

(*iii*) The type of cooling system (forced air cooling, natural cooling etc.)

The value of the thermal resistance θ depends on the size of the transistor, on convection or radiation to the surroundings, on forced air cooling (if used) and on the thermal connection of the device to a metal chassis, or to a heat sink. Also, θ is expressed as °C/W. Power transistor have lower values of θ as compared to the logic level transistors.

6.17.2 The condition for thermal stability

Here, we mention the condition to be met if the thermal runaway is to be avoided. The condition states that the rate at which heat is released at the collector junction must not exceed the rate at which heat can be dissipated to the atmosphere.

i.e., $$\frac{\partial P_C}{\partial T_j} < \frac{\partial P_D}{\partial T_j} \qquad ...(6.60)$$

where P_C = Maximum collector power

and P_D = Maximum power dissipated

Equation (6.59) states that

$$(T_j - T_A) = \theta \, P_D$$

Differentiating this with respect to T_j, we get,

$$1 = \theta \times \frac{\partial P_D}{\partial T_j},$$

Therefore, $$\frac{1}{\theta} = \frac{\partial P_D}{\partial T_j}$$

Substituting this into equation (6.60), we get,

$$\frac{\partial P_C}{\partial T_j} < \frac{1}{\theta} \qquad ...(6.61)$$

This condition must be satisfied to prevent thermal runaway.

Example 6.13. Prove that for a voltage divider circuit for common emitter (CE) amplifier, the condition for external stability is $V_{CE} < \frac{V_{CC}}{2}$.

Solution :

Let us consider figure 6.34 and assume that the transistor is in the active region. By applying KVL around the collector loop, we get,

$$V_{CE} = V_{CC} - I_C (R_C + R_E) - I_B R_E$$

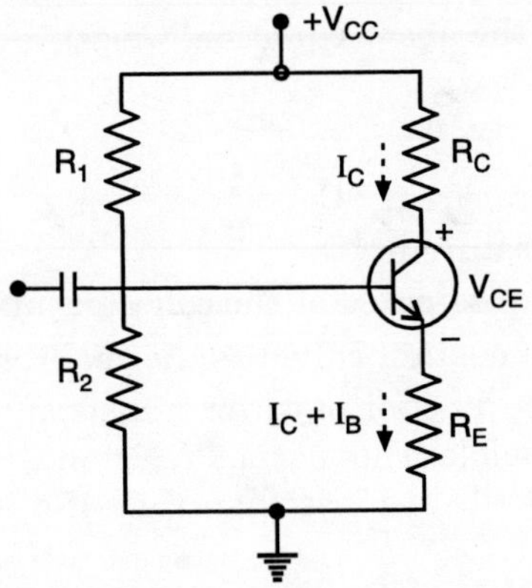

Fig. 6.34 Voltage divider bias circuit

Neglecting the last term, we get

$$V_{CE} = V_{CC} - I_C (R_C + R_E) \quad ...(i)$$

Therefore, $$I_C = \frac{V_{CC} - V_{CE}}{(R_C + R_E)}$$

The power developed at the collector junction in the absence of ac input signal, is given by,

$$P_C = V_{CB} I_C \cong V_{CE} I_C \quad ...(ii)$$

Substituting equation (*i*) in equation (*ii*), we get,

$$P_C = V_{CC} I_C - I_C^2 (R_C + R_E).$$

Now, differentiating this equation with respect to I_C, we obtain

$$\frac{\partial P_C}{\partial I_C} = V_{CC} - 2 I_C (R_C + R_E) \quad ...(iii)$$

The condition to avoid the thermal runaway can be rewritten as :

$$\frac{\partial P_C}{\partial I_C} \times \frac{\partial I_C}{\partial T_j} < \frac{1}{\theta}$$

Therefore, $$\frac{\partial P_C}{\partial I_C} \times \frac{\partial I_C}{\partial T_j} \times \theta < 1 \quad ...(iv)$$

In this expression, $\frac{\partial I_C}{\partial T_j}$ and θ will always be positive. Therefore, this equation will always be satisfied if $\frac{\partial P_C}{\partial I_C}$ is negative. Hence, to avoid thermal runaway, we have

$$\frac{\partial P_C}{\partial I_C} < 0 \quad ...(v)$$

Substituting equation (*iii*) into equation (*v*), we obtain,

$$V_{CC} - 2I_C (R_C + R_E) < 0$$

i.e., $$I_C > \frac{V_{CC}}{2(R_C + R_E)} \quad ...(vi)$$

But from equation (*i*), the collector current is expressed as:

$$I_C = \frac{V_{CC} - V_{CE}}{(R_C + R_E)} \quad ...(vii)$$

Substituting this in equation (*vi*), we obtain,

$$\frac{V_{CC} - V_{CE}}{(R_C + R_E)} > \frac{V_{CC}}{2(R_C + R_E)}$$

or $$(V_{CC} - V_{CE}) > \frac{V_{CC}}{2}$$

or $$V_{CE} < V_{CC}/2$$ **Hence Proved.** ...(8)

Important Conclusion : If V_{CE} is less than V_{CC} / 2, then the operating point will be in the safe region, where thermal runaway will not take place. But, if the *Q* point is located such that $V_{CE} > V_{CC} / 2$, then the transistor is likely to get damaged due to thermal runaway.

SOME IMPORTANT SOLVED EXAMPLES

EXAMPLE 6.14 Figure 6.35 shows the circuit of fixed bias using silicon transistor. Determine the following:

(*i*) base current I_B **(*ii*) collector current I_C**
(*iii*) V_{CE} **(*iv*) V_C and V_B**
(*v*) V_{CB}

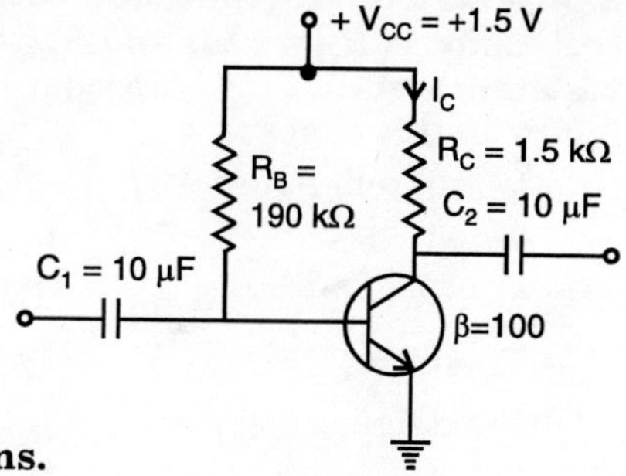

Fig. 6.35.

Solution: (*i*) We know that

$$V_{CC} = V_{BE} + I_B.R_B$$

Therefore, base current $I_B = \frac{V_{CC} - V_{BE}}{R_B}$

$$= \frac{15 - 0.7}{180 \times 10^3} = 79.44\ \mu A \quad \textbf{Ans.}$$

(*ii*) Collector Current , $I_C = \beta I_B$

$$I_C = 100 \times 79.44\ mA = 7.94\ mA \quad \textbf{Ans.}$$

(*iii*) Since $V_{CC} = V_{CE} + I_C R_C$, therefore

$$V_{CE} = V_{CC} - I_C.R_C = 15 - 7.94 \times 10^{-3} \times 1.5 \times 10^3$$

$$V_{CE} = 15 - 11.9 = 3.1 \text{ Volt}$$

(*iv*) $$V_C = V_{CE} = 3.1 \text{ Volt} \quad \textbf{Ans.}$$

$$V_B = V_{BE} = 0.7 \text{ Volt} \quad \textbf{Ans.}$$

(*v*) $$V_{CB} = V_C - V_B = 3.1 - 0.7$$

$$V_{CB} = 2.4 \text{ Volt} \quad \textbf{Ans.}$$

EXAMPLE 6.15 Figure 6.36 shows the circuit of fixed bias using a silicon transistor with β = 100. Determine the following:

(*i*) Operating point
(*ii*) Stability factor

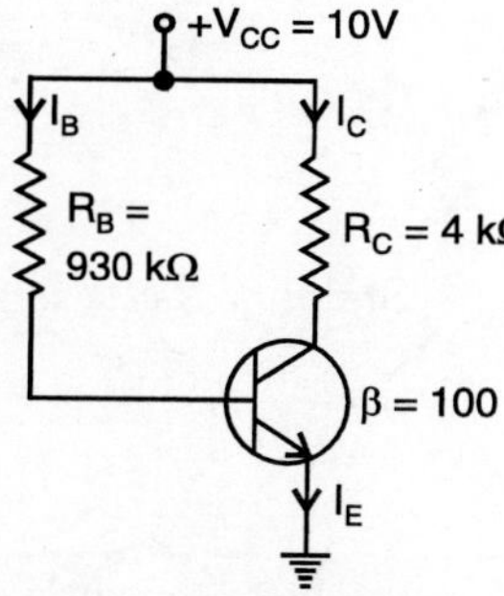

Fig. 6.36.

Solution: (*i*) Given that

$$V_{CC} = 10 \text{ Volt}$$
$$R_B = 930\ k\Omega$$
$$R_C = 4\ k\Omega$$
$$\beta = 100$$

Since, transistor is silicon transistor

Therefore, $V_{BE} = 0.7$ Volts

Applying KVL for the input side, we have

$$V_{CC} = I_B R_B + V_{BE}$$

or
$$I_B = \frac{V_{CC} - V_{BE}}{R_B} = \frac{10 - 0.7}{930 \times 10^3}$$

$$= \frac{9.3 \times 10^{-3}}{930} = \frac{93 \times 10^{-5}}{93} = 10 \text{ mA}$$

We know that collector current is given by

$$I_C = \beta I_B = 100 \times 10 \text{ mA} = 1000 \times 10^{-6} \text{ A} = 10^{-3} = 1 \text{ mA}$$

Now, Applying KVL to the output side, we have

$$V_{CE} = V_{CC} - I_C.R_C = 10 - 1 \times 10^{-3} \times 4 \times 10^3 = 10 - 4 = 6 \text{ Volt}$$

Therefore, operating point $Q \equiv (V_{CE}, I_C) \equiv (6V, 1mA)$ **Ans.**

(*ii*) For fixed-bias circuit, stability factor is given by

$$S = \beta + 1 = 100 + 1 = 101 \quad \textbf{Ans.}$$

EXAMPLE 6.16 A transistor with β = 100 is used in *CE* configuration. The collector circuit resistance is R_C = 1 kΩ and V_{CC} = 20 V. Assuming V_{BE} = 0, find the value of collector to base resistance, such that quiescent collector emitter voltage, is 4V. Also determine the stability factor in this case.

Solution: From figure 6.37, we observe that

$$I'_C = I_C + I_B$$

∵
$$I_C >> I_B$$

Therefore,
$$I'_C = I_C \quad (i)$$

From figure 6.37,
$$V_{CC} = I_C' R_C + V_{CE}$$

Using equation (*i*), we get

$$V_{CC} = I_C R_C + V_{CE}$$
$$20 = I_C . 1 \times 10^3 + 4$$
$$16 = I_C . 10^3$$
$$I_C = 16 \text{ mA}$$

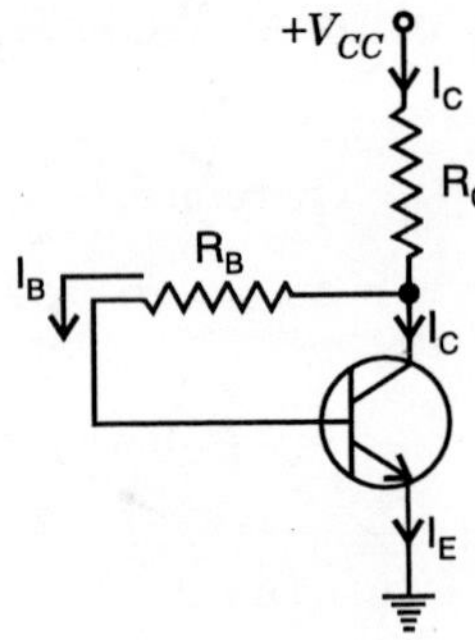

Fig. 6.37.

We know that
$$I_C = \beta I_B$$

or
$$I_B = \frac{I_C}{\beta} = \frac{16 \times 10^{-3}}{100} = 160 \, \mu\text{A}$$

From figure,
$$V_{CE} = I_B R_B + V_{BE} = I_B R_B + 0$$

$$R_B = \frac{V_{CE}}{I_B} = \frac{4}{160 \times 10^{-6}} = \frac{4}{16 \times 10^{-5}}$$

$$= \frac{1}{4} \times 10^{-5} = 25 \times 10^3 = 25 \text{ k}\Omega \quad \textbf{Ans.}$$

Stability factor is given as

$$S = \frac{\beta + 1}{1 + \frac{\beta R_C}{R_B + R_C}} = \frac{101}{1 + \frac{100.1}{25 + 1}} = \frac{101}{1 + \frac{100}{26}} = \frac{101}{1 + 3.84} = \frac{101}{4.84} = 20.86 \quad \textbf{Ans.}$$

The intelligence is proved not by ease of learning but by undeerstanding what we learn.
— **Joseph Whitney**

EXAMPLE 6.17 In a common emitter circuit an NPN transistor having a value of $\beta = 50$ is used with $V_{CC} = 10$ V and $R_C = 2$ kΩ. If a 100 K resistor is connected between collector and base and $V_{BE} = 0$, determine

(i) the position of quiescent point, and

(ii) stability factor, S.

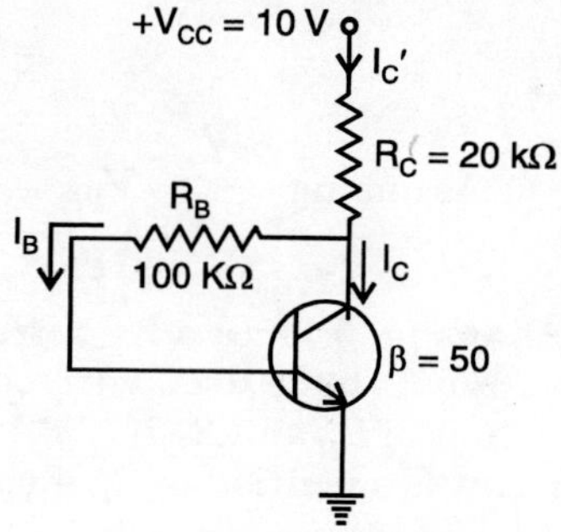

Fig. 6.38.

Solution: From figure 6.38, we observe that

$$I'_C = I_C + I_B$$

$\because$ $\quad I_B << I_C$

Therefore, $\quad I'_C \cong I_C \quad$ ....(i)

Writing loop equation

$$V_{CC} = I_C'R_C + I_BR_B + V_{BE}$$

Using (i) we get $\quad V_{CC} = I_CR_C + I_BR_B + V_{BE} \quad$ ...(ii)

$\because$ $\quad I_C = \beta I_B$

and $\quad V_{BE} = 0$

Therefore, $\quad V_{CC} = I_CR_C + \frac{I_C}{\beta}.R_B + 0$

$$V_{CC} = I_CR_C + \frac{I_C}{\beta}.R_B$$

$$V_{CC} = I_C\left(R_C + \frac{R_B}{\beta}\right)$$

or $\quad I_C = \dfrac{V_{CC}}{R_C + \dfrac{R_B}{\beta}} = \dfrac{10}{2 + \dfrac{100}{50}} = \dfrac{10}{2+2} = \dfrac{10}{4} = 2.5 \text{ mA}$

Again $\quad I_C = \beta I_B$

Therefore, $\quad I_B = \dfrac{I_C}{\beta} = \dfrac{2.5}{50} \text{ mA} = 0.05 \text{ mA} = 50 \text{ mA}$

Again from the figure, we may write

$$V_{CE} = I_BR_B + V_{BE}$$

But $\quad V_{BE} = 0$

Therefore, $\quad V_{CE} = I_BR_B = 50 \times 10^{-6} \times 10^6$

$$V_{CE} = 5 \text{ Volt}$$

$$Q \equiv (5 \text{ V}, 2.5 \text{ mA}) \quad \textbf{Ans.}$$

Stability factor $\quad S = \dfrac{\beta + 1}{\dfrac{\beta R_C}{R_B + R_C} + 1}$

$\because$ $\quad S = \dfrac{\beta + 1}{\dfrac{\beta R_C}{R_B + R_C} + 1} = \dfrac{51 \times 102}{100 + 102} = \dfrac{5202}{202} = 25.75 \quad$ **Ans.**

EXAMPLE 6.18 Determine the value of collector to emitter bias voltage for the given circuit if $\beta = 100$, and $I_C = 10$ mA.

Solution: Collector current is given as

$$I_C = \beta I_B$$

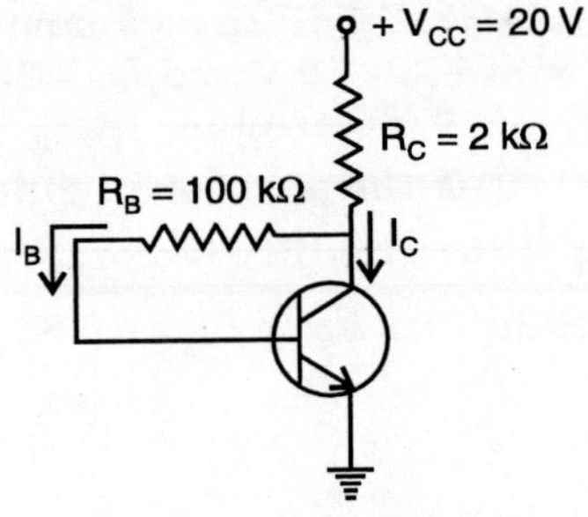

Fig. 6.39.

$$I_B = \frac{I_C}{\beta} = \frac{10\times10^{-3}}{100}$$

$$= 10 \times 10^{-5} = 100\ \mu A = 0.1\ mA$$

$\because$ $V_{CE} = I_B R_B + V_{BE}$

Assuming $V_{BE} = 0.7\ V = 100 \times 10^{-6} \times 100 \times 10^3 + 0.7$

$V_{CE} = 10 + 0.7 = 10.7$ Volt **Ans.**

EXAMPLE 6.19 In a NPN transistor amplifier stage, a 9 V battery supply is to be used with collector to base biasing. Determine the values of R_B and R_C if I_{CBO} is neglibible, β = 100 and if the quiescent point is specified by I_C = 0.2 A and V_{CE} = 5 V.

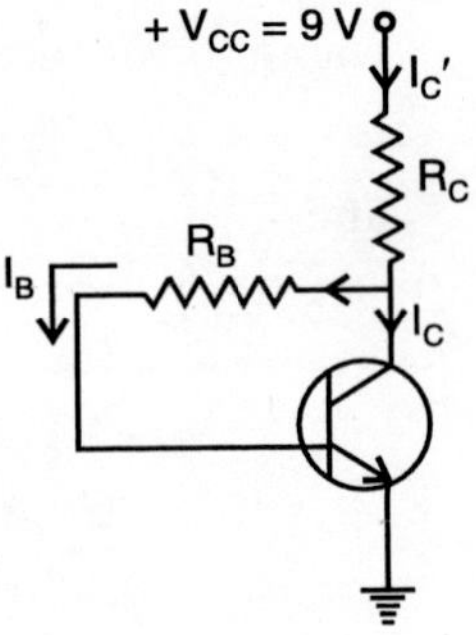

Fig. 6.40.

Solution: Given that $I_{CBO} \approx 0$

$I_C = 0.2A$ and

$V_{CE} = 5V$

From figure 6.55, we have

$$V_{CC} = I'_C R_C + V_{CE}$$

$$V_{CC} = (I_C + I_B) R_C + V_{CE}$$

But $I_C >> I_B$

Therefore, $V_{CC} = I_C R_C + V_{CE}$

$$I_C R_C = V_{CC} - V_{CE} = 9 - 5 = 4$$

$$R_C = \frac{4}{I_C} = \frac{40}{0.2} = 20\ \Omega$$

We know that $I_C = \beta I_B$

Therefore, $I_B = \frac{I_C}{\beta} = \frac{0.2}{100} = \frac{2}{1000} = 2$ mA

From figure 6.29 $V_{CE} = I_B R_B + V_{BE}$

Assuming $V_{BE} = 0$, we have

$$V_{CE} = I_B R_B$$

$$R_B = \frac{5\ V}{2\ mA} = 2.5\ k\Omega \quad \textbf{Ans.}$$

EXAMPLE 6.20 A transistor with β = 45 is used with collector to base resistor (*R*) biasing, with a quiscent value of 5 V for V_{CE}. If V_{CC} = 24 V, R_L = 10 kΩ and R_E = 270 Ω. Find the value of

(*i*) *R* and, (*ii*) Stability factor.

Solution: (*i*) Applying KVL to collector emitter circuit

$$V_{CC} = (I_C + I_B) R_L + V_{CE} + I_E R_E$$

$$V_{CC} = (I_C + I_B) R_L + V_{CE} + (I_C + I_B) R_E$$

$\because$ $I_E = I_B + I_C$

$$V_{CC} = (I_C + I_B) [R_L + R_E] + V_{CE}$$

$$V_{CC} = (\beta I_B + I_B) [R_L + R_E] + V_{CE}$$

$\because$ $I_C = \beta I_B$

Therefore, $V_{CC} = (\beta + 1)\, I_B\, [R_L + R_E] + V_{CE}$

$$24 = (45 + 1).I_B\, [10 \times 10^3 + 270] + 5$$

or $19 = 46 \times I_B\, [10.27 \times 10^3]$

or $$46\, I_B = \frac{19}{10.27 \times 10^3} = \frac{19 \times 10^{-3}}{10.27} = 1.85 \text{ mA}$$

$$I_B = \frac{1.85 \text{ mA}}{46} = 40\ \mu\text{A}$$

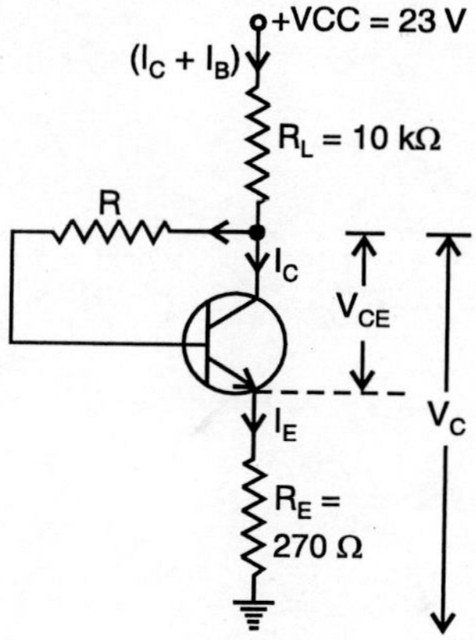

Fig. 6.41.

From circuit, we write

$$V_{CE} = V_{BE} + RI_B$$

$$R = \frac{V_{CE} - V_{BE}}{I_B} = \frac{5 - 0.6}{40 \times 10^{-6}} = 110 \text{ k}\Omega \qquad \textbf{Ans.}$$

(*ii*) Stability factor $$S = \frac{\beta + 1}{1 + \frac{\beta . R_E}{R_E + R}} = \frac{45 + 1}{1 + \frac{45 \times 270 \times 2}{270 + 110 \times 10^3}}$$

$$= \frac{46}{1 + \frac{12150}{110270}} = \frac{46}{1 + 0.11} = \frac{46}{1.11}$$

$$S = 41.44 \qquad \textbf{Ans.}$$

EXAMPLE 6.21 In a *CE* germanium transistor amplifier, self-bias is used. The various parameters are: V_{CC} = 16 Volt, R_C = 3 kΩ, R_E = 2 kΩ, R_1 = 56 kΩ, R_2 = 20 kΩ and α = 0.985. Determine the following:

(*i*) the operating point

(*ii*) the stability factor, *S*.

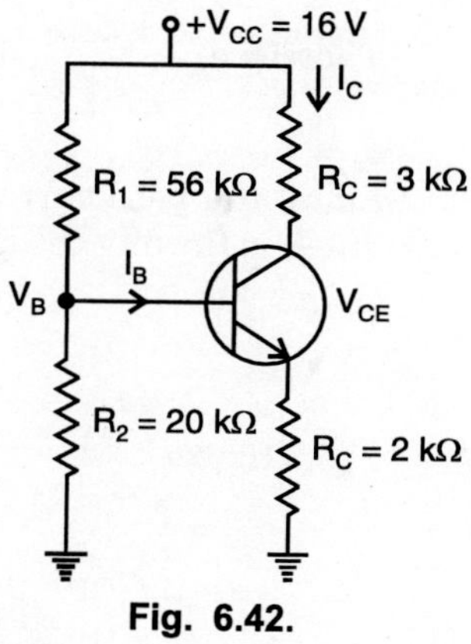

Fig. 6.42.

Solution: Given that $\alpha = 0.985$

So $$\beta = \frac{\alpha}{1 - \alpha} = \frac{0.985}{1 - 0.985} = 66$$

(*i*) Since, given transistor is a germanium transistor, therefore, V_{BE} = 0.3 Volt.

Value of I_C is given by

$$I_C = \frac{V_B - V_{BE}}{R_E} \qquad ...(i)$$

But $$V_B = V_{CC} \frac{R_2}{R_1 + R_2} = \frac{16 \times 20 \times 10^3}{(56 + 20) \times 10^3} = \frac{320}{76} = 4.21 \text{ Volt}$$

Using equation (*i*) $$I_C = \frac{V_B - V_{BE}}{R_E} = \frac{4.21 - 0.3}{2 \times 10^3}$$

$$I_C = 0.63 \times 10^{-3} \text{ A} = 0.63 \text{ mA}$$

Also from figure 6.57 we have

$$V_{CE} = V_{CC} - I_C.(R_C + R_E)$$

$$V_{CE} = 16 - 0.63 \times 10^{-3}.\ (3 + 2) \times 10^3$$

$$V_{CE} = 16 - 0.63 \times 5$$

$$V_{CE} = 12.85 \text{ Volt}$$

Therefore, operating point

$$Q \equiv (V_{CE}, I_C) \equiv (12.85 \text{ V}, 0.63 \text{ mA})$$

(*ii*) The stability factor,

$$S = \frac{1+\beta\left(1+\frac{R_{TH}}{R_E}\right)}{1+\beta+\frac{R_{TH}}{R_E}} \qquad (ii)$$

But $\quad R_{TH} = \frac{R_1 . R_2}{R_1+R_2} = \frac{56\times10^3\times20\times10^3}{(56+20)\times10^3} = \frac{1120\times10^3}{76} = 14.73 \times 10^3 = 14.73 \text{ k}\Omega$

Using equation (*ii*)

$$S = \frac{1+66\left(1+\frac{14.73}{2}\right)}{1+66+\frac{14.73}{2}} = \frac{1+66(1+7.36)}{1+66+7.36}$$

$$S = \frac{1+(66+8.36)}{67+7.36} = \frac{1+551.76}{74.36} = \frac{552.76}{74.36} = 7.40 \qquad \textbf{Ans.}$$

EXAMPLE 6.22 A potential divider biasing circuit uses a germanium transistor. The operating point is fixed such that I_C = 2 mA, V_{CE} = 4 volt. If R_C = 2 kΩ, V_{CC} = 10 V and β = 50, determine the values of R_1, R_2 and R_E. Assume $I_1 = 10.I_B$.

Solution: $\quad \because \quad I_C = \beta I_B$

Therefore, $\quad I_B = \frac{I_C}{\beta} = \frac{2\times10^{-3}}{50}$

$$I_B = 0.04 \text{ mA} = 40 \ \mu\text{A}$$

Since, the base current I_B is very small compared to I_1, therefore, we can assume that same current I_1 is flowing through R_1 and R_2.

Given that $\quad I_1 = 10\ I_B$

$$= 0.04 \times 10^{-3} \times 10$$

$$= 0.4 \text{ mA}$$

From figure $\quad V_{CC} = I_1.(R_1 + R_2)$

$$R_1 + R_2 = \frac{V_{CC}}{I_1}$$

$$R_1 + R_2 = \frac{10}{0.4 \text{ mA}}$$

$$R_1 + R_2 = 25 \text{ k}\Omega \qquad ...(i)$$

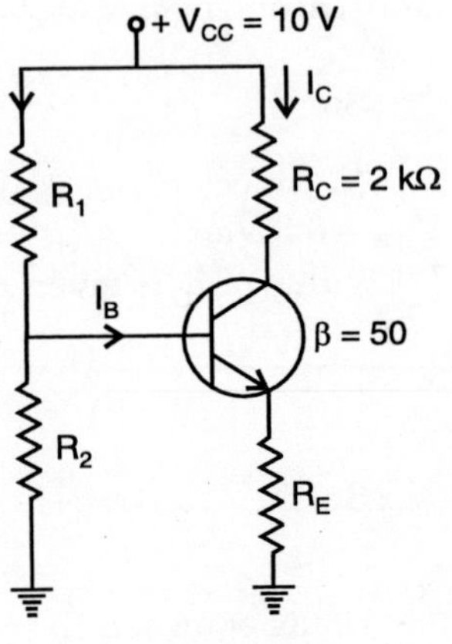

Fig. 6.43.

Applying KVL to the output side

$$V_{CC} = I_C.R_C + V_{CE} + I_E R_E$$

But $\quad I_E \cong I_C$

Therefore, $\quad V_{CC} = I_C.R_C + V_{CE} + I_C R_E$

$$10 = 2 \times 10^{-3} \times 2 \times 10^{-3} \times 4 + 2 \times 10^{-3}\ R_E$$

$$10 - 4 - 4 = 2 \times 10^{-3}.\ R_E$$

$$R_E = \frac{2 \text{ V}}{2 \text{ mA}} = 1 \text{ k}\Omega \qquad ...(ii)$$

The voltage V_2 across R_2 is given by

$$V_2 = V_{BE} + V_E = V_{BE} + R_E.I_C$$

$$V_2 = 0.3 + 1 \times 10^3 \times 2 \times 10^{-3}$$

$$V_2 = 0.3 + 2 = 2.3 \text{ Volt}$$

Value of $$R_2 = \frac{V_2}{I_1} = \frac{2.3}{0.4 \times 10^{-3}} = 5.75 \text{ k}\Omega$$

Using equation (*i*) $$R_1 + R_2 = 25 \text{ k}\Omega$$

$$R_1 = 25 \text{ k}\Omega - 5.75 \text{ k}\Omega = 19.25 \text{ k}\Omega \quad \textbf{Ans.}$$

EXAMPLE 6.23 In the circuit shown (Figure 6.44) V_{CC} = 24 V, R_C = 10 kΩ and R_E = 270 ohm. If the Silicon transistor is used with β = 45 and if V_{CE} = 5 V, find R_B. Neglect the reverse saturation current.

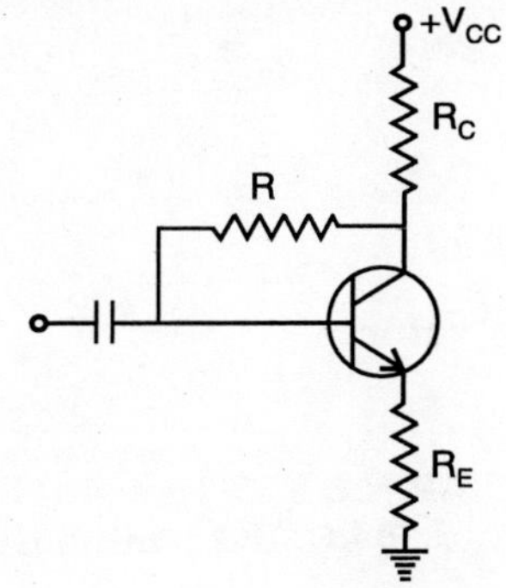

Fig. 6.44.

Solution: Let us redraw the circuit. (figure 6.45)

Applying KVL to collector-emitter circuit

$$V_{CC} = (I_C + I_B)R_C + V_{CE} + I_E R_E$$

But, we know that $$I_E = I_C + I_B$$

∴ $$V_{CC} = (I_C + I_B)\, R_C + V_{CE} + (I_C + I_B)R_E$$

$$V_{CC} = (I_C + I_B)\, [R_C + R_E] + V_{CE}$$

But $$I_C = \beta I_B + I_{CEO}$$

Neglecting I_{CEO}

$$I_C = \beta I_B$$

∴ $$V_{CC} = (\beta I_B + I_B)\, [R_C + R_E] + V_{CE}$$

$$V_{CC} = I_B\, (\beta + 1)\, (R_C + R_E) + V_{CE}$$

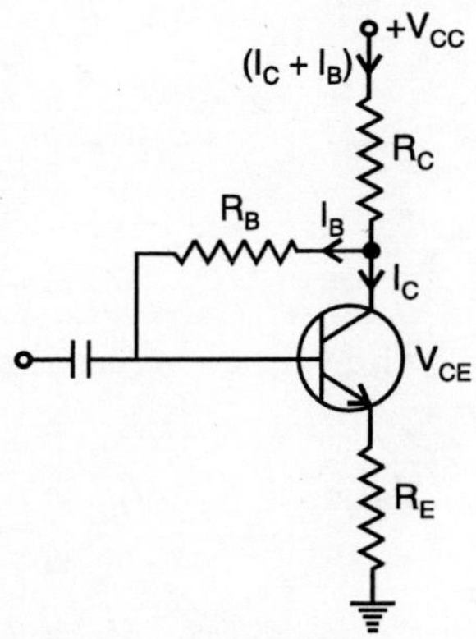

Fig. 6.45.

Putting all the given values

$$24 = (45 + 1).I_B\, (10 \times 10^3 + 270) + 5$$

$$19 = 46 I_B.(10.27 \times 10^3)$$

or $$46 I_B = \frac{19}{10.27 \times 10^3} = 1.85 \text{ mA}$$

or $$I_B = \frac{1.85 \text{ mA}}{46} = 40\ \mu\text{A}$$

From circuit,

$$V_{CE} = V_{BE} + R.I_B$$

$$R = \frac{V_{CE} - V_{BE}}{I_B} = \frac{5 - 0.6}{40 \times 10^{-6}}$$

$$= \frac{4.4}{40} \times 10^{-6} = 1.1 \times 10^5 = 110 \text{ k}\Omega \quad \textbf{Ans.}$$

EXAMPLE 6.24 Calculate the collector current and collector to emitter voltage of the circuit given in figure 6.46 assuming the following circuit components and transistor specifications: R_1 = 40 kΩ, R_2 = 4 kΩ, R_C = 10 kΩ, R_E = 1.5 kΩ, V_{BE} = 0.5 V, β = 40, V_{CC} = 22 V.

Solution: For given self-bias circuit, the value of collector current I_C is given by

$$I_C = \frac{V_B - V_{BE}}{R_E} \qquad \text{... (i)}$$

But base voltage $V_B = V_{CC} \times \frac{R_2}{R_1 + R_2}$

$$V_B = 22 \times \frac{4 \times 10^3}{(40+4) \times 10^3} = \frac{22 \times 4}{44} = 2 \text{ Volt}$$

Using equation (*i*) $I_C = \frac{V_B - V_{BE}}{R_E}$

$$= \frac{2 - 0.5}{1.5 \times 10^3} = \frac{1.5}{1.5 \times 10^3}$$

$\therefore \quad I_C = 1 \text{ mA}$

Now, Collector-to-emitter voltage

$$V_{CE} = V_{CC} - I_C (R_C + R_E)$$

$$V_{CE} = 22 - 1 \times 10^{-3} (10 + 1.5) \times 10^3$$

$$V_{CE} = 22 - 11.5$$

$$= 8.5 \text{ Volt}$$

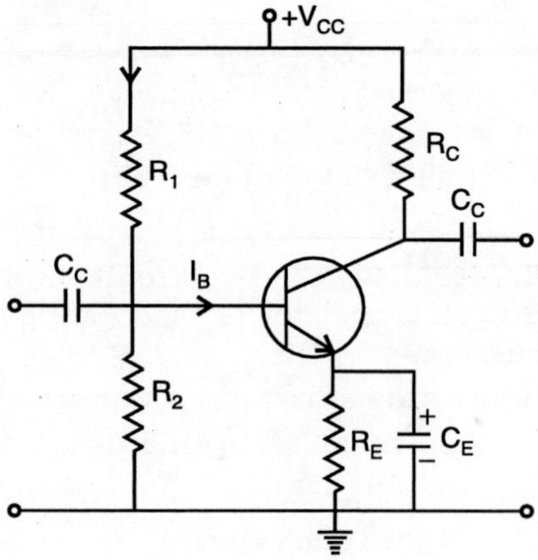

Fig. 6.46.

Hence, $I_C \equiv 1 \text{ mA}$

and $V_{CE} = 8.5 \text{ Volt}$ **Ans.**

EXAMPLE 6.25 In a single stage CE amplifier V_{CC} = 20 Volt, β = 50, R_E = 200 Ω, R_1 = 60 kΩ and R_2 = 30 kΩ. Determine the d.c. voltage across R_E.

Solution: The base voltage V_B (the voltage at base with respect to ground) is

$$V_B = V_{CC} \times \frac{R_2}{R_1 + R_2}$$

[$\because V_B$ is the voltage across R_2]

$$= 20 \times \frac{30 \times 10^3}{(60+30) \times 10^3} = \frac{20 \times 30}{90}$$

$$V_B = \frac{20}{3} = 6.67 \text{ Volt}$$

The voltage across the emitter resistor R_E is given by

$$V_E = V_B - V_{BE}$$

Taking $V_{BE} \cong 0.6$ Volt

$$V_E = 7.67 - 0.6$$

$$V_E = 7.07 \text{ Volt}$$ **Ans.**

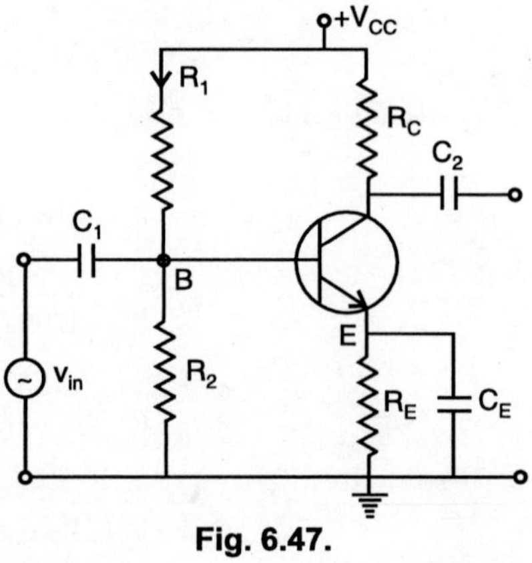

Fig. 6.47.

EXAMPLE 6.26 A Silicon transistor with V_{BE} = 0.8 V, h_{FE} = 100, $V_{CE\,sat}$ = 0.2 V, is used in the circuit shown in figure 6.48. Find the minimum value of R_C for which the transistor reaches in saturation.

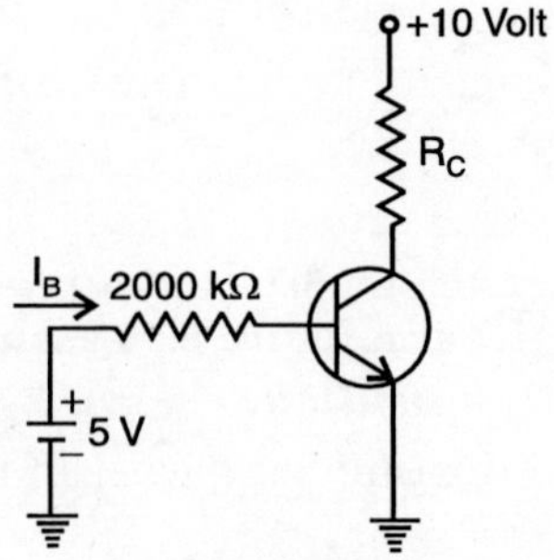

Fig. 6.48.

Solution: Base current I_B is given by

$$I_B = \frac{5 \text{ V}}{200 \text{ k}\Omega} = \frac{5}{200 \times 10^3}$$

$$I_B = \frac{5}{2} \times 10^{-5} = 25 \,\mu\text{A}$$

Collector current $I_C = \beta I_B = 100 \times 25 \times 10^{-6}$

$I_C = 25 \times 10^{-4} = 2.5$ mA

From the figure $V_{CC} = V_{CE} + I_C R_C$

$$10 = 0.2 + 2.5 \times 10^{-3} \times R_{C(\min)}$$

$$9.8 = 2.5 \times 10^{-3} \times R_{C(\min)}$$

$$R_{C(\min)} = \frac{9.8}{2.5} \text{ k}\Omega = 3.92 \text{ k}\Omega \quad \textbf{Ans.}$$

EXAMPLE 6.27 For the circuit shown in figure 6.49, assume $h_{FE} = 100$ and $V_{BE} = 0.8$ V.

(*a*) Find whether the silicon transistor is in cut-off, saturation or active region.

(*b*) Find V_C.

(*c*) Find the minimum value for the emitter resistor for which the transistor operates in active region.

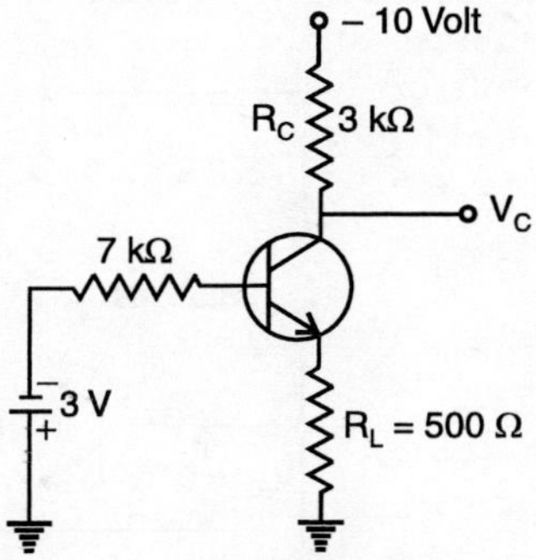

Fig. 6.49.

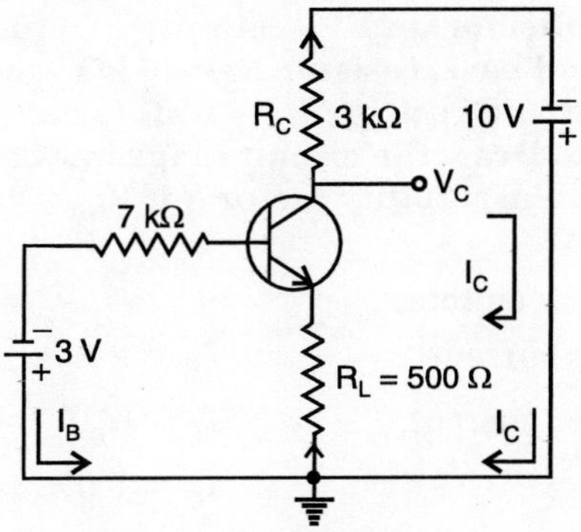

Fig. 6.50.

Solution: Redrawing the given circuit (figure 6.50)

Applying KVL to the input loop

$$7 \times 10^3 . I_B + 0.8 + 500\,(I_B + I_C) = 3$$

or $$7500\, I_B + 500\, I_C = 2.2 \qquad (i)$$

Applying KVL to the output loop

$$3 \times 10^3 . I_C + V_{CE} + 500\,(I_B + I_C) = 10$$

Taking limiting value of $V_{CE} = 0.2$ for saturation as in last question

$$3000\, I_C + 0.2 + 500\, I_B + 500\, I_C = 10$$

or $$500\, I_B + 3500\, I_C = 9.8 \qquad ....(ii)$$

Solving equations (*i*) and (*ii*), we have

$$I_C = 2.78 \text{ mA}, \quad I_B = 0.1 \text{ mA}$$

For saturation

$$I_{B\,(min)} = \frac{I_C}{h_{FE}} = -\frac{2.78}{100} \text{ mA}$$

$$I_{B\,(min)} = 0.0278 \text{ mA}$$

But $I_B = 0.1$ mA, which is greater than 0.0278 mA

$$I_B = 0.1 \text{ mA} > I_{B\,(min)} = 0.0278 \text{ mA}$$

Therefore, the transistor is in **saturation region.**

(*b*) Since, transistor is in saturation,

$$V_C = V_{BE} = 0.8 \text{ Volt} \quad \textbf{Ans.}$$

(*c*) From circuit, we have

$$V_{CE} = V_C - V_E = V_C - 500\,(I_C + I_B)$$

$$= 0.8 - 500(2.78 \times 10^{-3} + 0.1 \times 10^{-3})$$

$$V_{CE} = 0.8 - 500 \times 2.88 \times 10^{-3} = -\,0.64 \text{ Volt}$$

Again from circuit $\quad V_{CE} = V_{CC} - I_C \cdot (R_C + R_E)$

$$-\,0.64 = 10 - 2.78 \times 10^{-3}(3 \times 10^3 + R_E)$$

Solving above equation, we get

$$R_E = 827\ \Omega \quad \textbf{Ans.}$$

EXAMPLE 6.28 A CE amplifier employing an NPN transistor has load resistor R_C connected between collector and V_{CC} supply of +16 V. For biasing, a resistor R_1 is connected between collector and base, resistor R_2 = 30 kΩ is connected between base and ground and resistor R_E = 1 kΩ is connected between emitter and ground. Draw the circuit diagram. Calculate the values of R_1 and R_C and the stability factor S if V_{BE} = 0.2V, I_E = 2 mA, α_0 = 0.985 and V_{CE} = 6V.

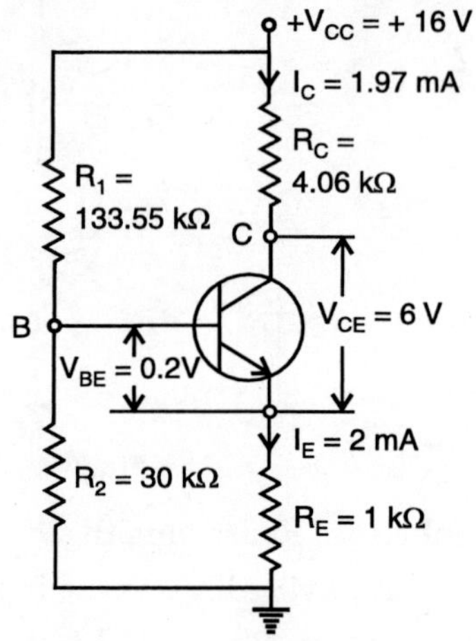

Fig. 6.51.

Solution: Given that:

Emitter current, $\quad I_E = 2$ mA

Collector current, $\quad I_C = \alpha_0 I_E = 0.985 \times 2$

or $\quad I_C = 1.97$ mA

Also, base current, $\quad I_B = I_E - I_C = 2 - 1.97 = 0.03$ mA

We know that

$$\beta = \frac{\alpha}{1-\alpha}$$

Therefore, $$\beta = \frac{0.985}{1-0.985} = 65.667$$

Collector resistance, $$R_C = \frac{V_{CC} - V_{CE} - I_E R_E}{I_C} = \frac{16 - 6 - 2 \text{ mA} \times 1 \text{ k}\Omega}{1.97 \text{ mA}} = 4.06 \text{ k}\Omega \quad \textbf{Ans.}$$

Voltage drop across resistor R_2

$$V_{Th} = \frac{R_2}{R_1 + R_2} V_{CC} = \frac{30 \times 16}{30 + R_1} = \frac{480}{30 + R_1} \quad \ldots(i)$$

Again, we know that

$$R_{Th} = \frac{R_1 R_2}{R_1 + R_2} = \frac{30 R_1}{30 + R_1}$$

and $$V_{Th} = I_B R_{Th} + V_{BE} + I_E R_E$$

or $$\frac{480}{30 + R_1} = 0.03 \times \frac{30 R_1}{30 + R_1} + 0.2 + 2 \text{ mA} \times 1 \text{ k}\Omega$$

or $$480 = 0.9\, R_1 + 2.2\,(20 + R_1)$$

Solving, we get $\quad R_1 = 133.55$ kΩ **Ans.**

For voltage-divider biasing, stability factor S is expressed as

$$\text{Stability factor, } S = \frac{1+\beta}{1+\beta\dfrac{R_E}{R_{Th}+R_E}}$$

Substituting all the values, we get

$$S = \frac{1+65.667}{1+65.667\dfrac{1}{\dfrac{133.55\times 30}{133.55+30}+1}} = 18.65 \quad \textbf{Ans.}$$

EXAMPLE 6.29 Assume that a silicon transistor with $\beta_0 = 50$, $V_{BE} = 0.6$ V and $V_{CC} = 20$ V and $R_C = 4.7$ kΩ is used in a self-biased circuit. It is designed to establish a Q-point at $V_{CE} = 8$ V and $I_C = 2$ mA and stability factor $S \le 5.0$. Design the circuit with all component values.

Solution: Given that

Collector current, $I_C = 2$ mA

and $\beta_0 = 50$

Base current, $$I_B = \frac{I_C}{\beta_0} = \frac{2}{50} = 0.04 \text{ mA}$$

Emitter current is given as

$$I_E = I_B + I_C = 0.04 + 2 = 2.04 \text{ mA}$$

$$V_{CE} = 8V$$

$$V_{BE} = 0.6 \text{ V}$$

$$V_{CC} = 20 \text{ V}$$

We know that for a self-biased circuit

Emitter resistance, $$R_E = \frac{V_{CC}-V_{CE}}{I_C} - R_C$$

Substituting, all the values, we get

$$R_E = \frac{20-8}{2\text{ mA}} - 4.7 = 1.3 \text{ k}\Omega \quad \textbf{Ans.}$$

Also, stability factor for self-bias circuit is given as

$$S = 1+\beta\frac{1+R_{Th}/R_E}{1+\beta+\dfrac{R_{Th}}{R_E}}$$

Substituting all the values, we get

$$S = (1+50)\frac{1+R_{Th}/1.3}{1+50+\dfrac{R_{Th}}{R_1}}$$

Solving, we get $R_{Th} = 5.765$ kΩ

Also, $$V_{Th} = I_B R_{Th} + V_{BE} + I_E R_E$$

or $$V_{Th} = 0.04 \times 5.765 + 0.6 + 2.04 \times 1.3 = 3.486 \text{ V}$$

Again, $$V_{Th} = \frac{R_2}{R_1+R_2}V_{CC} = \frac{R_{Th}}{R_1}V_{CC}$$

or $$R_1 = R_{Th}\cdot\frac{V_{CC}}{V_{Th}}$$

Substituting given values, we get

$$R_1 = 5.765 \times \frac{20}{3.4826} = 331.3 \text{ k}\Omega \qquad \textbf{Ans.}$$

Now since $$V_{TH} = \frac{R_2}{R_1 + R_2} V_{CC}$$

Therefore, $$R_2 = \frac{V_{Th} \times R_1}{V_{CC} - V_{Th}} = \frac{3.4826 \times 33.1}{20 - 3.4826} = 7.98 \text{ k}\Omega \qquad \textbf{Ans.}$$

EXAMPLE 6.30 Figure 6.52 shows a self-biased transistor amplifer using a silicon transistor with V_{CC} = 20 V, h_{FE} = 400 and V_{BE} = 0.65 V. The transistor should be biased at V_{CE} = 10 V and I_C = 0.6 mA. Find the values of R_C, R_E, R_1 and R_2 such that it meets the following specifications over the temperature range of 25°C to 145°C.

$\frac{\Delta I_C}{I_C} \leq 10°$, V_{BE} at 25°C = 650 ± 50 mV, I_{CO} at 25°C = 5 μA max, I_{CO} at 145° = 3.0 μA max.

Assume that percentage change in I_C due to V_{BE} and I_{CO} is same (5%).

Solution: Change in I_C due to V_{BE} and I_{CO} is 5%

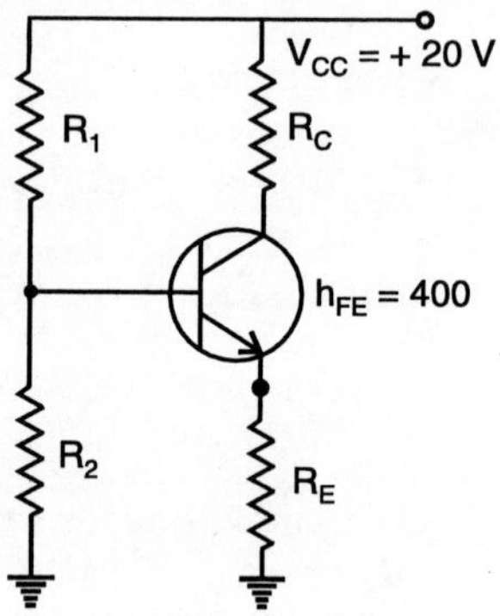

Fig. 6.52.

Change in collector current, $$\Delta I_C = \frac{5}{100} \times 0.6 = 0.03 \text{ mA}$$

Change in reverse saturation current,

$$\Delta I_{CO} = 3 \times 10^{-6} - 0.005 \times 10^{-6}$$
$$= 2.995 \text{ μA}$$

Stability factor, $$S = \frac{\Delta I_C}{\Delta I_{CO}}$$

or $$S = \frac{0.03 \times 10^{-3}}{2.995 \times 10^{-6}} = 10$$

For a silicon transistor, V_{BE} decreases at the rate of –2.5 mV/°C

Therefore, $$\Delta V_{BE} = -2.5\,(145 - 25) = -300 \text{ mV}$$

Now $$S_V = \frac{\Delta I_C}{\Delta V_{BE}} = \frac{0.03 \times 10^{-3}}{-300 \times 10^{-3}} = -0.0001$$

$$\beta = h_{FE} = 400$$

and $$S_V = \frac{-S}{R_{Th} + R_E} \times \frac{\beta}{\beta + 1}$$

or $$-0.0001 = \frac{-10}{R_{Th} + R_E} \times \frac{400}{401}$$

or $$R_{Th} + R_E = 99{,}750.6 \qquad ...(i)$$

Again, $$S = \frac{R_E + R_{Th}}{R_E + \dfrac{R_{Th}}{\beta + 1}}$$

or $$10 = \frac{R_E + R_{Th}}{R_E + \dfrac{R_{Th}}{\beta + 1}}$$

or $$R_E = \frac{391}{3{,}609} R_{Th} \qquad ...(ii)$$

Solving equations (*i*) and (*ii*), we get

$$R_E = 9.75 \text{ k}\Omega \quad \textbf{Ans.}$$

and

$$R_{Th} = 90 \text{ k}\Omega$$

$$\Delta I_C = S\, I_{CO} + S_V\, \Delta V_{BE} = 0.03 \times 10^{-3} + 0.03 \times 10^{-3} = 0.06 \text{ mA}$$

For the given circuit, we have

$$V_{CC} = I_C\, R_C + V_{CE} + I_E\, R_E$$

$$20 = 0.66\, R_C + 10 + \left(0.66 + \frac{0.66}{400}\right) \times 9.75$$

or

$$R_C = 5.377 \text{ k}\Omega \quad \textbf{Ans.}$$

Now

$$V_{Th} = I_B\, R_{Th} + V_{BE} + I_E\, R_E$$

Substituting all the values, we get

$$V_{Th} = \frac{0.66}{400} \times 90 + 0.65 + \left(0.66 + \frac{0.66}{400}\right) \times 9.75 = 7.25 \text{ V}$$

Now, since

$$R_1 = R_{Th} \frac{V_{CC}}{V_{Th}}$$

Therefore,

$$R_1 = 90 \times \frac{20}{7.25} = 248 \text{ k}\Omega \quad \textbf{Ans.}$$

Also,

$$R_2 = \frac{V_{Th} \times R_1}{V_{CC} - V_{Th}}$$

Substituting values, we get

$$R_2 = \frac{7.25 \times 2.48}{20 - 7.25} = 141 \text{ k}\Omega \quad \textbf{Ans.}$$

As R_1, R_2, R_C and R_E are designed for variation upto 145°C, the variation of V_{BE} at 25°C of 650 ± 50 mV will be absorbed in the overall variation.

EXAMPLE 6.31 Find the value of β, V_{CC} and R_B of the following circuit (Figure 6.53):

Solution: Given, voltage across emitter resistor R_E, $V_E = 2.1$ Volt.

But $V_E = I_E . R_E$

$$I_E = \frac{V_E}{R_E} = \frac{2.1 \text{ V}}{0.68 \text{ k}\Omega} = \frac{2.1}{0.68} \times 10^{-3} \text{ A} = 3.08 \text{ mA}$$

But, we know that $I_C \cong I_E$

∴ $I_C = 3.08$ mA

Also, given that $I_B = 20\ \mu\text{A}$

∵ $I_C = \beta . I_B$

$$\beta = \frac{I_C}{I_B} = \frac{3.08 \times 10^2}{2}$$

$$= 1.54 \times 10^2 = 154$$

From figure,

$$V_{CC} = V_{CE} + I_C . R_C + I_E . R_E$$

$$= V_{CE} + I_C(R_C + R_E) \quad [\because I_C \cong I_E]$$

$$V_{CC} = 7.3 + 3.08 \times 10^{-3}\, (2.7 + 0.68) \times 10^3$$

$$= 7.3 + 3.08 \times 3.38$$

$$V_{CC} = 7.3 + 10.4 = 17.7 \text{ Volt}$$

Again from figure, $V_{CC} = I_B.R_B + V_{BE} + V_E$

$= I_B.R_B + 0 + 2.1$ (neglecting V_{BE})

$17.7 = I_B R_B + 2.1$

or $I_B R_B = 17.7 - 2.1 = 15.6$

$$R_B = \frac{15.6}{I_B} = \frac{15.6}{20 \times 10^{-6}} = \frac{15.6}{2} \times 10^5$$

$= 7.8 \times 10^5 = 780$ kΩ **Ans.**

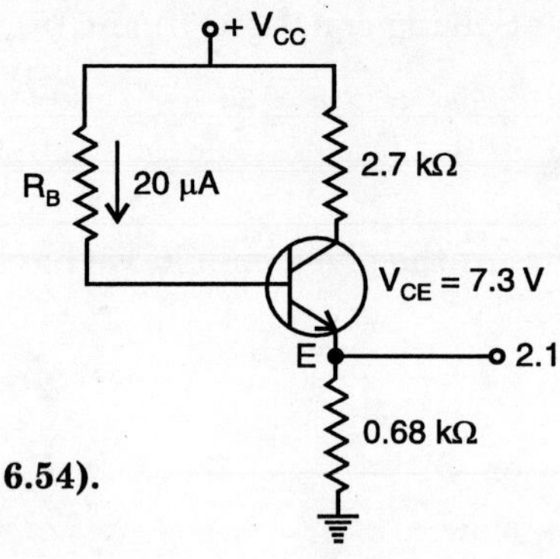

Fig. 6.53.

EXAMPLE 6.32 Find I_C and V_{CE} for the following circuit (figur 6.54). What will happen to V_{CE} if β increases due to temperature?

Solution: Applying KVL to the base-emitter loop

$$V_{CC} = (I_C + I_B) R_C + I_B.R_B + V_{BE} + I_E R_E$$

$$V_{CC} = I_C R_C + I_B.R_B + V_{BE} + I_E R_E$$

∵ $(I_C >> I_B)$

But $I_E \cong I_C$ and neglecting V_{BE}

We have $V_{CC} = I_C R_C + I_B R_B + 0 + I_C R_E$

$= I_C (R_C + R_E) + I_B R_B$

But $I_C = \beta I_B$

Therefore, $V_{CC} = \beta I_B (R_C + R_E) + I_B R_B$

$= I_B [\beta(R_C + R_E) + R_B]$

$$I_B = \frac{V_{CE}}{R_B + \beta(R_C + R_E)} = \frac{18}{510 \times 10^3 + 90(2.2 + 1.8)10^3}$$

$$= \frac{18 \times 10^{-3}}{870} = 0.02 \times 10^{-3}$$

$I_B = 20 \times 10^{-6}$ A $= 20$ μA

∵ $I_C = \beta I_B = 90 \times 20 \times 10^{-6}$

Hence, $I_C = 1800 \times 10^{-6} = 1.8 \times 10^{-3} = 1.8$ mA

Now, $V_{CE} = I_B R_B + V_{BE}$

Taking $V_{BE} \cong 0$

$V_{CE} \cong I_B.R_B = 20 \times 10^{-6} \times 510 \times 10^3 = 20 \times 510 \times 10^{-3} = 102 \times 10^{-1}$

$V_{CE} = 10.2$ Volt **Ans.**

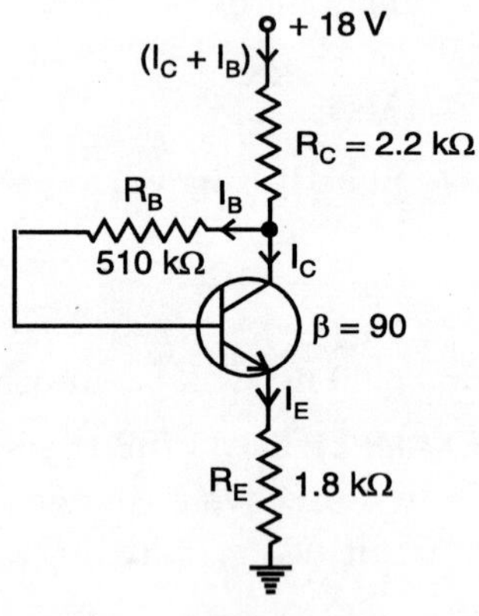

Fig. 6.54.

EXAMPLE 6.33 Consider the circuit shown in figure 6.55, where the transistor (silicon) has a β_{dc} = 80. Calculate the voltage between collector and ground. For what approximate value of β_{dc} does the transistor saturate?

Solution: From the given circuit, we have

$$I_B = \frac{V_{CC} - V_{BE}}{R_B + (\beta + 1) R_E}$$

Substituting all the given values, we get

$$I_B = \frac{25 - 0.7}{180 + (81)(0.2)} \text{ mA}$$

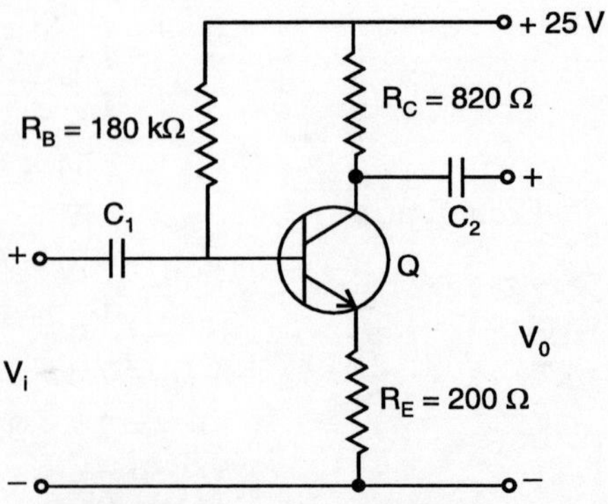

Fig. 6.55. Circuit for example 6.33.

Simplifying, we get $I_B = 0.124$ mA

Also, $I_C = \beta . I_B$

$= 80 \times 0.124 = 9.91$ mA

Therefore, $V_{CG} = V_{CC} - I_C R_C = 25 - 9.91 \times 0.82 = 17.88$ Volts **Ans.**

Under saturation, we have

$$V_{CE_{sat}} = 0.3 \text{ Volt}$$

Therefore, $$V_{CC} - V_{CE_{Sat}} = I_{C_{Sat}} . R_C + \frac{I_{C_{Sat}}}{\beta}(\beta + 1) R_E$$

Since, $\beta >> 1$, therefore, we write

$$V_{CC} - V_{CE_{Sat}} \cong I_{C_{Sat}} (R_C + R_E)$$

or $$I_{C_{Sat}} \cong I_C = \frac{V_{CC} - V_{CE_{Sat}}}{R_C + R_E} = \frac{25 - 0.3}{0.82 + 0.2} = 24.2 \text{ mA}$$

Hence, $$I_B = \frac{V_{CC} - (I_E R_E + V_{BE})}{R_B} = \frac{V_{CC} - (I_C R_E + V_{BE})}{R_B} = 0.11 \text{ mA}$$

Therefore, required $$\beta_{dc} \cong \frac{I_C}{I_B} = \frac{24.2}{0.11} = 220 \quad \textbf{Ans.}$$

EXAMPLE 6.34 Consider the voltage-divider biasing circuit for transistor shown in figure 6.56 and assume that the transistor has a nominal dc forward current transfer ratio h_{FE}. Determine the condition that R_1, R_2 and R_E should satisfy so that I_c becomes practically insensitive to transistor replacement. Indicate why R_C does not play any significant role in this regard.

Solution: From the given circuit, we have

$$I_B = \left[\frac{V_{CC} \times R_2}{R_1 + R_2} - V_{BE}\right] \frac{1}{[R_1 R_2/(R_1 + R_2) + (1 + h_{FE}) R_E]}$$

$$I_B = \left[V_{CC} \cdot \frac{R_B}{R_1} - V_{BE}\right] \frac{1}{R_B + (1 + h_{FE}) R_E}$$

and $I_C = h_{FE} I_B + (1 + h_{FE}) I_{CO}$

Assuming $h_{FE} >> 1$, we get

$$I_C = \frac{\left[V_{CC}(R_B/R_1) - V_{BE}\right] h_{FE}}{R_B + h_{FE} R_E} + h_{FE} I_{CO}$$

or $$I_C \cong \frac{V_{CC}(R_B/R_1) - V_{BE}}{(R_B/h_{FE}) + R_E}$$

Further, if we choose, $R_B << h_{FE} R_E$, we obtain

$$I_C \cong \frac{V_{CC}(R_B/R_1) - V_{BE}}{R_E}$$

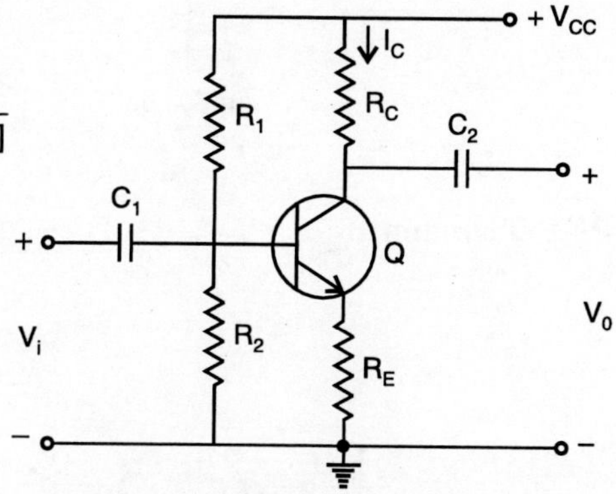

Fig. 6.56. Circuit for example 6.34.

Four steps to achievement : plan purposefully, prepare prayerfully, proceed positively, pursue persistently.

— **William Arthur Ward**

This shows that I_C is independent of h_{FE}. Since transistor replacement may result into significant change in h_{FE}, therefore, the desired condition is

$$R_B << h_{FE} R_E.$$

Again, since in a transistor, I_C is controlled by I_B and I_B is not dependent on R_C, therefore, R_C will not have any control over I_C. **Hence, proved.**

EXAMPLE 6.35 Differentiate between I_{CO} and I_{CBO}. What is the effect of temperature on I_{CBO}?

Solution: I_{CO} is a reverse saturation current in a P-N-junction when it is reverse biased. I_{CBO} is a leakage current in a common emitter transistor which flows from collector to base due to minority charge carriers, when emitter is common to both input and output circuit.

Effect of Temperature on I_{CBO}

When the temperature of the junction increases, the number of minority carriers increases and hence, the current I_{CBO} increases. Therefore, it can be concluded that on increasing the temperature, the current I_{CBO} increases.

EXAMPLE 6.36 Draw N-P-N and P-N-P transistors. Label all the currents and show the direction of flow. How are all the currents of a transistor related?

Solution: NPN Transistor

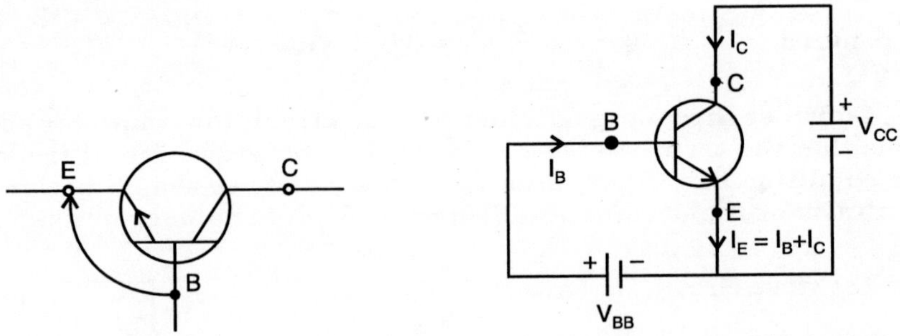

(a) Symbol (b) Common emitter NPN transistor

Fig. 6.57.

PNP Transistor

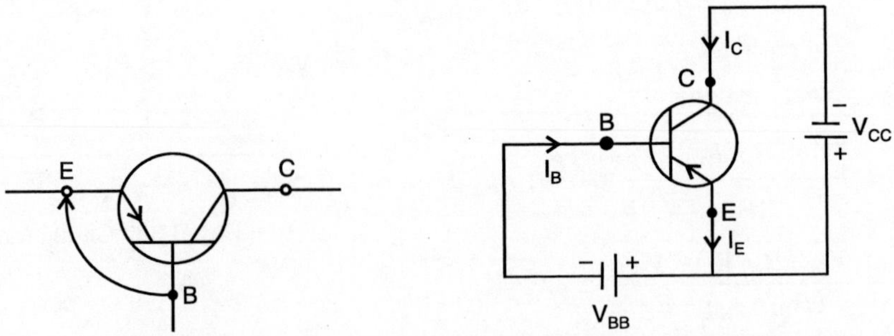

(a) Symbol (b) Common emitter PNP transistor

Fig. 6.58.

The currents of a transistor are related by the following relation:

$$I_E = I_B + I_C$$

where I_E = emitter current

I_C = collector current
I_B = base current.

EXAMPLE 6.37 **Consider a d.c. bias circuit with voltage feedback as shown in figure 6.59. Determine the quiescent levels of I_{CQ} and V_{CEQ}. The β of the transistor is 90, and cut-in voltage is 0.7 V.**

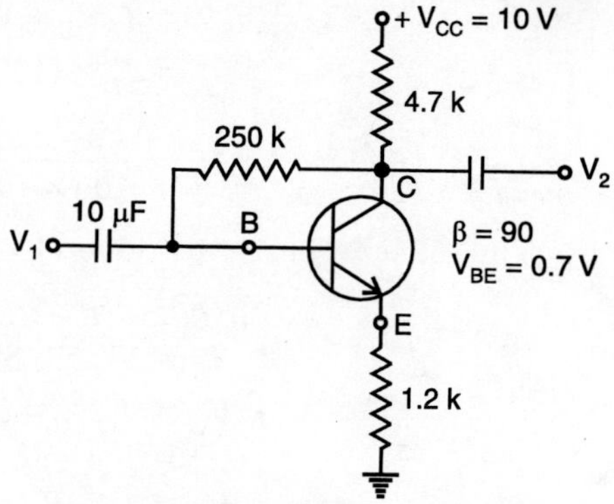

Fig. 6.59. Circuit for example 6.37.

Solution: Figure 6.60 shows the d.c. equivalent circuit for the given circuit.

Now, applying KVL to the input side, we get

$$V_{CC} = (I_C + I_B)\,R_C + I_B R_B + V_{BE} + I_E R_E$$

or $$V_{CC} = (I_C + I_B)R_C + I_B R_B + V_{BE} + (I_C + I_B)\,R_E$$

or $$V_{CC} = (I_C + I_B)\,(R_C + R_B) + I_B R_B + V_{BE}$$

But, $I_B = \dfrac{I_C}{\beta}$

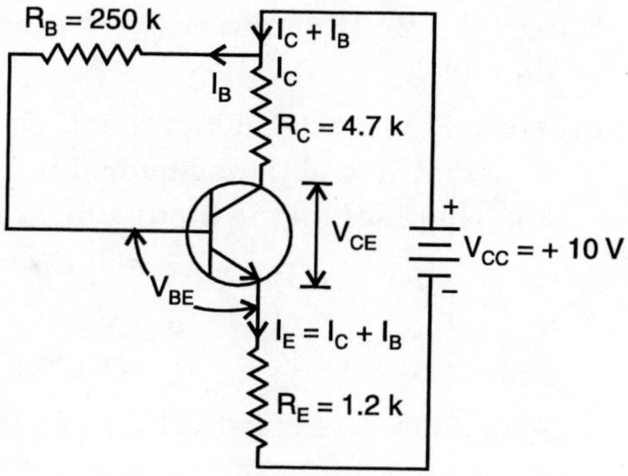

Fig. 6.60.

Therefore, we have

$$V_{CC} = \left(I_C + \frac{I_C}{\beta}\right)(R_C + R_B) + \frac{I_C}{\beta}\cdot R_B + V_{BE}$$

or $$V_{CC} = \left(\frac{I_C}{\beta}\right)(R_C + R_E) + \frac{I_C}{\beta}\cdot R_B + V_{BE}$$

or $$V_{CC} - V_{BE} = I_C\left[\left(1 + \frac{1}{\beta}\right)(R_C + R_E) + \frac{R_B}{\beta}\right]$$

Thus, we have

$$I_{CQ} = \frac{\beta(V_{CC} - V_{BE})}{(1+\beta)\,(R_C + R_E) + R_B} \qquad ...(i)$$

Given $\beta = 90$

$V_{CC} = 10$ Volt

$V_{BE} = 0.7$ Volt

$R_C = 4.7$ k = 4700 Ohm

$R_B = 250$ k = 250000 Ohm

Substituting all these values in equation (*i*), we get

$$I_{CQ} = \frac{90(10 - 0.7)}{(1 + 90)\,(4700 + 100) + 250000}$$

$$= \frac{90 \times 9.3}{91 \times 5900 + 250000} = 1.06 \text{ mA} \quad \textbf{Ans.}$$

Further, applying KVL to the output side, we get

$$V_{\text{CEQ}} = V_{CC} - I_C + \frac{I_C}{\beta}\,(R_C + R_E)$$

Again, substituting given values, we obtain

$$V_{CEQ} = 3.68 \text{ Volts} \quad \textbf{Ans.}$$

EXAMPLE 6.38 **Determine V_C and V_B for the circuit shown in figure 6.61. Given that β = 45 and V_{BE} = 0.7 Volt.**

Solution: Applying KVL to the input side, we get

$$V_{EE} = I_B R_B + V_{BE}$$

or $$V_{EE} = I_B \times (100\text{ k}) + 0.7$$

or $$I_B = \frac{9 - 0.7}{100 \times 10^3} = -83 \times 10^{-6} \text{ amp}$$

The collector current is given by

$$I_C = \beta I_B$$

or $$I_C = -45 \times 83 \times 10^{-6} = -3.735 \text{ mA}$$

$\because$ $$V_C = I_C R_C = (-3.735 \times 10^{-3}) \times 1.2 \times 10^3$$
$$= -4.482 \text{ Volt} \quad \textbf{Ans.}$$

Also, $$V_B = -I_B R_B = -(83 \times 10^{-6} \times 100 \times 10^3)$$
$$= -8.3 \text{ Volts} \quad \textbf{Ans.}$$

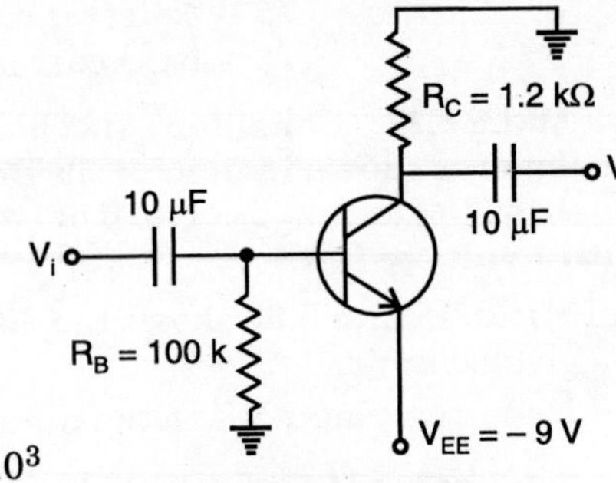

Fig. 6.61. Circuit for example 6.40.

EXAMPLE 6.39 **For the emitter bias circuit shown in Figure 6.62, determine I_B, I_C, V_{CE}, V_E, V_B and V_C.**

Solution: Figure 6.63 shows the dc equivalent circuit for the given circuit.

Note that it is obtained by making the capacitors open circuited.

Applying KVL to the input side, we get

$$V_{CC} = I_B R_B + V_{BE} + I_E R_E \quad ...(i)$$

$\because$ $$I_E = I_B + I_C$$

and $$I_C = \beta I_B$$

With above two substitutions, equation (*i*) becomes

$$V_{CC} = I_B R_B + V_{BE} + (I_B + \beta I_B) R_E$$

Simplifying, we get

$$I_B = \frac{V_{CC} - V_{BE}}{R_B + (\beta + 1) R_E}$$

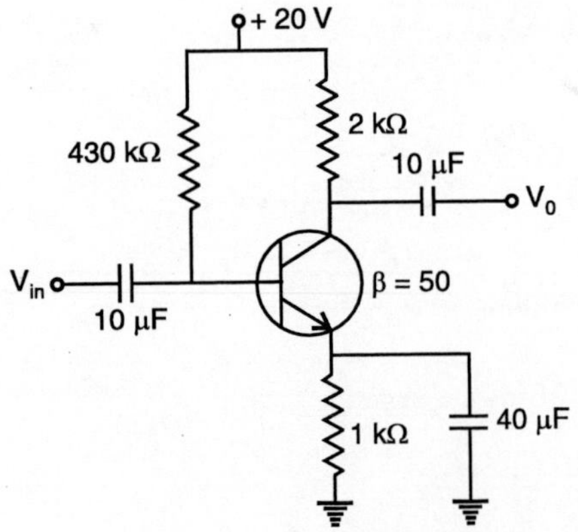

Fig. 6.62.

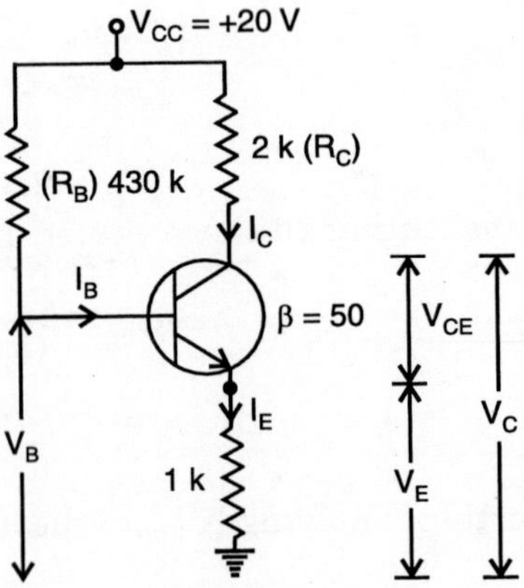

Fig. 6.63.

$V_{BE} \cong 0.7$ volts for Si transistor

Substituting all the given values, we obtain

$$I_B = \frac{20 - 0.7}{430 \times 10^3 + (50 + 1) \times 1 \times 10^3}$$

or $$I_B = 4 \times 10^{-5} \text{ Amp.} \quad \textbf{Ans.}$$

Also, $I_C = \beta I_B = 50 \times 4 \times 10^{-5} = 2$ mA **Ans.**

Now, applying KVL to the output side, we get

$$V_{CC} = I_C R_C + V_{CE} + I_E R_E$$

or $$V_{CC} = I_C R_C + V_{CE} + (I_C + I_B) R_E$$

or $$V_{CE} = V_{CC} - I_C R_C - (I_C + I_B) R_E$$

or $$V_{CE} = 20 - 2 \times 10^{-3} \times 2 \times 10^3 - (2 \times 10^{-3} + 4 \times 10^{-6}) \times 1 \times 10^3$$

Solving, we get

$V_{CE} = 13.96$ Volts **Ans.**

$\because$ $V_E = I_E R_E$

or $V_E = (I_C + I_B) R_E = (2 \times 10^{-3} + 40 \times 10^{-6}) \times 1 \times 10^3$

or $V_E = 2.04$ volts **Ans.**

Again, $V_B = V_{CC} - I_B R_B = 20 - 40 - 40 \times 10^{-6} \times 430 \times 10^3$

or $V_B = 2.8$ Volts **Ans.**

$\because$ $V_C = V_{CE} + V_E = 13.96 + 2.04$

or $V_C = 16$ Volts **Ans.**

EXAMPLE 6.40 **In the circuit shown in figure 6.64, $h_{FE} = 100$, $V_{BE} = 0.8$ Volt, $V_{CE} = 0.2$ volt. Determine whether or not the silicon transistor is in saturation and find I_B and I_C.**

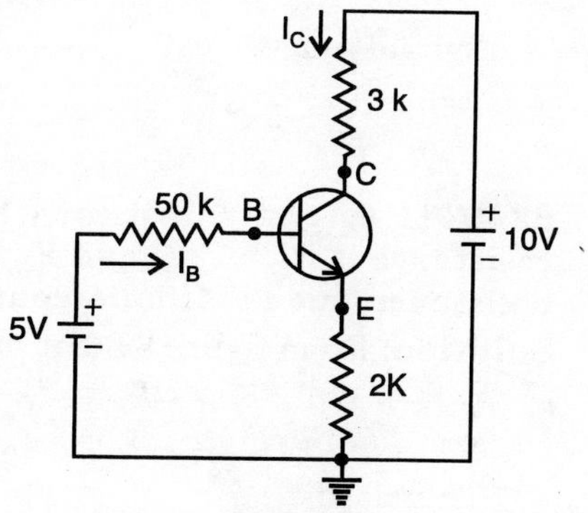

Fig. 6.64.

Solution: As $V_{CE} = 0.2$ volt for the S_i transistor which is in saturation state.

Applying KVL to the input side, we have

$$5 = 5 \times 10^3 I_B + V_{BE} + (I_C + I_B) \times 2 \times 10^3 \quad ...(i)$$

But, $I_C = \beta I_B$

or $I_C = h_{FE} \times I_B$

or $$I_C = 100\ I_B \quad ...(ii)$$

Using equations (*i*) and (*ii*), we get

$$5 = 50{,}000\ I_B + 0.8 + 101 \times 2 \times 10^3 I_B$$

Solving, we get

$$I_B = \frac{5 - 0.8}{5000 + 101 \times 2 \times 10^3} = 16.67\ \mu A \quad \textbf{Ans.}$$

Also, $I_C = 100\ I_B = 100 \times 16.67 \times 10^{-6} = 1.67$ mA **Ans.**

Now, applying KVL to the output side, we get

$$10 = I_C \times 3000 + V_{CE} + (I_C + I_B) \times 2 \times 10^3$$

or $$V_{CE} = 10 - 3000 \times 1.67 \times 10^{-3} - (1.67 \times 10^{-3} + 16.67 \times 10^{-6}) \times 2 \times 10^3$$

Solving, we get

$$V_{CE} = 1.617 \text{ Volts}$$

Thus, in the given circuit, we have

$$V_{CE} = 1.617 \text{ Volts} > 0.2 \text{ Volts}$$

Therefore, the transistor is not working in saturation region, rather, it is in active region.

Ans.

EXAMPLE 6.41 **A Silicon transistor with $(V_{BE})_{sat} = 0.8$ V, $\beta = h_{FE} = 100$, $(V_{CE})_{sat} = 0.2$ V is used**

in the circuit shown in figure 6.65. Find the minimum value of R_C for which the transistor remains in saturation.

Solution: Using KVL in base circuit for saturation, we have

$$V_{BB} = (I_B)_{sat} R_B + (V_{BE})_{sat}$$

$$5 = (I_B)_{sat} \times 200 \times 10^3 + 0.8$$

$$(I_B)_{sat} = \frac{5-0.8}{200 \times 10^3} = \frac{4.2}{200 \times 10^3} = 21 \times 10^{-6} \text{ Amp}$$

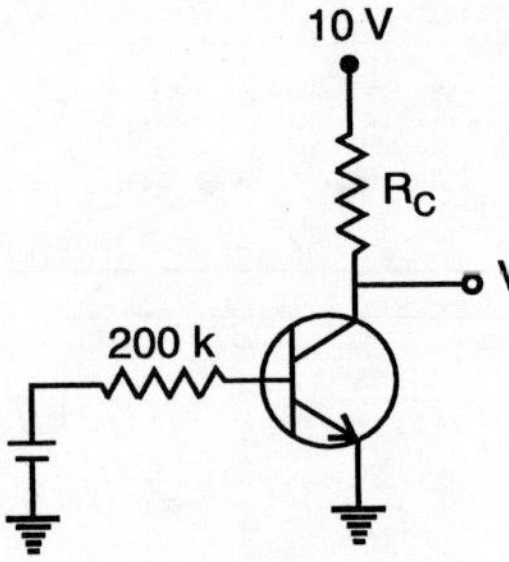

Fig. 6.65.

Now, we have

$$(I_C)_{sat} = \beta(I_B)_{sat}$$

$$= 100 \times 21 \times 10^{-6} = 2.1 \text{ m Amp}$$

Again, using KVL in collector circuit, we get

$$V_{CC} = (I_C)_{sat} \times R_C + (V_{CE})_{sat}$$

$$10 = 2.1 \times 10^{-3} \times R_C + 0.2$$

Simplifying, we get

$$R_C = \frac{10-0.2}{2.1 \times 10^{-3}} = \frac{9.8 \times 10^3}{2.1} = 4.66 \text{ k}\Omega. \textbf{ Ans.}$$

EXAMPLE 6.42 **A transistor with $\beta = 100$ is used in *CE* configuration. The collector circuit resistance is $R_C = 1$ kΩ and $V_{CC} = 20$ V. Assuming $V_{BE} = 0$, find the value of collector to base resistance, such that quiescent collector emitter voltage, is 4 V.**

Solution: From figure 6.66, it may be observed that

$$I'_C = I_C + I_B$$

$\because$ $\quad I_C >> I_B$

Therefore, $\quad I'_C = I_C \quad$ ...(*i*)

Again, from figure 6.66, we have

$$V_{CC} = I'_C R_C + V_{CE}$$

Using equation (*i*), we get

$$V_{CC} = I_C R_C + V_{CE}$$

Substituting values, we shall have

$$20 = I_C . 1 \times 10^3 + 4$$

or $\quad 16 = I_C . 10^3$

or $\quad I_C = 16$ mA

We know that $\quad I_C = \beta I_B$

or $\quad I_B = \frac{I_C}{\beta} = \frac{16 \times 10^{-3}}{100} = 160\ \mu\text{A}$

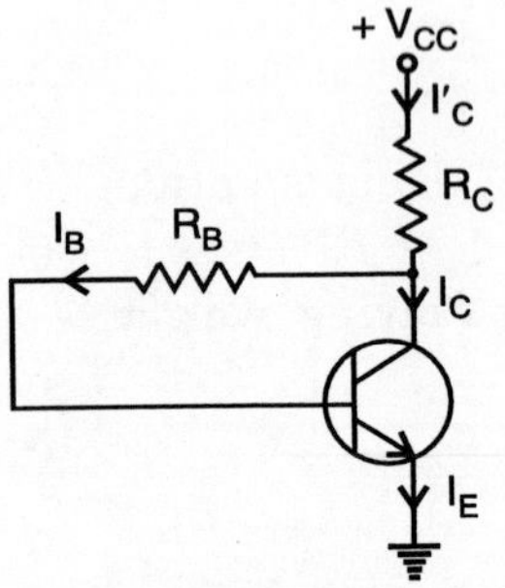

Fig. 6.66.

From figure 6.66, we have

$$V_{CE} = I_B R_B + V_{BE}$$

$$V_{CE} = I_B R_B + 0$$

$$R_B = \frac{V_{CE}}{I_B} = \frac{4}{160 \times 10^{-6}} = \frac{4}{16 \times 10^{-5}}$$

$$R_B = \frac{1}{4} \times 10^5 = 0.25 \times 10^5$$

or $\quad R_B = 25 \times 10^3$

or $\quad R_B = 25$ kΩ **Ans.**

EXAMPLE 6.43 Find I_C and V_{CE} for the following circuit. What will happen to V_{CE} if β increases due to temperature?

Solution: Applying KVL to the input side, we get

$$V_{CC} = I_C R_C + I_B R_B + V_{BE} + I_E R_E$$

Substituting values, we get

$$18 = 2.2 \times 10^3 I_C + 510 \times 10^3 I_B + 0 + 1.8 \times 10^3 I_E$$

or $$\frac{18}{1000} = 4I_E + 510\ I_B + 0$$

∴ $$I_E \cong I_C \text{ and } V_{BE} = 0 \text{ (Assuming)}$$

$$\frac{18}{1000} = 4\ (I_C + I_B) + 510\ I_B \qquad ...(i)$$

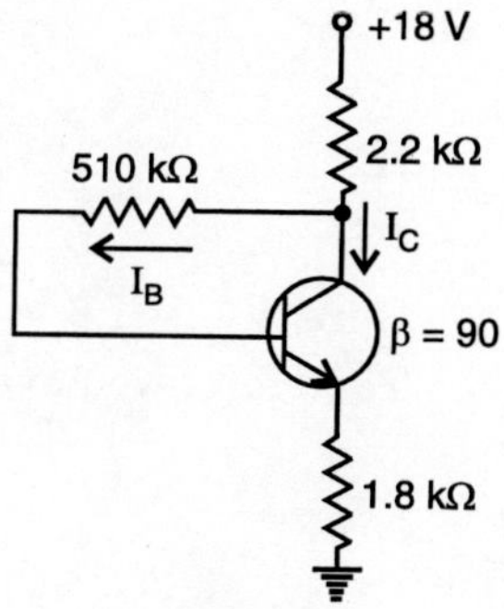

Fig. 6.67.

But $$\beta = \frac{I_C}{I_B} = 90$$

or $$I_C = 90\ I_B$$

Substituting, the value of I_C in equation (*i*), we have

$$\frac{18}{1000} = 4 \times 90\ I_B + 514\ I_B$$

$$\frac{18}{1000} = (360 + 514)I_B$$

$$I_B = \frac{18}{1000 \times 874} = 20.595\ \mu\text{A}$$

Now, value of I_C will be (using equation *ii*)

$$I_C = 90 \times 20.595\ \mu\ \text{Amp}$$

$$I_C = 1.85 \times 10^{-8}\ \text{Amp} = 1.85\ \text{mA}$$

Again applying KVL to the output side, we get

$$V_{CC} = 18 = 2.2 \times 10^3 I_E + V_{CE} + 1.8\ I_E$$

or $$18 = 4 \times 10^3 I_E + V_{CE}$$

or $$18 - 4 \times 10^3 (1.85 \times 10^{-3} + 20.595 \times 10^{-6}) = V_{CE}$$

or $$18 - 7.4824 = V_{CE}$$

or $$V_{CE} = 10.52\ \text{Volt}$$

If β increases due to temperature then V_{CE} will decrease.

EXAMPLE 6.44 Find V_{CE} and I_E in figure 6.68.

Solution: This is potential divider biasing of transistor so first let us calculate the R_{TH} and V_{TH} in input circuit; where

$$R_{TH} = \frac{R_1 \times R_2}{R_1 + R_2} \quad \text{and} \quad V_{TH} = \frac{V_{CC} \times R_2}{R_1 + R_2}$$

Now, $$R_{TH} = \frac{10 \times 2.2}{12.2} = 1.8\ \text{k}\Omega$$

One who thinks in terms of silver, cannot act in terms of gold. — Henry G. Weaver and $$V_{TH} = \frac{10 \times 2.2}{12.2} = 1.8 \text{ V}$$

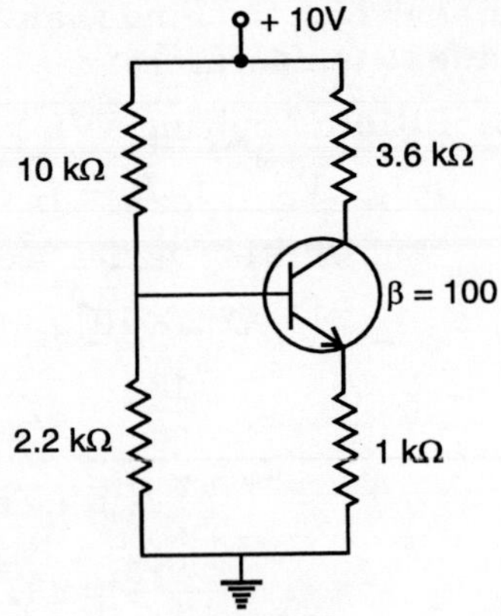

Fig. 6.68.

Applying KVL to input side, we get

$$V_{TH} = I_B R_{TH} + V_{BE} + I_E R_E$$

We know that $$I_E = (\beta + 1)I_B$$

$$V_{TH} = \frac{I_E}{\beta+1} R_{TH} + V_{BE} + I_E R_E$$

or $$I_E = \frac{V_{TH} - V_{BE}}{R_E + \dfrac{R_{TH}}{\beta+1}}$$

Here, $V_{TH} = 1.8$ V, $V_{BE} = 0.7$ V, $R_E = 1$ K, $R_{TH} = 1.8$ K, $\beta = 100$.
Therefore, we have

$$I_E = \frac{1.8 - 0.7}{1000 + \dfrac{1800}{101}} = 1.081 \text{ mA} \; ; I_B = \frac{I_E}{\beta + 1} = \frac{1.081}{100 + 1} = 0.0107 \text{ mA}$$

Now, $$I_C = \beta I_B = 100 \times 0.0107 \text{ mA} = 1.07 \text{ mA} \quad \textbf{Ans.}$$

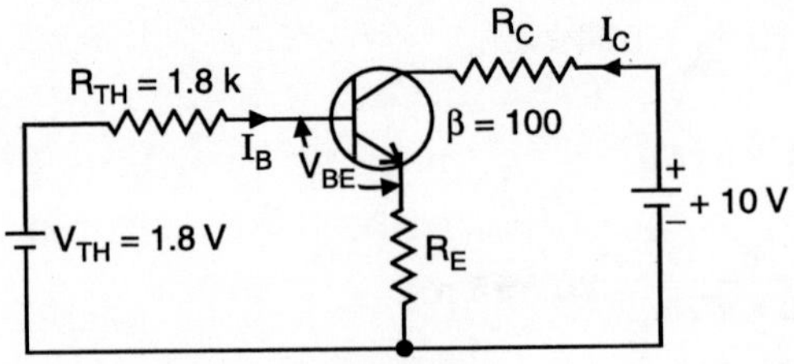

Fig. 6.69.

Further, KVL in output loop will yield

$$V_{CC} = I_C R_C - V_{CE} + I_E R_E$$

or $$V_{CE} = V_{CC} - I_C R_C - I_E R_E$$

$$= 10 - 1.07 \times 10^{-3} \times 3.6 \times 10^3 - 1.081 \times 10^{-3}$$

or $$V_{CE} = 5.067 \text{ Volts.} \quad \textbf{Ans.}$$

EXAMPLE 6.45 Determine V_C and I_B for the following network in figure 6.70.

Solution: Thevenin's equivalent circuit has been drawn in figure 6.71.

Here $$V_{TH} = \frac{R_2}{R_1 + R_2} \times V_{CC} = \frac{16 \text{ k}}{82 \text{ k} + 16 \text{ k}} \times (+22)$$

or $$V_{TH} = + 3.59 \text{ Volts}$$

and $$R_{TH} = \frac{R_1 R_2}{R_1 + R_2} = \frac{16 \times 82}{16 + 32}$$

$$= 13.39 \text{ k}\Omega$$

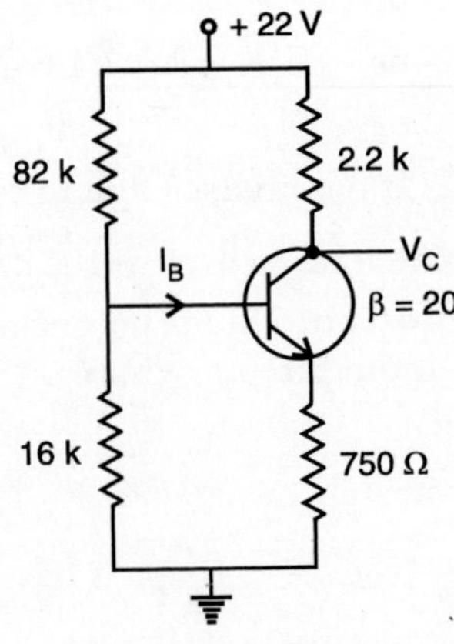

Fig. 6.70.

KVL in input loop will yield

$$V_{TH} = I_E R_E + V_{EB} + I_B R_{TH}$$

We know that

$$I_E = (\beta + 1)\, I_B$$

and V_{EB} for S_i transistor = 0.7 V.

So, $$V_{TH} = (\beta + 1)\, I_B\, R_E + V_{EB} + I_B\, R_{TH}$$

or $$I_B = \frac{V_{TH} - V_{EB}}{R_{TH} + (\beta+1)\, R_E}$$

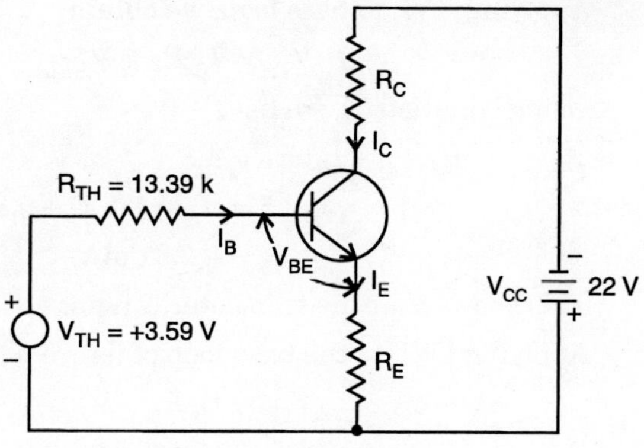

Fig. 6.71.

Here $V_{TH} = +\,3.59$ V, $R_{TH} = 13.39$ k

$$R_E = 750 \text{ Ohm}$$
$$\beta = 200$$
$$V_{EB} = 0.7 \text{ Volts}$$

so $$I_B = \frac{3.59 - 0.7}{13.39 \times 10^3 + (200 + 1)\, 750}$$

or $I_B = 17.61\ \mu$A. (in the shown direction) **Ans.**

We know that $I_C = \beta\, I_B = 200 \times 17.61 \times 10^{-6} = 3.52$ mA (in the shown direction)

KVL in the output loop will yield

$$V_{CC} = -\, I_C\, R_C + V_C$$

So $$V_C = V_{CC} + I_C\, R_C$$

Here $V_{CC} = -\,22$ Volts, $R_C = 2.2$ K, $I_C = 3.52$ mA

So $V_C = -\,22 + 3.52 \times 10^{-3} \times 2.2 \times 10^3 = -\,14.256$ Volts. **Ans.**

EXAMPLE 6.46 In the circuit diagram shown in figure 6.72, the transistor is made of Silicon. Will the biasing circuit shown place it in active region? If no, explain the situation, the transistor will be in. If yes, complete the bias collector current I_{CQ} and collector emitter voltage V_{CEQ} assuming β = 100. Appropriately assume missing data.

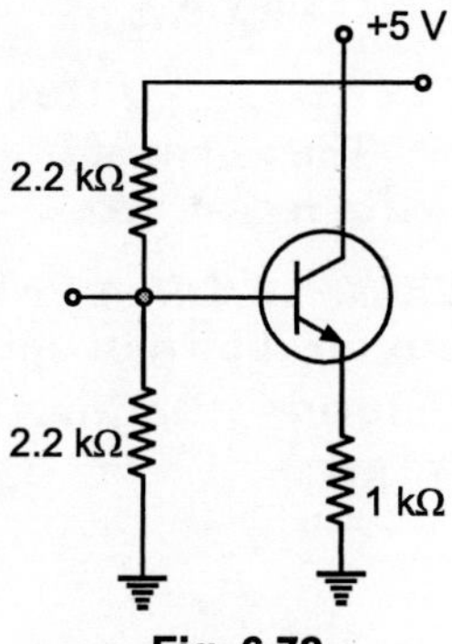

Fig. 6.72.

Solution : Drawing the Thevenin's equivalent circuit as shown in figure 6.73, we have

$$R_B = R_1 \mid\mid R_2$$
$$= 2.2 \text{ k} \mid\mid 2.2 \text{ k} = 1.1 \text{ k}\Omega$$

Also, $$V_{TH} = \frac{R_2}{(R_1 + R_2)} \times V_{CC}$$

or $$V_{TH} = \frac{2.2}{2.2 + 2.2} \times 5 = 2.5 \text{ V}$$

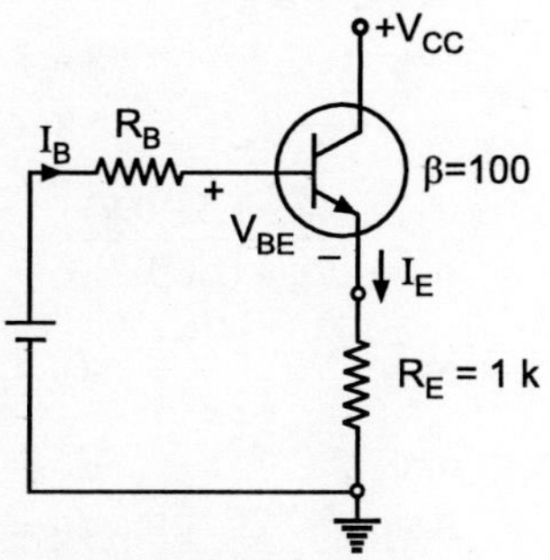

Fig. 7.73. Thevenin's equivalent circuit

Assuming the transistor to be in saturation, we have

$$V_{BE(sat)} = 0.8 \text{ V and } V_{CE(sat)} = 0.2$$

Now, Let us apply KVL to collector loop to get,

$$V_{CC} = [I_{C(sat)} + I_B]\, R_E + V_{CE(sat)}$$

or $$5 = I_{C(sat)} \times 1 + I_B + 0.2$$

or $$I_{C(sat)} + I_B = 4.8 \qquad \text{...}(i)$$

Applying KVL to base loop, we obtain

$$V_{TH} = I_B R_B + V_{BE(sat)} + [I_{C(sat)} + I_B] R_E$$

Using equation (*i*), we have

$$[I_{C(sat)} + I_B] R_E = 4.8 \text{ V}$$

or $$2.5 = (I_B \times 1.1 \text{ k}) + 0.8 + 4.8$$

or $$I_B = -2.81 \text{ mA}$$

This shows that the transistor can not go into saturation. So, it is in the active region.

Applying KVL to the base loop of figure 6.92, we have

$$V_{TH} = I_B R_B + V_{BE(act)} + I_E R_E$$

But $$I_E = (1 + \beta) I_B .$$ (Since, the transistor is in active region)

Therefore, $$V_{TH} = I_B R_B + 0.7 + (1 + \beta) I_B R_E$$

or $$I_B = \frac{V_{TH} - 0.7}{R_B + (1+\beta) R_E} = \frac{2.5 - 0.7}{1.1 \text{ k} + (101 \times 1 \text{ k})}$$

or $$I_B = 17.63 \ \mu\text{A}$$

Now, collector current is given by

$$I_{CQ} = \beta I_B = 100 \times 17.63 \times 10^{-6} = 1.763 \text{ mA}$$

Also, we can calculate value of I_E as under:

$$I_E = (1 + \beta) I_B = 101 \times 17.63 \times 10^{-6} = 1.78 \text{ mA}$$

Finally, V_{CEQ} can be calculated as under:

$$V_{CEQ} = V_{CC} - I_E R_E = 5 - (1.78 \times 1) = 3.22 \text{ Volts.}$$

The values of I_{CQ} and V_{CEQ} indicate that the transistor is indeed in the active region. **Ans.**

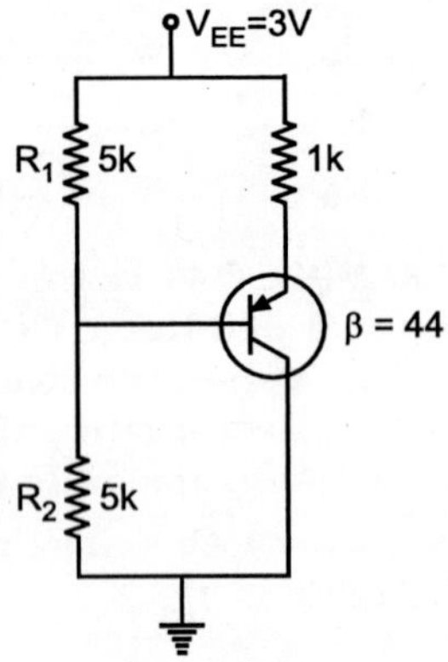

Fig. 6.74.

EXAMPLE 6.47 A PNP silicon transistor is used in a common collector circuit shown in figure 6.74. Find quiescient point.

Solution : Drawing Thevenin's equivalent circuit as shown in figure 6.75, we get

$$R_B = R_1 \parallel R_2 = 5 \text{ k} \parallel 5 \text{ k} = 2.5 \text{ k}\Omega$$

Also, $$V_{TH} = \frac{R_2}{R_1 + R_2} V_{CC} = \frac{5}{5+5} \times 3 = 1.5 \text{ V}$$

Applying KVL to base loop, we obtain

$$V_{EE} = I_E R_E + V_{EB} + I_B R_B + V_{TH}$$

or $$3 = (1 + \beta) I_B R_E + 0.7 + I_B R_B + 1.5$$

or $$\frac{(3 - 1.5 - 0.7)}{R_B + (1+\beta) R_E} = I_B$$

or $$I_B = \frac{0.8}{2.5 \text{ k} + (45 \times 1 \text{ k})} = 16.84 \ \mu\text{A}$$

or $$I_C = \beta I_B = 44 \times 16.84 \times 10^{-6} = 0.74 \text{ mA}$$

Also, $$I_E = (1 + \beta) I_B = 45 \times 16.84 \times 10^{-6} = 0.757 \text{ mA}$$

Now, applying KVL to collector loop, we get

$$V_{EE} = I_E R_E + V_{EC}$$

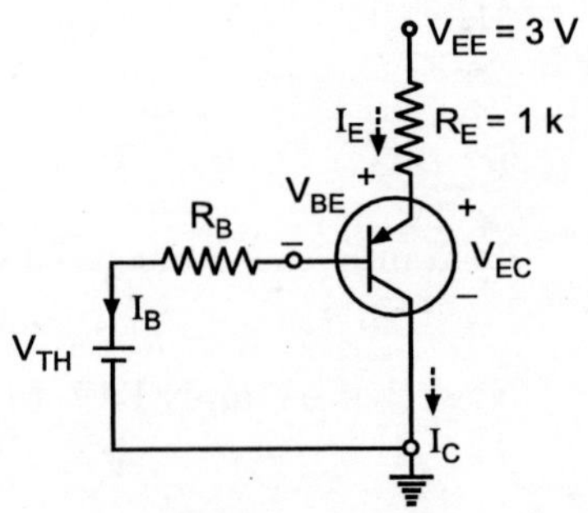

Fig. 6.75. (a) Thevenin's equivalent circuit.

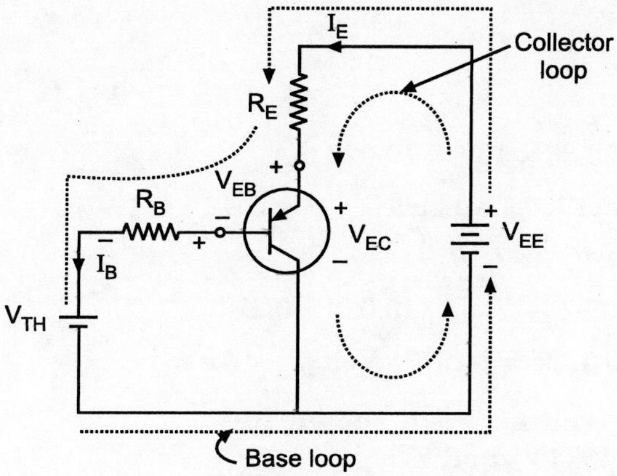

Fig. 6.76. (b)

or $$V_{EC} = V_{EE} - I_E R_E = 3 - (0.757 \times 1) = 2.243 \text{ Volts}$$

Also, $$V_{CE} = -2.243 \text{ Volts}$$

Hence, $$Q \text{ point} = (-2.243 \text{ V}, 0.74 \text{ mA}). \textbf{ Ans.}$$

EXAMPLE 6.48 Find I_C and V_{CE} for the following circuit if $\beta = 80$ for the BJT.

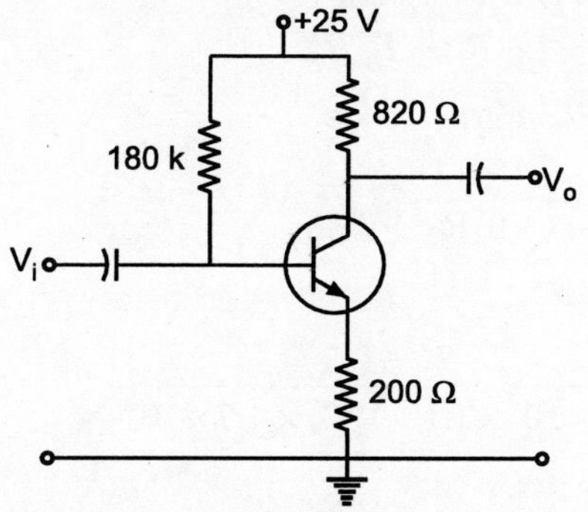

Fig. 6.77.

Solution: Given that $\beta = 80$, $R_B = 180 \text{ k}\Omega$, $R_C = 820\ \Omega$, $R_E = 200\ \Omega$

Applying KVL to the loop, we have

$$V_{CC} = I_B R_B + V_{BE} + I_E R_E$$

$$= I_B R_B + V_{BE} + (1 + \beta) I_B R_E$$

or $$I_B = \frac{V_{CC} - V_{BE}}{R_B + (1+\beta) R_E}$$

or $$I_B = \frac{25 - 0.7}{180 \text{ k} + (81 \times 0.2 \text{ k})}$$

or $$I_B = 123.85\ \mu\text{A}$$

Also, it can be calculated as under:

$$I_C = \beta I_B = 80 \times 123.85 \times 10^{-6}$$

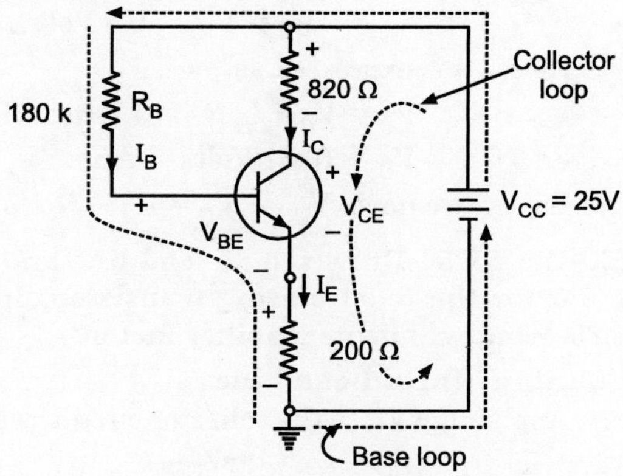

Fig. 6.78.

$I_C = 9.9$ mA. **Ans.**

Also, we have

$$I_E = (1 + \beta)\, I_B$$
$$= 81 \times 123.85 \times 10^{-6} = 10.023 \text{ mA}$$

Applying KVL to collector loop, we have

$$V_{CC} = I_C R_C + V_{CE} + I_E R_E$$

or $$25 = (9.9 \times 0.82) + V_{CE} + (10.023 \times 0.2)$$

or $$V_{CE} = 25 - 10.1226 = 14.8774 \text{ Volts.} \quad \textbf{Ans.}$$

EXAMPLE 6.49 For the given network shown in figure 6.79, determine I_C, V_{CE}, V_B and V_C.

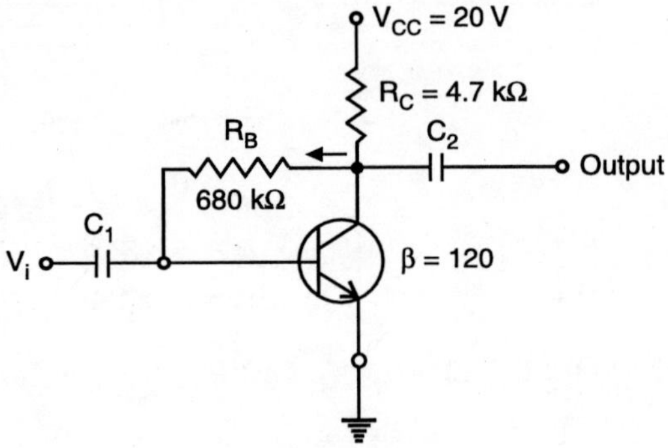

Fig. 6.79. Given circuit.

Solution: Given that

$$V_{CC} = 20 \text{ V}, R_C = 4.7 \text{ k}\Omega,$$
$$R_B = 680 \text{ k}\Omega, \beta = 120$$

First, let us find the value of I_B. For this, we shall apply KVL to the base-emitter loop of the circuit in figure 6.79.

$$V_{CC} = (I_C + I_B)\, R_C + I_B R_B + V_{BE}$$

or $$V_{CC} - V_{BE} = (\beta I_B + I_B)\, R_C + I_B R_B$$

or $$V_{CC} - V_{BE} = (1 + \beta)\, I_B R_C + I_B R_B$$

or $$I_B = \frac{V_{CC} - V_{BE}}{R_B + (1+\beta)R_C}$$

Subtituting values, we get

$$I_B = \frac{20 - 0.7}{680 \times 10^3 + [(1+120) \times 4.7 \times 10^3]}$$

Solving, we get $I_B = 15.46\ \mu$A

We know that I_C is given by

$$I_C = \beta I_B = 120 \times 15.46 \times 10^{-6} = 1.854 \text{ mA} \quad \textbf{Ans.}$$

Now, V_B is given by

$$V_B = V_{BE} = 0.7 \text{ Volt} \quad \textbf{Ans.}$$

Also, V_C is expressed as

$$V_C = V_B + I_B R_B = 0.7 + (15.46 \times 10^{-6} \times 680 \times 10^3)$$

or $$V_C = 10.51 \text{ Volts} \quad \textbf{Ans.}$$

Lastly, we have $V_{CE} = V_C = 10.51$ Volts **Ans.**

EXAMPLE 6.50 Draw the dc load line and locate the operating point for the fixed biasing transistor circuit shown in figure 6.80. What will be its stability factor?

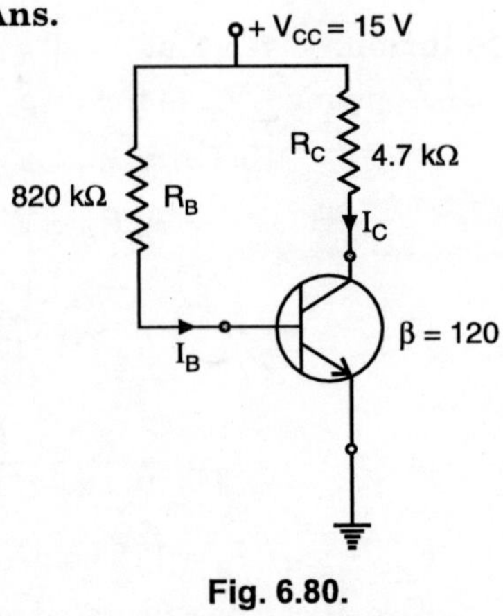

Fig. 6.80.

Solution: (*i*) DC Load Line

Applying KVL to the collector circuit, we get

$$V_{CC} = I_C R_C + V_{CE}$$

Argument is the worst sort of conversation. — **Jonathan Swift**

or $$I_C = -\frac{1}{R_C}V_{CE} + \frac{V_{CC}}{R_C} = -\frac{1}{4700}V_{CE} + \frac{15}{4700}$$

or $$I_C = -\frac{1}{4700}V_{CE} + 3.19 \text{ mA} \quad ...(i)$$

Equation (*i*) represents the expression for DC load line.

Now, let us substitute $I_C = 0$ in equation (*i*).

We have $V_{CE} = V_{CC} = 15$ V

Thus, point '*A*' on the dc load line will have coordinates (15 V, 0 mA) *i.e.*,

$$A \equiv (15 \text{ V}, 0 \text{ mA})$$

Next, let us substitute $V_{CE} = 0$ in equation (*i*).

We have $I_C = 3.19$ mA

Hence, point '*B*' on the load line will have coordinates (0, 3.19 mA). Figure 6.81 shows the dc load line.

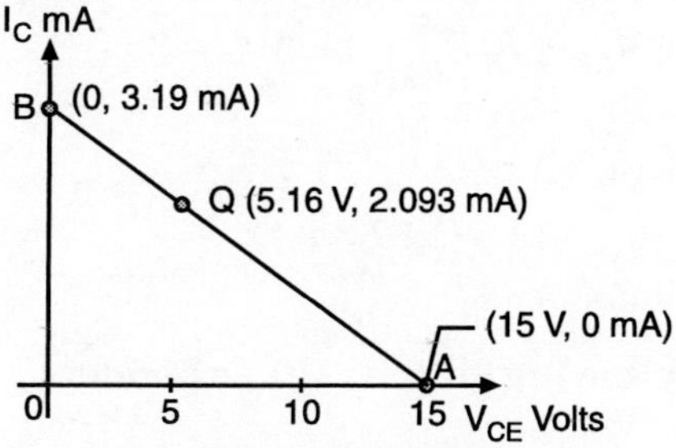

FIG. 6.81. DC load line

***(ii)* Q-point**

We have $$I_B = \frac{V_{CC} - V_{BE}}{R_B} = \frac{15 - 0.7}{820 \times 10^3} = 17.44 \ \mu\text{A}$$

Also, $I_C = \beta I_B = 120 \times 17.44 \times 10^{-6} = 2.093$ mA

We know that $V_{CE} = V_{CC} - I_C R_C = 15 - (2.093 \times 4.7)$

or $V_{CE} = 5.16$ Volts

Therefore, Q-point coordinates are : (5.16 V, 2.093 mA)

Ans.

EXAMPLE 6.51 Calculate I_B, I_C, V_E and V_{CE} for the following circuit, if $\beta = 130$.

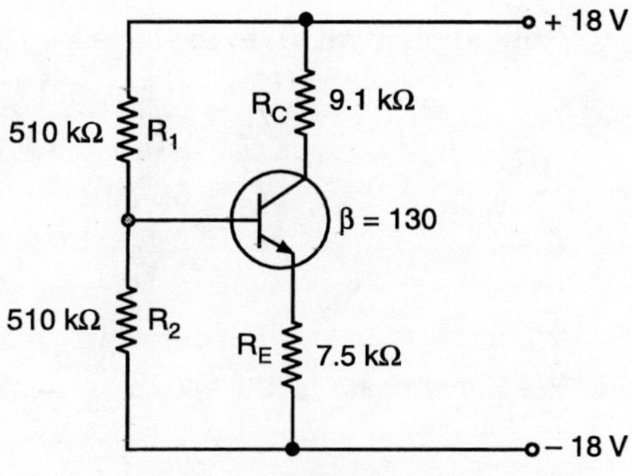

Fig. 6.82

Solution: First, let us obtain Thevenin's equivalent circuit. For this, we have

$$R_{TH} \text{ or } R_B = \frac{R_1 R_2}{R_1 + R_2}$$

or $$R_{TH} \text{ or } R_B = \frac{510 \times 510}{(510 + 510)} = 255 \text{ k}\Omega$$

and $$V_{TH} = \left[\frac{R_2}{R_1 + R_2} \times 36\right]^{-18} = \left[\frac{510}{510 + 510} \times 36\right]^{-18}$$

Thus, $V_{TH} = 0$ Volts with respect to GND. Figure 6.83 shows the Thevenin's equivalent circuit.

Now, applying KVL to the base loop of figure 6.83, we get

$$V_{TH} + 18 = I_B R_B + V_{BE} + I_E R_E$$

or $$0 + 18 = I_B R_B + 0.7 + (1 + \beta) I_B R_E$$

or $$I_B = \frac{18 - 0.7}{R_B + (1+\beta)R_E}$$

or $$I_B = \frac{17.3}{255\text{k} + (131 \times 7.5\text{k})}$$

$$= 13.98\ \mu\text{A} \cong 14\ \mu\text{A}$$

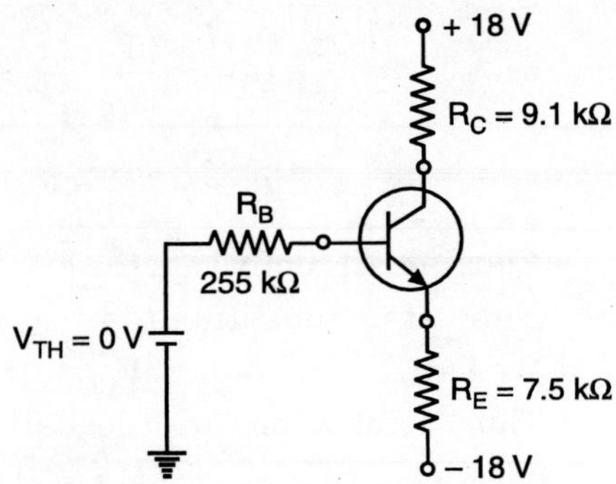

Fig. 6.83. Thevenin's equivalent circuit

We know that

$$I_C = \beta I_B = 130 \times 14 \times 10^{-6} = 1.82 \text{ mA} \quad \textbf{Ans.}$$

Also, $$I_E = (1 + \beta) I_B = 131 \times 14 \times 10^{-6} = 1.834 \text{ mA}$$

Next, appling KVL to the collector loop of figure 6.83, we obtain

$$18 + 18 = I_C R_C + V_{CE} + I_E R_E \quad \text{or} \quad V_{CE} = 36 - (1.82 \times 9.1) - (1.834 \times 7.5)$$

or $$V_{CE} = 5.683 \text{ Volt} \quad \textbf{Ans.}$$

Further from figure 6.83, we have

$$V_C = 18 - I_C R_C = 18 - (1.82 \times 9.1)$$

or $$V_C = 1.438 \text{ Volts} \quad \textbf{Ans.}$$

EXAMPLE 6.52 Determine the following for the voltage divider configuration :

***(i)* I_C *(ii)* V_E *(iii)* V_B *(iv)* V_C *(v)* R_i**

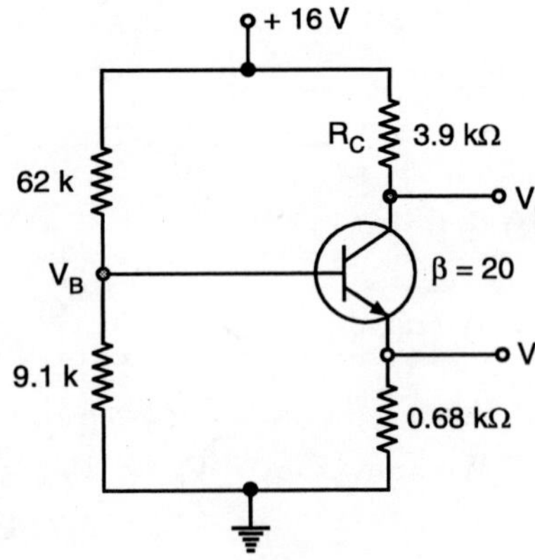

Fig. 6.84

Solution: Let us first obtain the Thevenin's equivalent circuit as under:

$$R_{TH} = R_B = \frac{R_1 R_2}{R_1 + R_2} = \frac{62 \times 9.1}{62 + 9.1} = 7.94 \text{ k}\Omega$$

and $$V_{TH} = \frac{R_2}{R_1 + R_2} \times V_{CC} = \frac{9.1}{62 + 9.1} \times 16 = 2 \text{ V}$$

Now, applying KVL to the base loop and simplifying, we obtain

$$I_B = \frac{V_{TH} - V_{BE}}{R_B + (1+\beta)R_E} = \frac{2 - 0.7}{7.94\text{k} + (21 \times 0.68\text{k})}$$

or $$I_B = 58.5\ \mu\text{A}$$

Also, $$I_C = \beta I_B = 20 \times 58.5 \times 10^{-6}$$

$$= 1.17 \text{ mA} \quad \textbf{Ans.}$$

We have $$I_E = (1 + \beta) I_B = 21 \times 58.5 \times 10^{-6} = 1.23 \text{ mA}$$

We know that $$V_E = I_E R_E = 1.23 \times 0.68$$

$$= 0.8364 \text{ V}$$

Also, $$V_B = V_E + V_{BE} = 0.8364 + 0.7 = 1.5364 \text{ V}$$

V_C is given by $$V_C = V_{CC} - I_C R_C = 16 - (1.17 \times 3.9) = 11.437 \quad \textbf{Ans.}$$

For voltage divider bias circuit, we have

$$R_i = [h_{ie} + (1 + h_{fe}) R_E] || R_B \quad \textbf{Ans.}$$

SUMMARY

- In *CB* configuration the d.c. current gain is given as

$$\alpha = \frac{I_C}{I_E}$$

- In *CB* configuration, the a.c. current gain is given as

$$\alpha_0 = \left.\frac{\Delta I_C}{\Delta I_E}\right|_{hfb = \text{constant}}$$

- In *CE* configuration, the d.c. current gain is given as

$$\beta_{dc} = \beta = \frac{I_C}{I_B}$$

- In *CE* configuration, the a.c. current gain is given as

$$\beta_0 = \frac{\Delta I_C}{\Delta I_B}$$

- The relationship between current gain α and β is

$$\beta = \frac{\alpha}{1 - \alpha}$$

- The collector current, in terms of leakage current in *CE* configuration is given as

$$I_C = \beta I_B + I_{CEO}$$

where $I_{CEO} = (1 + \beta)\, I_{CBO}$

or $I_C = \beta I_B + (1 + \beta)\, I_{CBO}$

- There are two types of transistor characteristics
  (*i*) input characteristics
  (*ii*) output characteristics.
- In *CB*-configuration, the a.c. or dynamic input resistance is given as

$$r_i = \left.\frac{\Delta V_{EB}}{\Delta I_E}\right|_{V_{CB} = \text{constant}}$$

- In *CB*-configuration, the dynamic output resistance is given as

$$r_0 = \left.\frac{\Delta V_{CB}}{\Delta I_C}\right|_{I_E = \text{constant}}$$

- In *CE*-configuration, the dynamic input resistance is given as

$$r_i = \left.\frac{\Delta V_{BE}}{\Delta I_B}\right|_{V_{CE} = \text{constant}}$$

- In *CE*-configuration, the dynamic output resistance is given as

$$r_0 = \left.\frac{\Delta V_{CE}}{\Delta I_C}\right|_{I_B = \text{constant}}$$

- *CE* configuration is most widely used configuration in amplifier circuits.

SHORT QUESTIONS WITH ANSWERS

Q. 1. Why biasing is required? Explain.

Ans. The transistor is used in a wide number of applications. The most basic application of a transistor is amplification. Amplification means to raise the strength or amplitude of a weak signal without any change in its original shape. However, for faithful amplification two things are necessary. Firstly, the transistor must operate in active region. It means that the emitter-base junction is forward-bias and collector-base junction is reverse bias. Active region for amplification is must because amplification is a linear process and transistor operates linearly only in active region. Secondly, we should get a fixed dc collector current at a fixed dc collector voltage. These fixed values of dc collector current and dc collector voltage are expressed by a term called operating point or quiescent point or *Q*-point.

Therefore, the primary aim of biasing is to obtain a fixed collector current at a fixed collector voltage. The circuits used for transistor biasing are simply called biasing circuits.

Q. 2. What is bias stabilization? Explain.

Ans. We know that for proper operation of transistor, the operating point should be properly fixed. However, only the fixing of operating point is not sufficient. The operating point must remain stable throughout the operation where it was fixed originally. This type of shift of operating point may drive the transistor into an unwanted region. Therefore, bias stabilization is of utmost importance. Following are the factors which are responsible for the shift of operating point.

Q. 3. What is stability factor? Explain.

The stability factor is expressed as the rate of change of collector current with respect to the reverse saturation current while keeping the common-emitter current gain β and base current I_B as constant.

Mathematically, the stability factor may be stated as

$$S = \frac{dI_C}{dI_{CO}}$$

The stability factor S is a measure of bias stability of a transistor circuit. It may be observed that higher the value of stability factor, poor is the stability whereas lower is the value of stability factor, good is the stability.

Q. 4. What are the requirements of biasing circuits? Explain.

Ans. Following are the requirements of a biasing circuit:

(*a*) A biasing circuit must set the operating point in the middle of the active region of the transistor characteristics.

(*b*) A biasing circuit must stabilize the collector current against temperature variations.

(*c*) A biasing circuit must ensure that the operating point is independent of the transistor parameters such as β so that operating point is not shifted if the transistor is replaced by another transistor of same type.

Q. 5. What are the methods of transistor biasing? Explain.

Ans. There are following four methods of transistor biasing which are most widely used:

1. Fixed bias
2. Collector to base bias
3. Self bias or voltage divider bias
4. Emitter bias

Q. 6. Define self bias or voltage divider bias?

Ans. For fixed bias circuit and collector-to-base bias circuit, the operating point or the values of d.c. collector current and collector voltage depend upon the value of current gain β of transistor. But the value of β is temperature dependent particularly for silicon transistors. Also, the value of β changes from transistor to transistor. This means that for same type of two transistors the values of β are different.

Q. 7. Explain why an ordinary transistor is called bipolar?

Ans. Because the transistor operation is carried out by two types of charge carriers (majority and minority carriers), an ordinary transistor is called bipolar.

Q. 8. What is done to the base region of a transistor to improve its operation ?

Ans. Base is made thin and very lightly doped in comparison to either emitter or collector so that it may pass most of the injected charge carriers to the collector.

Q. 9. Why transistor is called current controlled device ?

Ans. The output voltage, current or power is controlled by the input current in a transistor so it is called the current controlled device.

Q. 10. Why silicon type transistor are more often used than germanium type ?

Ans. Because, silicon has smaller cut-off current I_{CBO}, small variations in I_{CBO} due to variations in temperature as compared to those in case of germanium.

Q. 11. Why the width of the base region of a transistor is kept very small compared to other regions ?

Ans. Base region of a transistor is kept very small and very lightly doped so as to pass most of the injected charge carriers to the collector.

Q. 12. Why is there a maximum limit of collector supply voltage for a transistor ?

Ans. Although collector current is practically independent of collector supply voltage over the transistor operating range, but if V_{CB} is increased beyond a certain value, collector current I_C eventually increases rapidly and possibly destroys the device.

Q. 13. Why CE configuration is most popular in amplifier circuits ?

Ans. CE configuration is mainly used because its current, voltage and power gains are quite high and the ratio of output impedance and input impedance are quite moderate.

Q. 14. Why is CC configuration seldom used ?

Ans. CC configuration is seldom used because its voltage gain is always less than unity.

REVIEW QUESTIONS

1. What is a Bipolar junction transistor?
2. Explain the different configurations of BJT.
3. Explain the input and output characteristics of a transistor in *CB* configuration.
4. Define α and β for a transistor.
5. Derive the relationship between α and β.
6. What is meant by *Q*-point?

7. What is the need for biasing a transistor?
8. Explain the requirements of a biasing circuit.
9. What are the factors responsible for the instability of operating point.
10. Draw a fixed bias circuit and obtain the value of d.c. Voltage and currents in the circuit.
11. Draw a self-bias circuit and derive an expression for stability factor.
12. Define, stability factor. Derive general expression for stability factor of a biasing circuit.
13. State various methods of improving stability.
14. Define Stability factor with respect to transistor biasing.
15. Explain the function of emitter in the operation of a junction transistor.

NUMERICAL PROBLEMS

1. For a fixed bias circuit using a germanium NPN transistor, the value of β is 100. If V_{BE} = 0.2 V, V_{CC} = 16 V, R_C = 5 kΩ and R_B = 790 kΩ then determine its operating point.
2. Find I_C and V_{CE} for the following circuit. What happens to V_{CE} if β increases due to temperature?

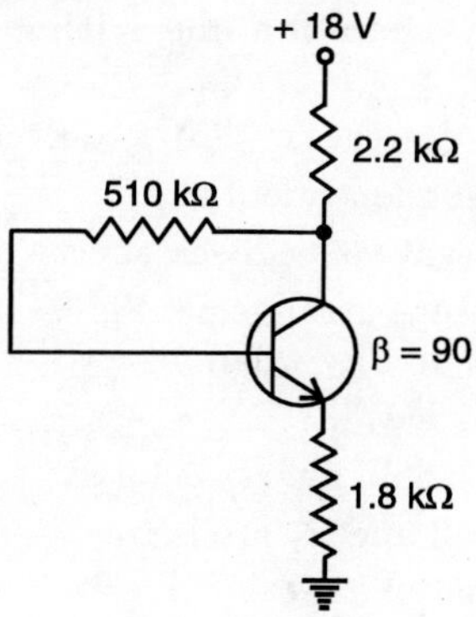

Fig. 6.85.

3. Find the value of operating point for the circuit of a potential divider bias arrangement with $R_2 = R_C$ = 5 kΩ, R_E = 1 kΩ and R_1 = 40 kΩ.
(Ans. V_{CE} = 6 V, I_C = 1 mA)
4. Consider a DC bias circuit with voltage feedback, given below. Determine the quiescient levels of I_{CQ} and V_{CEQ}. The β of the transistor is 90 and cut-in voltage is 0.7 V. What will be the effect on I_{CQ} and V_{CEQ} if β is increased by 50%.

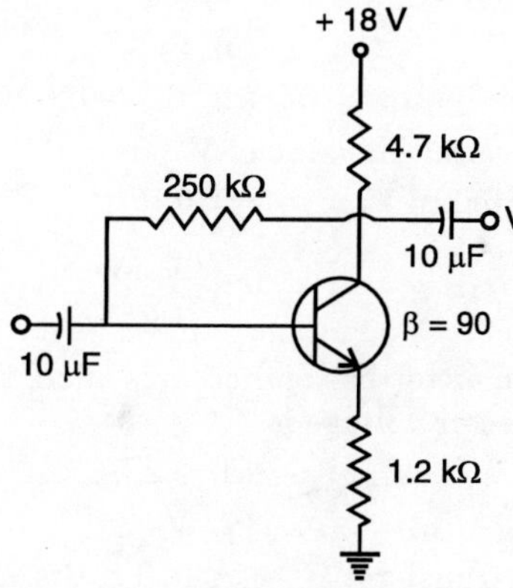

Fig. 6.86.

5. The various parameters of a *CE* amplifier which uses the self-bias method are V_{CC} = 12 V, R_1=10 kΩ, R_2 = 5 kΩ, R_C = 1 kΩ, R_E = 2 kΩ and β = 100. Determine the following:
(*i*) operating point (*ii*) the stability factor.
(Ans. V_{CE} = 7.05 V, I_C = 1.65 mA, S = 2.62)
6. Determine V_C and V_B for the network given below with β = 45 and V_{BE} = 0.7 V.

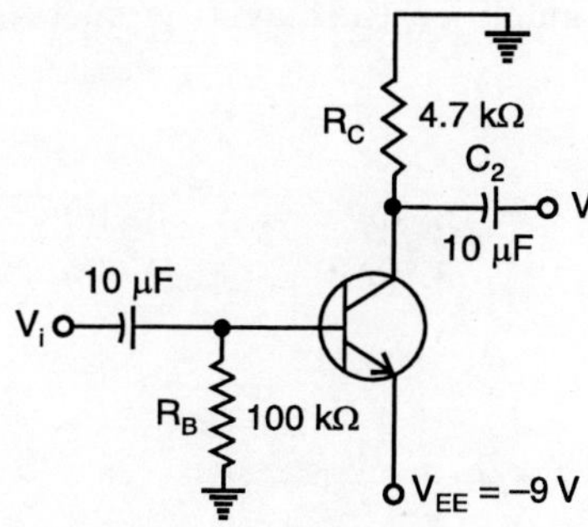

Fig. 6.87.

OBJECTIVE TYPE QUESTIONS

Choose one correct answer in the following :

1. The relationship between α_{dc} and β_{dc} is given by

   (a) $\beta_{dc} = \dfrac{\alpha_{dc}}{1+\alpha_{dc}}$ (b) $\beta_{dc} = \dfrac{\alpha_{dc}}{1-\alpha_{dc}}$

   (c) $\alpha_{dc} = \dfrac{1+\beta_{dc}}{\beta_{dc}}$ (d) $\alpha_{dc} = \dfrac{\beta_{dc}}{\beta_{dc}-1}$

2. In cut-off, the value of V_{CE} is

   (a) 0 V (b) V_{CC}

   (c) minimum (d) $V_{BE} = 0.6$ V

3. The saturation value of V_{CE} is

   (a) equal to V_{CC} (b) 0 V

   (c) 0.7 V (d) $\dfrac{V_{CC}}{2}$

4. If the emitter-base junction is open, the value of collector voltage is

   (a) 0 V (b) 0.2 V

   (c) Floating (d) V_{CC}

5. If in a transistor $(I_C)_{sat} < I_B \beta_{dc}$, the transistor is in

   (a) cut off

   (b) saturation

   (c) amplification state

   (d) none

6. As temperature is increased, the h-parameters h_{ie}, h_{re}, h_{fe} and h_{oe} of a transistor

   (a) decreases (b) increases

   (c) remains constant (d) none

7. For operation of a bipolar junction transistor as an amplifier, the transistor operates in

   (a) cut off region (b) saturation region

   (c) active region (d) deep in saturation

8. The load line shifts parallel to itself on the CE output characteristics of a transistor when

   (a) R_L is changed

   (b) V_{CC} is changed

   (c) both R_L and V_{CC} are changed

   (d) base current is changed

9. The dc load line of a transistor

   (a) has a positive slope

   (b) has a negative slope

   (c) does not contain the Q-point

   (d) is a plot of I_C versus I_B

10. A CE transistor circuit employing base-bias with collector feedback possesses greater stability than the one without feedback because

    (a) V_{BE} decreases (b) I_C decreases

    (c) I_C is independent of β

    (d) of negative feedback effect

11. The self bias arrangement gives an improved Q-point stability when

    (a) R_E is low

    (b) β is small, but R_E is large

    (c) both β and R_E are large

    (d) none

12. The biasing method which is considered independent of transistor β_{dc} is

    (a) fixed biasing

    (b) collector feedback bias

    (c) voltage divider bias

    (d) base bias with collector feedback

Answers

1. (b)	2. (b)	3. (a)	4. (d)	5. (b)	6. (b)
7. (c)	8. (b)	9. (b)	10. (d)	11. (a)	12. (c)

UNIT—V

Small Signal Low Frequency BJT Models

BJT Hybid Model, Determination of h-parameters from Transistor Characteristics, Analysis of a Transistor Amplifier Circuit using h-Parameters, Comparison of CB, CE, and CC Amplifier Configurations.

Above Syllabus has been covered in following Chapter:

CHAPTER 7 : Small Signal Low Frequency BJT Models

Chapter 7

Small Signal Low Frequency BJT Models

SYLLABUS

BJT Hybid Model, Determination of h-parameters from Transistor Characteristics, Analysis of a Transistor Amplifier Circuit using h-Parameters, Comparison of CB, CE, and CC Amplifier Configurations.

Chapter Outline

- Introduction
- Classification of Amplifiers
- Transistor Amplifier Circuit
- Analysis of Transistor Amplifier
- Equivalent Circuit of Transistor Amplifier
- Transistor a.c. Equivalent Circuit
- Hybrid Parameters (*h*-Parameters)
- *h*-Parameter Representation for Transistor
- General Hybrid Equivalent Circuit
- Hybrid Equivalent Circuit for Common Emitter Configuration
- Hybrid Equivalent Circuit for Common Base Configuration
- Hybrid Equivalent Circuit for Common Collector Configuration
- Analysis of a Transistor Amplifier Using *h*-Parameters

7.1. INTRODUCTION

The process of raising the strength of a weak signal without any change in its shape is called amplification. A device which amplifies the input signal is called the amplifier.

Almost no electronic system can work without an amplifier. For example, radio receivers amplify very weak signals until they are strong enough to make the sound audible. In television, both the sound and picture signals should be suitably amplified. Therefore, electronic amplifiers are a constant and important part of different electronic equipments and systems.

We know that a properly biased transistor increases the amplitude of a weak signal and hence works as an amplifier. Therefore, an amplifier may also be defined as a device which increases the current, voltage or power of an input signal with the help of a transistor by providing additional power from a separate source of power supply or battery.

An amplifier, in which the instantaneous output signal is directly proportional to the corresponding input signal is known as a linear amplifier. On the other hand, the amplifier, in which the output signal is not directly proportional to the input signal, is known as a non-linear amplifier.

7.2. CLASSIFICATION OF AMPLIFIERS

We may classify the linear amplifiers according to their mode of operation as listed below:

1. based on the input signal
   (*i*) small-signal amplifier
   (*ii*) large-signal amplifier
2. based on the output signal
   (*i*) voltage amplifier
   (*ii*) power amplifier
3. based on transistor configuration
   (*i*) common-emitter (CE) amplifier
   (*ii*) common-base (CB) amplifier
   (*iii*) common-collector (CC) amplifier
4. based on biasing conditions
   (*i*) class–A
   (*ii*) class–B
   (*iii*) class–AB
   (*iv*) class–C
5. based on nature of load resistance
   (*i*) untuned amplifier or wideband amplifier
   (*ii*) tuned amplifier or narrowband amplifier
6. based on frequency response
   (*i*) direct- coupled (DC) amplifier
   (*ii*) audio frequency (AF) amplifier
   (*iii*) radio frequency (RF) amplifier
   (*iv*) ultra high frequency (UHF) and microwave frequency amplifier
7. based on number of stages
   (*i*) single-stage amplifier
   (*ii*) multistage amplifier
8. based on the method of coupling between the stages
   (*i*) direct-coupled (DC) amplifier
   (*ii*) R-C coupled amplifier
   (*iii*) transformer coupled amplifier

In this chapter, we shall discuss the small-signal, single-stage common-emiiter (CE), common-base (CB) and common-collector (CC) amplifiers.

7.3. TRANSISTOR AMPLIFIER CIRCUIT

In the last chapter, we have studied that the self-bias or voltage divider biasing method is the best. Figure 7.1 shows the circuit of voltage-divider or self-bias. This biasing circuit is most widely used in almost all amplifier circuits because this circuit can be designed very easily and also it provides very good stabilization of the operating point. This self-bias or voltage divider bias circuit can be used as an amplifier if we add input signal source and few components to it.

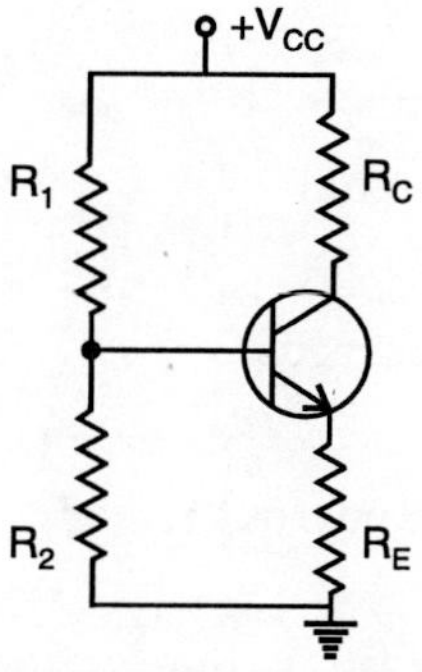

Fig. 7.1. Self-bias or voltage divider bias

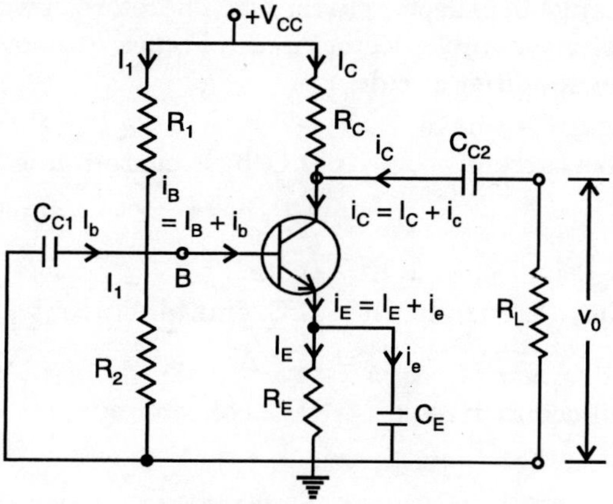

Fig. 7.2. Transistor amplifier circuit

Figure 7.2 shows such an amplifier circuit.

As discussed in last article, there are three types of transistor configurations, *i.e.*, common-base (CB), common-emitter (CE) and common collector (CC). Out of these three, common-emitter configuration is the best and hence this is most widely used in amplifier circuits.

In Figure 7.2, we are also considering a common-emitter amplifier.

In this circuit, the resistors R_1, R_2 and R_E form the biasing and stabilization circuit arrangement. This biasing circuit must fix a proper operating point. The resistor R_L is the load resistance which is connected across the output terminals.

The capacitors C_{C1} and C_{C2} are called the coupling capacitors. Capacitor C_{C1} is used to couple the input a.c. signal to the base of the transistor. A coupling capacitor is also called a blocking capacitor because it blocks d.c. $X_C = \dfrac{1}{2\pi fC}$ and allows a.c. to pass through.* Capacitor C_{C1} prevents the d.c.current from flowing to the signal source. Hence C_{C1} isolates the a.c. signal source from R_2.

Similarly, coupling capacitor C_{C2} blocks the d.c. and allows a.c. to pass through. Due to capacitor C_{C2} the output across the load resistor R_L is free from the collector's d.c. voltage.

Another capacitor is C_E. It is also called a **by pass capacitor** because it bypasses all the a.c. currents from the emitter to the ground. If capacitor C_E is not connected in the circuit, the a.c. voltage developed across R_E will affect the a.c. input voltage. This feedback of a.c. signal is reduced by connecting the capacitor C_E in the circuit.

In this amplifier circuit, when a weak a.c. signal is fed to the base terminal of the transistor, a base current i_B flows. Because of transistor action, a much larger current βi_B flows in the collector circuit. Since the value of load resistor R_L is large, a very high value of voltage is developed at the output. Hence, a small change in the base current i_B gives rise to a

DO YOU KNOW?

The voltage gain and output voltage of an amplifier are affected by the following capacitances: coupling capacitors, bypass capacitors, internal transistor capacitance and stray wiring capacitance. The coupling and bypass capacitors affect the output voltage at low frequencies because their X_C begins to increase substantially. The internal transistor capacitance and stray wiring capacitance affect the output voltage at higher frequencies because their X_C begins to decrease substantially. A graph of output voltage versus frequency for an amplifier is called its frequency response.

* Since a capacitor acts as an open-circuit for d.c. (*i.e.* blocks it) and a short-circuit for a.c. (*i.e.* allows it to pass easily).

very large change in collector current i_C. Therefore, a weak a.c. signal in the base circuit appears as a large a.c. voltage in the collector circuit. Figure 7.2 shows the different currents and their directions flowing in the amplifier circuit.

From Figure, we have

Total Base current = D.C. base current due to biasing ciruit + A.C. base current due to a.c. input signal

or $i_B = I_B + i_b$

Total Emitter current = D.C. emitter current + a.c. emitter current

or $i_E = I_E + i_e$

Total collector current = D.C. collector current + a.c. collector current

or $i_C = I_C + i_c$

Thus, we conclude that if a self-bias circuit is provided with an input a.c. signal and some components, then an important device known as amplifier is formed.

7.3.1. Phase Reversal

As discussed earlier, a common-emitter (CE) amplifier is the most widely used amplifier circuit. This common-emitter amplifier has one important property.

This property is that a phase difference of 180° occurs between the input signal voltage and the output voltage in a common emitter amplifier. This is also known as phase reversal.

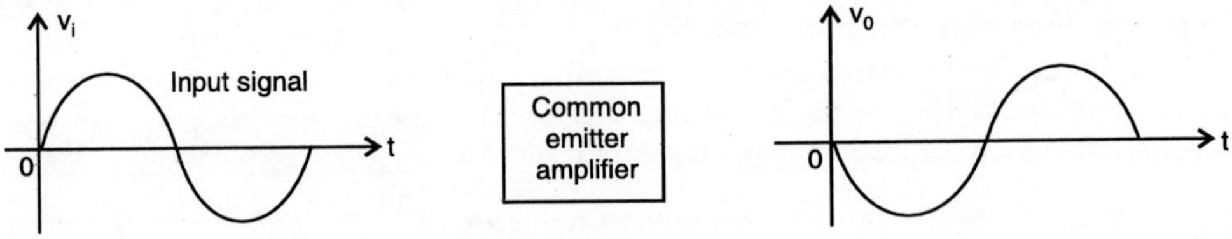

Fig. 7.3. Phase reversal in common emitter amplifier

7.3.2. Graphical Representation of Common Emitter Amplifier

Figure 7.4 shows the graphical representation of a common emitter amplifier. When input signal is applied, base current increases. When base current increases, collector current is also increased ($\because I_C = \beta I_B$). This increased collector current produces a large output voltage across the load resistor. When base current is varied, the corresponding variations in collector current and collector to emitter voltage can be observed from the characteristics shown in figure 7.4.

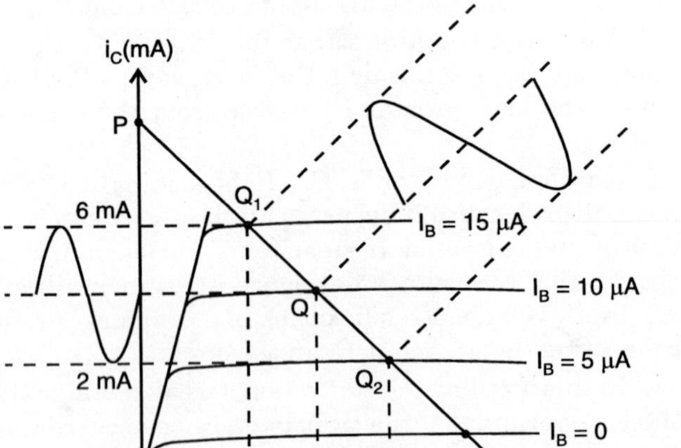

Fig. 7.4. Graphical representation of common emitter amplifier

The common emitter amplifier also produces 180° phase difference between the input and the output. This phase reversal can be explained as under:

When input signal voltage is increased during the positive half cycle, the base current i_B increases. When i_B increases, the collector current i_C also increases, with increase in i_C, the voltage drop $i_C R_C$ also increases. But V_{CC} is constant, therefore, the output voltage V_{CE} decreases. This implies that when the signal voltage is increasing during the positive half cycle, the output voltage is decreasing. This means that the output is 180° out of phase with the input.

7.4. ANALYSIS OF TRANSISTOR AMPLIFIER

As discussed earlier, in a transistor amplifier circuit, the d.c. sources established d.c. biasing currents and voltages, also known as quiescent operating currents and voltages. After this, when a.c. input signal source is applied, it produces fluctuations in these currents and voltages. As a result of this, an amplified signal at the output is obtained.

Sometimes, we need to analyse an amplifier circuit in order to determine the a.c. and d.c current gain, voltage gain, input resistance and output resistance of an amplifier circuit. As a simplest way, to analyze an amplifier circuit, we break the analysis into following two parts:

(*i*) D.C. analysis

(*ii*) A.C. analysis

(*i*) D.C. analysis

We adopt the following procedure to carry-out the a.c. analysis.

(*a*) Firstly, reduce a.c source to zero. This means that a voltage source is replaced by a short-circuit and a current-source by an open-circuit.

(*b*) After this, open all the capacitors as they block direct current.

Thus, the resulting circuit, after applying above two steps is known as a d.c. equivalent circuit. This circuit is utilized to determine the d.c. currents and voltages, which are needed for the analysis.

(*ii*) A.C. analysis

We adopt the following procedure to carry-out the a.c. analysis:

(*a*) Firstly, reduce the d.c. source to zero. This means that a voltage source is replaced by a short-circuit and a current-source by an open-circuit.

(*b*) After this, short all the capacitors as they pass alternating current.

(*c*) Then, replace the transistor with its small-signal equivalent circuit.

Thus, the resulting circuit, after applying the above three steps is known as an a.c. equivalent circuit. This circuit is utilized to determine the a.c. currents and voltages which are needed for analysis.

This method is valid for single-stage Common-Emitter (CE), Common-Base (CB) and Common-Collector (CC) amplifiers. This is also valid to analyse multistage amplifiers.

7.5. EQUIVALENT CIRCUIT OF TRANSISTOR AMPLIFIER

Figure 7.5 shows the circuit of a common emitter transistor amplifier using self-bias or voltage divider bias.

This circuit can be analyzed for dc and ac behaviour separately. For this, we obtain d.c. and a.c. equivalent circuit of amplifier.

(*i*) D.C. equivalent circuit: To obtain d.c. equivalent circuit, we reduce signal voltage to zero and open all the capacitors because for d.c., capacitors behave as open circuit.

It is the weak man who urges compromise, never the strong man. — **Elbert Hubbard**

$$X_C = \frac{1}{2\pi fC}$$

$$f = 0, \text{ for d.c.}$$

$$X_C = \frac{1}{0} = \infty = \text{open circuit}$$

Therefore, the d.c. equivalent circuit will be as shown in figure 7.6 (*a*).

***(ii)* A.C. equivalent circuit:** To obtain a.c. equivalent circuit, first we reduce d.c. supply to zero and then replace the capacitors with short-circuits because for a.c. the capacitors act almost as short-circuit. If we do this, we get the a.c. equivalent circuit shown in figure 7.6 (*b*).

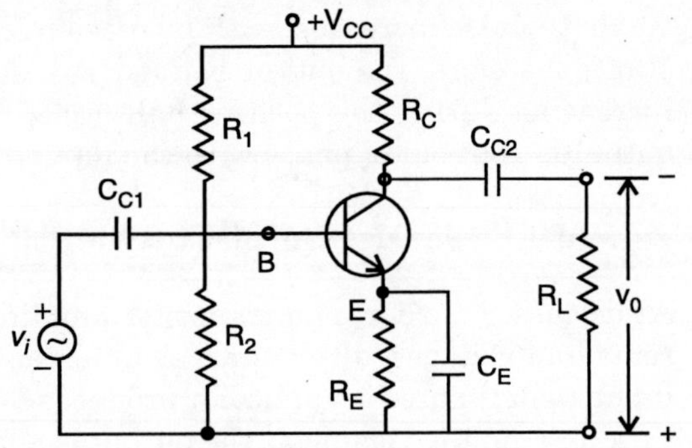

Fig. 7.5. Transistor amplifier using Self-bias.

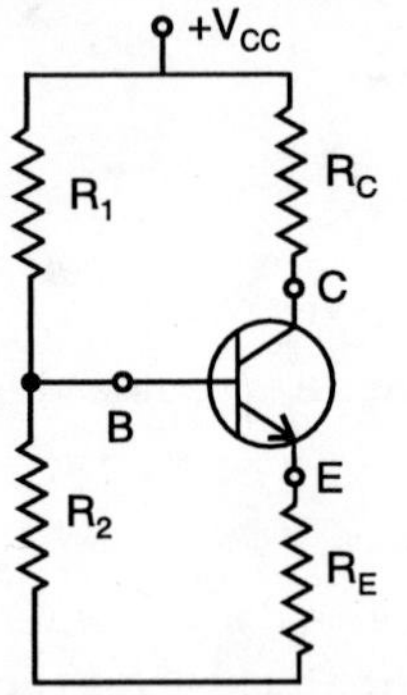

Fig. 7.6. (*a*) D.C. equivalent circuit.

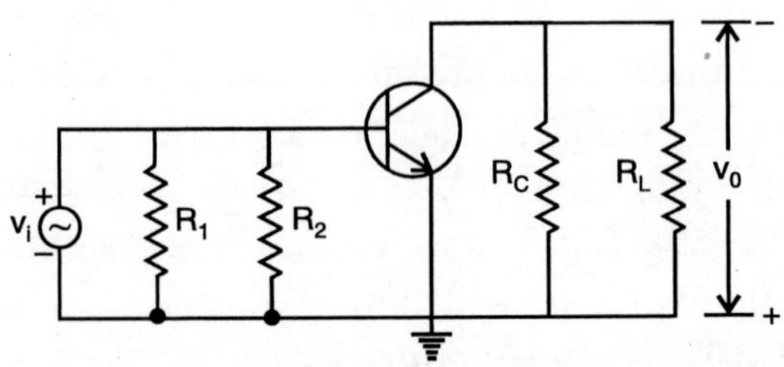

Fig. 7.6. (*b*) A.C. equivalent circuit.

From, the above, a.c. equivalent circuit, it may be observed that the resistor R_C comes in parallel with load resistor R_L. This combination forms the a.c. load for the amplifier.

$$R_{ac} = R_C \| R_L$$

Also, the resistors R_1 and R_2 come in parallel. The a.c. equivalent circuit shown in the figure describes the complete a.c. behaviour.

7.6. TRANSISTOR A.C. EQUIVALENT CIRCUIT *(Important)*

We can obtain transistor a.c. equivalent circuit with the help of output characteristics of transistor in *CE* cofiguration.

Figure 7.7 shows the output characteristics of a transistor in CE configuration. We observe that characteristics curves are almost horizontal. This means that the collector current I_C hardly depends on the value of V_{CE} for a given I_B. In other words, for a constant I_B, if V_{CE} is changed then the change in I_C is very small. This shows that the output resistance is very high. Hence, we can replace the transistor with a current source between output terminals. From this, we can draw a transistor a.c. equivalent circuit for output portion as shown in figure 7.8.

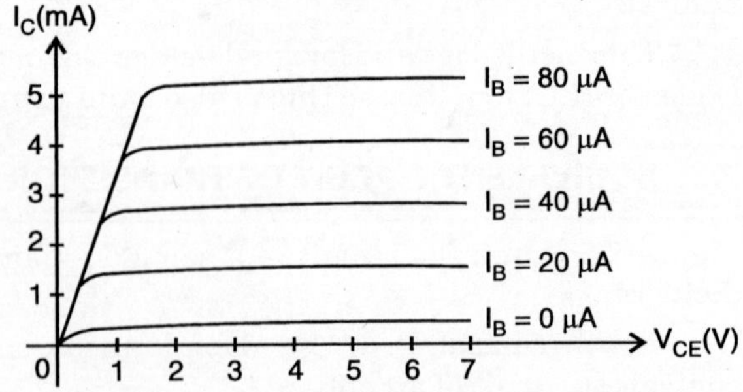

Fig. 7.7. Typical output CE characteristics of a transistor

In figure 7.8, r_o is output resistance. The value of current source βi_b depends upon a.c. input current i_b and current gain β.

The input section of transistor may be replaced by a resistor r_i because the emitter-base junction of the characteristic is like that of a forward-biased diode. Hence equivalent circuit for input side is shown in figure 7.9.

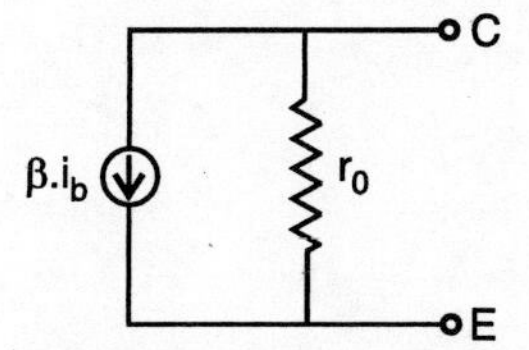

Fig. 7.8. Transistor a.c. equivalent circuit for output portion

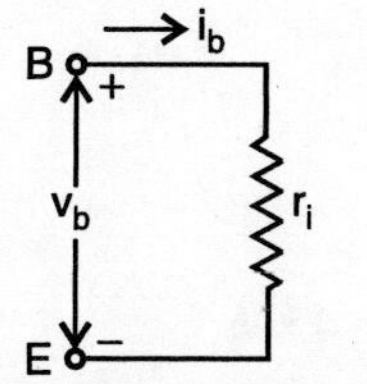

Fig. 7.9. Transistor A.C. equivalent circuit for input section

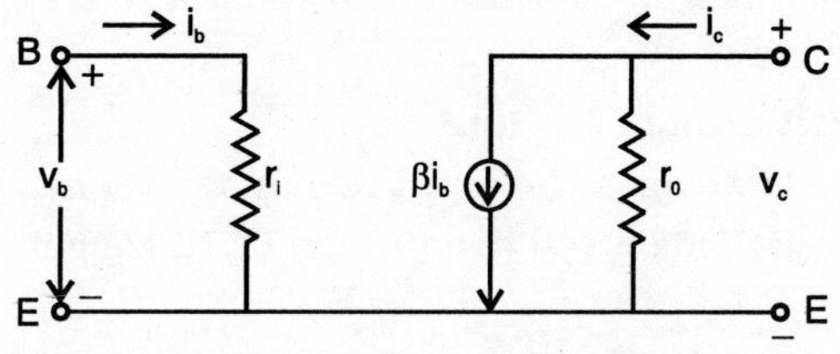

Fig. 7.10. Transistor a.c. equivalent circuit

To obtain complete a.c. equivalent circuit of a transistor, we combine the input and output section as shown in figure 7.10.

7.7. HYBRID PARAMETERS (h-PARAMETERS)

Hybrid parameters are also called as h-parameters. Characteristics of amplifier can be specified easily in terms of h-parameters. The hybrid or h-parameters are very popular and widely used because they give accurate result and can be measured very easily.

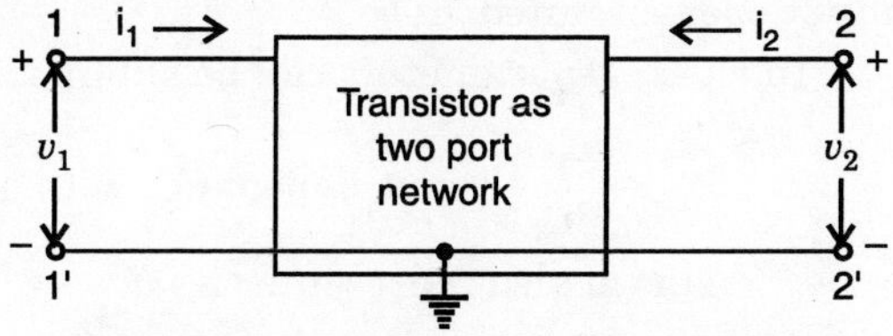

Fig. 7.11. Transistor as two port network.

We know that a transistor is a three terminals device. If one terminal is made common to the input and output, then the transistor can be represented as a two-part network as shown in figure 7.11.

There are two ports, input port and output port. For each port there are two variables, currents and voltages. Figure 7.11 shows the standard directions of currents and polarities of voltages for a two-port network. For a two-port network, the four variables are related in terms of h-parameters as

$$v_1 = h_{11}i_1 + h_{12}v_2 \quad ...(7.1)$$

$$i_2 = h_{21}i_1 + h_{22}v_2 \quad ...(7.2)$$

The parameters h_{11}, h_{12}, h_{21} and h_{22} are called hybrid parameters or h-parameters.

> **DO YOU KNOW?**
>
> The four parameters associated with hybrid equivalent model are input impedance, voltage ratio, current gain and output conductance. Since their units are completely different from each other, this set of parameters is called as hybrid parameters.

Determination of h-Parameters

We can derive the expressions for different h-parameters as under:

(*i*) Value of h_{11} and h_{21}

If we short-circuit the output terminals, then $v_2 = 0$

Putting this value $v_2 = 0$ in equation (7.1), we get

$$v_1 = h_{11}i_1 + h_{12}v_2 = h_{11}i_1 + 0$$

or $$h_{11} = \frac{v_1}{i_1}$$

Since h_{11} is the ratio of input voltage and current, it is called **input impedance** with output short-circuit.

Now, putting $v_2 = 0$ in equation (7.2), we obtain

$$i_2 = h_{21}i_1 + h_{22}v_2 = h_{21}i_1 + 0$$

$$h_{21} = \frac{i_2}{i_1}$$

Since, h_{21} is ratio of output and input currents, it is called **forward current gain** with output short-circuit.

***(ii)* Value of h_{12} and h_{22}**

If we open the input terminals, then $i_1 = 0$

putting $i_1 = 0$ in equation (7.1), we get

$$v_1 = h_{11}i_1 + h_{12}v_2 = 0 + h_{12}\,v_2$$

$$h_{12} = \frac{v_1}{v_2}$$

Since h_{12} is the ratio of input and output voltages, it is called **reverse voltage gain** with input open- circuited.

Now, putting $i_1 = 0$ in equation (7.2), we get

$$i_2 = h_{21}i_1 + h_{22}v_2 = 0 + h_{22}v_2$$

$$h_{22} = \frac{i_1}{v_2}$$

Since h_{22} is ratio of output current and output voltage, it is called **output admittance** with input open-circuited.

In short, h-parameters can be summarized as

1. $h_{11} = \frac{v_1}{i_1}$ = input impedance with output shorted and unit is ohm.

   This is also represented as h_i

   $h_{11} = h_i$

2. $h_{21} = \frac{i_2}{i_1}$ = forward current gain with output shorted and unit is none.

   This is also represented as h_f

   $h_{21} = h_f$

3. $h_{12} = \frac{v_1}{v_2}$ = reverse voltage gain with input open circuit and has no unit.

   This is also represented as h_r

   $h_{12} = h_r$

4. $h_{22} = \frac{i_2}{v_2}$ = output admittance with input open circuit and has unit ohm^{-1} or mho.

   This is also represented as h_o

   $h_{22} = h_o$

DO YOU KNOW?

The symbol h_{FE} represents the forward current transfer ratio in the common-emitter configuration. The symbol h_{FE} is a hybrid h-parameter symbol. The h-parameter system is the most common system in use today for specifying transistor parameters.

From above, it is clear that all the four parameters have different units and thus the parameters are hybrid (mixed) in nature. Therefore, they are called **hybrid or *h*-parameters.**

Adopting the right attitude can convert a negative stress into a positive one.

— **Hans Selye**

Table 7.1: Also Summaries h-parameters and Their Meaning

S.No.	*h*-Parameter	Meaning	Condition
1.	h_{11}	Input impedence	Output is shorted
2.	h_{21}	Forward current gain	Output is shorted
3.	h_{12}	Reverse voltage gain	Input open
4.	h_{22}	Output admittance	Input open

7.8. h-PARAMETER REPRESENTATION FOR TRANSISTOR

The *h*-parameter representation for transistor depends upon the type of configuration (*CE*, *CB*, or *CC*). Therefore, each *h*-parameter is assigned a second subscript letter *e*, *b* or *c*, depending upon the type of configuration. The letter '*e*' is used when the configuration is common emitter, the letter '*b*' is used if the configuration is common base and letter '*c*' is used if the configuration is common collector.

Table 7.2: Shows the h-parameter Representation for Three Types of Configuration

S.No.	*h*-Parameter	Types of Configuration		
		Common emitter	Common base	Common collector
1.	h_{11}	h_{ie}	h_{ib}	h_{ic}
2.	h_{21}	h_{fe}	h_{fb}	h_{fc}
3.	h_{12}	h_{re}	h_{rb}	h_{rc}
4.	h_{22}	h_{oe}	h_{ob}	h_{oc}

7.9. BJT HYBRID MODEL : GENERAL HYBRID EQUIVALENT CIRCUIT

There are three transistor configurations. *CB*, *CE* and *CC*. Each configuration will have its separate hybrid or *h*-parameter equivalent circuit. But, first let us develop a general *h*-parameter equivalent circuit which can be applied for any of three configurations. The equivalent circuit thus obtained for any type of configuration is also called as *small-signal, low-frequency hybrid model of a transistor.*

h-parameter equations are

$$v_1 = h_{11}i_1 + h_{12}v_2$$

$$i_2 = h_{21}i_1 + h_{22}v_2$$

or

$$v_1 = h_i.i_1 + h_r.v_2 \quad ...(7.3)$$

$$i_2 = h_f.i_1 + h_o.v_2 \quad ...(7.4)$$

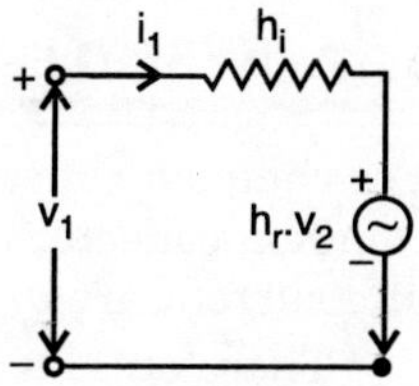

Fig. 7.12.

Since, each term of equation (7.3) has the units of volts, we can use Kirchoff's voltage law to find a circuit that fits this equation. Figure 7.12 shows this circuit.

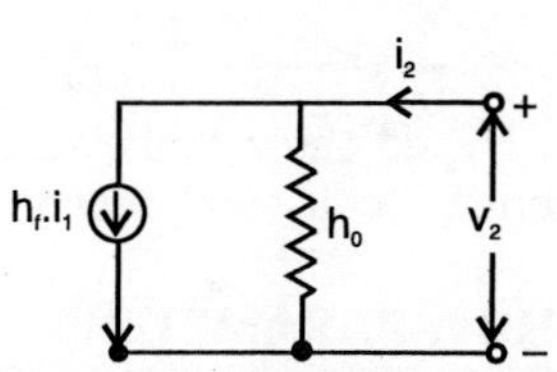

Fig. 7.13.

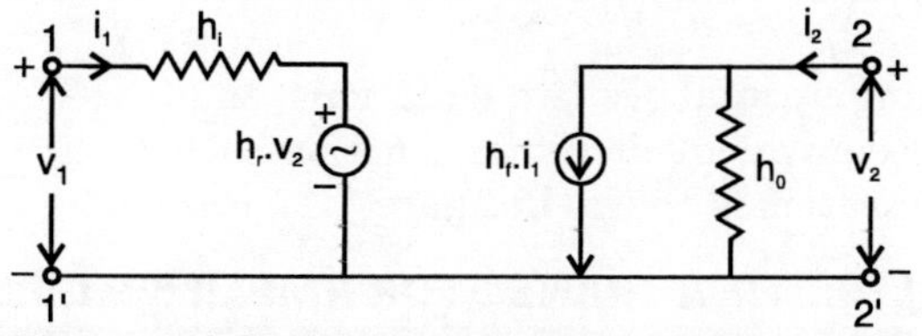

Fig. 7.14. General h-parameter equivalent circuit.

Similarly, each term of equation (7.4) has the units of currents, we can use Kirchoff's current law to find a circuit that fits this equation. Figure 7.13 shows this circuit.

Combining these two figures, we get the general hybrid (h-parameter) equivalent circuit as shown in figure 7.14.

7.10. HYBRID EQUIVALENT CIRCUIT FOR COMMON EMITTER CONFIGURATION

We have already discussed that for common emitter configuration, an additional suffix '*e*' is added to the h-parameters. Also, for common-emitter configuration, input voltage and current are v_{be} (base voltage) and i_b (base current). The output voltage and current are v_{ce} and i_c.

Therefore, for common emitter configuration, the h-parameter equations are

$$v_{be} = h_{ie}.i_b + h_{re}.v_{ce} \quad ...(7.5)$$

$$i_c = h_{fe}.i_b + v_{ce}.h_{oe} \quad ...(7.6)$$

Depending upon above two equations, the h-parameter equivalent circuit for common-emitter configuration is

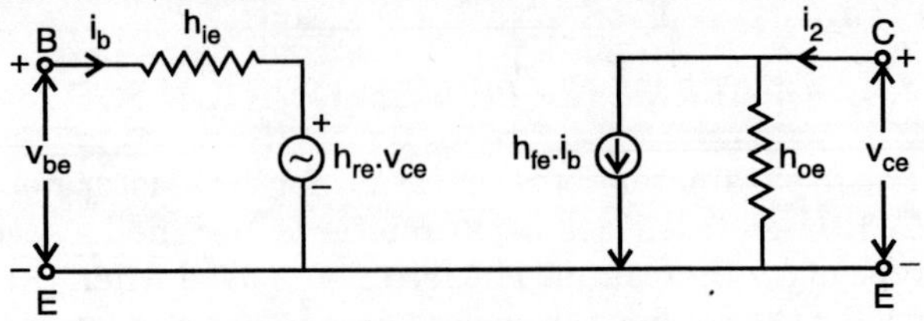

Fig. 7.15. Hybrid equivalent circuit for common-emitter configuration.

7.11. HYBRID EQUIVALENT CIRCUIT FOR COMMON BASE CONFIGURATION

For common-base configuration, an additional suffix '*b*' is added to the h-parameters. Also, for common-base configuration, input voltage and input current are i_e and v_{eb}. The output voltage and current are v_{cb} and i_c.

Therefore, for common-base configuration, the h-parameter equations are

$$v_{eb} = h_{ib}.i_e + h_{rb}.\ v_{cb} \quad ...(7.7)$$

$$i_c = h_{fb}.i_e + v_{cb}.h_{ob} \quad ...(7.8)$$

Depending upon above two equations, the h-parameter equivalent circuit for common-base configuration is shown in figure 7.16.

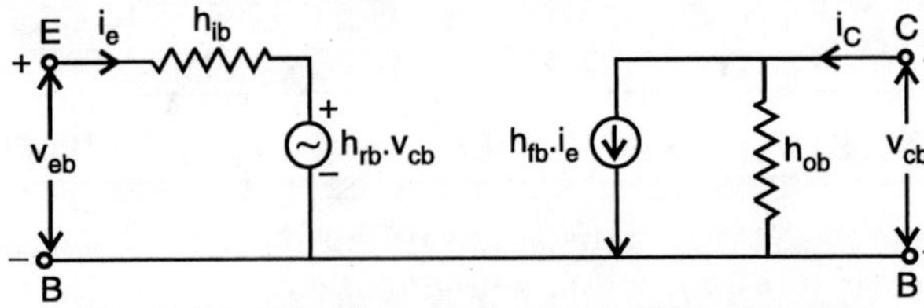

Fig. 7.16. Hybrid equivalent circuit for common-base configuration.

7.12. HYBRID EQUIVALENT CIRCUIT FOR COMMON COLLECTOR CONFIGURATION

For common-collector configuration an additional suffix '*c*' is added to the h-parameters. Also, for common-collector configuration, input voltage and input current are v_{bc} and i_b. The output voltage and current are v_{ec} and i_e.

Therefore, for common-collector configuration, the h-parameter equations are

$$v_{bc} = h_{ic}.i_b + h_{rc}.\ v_{ec}$$

$$i_e = h_{fc}.i_b + v_{ec}.h_{oc}$$

Depending upon above two equations, the h-parameter equivalent circuit for common-collector configuration is shown in figure 7.17.

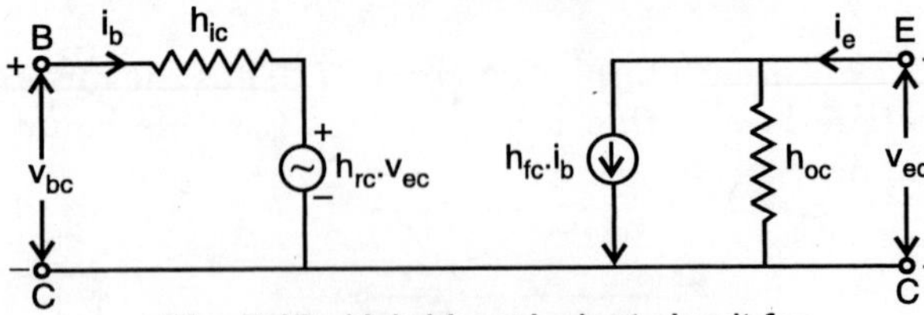

Fig. 7.17. Hybrid equivalent circuit for common-collector configuration.

7.13. ANALYSIS OF A TRANSISTOR AMPLIFIER CIRCUIT USING h-PARAMETERS*

(Important)

Figure 7.18 shows the circuit of a general amplifier. In this circuit, the transistor may be connected

* Draw the hybrid equivalent circuit for common base configuration and find the expression for current gain, voltage gain and input impedance.

in any one of three configurations, *CE*, *CB* or *CC*. Figure 7.19 shows the hybrid equivalent circuit for this general amplifier. To obtain this circuit, we replace the transistor by its small-signal hybrid model.

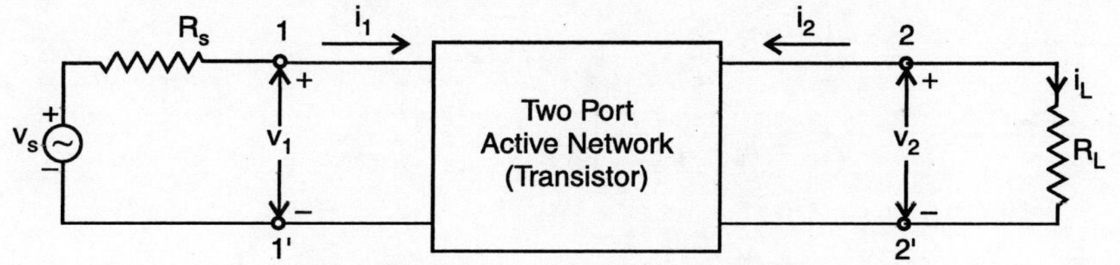

Fig. 7.18. General amplifier circuit.

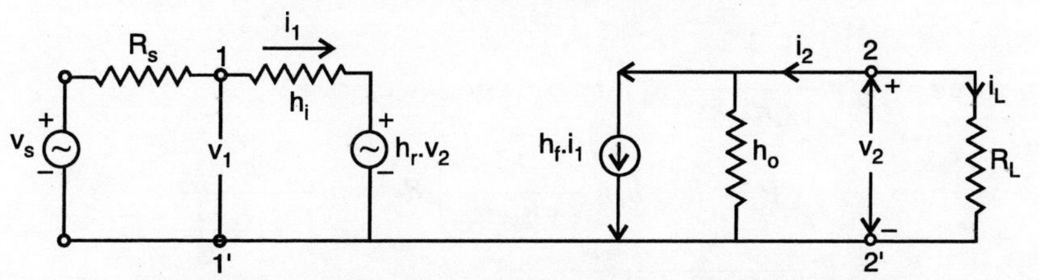

Fig. 7.19. Hybrid equivalent circuit for general amplifier.

We shall derive expressions for current gain, voltage gain, input resistance, output resistance for amplifier in terms of *h*-parameters.

First writing the *h*-parameter equations

$$v_1 = h_i . i_1 + h_r . v_2 \quad \text{... (7.9)}$$

$$i_2 = h_f . i_1 + h_o . v_2 \quad \text{... (7.10)}$$

1. The current gain: The current gain A_i for transistor amplifier is defined as the ratio of output to input currents. Mathematically,

$$A_i = \frac{i_L}{i_1} = -\frac{i_2}{i_1} \qquad (\because i_L = -i_2)$$

From figure, voltage across the output terminals is v_2

$$v_2 = i_L . R_L = -i_2 R_L$$

Putting value of v_2 in equation (7.10), we get

$$i_2 = h_f . i_1 - h_o i_2 R_L$$

$$i_2 (1 + h_o . R_L) = h_f . i_1$$

$$\frac{i_2}{i_1} = \frac{h_f}{1 + h_o . R_L}$$

or

$$A_i = \frac{-i_2}{i_1} = \frac{-h_f}{1 + h_o . R_L} \quad \text{...(7.11)}$$

2. The input resistance R_i: The input resistance R_i is defined as the resistance we see looking into the amplifier input terminals (1,1′). Mathematically.

$$R_i = \frac{v_1}{i_1}$$

Putting the value of $v_2 = -i_2 . R_L$ in equation (7.9), we get

$$v_1 = h_i . i_1 + h_r . v_2 = h_i . i_1 + h_r . (-i_2 R_L) = h_i . i_1 - h_r . i_2 . R_L$$

Dividing the above equation on the both sides by i_1, we have

$$\frac{v_1}{i_1} = h_i - h_r.\left(\frac{i_2}{i_1}\right)R_L$$

But $$\frac{v_1}{i_1} = R_i$$

Therefore, $$R_i = h_i - h_r\left(\frac{i_2}{i_1}\right).R_L$$

We know that $$A_i = \frac{-i_2}{i_1}$$

Thus, $$R_i = h_i + h_r.A_i.R_L$$

Putting the value of $A_i = \dfrac{-h_f}{1+h_o.R_L}$ in last equation, we have

$$R_i = h_i + h_r.\left(\frac{-h_f}{1+h_o.R_L}\right).R_L = h_i - \frac{h_r.h_f}{h_o + \frac{1}{R_L}}$$

or $$\boxed{R_i = h_i - \frac{h_r.h_f}{h_o + Y_L}} \quad ...(7.12)$$

where $\dfrac{1}{R_L} = Y_L$ = Load admittance.

Therefore, input resistance is a function of the load resistance.

3. The voltage gain: The voltage gain is defined as the ratio of output voltage v_2 to input voltage v_1.

$$A_v = \frac{v_2}{v_1} = -\frac{i_2.R_L}{v_1} \quad ...(7.13)$$

Since we know that $$A_i = -\frac{i_2}{i_1} \quad \text{or} \quad i_2 = -A_i.i_1$$

Therefore, $$A_v = \frac{A_i.i_1.R_L}{v_1} = A_i.R_L\left(\frac{i_1}{v_1}\right)$$

But $$\frac{v_1}{i_1} = R_i$$

$$A_v = \frac{A_i R_L}{R_i} \quad ...(7.14)$$

Again, since $$A_i = \frac{-h_f}{1+h_o.R_L}$$

and $$R_i = h_i - \frac{h_r.h_f}{h_o + \frac{1}{R_L}}$$

Putting the values of A_i and R_i in equation (7.14) and rearranging the terms, we have

$$A_v = -\frac{h_f.R_L}{h_i + (h_i.h_o. - h_r.h_f)R_L}$$

Replacing $$\Delta h = h_i h_o - h_r.h_f$$

Hence, $$A_v = -\frac{h_f . R_L}{h_i + \Delta h . R_L} \qquad ...(7.15)$$

4. Output resistance: The output resistance may be calculated by reducing the source voltage v_s to zero and load resistance R_L to infinity* and driving the output terminals by a voltage generator v_2. This arrangement is shown in figure 7.20.

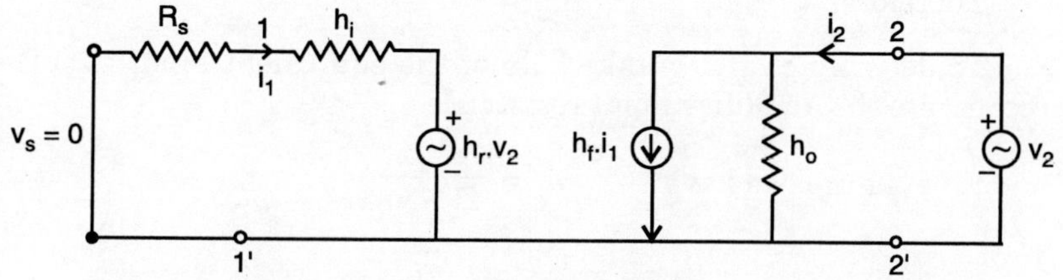

Fig. 7.20. Hybrid equivalent circuit to calculate output resistance.

From figure, output resistance R_o is the ratio of voltage v_2 and the current drawn from the voltage source, i_2.

Mathematically, $$R_0 = \frac{v_2}{i_2}$$

But $$i_2 = h_f . i_1 + h_o . v_2$$

Hence $$R_0 = \frac{v_2}{h_f . i_1 + h_o . v_2} \qquad ...(7.16)$$

Applying KVL to the input side of the circuit, we have

$$R_s . i_1 + h_r . v_2 + h_i . i_1 = 0$$

or $$i_1 (R_s + h_i) = - h_r . v_2$$

or $$i_1 = -\frac{h_r . v_2}{R_s + h_i} \qquad ...(7.17)$$

Now, putting the value of i_1 in equation (7.16), we get

$$R_0 = \frac{v_2}{h_f . \left(\frac{-h_r . v_2}{R_s + h_i}\right) + h_o . v_2} = \frac{v_2}{\frac{-h_f . h_r . v_2}{R_s + h_i} + h_o . v_2}$$

Rearranging, we get $$R_0 = \frac{R_s + h_i}{h_o . (R_s + h_i) - h_f . h_r} = \frac{R_s + h_i}{R_s . h_o + (h_i h_o - h_f . h_r)} = \frac{R_s + h_i}{R_s . h_o + \Delta h}$$

where $$\Delta h = h_i h_o - h_f h_r$$

If source resistance R_s is zero, then output resistance will be

$$\boxed{R_0 = \frac{h_i}{\Delta h}} \qquad ...(7.18)$$

5. The overall voltage gain: The overall voltage gain of the transistor amplifier is the ratio of output voltage v_2 to the source voltage v_s. It is denoted by A_{vs}. Hence, A_{vs} is the voltage amplification factor taking into account the source resistance R_s.

Mathematically, overall voltage gain

$$A_{vs} = \frac{v_2}{v_s} = \frac{v_2}{v_1} . \frac{v_1}{v_s}$$

* Load resistance R_L equal to infinity means open circuit.

But $$A_v = \frac{v_2}{v_1}$$

Therefore, $$A_{vs} = A_v . \frac{v_1}{v_s} \quad ...(7.19)$$

To determine A_{vs}, we take help of the equivalent input circuit shown in figure 7.21. Here Z_i represents the amplifier input resistance.

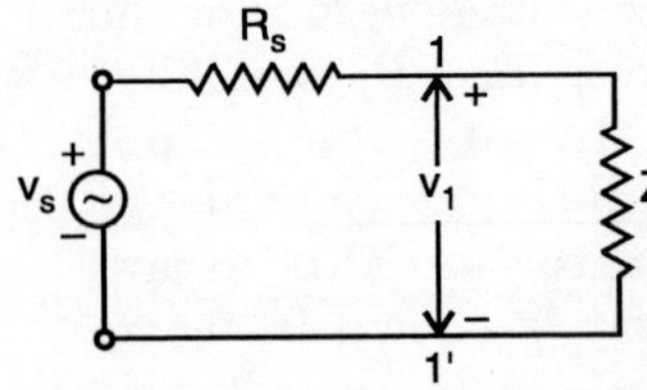

Fig. 7.21. Equivalent input circuit.

From figure, we have $$v_1 = \frac{v_s . Z_i}{Z_i + R_S}$$

or $$\frac{v_1}{v_s} = \frac{Z_i}{Z_i + R_S}$$

Putting the value of $\frac{v_1}{v_s}$ in equation (7.19), we get

$$A_{vs} = A_v \frac{Z_i}{Z_i + R_S} \quad ...(7.20)$$

Note that if $R_S = 0$.

Then $A_{vs} = A_v$

Therefore, A_v is the voltage gain with an ideal voltage source ($R_S = 0$).

6. Overall current gain: Overall current gain of the transistor amplifier is the ratio of output current (i_L) to the current delivered by the source (i_S).

Overall current gain $$A_{is} = \frac{i_L}{i_s} = \frac{-i_2}{i_s}$$

Modifying the above expression, we have

$$A_{is} = \frac{-i_2}{i_1} \times \frac{i_1}{i_s}$$

But, $$\frac{i_2}{i_1} = A_i$$

$$A_{is} = -A_i \times \frac{i_1}{i_s} \quad ...(7.21)$$

To determine A_{is}, we take help of modified input equivalent circuit shown in figure 7.22. Here Z_i represents the amplifier input resistance.

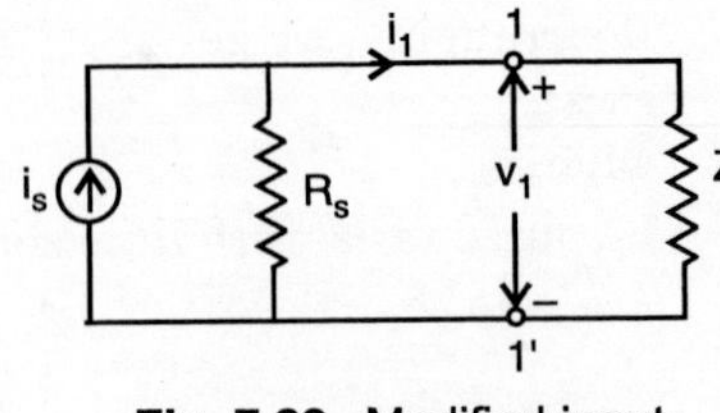

Fig. 7.22. Modified input equivalent circuit

From figure

$$i_1 = i_s \times \frac{R_s}{R_s + Z_i}$$

or $$\frac{i_1}{i_s} = \frac{R_s}{R_s + Z_i} \quad ...(7.22)$$

Putting the value of $\frac{i_1}{i_s}$ in equation (7.21), we get

$$A_{is} = -A_i \times \frac{R_S}{R_S + Z_i} \quad ...(7.23)$$

Now, if $R_s = \infty$

then $A_{is} = A_i$

Therefore, A_i is the current gain with an ideal curent source (infinite resistance).

7.14. h-PARAMETER (HYBRID) EXPRESSION FOR COMMON-EMITTER AMPLIFIER*

h-parameter expressions for CE amplifier will be obtained with the help of general h-parameter formulas discussed in last article.

1. Current gain: The general expression for current gain is

$$A_i = -\frac{h_f}{1+h_o.R_L} \qquad ...(7.24)$$

For common-emitter amplifier, we add an additional subscript 'e'.

Hence, current gain for CE amplifier

$$A_i = -\frac{h_{fe}}{1+h_{oe}.R_L} \qquad ...(7.25)$$

Here R_L is the a.c. load resistance.

2. Input resistance: The resistance looking into the amplifier input terminals (base of a transistor) is

General expression $$R_i = h_i - \frac{h_r.h_f}{h_o + \dfrac{1}{R_L}} \qquad ...(7.26)$$

Expression for common emitter amplifier

$$R_i = h_{ie} - \frac{h_{re}.h_{fe}}{h_{oe} + \dfrac{1}{R_L}} \qquad ...(7.27)$$

We know that input resistance of the amplifier stage, Z_i depends upon the type of biasing arrangement.

As an example, for a fixed bias circuit

$$Z_i = R_i \| R_B$$

If there is no biasing resistance then

$$Z_i = R_i \qquad ...(7.28)$$

3. Voltage gain: General expression

$$A_v = -\frac{h_f.R_L}{h_i + \Delta h.R_L} \qquad ...(7.29)$$

Expression for common emitter amplifier

$$A_v = -\frac{h_{fe}.R_L}{h_{ie} + \Delta h.R_L} \qquad ...(7.30)$$

Here, $$\Delta h = h_{ie}.h_{oe} - h_{re}.h_{fe}$$

4. Output resistance: The resistance looking into the amplifier output terminals is general expression

$$R_0 = \frac{R_s + h_i}{R_s.h_o + \Delta h} \qquad ...(7.31)$$

* Using h-parameter model, obtain the expression for a CE configuration circuit for (*i*) A_I, (*ii*) R_i, (*iii*) A_v (*iv*) R_o.

$$R_S = \text{source resistance}$$
$$\Delta h = h_{ie}.h_{oe} - h_{re}.h_{fe}$$

The output resistance of the amplifier stage will be

$$Z_0 = R_o \| R_L \qquad \text{...(7.32)}$$

5. Overall voltage gain: General expression

$$A_{vs} = A_v.\frac{Z_i}{Z_i + R_s} \qquad \text{...(7.33)}$$

Expression for common emitter amplifier is also same.

Thus,
$$A_{vs} = \frac{A_v.Z_i}{Z_i + R_s} \qquad \text{...(7.34)}$$

6. Overall current gain:

$$A_{is} = -\frac{A_i.R_s}{R_s + Z_i} \qquad \text{...(7.35)}$$

EXAMPLE 7.1 **A common emitter amplifier uses a voltage source having internal resistance R_S = 800 Ω and the load resistance R_L = 1000 Ω. The *h*-parameters are h_{ie} = 1 kΩ, $h_{re} = 2 \times 10^{-4}$, h_{fe} = 50 and h_{oe} = 25 μA/volt. Calculate the current gain A_i, input resistance R_i, voltage gain A_v.**

Solution: Given that

$$R_s = 800\ \Omega$$
$$R_L = 1000\ \Omega$$
$$h_{ie} = 1\ \text{k}\Omega$$
$$h_{re} = 2 \times 10^{-4}$$
$$h_{fe} = 50$$
$$h_{oe} = 25\ \mu\text{A/Volt} = 25 \times 10^{-6}\ \text{A/Volt}$$

***(i)* Current gain:**
$$A_i = -\frac{h_{fe}}{1 + h_{oe}.R_L}$$

Putting the given values, we get

$$A_i = -\frac{50}{1 + 25 \times 10^{-6} \times 10^3} = -48.78 \quad \textbf{Ans.}$$

***(ii)* Input resistance:**
$$R_i = h_{ie} - \frac{h_{re}.h_{fe}}{h_{oe} + \frac{1}{R_L}} = 1000 - \frac{50 \times 2 \times 10^{-4}}{25 \times 10^{-6} + \frac{1}{1000}} = 990.24\ \Omega \quad \textbf{Ans.}$$

***(iii)* Voltage gain:**
$$A_v = A_i.\frac{R_L}{R_i} = -48.78 \times \frac{1000}{990.24} = -49.26 \quad \textbf{Ans.}$$

7.15. h-PARAMETER (HYBRID) EXPRESSIONS FOR COMMON BASE AMPLIFIER

***(i)* Current gain:**

General expression
$$A_i = -\frac{h_f}{1 + h_o.R_L} \qquad \text{...(7.36)}$$

Expression for common base amplifier

$$A_i = \frac{h_{fb}}{1 + h_{ob}.R_L} \qquad \text{...(7.37)}$$

***(ii)* Input resistance:** The resistance looking into the input terminals is

General expression
$$R_i = h_i - \frac{h_r.h_f}{h_o + \frac{1}{R_L}} \quad ...(7.38)$$

Expression for common base amplifier
$$R_i = h_{ib} - \frac{h_{rb}.h_{fb}}{h_{ob} + \frac{1}{R_L}} \quad ...(7.39)$$

The input resistance of the amplifier stage depends upon the biasing arrangement. If there is no biasing resistance, then
$$Z_i = R_i \quad ...(7.40)$$

(iii) **Voltage gain:** General expression
$$A_v = -\frac{h_f.R_L}{h_i + \Delta h.R_L} \quad ...(7.41)$$

Expression for common base amplifier will be
$$A_v = -\frac{h_{fb}.R_L}{h_{ib} + \Delta h.R_L} \quad ...(7.42)$$

(iv) **Output resistance:** The resistance looking into amplifier output terminals is

General expression
$$R_0 = \frac{R_s + h_i}{R_S.h_o + \Delta h} \quad ...(7.43)$$

Expression for common-base amplifier
$$R_0 = \frac{R_s + h_{ib}}{R_S.h_{ob} + \Delta h} \quad ...(7.44)$$

The output resistance of the amplifier stage
$$Z_0 = R_o \| R_L \quad ...(7.45)$$

(v) **Overall voltage gain:**
$$A_{vs} = -\frac{A_v.Z_i}{Z_i + R_s} \quad ...(7.46)$$

(vi) **Overall current gain:**
$$A_{is} = -\frac{A_i.R_s}{R_s + Z_i} \quad ...(7.47)$$

EXAMPLE 7.2 For a common base amplifier, a transistor has the following values of *h*-parameters. h_{ib} = 28 Ω, h_{fb} = – 0.98, h_{rb} = 5 ×10^{-4} and h_{ob} = 0.34 × 10^{-6} *S*. Find the values of current gain, input resistance, voltage gain and output resistance. The load resistance is 1.2 kΩ. The source resistance is zero. ***(V. Important)***

Solution: (*i*) The current gain:
$$A_i = -\frac{h_{fb}}{1 + h_{ob}.R_L} = -\frac{(-0.98)}{1 + 0.34 \times 10^{-6} \times 1200} = 0.98 \textbf{ Ans.}$$

(*ii*) The input resistance:
$$R_i = h_{ib} + h_{rb}.A_i.R_L = 28 + 5 \times 10^{-4} \times 0.98 \times 1200$$
$$R_i = 28 + 0.6 = 28.6\ \Omega \quad \textbf{Ans.}$$

***(iii)* The voltage gain:** $A_v = \dfrac{A_i . R_L}{R_i} = \dfrac{0.98 \times 1200}{28.6} = 41$ **Ans.**

***(iv)* Output resistance:**

$$R_0 = \frac{R_s + h_{ib}}{R_s . h_{ib} + \Delta h} \cong \frac{h_{ib}}{\Delta h}$$

where $\Delta h = h_{ib} . h_{ob} - h_{rb} . h_{fb}$

Substituting all the values, we have

$$\Delta h = 28 \times 0.34 \times 10^{-6} - 5 \times 10^{-4} \times (-0.98) = 5 \times 10^{-4}$$

Thus, we have $R_0 = \dfrac{h_{ib}}{\Delta h} = \dfrac{28}{5 \times 10^{-4}}$

$= 5.6 \times 10^4$ Ohm $= 56$ KΩ **Ans.**

7.16. h-PARAMETER (HYBRID) EXPRESSION FOR COMMON COLLECTOR AMPLIFIER

***(i)* Current gain:** General expression

$$A_i = -\frac{h_f}{1 + h_o . R_L} \quad \text{...(7.48)}$$

Expression for common collector amplifier

$$A_i = -\frac{h_{fc}}{1 + h_{oc} . R_L} \quad \text{...(7.49)}$$

***(ii)* Input resistance:** The input resistance looking into the input terminals is

$$R_i = h_{ic} - \frac{h_{rc} . h_{fc}}{h_{oc} + \dfrac{1}{R_L}} \quad \text{...(7.50)}$$

The input resistance of the amplifier stage depends upon the biasing arrangement. If there is no biasing resistance, then

$$Z_i = R_i \quad \text{...(7.51)}$$

***(iii)* Voltage Gain:**

$$A_v = \frac{A_i . R_L}{R_i} \quad \text{...(7.52)}$$

$$A_v = -\frac{h_{fc} . R_L}{h_{ic} + \Delta h . R_L} \quad \text{...(7.53)}$$

***(iv)* Output resistance:**

$$R_0 = \frac{R_s + h_{ic}}{R_s . h_{oc} + \Delta h} \quad \text{...(7.54)}$$

***(v)* Overall voltage gain:**

$$A_{vs} = \frac{A_v . Z_i}{R_s + Z_i} \quad \text{...(7.55)}$$

***(vi)* Overall current gain:**

$$A_{is} = \frac{A_i + R_s}{R_i + Z_i} \quad \text{...(7.56)}$$

The happiness of every country depends upon the character of its people rather than the form of its government.
— **Thomas C. Haliburton**

7.17. COMPARISON OF CE, CB AND CC TRANSISTOR AMPLIFIER CONFIGURATIONS

S. No.	Quantity/Parameter	CE	CB	CC
1.	Configuration	Refer Fig. 7.23 (*a*)	Refer Fig. 7.23 (*b*)	Refer Fig. 7.23 (*c*)
2.	Input terminal	Base	Emitter	Base
3.	Output terminal	Collector	Collector	Emitter
4.	Common terminal	Emitter	Base	Collector
5.	Input resistance (R_i)	Medium	Low	High
6.	Output resistance (R_0)	Medium	High	Low
7.	Current Gain (A_I)	High	Less than 1	High
8.	Voltage Gain (A_V)	High	High	Less than 1
9.	Applications	AF voltage amplifiers	Low noise pre-amplifiers	Buffer amplifiers

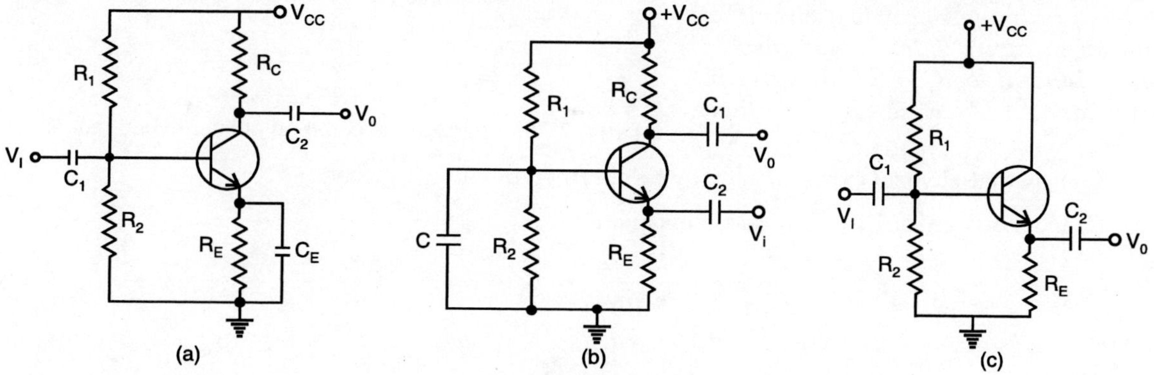

Fig. 7.23.

7.18. SIMPLIFIED OR APPROXIMATE MODEL

As a matter of fact, the *h*-parameters are widely used in amplifier analysis. To simplify the analysis, the exact *h*-parameters expressions are converted into approximated expressions. The complete behaviour of the transistor circuit can be obtained by using this simplified hybrid model.*

Figure 7.24 shows the *h*-parameter equivalent circuit of transistor in *CE* configuration.**

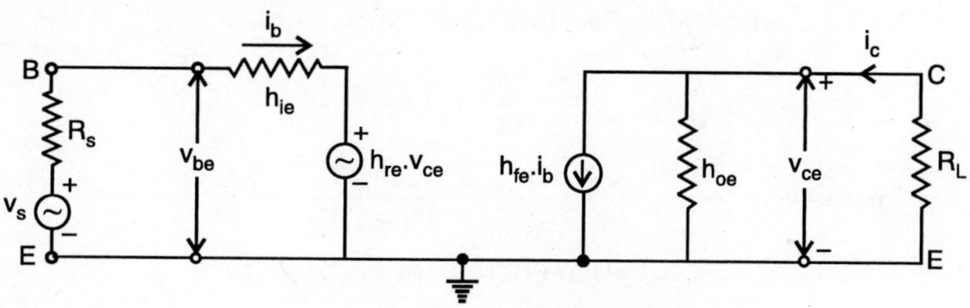

Fig. 7.24. Exact CE hybrid model

* Since the *h*-parameters themselves vary widely for the same type of transistor, it is justified to make approximations and simplify the expression for A_i, A_v, R_i and R_o. In addition, a better understanding of the behaviour of the transistor circuit can be obtained by using the simplified hybrid model.

** Since *CE* configuration is more useful and general, it is taken for consideration.

In exact CE hybrid model, $\frac{1}{h_{oe}}$ and R_L are in parallel. The parallel combination of these two is approximately equal to lower value, *i.e.* R_L. Hence, If $1/h_{oe} >> R_L$, then the term h_{oe} may be neglected provided that $h_{oe} R_L << 1$.

Further, if h_{oe} is omitted, the collector current i_c is given by $i_c = h_{fe}.i_b$.
Under this condition, the magnitude of voltage in the emitter circuit is

$$h_{re}|v_{ce}| = h_{re}.i_c\, R_L = h_{re}.h_{fe}.i_b.R_L$$

But, since, $h_{re}.h_{fe} \cong 0.01$

Therefore, the voltage $h_{re}.v_{ce}$ may be neglected in comparison to the voltage drop across $h_{ie} = h_{ie}.i_b$ provided that R_L is not too large.

To summarize, if the load resistance R_L is small, it is possible to neglect the parameter h_{re} and h_{oe} and obtain the approximate equivalent circuit as shown in figure 7.24. It may also be proved that if $h_{oe}R_L \leq 0.1$, the error in calculating A_i, A_v, R_i and R_o is less than 10%.

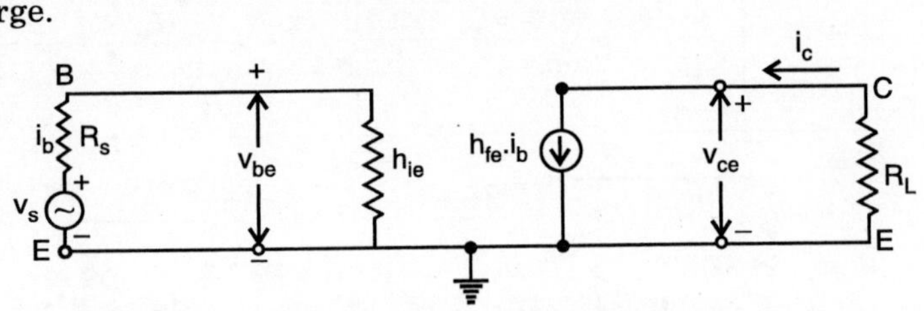

Fig. 7.25. Approximate common emitter hybrid model

Approximate hybrid expressions may be obtained in following way:

***(i)* Current gain:** Exact expression is given by

$$A_i = -\frac{h_{fe}}{1+h_{oe}R_L} \quad ...(7.57)$$

If $h_{oe}.R_L < 0.1$ ...(7.58)

Then $A_i = -h_{fe}$

***(ii)* Input resistance:** Exact expression is given by

$$R_i = h_{ie} + h_{re}A_i.R_L \quad ...(7.59)$$

It can be written as

$$R_i = h_{ie}\left[1-\frac{h_{re}\,h_{fe}|A_i|h_{oe}\,R_L}{h_{ie}\,h_{oe}\,h_{fe}}\right] \quad ...(7.60)$$

Using the typical values for the h-parameters,

$$\frac{h_{re}.h_{fe}}{h_{ie}.h_{oe}} \cong 0.5 \quad ...(7.61)$$

Further,

$$|A_i| = \frac{h_{fe}}{1+h_{oe}.R_L} \cong h_{fe} \quad ...(7.62)$$

Therefore, the equation (7.60) approximates to

$$R_i \cong h_{ie}\left[1-\frac{0.5\,h_{fe}.h_{oe}.R_L}{h_{fe}}\right]$$

If $h_{oe}.R_L < 0.1$

then

$$R_i \cong h_{ie} = \frac{v_b}{i_b}\,^* \quad ...(7.63)$$

* This over estimates the input resistance by less than 5%.

***(iii)* Voltage gain:** We know that the voltage gain is given by

$$A_v = A_i . \frac{R_L}{R_i} \qquad \text{...(7.64)}$$

But $\quad A_i = -h_{fe}$

and $\quad R_i = h_{ie}$

Therefore, we have $\quad A_v = -\frac{h_{fe}.R_L}{h_{ie}}$

(iv) Similarly $\quad R_o = \frac{1}{h_{oe}}$

Approximate hybrid expressions for *CB* and *CC* configurations may be found in a similar way.

Table 7.3 : List of Approximate Hybrid Expressions for CE, CB and CC Configuration

S. No.	Parameter	Symbol	Approximate hybrid expressions for		
			Common Emitter	Common Base	Common Collector
1.	Current gain	A_i	$-h_{fe}$	$h_{fb} = 1$	h_{fc}
2.	Input resistance	R_i	$-h_{ie}$	h_{ib}	$h_{ic}+h_{fc}.R_L$
3.	Voltage gain	A_v	$\frac{h_{fe}.R_L}{h_{ie}}$	$\frac{h_{fb}.R_L}{h_{ib}}$	1
4.	Output resistance	R_o	$\frac{1}{h_{oe}}$	$\frac{1}{h_{ob}}$	$\frac{R_s + h_{ic}}{h_{fc}}$

Table 7.4 : Typical h-parameter Values for Transistor

S.No.	Parameter	Values of parameter for		
		CE	CB	CC
1.	h_i	1100 Ω	1100 Ω	22 Ω
2.	h_r	2.5×10^{-4}	1	3×10^{-4}
3.	h_f	50	–51	–0.98
4.	h_o	25 μA/V	25 μA/V	0.49 μA/V

Table 7.5: Conversion Expression (Formulae) for h-parameters

S. No.	Types of Configuration	
	CB	CC
1.	$h_{ib} = \frac{h_{ie}}{1+h_{fe}}$	$h_{ic} = h_{ie}$
2.	$h_{rb} = \frac{h_{ie}\,h_{oe}}{1+h_{fe}} - h_{re}$	$h_{rc} = 1$
3.	$h_{fb} = \frac{-h_{fe}}{1+h_{fe}}$	$h_{fc} = -(1 + h_{fe})$
4.	$h_{ob} = \frac{h_{oe}}{1+h_{fe}}$	$h_{oc} = h_{oe}$

7.19. IMPORTANT POINTS ABOUT h-PARAMETERS

(*i*) We can obtain the values of *h*-parameters for the other two configurations (CB and CC) by following the same procedure as for CE configuration.

(*ii*) It is important to note that the *h*-parameters are calculated at the quiescent condition. Hence, any variation in the *Q* point will change the values of *h*-parameters.

(*iii*) As the *Q* point changes its position with the changes in temperature, the *h*-parameter values will also change with the changes in temperature.

7.19.1. Merits of Hybrid Model

(*i*) *h*-parameters can be easily measured.

(*ii*) They can be calculated from the static characteristics of transistor.

(*iii*) Manufacturers provide *h*-parameters in their data sheets.

(*iv*) *h*-parameters can be used easily and conveniently in circuit analysis and design.

7.19.2. Demerits of the h-parameters

The hybrid parameters suffer from a major disadvantage that they are defined only for a particular set of operating conditions. The parameters of other equivalent circuit (r_e model) can be determined for any region of operation within the active region and are not limited by the single set of parameters provided by the specification sheet.

7.20. DETERMINATION OF h-PARAMETERS FROM TRANSISTOR CHARACTERISTICS

Let us consider the common emitter (CE) configuration to find the *h*-parameters graphically. These are static input and output characteristics and the voltages and currents plotted in them are all dc quantities. We define the four *h*-parameters for CE configuration in a different way as under:

(*i*) $h_{ie} = \left.\dfrac{\Delta V_{BE}}{\Delta I_B}\right|_{\text{constant } V_{CE}}$

(*ii*) $h_{re} = \left.\dfrac{\Delta V_{BE}}{\Delta V_{CE}}\right|_{\text{constant } I_B}$

(*iii*) $h_{fe} = \left.\dfrac{\Delta I_C}{\Delta I_B}\right|_{\text{constant } V_{CE}}$

(*iv*) $h_{oe} = \left.\dfrac{\Delta I_C}{\Delta I_{CE}}\right|_{\text{constant } I_B}$

In each case, symbol Δ refers to a small change in that quantity around the quiescent point of operation. For the greatest accuracy, these changes should be as small as possible.

The first two *h*-parameters h_{ie} and h_{re} can be calculated from the input characteristics of the CE configuration while the other two can be obtained from the output characteristics.

7.20.1. To Calculate h_{ie} from the Input Characteristics

To determine the value of h_{ie}, let us locate the *Q* point on the characteristics as shown in figure 7.26 (*a*).

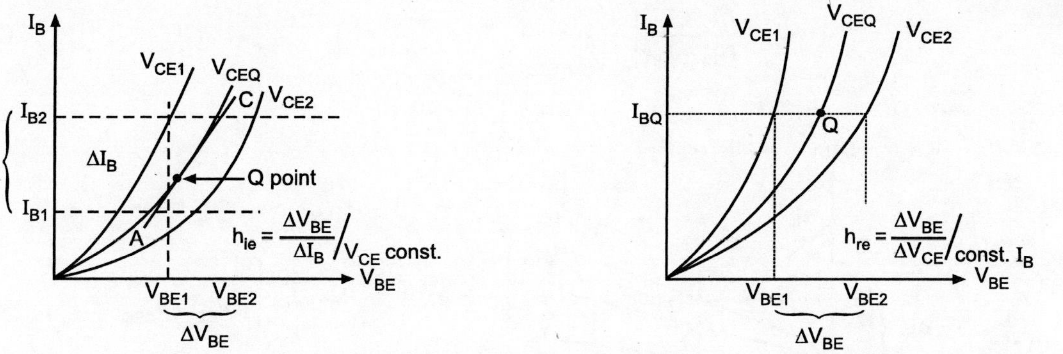

(a) Determining hie from the input characteristics (b) Determining hre from the input characteristics

Fig. 7.26.

We draw a line AC tangent to the input characteristics for $V_{CE} = V_{CEQ}$. The slope of this line gives us the value of h_{ie}. A small change in V_{BE} is chosen and the resultant change in I_B is recorded as ΔI_B. Now, h_{ie} is practically calculated by using the values of ΔV_{BE} and ΔI_B as under :

$$h_{ie} = \left.\frac{\Delta V_{BE}}{\Delta I_B}\right|_{\text{constant } V_{CE}} \quad \text{... (7.65)}$$

or

$$h_{ie} = \frac{V_{BE2} - V_{BE1}}{I_{B2} - I_{B1}} \quad \text{... (7.66)}$$

This is how we get the value of h_{ie} from the input characteristics.

7.20.2. Graphical Measurement of h_{re}

We know that

h_{re} is defined as under :

$$h_{re} = \left.\frac{\Delta V_{BE}}{\Delta V_{CE}}\right|_{\text{constant } I_B} \quad \text{... (7.67)}$$

We draw a horizontal line through the Q point as shown in figure 7.26 (*b*). Hence, I_B remains constant at I_{BQ}. The change in V_{CE} is then decided. Here $\Delta V_{CE} = V_{CE2} - V_{CE1}$. The points of intersection of the input characteristics corresponding to V_{CE1} and V_{CE2} with the horizontal line at I_{BQ} are then projected on the X-axis to calculate ΔV_{BE}.

Substituting the values of ΔV_{BE} and ΔV_{CE}, we get the value of h_{re} as under :

$$h_{re} = \left.\frac{V_{BE2} - V_{BE1}}{V_{CE2} - V_{CE1}}\right|_{I_B = I_{BQ}} \quad \text{... (7.68)}$$

7.20.3. To obtain h_{fe} from the Output Characteristics

We draw a perpendicular line at $V_{CE} = V_{CEQ}$ as shwon in figure 7.27.

Then, let us choose ΔI_B to be from I_{B2} to I_{B1}, about the Q point base current I_{BQ}.

Note the corresponding change in collector current as $I_{C2} - I_{C1}$. Then, we calculate h_{fe} as under:

$$h_{fe} = \left.\frac{\Delta I_C}{\Delta I_B}\right|_{\text{constant } V_{CE}} \quad \text{... (7.69)}$$

$$h_{fe} = \left.\frac{I_{C2} - I_{C1}}{I_{B2} - I_{B1}}\right|_{\text{At } V_{CE} = V_{CEQ}} \quad \text{... (7.70)}$$

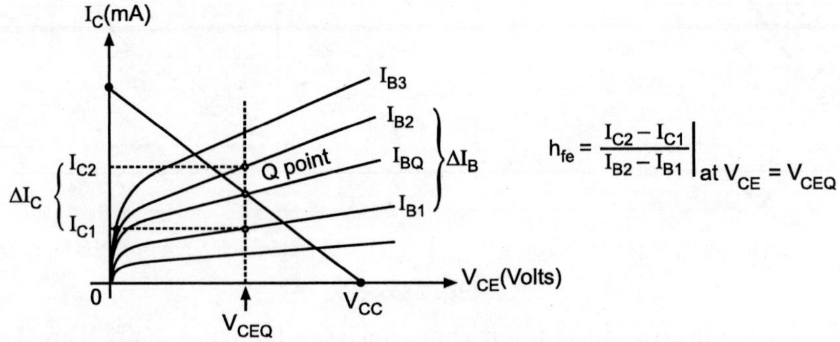

Fig. 7.27. Measurement of h_{fe} from the output characteristics of CE configuration.

7.20.4. To Obtain h_{oe} from the Output Characteristics

Here, h_{oe} is calculated at a constant value of $I_B = I_{BQ}$. Hence, we will make all the measurements on the characteristics corresponding to $I_B = I_{BQ}$. Corresponding to the characteristics of $I_B = I_{BQ}$, a change in V_{CE} is first decided as $V_{CE2} - V_{CE1}$. Note the corresponding change in I_C as $I_{C2} - I_{C1}$.

We calculate h_{oe} as under :

$$h_{oe} = \left.\frac{\Delta I_C}{\Delta V_{CE}}\right|_{\text{constant } I_B} \quad \text{... (7.71)}$$

or

$$= \left.\frac{I_{C2} - I_{C1}}{V_{CE2} - V_{CE1}}\right|_{\text{At } I_B = I_{BQ}} \quad \text{... (7.72)}$$

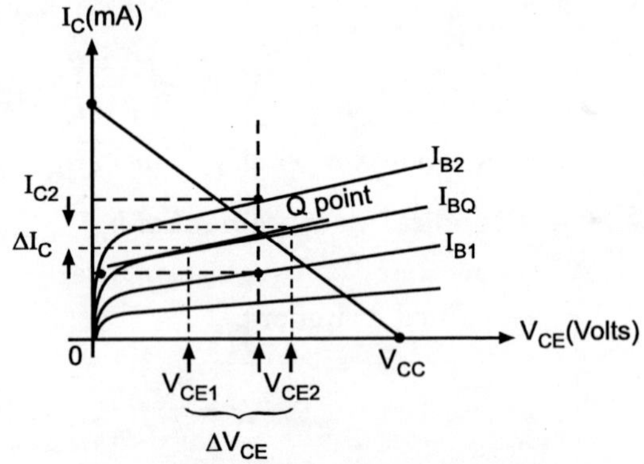

Fig. 7.28. Measurement of h_{oe} from output characteristics.

If a tangent *PR* which passes through the *Q* point is drawn as shown in figure 7.28, then the slope of line *PR* at *Q* point gives us the value of h_{oe}.

EXAMPLE 7.3 **A *CB* transistor amplifier uses a voltage source of internal reistance $R_s = 1200\ \Omega$ and the load resistance is $R_L = 1000\ \Omega$. The *h*-parameters are $h_{ib} = 22\ \Omega$, $h_{rb} = 3 \times 10^{-4}$, $h_{fb} = -0.98$ and $h_{ob} = 0.5\ \mu A/V$. Calculate the following:**

(*i*) Current gain

(*ii*) Input impedance

(*iii*) Voltage gain

(*iv*) Overall voltage gain

(*v*) Overall current gain.

Solution:

(*i*) Current gain: $$A_i = \frac{-h_{fb}}{1 + h_{ob} \cdot R_L} = \frac{(0.98)}{1 + 0.5 \times 10^{-6} \times 1000} = 0.98 \quad \textbf{Ans.}$$

***(ii)* Input resistance:** $R_i = h_{ib} + h_{rb}.A_i R_L$

$= 22 + (3 \times 10^{-4}) \times 0.98 \times 1000 = 22.3\ \Omega$. **Ans.**

***(iii)* Voltage gain:** $A_v = \dfrac{A_i.R_L}{R_i} = \dfrac{0.98 \times 1000}{22.3} = 43.946$ **Ans.**

***(iv)* Overall voltage gain:** $A_{vs} = \dfrac{A_v.R_i}{R_i + R_s} = \dfrac{43.946 \times 22.3}{22.3 + 1200} = 0.802$ **Ans.**

***(v)* Overall current gain:** $A_{is} = \dfrac{A_i.R_s}{R_i + R_s} = \dfrac{0.98 \times 1200}{22.3 + 1200} = 0.962$ **Ans.**

EXAMPLE 7.4 **Figure 7.29 shows the circuit of common collector amplifier or emitter follower. The *h*-parameters are $h_{ic} = 2$ kΩ, $h_{fc} = -51$, $h_{rc} = 1$ and $h_{oc} = 25 \times 10^{-6}$ mhos. Determine the following:**

(*i*) Current gain
(*ii*) Input resistance
(*iii*) Voltage gain
(*iv*) Output resistance ***(Important)***

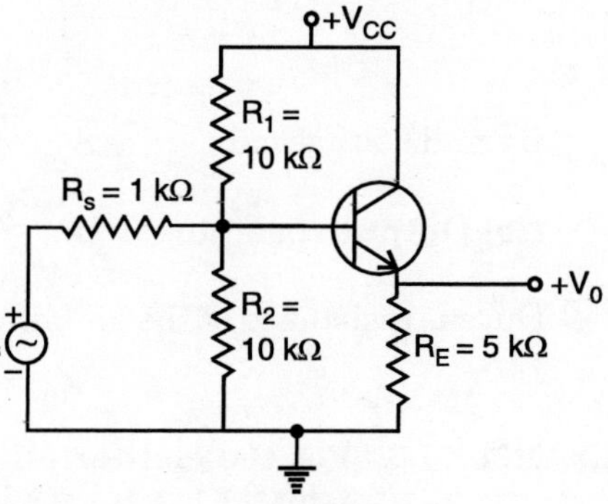

Fig. 7.29.

Solution:

***(i)* Current gain:** $A_i = \dfrac{-h_{fc}}{1 + h_{oc}.R_L}$

$A_i = \dfrac{51}{1 + 25 \times 10^{-6} \times 5 \times 10^3}$

$A_i = 45.3$ **Ans.**

***(ii)* Input resistance:** $R_i = h_{ic} + h_{rc}.A_i R_L$

$= 2000 + 1 \times 45.3 \times 5 \times 10^3 = 228$ kΩ. **Ans.**

Input resistance of amplifier stage

$$Z_i = R_i \,\|\, (R_1 \| R_2) = 228 \,\|\, (10 \,\|\, 10) = 4.9 \text{ k}\Omega \quad \textbf{Ans.}$$

***(iii)* Voltage gain:** $A_v = \dfrac{A_i.R_L}{R_i} = \dfrac{45.3 \times 5 \times 10^3}{228 \times 10^3} = 1$ **Ans.**

***(iv)* Output resistance:**

$$R_0 = -\frac{R_s + h_{ic}}{h_{fc}} = -\frac{1000 + 2000}{-51} = 58.8\ \Omega \quad \textbf{Ans.}$$

Output resistance of the amplifier stage

$$Z_o = R_O \,\|\, R_E = 58.8 \,\|\, 5 \times 10^3 = 581.1\ \Omega \quad \textbf{Ans.}$$

EXAMPLE 7.5 **Figure 7.30 shows the circuit of a single stage *CE* amplifier. The values of *h*-parameters are $h_{ie} = 1.5$ kΩ, $h_{re} = 5 \times 10^{-3}$, $h_{fe} = 50$, $h_{oe} = 2 \times 10^{-5}$ μA/V.***

Determine the following:

(*i*) Current gain
(*ii*) Input resistance
(*iii*) Voltage gain
(*iv*) Output resistance ***(Use Approximate Analysis)***

* Derive the voltage gain, current gain expressions for a potential divider biased common emitter amplifier using *h* parameters.

Solution: In this case, value of a.c. load resistance

$$R_L = R_C \parallel R = 5 \parallel 10 = 3.3\ k\Omega$$

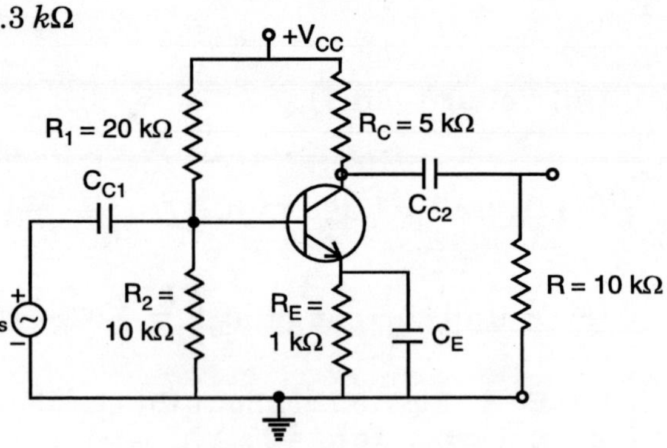

Fig. 7.30.

(*i*) **Current gain** $A_i = -h_{fe} = -50$ **Ans.**

(*ii*) **Input resistance** $R_i = h_{ie} = 1.5\ \text{k}\Omega$ **Ans.**

Input resistance of amplifier stage

$$Z_i = R_i \parallel (R_1 \parallel R_2)$$

$$= 1.5 \parallel (20 \parallel 10)$$

$$Z_i = 1.22\ \text{k}\Omega \quad \textbf{Ans.}$$

***(iii)* Voltage gain:** $A_v = \dfrac{A_i . R_L}{R_i}$

$$A_v = \frac{(-50)\times(3.3\times10^3)}{1500} = -110 \quad \textbf{Ans.}$$

Overall voltage gain $A_{vs} = A_v = -110$ **Ans.**

***(iv)* Output resistance:** $R_0 = \dfrac{1}{h_{oe}} = \dfrac{1}{20\times10^{-6}} = 0.05 \times 10^6 \text{ ohm} = 50\ \text{k}\Omega.$

Output resistance of the amplifier

$$Z_0 = R_o \parallel R_L = 50 \parallel 3.3 = 3.1\ \text{k}\Omega \quad \textbf{Ans.}$$

EXAMPLE 7.6 **For the emitter follower with $R_S = 0.5\ \text{k}\Omega$ and $R_L = 5\ \text{k}\Omega$, calculate A_I, R_i, A_v. Assume $h_{fe} = 50$, $h_{ie} = 1\ \text{k}\Omega$, $h_{oe} = 25$ micro amp/volt.**

Solution: For emitter follower,

(*i*) **The current gain:** $A_I = \dfrac{1+h_{fe}}{1+h_{oe}.R_L} = \dfrac{1+50}{1+25\times10^{-6}\times5\times10^3} = \dfrac{51}{1.125} = 45.33$ **Ans.**

***(ii)* Input resistance:** $R_i = h_{ic} + h_{rc}.A_I.R_L = h_{ie} + 1.A_I.R_L = h_{ie} + A_I.R_L$

Putting all the values $R_i = 1 \times 10^3 + 45.33 \times 5 \times 10^3$

$$R_i = (1+226.6)10^3 = 228.6\ \text{k}\Omega \quad \textbf{Ans.}$$

(iii) $A_v = \dfrac{v_O}{v_i} = \dfrac{A_I.R_L}{R_i} = \dfrac{45.33\times5}{227.6} = \dfrac{226.6}{227.6} = 0.9958$ **Ans.**

EXAMPLE 7.7 **Find the values of voltage gain, current gain, input resistance and power gain for a common emitter transistor amplifier with $R_L = 1600$ ohm and $R_S = 1\ \text{k}\Omega$. The transistor has $h_{ie} = 1100$ ohm, $h_{fe} = 2.5 \times 10^{-4}$, $h_{oe} = 25$ micro amp/V.**

Solution: For common emitter transistor amplifier

(*i*) **Current gain:** $A_I = -\dfrac{h_{fe}}{1+h_{oe}.R_L} = -\dfrac{2.5\times10^{-4}}{1+25\times10^{-6}\times1600}$

$$A_I = -\frac{2.5\times10^{-4}}{1.04} = -2.4 \times 10^{-4} \quad \textbf{Ans.}$$

***(ii)* Input resistance** $R_i = h_{ie} + h_{re}\,A_I.R_L$

$$\equiv h_{ie} \text{ (neglecting the factor } h_{re}A_I.R_L) \equiv 1100 \text{ ohm} \quad \textbf{Ans.}$$

***(iii)* Voltage gain:** $A_v = A_I \dfrac{R_L}{R_i} = -2.4\times10^{-4}\dfrac{1600}{1100}$

$$A_v = -\frac{38.4\times10^{-4}}{11} = -3.49\times10^{-4} \quad \textbf{Ans.}$$

(iv) Power gain: $= A_I . A_v = (-2.4 \times 10^{-4}) \times (-3.49 \times 10^{-4}) = 8.37 \times 10^{-8}$ **Ans.**

EXAMPLE 7.8 For the circuit shown in figure 7.31, given $\beta_{dc} = h_{FE} = 130$.

(a) Find I_{CQ} and V_{CEQ}

(b) Find A_v and R_{in} for the circuit of part (a) if $h_{fe} = 50$, $h_{ie} = 1$ kΩ, $h_{re} = 0$ and $h_{oe} = 0$.

Solution: (a) ∵ Base Voltage $V_B = V_{CC} \times \frac{R_2}{R_1 + R_2} = 18 \times \frac{510}{510+510}$

$$V_B = \frac{18}{2} = 9 \text{ Volt}$$

Value of I_C is given by

$$I_C = \frac{V_B - V_{BE}}{R_E} = \frac{V_B - 0}{R_E} \text{ (neglecting } V_{BE})$$

$$I_C = \frac{V_B}{R_E} = \frac{9}{7.5\times10^3} = \frac{9\times10^{-3}}{7.5}$$

$$I_C = 1.2 \text{ mA}$$

or $\quad I_{CQ} \equiv 1.2$ mA **Ans.**

Now, $\quad V_{CE} = V_{CC} - I_C . (R_C + R_E)$

$$= 18 - 1.2 \times 10^{-3} (9.1 + 8.5) \times 10^3$$

$$V_{CE} = 18 - 1.2 \times 16.6 = 18 - 19.9$$

$$V_{CE} = -1.9 \text{ Volt}$$

or $\quad V_{CEQ} \equiv -1.9$ Volt **Ans.**

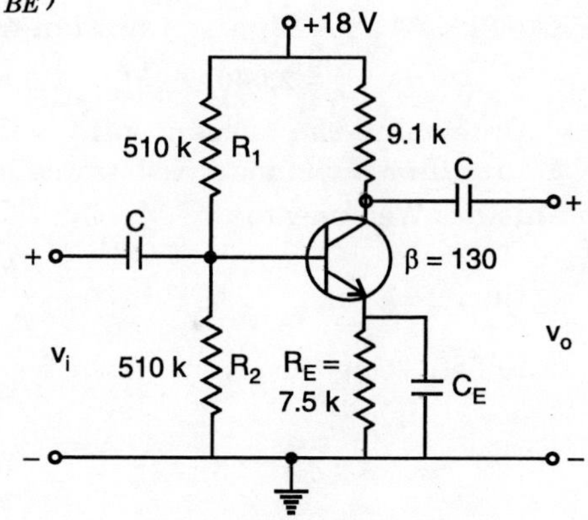

Fig. 7.31.

(b) Input resistance

$$R_i \cong h_{ie} \cong 1 \text{ k}\Omega \quad \textbf{Ans.}$$

Input resistance of amplifier stage

$$R_{in} = R_i \,\|\, (R_1 \,\|\, R_2) = 1 \,\|\, (510 \,\|\, 510)$$

$$R_{in} = 1 \,\|\, 255 = 0.99 \text{ K}\Omega \quad \textbf{Ans.}$$

Current gain $\quad A_I \cong -h_{fe} = -50$ **Ans.**

Voltage gain $\quad A_v = \frac{A_I . R_L}{R_i} = \frac{A_I . R_C}{R_i} = \frac{-50\times9.1}{1} = -455$ **Ans.**

EXAMPLE 7.9 Given that $h_{fe} = 50$, $h_{ie} = 0.83$ kΩ. Find out the current gain (h_{fb}) and input impedance (h_{ib}) for a transistor in *CB* configuration.

Solution: We know that

$$h_{fb} = -\frac{h_{fe}}{1 + h_{fe}}$$

Substituting the given values, we have

$$h_{fb} = -\frac{50}{1+50} = -0.98 \quad \textbf{Ans.}$$

Also, $$h_{ib} = \frac{h_{ie}}{1+h_{fe}} = \frac{0.83 \times 10^3}{1+50} = 16.27\ \Omega \quad \textbf{Ans.}$$

EXAMPLE 7.10 The h-parameters for a *CE* configuration are $h_{ie} = 2600\ \Omega$, $h_{fe}=100$, $h_{re}= 0.02 \times 10^{-2}$ and $h_{oe} = 5 \times 10^{-6}\ S$. Find h-parameters for *CC* configuration.

Solution: We know that

$$h_{ic} = h_{ie}$$

But given that $$h_{ie} = 2600\ \Omega$$

Therefore, $$h_{ic} = h_{ie} = 2600 \text{ ohms} \quad \textbf{Ans.}$$

Also, $$h_{fc} = -(1 + h_{fe}) = -(1 + 100)$$

or $$h_{fc} = -101 \quad \textbf{Ans.}$$

and $$h_{rc} = 1 - h_{re} \cong 1 \quad \textbf{Ans.}$$

$$h_{oc} = h_{oe} = 5 \times 10^{-6}\ S \quad \textbf{Ans.}$$

EXAMPLE 7.11 A Bipolar junction transistor has the following h-parameters:

$$h_{ie} = 2{,}000\ \Omega;\ h_{re} = 1.6 \times 10^{-4},\ h_{fe} = 49;\ h_{oe} = 50\ \mu A/V$$

Determine the current gain, voltage gain, input resistance and output resistance of the *CE* amplifier if the load resistance is 30 KΩ and the source resistance is 600 Ω.

Solution: We know that

Current gain, $$A_i = \frac{-h_{fe}}{1+h_{oe} R_L}$$

Substituting the given values, we get

$$A_i = \frac{-49}{1+50\times 10^{-6} \times 30 \times 10^3} = -19.6$$

Input resistance, $$R_{in} = h_{ie} - \frac{h_{re}\, h_{fe}}{h_{oe} + \dfrac{1}{R_L}}$$

Substituting the given values, we get

$$R_{in} = 2{,}000 - \frac{1.6 \times 10^{-4} \times 49}{50 \times 10^{-6} + \dfrac{1}{30 \times 10^3}} = 1.906\ \Omega \quad \textbf{Ans.}$$

Also, voltage gain, $$A_v = \frac{-h_{fe}}{\left(h_{oe} + \dfrac{1}{R_L}\right) R_{in}} = \frac{-h_{fe}\, R_L}{(1+h_{oe}\, R_L)\, R_{in}} = \frac{A_i\, R_L}{R_{in}}$$

Substituting the given values, we get

$$A_v = \frac{-19.6 \times 30 \times 10^3}{1.906} = -308.5 \quad \textbf{Ans.}$$

Overall voltage gain, $$A_{vs} = A_v \frac{R_{in}}{R_{in} + R_s} = -308.5 \times \frac{1.906}{1.906 + 600} = -235 \quad \textbf{Ans.}$$

Overall current gain, $$A_{is} = A_i \frac{R_s}{R_{in} + R_s} = -19.6 \times \frac{600}{1.906 + 600} = -4.7 \quad \textbf{Ans.}$$

Output conductance, $G_{out} = h_{oe} - \dfrac{h_{fe}\,h_{re}}{h_{ie} + R_s}$

$$= 50\times10^{-6} - \frac{49\times1.6\times10^{-4}}{2{,}000+600} = 46.985\times10^{-6}\ \text{S} \quad \textbf{Ans.}$$

Output resistance, $R_{out} = \dfrac{1}{G_{out}} = \dfrac{1}{46.985\times10^{-6}} = 21{,}283\ \Omega$ or $21.283\ \text{k}\Omega$ **Ans.**

EXAMPLE 7.12 **A BJT has $h_{ie} = 2\ \text{k}\Omega$, $h_{fe} = 100$, $h_{re} = 2.5 \times 10^{-4}$ and $h_{oe} = 25$ mA/V as parameters in *CE* configuration. It is used as an emitter follower amplifier with $R_s = 1\ \text{k}\Omega$ and $R_L = 500\ \Omega$. Determine for the amplifier, the voltage gain $A_{Vs} = V_o/V_s$, the current gain $A_{is} = I_o/I_s$, the input resistance R_i and output resistance R_o.**

Solution: For emitter follower (*i.e.* common-collector amplifier), transistor parameters are given as under:

$$h_{ic} = h_{ie} = 2\ \text{K}\Omega = -(1 - h_{fe}) = -(1 + 100) = -101$$

$$h_{rc} = 1 - h_{re} = 1 - 2.5\times10^{-4} = 0.99975 \simeq 1$$

$$h_{oc} = h_{oe} = 25\times10^{-6}\ \text{S}$$

Current gain, $A_i = \dfrac{-h_{fc}}{1 + h_{oc}R_L} = \dfrac{-(-101)}{1 + 25\times10^{-6}\times500} = 99.75$

Input resistance, $R_{in} = h_{ic} - \dfrac{h_{rc}\,h_{fc}}{h_{oc} + \dfrac{1}{R_L}}$

or $R_{in} = 2\times10^3 - \dfrac{1\times(-101)}{25\times10^{-6} + \dfrac{1}{500}} = 51.876\ \text{k}\Omega$ **Ans.**

Voltage gain, $A_v = \dfrac{-h_{fc}}{\left(h_{oc} + \dfrac{1}{R_L}\right)R_{in}} = \dfrac{-(-101)}{\left(25\times10^{-6} + \dfrac{1}{500}\right)\times51.876\times10^3} = 0.9614$

Overall voltage gain, $A_{vs} = A_v\dfrac{R_{in}}{R_{in} + R_s} = 0.9614\times\dfrac{51.876}{51.876+1} = 0.9432$ **Ans.**

Overall current gain, $A_{is} = A_i\dfrac{R_s}{R_{in} + R_s}$

or $A_{vs} = 99.75\times\dfrac{1}{51.876+1} = 1.886$ **Ans.**

Output conductance, $G_o = h_{oc} - \dfrac{h_{fc}\,h_{rc}}{h_{ic} + R_s} = 25\times10^{-6} - \dfrac{(-101\times1)}{2\times10^3 + 1\times10^3} = 33.69\times10^{-3}\ \text{S}$ **Ans.**

Output resistance, $R_o = \dfrac{1}{G_o} = \dfrac{1}{33.69\times10^{-3}} = 29.68\ \Omega$ **Ans.**

EXAMPLE 7.13 **For the *CE* amplifier shown in figure 7.32, following data are given:**

$h_{ie} = 1.1\ \text{k}\Omega$, $h_{re} = 2.5\times10^{-4}$, $h_{fe} = 50$, $h_{oe} = 24\ \mu$A/V. Calculate A_i, A_v, R_i, A_{is} and A_{vs} for $R_L = 10\ \text{k}\Omega$.

Solution: We have

$$A_i = -\frac{h_{fe}}{1+h_{oe}R_L} = -40$$

$$A_v = \frac{A_i R_L}{R_i} = -400$$

$$R_i = h_{ie} + h_{re} A_i R_L = 1 \text{ k}\Omega$$

$$A_{vs} = \frac{A_v . R_i}{R_i + R_S} = -200$$

$$A_{is} = \frac{A_i R_S}{R_i + R_S} = -26 \quad \textbf{Ans.}$$

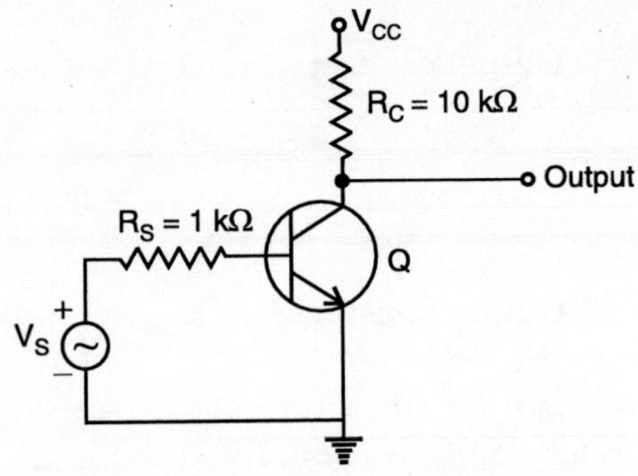

Fig. 7.32. Circuit for example 7.13.

EXAMPLE 7.14 **Find A_v and R_{in} for the circuit shown in figure 7.33, if h_{fe} = 50, h_{ie} = 1 kΩ, h_{re} = 0 and h_{oe} = 0.**

Solution: From figure 7.33, we have

Base voltage, $$V_B = V_{CC} \times \frac{R_2}{R_1 + R_2}$$

or $$V_B = 18 \times \frac{510}{510 + 510}$$

or $$V_B = \frac{18}{2} = 9 \text{ Volt}$$

Value of current I_C is given by

$$I_C = \frac{V_B - V_{BE}}{R_E} = \frac{V_{B-0}}{R_E} \quad \text{(neglecting } V_{BE}\text{)}$$

or $$I_C = \frac{V_B}{R_E} = \frac{9}{7.5 \times 10^3} = \frac{9 \times 10^{-3}}{7.5}$$

or $$I_C = 1.2 \text{ mA}$$

$$I_{CQ} = 1.2 \text{ mA} \quad \textbf{Ans.}$$

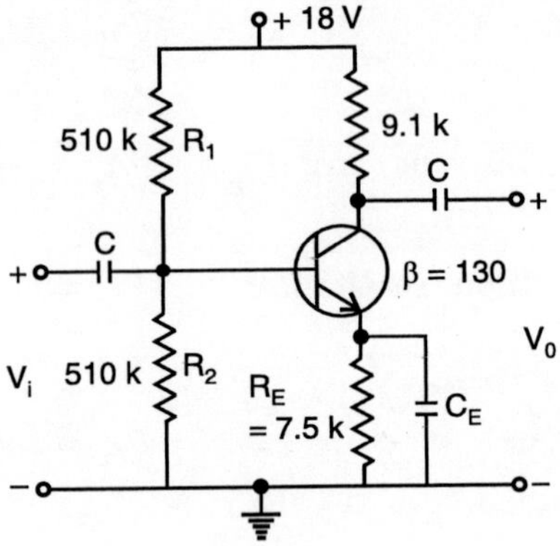

Fig. 7.33.

Again, from figure 7.33, we have

$$V_{CE} = V_{CC} - I_C \cdot (R_C + R_E)$$

Substituting values, we get

$$V_{CE} = 18 - 1.2 \times 10^{-3} (9.1 + 7.5) \times 10^3$$

or $$V_{CE} = 18 - 1.2 \times 16.6 = 18 - 19.9 = -1.9 \text{ volt}$$

or $$V_{CEQ} = -19 \text{ volt} \quad \textbf{Ans.}$$

Also, input resistance

$$R_i \cong h_{ie} \cong 1 \text{ k}\Omega \quad \textbf{Ans.}$$

Further, input resistance of amplifier stage is given by

$$R_{in} = R_i || (R_1 || R_2) = 1 || (510 || 510)$$

or $$R_{in} = 1 || 255 = 0.99 \text{ k}\Omega \quad \textbf{Ans.}$$

Current gain is given by

$$A_i = -h_{fe} = -50 \quad \textbf{Ans.}$$

Voltage gain

$$A_v = \frac{A_i \cdot R_L}{R_i}$$

But $R_L = R_C$

Therefore, $A_v = \frac{A_i\ R_C}{R_i} = \frac{-50 \times 9.1}{1}$

or $A_v = -455$ **Ans.**

EXAMPLE 7.15 **Find I_{CQ} and I_{CEQ} for the circuit shown in figure 7.34. Given that $\beta_{dc} = h_{FE} = 130$.**

Solution: From figure 7.34, we have

$$R_{th} = \frac{R_1 \times R_2}{R_1 + R_2} = \frac{510 \times 510}{510 + 510}$$ [∵ R_1 and R_2 are in parallel]

or $R_{th} = 255$ k

and $V_{th} = \frac{R_2 \times V_{CC}}{R_1 + R_2} = \frac{510 \times 18}{510 + 510} = 9$ Volt

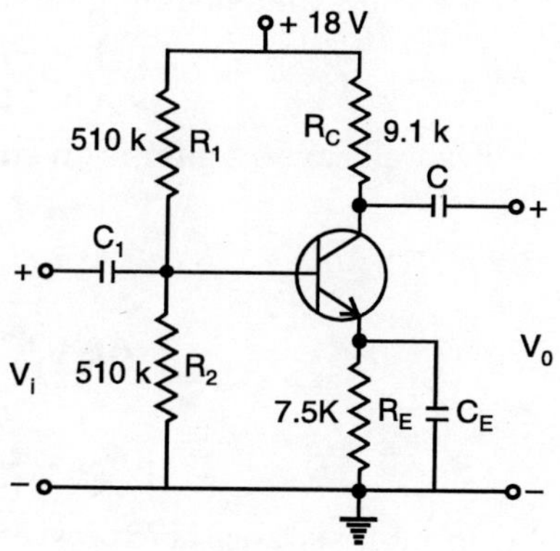

Fig. 7.34.

Therefore, we have

$$I_B = \frac{9 - 0.7}{255 + (130 + 1) \times 7.5} = \frac{8.3}{255 + 131 \times 7.5}$$

or $I_B = \frac{8.3}{1237.5}$

or $I_B = 0.00670$ mA

$= 6.7 \times 10^{-3}$ mA

We know that

$$I_C = \beta I_B$$

or $I_C = 130 \times 6.7 \times 10^{-3}$

or $I_C = 0.8719$ mA

Again, from figure 7.34, we get

$$V_{CC} = V_{CC} - I_C R_C - I_E R_E$$

or $V_{CE} = 18 - 0.8719 \times 91 - (0.8719 + 0.0067) \times 7.5$

or $V_{CE} = 3.476$ Volt **Ans.**

EXAMPLE 7.16 **Given $h_{ie} = 2.4\ k\Omega$, $h_{fe} = 100$, $h_{re} = 4 \times 10^{-4}$ and $h_{oe} = 25\ \mu s$. Sketch the common emitter hybrid equivalent model.**

Solution: The required model is as shown in figure 7.35.

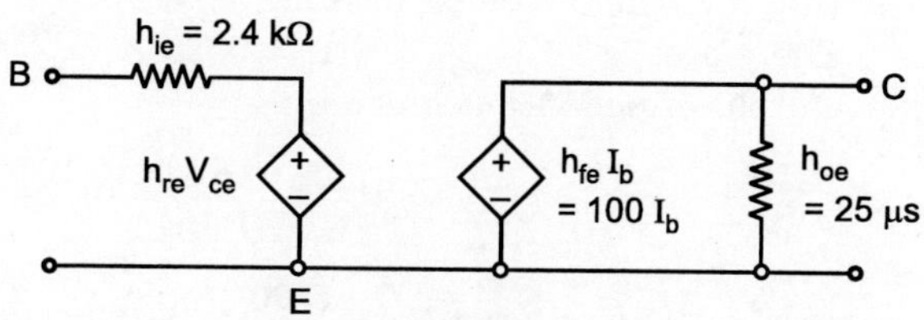

Fig. 7.35.

EXAMPLE 7.17 **For the circuit shown in figure 7.36, find A_v, A_i, Z_i and Z_o.**

$$V_{CC} = 15\ V,\quad h_{ie} = 1100\ \Omega,\quad R_B = 100\ k\Omega,\quad h_{fe} = 50,\quad R_E = 11\ k\Omega.$$

Solution :

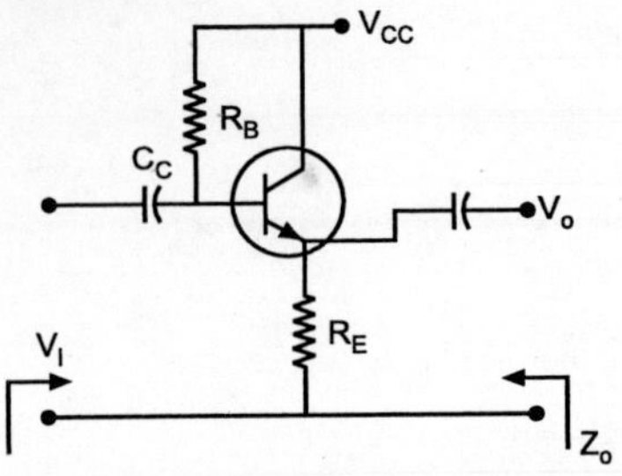

Fig. 7.36.

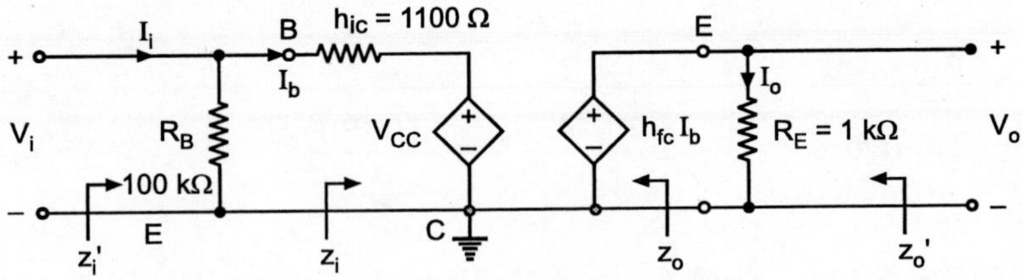

Fig. 7.37.

Type of Amplifier : Type of analysis : Approximate CC with fixed bias.

First, let us draw the h parameter equivalent circuit as shown in figure 7.37.

We know that current gain of transistor (A_i) is given by

$$A_i = -h_{fc} = -[-(1+h_{fe})] = -[-(1+50)]$$

or $$A_i = 51$$

Overall current gain is given by

$$A_{IS} = \frac{I_o}{I_i} = \frac{I_o}{I_b} \times \frac{I_b}{I_i} = A_I \times \frac{I_b}{I_i}$$

or $$A_{IS} = 51 \times \frac{Z_i}{Z_i + R_B} = 51 \times \frac{[h_{ie} + (1+h_{fe})\,R_E]}{100\text{ k}\Omega + [h_{ie} + (1+h_{fe})R_E]}$$

or $$A_{IS} = \frac{51 \times [1.1\text{ k} + (51 \times 1\text{ k})]}{100\text{ k} + [1.1\text{ k} + (51 \times 1\text{ k})]} = 17.49. \quad \textbf{Ans.}$$

Input resistance is expressed as

$$Z_i = h_{ie} + (1+h_{fe})\,R_E = 1.1\text{ k}\Omega + (51 \times 1\text{ k}\Omega)$$

or $$Z_i = 52.1\text{ k}\Omega$$

and $$Z_1' = Z_i \parallel R_B = 52.1\text{ k} \parallel 100\text{ k} = 34.25\text{ k}\Omega$$

Also, voltage gain is given by

$$A_V = A_i \times \frac{R_L}{Z_i} = 51 \times \frac{1\text{ k}}{52.1\text{ k}} = 0.9788.$$

$$A_{VS} = \frac{A_V \times Z_i}{Z_i + R_S}$$

But, R_S is not given so, assume $R_S = 0$

Therefore, $A_{VS} = A_V = 0.9788$

Output impedance is given by

$$Z_o = \frac{h_{ie} + (R_S \parallel R_B)}{(1+h_{fe})} = \frac{h_{ie} + (0 \parallel 100\text{ k})}{(1+h_{fe})}$$

or $$Z_o = \frac{1100}{51} = 21.58\ \Omega$$

and $$Z'_o = Z_o \parallel R_C = \frac{21.58 \times 1000}{(21.58 + 1000)} = 21.12\ \Omega. \quad \textbf{Ans.}$$

EXAMPLE 7.18 A CE amplifier has the following h-parameters :

$$h_{ie} = 1.1\text{ k}\Omega,\quad h_{re} = 0.25 \times 10^{-3},\ h_{fe} = 50,\ h_{oe} = 25\ \mu\text{ mho}$$

If the load resistance is 1 kΩ and source resistance is also 1 kΩ , find :

(i) Current gain,

(ii) Voltage gain.

Solution : Exact or Approximate analysis :

$$h_{oe}\,R_L = 25 \times 10^{-6} \times 1 \times 10^{3} = 0.025$$

Since $h_{oe}\,R_L$ is less than 0.1, therefore, we shall use the approximate analysis.

First, let us draw the *h*-parameter equivalent circuit as shown in figure 7.38.

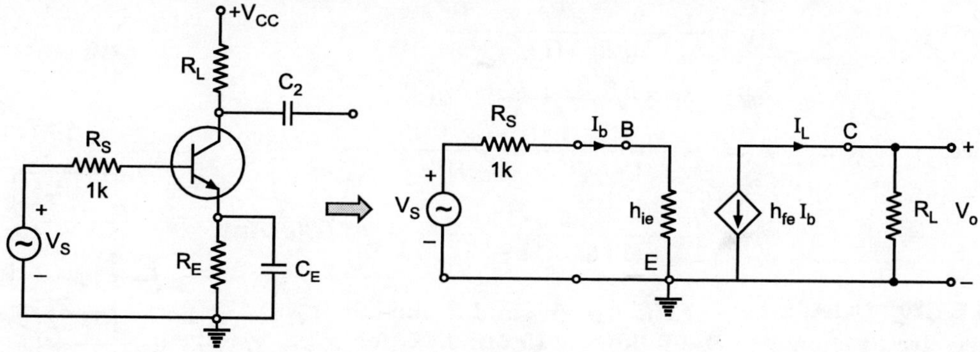

Fig. 7.38.

Now, let us calculate current gain and voltage gain as under:

$$A_I = \frac{I_L}{I_b} = \frac{I_L}{I_s} = \frac{-h_{fe}\,I_b}{I_b} = -\,h_{fe}$$

or $$A_I = -\,50$$

Also, $$A_V = \frac{V_o}{V_i} = \frac{-h_{fe}\,I_b R_L}{I_b\,h_{ie}} = \frac{-h_{fe}\,R_L}{h_{ie}}$$

or $$A_V = \frac{-50 \times 1\text{ k}}{1.1\text{ k}} = -\,45.45.$$ **Ans.**

EXAMPLE 7.19 In the following given circuit, explain with reasoning, in what configuration this circuit is working. Also calculate the voltage-gain.

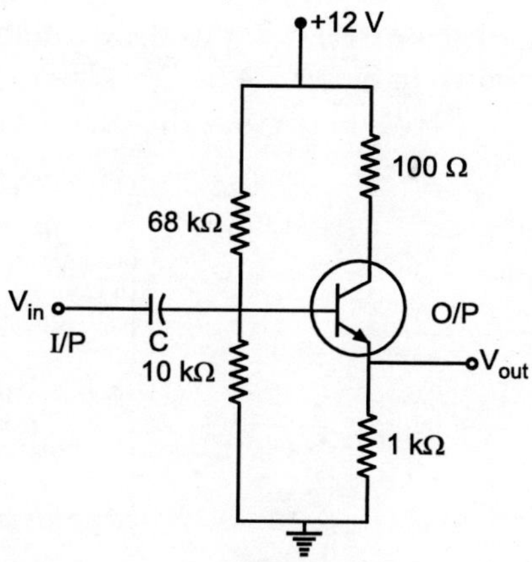

Fig. 7.39.

Solution : The given circuit is in the common collector configuration or emitter follower. Let us assume h_{fe} = 50, h_{ie} = 1 kΩ, $h_{re} = h_{oe} = 0$.

From figure 7.40, the voltage gain is given by

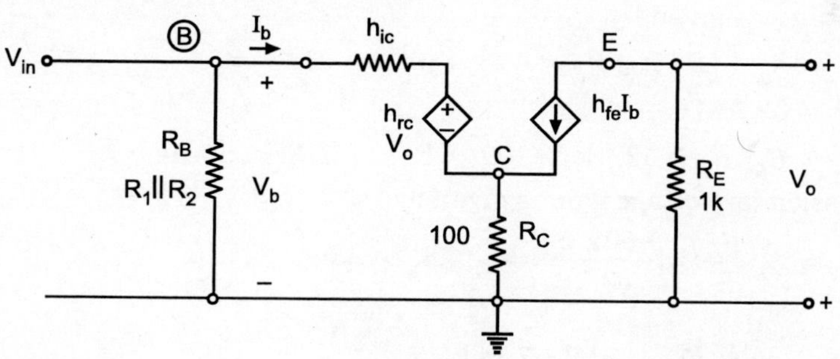

Fig. 7.40.

$$A_V = \frac{V_o}{V_b} = \frac{-h_{fc}\, I_b\, R_E}{I_b\, h_{ic} + h_{rc} V_o + (1 + h_{fc}) I_b\, R_C}$$

But $V_o = -\,h_{fc}\, I_b\, R_E$

Therefore, $$A_V = \frac{-h_{fc}\, I_b\, R_E}{h_{ic}\, I_b - h_{rc}\, h_{fc}\, I_b\, R_E + (1 + h_{fc}) I_b\, R_C}$$

But $h_{rc} \cong 1$

Therefore, $$A_V = \frac{-h_{fc}\, R_E}{h_{ic} - h_{fc}\, R_E + (1 + h_{fc}) R_C}$$

But $h_{fc} = -\,(1 + h_{fe})$ and $h_{ic} = h_{ie}$

Hence, $$A_V = \frac{(1 + h_{fe}) R_E}{h_{ie} + (1 + h_{fe}) R_E + R_C - (1 + h_{fe}) R_C}$$

$$A_V = \frac{(1+50)\times 1\text{ k}}{1\text{ k} + (1+50)\times 1\text{k} - 50\times 0.1\text{ k}} = 1.085. \textbf{ Ans.}$$

EXAMPLE 7.20 Calculate $A_i = I_o/I_i$, A_V, A_{VS} and R_i for an amplifier circuit shown in the following figure 7.41 for which the h-parameters are:

(JNTU, Hyderabad, Sem. Exam, 2002-03) (04 marks)

h_{ie} = **1100 Ω**, h_{re} = **0,**

h_{fe} = **50 and** $1/h_{oe} = \infty$

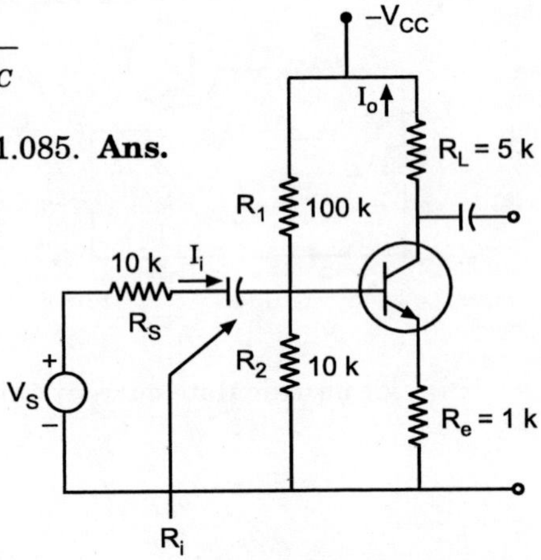

Fig. 7.41.

Solution: First, let us draw small signal and equivalent as shown in figure 7.42.

(*i*) We know that the current gain is given by

$$A_i = -\,h_{fe} = -\,50$$

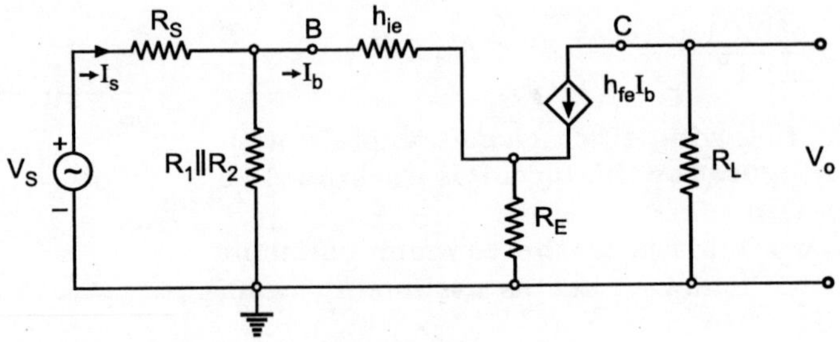

Fig. 7.42.

(*ii*) Input resistance is given by

$$R_i = h_{ie} + (1 + h_{fe}) R_E = 1.1\text{ k} + (51 \times 1\text{ k})$$

or $R_i = 52.1\text{ k}\Omega$

Also, $R_i' = R_i \,||\, R_B = 52.1\text{ k} \,||\, (100 \,||\, 10\text{ k}) = 7.74\text{ k}\Omega$. **Ans.**

(*iii*) The expression for voltage gain is given by

$$A_V = A_i \frac{R_L}{R_i} = \frac{-50\times 5\text{ k}}{52.1\text{ k}} = -\,4.8. \quad \textbf{Ans.}$$

Also, we have

$$A_{VS} = \frac{A_V \times R_i'}{(R_i' + R_s)} = \frac{-4.8\times 7.74\text{ k}\Omega}{(7.74\text{ k} + 10\text{ k})} = -\,2.094 \quad \textbf{Ans.}$$

EXAMPLE 7.21 **Input is given to the circuit in figure 7.43 through the base and output taken from emitter. Draw the small signal equivalent circuit, using common emitter hybrid parameters. Hence, compute the small signal voltage gain and input impedance. Given that h_{ie} = 1 kΩ, h_{fe} = 100, h_{re} = 0 and h_{oe} = 0. What is the configuration of the amplifier?**

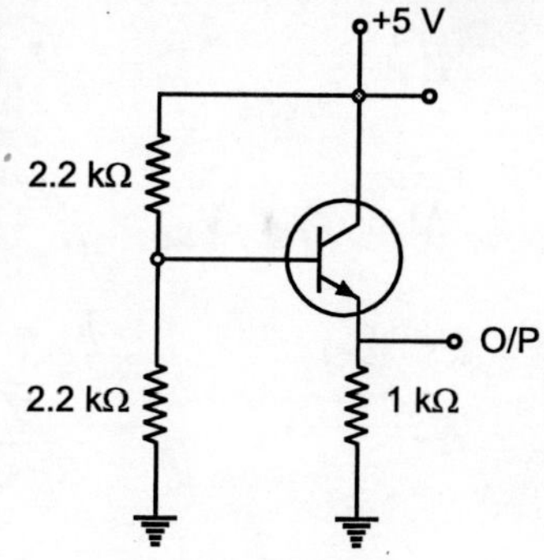

Fig. 7.43.

Solution: We have to find the following parameters:

A_V, R_i and R_i'

Let us first draw small signal equivalent circuit as shown in figure 7.44.

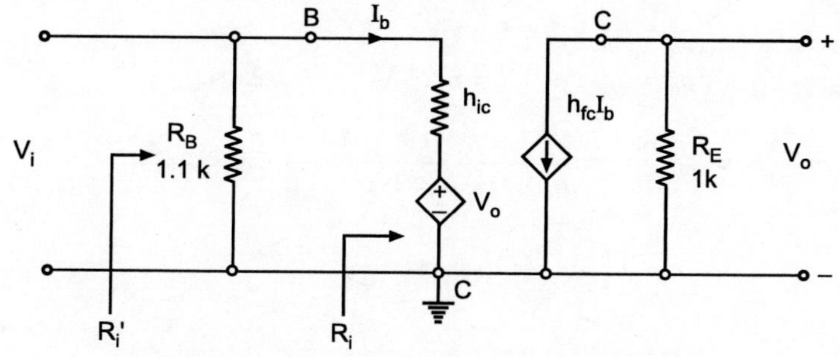

Fig. 7.44. Small signal equivalent circuit.

We can calculate R_i and R_i' as under:

$$R_i = h_{ie} + (1 \,.\, h_{fe})\, R_E = 1\text{ k} + (51 \times 1\text{ k}) = 52\text{ k}\Omega.$$

$$R'_i = R_i \,||\, R_B = R_i \,||\, (R_1 \,||\, R_2) = 52\text{ k} \,||\, (2.2\text{ k} \,||\, 2.2\text{ k}) = 52\text{ k} \,||\, (1.1\text{ k})$$

or $$R_i' = 1.077\text{ k}\Omega.$$

We know that A_V is given by

$$A_V = \frac{V_o}{V_i} = \frac{-h_{fe} I_b R_E}{I_b R_i} = \frac{-h_{fe} R_E}{h_{ie}} = \frac{(1+h_{fe})\, R_E}{R_i} = \frac{51 \times 1\text{ k}}{52\text{ k}} = 0.98.$$

EXAMPLE 7.22 **For the common collector circuit shown in figure 7.45 find I_E, V_E, V_B and V_o/V_i. For the transistor h_{ie} = 1 kΩ and h_{fe} = 200.**

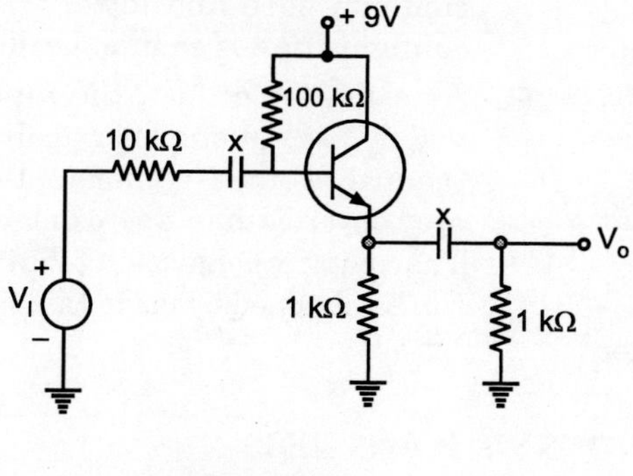

Fig. 7.45.

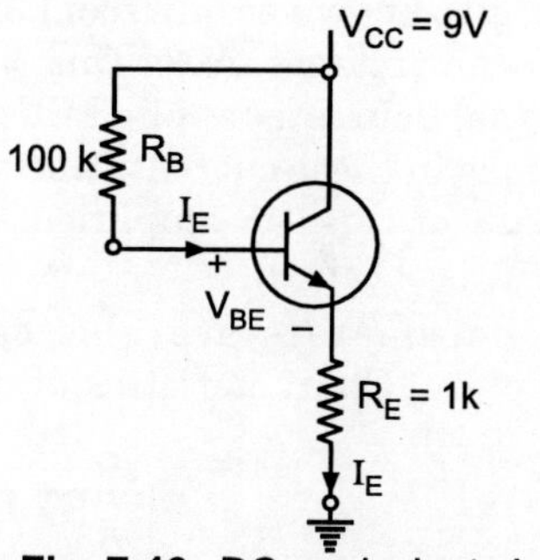

Fig. 7.46. DC equivalent circuit.

First, let us draw the dc equivalent circuit as shown in figure 7.46.

Applying KVL to the base loop, we obtain

$$V_{CC} = I_B\, R_B + V_{BE} + I_E\, R_E$$

But $$I_B = \frac{I_E}{(1+\beta)}$$

Also, $$V_{CC} - V_{BE} = \frac{R_B}{(1+\beta)} I_E + I_E R_E$$

Let $$\beta = h_{fe} = 200$$

or $$I_E = \frac{V_{CC} - V_{BE}}{R_E + [R_B/(1+\beta)]} = \frac{9-0.7}{1\text{ k} + [100\text{ k}/201]}$$

$\therefore$ $$I_E = 5.54 \text{ mA} \qquad \textbf{Ans.}$$

$$V_E = I_E \times R_E = 5.54 \times 1 = 5.54 \text{ V} \qquad \textbf{Ans.}$$

$$V_B = V_E + V_{BE} = 5.54 + 0.7 = 6.24 \text{ V} \qquad \textbf{Ans.}$$

$$A_V = \frac{V_o}{V_i} = A_I \frac{R_L}{R_i} = \frac{(1+h_{fe})(1\text{ k} \parallel 1\text{ k})}{h_{ie} + (1+h_{fe})(1\text{ k} \parallel 1\text{ k})} = \frac{201 \times 0.5\text{ k}}{1\text{ k} + (201 \times 0.5\text{ k})} = 0.99 \qquad \textbf{Ans.}$$

SUMMARY

- The self-bias or voltage divider bias circuit can be used as an amplifier if we add input signal source and few components to it.
- A Common-Emitter (CE) amplifier is the most widely used amplifier circuit. This common-emitter amplifier has one important property. This property is that a phase difference of 180° occurs between the input signal voltage and the output voltage in a common emitter amplifier. This is also known as phase reversal.
- In a transistor amplifier circuit, the d.c. sources established d.c. biasing currents and voltages, also known as quiescent operating currents and voltages. After this, when a.c. input signal source is applied, it produces fluctuations in these currents and voltages. As a result of this, an amplified signal at the output is obtained.
- Hybrid parameters are also called as h-parameters. Characteristics of amplifier can be specified easily in terms of h-parameters. The hybrid or h-parameters are very popular and widely used because they give accurate result and can be measured very easily.
- The h-parameter representation for transistor depends upon the type of configuration (*CE*, *CB* or *CC*). Therefore, each h-parameter is assigned a second subscript letter *e*, *b* or *c*, depending upon the type of configuration. The letter '*e*' is used when the configuration is common emitter, the letter '*b*' is used if the configuration is common base and letter '*c*' is used if the configuration is common collector.
- As a matter of fact, the h-parameters are widely used in amplifier analysis. To simplify the analysis, the exact h-parameters expressions are converted into approximated expressions. The complete behaviour of the transistor circuit can be obtained by using this simplified hybrid model.

SHORT QUESTIONS WITH ANSWERS

Q. 1. What is phase reversal? Explain.

Ans. As discussed earlier, a common-emitter (CE) amplifier is the most widely used amplifier circuit. This common-emitter amplifier has one important property.

This property is that a phase difference of 180° occurs between the input signal voltage and the output voltage in a common emitter amplifier. This is also known as phase reversal.

Q. 2. What are hybrid parameters (h-parameters)? Explain.

Ans. Hybrid parameters are also called as h-parameters. Characteristics of amplifier can be specified easily in terms of h-parameters. The hybrid or h-parameters are very popular and widely used because they give accurate result and can be measured very easily.

Q. 3. What is meant by small signal amplifier ?

Ans. When the input signal is quite weak and produces small fluctuations in the output current in comparison to its quiescent value, the amplifier is called the small signal or voltage amplifier.

Q. 4. What is meant by phase reversal ?

Ans. In a CE configuration, the output voltage increases in the negative direction when the input signal voltage increases in the positive direction and vice-versa. This is called the phase reversal and causes a phase difference of 180° between the input signal voltage and output voltage.

Q. 5. What is the effect of removal of emitter bypass capacitor in a CE amplifier circuit ?

Ans. Removal of bypass capacitor in a CE amplifier circuit causes excessive degeneration in the amplifier circuit.

Q. 6. What is the effect of source resistance on voltage gain of a common base transistor amplifier ?

Ans. The voltage gain of a CB transistor amplifier will decrease if source resistance is considered because in such a case, there will be a voltage drop across the source resistance and output voltage will decrease.

Q. 7. What are the main purposes for which a common-collector amplifier may be used ?

Ans. For a common collector amplifier, current gain is as high as for CE amplifier, voltage gain is less than unity, input resistance is the highest and the output resistance is the lowest of all the three (CE, CC and CB) configurations. This circuit finds wide application as a buffer amplifier between a high impedance source and a low impedance load.

Q. 8. What is meant by distortion in an amplifier ?

Ans. Change in amplified output signal waveform is called the distortion.

Q. 9. Why distortion in amplifier is considered harmful ?

Ans. Because distortion may change the intelligence (useful information) carried by the signal.

Q. 10. Write merits of Hybrid model.

Ans. **Merits of Hybrid Model**

(*i*) h-parameters can be easily measured.

(*ii*) They can be calculated from the static characteristics of transistor.

(*iii*) Manufacturers provide h-parameters in their data sheets.

(*iv*) h-parameters can be used easily and conveniently in circuit analysis and design.

Q. 11. What are Demerits of the h-parameters?

Ans. The hybrid parameters suffer from a major disadvantage that they are defined only for a particular set of operating conditions. The parameters of other equivalent circuit (r_e model) can be determined for any region of operation within the active region and are not limited by the single set of parameters provided by the specification sheet.

REVIEW QUESTIONS

1. Draw and explain the transistor a.c. equivalent circuit.
2. Derive the hybrid or h-parameters for a two port network.
3. Why hybrid parameters are called so? Define them.
4. Draw the circuit diagram of a single-stage transistor amplifier. State the function of each component used in this circuit.
5. Define the h-parameters of a *CE*-transistor.
6. Derive the general expressions in terms of h-parameters and the load for:
   (*i*) Input impedence
   (*ii*) Current gain
   (*iii*) Voltage gain
   (*iv*) Output resistance
7. Find the h_{ie} in terms of the *CB* h-parameter.

NUMERICAL PROBLEMS

1. A *CE* amplifier uses a voltage source having internal resistance R_s = 1000 W and the load resistance is R_L = 1200 W. The values of *h*-parameters are h_{ie} = 1.2 kW, $h_{re} = 2 \times 10^{-4}$, h_{fe} = 60 and h_{oe} = 25 mA/V. Calculate the current gain, input resistance, voltage gain and output resistance.

   (Ans. A_i = –58.25, R_i = 1.186 kW, A_v = –58.937, R_0 = 51.162 kW)

2. A *CB* amplifier uses a voltage source having internal resistance R_s = 600 W and the load resistance is R_L = 1200 W. The *h*-parameters are h_{ib} = 22 W, $h_{rb} = 4 \times 10^{-4}$, h_{fb} = –0.98 and h_{ob} = 0.25 mA/V. Calculate the current gain, input resistance, voltage gain and output resistance.

   (Ans. A_i = 0.9797, R_i = 22.47 W, A_v = 52.32 and R_0 = 1.13607 mW)

3. The following test results were obtained in a *CE* amplifier circuit while measuring *h*-parameters experimentally:

   (*i*) With a.c. output shorted I_b = 25 mA, I_c = 1.2 mA, V_{bc} = 30 mV and V_{oe} = 0.

   (*ii*) With a.c. circuit open-circuited I_b = 0, I_c = 32 mA, V_{bc} = 0.3 mV and V_{ce} = 1.2 V.

   Determine hybrid parameters of the given transistor.

   (Ans. h_{oe} = 1.2 kΩ, h_{fe} = 48 ; $h_{re} = 2.5 \times 10^{-4}$, h_{oe} = 25 μS)

4. A BJT has the following *h*-parameters:

   h_{ie} =2,000 Ω; $h_{ie} = 16 \times 10^{-5}$; h_{fe} = 49 and h_{oe} = 50 μA/V

   Determine the current gain, voltage gain, input resistance and output resistance of the *CE* amplifier if the load resistance is 30 kΩ. Neglect source resistance.

   [Ans. A_i = –19.6, A_v = – 308.5; Z_{in} = 1.966 Ω; Z_{out} = 21.7 kΩ]

5. Define (*h*) parameters of a common emitter amplifier. How can these parameters be determined from the characteristic curves?

6. A CE amplifier has the following *h*-parameters:

   h_{ie} = 1.1 kΩ, $h_{re} = 0.25 \times 10^{-3}$, h_{fe} = 50, h_{oe} = 25 μmho

   If the load resistance is 1 kΩ and source resistance is also 1 kΩ, find:

   (*i*) current gain, (*ii*) voltage gain,

7. For the circuit shown in figure 7.48 below find A_v, A_i, Z_i and Z_0

   V_{CC} = 15 V, h_{ie} = 1100 Ω, R_B = 100 kΩ, h_{fe} = 50, R_E = 1 kΩ

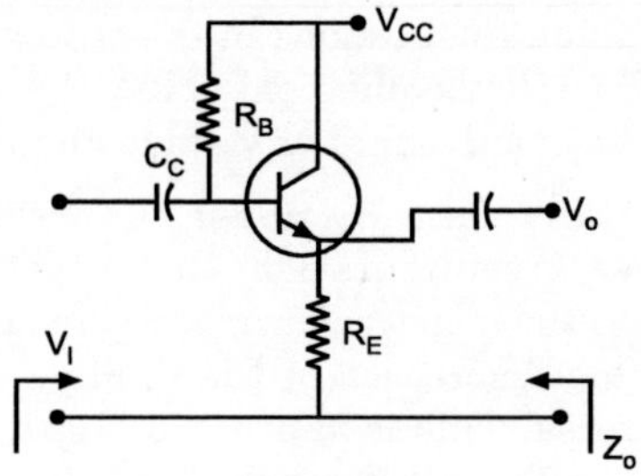

Fig. 7.48.

8. In the following given circuit, explain with reasoning, in what configuration this circuit is working.
   Also calculate the voltage-gain.

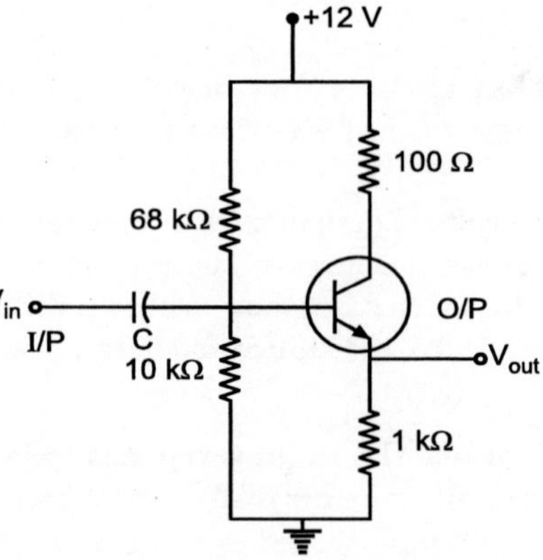

Fig. 7.49.

9. Compare CE, CB and CC transistor amplifiers for their parameters *e.g.* Z_1, Z_0, A_i and A_v.

   (U.P. Tech. Sem. Exam. 2001-02) (04 marks)

10. Draw and describe the *h*-parameter equivalent circuit for a common collector transistor amplifier. Show how the values of h_{ic}, h_{rc}, h_{fc} and h_{oc} are obtained from the common collector input characteristic curves and CC output characteristic curves.

OBJECTIVE TYPE QUESTIONS

Choose one correct answer in the following :

1. For a transistor, the h_{fe} parameter is the same as
   (*a*) β_{dc} (*b*) β_{ac}
   (*c*) α_{dc} (*d*) r_e
2. If the dc emitter current in a transistor amplifier is 4 mA, the approximate value of r_e is
   (*a*) 4 Ω (b) 6.25 Ω
   (*c*) 22 Ω (*d*) 26 Ω
3. In a common emitter amplifier, the voltage gain = 90. If the emitter bypass capacitor is removed
   (*a*) the voltage gain will decrease
   (*b*) the voltage gain will increase
   (*c*) the Q-point will shift
   (*d*) the circuit will be unstable
4. In a common emitter amplifier, R_C = 1 kΩ, R_E = 200 Ω, r_e = 16 Ω and β_{ac} = 75. Assuming that R_E is completely bypassed at the operating frequency, the voltage gain will be
   (*a*) 4 (*b*) 5
   (*c*) 63 (*d*) 75
5. For a common collector amplifier, R_E = 90 Ω, r_e = 10 Ω and β_{ac} = 130. The ac input resistance at the base will be
   (*a*) 100 Ω (*b*) 1080 Ω
   (*c*) 1.3 kΩ (*d*) 1170 Ω
6. For a common collector amplifier, R_E = 90 Ω r_e = 10 Ω and β_{ac} = 10. The voltage gain of the amplifier will be
   (*a*) 1.2 (*b*) 120
   (*c*) 0.85 (*d*) 0.9
7. In a common emitter amplifier, the value of R_C = 1 kΩ, R_L = 10 kΩ, r_e = 13 Ω. The voltage gain of the transistor amplifier will be (when R_E is bypassed)
   (*a*) 10 (*b*) 69.9
   (*c*) 76.9 (*d*) 84.6
8. The input resistance of a common base amplifier is
   (*a*) very low (*b*) very high
   (*c*) the same as common emitter
   (*d*) the same as common collector amplifier
9. For an amplifier, the current gain ≅ 1, power gain = voltage gain, input resistance is very low (≅ r_e) and output resistance ≅ R_C. The amplifier is
   (*a*) common base (*b*) common emitter
   (*c*) common collector
   (*d*) Darlington pair
10. For an amplifier, the input resistance is high, the current gain is high and voltage gain is near unity. It is a
    (*a*) common base (*b*) common emitter
    (*c*) common collector
    (*d*) Darlington amplifier
11. An amplifier has good voltage, current and power gains and the input resistance is low. It is a
    (*a*) common base (*b*) common emitter
    (*c*) common collector amplifier
    (*d*) None
12. The maximum current gain of an amplifier is unity. It is a
    (*a*) common base (*b*) common collector
    (*c*) common emitter amplifier
    (*d*) None.

Answers:

1. (*b*)	2. (*b*)	3. (*a*)	4. (*c*)	5. (*c*)	6. (*d*)
7. (*b*)	8. (*a*)	9. (*b*)	10. (*c*)	11. (*b*)	12. (*a*)

UNIT—VI

Field Effect Transistor

The Junction Field Effect Transistor (Construction, principle of operation, symbol)—Pinch-off Voltage, Volt-Ampere characteristics, The JFET Small Signal Model, MOSFET (Construction, principle of operation, symbol).

Above Syllabus has been covered in following Chapter:

CHAPTER 8 : Field Effect Transistor

Chapter 8

Field-Effect Transistor

SYLLABUS

The Junction Field Effect Transistor (Construction, principle of operation, symbol)—Pinch-off Voltage, Volt-Ampere characteristics, The JFET Small Signal Model, MOSFET (Construction, principle of operation, symbol).

Chapter Outline

8.1. INTRODUCTION

We have already discussed in Chapter 5 that a bipolar junction transistor (BJT) is a current controlled device. Due to this reason, the output characteristics of BJT are controlled by the base current and not the base voltage. Another type of transistor, known as the field-effect transistor (FET), is a voltage-controlled device. This means that the output characteristics of FET are controlled by the applied input voltage and not by the input current. In addition to this, FET is an unipolar device. This means that the current flowing through it is only due to one type of charge particles *i.e.*, holes or electrons. BJT, on the other hand, is a bipolar device because holes and electrons both contribute to the flow of current.

In a broad sense, there are two main types of field-effect transistors as under:

(*i*) Junction field-effect transistors (JFET).

(*ii*) Metal oxide semiconductor field-effect transistor (MOSFET) or insulated gate field-effect transistor (IGFET).

Both of these types of FETs may be further sub-divided as under:

Now, let us discuss the construction, operation and characteristics of all the above-mentioned FET in the sections to follow.

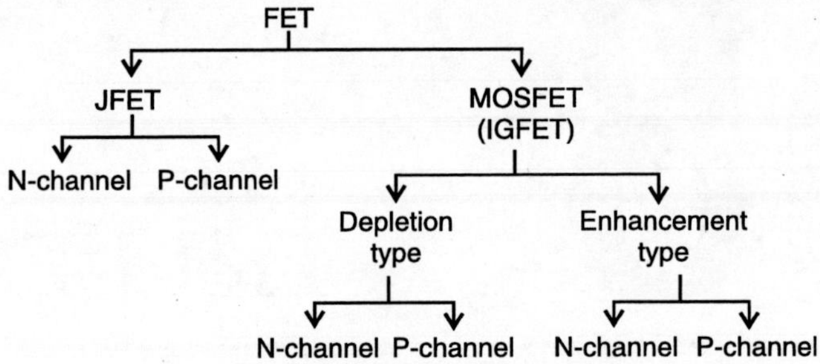

Fig. 8.1. Different types of FETs.

8.2. JUNCTION FIELD-EFFECT TRANSISTOR (JFET)

As shown in figure 8.1, the junction field-effect transistor (JFET) may be divided, depending on their structure, into the following two categories:

(*i*) N-channel JFET.

(*ii*) P-channel JFET.

Figure 8.2 shows the basic construction of an *N*-channel JFET. It simply consists of an N-type semiconductor bar with two P-type heavily doped regions diffused on opposite sides of its middle part. The P-type regions form two PN-junctions. The space between the junctions (*i.e.,* N-type regions) is known as a **channel**. Both the P-type regions are connected internally and a single wire is taken out in the form of a terminal known as the gate (*G*). The electrical connections (known as ohmic contacts) are made to both ends of the N-type semiconductor and are taken out in the form of two terminals known as drain (*D*) and source (*S*). The drain (*D*) is a terminal through which electrons leave the semiconductor bar and source (*S*) is a terminal through which the electrons enter the semiconductor.

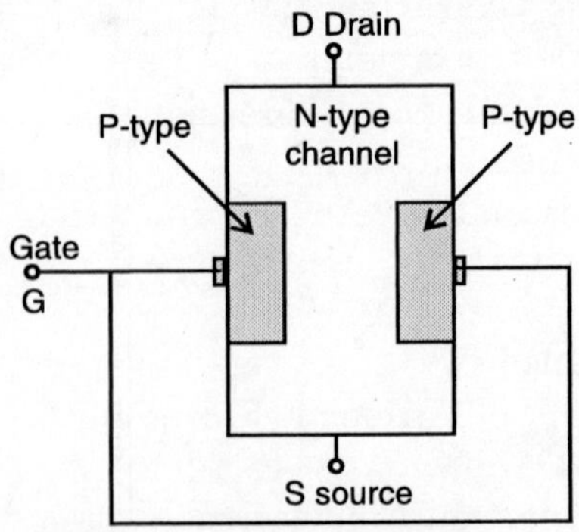

Fig. 8.2. A N-channel JFET

The more you praise and celebrate you life, the more there is in life to celebrate.

— **Opran Winfrey**

Whenever a voltage is applied across the drain and source terminals, a current flows through the N-channel. The current consists of only one type of carriers (*i.e.* electrons). Therefore, the field-effect transistor (FET) is known as an unipolar device. In fact, this quality distinguishes an FET from a BJT (*i.e.*, a bipolar Junction transistor) where the current consists of the flow of both the electrons and holes.

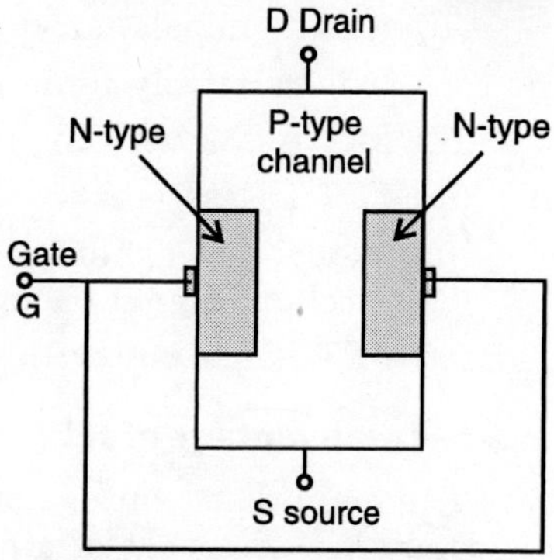

Fig. 8.3. A P-channel JFET

Figure 8.3 shows a *P*-channel JFET. Its construction is similar to that of N-channel JFET, except that it consists of a P-channel and N-type junctions. The current carriers in JFET are the holes which flow through the *P*-type channel.

8.2.1. Symbol for JFET

Figure 8.4(*a*) shows the schematic symbol for an N-channel JFET. Similarly, figure 8.4(*b*) shows the schematic symbol for a P-channel JFET.

> **DO YOU KNOW?**
>
> Because of the extremely high input impedance of a JFET, the input current is generally assumed to be 0 μA, and the current gain of a JFET amplifier is an undefined quantity.

It may be noted that in an N-channel JFET, the arrow points towards the vertical line. The vertical line represents the N-channel. On the other hand, in a P-channel symbol, the arrow points away from the vertical line. In this case, the vertical line represents the P-channel.

Now let us discuss the operation and characteristics of JFET with reference to N-channel JFET. In fact, the operation of a P-channel JFET is similar to N-channel except that all voltages and currents are reversed.

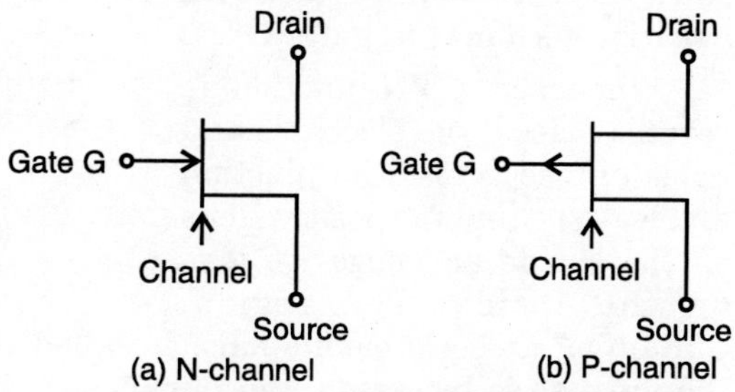

Fig. 8.4. Symbols for JFET

8.2.2. Voltage Controlled or Current Controlled?

The voltage applied between gate and source (V_{GS}) controls the drain current I_D. Therefore, FET is a voltage controlled device. The transistor on the contrary is a current controlled device because the output current I_C is controlled by varying the base current I_B. The name field effect is derived from the fact that the current flow in FET is controlled by an electric field set by an external voltage.

8.2.3. Unipolar or Bipolar?

FET is a unipolar device. This means that the current flowing through it is only due to one type of charge particles, holes or electrons. Transistor, on the other hand, is a bipolar device as holes and electrons both contribute to the flow of current. Just as there are NPN and PNP transistors, there are n-channel and p-channel field effect transistors. One of the most important characteristics of the FET is its very high input impedance. Typically, it is in the range of 1 to several megaohms, which is much higher than the input resistance of a BJT. FETs are more temperature stable as compared to the BJT and it requires less space than that for a BJT. Therefore, FETs are preferred in integrated circuits.

8.2.4. Advantages of JFET Over BJT

The FET enjoys several advantages over the conventional bipolar junction transistor (BJT). Some of them may be listed as under :

(*i*) It is a unipolar device so its operation depends on the flow of majority carriers only.

(*ii*) It is relatively immune to radiation.

(*iii*) FET has a very high input resistance (typically few megaohms)

(*iv*) FET is less noisy

(*v*) It does not exhibit any offset voltage at zero drain current hence it can be used as an excellent signal chopper.

(*vi*) FET has a better thermal stability.

8.2.5. Disadvantage of JFET

The main disadvantage of the JFET is that it has a relatively small gain-bandwidth product as compared with that of a conventional transistor (*i.e.* BJT).

8.3. FORMATION OF DEPLETION REGION IN JFET

Let us consider an N-channel JFET as shown in figure 8.5 (*a*). Here the P-type gate and N-type channel constitute a PN-junction. This PN-junction is always reverse-biased in JFET operation. The reverse bias is applied by a voltage V_{GG} connected between the gate and source terminals as shown in figure 8.5(*b*). Here, it can be noted that the positive terminal of the voltage V_{GG} is connected to source (*S*) and negative terminal to gate (*G*).

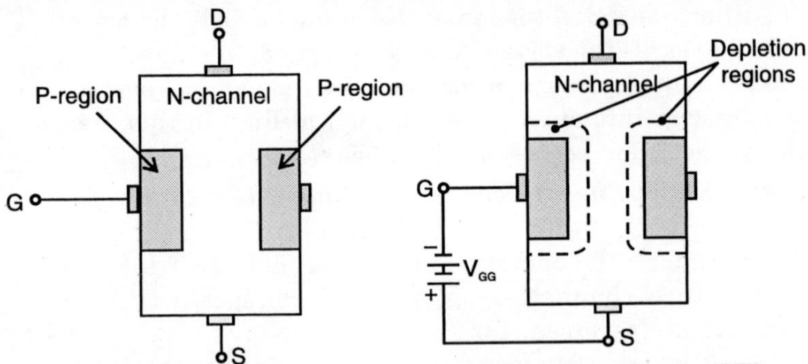

(a) N-channel JFET (b) Depletion regions in JFET

Fig. 8.5.

Whenever a PN-junction is reverse-biased, the electrons and holes diffuse across the junction and leave behind the positive ions on the N-side and negative ions on the P-side. The region, containing these immobile ions, is called **depletion region.** As the reverse bias voltage across the junction is increased, the thickness of depletion region is also increased. If both the P-side and *N*-side of the junctions are equally doped, the depletion region will extend equally in both the regions.

However, if one side of the junction is heavily doped as compared to the other side, the depletion region extends more into the region of lower doping.

A similar action takes place in JFET. Because the *P*-region of an *N*-channel JFET is heavily doped as compared to the N-channel, the depletion region extends less into the P-region and deeper into the N-channel as depicted in figure 8.5(*b*). Incidentally, when there is no applied voltage between the drain and the source, the depletion region is symmetrical around the junction. The conductivity of depletion region is zero since there are no mobile charge carriers in this region. Therefore, the effective width of the N-channel is reduced. It further reduces with the increased reverse bias voltage applied across the gate and source terminals of the JFET.

DO YOU KNOW?

In a transistor, the emitter-base depletion layer is narrower than the collector-base depletion layer. The reason can be attributed to the different doping levels of the emitter and collector regions. With much heavier doping in the emitter region, the penetration into the *n* material is minimal because of the availability of many more free electrons. However, on the collector side, fewer free electrons are available and the depletion layer must penetrate more deeply in order to set-up the barrier potential.

The reverse-bias across the gate-source junction of a JFET can also be achieved simply by applying a voltage across the drain and source terminals as shown in figure 8.6(*a*)

It may be noted that the drain (*D*) is connected to the positive terminal of the d.c. supply (V_{DD}) and source (*S*) is connected to the negative terminal.

Now, let us discuss as to how the drain-to-source voltage produces a reverse-bias across the gate-source junction in the same way as the gate-to-source voltage. In order to do so, we represent the channel resistances r_a and r_b as shown in figure 8.6(a). These resistances have been shown variable since their values depend on the magnitude of the drain-to-source voltage (V_{DS}) and gate-to-source voltage (V_{GS}). In the presence of positive supply voltage V_{DD} (with gate open) the electrons flow from source to drain through the *N*-channel and constitutes a current known as **drain current (I_D).** The conventional direction of the drain current is indicated from drain-to-source through the device. The drain current causes a voltage drop across the resistance r_b, which has the effect of reverse-biasing the gate-to-source junction. Hence, even if the gate is open, the gate-to-source junction is reverse biased by the drain-to-source voltage. It creates the depletion regions within the channel as shown in figure 8.6(*b*).

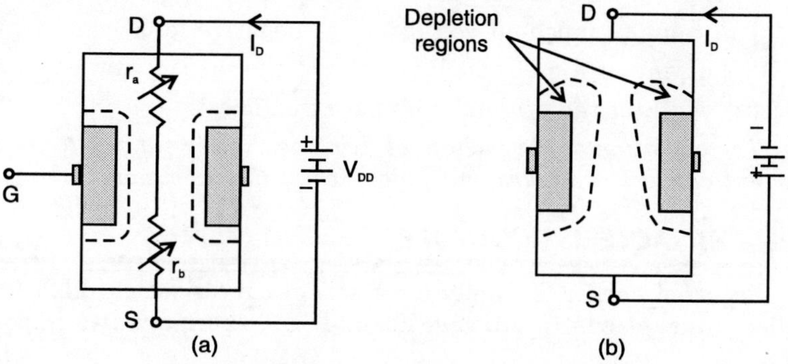

Fig. 8.6. Effect of drain-to-source supply

It may be noted from figure 8.6 (*b*) that the depletion region is not symmetrical around the gate-to-source junction. As a matter of fact, it is extended more deeper into the channel near the drain terminal and less on the source terminal. It is due to the fact that voltage drop across r_a is greater than that across r_b. Thus, because of this reverse bias voltage is higher near the drain end of the channel as compared to the source end.

8.4. PRINCIPLE OF OPERATION OF JFET*

(V. Important)

In last article we have discussed the application of negative gate voltage or positive drain voltage, with respect to source, reverse biases the gate-source junction of an N-channel JFET. The effect of reverse-bias voltage is to form the depletion regions within the channel. When a voltage is applied between the drain and source with a dc supply voltage (V_{DD}), the electrons tend to flow from source to drain side through the narrow channel existing between the depletion regions. This constitutes the drain current (I_D) and its conventional direction is indicated from drain-to-source. The value of this drain

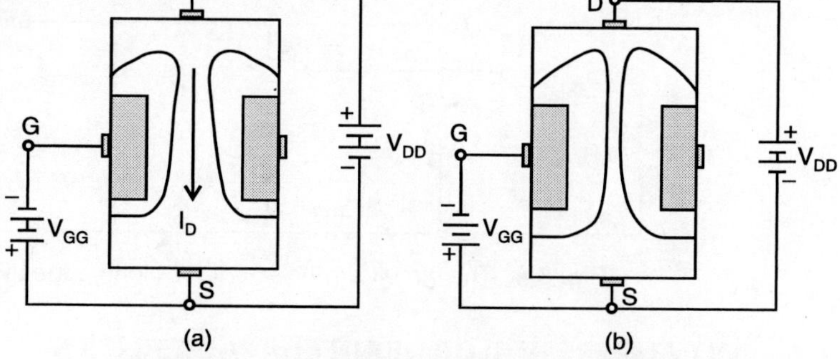

Fig. 8.7. Illustration of operation of JFET.

* Explain the working of a JFET along with its construction with neat sketch. Also draw its transfer characteristic curve with explanation.

current is maximum when no external voltage is applied between the gate and the source and is designated by the symbol I_{DSS}.

When the gate-to-source voltage (V_{GS}), is applied by a dc supply (V_{GG}) and increased above zero as depicted in figure 8.7(*a*), the reverse-bias voltage across the gate-source junction is increased. Because of this, the depletion regions are widened. This reduces the effective width of the channel and hence controls the flow of drain current through the channel.

When the gate-to-source voltage (V_{GS}) is further increased, a stage is reached at which two depletion regions touch each other as shown in figure 8.7(*b*). At this gate-to-source voltage, the channel is completely blocked or pinched off and drain current is reduced to zero. The gate-to-source voltage (V_{GS}) at which the drain current is zero (or completely cut-off) is known as **pinch off voltage**. It is denoted by the symbol V_P or V_{GS} (off).

The value of pinch-off voltage V_P is negative for N-channel JFET. It depends upon

(*i*) doping of the N and P regions of the device, and

(*ii*) width of the original channel structure.

The operation of the P-channel JFET is exactly similar to the N-channel JFET except that the current carriers are holes and polarities of the dc supply voltage V_{GG} and V_{DD} are reversed.

8.5. CHARACTERISTICS OF JFET

As discussed earlier, a family (or a set) of curves which relate device currents and voltages are called **characteristic curves**. For a JFET, there are two important characteristics as under:

(i) V-I or drain characteristics

These curves give relationship between the drain current (I_D) and drain-to-source voltage (V_{DS}) for different values of gate-to-source voltage (V_{GS}).

(ii) Transfer characteristics

These curves give relationship between drain current (I_D) and gate-to-source voltage (V_{GS}) for different values of drain-to-source voltage (V_{DS}).

The drain and transfer characteristics of JFET can be obtained by using an *N*-channel JFET connected in the common source mode as shown in figure 8.8. Here, the potentiometers R_1 and R_2 are used to vary the voltages V_{GS} and V_{DS} respectively. The voltages V_{DS} and V_{GS} can be measured by the voltmeters connected across the JFET terminals. The drain current (I_D) can be measured by the milli-ammeter connected in series with the JFET and the supply voltage V_{DD}.

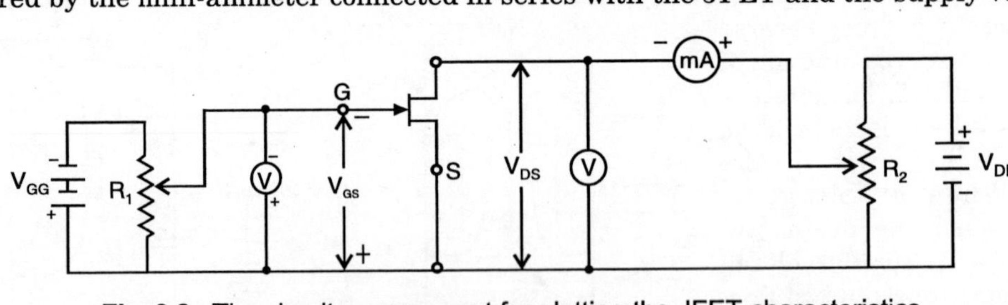

Fig. 8.8. The circuit arrangement for plotting the JFET characteristics.

8.6. VOLT AMPERE (V–I) OR DRAIN CHARACTERISTICS

The drain characteristics curves may be obtained by using the circuit arrangement shown in figure 8.8. Firstly, we adjust the gate-to-source voltage (V_{GS}) to zero volt. After this, we increase the drain-to-source voltage (V_{DS}) in small suitable steps and note the corresponding values of drain current (I_D) at each step. Now, if we plot a graph taking drain-to-source voltage along the horizontal axis and drain current along the vertical axis, then we shall obtain a curve marked $V_{GS} = 0$ as shown in figure 8.9. A similar procedure can be followed to obtain curves for different values of gate-to source voltage *i.e.*, $V_{GS} = 1, 2, 3$ and 4 volt *etc.*

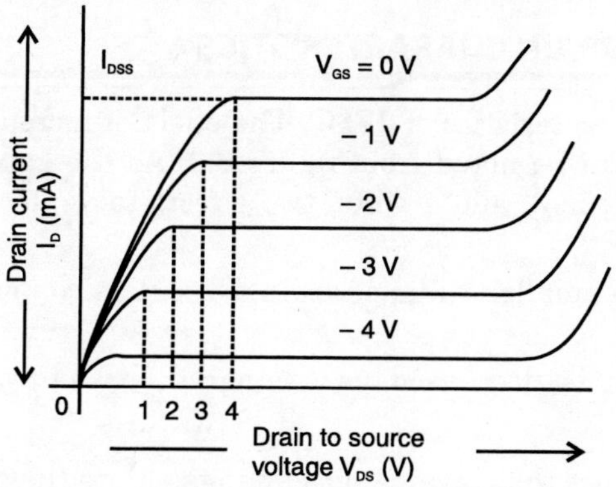

Fig. 8.9. The drain characteristics of JFET.

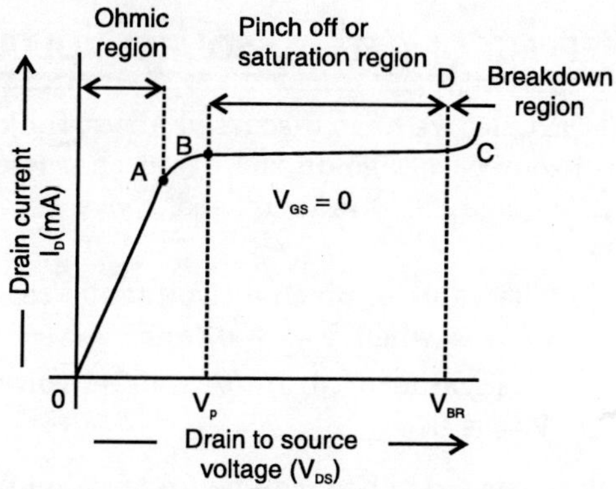

Fig. 8.10. The drain characteristics with $V_{GS} = 0$ *Volt.*

Now, in order to explain the typical shape of drain characteristics, let us select the curve with $V_{GS} = 0$ volt as shown in figure 8.10. The curve can be sub-divided into the following regions:

(*i*) Ohmic region

Ohmic region has been shown as a curve *OA* in the figure 8.10. In this region, the drain current increases linearly with the increase in drain-to-source voltage obeying Ohm's law. The linear increase in drain current is due to the fact that *N*-type semiconductor bar acts like a simple resistance.

(*ii*) Curve *AB*

In this region, the drain current increases at the reverse square law rate with the increase in drain-to-source voltage. It means that drain current increases slowly as compared to that in the ohmic region. It is due to the fact that with the increase in drain-to-source voltage, the drain current increases. This, in turn, increases the reverse-bias voltage across the gate -source junction. As a result of this, the depletion region grows in size thereby reducing the effective width of the channel. At the drain-to-source voltage, corresponding to point *B*, the channel width is reduced to a minimum value and is called **pinch off**. The drain-to-source voltage at which the channel pinch-off occurs, is called **pinch off voltage (V_P).**

(*iii*) Pinch-off region

This region has been shown by the curve *BC*. It is also known as the **saturation region** or **constant current region**. In this region, the drain current remains constant at its maximum value (*i.e.*, I_{DSS}). The drain current in the pinch off region depends on the gate-to-source voltage and is given by the following relation:

$$I_D = I_{DSS}\left(1 - \frac{V_{GS}}{V_P}\right)^2$$

This relation is known as ***Shockly's equation****. The pinch-off region is the normal operating region of JFET, when used as an amplifier.*

(*iv*) Breakdown region

This region has been shown by the curve *CD*. In this region the drain current increases rapidly as the drain-to-

DO YOU KNOW?

The pinch off voltage. V_p is the point at which further increases in V_{DS} are offset by a proportional increase in the channel's resistance. This means that if the channel resistance is increasing in direct proportion to V_{DS} above V_P, I_D must remain the same above V_P.

source voltage is also increased. It happens because of the breakdown of gate-to-source junction due to the avalanche effect. The drain-to-source voltage corresponding to point C is known as the **breakdown voltage**.

8.7. EFFECT OF GATE-TO-SOURCE VOLTAGE ON DRAIN CHARACTERISTICS

In last article, we have discussed about the drain characteristics of JFET. The effect of increasing gate-to-source voltage on the drain characteristics can be noted from figure 8.9. As the gate-to-source voltage (V_{GS}) is increased above zero to the 1, 2, 3, and 4 volts, two effects take place as under:

(*i*) The value of pinch-off voltage is reached at a smaller value of drain current as compared to that when $V_{GS} = 0$, and

(*ii*) The value of drain-to-source voltage (V_{DS}) is decreased as compared to that when $V_{GS} = 0$.

Now, the first effect can be understood from the fact that reverse bias voltage across the gate-to-source junction of JFET is the sum of gate-to-source voltage (V_{GS}) and drain-to-source voltage (V_{DS}). Hence, for fixed value of V_{DS}, the increased value of voltage V_{GS} would increase the reverse-bias voltage across the junction. For example, if V_{GS} of –1 volt is applied between the gate and the source of JFET, then the gate-to-source junction becomes reverse-biased even when the drain current is zero. This causes the depletion regions to be formed within the channel. As the voltage V_{DS} is increased above zero, the drain current increases. This produces voltage drop across channel resistance, which in turn, produces reverse-bias voltage across the gate-source junction. The amount of reverse-bias voltage required to be produced by drain current will obviously be decreased by one volt. In other words, a smaller voltage drop along the channel (*i.e.*, smaller than that when $V_{GS} = 0$) will increase the depletion regions to the point where they will pinch off the drain current. As a result of this the pinch-off voltage (V_P) is reached at a lower value of drain current than that when $V_{GS} = 0$.

Also the second effect is caused due to the fact that the increased value of gate-to-source voltage (V_{GS}) keeps adding to the reverse-bias voltage at the junction produced by the drain current. Hence, its lower value is required to reach the breakdown voltage.

It can be noted that with zero gate-to-source voltage the drain current saturates at I_{DSS} value and the characteristic indicates the pinch-off voltage (V_P) equal to 4 volt. However, if V_{GS} of – 1 volt is applied, the channel still requires –4 volt to achieve pinch-off condition. It means that now the channel requires only 3 volt drop (instead of previously 4 volt) from the drain-to-source voltage. This 3 volt drop can be achieved with the lower value of drain current. Similarly when V_{GS} is –2 and –3 volts, pinch-off is achieved with 2 volt and 1 volt drops respectively along the channel. These voltage drops are obtained with further reduced values of the drain current. It can be noted that when V_{GS} is –4 volt (*i.e.*, equal to pinch-off voltage), no channel drop is required and thus the drain current is zero.

8.8. TRANSFER CHARACTERISTICS *(V. Important)*

Transfer characteristics are also known as transconductance curves which give us the relationship between drain current (I_D) and gate-to-source voltage (V_{GS}) for a constant value of drain-to-source voltage (V_{DS}). The transfer characteristics can be obtained by using the circuit arrangement shown in figure 8.8. Firstly, we adjust the drain-to-source voltage to some suitable value and increase the gate-to-source voltage in small suitable steps. After this, note the corresponding values of drain current at each step. If we plot a graph with gate-to-source voltage (V_{GS}) along the horizontal axis and the drain current (I_D) along the vertical axis, we shall obtain a curve as shown in figure 8.11. A similar procedure can be used to obtain curves at different values of the gate-to-source voltage (V_{GS}).

It may be observed that the upper end of the curve has been shown by the drain current value equal to I_{DSS}, while the lower end is indicated by a voltage equal to V_{GS} (off) or V_P. It can be noted that the curve is part of a parabola and hence can be expressed by the following expression

$$I_D = I_{DSS}\left[1-\frac{V_{GS}}{V_{GS}(off)}\right]^2 = I_{DSS}\left(1-\frac{V_{GS}}{V_P}\right)^2$$

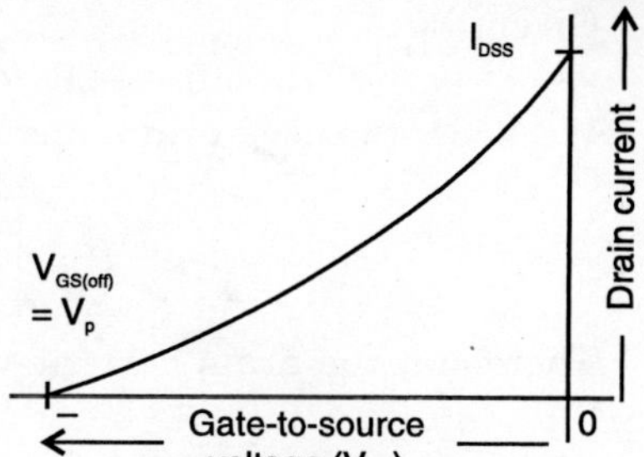

Fig. 8.11. The transfer characteristics of N-channel.

There is one important point which must be mentioned at this stage. Most JFET specifications sheet will list more than one value of V_{GS} (off) and I_{DSS}.

For example, the specifications sheet for the JFET 2*N* 5457 lists the following:

V_{GS} (off) (minimum) = – 0.5 volt

and I_{DSS} (min) = 1 mA

V_{GS}(off) (maximum) = – 6 volt

and I_{DSS} (max) = 5 mA

In such a situation, we must use the two minimum values to plot one curve and the two maximum values to plot another curve on the same graph.

> **DO YOU KNOW?**
>
> The transconductance curve of a JFET is unaffected by the circuit or configuration in which the JFET is used.

EXAMPLE 8.1 **The data sheet of a certain JFET indicates that I_{DSS} is equal to 15 mA and V_{GS} (off) is equal to –5 volt. Find the drain current for V_{GS} equal to 0 volt, –1 volt and – 4 volt.**

Solution: Given that

I_{DSS} = 15 mA and, V_{GS} (off) = –5 volt

We know that the drain current is given by

$$I_D = \left[1-\frac{V_{GS}}{V_{GS}(off)}\right]^2$$

Therefore, drain current when (V_{GS} = 0), will be

$$I_D = 15\left(1-\frac{0}{-5}\right)^2 = 15 \text{ mA} \quad \textbf{Ans.}$$

Similarly, drain current when V_{GS} – 1 V, will be

$$I_D = 15\left(1-\frac{-1}{-5}\right)^2 = 9.6 \text{ mA} \quad \textbf{Ans.}$$

and the drain current when V_{GS} = –4 *V*, will be

$$I_D = 15\left(1-\frac{-4}{-5}\right)^2 = 0.6 \text{ mA} \quad \textbf{Ans.}$$

EXAMPLE 8.2 **A JFET has parameters of V_{GS} (off) equal to –20 volt and I_{DSS} equal to 12 mA. Plot the transconductance curve for the device using V_{GS} values of 0 V, –5 V, –10 V, –15 V, and –20 V.**

Solution: Firstly, let us find the value of drain currents for the given values of V_{GS} (0 V , –5 V, –10 V, –15 V and –20 V) as discussed below:

Given that

$$V_{GS}\text{ (off)} = -20 \text{ volt} \quad \text{and,} \quad I_{DSS} = 12 \text{ mA}$$

We know that the drain current is given by

$$I_D = I_{DSS}\left[1 - \frac{V_{GS}}{V_{GS(off)}}\right]^2$$

Therefore, the drain current when $V_{GS} = -5$ V, will be

$$I_D = 12\left(1 - \frac{-5}{-20}\right)^2 = 6.75 \text{ mA}$$

Similarly, the drain current when $V_{GS} = -10$ V, will be

$$I_D = 12\left(1 - \frac{-10}{-20}\right)^2 = 3 \text{ mA}$$

and the drain current when $V_{GS} = -15$ V, will be

$$I_D = 12\left(1 - \frac{-15}{-20}\right)^2 = 0.75 \text{ mA}$$

The value of V_{GS} and the corresponding values of I_D obtained above have been shown in Table 8.1.

Table 8.1.

V_{GS}(V)	I_D(mA)
–20	0
–15	0.75
–10	0.3
–5	6.75
–0	12

These points are now plotted and joined together to form a curve shown in figure 8.12.

EXAMPLE 8.3 A 2N 5486 JFET has values of V_{GS}(off) equal to –2 volt to –6 volt and I_{DSS} is equal to 8 mA to 20 mA. Plot the minimum and maximum transconductance curves for the device.

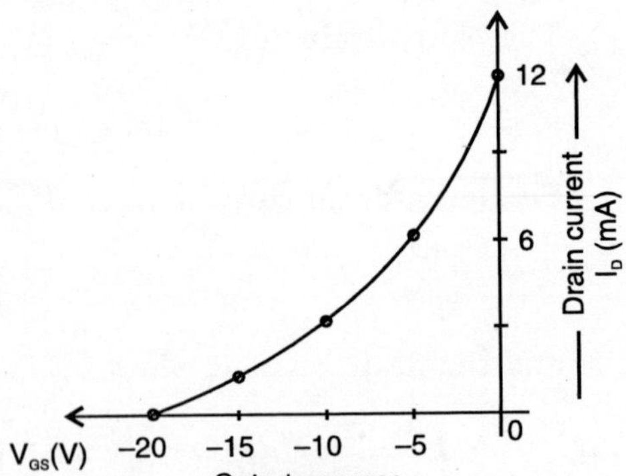

Fig. 8.12.

Solution: Given that

$$V_{GS}\text{ (off)} = -2 \text{ volt to } -6 \text{ volt}$$

and $$I_{DSS} = 8 \text{ mA to } 20 \text{ mA}$$

Firstly, let us find the values of drain currents for the given values of V_{GS}.

We know that the drain current is given by

$$I_D = I_{DSS}\left[1 - \frac{V_{GS}}{V_{GS(off)}}\right]^2 \quad ...(i)$$

In this equation, substituting the values of V_{GS} = – 0.5, –1.0 and –1.5 with V_{GS}(off) = – 2.0 volt, we get the corresponding values of I_D = 4.5 mA, 2 mA and 0.5 mA.

With these values, we have the set of V_{GS} versus I_D value as shown in Table 8.2.

Table 8.2. For Minimum Transconductance Curve

V_{GS} (V)	I_D (mA)
– 2.0	0
– 1.5	0.5
– 1.0	2.0
– 0.5	4.5
0	8

Similarly, using the maximum values of V_{GS} (off) = –6 Volt and I_{DSS} = 20 mA, we can use equation (*i*) again for V_{GS} = –4 V and –2 V to obtain the values of I_D in between the end points. The end points of the transcondutance curve are (–6 V, 0 mA) and (0 V, 20 mA). The values of I_D for in between points are 2.2 mA and 9.9 mA. The values obtained have been shown in Table 8.3.

Table 8.3. For Maximum Transconductance Curve

V_{GS}(V)	I_D (mA)
–6	0
–4	2.2
–2	3.9
0	20

Now, if we plot the values corresponding to the sets shown in Tables 8.2 and 8.3 and joining them, we get the maximum and minimum transconductance curve as shown in figure 8.13.

8.9. THE JFET PARAMETERS

The electrical behaviour of the JFET can be described in terms of certain parameters known as JFET parameters. Such parameters can be obtained from the characteristic curves. The important parameters of JFET have been discussed below:

(*i*) D.C. drain resistance (R_{DS})

It is also known as the static or ohmic resistance of the channel and is given by the ratio of voltage (V_{DS}) to the drain current (I_D). Mathematically, the d.c. drain resistance is expressed as

$$R_{DS} = \frac{V_{DS}}{I_D}$$

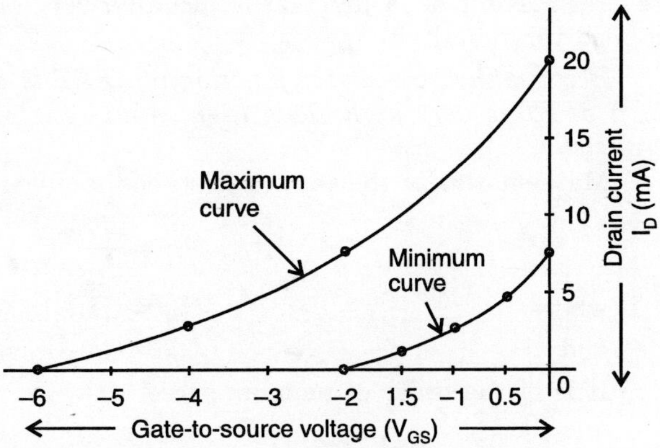

Fig. 8.13.

(*ii*) A.C. drain resistance (r_d)

It is also known as the dynamic drain resistance and is the a.c. resistance between the drain and the source terminal when the JFET is operating in the pinch-off or saturation region. It is given by the ratio of small change in drain-to-source voltage (ΔV_{DS}) to the corresponding change in drain current (ΔI_D) for a constant gate-to-source voltage (V_{GS}).

Mathematically, the a.c. drain resistance is expressed as

$$r_d = \frac{\Delta V_{DS}}{\Delta I_D}$$

It is, sometimes, written as r_{ds}, which indicates that it is the resistance from drain-to-source terminals. Since r_d is the output resistance of JFET, it can also be expressed as an output admittance y_{DS} (*i.e.*, equal to $1/r_d$).

(*iii*) Transconductance (g_m)

It is also known as the forward transconductance or forward transmittance. It is given by the ratio of small change in drain current (ΔI_D) to the corresponding change in gate-to-source voltage (ΔV_{GS}) for a constant drain-to-source voltage (V_{DS}).

Mathematically, the transconductance is expressed as

$$g_m = \frac{\Delta I_D}{\Delta V_{GS}}$$

The value of transconductance (g_m) is expressed in siemens (S) or mhos. We know that transfer characteristic of a JFET is a part of a parabola *i.e.*, it is non-linear. Therefore, the value of g_m varies depending upon the location on the curve as set by the gate-to-source voltage (V_{GS}). It has a greater value near the top of the curve than near the bottom.

(*iv*) Amplification factor (μ)

It is given by the ratio of small change in drain-to-source voltage (ΔV_{DS}) to the corresponding change in gate-to-sources voltage (ΔV_{GS}) for a constant drain current (I_D).

Mathematically, the amplification factor is expressed as

$$\mu = \frac{\Delta V_{DS}}{\Delta V_{GS}}$$

The amplification factor (μ) can also be expressed as a product of transconductance (g_m) and a.c. drain resistance (r_d)

Hence,

$$\mu = \frac{\Delta V_{DS}}{\Delta V_{GS}} = \frac{\Delta V_{DS}}{\Delta I_D} \times \frac{\Delta I_D}{\Delta V_{GS}} = r_d \times g_m$$

(*v*) Input resistance (R_i)

We know that a JFET operation with its gate-to-source junction reverse biased. As the reverse current of a junction is, usually, very small, therefore, gate-reverse current of a JFET is also very small.

Because the gate-source junction of JFET is used as an input, therefore, the input resistance of a JFET is very high. This high input resistance is an advantage of JFET over the bipolar transistor.

Mathematically the input resistance is given by

$$R_i = \frac{V_{GS}}{I_{GSS}}$$

where V_{GS} = Gate-to-source voltage

and I_{GSS} = Gate reverse current

It is of the order of nano-amperes.

8.10. THE MATHEMATICAL EXPRESSION FOR TRANSCONDUCTANCE *(V. Important)*

As discussed earlier, the drain current is expressed by the relation

$$I_D = I_{DSS}\left(1 - \frac{V_{GS}}{V_P}\right)^2$$

Differentiating the above equation on both sides with respect to V_{GS}, we have

$$\frac{dI_D}{dV_{GS}} = 2I_{DSS}\left(1-\frac{V_{GS}}{V_P}\right)\times\left(\frac{0-1}{V_P}\right)$$

or
$$g_m = -\frac{2I_{DSS}}{V_P}\left(1-\frac{V_{GS}}{V_P}\right) \quad ...(i)$$

Substituting $V_{GS} = 0$ in the above expression, we get

$$g_m = -\frac{2I_{DSS}}{V_P} = g_{mo}$$

The value of g_m at $V_{GS} = 0$ is the maximum and is equal to $-\frac{2I_{DSS}}{V_P}$. This value of g_m is designated as g_{mo}. The equation (*i*) can now be rewritten as

$$g_m = g_{mo}\left(1-\frac{V_{GS}}{V_P}\right)$$

EXAMPLE 8.4 The following information is included on the data sheet for an N-channel JFET:

$$I_{DSS} = 20 \text{ mA} \qquad V_P = -8 \text{ Volt}$$

and, $g_{mo} = 5000$ **µs**

Find the values of the drain current and transconductance at V_{GS} = –4 Volts.

Solution: Given that: $I_{DSS} = 20$ mA

$V_P = -8$ Volts $\qquad g_{mo} = 5000$ µs

and $V_{GS} = -4$ Volts

Now, we know that the drain current is given by

$$I_D = I_{DSS}\left(1-\frac{V_{GS}}{V_P}\right)^2$$

Substituting values, we get

or
$$I_D = 20\left(1-\frac{-4}{-8}\right)^2 = 5 \text{ mA} \qquad \textbf{Ans.}$$

Also the transconductance is given by

$$g_m = g_{mo}\left(1-\frac{V_{GS}}{V_P}\right) = 5000\left(1-\frac{-4}{-8}\right) = 2500 \text{ µs} \qquad \textbf{Ans.}$$

EXAMPLE 8.5 The data sheet for a certain enhancement-type MOSFET reveals that $I_{D(on)}$ = 10 mA at V_{GS} = – 12 V and $V_{GS(th)}$ = – 3 V. Is this device P-channel or *N*-channel? Find the value of I_D, when V_{GS} = – 6 V.

Solution: Given that: $I_{D(on)} = 10$ mA; $V_{GS} = -12$ volts and $V_{GS(th)} = -3$ Volts.

Since the value of V_{GS} is negative for the enhancement-type MOSFET, this indicated that the device is P-channel.

We know that the drain current is given by

$$I_D = K\,(V_{GS} - V_{GS(th)})^2 \quad ...(i)$$

or
$$I_{D(on)} = K\,[(V_{GS} - V_{GS(th)})]^2$$

$$10 = K\,[-12-(-3)]^2 = 81 \text{ k}$$

$$K = 10/81 = 0.12 \text{ mA/V}$$

Substituting this value of k and V_{GS} (equal to – 6) in the equation (*i*),

$$I_D = 0.12\,[-6-(-3)]^2 = 1.08 \text{ mA} \qquad \textbf{Ans.}$$

EXAMPLE 8.6 **Show that if $|V_{GS}| << |V_P|$, then drain current can be approximated as**

$$I_{DS} = I_{DSS} - g_{mo} V_{GS}$$

Solution: We know that

$$I_{DS} = I_{DSS}\left[1-\frac{V_{GS}}{V_P}\right]^2$$

Simplifying, we get $$I_{DS} = I_{DSS}\left[1-2\frac{V_{GS}}{V_P}+\left(\frac{V_{GS}}{V_P}\right)^2\right]$$

Neglecting the square term, we get

$$I_{DS} = I_{DSS} - \frac{2I_{DSS}V_{GS}}{V_P}$$

Putting $$g_{mo} = -\frac{2I_{DSS}}{V_P}, \text{ we get}$$

$$I_{DS} = I_{DSS} + g_{mo} V_{GS}$$ **Proved. Ans.**

EXAMPLE 8.7 **An N-channel JFET has I_{DSS} = 8 mA and V_P = – 5 V. Find the minimum value of V_{DS} for pinch-off region and the drain current I_{DS}, for V_{GS} = – 2 V in the pinch-off region.**

Solution: The minimum value of V_{DS} for pinch-off to occur for V_{GS} = – 2 V is

$$V_{DS(min)} = V_{GS} - V_P = -2 - (-5) = 3 \text{ V}$$

$$I_{DS} = I_{DSS}\left[1-\frac{V_{GS}}{V_P}\right]^2$$

Substituting values, we get

$$I_{DS} = 8 \times 10^{-3} [1 - (-2)/(-5)]^2$$

Solving, we get $$I_{DS} = 2.88 \text{ mA}$$ **Ans.**

EXAMPLE 8.8 **V_P = – 2 V, I_{DSS} = 1.55 mA for the circuit in figure 8.14. It is desired to bias the circuit at I_D = 0.8 mA, V_{DD} = 24 V. Find: (*i*) V_{GS} (*ii*) g_m.**

Solution: (*i*) We know that

$$V_{GS} = V_P\left[1-\sqrt{\frac{I_D}{I_{DSS}}}\right]$$

Substituting values we get

$$V_{GS} = -2\left[1-\sqrt{\frac{0.8\times10^{-3}}{1.65\times10^{-3}}}\right]$$

Simplifying, we get $$V_{GS} = -0.6074 \text{ V}$$

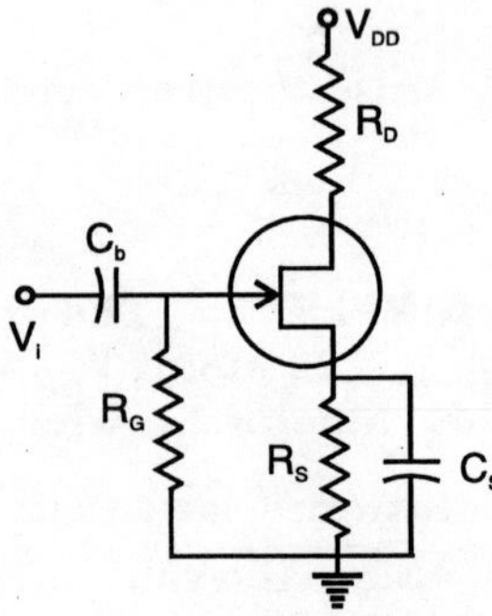

Fig. 8.14.

(*ii*) $$g_{mo} = -\frac{2I_{DSS}}{V_P}$$

or $$g_{mo} = -\frac{2\times1.65}{-2} = 1.65 \text{ ms} = g_{mo}\left(1-\frac{V_{GS}}{V_P}\right)$$

or $$g_m = 1.65\left(1-\frac{0.6074}{2}\right)\times10^{-3} = 1.65\times10^{-3}\times0.6963 = 1.15 \text{ ms}$$ **Ans.**

EXAMPLE 8.9 **The pinch-down voltage of a P-channel junction FET is V_P = 5 V and the drain-to-source saturation current I_{DSS} = – 40 mA. The value of drain-source voltage V_{DS} is such that the transistor is operating in the saturated region. The drain current is given as I_D = – 15 mA. Determine the gate-source voltage V_{GS}.**

Solution: It is found experimentally that a square-law characteristic closely approximate the drain current in saturation:

$$I_{D(\text{sat})} \approx I_{DSS}\left(1+\frac{V_{GS}}{V_P}\right)^2$$

Givan that

$$V_P = 5 \text{ V}$$
$$I_{DSS} = -40 \text{ mA}$$
$$I_D = -15 \text{ mA}$$

$$-15 \text{ mA} = -40 \text{ mA}\left(1+\frac{V_{GS}}{5}\right)^2$$

or

$$\sqrt{\frac{15}{40}} = 1+\frac{V_{GS}}{5}$$

$\therefore$

$$1+\frac{V_{GS}}{5} = 0.612$$

$$\frac{V_{GS}}{5} = 0.612 - 1 = -0.3876$$

$$V_{GS} = 1.938 \text{ volt.} \quad \textbf{Ans.}$$

EXAMPLE 8.10 Determine the value of transconductance of a FET, when the drain current changes from 1 mA to 1.9 mA with a change in gate-source voltage from –3.3 V to –3 V.

Solution: Change in drain current,

$$\Delta I_D = 1.9 - 1.0 = 0.9 \text{ mA}$$

Change in gate-source voltage,

$$\Delta V_{GS} = 3.3 - 3.0 = 0.3 \text{ V}$$

Now, transconductance is given by

$$g_m = \frac{\Delta I_D}{\Delta V_{GS}} = \frac{0.9 \text{ mA}}{0.3 \text{ V}} = 3 \text{ mA/V} = 3000 \ \mu \text{ mhos} \quad \textbf{Ans.}$$

EXAMPLE 8.11 The following readings were obtained experimentally from a FET:

V_{GS}	**– 0.1 V**	**– 0.1 V**	**– 0.4 V**
V_{DS}	**5 V**	**14 V**	**14 V**
I_D	**3 mA**	**3.3 mA**	**0.71 mA**

Obtain (*i*) a.c. drain resistance, (*ii*) transconductance and (*iii*) amplification factor.

Solution: (*i*) At constant V_{GS} (*i.e.*, at $V_{GS} = -0.1$ V), when V_{DS} is increased from 5 V to 14 V the drain current I_D increases from 3 mA to 3.3 mA.

Therefore, change in drain-source voltage, $\Delta V_{DS} = 14 - 5 = 9$ V

Thus, change in drain current, $\Delta I_D = 3.3 - 3 = 0.3$ mA

Hence, a.c. drain resistance will be

$$r_d = \frac{\Delta V_{DS}}{\Delta I_D} \text{ at constant } V_{GS}$$

$$r_d = \frac{9 \text{ V}}{0.3 \text{ mA}} = \textbf{30 K ohms}$$

(*ii*) At constant V_{DS} (*i.e.*, at $V_{DS} = 14$ V), when V_{GS} is changed from – 0.1 V to – 0.4 V, the drain current I_D changes from 3.3 mA to 0.71 mA.

The change in gate-source voltage will be

$$\Delta V_{GS} = 0.4 - 0.1 = 0.3 \text{ V}$$

The change in drain current will be

$$\Delta I_D = 3.3 - 0.71 = 2.59 \text{ mA}$$

Therefore, transconductance will be

$$g_m = \frac{\Delta I_D}{\Delta V_{GS}} \text{ at constant } V_{DS}$$

or
$$g_m = \frac{2.59 \text{ mA}}{0.3 \text{V}} = 8.63 \text{ mA/V} = 4000 \ \mu \text{ mhos}$$

(*iii*) Amplification factor is given by

$$\mu = r_d \times g_m = (30 \times 10^3) \times (8630 \times 10^{-6}) = 258.9 \quad \textbf{Ans.}$$

8.11. COMPARISON BETWEEN FET AND BJT

Table 8.4 shows some of the important points of comparison between the field effect transistor (FET) and bipolar junction transistor (BJT).

Table 8.4. Comparison Between Field-effect Transistor (FET) and Bipolar Junction Transistor (BJT) ***(Important)***

S.No.	Field-Effect Transistor (FET)	Bipolar junction transistor (BJT)
1.	It is an unipolar device *i.e.*, current in the device is carried either by electrons or holes.	It is bipolar device, *i.e.*, current in the device is carried by both electrons and holes.
2.	It is a voltage-controlled device, *i.e.*, voltage at the gate (or drain) terminal controls the amount of current flowing through the device.	It is a current-controlled device, *i.e.*, the base current controls the amount of collector current.
3.	Its input resistance is very high and is of the order of several megaohms.	Its input resistance is very low as compared to FET and is of the order of few kilo-ohms.
4.	It has a negative temperature coefficient at high current levels. It means that current decreases as the temperature increases. This characteristic prevents the FET from thermal breakdown.	It has a positive temperature coefficient at high current level. It means that collector current increases with the increase in temperature. This characteristic leads the BJT to thermal breakdown.
5.	It does not suffer from minority-carrier storage effects and therefore, has higher switching speeds and cut-off frequencies.	It suffers from minority carrier storage effects and therefore, has lower switching speed and cut-off frequencies than that of FET's.
6.	It is less noisy than a BJT or vacuum tube and is thus more suitable as an input amplifier for low-level signals. It is used extensively in high fidelity frequency modulated receivers.	It is comparatively more noisy than a field-effect transistor.
7.	It is much simpler to fabricate as an integrated circuit (IC) and occupies a less space on *IC* chip than that of BJT.	It is comparatively difficult to fabricate as an integrated circuit (IC) and occupies more space on IC chip than that of FET.

8.12. FET AS VOLTAGE VARIABLE RESISTANCE (VVR)

In almost all the linear applications of FET, it is operated in the constant current portion of its output characteristics (*i.e.* in the saturation region).

Now, let us observe the operation of JFET in the region just before pinch off. In this region, V_{DS} is small but I_D is large. This region is marked as ohmic region in the output characteristics of a JFET. In this region, the FET can be used as a voltage variable resistor (VVR) or voltage dependent resistor (VDR). The drain to source resistance r_{DS} in this region is dependent on the gate to source voltage. As V_{GS} becomes less negative, more drain current flows and r_{DS} decreases. For higher negative gate voltage, I_D decreases and r_{DS} will increase. Thus, the JFET acts as a voltage variable resistor (VVR).

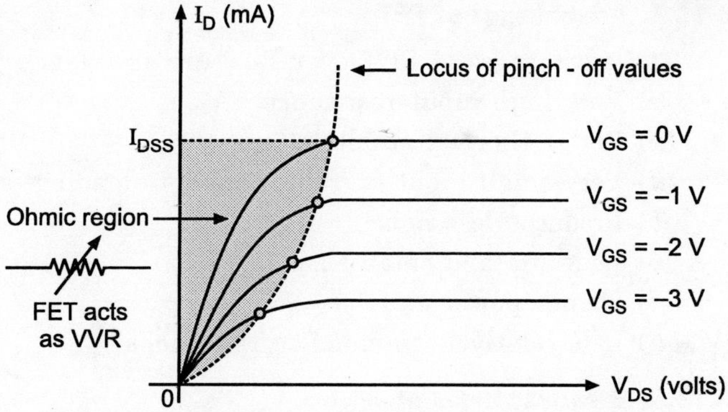

Fig. 8.15. Ohmic region in which FET acts as VVR

8.12.1. Application of VVR

(*i*) The FET used as VVR finds applications in many areas where this property can be used.

(*ii*) The VVR, for example, can be used to vary the voltage gain of a multistage amplifier A as the signal level is increased. This is called as **automatic gain control**.

(*iii*) A typical arrangement for controlling the gain of an amplifier automatically is shown in figure 8.16.

Working Operation

The gain of the common emitter amplifier Q_1 is controlled. The output signal v_0 is rectified and filtered and the dc voltage proportional to v_0 is applied to the gate of FET operating in the ohmic region. Depending on the dc gate voltage, the resistance r_{DS} of FET will change, this will change the emitter resistance of Q_1 since R_E is in parallel with r_{DS}.

The gain of CE amplifier is approximately equal to $-\dfrac{R_C}{R_E}$. Hence, change in R_E will change the gain automatically. If v_0 increases, then dc voltage at the rectifier filter output will increase. Hence, V_{GS} becomes more negative. This decreases I_D or increases r_{DS} of FET. Hence, the parallel combination ($R_E \,||\, r_{DS}$) will increase and the gain of Q_1 will increase. Exactly opposite action takes place to increase gain of Q_1 if v_0 decreases. This is how the automatic gain control is achieved.

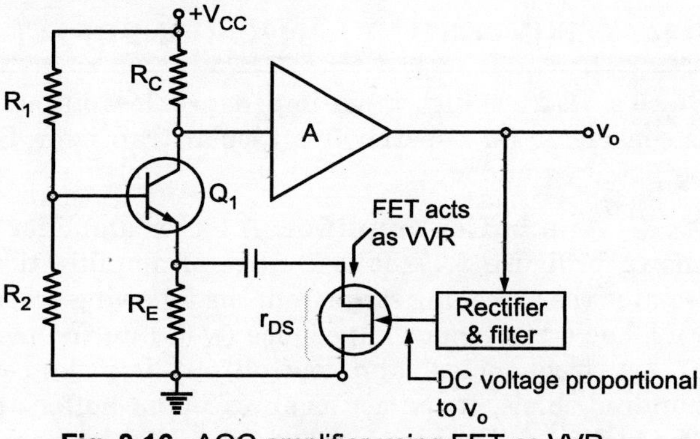

Fig. 8.16. AGC amplifier using FET as VVR.

8.13. SUMMARY OF VARIOUS FEATURES OF JFET

8.13.1. Peculiarities

(*i*) It is a voltage controlled device. (V_{GS} controls I_D).
(*ii*) It is equivalent to a voltage controlled current source.
(*iii*) The gate-source junction is always reverse biased.
(*iv*) Very small gate current (I_G = 0).
(*v*) High input resistance and input capacitance.
(*vi*) Can be used as switch (operated in ohmic region) or as an amplifier (operated in saturation region).
(*vii*) It can be used as voltage variable resistance VVR.

8.13.2. Advantages of FET

The advantages of FET over BJT are as under :

(*i*) Very high input resistance.

(*ii*) Better thermal stability.

(*iii*) Very small input current, hence, to loading on the input source.

(*iv*) Produces less noise.

(*v*) Long life and small size.

(*vi*) Higher power gain.

(*vii*) It is relatively immune to radiations.

8.13.3. Disadvantages of JFET

(*i*) It can not be used as a current amplifier.

(*ii*) The value of transconductance g_m is small.

(*iii*) The voltage gain of JFET amplifier is lower than that of the transistor amplifier.

(*iv*) Special handling precautions are required to be taken.

8.13.4. Applications of JFET

(*i*) JFET can be used as an amplifier.

(*ii*) JFET can be used as a switch.

(*iii*) It can be used as analog switch in circuits like sample and hold, amplitude modulation, ADC/DAC (analog to digital or digital to analog) converters.

(*iv*) As a voltage variable resistor (VVR)

(*v*) In digital circuits.

8.14. FEW IMPORTANT APPLICATIONS OF FET *(Expected)*

Since a FET has high input impedance, low output impedance and low noise level, therefore, FET is considered far superior to a bipolar transistor. Some of the important applications of FET can be listed as under:

(*i*) As a buffer amplifier: A buffer amplifier stage is shown in figure 8.17. It is a stage of amplification which isolates the preceding stage from the following stage. Since, FET has a high input impedance (well into the megaohms at low frequencies) and low output impedance (a few hundred ohms), it can act as an excellent buffer amplifier. The high input impedance means high load on the preceding stage. This permits almost the entire output from stage A to appear at the buffer input. The low output impedance means that the buffer (FET) can drive heavy loads. This ensures that the output from buffer reaches at the input of stage B.

DO YOU KNOW?

In general, JFETs are more temperature stable than bipolar transistors. Furthermore, JFETs are typically much smaller than bipolar transistors. This size difference makes them particularly suitable for use in ICs, where the size of each component is very critical.

(*ii*) As a low-noise amplifier: Noise is an unwanted disturbance superimposed upon the useful signal. The useful signal contains the required information. As the noise level increases the information becomes inferior.

For instance, the noise in radio receivers produces hissing; severe noise may completely mask the voice or music. Similarly, the noise in television receivers produces white spots (snow) on the picture; a heavy noise may wipe off the picture. Noise is independent of the signal strength because it exists even when the signal is off.

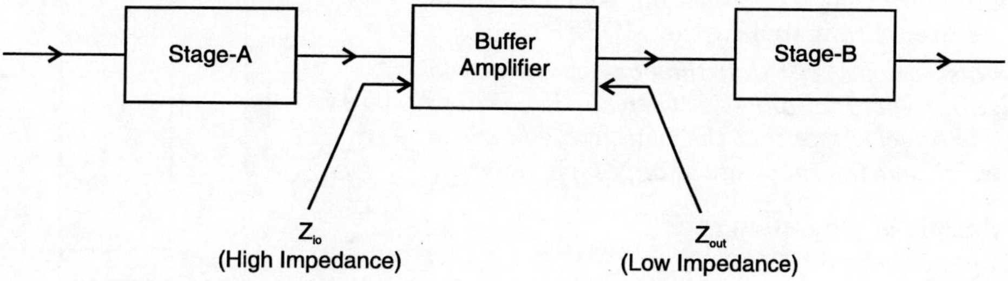

Fig. 8.17. The buffer amplifier stages.

Since FET is an outstanding low-noise device, therefore, it is preferred to apply it near the front end of the receivers and other electronic equipment. Hence, when we use FET amplifier at the front end, we get less amplified noise (disturbance) at the final output.

***(iii)* As phase-shift oscillator:** Although all the oscillators can also work with FETs, however, the high input impedance of FET is especially valuable in a phase-shift oscillator to minimise loading. The circuit diagram of a phase-shift oscillator using N-channel FET is shown in figure 8.18.

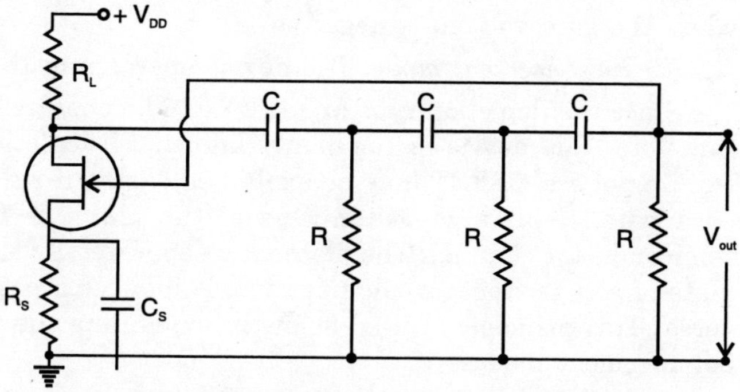

Fig. 8.18. Phase-shift oscillator using N-channel FET

8.15. MOSFETs

The MOSFET is an abbreviation for metal-oxide semiconductor field-effect transistor Like JFET, it has a source, gate and drain. However, unlike JFET, the gate of a MOSFET is insulated from the channel. Due to this, the MOSFET is sometimes called an IGFET which stands for insulated-gate field effect transistor.

Basically, the MOSFETs are of two types namely depletion type MOSFET and enhance-type MOSFET.

8.16. DEPLETION-TYPE MOSFETs*

Figure 8.19 illustrates the basic structure of an N-channel depletion type MOSFET. It simply consists of a conducting bar of N-type material with an insulated gate on the left and P-region on the right. Free electrons can flow from source to drain through the N-type material. The P-region is called substrate (or body). It physically reduces the conducting path a narrow channel. A thin layer of silicon dioxide S_iO_2 is deposited on the left side of the channel. This layer insulates the gate from the

DO YOU KNOW?

In many cases, bipolar transistors and MOSFETs are used in the same electronic circuit.

* Write notes on the following:
(*i*) depletion type MOSFET (*ii*) pinch-off voltage (*iii*) transfer characteristics of JFET
(*iv*) Transconductance

channel. Because of this, a negligible gate current flows even when this gate voltage is positive. It will be interesting to know that a PN junction, which exists in a JFET, has been eliminated in the MOSFET.

It may also be observed that the basic construction of a depletion-type P-channel MOSFET is similar to that of N-channel except that the conducting bar is of P-type material and the substrate is of N-type material.

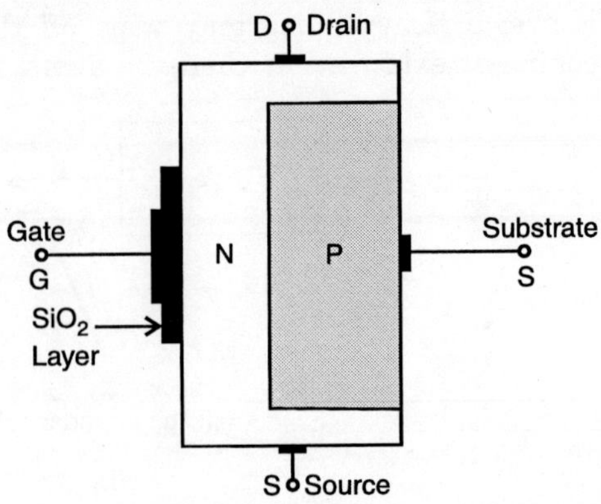

Fig. 8.19. Basic structure of an N-channel depletion-type MOSFET

8.16.1. Principle of Operation of a Depletion-Type MOSFET

The depletion-type MOSFET can be operated in two different modes as given below:

(*i*) *Depletion mode.* The device operates in this mode, when the gate voltage is negative.

(*ii*) *Enhancement mode.* The device operates in this mode, when the gate voltage is positive.

Since the depletion-type MOSFET can be operated in either depletion or enhancement mode, therefore, this device is commonly known as depletion-enhancement (DE) type MOSFET. The working of a MOSFET may be explained easily, if we visualize the entire structure of the device as a parallel plate capacitor. One of the plates is formed by the gate and the other by the semiconductor channel. The plates are separated by a dielectric (S_iO_2) layer. We know that if one plate of a capacitor is made negative, it induces a positive charge on the opposite plate and vice versa. This principle is used below in explaining the working of MOSFETs in the depletion and enhancement modes.

(*i*) Depletion mode: Figure 8.18(*a*) shows a MOSFET with a negative gate-to-source voltage. The negative voltage, on the gate, induces a positive charge in the channel. Because of this, free electrons in the vicinity of positive charge are repelled away in the channel. As a result of this, the channel is depleted of free electrons. This reduces the number of free electrons (which constitute the drain current) passing through the channel. Thus, as the value of negative gate-to-source voltage is increased, the value of drain voltage, called $V_{GS(\text{off})}$, the channel is totally depleted of free electrons and, therefore, the drain current reduces to zero. Thus, with the negative gate voltage, the operation of MOSFET is similar to that of a JFET.

> **DO YOU KNOW?**
>
> Like a JFET, a depletion-mode MOSFET is considered a normally on device. This is because both devices have drain current when $V_{GS} = 0$ V. Recall that for a JFET, I_{DSS} is the maximum possible drain current. With a depletion-mode MOSFET, the drain current can exceed I_{DSS} if the gate voltage is of the correct polarity to increase the number of charge carriers in the channel. For an n-channel D-MOSFET, I_D is greater than I_{DSS} when V_{GS} is positive.

It is evident from the above discussion that negative gate voltage depletes the channel of free electrons. It is due to this fact that the working of a MOSFET, with a negative gate voltage, is called depletion mode.

(*ii*) Enhancement mode: Figure 8.20(*b*) shows a MOSFET with a positive gate-to-source voltage. The positive gate voltage increases the number of free electrons passing through the channel. The greater the gate voltage, greater is the number of free electrons passing through the channel. This increases *i.e.*, enhances the conducting of the channel. Because of this fact, positive gate operation is called **enhancement mode**.

Man is not divine. His divine is something to be built up by good thoughts, good words and good deads.
— **Swami Dayanand**

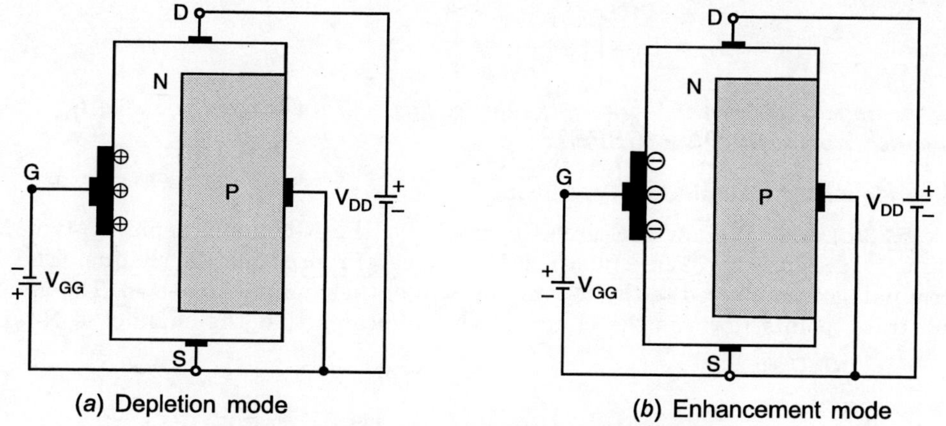

Fig. 8.20. Working of a depletion-type MOSFET

*It will be interesting to know that depletion-type MOSFET can conduct even if the gate-to-source voltage (V_{GS}) is zero. Because of this, it is commonly known as **Normally-ON MOSFET**.*

8.16.2. Drain Characteristic of Depletion-Type MOSFET

Figure 8.20 shows the drain characteristic for the N-channel depletion-type MOSFET in the common source configuration. These curves are plotted for both negative and positive values of gate-to-source voltage (V_{GS}). The curves shown above the curve for $V_{GS} = 0$ have a positive zero whereas those below it have a negative value of V_{GS}. When V_{GS} is zero and negative, the MOSFET operates in the depletion-mode. On the other hand, if V_{GS} is zero and positive, the MOSFET operates in the enhancement-mode.

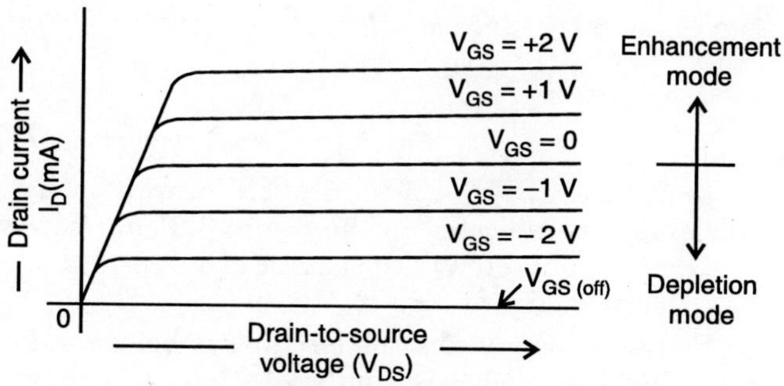

Fig. 8.21. Drain characteristic of N-channel depletion-type MOSFET

It may be noted that the drain characteristics of depletion-type MOSFETs are similar to that of JFET. The only difference is that JFET does not operate for positive values of gate-to-source voltage (V_{GS}).

8.16.3. Transfer Characteristic of Depletion-Type MOSFET

Figure 8.22 shows the transfer characteristic (also called transconductance curve) for an N-channel depletion-type MOSFET. It may be noted from this curve that the region AB of the characteristic is similar to that of JFET. But here, this curve extends for the positive values of gate-to-source voltage (V_{GS}) also. The value I_{DSS} represents the current from drain-to-source with V_{GS} = 0. The drain current at any point along the transfer characteristic (*i.e.*, the curve *ABC*) is given by the relation,

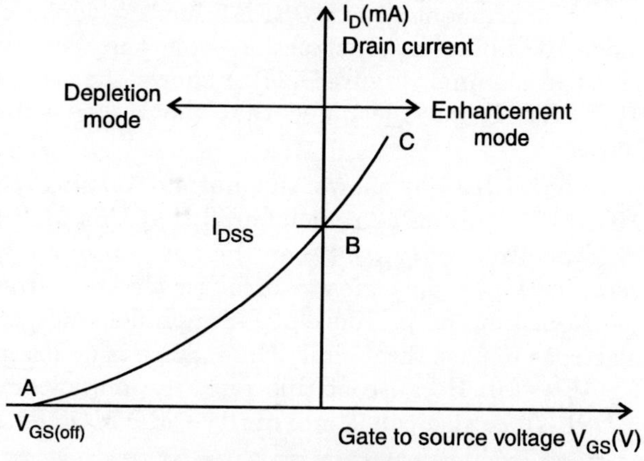

Fig. 8.22. Transfer characteristics of a N-channel depletion-type MOSFET

$$I_D = I_{DSS}\left[1 - \frac{V_{GS}}{V_{GS(off)}}\right]^2.$$

*It may be noted that even if V_{GS} = 0, the device has a drain current equal to I_{DSS}. Due to this fact, it is called **normally-ON MOSFET.***

8.16.4. Circuit Symbol for Depletion-Type MOSFET

Figure 8.23(*a*) and (*b*) show the circuit symbols for the N-channel depletion-type MOSFET. In these figures, a thin vertical line (just right to the gate) represents the channel. The drain and source terminals are connected to the top and bottom of the channel as shown. The arrow, on the P-type substrate, points towards the channel. This indicates that the channel is N-type.

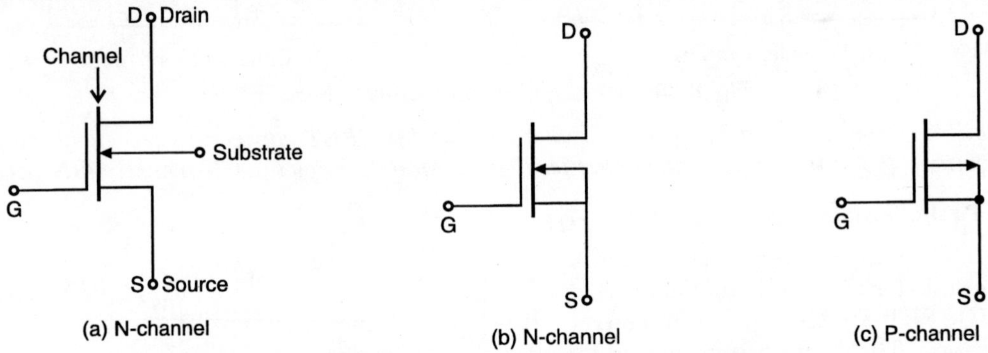

Fig. 8.23. Circuit symbols for depletion type MOSFETs

In some MOSFETs, a connection from the substrate is also taken out. Such MOSFETs have 4-terminals as indicated in the figure 8.21(*a*). However, in most of the MOSFETs, the substrate is internally connected to the source. This results in a three terminal device, whose circuit symbol is as shown in figure 8.23(*b*).

Figure 8.23(*c*) shows the circuit symbol for a P-channel depletion type MOSFET. It may be noted that the symbol is similar to that of N-channel, except the direction of the arrow on the substrate. Its direction is away from the channel, which indicates that the channel is of P-type material.

8.17. ENHANCEMENT-TYPE MOSFET*

The enhancement-type MOSFET has no depletion mode and it operates only in enhancement mode. It differs in construction from the depletion-type MOSFET in the sense that it has no physical channel. Figure 8.24(*a*) shows the basic structure of the N-channel enhancement-type MOSFET. It may be noted that the P-type substrate extends the silicon dioxide (S_iO_2) layer completely.

Figure 8.24(*b*) shows the normal biasing polarities for the N-channel enhancement-type MOSFET. It must be remembered that this MOSFET is always operated with the positive gate-to-source voltage (V_{GS}). When the gate-to-source voltage is zero, the V_{DD} supply tries to force free electrons from source-to-drain. But the presence of P-region does not permit the electrons to pass through it. Thus, there is no drain current for V_{GS} = 0. Because of this fact, the enhancement type MOSFET is also called **normally-OFF MOSFET**.

> **DO YOU KNOW?**
>
> With the E-MOSFET, V_{GS} has to be greater than $V_{GS(th)}$ to get any drain current at all.

* What is the significant difference between the construction of an enhancement type MOSFET and depletion type MOSFET ? Explain with suitable diagram ?

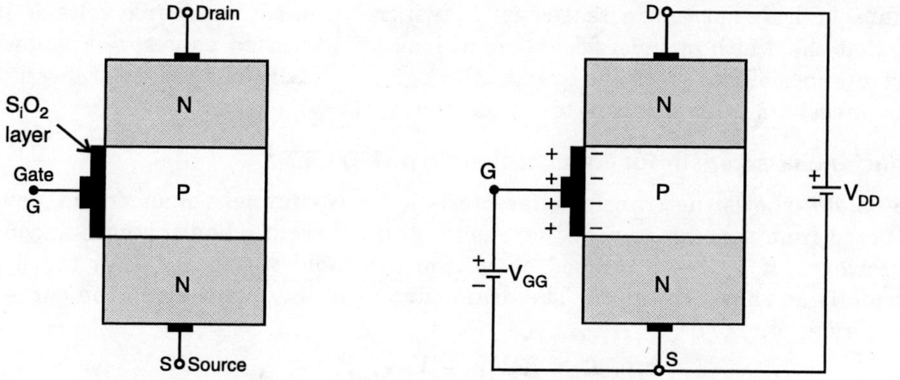

(a) Basic construction, (b) Formation of invertion layer.

Fig. 8.24. Enhancement type MOSFET

Now, if some positive voltage is applied at the gate, it induces a negative charge in the P-type substrate just adjacent to the silicon dioxide (S_iO_2) layer. The induced negative charge is produced by attracting the free electrons from the source. When the gate is positive enough, it can attract a number of free electrons. This forms a thin layer of electrons, which stretches from source to drain. This effect is equivalent to producing a thin layer of N-type channel in the P-type substrate. This layer of free electrons is called **N-type inversion layer.**

The minimum gate-to-source voltage (V_{GS}), which produces inversion layer, is called **threshold voltage** and is designated by the symbol $V_{GS(th)}$. When the voltage V_{GS} is less than $V_{GS(th)}$, no current flows from drain to source. However, when the voltage V_{GS} is greater than $V_{GS(th)}$, the inversion layer connects the drain and source and we get a significant value of current.

8.17.1. Drain Characteristics for Enhancement-Type MOSFET

Figure 8.25(*a*) shows the drain characteristics for N-channel enhancement-type MOSFET. It may be noted from this figure, that the gate-to-source voltage (V_{GS}) is less than threshold voltage, $V_{GS(th)}$, there is no drain current. However, in actual practice, an extremely small value of drain current does flow through the MOSFET. This current flow is due to the pressure of thermally generated electrons in the P-type substrate. When the value of V_{GS} is kept above $V_{GS(th)}$, a significant amount of drain current flows.

DO YOU KNOW?

The depletion-mode MOSFET was part of the evolution toward the enhancement-mode MOSFET, abbreviated E-MOSFET. Without the E-MOSFET, the personal computers that are now so widespread would not exist.

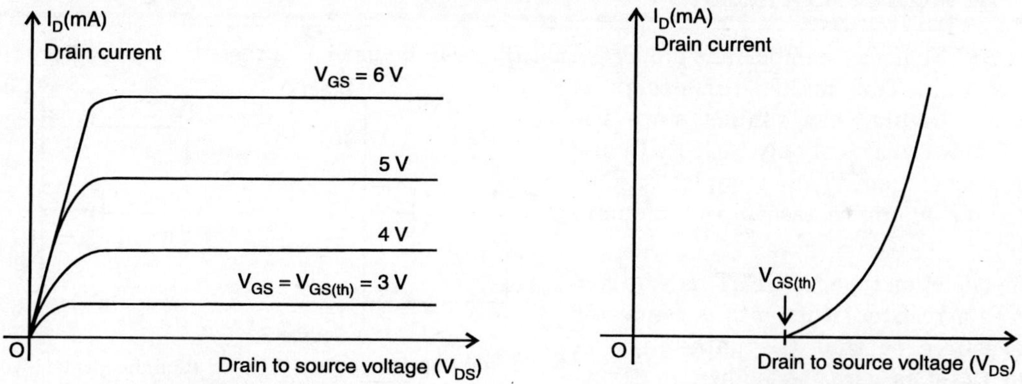

(a) Drain characteristic (b) Transfer characteristic

Fig. 8.25. Characteristics for N-channel enhancement-type MOSFET

The values of drain current increases with the increase in gate-to-source voltage. It is because of the fact that the width of inversion layer widens for increased values of V_{GS} and, therefore, allows more number of free electrons to pass through it. The drain current reaches its saturation value above a certain value of drain-to-source voltage (V_{DS}).

8.17.2. Transfer Characteristic for Enhancement-Type MOSFET

Figure 8.25(*b*) shows the transfer characteristic for N-channel enhancement type MOSFET. It may be noted from this figure that there is no drain current when the gate-to-source voltage, V_{GS} = 0. However, if V_{GS} is increased above the threshold voltage, $V_{GS(th)}$, the drain current increases rapidly as shown in figure. The drain current at any point along the curve is given by the following relation:

$$I_D = K[V_{GS} - V_{GS(th)}]^2$$

where K is a constant, whose value depends on the type of MOSFET. Its value can be determined from the data sheet by taking specified value of drain current called $I_{D(ON)}$ at the given value of V_{GS} and then substituting these values in the above equation. Incidently, it may be noted that enhancement-type MOSFET does not have an I_{DSS} parameter like JFET and depletion-type MOSFET.

8.17.3. Circuit Symbol for Enhancement-Type MOSFET

Figure 8.26(*a*) shows the circuit symbol for N-channel enhancement-type MOSFET. In this figure, the broken line indicates that there is no conducting channel between drain and source, when V_{GS} = 0. Because of this fact, this device is also known as "**Normally-OFF MOSFET**". The drain and source terminals are shown at the top and bottom end of the broken line. The substrate is internally connected to the source as shown. The arrow points in the direction of channel (or inversion layer), which is created when V_{GS} is increased above $V_{GS(th)}$.

Figure 8.26(*b*) shows the circuit symbol for the P-channel enhancement-type MOSFET. This symbol is similar to N-channel except that the arrow points outwards. This indicates that the channel (or inversion layer) created in the substrate is of P-type. In actual practice, the N-channel and P-channel enhancement-type MOSFETs are known as **NMOS and PMOS field-effect transistors respectively.**

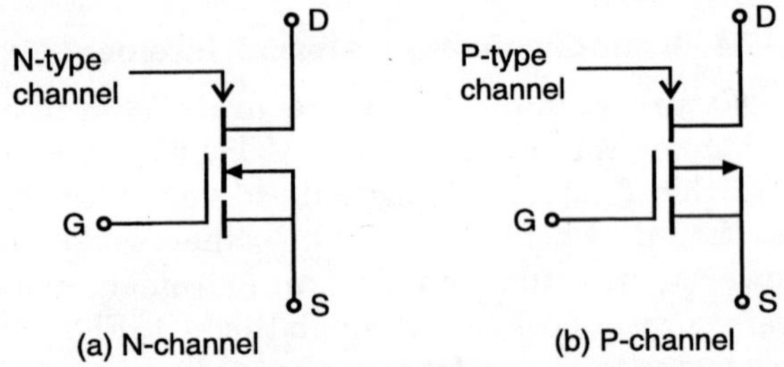

Fig. 8.26. Circuit symbols for enhancement-type MOSFET

8.18. THE MOSFET AS A RESISTOR

The MOSFETs have an important property that they can be used as a resistor capacitor, amplifier and a switch. This makes the design of electronic circuits very simple, since the entire circuit consists of only MOSFETs and no other component. The examples of such circuits are microprocessors and memory circuits.

The depletion type MOSFET may be connected as a resistor, if its gate is connected to the source so that the gate-to-source voltage (V_{GS}) is zero, as shown in figure 8.27(*a*).

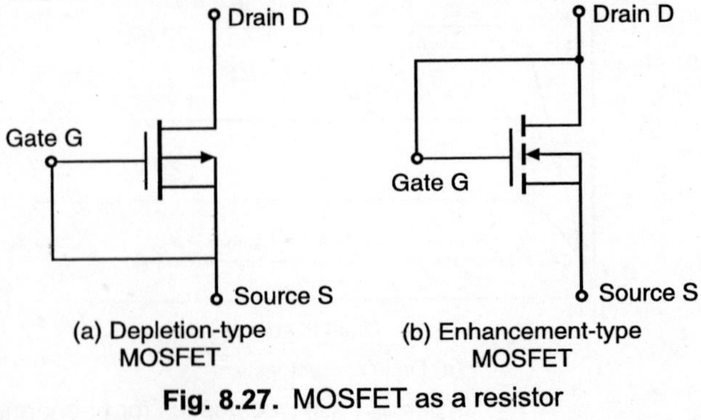

Fig. 8.27. MOSFET as a resistor

The enhancement-type MOSFET may be connected as a resistor, if its gate is connected to drain so that $V_{GS} = V_{DS}$ as shown in figure 8.27(*b*).

8.19. ADVANTAGES OF N-CHANNEL MOSFETs OVER P-CHANNEL

These days N-channel MOSFETs have become much popular than P-channel for fabricating MOS circuits and systems. The P-channel MOSFETs find their use only in complementary metal-oxide semiconductor (CMOS) integrated circuits. The main advantage of N-channel MOSFET, over P-channel, is due to the fact that charge carriers in N-channel devices are the electrons, which have a mobility of about 1300 cm^2/V.S. On the other hand, the charge carriers in P-channel devices are the holes which have a mobility of about 500 cm^2/V.S. Since the current, in a semiconductor, is directly proportional to the mobility, therefore, the current in the N-channel MOSFET is more than two times that of P-channel for the same dimension.

The ON resistance of the N-channel MOSFET is one-third of that for the P-channel. It means that in order to achieve the same value of current and ON resistance, the P-channel MOSFET requires three times the area of an equivalent N-channel MOSFET. Thus, electronic currents using N-channel MOSFETs are much smaller in size than those of P-channel. This results in higher packing density.

8.20. COMPARISON OF JFET AND MOSFET

Table 8.5. Comparison of JFET and MOSFET

S.No.	JFET	MOSFET
1.	JFET are of two types, P-channel and N-channel JFETs.	MOSFETs can be of depletion type or enhancement type.
2.	The symbols of JFETs are as under : 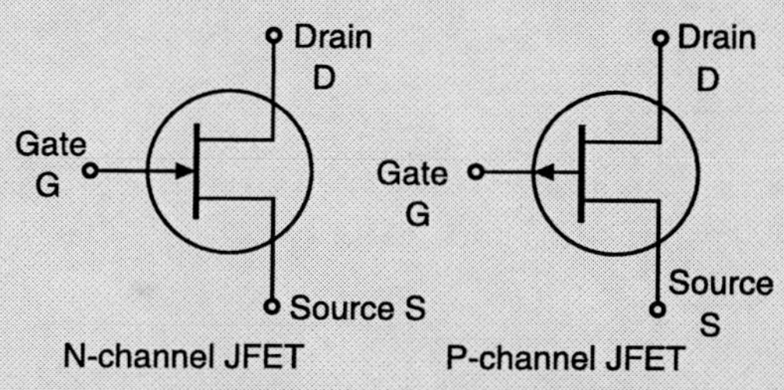	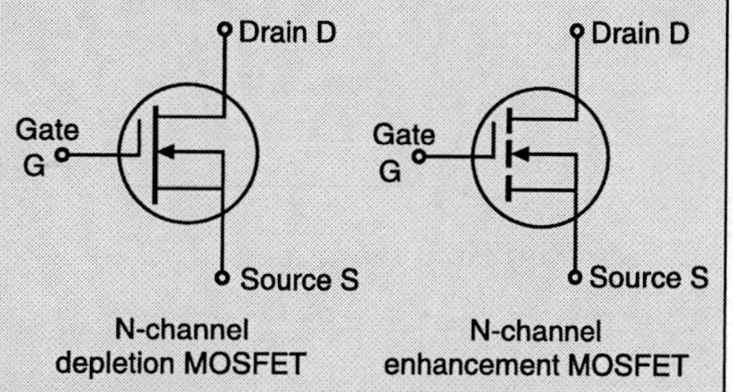
3.	JFETs do not have the insulated gate.	MOSFETs have the insulated gate structure.
4.	Input impedance is lower than that of the MOSFETs.	Input impedance is higher than that of JFET due to the insulated gate structure.
5.	Drain resistance is lower than that of a MOSFET.	Drain resistance is higher than that of a JFET.
6.	The input resistance r_{gs} and the feedback resistance r_{gd} are very much smaller than that of MOSFET.	r_{gs} and r_{gd} are much larger than that of a JFET.
7.	Biasing methods used are self bias, fixed bias and voltage divider bias.	Biasing circuits of deplection MOSFET are same as those used for JFET but biasing circuits for enhancement MOSFET are, feedback biasing and voltage divider biasing.

EXAMPLE 8.12 A certain JFET has a g_m of 4 mS. With an external drain resistance of 1.5 kΩ, determine the value of ideal voltage gain.

Solution: Given that $g_m = 4 \text{ mS} = 4 \times 10^{-3}$ S

and $R_D = 1.5 \text{ k}\Omega = 1.5 \times 10^3\ \Omega$

We know that the voltage gain is given by

$$A_v = -g_m \cdot R_D = -(4 \times 10^{-3}) \times (1.5 \times 10^3) = -6 \quad \textbf{Ans.}$$

EXAMPLE 8.13 A JFET amplifier has g_m = 2.5 mA/V and r_d = 500 kΩ. The load resistance is 10 kΩ. Find the value of voltage gain.

Solution: Given that $g_m = 2.5 \text{ mA/V} = 2.5 \times 10^{-3}$ A/V;

$r_d = 500$ kΩ

and $R_D = 10$ kΩ.

We know that the a.c. equivalent resistance is given by

$$r_L = \frac{R_D \times r_d}{R_D + r_d} = \frac{10 \times 500}{10 + 500} \text{ k}\Omega = 9.8 \text{ k}\Omega = 9.8 \times 10^3\ \Omega$$

Now, voltage gain, $A_v = -g_m \cdot r_L = -(2.5 \times 10^{-3}) \times (9.8 \times 10^3) = -24.5$ **Ans.**

EXAMPLE 8.14 A FET follows the following relation :

$$I_D = I_{DSS}\left[1 - \frac{V_{GS}}{V_P}\right]^2$$

I_{DSS} = 8.4 mA, V_P = – 3 V. What is the value of I_D for V_{GS} = – 1.5 V? Find g_m at this point.

Solution: Drain-source saturation current,

$I_{DSS} = 8.4$ mA

Pinch-off voltage, $V_P = -3$ V

Gate-source voltage, $V_{GS} = -1.5$ V

We know that the drain current is given by

$$I_D = I_{DSS}\left[1 - \frac{V_{GS}}{V_P}\right]^2$$

Substituting all the values, $I_D = 8.4\left(1 - \frac{-1.5}{-3}\right)^2 = 2.1$ mA **Ans.**

Transconductance for $V_{GS} = 0$ is given by

$$g_{mo} = \frac{-2I_{DSS}}{V_P} = \frac{-2 \times 8.4}{-3} = 5.6 \text{ mA/V or } 5.6 \text{ mS}$$

Transconductance, $g_m = g_{mo}\left(1 - \frac{V_{GS}}{V_P}\right) = 5.6\left(1 - \frac{-1.5}{-3}\right) = 2.8$ mS **Ans.**

EXMAPLE 8.15 An *N*-channel JFET has a pinch off voltage of – 4.5 V and I_{DSS} = 9 mA. At what value of V_{GS} will I_{DS} be equal to 3 mA? What is its g_m at this I_{DS}?

Solution: Pinch-off voltage,

$V_P = -4.5$ V

Drain-source saturation current,

$I_{DSS} = 9 \text{ mA} = 9 \times 10^{-3}$ A

Drain-source current, $I_{DS} = 3 \text{ mA} = 3 \times 10^{-3}$ A

From Shockley's equation (equation (8.1))

$$V_{GS} = V_P\left[1-\sqrt{\frac{I_{DS}}{I_{DSS}}}\right] = -4.5 \times \left[1-\sqrt{\frac{3\times 10^{-3}}{9\times 10^{-3}}}\right] = 1.902 \text{ V} \qquad \textbf{Ans.}$$

Transconductance g_m for $I_{DS} = 3$ mA for which $V_{GS} = -1.902$ V

$$g_m = \frac{-2I_{DSS}}{V_P}\left[1-\frac{V_{GS}}{V_P}\right] = \frac{-2\times 9\times 10^{-3}}{-4.5}\left(1-\frac{-1.902}{-4.5}\right)$$

$$g_m = 2.31 \text{ mA/V} = 2.31 \text{ mS} \qquad \textbf{Ans.}$$

EXAMPLE 8.16 **When drain-source voltage is changed by1.5 volts, the change in drain current is of 120 μA, the gate-source voltage remaining unchanged. Determine the a.c. drain resistance of the JFET.**

Solution: Change in drain-source voltage,

$$\Delta V_{DS} = 1.5 \text{ V}$$

Change in drain-current, $\Delta I_o = 120 \times 10^{-6}$ A

AC drain resistance of the JFET,

$$r_d = \frac{\Delta V_{DS}}{\Delta I_D} = \frac{1.5}{120\times 10^{-6}} = 12.5 \text{ k}\Omega \qquad \textbf{Ans.}$$

EXAMPLE 8.17 **The readings obtained experimentally from a JFET are as follows:**

Drain-source voltage, V_{DS} in volts	**5**	**12**	**12**
Gate-source voltage V_{GS} in volts	**0**	**0**	**– 0.25**
Drain current, I_D in mA	**8**	**8.2**	**7.5**

Determine (*i*) a.c. drain resistance (*ii*) transconductance and (*iii*) amplification factor.

Solution: (*i*) The drain current increases from 8.0 mA to 8.2 mA when drain-source voltage is increased from 5 V to 12 V, keeping gate-source voltage constant at 0.V.

$$\therefore \text{ A.C. drain resistance, } r_d = \frac{\Delta V_{DS}}{\Delta I_D} = \frac{(12-5)V}{(8.2-8)\text{mA}}$$

or

$$r_d = \frac{7}{0.2\times 10^{-3}} = 35 \text{ k}\Omega \qquad \textbf{Ans.}$$

(*ii*) The drain current changes from 8.2 mA to 7.5 mA when gate-source voltage is changed from 0 to – 0.25 V, keeping drain-source voltage constant at 12 V.

$$\therefore \text{ Transconductance, } g_m = \frac{\Delta I_D}{\Delta V_{GS}} = \frac{(7.5-8.2)\text{mA}}{(-0.25-0)\text{V}}$$

$$g_m = 2.8 \text{ mA/V} = 2.8 \text{ mS} \qquad \textbf{Ans.}$$

(*iii*) Amplification factor, $\mu = r_d \times g_m = 35 \times 1{,}000 \times 2.8 \times 10^{-3} = 98$ **Ans.**

EXAMPLE 8.18 **A JFET has $V_P = -4.5$ V, $I_{DSS} = 10$ mA and $I_{DS} = 2.5$ mA. Determine the transconductance.**

Solution: Drain-source saturation current,

$$I_{DSS} = 10 \text{ mA}$$

Pinch-off voltage, $V_P = -4.5$ V

Drain-source current, $I_{DS} = 2.5$ mA

From Shockley's equation, drain-source current,

$$I_{DS} = I_{DSS}\left[1-\frac{V_{GS}}{V_P}\right]^2$$

or $$V_{GS} = V_P\left[1-\sqrt{\frac{I_{DS}}{I_{DSS}}}\right] = -4.5\left[1-\sqrt{\frac{2.5}{10}}\right] = -2.25 \text{ V}$$

From equation (8.8), transconductance,

$$g_m = \frac{-2I_{DSS}}{V_P}\left[1-\frac{V_{GS}}{V_P}\right] = \frac{-2\times 10\times 10^{-3}}{-4.5}\left[1-\frac{-2.25}{-4.5}\right]$$

$$= 2.22 \text{ m A/V} \quad \textbf{Ans.}$$

EXAMPLE 8.19 A JFET has g_m = 10 mS, I_{DSS} = 10 mA. Calculate $V_{GS(OFF)}$

Solution: Transconductance, $g_m = 10 \text{ mS} = 10\times 10^{-3}$ S

Drain-source saturation current,

$$I_{DSS} = 10 \text{ mA} = 10\times 10^{-6} \text{ A}$$

$$V_{GS(OFF)} \rightarrow V_{GS} = V_P$$

$$g_m = g_{mo} = \frac{-2I_{DSS}}{V_P} = \frac{-2I_{DSS}}{V_{GS(OFF)}}$$

or $$10\times 10^{-3} = \frac{-2\times 10\times 10^{-6}}{V_{GS(OFF)}}$$

or $$V_{GS(OFF)} = \frac{-2\times 10\times 10^{-6}}{10\times 10^{-3}} = 2\text{mV} \quad \textbf{Ans.}$$

EXAMPLE 8.20 An N-channel JFET has I_{DSS} = 10 mA and V_P = – 4 V. Determine the minimum value of V_{DS} for pinch-off region and drain current I_D for V_{GS} = – 2 V in pinch-off region.

Solution: Pinch-off voltage, $V_P = -4$ V

Gate-source voltage, $V_{GS} = -2$ V

Drain-source saturation current,

$$I_{DSS} = 10 \text{ mA} = 10\times 10^{-3} \text{ A}$$

Drain current, $$I_D = I_{DSS}\left[1-\frac{V_{GS}}{V_P}\right]^2 = 10\times 10^{-3}\left[1-\frac{-2}{-4}\right]^2 = 2.5 \text{ mA} \quad \textbf{Ans.}$$

The minimum value of V_{DS} for pinch-off region is equal to V_P. Thus the minimum value of V_{DS}

$$V_{DS(\min)} = V_P = -4 \text{ V} \quad \textbf{Ans.}$$

EXAMLE 8.21 For an N-channel JFET

$$\boldsymbol{I_{DSS} = 8.7 \text{ mA}}$$

$$\boldsymbol{V_P = -3 \text{ V}}$$

$$\boldsymbol{V_{GS} = -1 \text{ V.}}$$

Find the values of (*i*) I_D (*ii*) g_{mo} (*iii*) g_m

Solution: Drain-source saturation current,

$$I_{DSS} = 8.7 \text{ mA} = 0.0087 \text{ A}$$

Pinch-off voltage, $V_{GS} = -1$ V

(*i*) Drain current, $$I_D = I_{DSS}\left(1-\frac{V_{GS}}{V_P}\right)^2 = 0.0087\left(1-\frac{-1}{-3}\right)^2 = 3.8667 \text{ mA} \quad \textbf{Ans.}$$

(*ii*) Transconductance for $V_{GS} = 0$, is given by

$$g_{mo} = \frac{-2I_{DSS}}{V_P} = \frac{-2 \times 8.7 \text{ mA}}{-3}$$

or $$g_{mo} = 5.8 \text{ mA/V} = 5.8 \text{ mS} \quad \textbf{Ans.}$$

(*iii*) Now, transconductance, $g_m = g_{mo}\left(1 - \frac{V_{GS}}{V_P}\right)$

$$g_m = 5.8\left(1 - \frac{-1}{-3}\right) = 3.867 \text{ mA/V} = 3.867 \text{ mS} \quad \textbf{Ans.}$$

EXAMPLE 8.22 **The maximum transconductance of a certain N-channel JFET is 3.4 ms. If I_{DSS} is 20 mA, what is the pinch off voltage ?**

Solution: The maximum transconductance is given by

$$g_{mo} = \frac{-2I_{DSS}}{V_P} \quad \text{or} \quad V_P = \frac{-2I_{DSS}}{g_{mo}}$$

Substituting the values, we get

$$V_P = \frac{-2 \times 20 \times 10^{-3}}{9.4 \times 10^{-3}}$$

or Pinch off voltage $V_P = -4.255$ Volts **Ans.**

EXAMPLE 8.23 **The maximum transconductance of a certain N-channel JFET is 3.8×10^{-3} *S*. If $I_{DSS} = 18$ mA. What would be the pinch off voltage?**

Solution: We know that

Maximum transconductance,

$$g_{mo} = \left|\frac{2I_{DSS}}{V_P}\right|$$

or $$V_P = \frac{2I_{DSS}}{g_{mo}} = \frac{2 \times 18 \text{mA}}{9.8 \times 10^{-3}} = 3.6 \text{ Volts} \quad \textbf{Ans.}$$

SUMMARY

- The biasing deals with setting a fixed level of current which should flow through the FET with a desired fixed voltage drop across the FET junctions.
- The commonly used biasing circuits for FET are as under:
  (*i*) Fixed-bias circuit.
  (*ii*) Self-bias circuit.
  (*iii*) Voltage divider bias circuit.
- The *Q*-point (*i.e.*, quiescent or operating point or bias point) for a self-biased JFET is established by determining the value of drain current (I_D) for a desired value of gate-to-source voltage (V_{GS}) or vice-versa.
- In analytical method, we note down the values of the maximum drain current (I_{DSS}) and the gate-to-source cut-off voltage [$V_{GS(off)}$] from the data sheets of JFET supplied by the manufacturer.
- In graphical method, we draw a self-bias line such that it intersects the transfer characteristic curve near its mid-point.
- The *Q*-point for the self-biased JFET circuit can also be obtained graphically by using d.c load line.
- However, it has been observed that the JFET parameters like maximum drain current (I_{DSS}) and the gate-to-source cut-off [$V_{GS(off)}$] vary with the temperature.
- In order to provide compensation for the device parameter, variation the manufacturers provide two transistor characteristic curves.

- The first current corresponds to the maximum values of I_{DSS} and $V_{GS\ (off)}$.
- The self bias circuit is not suitable since it cannot provide the positive V_{GS} required for the operation of an N-channel depletion type MOSFET in the enhancement region.
- The transfer characteristics of the enhancement type MOSFET are completely different from that of a JFET.
- The other most popular biasing arrangement for the enhancement type MOSFET is voltage divider biasing circuit.
- The field-effect transistor (FET), is a voltage-controlled device.
- The output characteristics of FET are controlled by the applied input voltage and not by the input current.
- FET is an unipolar device.
- The current flowing through it is only due to one type of charge particles *i.e.*, holes or electrons.
- In a broad sense, there are two main types of field-effect transistors as under:
  (*i*) Junction field-effect transistors (JFET).
  (*ii*) Metal oxide semiconductor field-effect transistor (MOSFET) or insulated gate field-effect transistor (IGFET).
- The junction field-effect transistor (JFET) may be divided, depending on their structure, into the following two categories:
  (*i*) N-channel JFET.
  (*ii*) P-channel JFET.
- The space between the junction (*i.e.*, N-type regions) is known as channel.
- Whenever a voltage is applied across the drain and source terminals a current flows through the N-channel.
- The current carriers in JFET are the holes which flow through the P-type channel.
- This PN-junction is always reverse-bias in JFET operation.
- The reverse bias is applied by a voltage V_{GG} connected between the gate and source terminals.
- Whenever a *PN*-junction is reverse-biased, the electrons and holes diffuse across the junction and leave behind the positive ions on the N-side and negative ions on the *P*-side. The region containing these immobile ions is called depletion region.
- The reverse-bias across the gate-source junction of a JFET can also be achieved simply by applying a voltage across the drain and source terminals.
- In the presence of positive supply voltage V_{DD} (with gate open) the electrons flow from source to drain through the N-channel and constitutes a current known as drain current (I_D).
- The gate-to-source voltage (V_{GS}) at which the drain current is zero (or completely cut-off) is known as pinch off voltage.
- The value of pinch-off voltage V_P is negative for N-channel JFET. It depends upon
  (*i*) doping of the *N* and *P* regions of the device,
  (*ii*) width of the original channel structure.

SHORT QUESTIONS WITH ANSWERS

Q. 1. What is junction field-effect transistor (JFET)? Explain.

Ans. The junction field-effect transistor (JFET) may be divided, depending on their structure, into the following two categories:

(*i*) N-channel JFET.
(*ii*) P-channel JFET.

Q. 2. What is depletion region in JFET? Expain.

Ans. Whenever a *PN*-junction is reverse-biased, the electrons and holes diffuse across the junction and leave behind the positive ions on the *N*-side and negative ions on the *P*-side. The region, containing these immobile ions, is called depletion region. As the reverse bias voltage across the junction is increased, the thickness of depletion region is also increased. If both the *P*-side and *N*-side of the junction are equally doped, the depletion region will extend equally in both the regions.

Q. 3. What is pinch-off region? Explain.

Ans. This region has been shown by the curve *BC*. It is also known as the saturation region or constant current region. In this region, the drain current remains constant at its maximum value (*i.e.*, I_{DSS}). The drain current in the pinch off region depends on the gate-to-source voltage and is given by the following relation:

$$I_D = I_{DSS}\left(1 - \frac{V_{GS}}{V_P}\right)^2$$

This relation is known as Shockley's equation. The pinch-off region is the normal operating region of JFET, when used as an amplifier.

Q. 4. Define breakdown region for JFET.

Ans. This region has been shown by the curve *CD*. In this region the drain current increases rapidly as the drain-to-source voltage is also increased. It happens because of the breakdown of gate-to-source junction due to the avalanche effect. The drain-to-source voltage corresponding to point *C* is known as the breakdown voltage.

Q. 5. How MOSFET works as a resistor?

Ans. The MOSFETs have an important property that they can be used as a resistor capacitor, amplifier and a switch. This makes the design of electronic circuits very simple, since the entire circuit consists of only MOSFETs and no other component. The examples of such circuits are microprocessors and memory circuits.

Q. 6. What is FET biasing? Explain.

Ans. We know the meaning and the need of biasing a device. We have already studied different biasing circuits for a bipolar junction transistor (BJT) in previous chapters. As a matter of fact, the biasing deals with setting a fixed level of current which should flow through the FET with a desired fixed voltage drop across the FET junctions.

Q. 7. What are the FET biasing circuits?

Ans. The commonly used biasing circuits for FET are as under:

(*i*) Fixed-bias circuit.
(*ii*) Self-bias circuit.
(*iii*) Voltage divider bias circuit.

Q. 8. What do you mean by field-effect transistor amplifier?

Ans. The field-effect transistor (FET) has a capability to amplify a.c. signals like a bipolar junction transistor (BJT). Depending upon the types of configuration, the FET amplifiers may be classified under the following three heads:

(*i*) Common source amplifier
(*ii*) Common drain amplifier, and
(*iii*) Common gate amplifier.

Q. 9. Why are field-effect transistors called unipolar transistors ?

Ans. In field-effect transistors, current conduction is by only one of type of majority carriers (either by electrons or by holes) and, therefore are called unipolar transistors.

Q. 10. Why the channel of a JFET is never completely closed at the drain end ?

Ans. If the channel is completely closed in JFET, then there will be no drain current, so there will be no voltage drop along the channel length and amount of reverse bias will become uniform and wedge-shaped depletion region will become rectangular one.

Q. 11. How is drain current controlled in a JFET ?

Ans. In a JFET, drain current is controlled by controlling the reverse bias given to its gate (i.e. V_{GS}).

Q. 12. What is meant by drain characteristics of FETs ?

Ans. The curve drawn between drain current and drain-source voltage with gate-to-source voltage as the parameter is called the drain characteristic.

Q. 13. What is meant by transfer characteristic of FETs ?

Ans. The curve drawn between drain current and gate-source voltage for a given value of drain-source voltage is called the transfer characteristic.

Q. 14. What is pinch-off voltage in a JFET ?

Ans. The value of drain-source voltage at which channel is pinched-off (*i.e.*, all the free charges from the channel get removed) is called the pinch-off voltage.

Q. 15. What is meant by saturation region ?

Ans. The region of drain characteristic of a FET in which drain current remains fairly constant is called the saturation or pinch-off region.

Q. 16. Why is input impedance of the FET very high ?

Ans. FET has very high input impedance because its input circuit (gate-to-source) is reverse biased and the input gate current is very small (of the order of few nano-amperes)

Q. 17. What is the significant difference between the construction of an enhancement type MOSFET and a depletion type MOSFET ?

Ans. In the depletion type MOSFET, a channel is physically constructed and a current between drain and source is due to voltage applied across the drain-source terminals while in enhancement type construction, no channel is formed during its construction. Voltage is applied to the gate in case of enhancement type MOSFET, to develop a channel of charge carriers so that a current results when a voltage is applied across the drain-source terminals.

Q. 18. Why E-MOSEFT is called sometimes normally-off MOSFET ?

Ans. E-MOSFET operates with large positive gate voltages only and does not conduct when gate-source voltage $V_{GS} = 0$, so it is called normally-off MOSFET.

Q. 19. Why MOSFETs are never connected or disconnected in the circuit when power is ON ?

Ans. If a MOSFET is connected or disconnected in circuit while the power is ON, transient voltages caused by inductive kickback and other effects may exceed $V_{GS(max)}$ and thus wipe out the MOSFET.

Q. 20. Name the factors which make the JFET superior to BJT?

Ans. The high input impedance, low output impedance and low noise level make JFET far superior to the BJT.

Q. 21. Explain FET is a voltage controlled device or current controlled device.

Ans. The voltage applied between gate and source (V_{GS}) controls the drain current I_D. Therefore, FET is a voltage controlled device. The transistor on the contrary is a current controlled device because the output current I_C is controlled by varying the base current I_B. The name field effect is derived from the fact that the current flow in FET is controlled by an electric field set by an external voltage.

Q.22. Explain FET is an unipolar device or a bipolar device.

Ans. FET is a unipolar device. This means that the current flowing through it is only due to one type of charge particles, holes or electrons. Transistor, on the other hand, is a bipolar device as holes and electrons both contribute to the flow of current. Just as there are NPN and PNP transistors, there are n-channel and p-channel field effect transistors. One of the most important characteristics of the FET is its very high input impedance. Typically, it is in the range of 1 to several megaohms, which is much higher than the input resistance of a BJT. FETs are more temperature stable as compared to the BJT and it requires less space than that for a BJT. Therefore, FETs are preferred in integrated circuits.

Q.23. Write the Advantages of JFET Over BJT

Ans. The FET enjoys several advantages over the conventional bipolar junction transistor (BJT). Some of them may be listed as under :

(*i*) It is a unipolar device so its operation depends on the flow of majority carriers only.

(*ii*) It is relatively immune to radiation.

(*iii*) FET has a very high input resistance (typically few megaohms)

(*iv*) FET is less noisy

(*v*) It does not exhibit any offset voltage at zero drain current hence it can be used as an excellent signal chopper.

(*vi*) FET has a better thermal stability.

Q.24. Write Disadvantage of JFET.

Ans. The main disadvantage of the JFET is that it has a relatively small gain-bandwidth product as compared with that of a conventional transistor (*i.e.* BJT).

REVIEW QUESTIONS

1. Illustrate with neat circuit diagrams the technique of biasing an enhancement MOSFET.
2. Explain giving necessary diagrams how JFET can be biased suitable to overcome the effect of unit to unit variation.
3. Explain briefly the technique of biasing a JFET for zero current drift with temperature.
4. Write notes on: Biasing techniques of MOSFET.
5. Explain any one method of biasing a single stage JFET amplifier.
6. With the help of transfer curve and bias line, explain the biasing techniques of N-channel FET.
7. With the help of neat diagrams explain the operation and drain characteristics of N-channel MOSFET.
8. Draw circuit diagram and describe the following applications of MOSFET:
   (*i*) Active load switching
   (*ii*) CMOS inverter.
9. Write a short note on: Depletion mode MOSFET.

10. Explain giving illustrative diagrams, how the pinch off condition occurs in a MOSFET.
11. Explain the difference between the operation of a JFET in ohmic and current saturation regions. Also explain the cut off condition.
12. What is thermal runway in BJT ? How can it be avoided? Is the thermal runway possible in JFET?
13. Draw the laboratory experimental setup for plotting output static characteristics of a JFET. Explain clearly, different operating regions in characteristics with proper reasoning.
14. Compare BJT and FET.
15. Define the parameters: Amplification factor, drain resistance, and mutual conductance for JFET and obtain the expression relating them.
16. Write a short note on MOSFET.
17. Compare JFET and MOSFET.
18. Give reason: FET behaves as a voltage controlled current source.
19. Sketch the drain characteristics of a JFET and explain its behaviour before and after pinch off.
20. Write a short note on: FET as *VVR*.
21. State true or false with reason.
    (*i*) The relationship between I_D and V_{GS} is non-linear.
    (*ii*) Gate current in FET is always neglected.
    (*ii*) In JFET, the P-N junction between gate and source is always kept in reverse biased condition.
    (*iv*) FET is a current controlled device.
22. What do you understand by "pinch off" voltage and cut-off voltage as applied to FET?
23. Draw the drain characteristics of depletion type MOSFET. Explain different operating regions in the characteristics with proper reasoning.
24. Draw the circuit diagram of JFET as an amplifier and explain its working.
25. Draw the a.c. equivalent circuit of JFET and mention important points.
26. Obtain the approximate equivalent circuit for JFET as an amplifier.
27. Draw the circuit diagram of JFET amplifier using fixed-bias configuration and explain the process.
28. What is the significant difference between the construction of an enhancement type MOSFET and a depletion type MOSFET ?
29. What is meant by gate-to-source threshold voltage in E-MOSFET ?

NUMERICAL PROBLEMS

1. An *n*-channel depletion type MOSFET is operated in the pinch off region. If I_{DSS} = 15 mA and V_P = – 4 V, calculate I_D when: (*a*) V_{GS} = – 3 V and (*b*) V_{GS} = + 2.5 V.

   [Ans. I_D = 3.75 mA, 3961 mA]

2. An *N*-channel JFET having a pinch off voltage of – 3 volts has a saturation current of 2.5 mA; when V_{GS} = – 1 V. What is its saturation current when,
   (*a*) V_{GS} = 0 Volts
   (*b*) V_{GS} = – 2 Volts **[Ans. (*i*) I_{DSS} = 5.625 mA at V_{GS} = 0 V, (*ii*) I_{DSS} = 0.625 mA at V_{GS} = –2 V]**

3. A potential divider bias circuit of FET is shown in figure 8.28. Find the quiescent values of : (*i*) I_D, (*ii*) V_{GS}, (*iii*) V_{DS}, (*iv*) V_S.

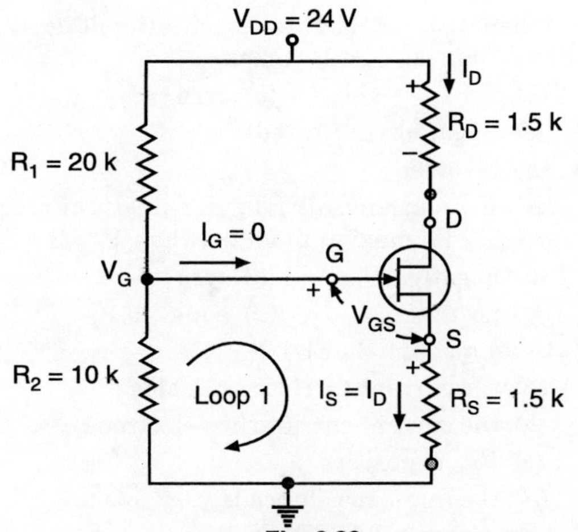

Fig. 8.28.

[Ans. (*i*) I_{DQ} = 3 mA, (*ii*) V_{GSQ} = –2 Volt (*iii*) V_{DS} = 1.5 Volt (*iv*) V_S = 6 Volt]

4. Given that for a certain JFET, $I_{DSS} = 15$ mA, $V_P = -5$ V, and $g_{mo} = 3$ ms. *Calculate:* Transconductance and drain current I_D for $V_{GS} = -3.5$ Volts.
[Ans. (*i*) $g_m = 0.9 \times 10^{-3}$ *AV*, (*ii*) $I_D = 1.35$ mA]

5. When a reverse gate voltage of 18 V is applied to a FET, the gate current is 0.9×10^{-3} mA. Determine the resistance between gate and source. **[Ans. 20,000 M]**

6. The following readings were obtained experimentally from a FET:

V_{GS}	0 V	0.V	– 0.3 V
V_{DS}	7 V	14.5 V	14.5 V
I_D	3.25 mA	10 mA	9.2 mA

Determine r_d, g_m and μ.
[Ans. 30 kΩ, 3500 μS; 10^5]

7. Determine the value of transconductance of a FET, when the drain current changes from 1 mA to 1.5 mA with a change in gate voltage from – 2.125 V to – 2 V. **[Ans. 4,000 μ mhos]**

OBJECTIVE TYPE QUESTIONS

Choose one correct answer in the following:

1. The JFET is
(*a*) a bipolar device (*b*) unipolar device
(*c*) current controlled device
(*d*) voltage controlled device
(*e*) *b* and *d* (*f*) *a* and *b*

2. The channel of a JFET exists between
(*a*) gate and source (*b*) drain and source
(*c*) gate and drain (*d*) input and output

3. For low values of V_{DS}, the JFET behaves like a
(*a*) resistance
(*b*) constant voltage device
(*c*) constant current device
(*d*) negative resistor

4. When V_{DS} reaches the pinch off voltage V_P, the drain current becomes
(*a*) low (*b*) zero
(*c*) constant (saturated)
(*d*) reversed

5. In an *n*-channel JFET, the drain current reaches its maximum value when V_{GS} is
(*a*) negative (*b*) zero
(*c*) positive (*d*) equal to V_P

6. In an *n*-channel JFET
(*a*) the current carriers are holes
(*b*) the current carriers are electrons
(*c*) V_{GS} is positive
(*d*) the input resistance is very low

7. In a p-channel JFET
(*a*) the current carriers are electrons
(*b*) the current carriers are holes
(*c*) V_{GS} is negative
(*d*) the input resistance is very small

8. For an *n*-channel JFET
(*a*) V_{GS} can vary between zero negatively to V_{GSO}
(*b*) V_{GS} can vary between zero positively to V_{GSO}
(*c*) pinch off occurs for positive V_{GS}
(*d*) V_{DD} is negative

9. An FET cannot operate at $V_{GS} = 0$ V. The FET is
(*a*) JFET (*b*) D-MOSFET
(*c*) E-MOSFET (*d*) both (*a*) and (*b*)

10. The transconductance g_m of JFET is defined as
(*a*) $\frac{\Delta I_D}{\Delta V_{GS}}$ (*b*) $\frac{\Delta I_D}{\Delta V_{DS}}$
(*c*) $\frac{\Delta V_{GS}}{\Delta I_D}$ (*d*) $\frac{\Delta I_{DSS}}{\Delta I_D}$

11. The amplification factor μ of JFET is given by
(*a*) $g_m \times r_d$ (*b*) $\frac{g_m}{r_d}$
(*c*) $g_m \cdot I_{DSS}$ (*d*) $g_m\left(1-\frac{V_{GS}}{V_P}\right)$

12. A channel is induced in a E-MOSFET by the application of a
(*a*) $V_{GS} < (V_{GS})$th (*b*) $V_{GS} > V_P$
(*c*) $V_{GS} > (V_{GS})$th (*d*) $V_{DS} > V_P$

13. The Shockley equation is given by
(*a*) $I_D = I_{DSS}\left[1-\frac{V_{GS}}{V_P}\right]^2$
(*b*) $I_D = I_{DSS}\left[1-\frac{V_{GS}}{V_P}\right]$

(c) $g_m = g_{mo}\left[1 - \frac{V_{GS}}{V_P}\right]$

(d) $I_D = C[V_{GS} - V_{GS}(\text{th})]^2$

14. Ideally, the equivalent circuit of an FET consists of
    (a) a resistance between drain and source
    (b) a current source between the gate and the source
    (c) a current source between the drain and the source
    (d) a current source between the gate and the drain
15. The magnitude of the current source in the ac equivalent circuit of an FET depends on
    (a) the dc supply voltage
    (b) V_{ds}
    (c) external drain resistance
    (d) transconductance and gate to source voltage
16. A CS amplifier has a drain load $R_D = 1\ \text{k}\Omega$ and an external load resistance of $R_L = 1\ \text{k}\Omega$. If $g_m = 10$ mS, the voltage gain will be
    (a) 10 (b) 5
    (c) 20 (d) 1
17. If the external load resistance is removed in problem 16, the voltage gain will be
    (a) decrease (b) increase
    (c) remain the same (d) become zero
18. In a *CD* amplifier, $R_S = 500\ \Omega$ and the transconductance is 9000 μS. The voltage gain will be
    (a) 0.82 (b) 1
    (c) 1.22 (d) 9
19. In a common gate amplifier, the transconductance is $g_m = 6000$ μS, $R_G = 1\ \text{M}\Omega$. The total input resistance will be
    (a) 1666 Ω (b) 166.6 Ω
    (c) 1 MΩ (d) 100166.6 Ω
20. In a *CD* amplifier, $I_{DSS} = 5$ mA at $V_{GS} = 10$ V. If the resistor R_G is 20 *M*Ω, total input the resistance will be
    (a) 20 MΩ (b) 200 MΩ
    (c) 40 MΩ (d) 100 MΩ
21. If there is an internal open between the source and the drain in a CS amplifier, the drain voltage will be
    (a) 0 V (b) V_{GS}
    (c) V_{DD} (d) $\frac{V_{DD}}{2}$

Answers :

1. (e)	2. (b)	3. (a)	4. (c)	5. (d)	6. (b)
7. (b)	8. (a)	9. (c)	10. (a)	11. (a)	12. (c)
13. (a)	14. (c)	15. (c)	16. (d)	17. (b)	18. (a)
19. (b)	20. (a)	21. (c)			

■ ■ ■

UNIT—VII

FET Amplifiers

FET Common Source Amplifier, Common Drain Amplifier, Generalized FET Amplifier, Biasing FET, FET as Voltage Variable Resistor, Comparison of BJT and FET, The Uni Junction Transistor (UJT).

Above Syllabus has been covered in following Chapter:

CHAPTER 9 : FET Amplifiers

Chapter 9

FET Amplifiers

SYLLABUS

ET Common Source Amplifier, Common Drain Amplifier, Generalized FET Amplifier, Biasing FET, FET as Voltage Variable Resistor, Comparison of BJT and FET, The Uni Junction Transistor (UJT).

Chapter Outline

9.1. INTRODUCITON

Like BJT, the FET can also be used for the amplification of a small time varying signal. In this chapter, we shall discuss various FET amplifier configurations. However, before discussing various FET amplifier configurations in detail, let us first discuss various FET biasing techniques.

9.1. FET BIASING

We know the meaning and the need of biasing a device. We have already studied different biasing circuits for a bipolar junction transistor (BJT) in previous chapters. As a matter of fact, the biasing deals with setting a fixed level of current which should flow through the FET with a desired fixed voltage drop across the FET junctions.

Now, let us study the different biasing circuits for FET. The commonly used biasing circuits for FET are as under:

(i) Fixed-bias circuit.

(ii) Self-bias circuit.

(iii) Voltage divider bias circuit.

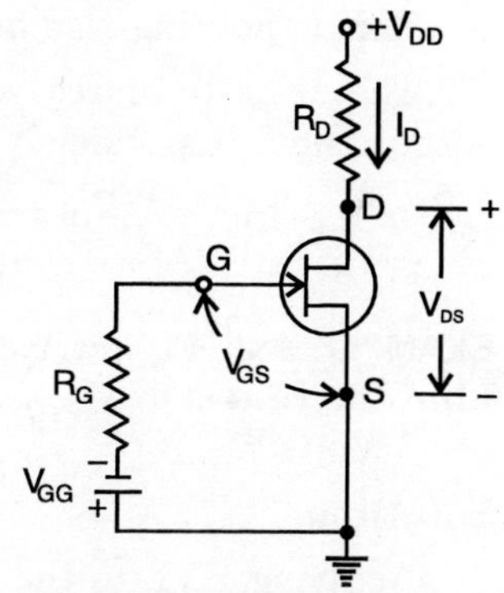

Fig. 9.1. Fixed-bias circuit for JFET

Whatever be the biasing method, the gate-source junction of a JFET will always be reverse-biased. Even when the gate and the source terminals are shorted, the drain current (I_D) will cause the junction to be reverse biased. As a matter of fact, the gate-source junction of a JFET must never be allowed to become forward-biased. It is because of the reason that the gate material is not constructed to handle any significant amount of current. If the junction is allowed to become forward-biased, current will start to flow in the gate circuit. Under most circumstances, this current will destroy the JFET.

DO YOU KNOW?

When biased in the active region, the JFET is equivalent to a current source.

The general relationships which can be applied to the d.c. analysis of all FET amplifiers are:

$$I_G \cong 0 \quad ...(9.1)$$

and

$$I_D = I_S \quad ...(9.2)$$

9.2. FIXED-BIAS CIRCUIT

Fixed-bias circuit for JFET is also known as gate-bias circuit. Figure 9.2 shows the fixed-bias circuit or gate-bias circuit for the N-channel JFET. In this circuit, the gate supply voltage ($-V_{GG}$) is used to ensure that the gate-source junction is reverse-biased. Since, there is no gate current, therefore, there is no voltage drop across resistor R_G.

It will be interesting to know that the fixed bias circuit cannot provide a stable Q-point from one JFET to another. The reason for this is that for a given JFET, there is a range of values for $V_{GS\,(off)}$ and I_{DSS} which can lead to a major problem with the fixed bias circuit.

It can also be noted that even though there is no gate current or any voltage drop across R_G, but still the circuit contains R_G. Actually, R_G is used for ac operation purpose. Thus R_G has little to do with the d.c operation of the amplifier, but it is a vital part of the ac operation of the circuit.

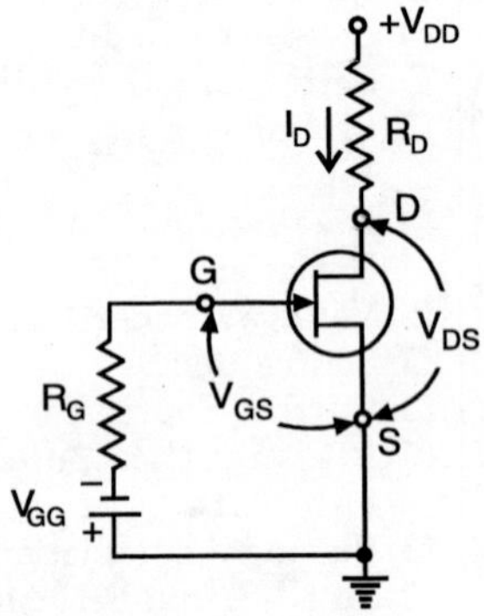

Fig. 9.2.

We know that for dc analysis,

$$I_G = 0$$

Applying KVL to the input circuit or gate circuit, we have

$$V_{GG} + V_{GS} = 0$$

or

$$V_{GS} = -\,V_{GG} \quad ..(9.3)$$

The expression for the drain current I_D is as under:

$$I_D = I_{DSS}\left[1 - \frac{V_{GS}}{V_P}\right]^2 \quad ...(9.4)$$

This is nothing else but the Shockly's equation as discussed earlier.

The drain-to-source voltage can be obtained by applying the KVL to the output circuit as under :

$$V_{DD} - I_D R_D - V_{DS} = 0$$

or

$$V_{DS} = V_{DD} - I_D R_D \quad ...(9.5)$$

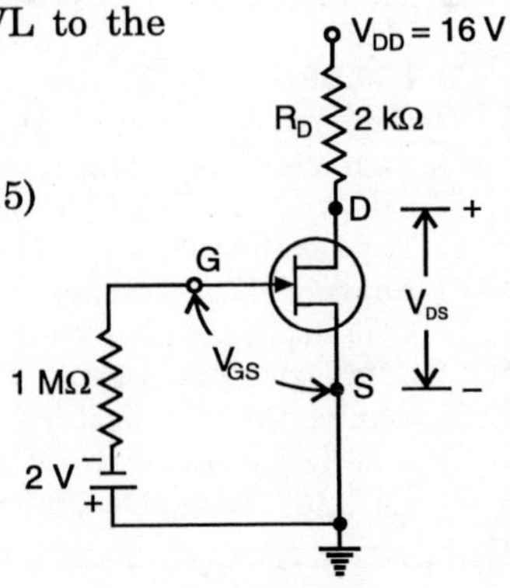

Fig. 9.3.

EXAMPLE 9.1 Figure 9.3 shows the fixed-bias circuit for JFET. Find the values of V_{GSQ}, I_{DQ} and V_{DS}. Assume I_{DSS} = 8 mA and V_P = – 8 Volts.

Solution:

Applying KVL to the gate circuit, we have

$$V_{GG} + I_G R_G + V_{GS} = 0$$

Substituting all the given values, we get

$$2 + (I_G \times 1 \times 10^6) + V_{GS} = 0$$

But for dc analysis, we have

$$I_G = 0$$

Therefore, $2 + V_{GS} = 0$

$$V_{GS} = -2 \text{ Volts} \qquad \textbf{Ans.}$$

Now, using Shocky's equation, we get

$$I_{DQ} = I_{DSS}\left[1 - \frac{V_{GS}}{V_P}\right]^2$$

Substituting values, we get

$$I_{DQ} = 8 \times 10^{-3}\left[1 - \frac{-2}{-8}\right]^2$$

Solving, we get $I_{DQ} = 4.5$ mA **Ans.**

For drain-to-source voltage V_{DS}, applying KVL to the drain circuit, we have

$$V_{DD} = I_D R_D + V_{DS} = V_{DD} - I_D R_D$$

Substituting the values we get

$$V_{DS} = 16 - (4.5 \times 2) = 16 - 9 = 7 \text{ Volt} \qquad \textbf{Ans.}$$

9.3. SELF-BIAS CIRCUIT

Figure 9.4 shows the self-bias circuit for an N-channel junction field-effect transistor (JFET). In this circuit, there is only one drain supply and no gate supply. The gate terminal is connected through a resistor (R_G) to the ground. The source terminal is connected through a resistor (R_S) to the ground. When the drain voltage is applied, a drain current flows even if there is no gate current. The drain current produces a voltage drop across resistor R_S (equal to $I_D.R_S$). The voltage drop produces the gate-to-source reverse voltage required for an FET operation. *The resistor R_S is known as a feedback resistor.* Its function is to prevent any *variation in the FET drain current, which can be understood from the following discussion:*

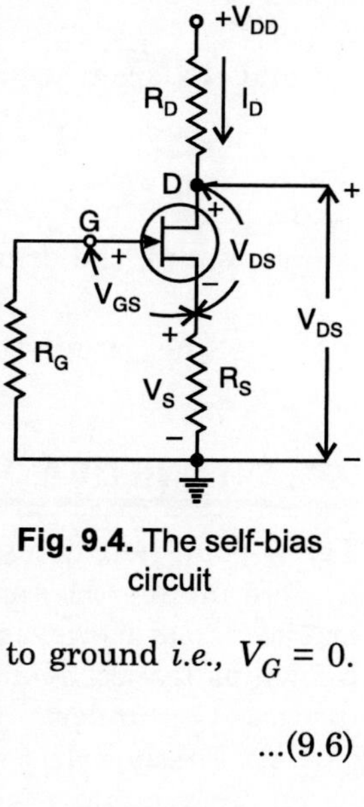

Fig. 9.4. The self-bias circuit

Firstly, let us consider an increase in the drain current, which will increase the voltage drop across the resistor (R_S). The increased voltage drop increases the reverse gate-to-source voltage, which decreases the effective width of the channel. It reduces the value of drain current. Now, if the drain current decreases, then the reverse action takes place *i.e.*, the reduced drain current decreases the gate-to-source voltage, which in turn increases the effective width of the channel and thereby increasing the value of drain current.

We know that in a JFET, no current flows through the gate terminal to the ground since the gate-source junction is reverse-biased. However, in actual practice, a small current known as reverse leakage current (I_{GSS}) does flow through the gate terminal. Usually, its value is very small and can be neglected. Therefore, we shall assume that the gate-terminal is at zero voltage. In other words, the gate voltage with respect to ground *i.e.*, $V_G = 0$.

We know that the source voltage with respect to ground is given by

$$V_S = I_D.R_S \qquad ...(9.6)$$

and the drain voltage is given by

$$V_D = V_{DD} - I_D.R_D \qquad ...(9.7)$$

The drain-to-source voltage (V_{DS}) is equal to the difference between the drain voltage (V_D) and the source voltage (V_S)

i.e, $V_{DS} = V_D - V_S$ ...(9.8)

or $V_{DS} = (V_{DD} - I_D.R_D) - I_D.R_S = V_{DD} - I_D\,(R_D + R_S)$ ...(9.9)

The gate-to-source voltage (V_{GS}) is equal to the difference between the gate voltage (V_G) and the source voltage (V_S)

i.e., $V_{GS} = V_G - V_S = 0 - I_D.R_S = -\,I_D.R_S$ ...(9.10)

It is evident from this expression that the gate-to-source voltage is equal to the negative of the voltage across the source resistor. Hence, it is obvious that greater the value of drain current, more negative will be the gate-to-source voltage.

Rearranging equation (9.10), we get

$$I_D = -\frac{V_{GS}}{R_S} \qquad ...(9.11)$$

Now, if we plot a graph using the value of drain current obtained from equation (9.11) for a constant value resistor (R_S) against the gate-to-source voltage, we get a straight line as shown in figure 9.5. This straight line is known as a **self-bias line**.

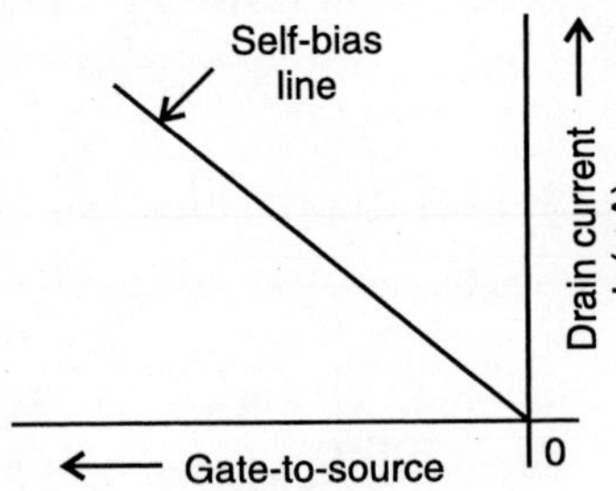

Fig. 9.5.

EXAMPLE 9.2 Determine the values of drain-to-source voltage (V_{DS}) and the gate-to-source voltage (V_{GS}) for the circuit shown in figure 9.6. Given I_D = 5 mA.

Solution: Given that

$$I_D = 5\text{ mA} = 5 \times 10^{-3}\text{ A}$$
$$V_{DD} = 10\text{ Volts}$$
$$R_D = 1\text{ k}\Omega = 1 \times 10^3\ \Omega \quad \text{and} \quad R_S = 500\ \Omega$$

We know that the source voltage is given by

$$V_S = I_D.R_S$$

or $V_S = 5 \times 10^{-3} \times 500 = 2.5$ Volt

and the drain voltage V_D is given by

$$V_{DD} = V_D + I_D.R_D = V_{DD} - I_D R_D$$

or $V_D = 10 - (5 \times 10^{-3}) \times (1 \times 10^3)$

or $V_D = 10 - 5 = 5$ Volt

Therefore, the drain-to-source voltage will be

$$V_{DS} = V_D - V_S = 5 - 2.5 = 2.5\text{ Volt}$$

Also, the gate-to-source voltage is given by

$$V_{GS} = -\,V_S = -\,2.5\text{ Volt} \qquad \textbf{Ans.}$$

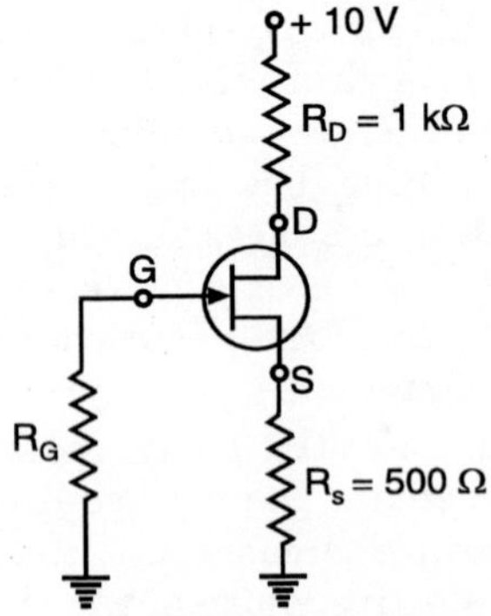

Fig. 9.6. Circuit for example 9.2

9.4. SETTING OF A Q-POINT

The *Q*-point (*i.e.,* quiescent or operating point or bias point) for a self-biased JFET is established by determining the value of drain current (I_D) for a desired value of gate-to-source voltage (V_{GS}) or vice-versa. However, if the data sheet of JFET includes a transfer characteristics curve (also known as transconductance curve), then we may determine the *Q*-point by using the procedure discussed as under:

(*i*) Firstly, select a convenient value of drain current. Generally, its value is taken half of the maximum possible value of drain current (*i.e.,* I_{DSS}). After this, find the voltage drop across source resistor (R_S) by the expression

$$V_S = I_D.R_S \qquad ...(9.12)$$

and the gate-to-source voltage from the expression

$$V_{GS} = -\,V_S \qquad ...(9.13)$$

(*ii*) Now, plot the assumed value of drain current (I_D) and the corresponding gate-to-source voltage (V_{GS}) on the transfer characteristics curve.

(*iii*) Draw a line through the plotted point and the origin. The point of intersection of the line and the curve gives the desired Q-point. Now, read the co-ordinates of Q-point.

As a matter of fact, it is required to set the Q-point near the mid-point of the transfer characteristic curve of a JFET. Under signal conditions, the mid-point bias allows a maximum amount of drain current swing between the values of I_{DSS} and O.

However, if we desire to design a self-bias circuit, then anyone of the following two methods can be used for this purpose:

(*i*) Analytical Method

In analytical method, we note down the values of the maximum drain current (I_{DSS}) and the gate-to-source cut-off voltage [$V_{GS(\text{off})}$] from the data sheets of JFET supplied by the manufacturer.

The value of the drain current is found by the expression

$$I_D = \left[1 - \frac{V_{GS}}{V_{GS(off)}}\right]^2 \qquad \text{...(9.14)}$$

For example, if we select the gate-to-source voltage (V_{GS}) equal to $V_{GS(\text{off})}/4$, then the value of the drain current would be

$$I_D = I_{DSS}\,[\,1 - 0.25\,]^2 = I_{DSS}\,(0.75)^2 = 0.56\,I_{DSS} \qquad \text{...(9.15)}$$

It shows that the drain current is slightly more than one-half of I_{DSS}. But it will bias the JFET close to the mid-point of the curve. The value of the drain resistor (R_D) is selected in such a way that the drain voltage (V_D) is equal to half the drain supply voltage (V_{DD}). The value of gate resistor (R_G) is chosen arbitrarily large, so that it prevents loading on the driving stages.

(*ii*) Graphical Method

In graphical method, we draw a self-bias line such that it intersects the transfer characteristic curve near its mid-point. The point of intersection of the self-bias line and the transfer characteristic curve gives us the required *Q*-point. After this, read the co-ordinates of the *Q*-point. The value of source resistance (R_S) is expressed by the ratio of gate-to-source voltage (V_{GS}) to the drain current (I_D).

Mathematically, the source resistance is given by

$$R_S = \frac{V_{GS}}{I_D} \qquad \text{...(9.16)}$$

However, a more accurate method is to draw a self-bias line through the co-ordinates of I_{DSS} and $V_{GS(\text{off})}$ as shown in figure 9.34. Then the point of intersection of self-bias line and the transfer characteristic curve locates the *Q*-point. The value of the source resistor is expressed by the relation

$$R_S = \frac{V_{GS}(off)}{I_{Dss}} \qquad \text{...(9.17)}$$

The value of drain resistor (R_D) and the gate resistor (R_G) are selected in the same way as discussed above for the analytical method.

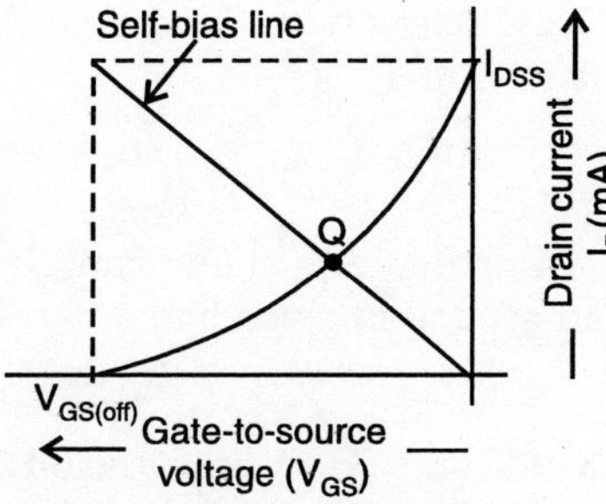

Fig. 9.7. Setting of a Q-Point in self-bias

EXAMPLE 9.3 Figure 9.8 (*a*) shows that circuit of a self-biased JFET amplifier and figure 9.8(*b*) indicates the transfer characteristic curve of the JFET.

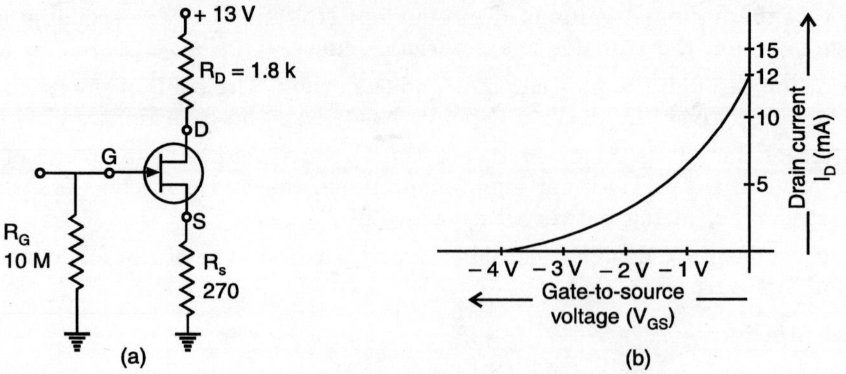

Fig. 9.8.

Determine the quiescent values of I_D and V_{GS}. Compute the values of d.c. voltage between the drain and the ground.

Solution: Given that

$$R_D = 1.9\text{ k}\Omega = 1.8 \times 103\ \Omega$$
$$R_S = 270\ \Omega$$
$$R_G = 10\text{ M}\Omega \text{ and}$$
$$I_{DSS} \text{ (from transfer characteristic curve)} = 12\text{ mA}$$

(*i*) Firstly, let us assume the drain current (I_D) equal to the half of I_{DSS} (*i.e.*, 6 mA or 6×10^{-3} A). Using this value of drain current, the voltage drop across resistor (R_S) is given by

$$V_S = I_D.R_S = (6 \times 10^{-3}) \times 270 = 1.62\text{ V}$$

and the gate-to-source voltage is given by

$$V_{GS} = -V_S = -1.62\text{ V}$$

(*ii*) Now, let us plot the values of drain current ($I_D = 6$ mA) and the corresponding value of voltage V_{GS} (*i.e.*, – 1.62 V) on the graph of transfer characteristic curve as shown in figure 9.9.

(*iii*) Now, let us draw a line through the plotted point (P) and the origin as shown in the figure and read the co-ordinates of Q-point.

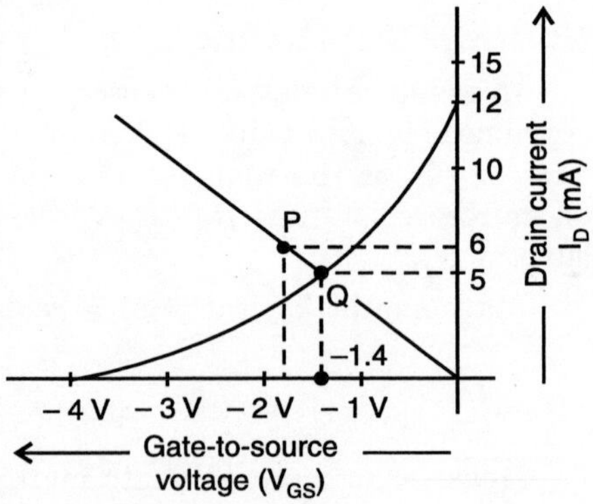

Fig. 9.9.

It is clear from this graph that the coordinates of Q-point are as under:

$$I_D = 5\text{ mA}$$
$$V_{GS} = -1.4\text{ V} \qquad \textbf{Ans.}$$

Further, we know that the d.c. voltage between drain and ground is given by

$$V_D = V_{DD} - I_D.R_D = 15 - (5 \times 10^{-3}) \times (1.8 \times 10^3)\text{ V} = 15 - 9 = 6\text{ V} \quad \textbf{Ans.}$$

EXAMPLE 9.4 **A self-biased P-channel JFET has a pinch off voltage (V_P) = 5 V and I_{DSS} = 12 mA. The supply voltage available is equal to 12 V. Find the values of resistors R_D and R_S, so that I = 5 mA and V_{DS} = 6 V.**

Solution: Given that : V_P, *i.e.*,

$$V_{GS\,(off)} = 5\text{ Volt}$$
$$I_{DSS} = 12 \times 10^{-3}\text{ A}$$
$$V_{DD} = 12\text{ volt}$$

$$V_{DS} = 6 \text{ volt}$$

We know that the drain current (I_D) is expressed as

$$I_D = I_{DSS}\left[1-\frac{V_{GS}}{V_{GS(off)}}\right]^2$$

or $$4 \times 10^{-3} = I_{DSS}\left[1-\frac{V_{GS}}{V_{GS(off)}}\right]^2 = 12\times10^{-3}\left[1-\frac{V_{GS}}{5}\right]^2 = \left(1-\frac{V_{GS}}{5}\right)^2$$

Taking square root, we have

$$\frac{1}{\sqrt{3}} = 1-\frac{V_{GS}}{5}$$

or $$V_{GS}\left(1-\frac{1}{\sqrt{3}}\right)\times 5 = (1 - 0.58) \times 5 \text{ V} = 2.1 \text{ V}$$

The source voltage is given by

$$V_S = V_{GS} = 2.1 \text{ V}$$

and voltage drop across the source resistor (V_S) is expressed by

$$2.1 = I_D.R_S = (4 \times 10^{-3})\ R_S$$

or $$R_S = \frac{2.1}{(4\times10^{-3})} = 525\ \Omega$$ **Ans.**

and drain-to-source voltage (V_{DS}) is given as under:

$$6 = V_{DS} - I_D.R_D = 12 - (4 \times 10^{-3}) \times R_D$$

or $$R_D = 1500\ \Omega = 1.5\text{ k}\Omega$$ **Ans.**

9.5. SETTING OF A Q-POINT USING D.C. LOAD LINE

The *Q*-point for the self-biased JFET circuit can also be obtained graphically by using d.c. load line. The d.c. load line can be drawn easily on the drain characteristics as shown in figure 9.10(*a*). The values of drain current (I_D) and drain-to-source voltage (V_{DS}) at the upper and lower ends of the d.c. load line can be obtained by using the expression

$$V_{DS} = V_{DD} - I_D.\ (R_D + R_S) \quad ...(9.18)$$

At the upper end (*i.e.*, point *A*) the value of drain-to-source voltage (V_{DS}) is equal to zero. Therefore, substituting $V_{DS} = 0$ in last equation (9.18), we have

$$0 = V_{DD} - I_D.\ (R_D + R_S) \quad ..(9.19)$$

or $$I_D = \frac{V_{DD}}{(R_D + R_S)} \quad ...(9.20)$$

Similarly, at the lower end (*i.e.*, point *B*) the value of drain current (I_D) is equal to zero. Therefore, substituting, $I_D = 0$ in equation (9.18) we have

$$V_{DS} = V_{DD} - 0(R_D + R_S) = V_{DD} \quad ...(9.21)$$

Hence the upper and lower ends of the d.c load line are given by the points,

$$I_D = V_{DD}/(R_D + R_S) \text{ and } V_{DS} = 0 \quad \text{(Upper end)} \quad ...(9.22)$$

$$I_D = 0 \text{ and } V_{DS} = V_{DD} \quad \text{(Lower end)} \quad ...(9.23)$$

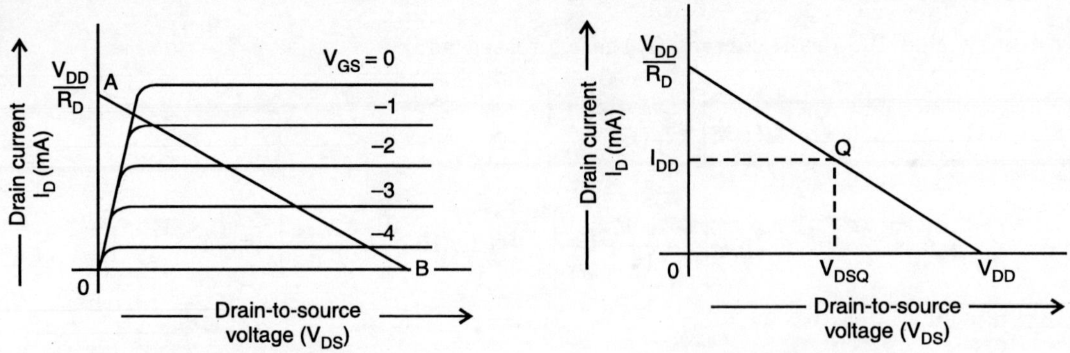

(a) Drawing d.c. load line (b) Setting a Q-point

Fig. 9.10.

The Q-point is set at mid-point of the d.c. load as shown in figure 9.10 (*b*). In this case, the value of drain current at *Q* point will be given by

$$I_{DQ} = \frac{V_{DD}}{2(R_D + R_S)} \qquad ...(9.24)$$

and the drain-to-source voltage at the *Q*-point is given by

$$V_{DSQ} = \frac{V_{DD}}{2} \qquad ...(9.25)$$

EXAMPLE 9.5 Calculate the self-bias operating point for the FET circuit shown in figure 9.11. Also find the values of resistors R_D and R_S to obtain this bias condition. Given the maximum value of drain current as 10 mA and V_{GS} = – 2.2 volt at I_D = 5 mA.

Solution: Given that I_{DSS} = 10 mA and V_{DD} = 20 volt.

We know that the values of drain current at *Q*-point is given by

$$I_{DQ} = \frac{I_{DSS}}{2} = \frac{10}{2} = 5 \text{ mA}$$

and the value of drain-to-source voltage is expressed as

$$V_{DSQ} = \frac{V_{DD}}{2} = \frac{20}{2} = 10 \text{ V}$$

Therefore, operating point will be given as

$$I_D = 5 \text{ mA}$$

and V_{DS} = 10 V **Ans.**

Now, V_{GS} = – 2.2 volt

and I_D = 5 mA = 5 × 10^{-3} A

The drain-to-source voltage (V_{DS}) is given by

$$10 = V_{DD} - I_D.R_D = 20 - (5 \times 10^{-3})R_D$$

or $$R_D = \frac{20 - 10}{5 \times 10^{-3}} = 2 \times 10^3 \text{ k}\Omega = 2 \text{ k}\Omega$$ **Ans.**

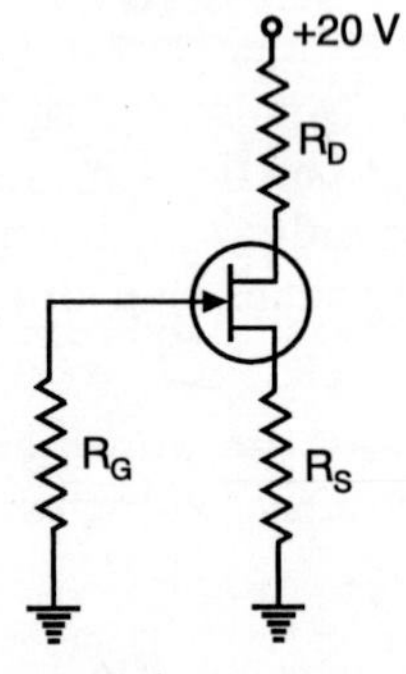

Fig. 9.11. Circuit for example 9.5.

The source voltage is given by

$$V_S = -V_{GS} = -(-2.2) = 2.2 \text{ V}$$

Little minds are tarned and sub dued by misfortune; but great minds rise about it.

— Washington Irving

and the value of voltage drop across the source resistor (V_S) is expressed by

$$2.2 = I_D.R_S = (5 \times 10^{-3}) \times R_S$$

or

$$R = 2.2/(5 \times 10^{-3}) = 440\ \Omega \quad \textbf{Ans.}$$

9.6. BIASING AGAINST DEVICE PARAMETER VARIATION

As already discussed that for self-biased JFET, the Q-point is fixed at mid-point of the transfer characteristic curve. However, it has been observed that the JFET parameters like maximum drain current (I_{DSS}) and the gate-to-source cut-off [$V_{GS(off)}$] vary with the temperature. In this case, the Q-point will also vary with the temperature and thus will not remain stable. This degrades the performance of the FET amplifier circuit.

In order to provide compensation for the device parameter variation the manufactures provide two transistor characteristic curves as shown in figure 9.12. The first current corresponds to the maximum values of I_{DSS} and $V_{GS\ (off)}$. In such a case, we draw a bias line through both the curves which is near the mid-point of either curve. Then we read the values of gate-to-source voltage (V_{GS}) and the drain current (I_D) for either Q-point. The optimum value of source resistor R_S is given by the ratio of gate-to-source voltage (V_{GS}) and drain current (I_D).

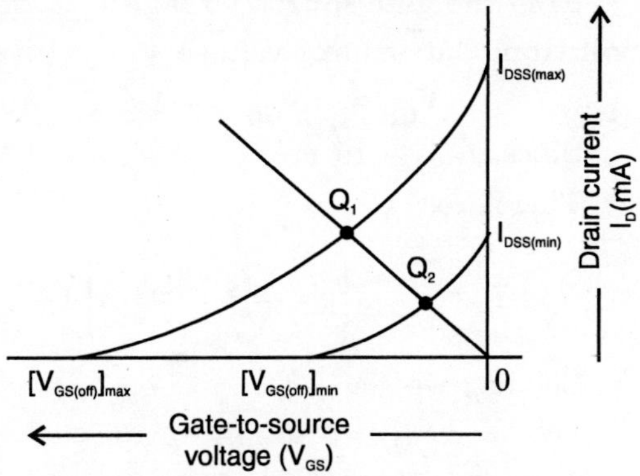

Fig. 9.12. Use of two transfer characteristic curves for setting a Q-point

Sometimes the manufacturers do not supply the transfer characteristic curves, but provide the maximum and minimum values of the I_{DSS} and $V_{GS(\text{off})}$. In this case, we calculate the maximum and minimum values of source resistor (R_S) and then take the average of these two values. This average value will provide a Q-point near the mid-point of the either transfer characteristic curve.

9.7. VOLTAGE DIVIDER BIAS

Figure 9.13 shows the voltage divider bias circuit. The name *'voltage divider'* is derived from the fact that resistor R_1 and R_2 are connected on the gate side (between the V_{DD} supply and ground) from a voltage divider. Assuming the current to be zero, the voltage is given by the expression

$$V_G = \left(\frac{R_2}{R_1 + R_2}\right) V_{DD} \qquad ...(9.26)$$

and the voltage from source to ground is given by,

$$V_S = V_G - V_{GS} \qquad ...(9.27)$$

The value of drain current will be

$$I_D = \frac{V_S}{R_S} = \frac{V_G - V_{GS}}{R_S} \qquad ...(9.28)$$

and the d.c. voltage from drain to ground is given as

$$V_D = V_{DD} - I_D.R_D \qquad ...(9.29)$$

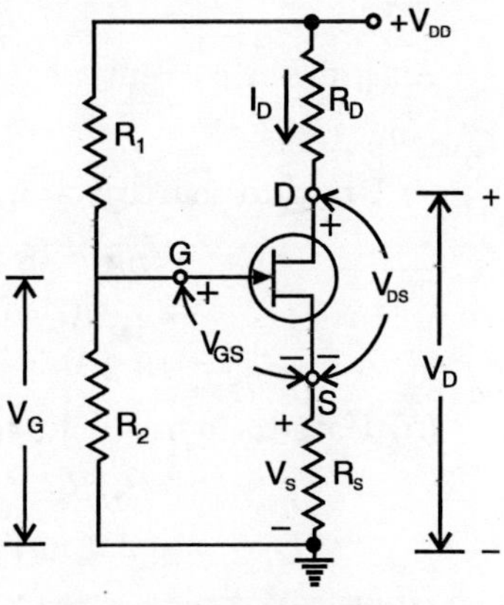

Fig. 9.13. Voltage divider bias circuit

Now, if the gate voltage is very large as compared to gate-to-source voltage, then the drain current is approximately constant for any JFET. It has been found that in actual practice the voltage divider bias is less effective with JFET than bipolar transistors. It is because of the fact that in bipolar transistors the base-to-emitter voltage (V_{BE}) is approximately 0.7 V with only minor variations from one transistor to next. However, in a JFET, the gate-to-source voltage (V_{GS}) can vary several volts from one JFET to another. Because of this, it is difficult to make gate voltage (V_G) large enough than the gate-to-source voltage (V_{GS}).

EXAMPLE 9.6 In the circuit shown in figure 9.14, the following datas are given:
V_{GG} = 1.5V, V_{DD} = 15V, R_D = 1.5 kΩ, R_G = 2 MΩ, I_{DSS} = 15 mA and V_p = – 4 V.
Find the gate-source voltage V_{GS}, the drain current I_D and drain-to-source voltage V_{DS}.

Solution: Gate-Source voltage V_{GS} is given by

$$V_{GS} = -V_{GG} = -1.5 \text{ V} \quad \textbf{Ans.}$$

Since, I_{DSS} = 15 mA and V_p = – 4 V

Therefore,

$$I_D = I_{DSS}\left(1-\frac{V_{GS}}{V_p}\right)^2 = 15\left[1-\frac{(-1.5)}{-4}\right]^2$$

Solving, we get

$$I_D = 5.86 \text{ mA} \quad \textbf{Ans.}$$

Also, $V_D = V_{DD} - I_D R_D$

$= 15 - 5.86 \times 10^{-3} \times 1.5 \times 10^3 = 6.2$ V

Lastly, the drain-to-source voltage V_{DS} is given by

$$V_{DS} = V_D - V_S = 6.2 - 0 = 6.2 \text{ V} \quad \textbf{Ans.}$$

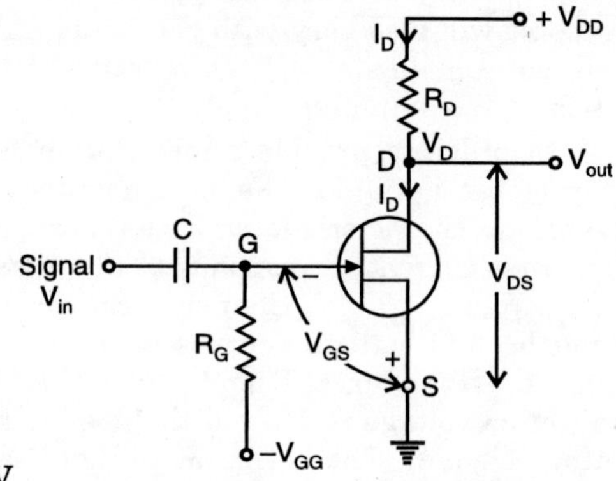

Fig. 9.14. Circuit for example 9.6.

EXAMPLE 9.7 In the JFET circuit shown in figure 9.15, the following data are given:
V_{DD} =25 V, R_D = 3 kΩ, R_S = 400 Ω, I_D = 2 mA.
Find: (i) V_{DS} (ii) V_{GS}.

Solution: Given that

Drain supply voltage, $V_{DD} = 25$ V

Drain circuit resistance, $R_D = 3$ kΩ

Source circuit resistance, $R_S = 400$ Ω $= 0.4$ kΩ

Also, drain current, $I_D = 2$ mA

Now,

(*i*) Drain-to-source voltage is given by

$$V_{DS} = V_{DD} - I_D(R_D + R_S)$$
$$= 25 - (2 \text{ mA}) [3 + 0.4) \text{ k}\Omega]$$

or $V_{DS} = 18.2$ V **Ans.**

(*ii*) Gate-to-source voltage, V_{GS}

$= -I_D R_S$

$= -2.0 \text{ mA} \times 0.4 \text{ k}\Omega = -0.8$ V **Ans.**

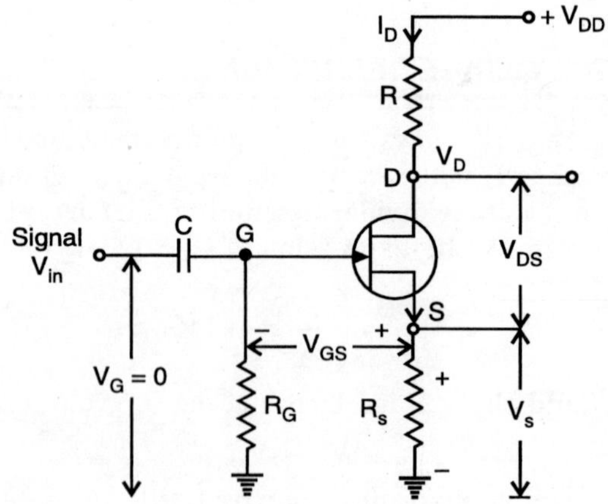

Fig. 9.15.

EXAMPLE 9.8 In an N-channel JFET biased by the voltage-divider method, determine the value of resistance R_S so that the operating point values are,

$$I_D = 4 \text{ mA and } V_{DS} = 8 \text{ V.}$$

Given that $V_{DD} = 25$ **V,** $R_{G1} = 1.2$ **MΩ,**

$R_{G2} = 0.6$ **MΩ.**

JFET parameters are $I_{DSS} = 12$ **mA and** $V_p = -4$ **V.**

Solution: Given that

$$V_{DD} = 25 \text{ V}$$
$$R_{G1} = 1.2 \text{ M}\Omega \text{ or } 1200 \text{ k}\Omega$$
$$R_{G2} = 0.6 \text{ M}\Omega \text{ or } 600 \text{ k}\Omega$$
$$I_D = 4 \text{ mA} \qquad I_{DSS} = 12 \text{ mA}$$
$$V_{DS} = 8 \text{ V} \qquad V_p = -4 \text{ V}$$

We know that
$$I_D = I_{DSS}\left[1-\frac{V_{GS}}{V_p}\right]^2$$

Therefore, we have
$$1-\frac{V_{GS}}{V_p} = \sqrt{\frac{I_D}{I_{DSS}}} = \sqrt{\frac{4}{12}} = 0.577$$

or
$$\frac{-V_{GS}}{-4} = 0.577 - 1 = 4(0.577 - 1) = -1.69 \text{ V}$$

Also, since,
$$V_G = \frac{V_{DD}}{R_{G1}+R_{G2}} \times R_{G2}$$

Therefore,
$$V_G = \frac{25 \times 600}{1200+600} = 8.33 \text{ Volt}$$

Further,
$$R_S = \frac{V_G - V_{GS}}{I_D} = \frac{8.33-(-1.69)}{4 \text{ mA}} = 2.5 \text{ k}\Omega$$

EXAMPLE 9.9. A JFET amplifier with stabilized biasing circuit shown in figure 9.16, has following parameters:

$V_p = -2$ **V,** $V_{DSS} = 5$ **mA,** $R_L = 910$ **Ω,** $R_F = 2.29$ **kΩ,** $R_1 = 12$ **MΩ,** $R_2 = 8.57$ **MΩ and** $V_{DD} = 24$ **V.**

Determine the value of drain current I_D **at the operating point. Also, verify that FET will operate in pinch-off region.**

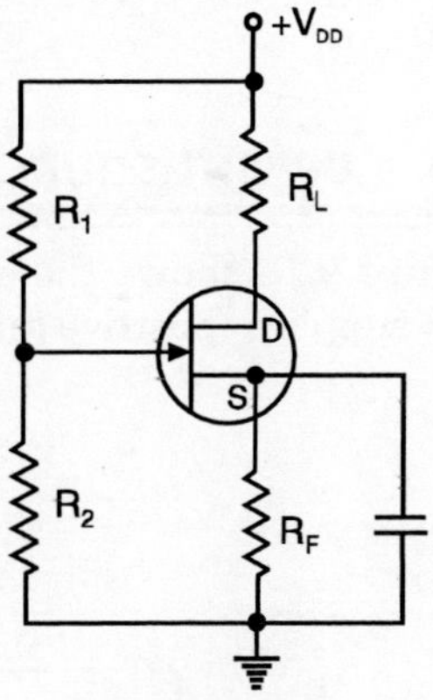

Fig. 9.16.

Solution: Given that

Drain supply voltage, $V_{DD} = 24$ Volt

Drain circuit or load resistance, $R_L = 910\ \Omega$

Source circuit resistance, $R_F = 2.29$ kΩ

Drain-source saturation current, $I_{DSS} = 5 \text{ mA} = 5 \times 10^{-3}$ A.

Pinch-off voltage, $V_p = -2$ V

$R_1 = 12$ MΩ

$R_2 = 8.57$ MΩ

Now, we know that the gate voltage is given by
$$V_G = \frac{V_{DD} \times R_2}{R_1 + R_2} = \frac{24 \times 8.57}{12+8.57} = 10 \text{ V}$$

Also, drain-current, I_D, is given by
$$I_D = I_{DSS}\left[1-\frac{V_{GS}}{V_p}\right]^2$$

Substituting values, we get $I_D = 5\times 10^{-3}\left[1-\frac{10-2.29\times 10^3 I_D}{-2}\right]^2$

or $I_D = 5 \times 10^{-3}\ [1 + 5 - 1,\ 145\ I_D]^2 = 0.005\ [6 - 1,\ 145\ I_D]^2$

Solving, we get $I_D = 4.46$ mA

Gate-source voltage, V_{GS} is given by

$$V_{GS} = V_G - I_D\ R_F = 10 - 2.29 \times 10^3 \times 4.46 \times 10^{-3}$$

or $V_{GS} = -\ 0.2134$ Volt

Current at operating point is given by

$$I_D = \frac{V_G - V_{GS}}{R_F} = \frac{10-0.2134}{2.29\times 10^3} = 4.274 \text{ mA} \qquad \textbf{Ans.}$$

It may be noted that the value of I_D at operating point is almost equal to previously computed value of I_D. Therefore, we can say that the FET is operating in pinch-off region.

9.8. SOURCE BIAS

Figure 9.17 shows the circuit of a source bias for JFET. It can be noted that this circuit is similar to the emitter bias circuit used for biasing bipolar transistors.

The value of drain current is given by the expression,

$$I_D = \frac{V_{SS} - V_{GS}}{R_S} \qquad \text{...(9.30)}$$

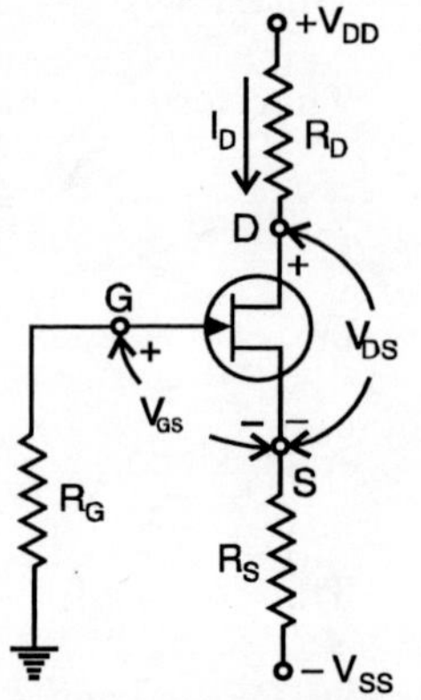

Fig. 9.17. Source bias

Because $V_{SS} >> V_{GS}$, therefore, drain current will be

$$I_D \cong \frac{V_{SS}}{R_S} \qquad \text{...(9.31)}$$

9.9. CURRENT SOURCE BIAS

Figure 9.18 shows the current-source bias circuit for a JFET. The circuit of figure 9.18 (*a*) uses two supplies (positive and negative), whereas figure 9.18(*b*) employs only one (*i.e.*, positive) supply.

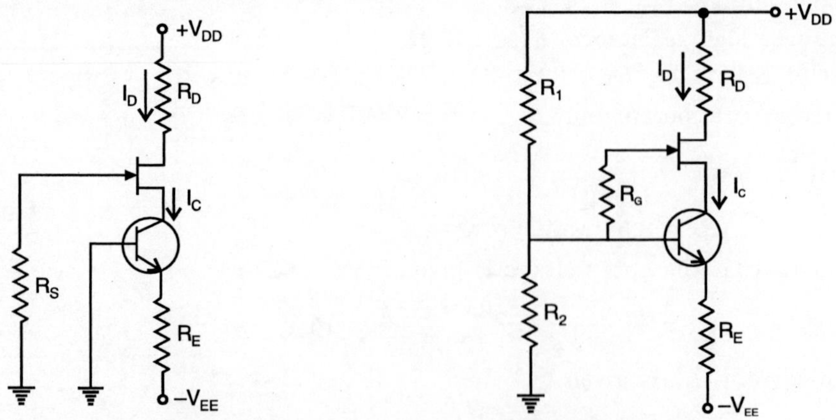

(a) Two supply circuits (b) One supply circuits

Fig. 9.18. Current-source bias circuit.

Firstly, let us consider the two supply current-source bias circuit. Here the bipolar transistor is emitter biased. Hence, its collector current is given by the expression

$$I_C = \frac{V_{EE} - V_{BE}}{R_E} \quad ...(9.32)$$

Now, if $V_{EE} >> V_{BE}$

then $$I_C = \frac{V_{EE}}{R_E}$$

The drain current of the JFET is equal to the collector current because the bipolar transistor acts like a d.c. current-source, *i.e.*,

$$I_D = I_C \quad ...(9.33)$$

Now, let us consider the one supply current-source bias circuit. Here, the bipolar transistor is voltage divider biased. Its collector current is given by the expression,

$$I_C = \frac{V_B - V_{BE}}{R_E} \quad ...(9.34)$$

and the drain current will be

$$I_D = I_C \quad ...(9.35)$$

The current-source bias circuit provides a solid *Q*-point with JFET. Thus, this circuit is very useful in actual practice.

9.10. BIASING THE ENHANCEMENT TYPE MOSFETs

We have already discussed in the chapter on field-effect transistors (FETs) that the operation of an enhancement types MOSFET requires a gate-to-source voltage greater than its threshold value (*i.e.*, $V_{GS} > V_{GS(th)}$). This eliminates the use of self-bias because it makes the gate voltage negative with respect to source. There are two methods for biasing the enhancement type MOSFETs namely the voltage divider bias and the drain feedback bias. The voltage divider bias has been already discussed. The drain feedback bias is similar to the collector feedback bias of bipolar transistors and is discussed as below:

Figure 9.19 shows the circuit of a drain feedback bias. It may be noted that this type of bias is used only for enhancement type MOSFETs. Since the value of gate current in FETs is approximately zero, therefore, there is no voltage drop across gate resistor (R_G). This makes the gate-to-source voltage equal to the drain-to-source voltage, *i.e.*,

$$V_{GS} = V_{DS}$$

The value of drain current is given by the following relation,

$$I_D = \frac{V_{DD} - V_{DS}}{R_D}$$

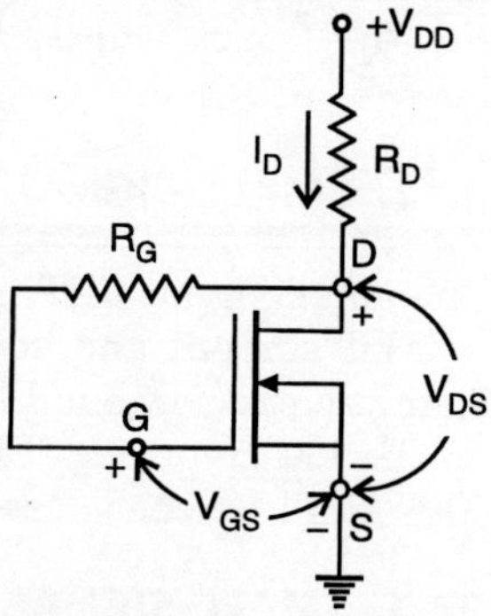

Fig. 9.19. Drain feedback bias

The drain feedback bias provides a stable *Q*-point, because it tends to compensate for any changes in device parameter variation.

9.11. BIASING THE DEPLETION TYPE MOSFETs

We have already discussed that the depletion type MOSFET can be operated with either positive or negative values of gate-to-source voltage (V_{GS}).

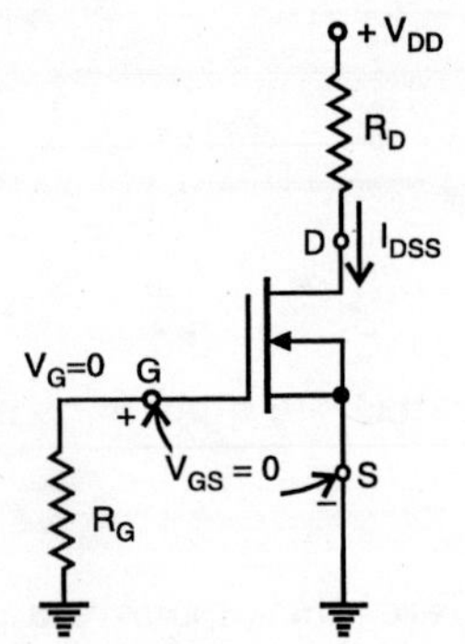

(a) Bias circuit for depletion type MOSFET

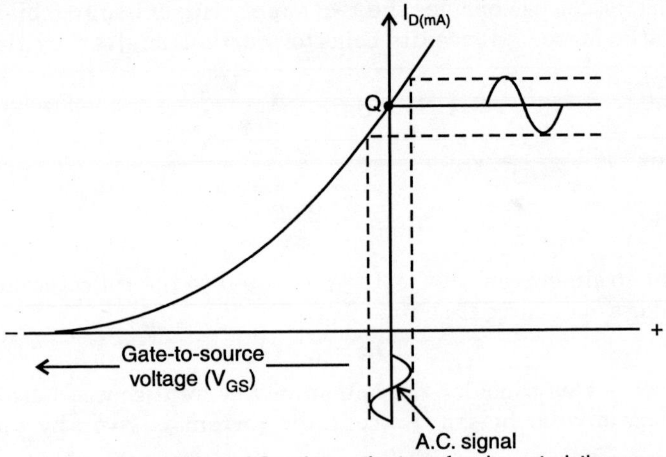

(b) Location of Q-point on the transfer characteristics curve

Fig. 9.20.

This means that the simplest method is to set the gate-to-source voltage equal to zero as shown in figure 9.20 (*a*). This will bias the MOSFET at a point which is lying on the vertical axis of the transfer characteristic curve as shown in figure 9.20 (*b*). If an a.c. signal is made to apply at the gate, it would vary the gate-to-source voltage above and below the *Q*-point.

Note that the biasing circuit of a depletion type MOSFET has no applied gate or source voltage. Hence, the gate-to-source voltage is equal to zero and the drain current is equal to its maximum value, (i.e., $I_D = I_{DSS}$*).*

The drain-to-source voltage is given by the following relation:

$$V_{DS} = V_{DD} - I_D \cdot R_D = V_{DD} - I_{DSS} \cdot R_D$$

It may be noted that the zero bias is possible only with depletion type MOSFET. It cannot be used for biasing the JFET or bipolar transistor. However, all the biasing methods, discussed for the JFET, can be used for depletion type MOSFET in the depletion mode but not in enhancement mode.

9.12. SMALL-SIGNAL FET MODEL

The small-signal FET model (valid both for JFET and MOSFET) is used to relate small changes in FET current and voltages about the quiescent operating point or simply *Q*-point. The model is different at low and high-frequencies. Therefore, we shall study the small-signal models separately as the low-frequency FET model and high-frequency model. However, in both these models, the FET shall be considered in common source configuration.

9.13. SMALL-SIGNAL LOW-FREQUENCY FET MODEL

Figure 9.21 shows the small-signal low-frequency model of a field-effect transistor (FET). In this model, the gate-to-source junction is represented by an open circuit and no current is drawn by the input terminal of the field-effect transistor (FET). It is because of the fact, that the input resistance (*i.e.*, the resistance of gate-source junction) is very high. Its value at d.c. or zero frequency is typically 10^8 to 10^{10} Ω for JFETs and 10^{10} or 10^{14} Ω for MOSFETs. However, for all practical purposes, these values can be considerd to be so high that the input of FET can be considered as an open circuit.

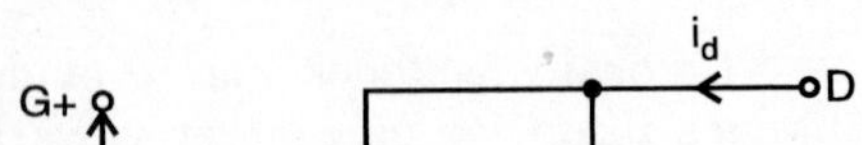

Fig. 9.21. Low-frequency FET model

It will be interesting to know that although the gate-source junction appears as an open circuit, yet the gate-to-source voltage affects the value of drain current. It is indicated by a voltage-controlled current source ($g_m \cdot v_{gs}$) whose value is proportional to the gate-to-source voltage. The FET transconductance (g_m) is measured in milliamperes per volt (mA/V), millisiemens (mS) or milli mhos (mΩ). Typical values of transconductance are from 0.5 mA/V to 10 mA/V for JFETs and 0.5 mA/V to 20 mA/V for MOSFETs.

The FET drain resistance (also FET output resistance) is represented by the resistance (r_d). Typical values of drain resistance are from 100 kΩ to 1 MΩ for JFETs and 1 kΩ to 50 kΩ for MOSFETs.

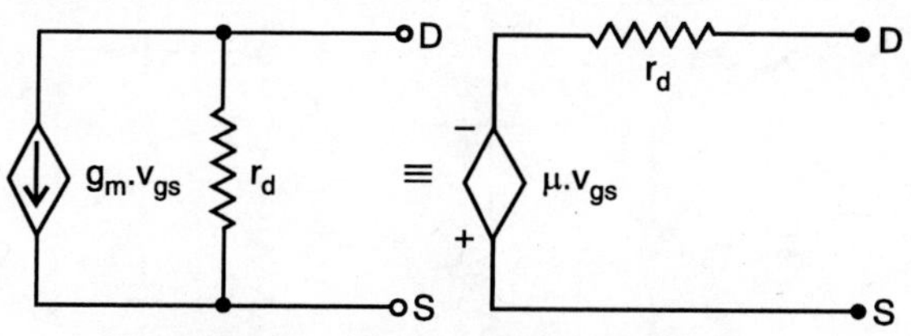

(a) Voltage-controlled current source with its equivalent voltage-controlled voltage source.

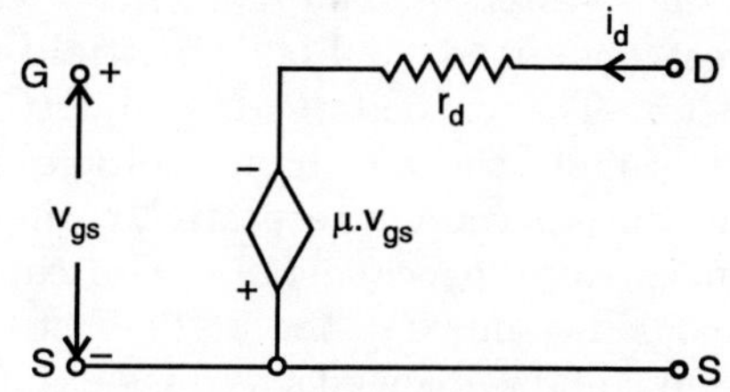

(b) Another form of low-frequecy FET model.

Fig. 9.22.

It is possible to convert the voltage-controlled current source ($g_m \cdot v_{gs}$) into its equivalent voltage-controlled voltage source as shown in figure 9.22 (a).

The equivalent circuit consists of a voltage-controlled voltage source ($\mu \cdot v_{gs}$) in series with the drain resistance (r_d). The parameter (μ) is the amplification factor of the field effect transistor (FET) and its value is equal to the product of transconductance (g_m) and the drain resistance (r_d). Mathematically, the amplification factor is given by

$$\mu = g_m \cdot r_d$$

Figure 9.22 (b) shows another form of the low frequency model of a field effect transistor (FET) using the voltage source representation. Sometimes this representation is also very useful in the analysis of the FET amplifiers.

9.14. SMALL-SIGNAL HIGH-FREQUENCY FET MODEL

Figure 9.23 shows the small-signal high-frequency model of a field-effect transistor (FET). It is identical to the low-frequency model, except the addition of capacitances between each pair of terminals. The capacitor, (C_{gs}) represents the barrier capacitance between the gate and source. Its typical value is from 1 pF to 10 pF for both JEFTs and MOSFETs. The capacitor (C_{gd}) represents the barrier capacitance between the gate and drain. Its typical value is also from 1 pF to 10 pF for both JFETs and MOSFETs. Similarly, the capacitor (C_{ds}) represents the drain-to-source capacitance. The typical value of C_{ds} is from 0.1 pF to 1 pF.

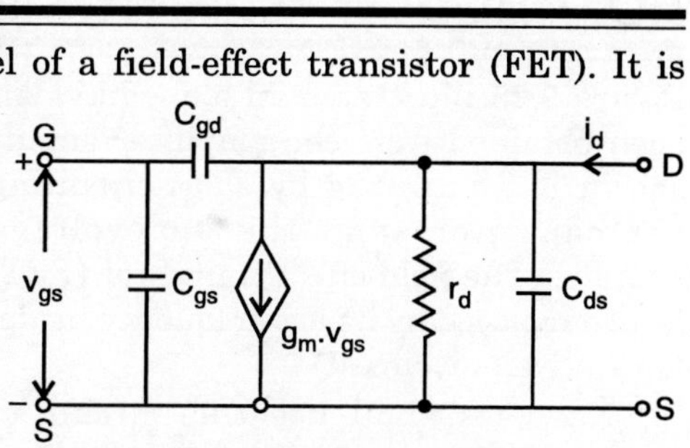

Fig. 9.23. High-frequency FET model.

9.15. FIELD-EFFECT TRANSISTOR AMPLIFIER

The field-effect transistor (FET) has a capability to amplify a.c. signals like a bipolar junction transistor (BJT). Depending open the types of configuration, the FET amplifiers may be classified under the following three heads:

(*i*) Common source amplifier
(*ii*) Common drain amplifier, and
(*iii*) Common gate amplifier.

9.16. COMMON SOURCE (CS) AMPLIFIER

(Very Very Important)

Figure 9.24 illustrates the circuit of a common source N-channel JFET amplifier. It is quite similar to a common emitter (CE) amplifier. Here, the resistors R_1 and R_2 (called a voltage divider) are used to bias the field-effect transistor. The capacitors (C_1) and (C_2) are used to couple the a.c. input voltage source and the output voltage respectively, these are known as *coupling capacitors*. The capacitor (C_s) keeps the source of the FET effectively at a.c. ground and is known as **bypass capacitor**.

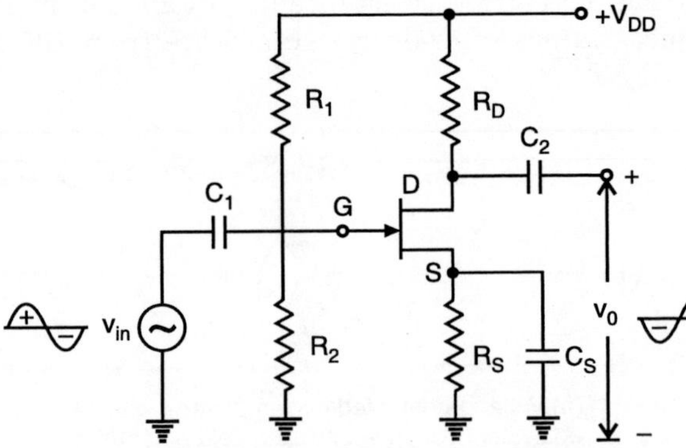

Fig. 9.24. A common source amplifier

The operation of the circuit may be understood from the assumption that when a small a.c. signal is made to apply to the gate, it produces variations in gate-to-source voltage which, in turn, produces variations in the drain current. As the gate-to-source voltage increases, the drain current also increases. Because of this, the voltage drop across the resistor (R_D) also increases. This causes the drain voltage to decrease. It means that the positive half cycle of the output voltage produces the negative half cycle of the output voltage. In other words, the output voltage (at the drain) is 180° out-of-phase with the input voltage (at the gate). This phenomenon of *phase inversion* is similar to that exhibited by a common emitter bipolar transistor amplifier.

> **DO YOU KNOW?**
>
> One of the most important applications of the JFET is the source follower, a circuit like the emitter follower, except that the input impedance is in the hundreds of megaohms for lower frequencies.

9.17. ANALYSIS OF COMMON SOURCE (CS) AMPLIFIER

Figure 9.25 illustrates an a.c. equivalent circuit of a common source amplifier. This circuit has been obtained from the amplifier circuit shown in figure 9.24 by short-circuiting the capacitors and the d.c. voltage supplies. The field-effect transistor (FET) is also replaced by its low-frequency model (or equivalent circuit).

Now, we shall use this circuit to derive the expressions for amplifier voltage gain, input resistance and output resistance.

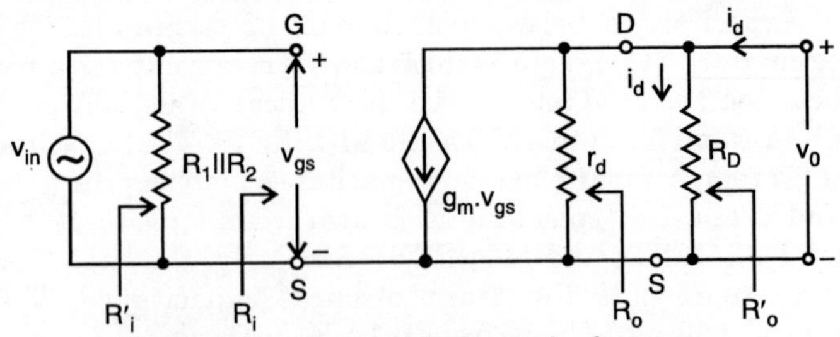

Fig. 9.25. A.C. equivalent circuit of a common source amplifier.

1. Voltage gain

It is the ratio of the output voltage (v_o) to the input voltage (v_{in}).

Mathematically the voltage gain,

$$A_v = \frac{v_o}{v_{in}}$$

It may be noted that the current from the current source divides between the resistors r_d and R_D. The current through resistor R_D (as per current divider rule) is given by the relation,

DO YOU KNOW?

For any JFET small-signal amplifier, the input signal that drives the gate should never reach a point at which the gate-source junction is forward biased.

$$i_d = \frac{r_d}{R_D + r_d}(g_m \cdot v_{gs})$$

and the output voltage will be

$$v_o = i_d \cdot R_D = -\left(\frac{r_d}{R_D + r_d}\right)(g_m \cdot v_{gs})\, R_D$$

or $$v_0 = -\, g_m\left(\frac{r_d \cdot R_D}{R_D + r_d}\right) v_{gs}$$

or $$v_0 = -\, g_m\,(r_d \,||\, R_D) \cdot v_{gs}$$

or $$v_0 = -\, g_m \cdot r_L \cdot v_{gs} \qquad \text{...}(\because r_L = r_d \,||\, R_D)$$

or $$v_0 = -\, g_m \cdot r_L\, v_{in} \qquad \text{...}(\because v_{in} = v_{gs})$$

Here, the minus sign indicates that the output voltage is 180° out-of-phase with the input voltage.

Now, voltage gain $$A_v = \frac{v_o}{v_{in}} = -\, g_m \cdot r_L$$

Note that if the drain resistance (r_d) is sufficiently large than R_D, (*i.e.*, $r_d \geq 10\ R_D$) then the equivalent resistance of the resistors r_d and R_D in parallel can be written as

$$r_L \cong R_D$$

and the voltage gain will be

$$A_v = -\, g_m \cdot R_D$$

Important Note: *If the resistance R_D is much larger than r_d (i.e., $R_D >> r_d$) then a.c. load resistance (r_L) is approximately equal to r_d. In that case, the amplifier gain is equal to the FET amplification factor (μ), which is the theoretical maximum voltage gain of the FET. However, in actual practice, we cannot select R_D to be much larger than r_d because it makes the d.c. bias current in the range of microamperes. The FET transconductances (i.e., g_m) is near zero at this low current value.*

2. Input resistance

It is the ratio of the input voltage (v_{in}) to the input current (i_{in}).

Mathematically, the input resistance will be given by

$$R_i = \frac{v_{in}}{i_{in}}$$

we know that resistance (R_i) of a field-effect transistor (FET) is very high and hence can be considered to be infinity (*i.e.*, open-circuit). However, the input resistance of the amplifier stage (R'_i) is equal to the parallel combination of resistors R_1 and R_2 and the FET input resistance (R_i). Thus, we have

$$R_i = (R_1 \,||\, R_2) \,||\, R_i$$
$$= R_1 \,||\, R_2 \qquad \text{... (when } R_i \text{ is infinite)}$$

It may be noted that if we use a self-bias circuit (instead of a voltage divider bias), then the input resistance of the amplifier stage will be

$$R'_i = R_G$$

3. Output resistance

It is the ratio of the output voltage (v_o) to the output current (i_d). Mathematically, the output resistance will be

$$R'_o = \frac{v_o}{i_d}$$

We know that the output resistance (R_o) of a field-effect transistor (FET) is r_d. But for the amplifier stage, the output resistance is the parallel combination of resistors R_D and r_d i.e.,

$$R'_0 = R_D || r_d$$

However, if $r_d >> R_D$, then the output resistance of amplifier stage will be

$$R'_0 = R_D$$

EXAMPLE 9.10 The input and output resistances of the FET amplifier have been shown in figure 9.26. Calculate the value of voltage gain. The FET amplifier has g_m = 2 mA/V and r_d = 40 kW.

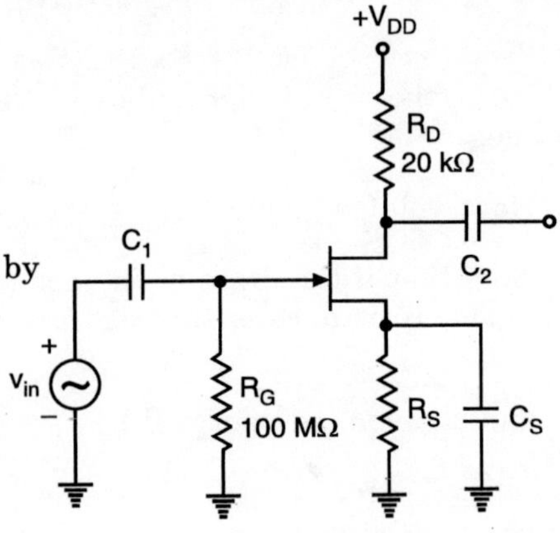

Fig. 9.26.

Solution: Given that $g_m = 2$ mA/V $= 2 \times 10^{-3}$ A/V;

$r_d = 40$ kΩ; $R_D = 20$ kΩ

and $R_G = 100$ MΩ.

Voltage gain

We know that the a.c. equivalent resistance is given by

$$r_L = \frac{R_D \times r_d}{R_D + r_d} = \frac{20 \times 10}{20 + 40}$$

or $r_L = 13.3$ kΩ $= 13.3 \times 10^3$ Ω

Therefore, voltage gain,

$$A_v = - g_m \cdot r_L$$
$$= - (2 \times 10^{-3}) \times (13.3 \times 10^3)$$
$$= -26.7 \quad \textbf{Ans.}$$

Input resistance

We know that the input resistance is given by

$R'_i = R_G = 100$ MΩ, **Ans.**

Output resistance

We know that the output resistance is given by

$R'_i = r_L = 13.3$ kΩ **Ans.**

9.18. EFFECT OF A.C. LOAD ON AMPLIFIER PARAMETERS

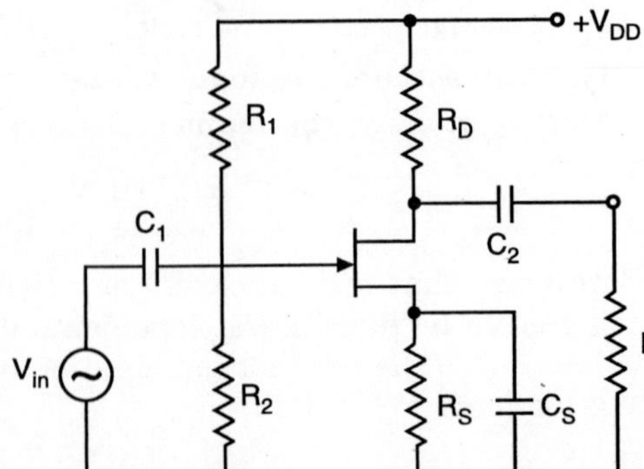

Fig. 9.27. Common source amplifier connected to a load resistor.

Let us consider a common source amplifier as shown in figure 9.27. Here, a load resistance (R_L) is connected to the amplifier's output through a coupling capacitor. In this case, the total or effective drain resistance is the parallel combination of resistors R_D and R_L i.e.,

$$r_L = R_D \parallel R_L$$

It may be noted that in the above expression, if we also consider the FET drain resistance (r_d), then the effective drain resistance will be

$$r_L = (R_D \parallel R_L) \parallel r_d$$

The effect of load resistor (R_L) is to reduce the voltage gain as well as the output resistance. The value of voltage gain with load resistance is given by the following relation:

$$A_v = -g_m \cdot r_L$$

$$= -g_m (R_D \parallel R_L) \parallel r_d$$

$$= -g_m (R_D \parallel R_L) \quad \text{... (if } r_d >> (R_D \,||\, R_L))$$

and the output resistance of the amplifier stage will be

$$R'_o = (R_D \parallel R_L) \parallel r_d$$

$$= R_D \parallel R_L \quad \text{...(if } r_d >> (R_D \,||\, R_L)$$

EXAMPLE 9.11 Figure 9.28 shows the circuit of a common source FET amplifier.

Determine the value of r.m.s output voltage. The g_m is 4500 μS and I_D is 2 mA. Neglect the FET drain resistance (r_d).

Solution: Given that $g_m = 4500\ \mu S = 4500 \times 10^{-6}$ S;

$R_D = 3$ kΩ; $v_{in} = 100$ mV $= 100 \times 10^{-3}$ V and $I_D = 2$ mA.

$R_L = 5$ kΩ

We know that the effective value of a.c. drain resistance is given by

$$r_L = R_D \,||\, R_L = \frac{R_D \times R_L}{R_D + R_L} = \frac{3 \times 5}{3+5} \text{ k}\Omega$$

$$= 1.87 \text{ k}\Omega = 1875\ \Omega$$

Therefore, output voltage,

$$v_o = -g_m \cdot r_L \times v_{in}$$

$$= -(4500 \times 10^{-6}) \times (1875) \times (100 \times 10^{-3})$$

$$= 0.844 \text{ V} \quad \textbf{Ans.}$$

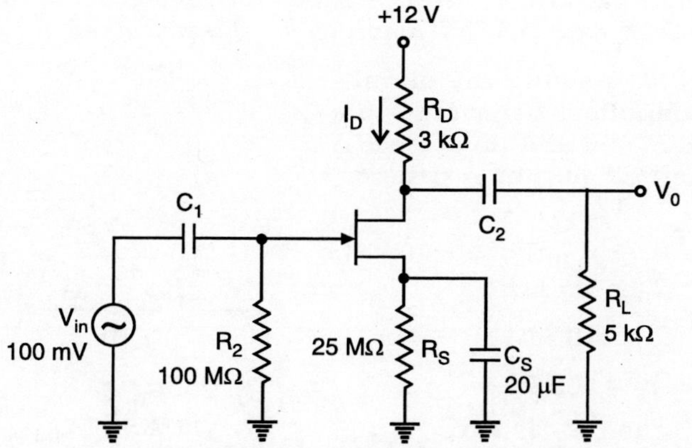

Fig. 9.28. Circuit for example 9.11

9.19. EFFECT OF EXTERNAL SOURCE RESISTANCE ON VOLTAGE GAIN

We have already discussed a field-effect transistor (FET) amplifier in which the source resistor is completely bypassed for a.c. signals. It means that for a.c. signals, the source is grounded. Now, let us consider the amplifier circuit in which we include an external resistor (r_s) to the source resistance (R_S) as shown in figure 9.29. The source is no longer at a.c. ground. The drain current through resistor (r_s) produced an a.c. voltage between the source and ground.

Note that the total input voltage between the gate and ground is given by

$$v_{in} = v_{gs} + i_d \cdot r_s$$

and the output voltage,

$$v_o = -\, i_d \cdot r_L$$

where r_L is the parallel combination of resistor R_D and the FET output resistance (r_d). Therefore, voltage gain is given by

$$A_v = \frac{v_o}{v_{in}} = -\frac{i_d \cdot r_L}{v_{gs} + i_d \cdot r_s}$$

Now, substituting the value of i_d (equal to $g_m \cdot v_{gs}$) in the above expression, we obtain

$$A_v = -\frac{g_m \cdot v_{gs} \cdot r_L}{v_{gs} + g_m \cdot v_{gs} \cdot r_s}$$

$$= -\frac{g_m \cdot v_{gs} \cdot r_L}{v_{gs}\,(1 + g_m \cdot r_s)} = -\frac{g_m \cdot r_L}{1 + g_m \cdot r_s} = \frac{r_L}{r_e + \dfrac{1}{g_m}}$$

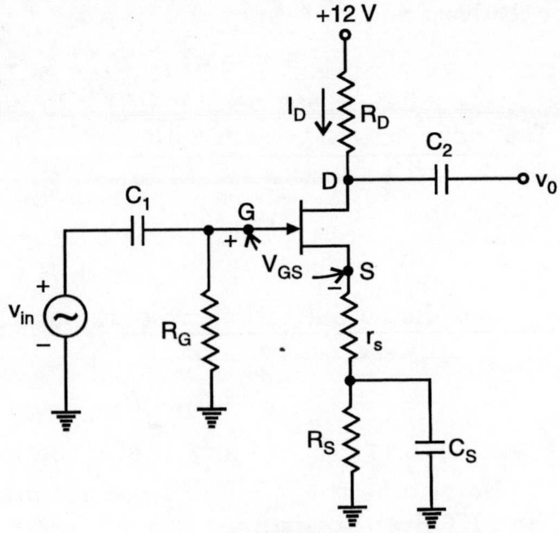

Fig. 9.29. JFET amplifier with external source resistance

It is obvious from the above expression that the external source resistance (r_s) reduces the voltage gain. If the value of resistor r_s is made sufficiently large as compared to $1/g_m$, then the voltage gain will be

$$A_v = -\frac{r_L}{r_s}$$

This means that the voltage gain is independent of the changes in the transconductance value. It is known as the ***swamping effect****. Because of this fact, the external resistance (r_s) is referred to as the swamping resistor and the FET amplifier as the* ***swamped amplifier****.*

EXAMPLE 9.12 Figure 9.30 shows the circuit of a swamped FET amplifier. Determine the voltage gain (*a*) when R_L = 0 and (*b*) when R_L is 100 kΩ. Neglect the FET output resistance (r_d). Take g_m = 4 mS.

Solution: Given that $g_m = 4 \text{ mS} = 4 \times 10^{-3}$ S

$R_D = 1.5$ kΩ;

$R_G = 10$ MΩ

and $r_s = 500$ Ω.

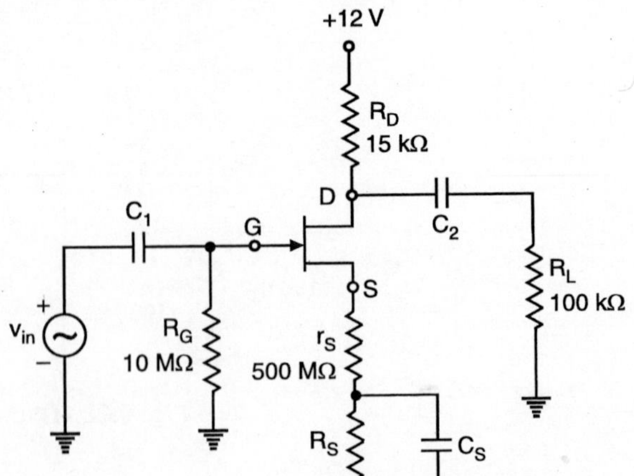

Fig. 9.30.

(*a*) Voltage gain when R_L is zero

We know that the effective a.c. drain resistance is given by

$$r_L = R_D = 1.5 \times 10^3 \ \Omega$$

and the voltage gain is given by

$$A_v = -\frac{g_m \cdot r_L}{1 + g_m \cdot r_s}$$

or $$A_v = -\frac{(4 \times 10^{-3}) \times (1.5 \times 10^{3})}{1 + (4 \times 10^{-3}) \times 500} = -2 \quad \textbf{Ans.}$$

(b) Voltage gain when R_L is 100 kΩ

We know that the effective a.c. drain resistance is given by

$$r_L = \frac{R_D \times R_L}{R_D + R_L} = \frac{1.5 \times 100}{1.5 + 100} = 1.48 \text{ k}\Omega = 1.48 \times 10^3 \ \Omega$$

and the voltage gain will be

$$A_v = -\frac{g_m \cdot r_L}{1 + g_m \cdot r_s} = -\frac{(4 \times 10^{-3}) \times (1.48 \times 10^{3})}{1 + (4 \times 10^{-3}) \times 500} = -1.97 \quad \textbf{Ans.}$$

9.20. THE COMMON DRAIN AMPLIFIER

Figure 9.31 shows the circuit of a common drain amplifier. It is similar to common collector (or emitter follower) amplifier. Self-biasing has been used in this circuit. The input signal is applied to the gate through a coupling capacitor (C_1). And the output is taken from the source terminal through the coupling capacitor (C_2).

DO YOU KNOW?

Because of the extremely high input impedance of a JFET, the input current is generally assumed to be 0 μA, and the current gain of a JFET amplifier is an undefined quantity.

The operation of the amplifier may be understood by assuming that when a small a.c. signal is applied to the gate, it produces variations in the gate-to-source voltage. This further produces the variations in drain current (I_D). As the gate-to-source voltage increases, the drain current also increases. Because of this, the voltage drop across the source resistor (R_S) also increases. Thus, the output voltage (v_o) increases.

It may be noted that the output voltage of a common drain amplifier is approximately equal to and in phase with the input voltage. Due to this fact, the circuit is known as ***source follower****.*

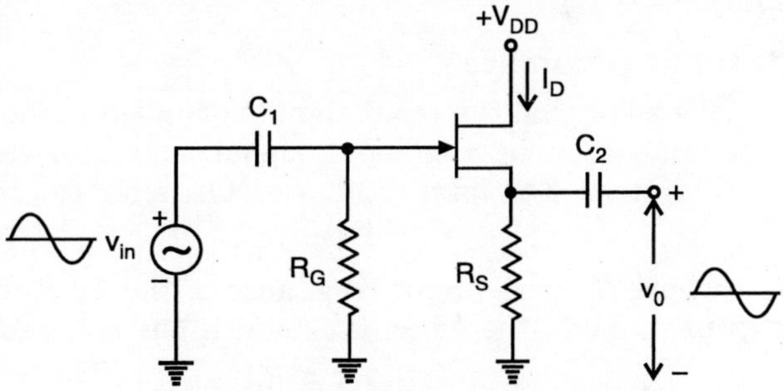

Fig. 9.31. Common drain amplifier

A common drain amplifier (or a source follower) has a very high value of input resistance. It is because of this property that a common drain amplifier is used at the front end of measuring instruments like electronic voltmeters and cathode ray oscilloscopes.

9.21. ANALYSIS OF A COMMON DRAIN AMPLIFIER

Figure 9.32 shows the a.c. equivalent circuit of a common drain amplifier. This circuit has been obtained from the amplifier circuit shown in figure 9.31 by short circuiting the capacitors and d.c. voltage supply. The field-effect transistor is also replaced by its low-frequency model. Now, we shall use this circuit to find the expressions for voltage gain, input resistance and output resistance.

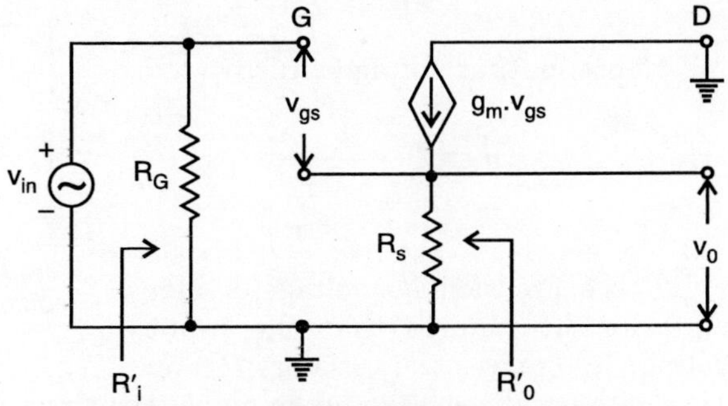

Fig. 9.32. A.C. equivalent circuit of a common drain amplifier.

1. Voltage gain

It is the ratio of output voltage (v_o) to the input voltage (v_{in}). Mathematically, the voltage gain is given by

$$A_v = \frac{v_o}{v_{in}}$$

For the common drain amplifier, the input voltage is expressed as

$$v_{in} = v_{gs} + i_d \cdot R_s$$

and the output voltage will be

$$v_o = i_d \cdot R_s$$

Therefore, voltage gain will be given by

$$A_v = \frac{v_o}{v_{in}} = \frac{i_d \cdot R_s}{v_{gs} + i_d \cdot R_s}$$

Now, substituting the value of i_d ($= g_m \cdot v_{gs}$) in the above expression, we get

$$A_v = \frac{(g_m \cdot v_{gs})\, R_s}{v_{gs} + g_m \cdot v_{gs} \cdot R_s} = \frac{g_m \cdot v_{gs} \cdot R_s}{v_{gs}\,(1 + g_m \cdot R_s)}$$

or

$$A_v = \frac{g_m \cdot R_s}{1 + g_m \cdot R_s} = \frac{R_s}{R_s + \dfrac{1}{g_m}}$$

It may be noted that when R_s is much greater than $1/g_m$, the value of voltage gain approaches unity.

2. Input resistance

We know that the input signal is applied to the gate of amplifier. As a result of this, the input resistance seen by the input signal source is extremely high just as in the common source configuration. The total input resistance (or the input resistance of the amplifier) will be

$$R'_i = R_G \,||\, R_i$$

where R_i is the input resistance of the field effect transistor (FET). Since, the value of R_i is extremely high, therefore, the total input resistance will be

$$R'_i = R_G$$

3. Output resistance

We know that voltage gain of the common drain amplifier is given by

$$A_v = \frac{v_o}{v_{in}} = \frac{R_s}{R_s + \dfrac{1}{g_m}}$$

Hence, output voltage will be

$$v_o = \left(\frac{R_s}{R_s + \dfrac{1}{g_m}}\right) \times v_{in}$$

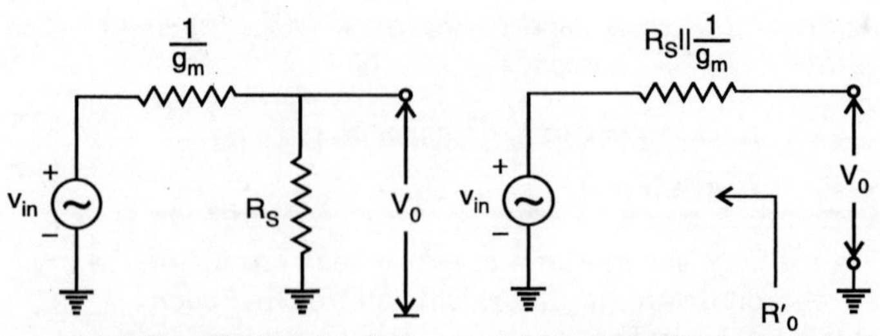

(a) A.C. equivalent circuit of output side of the amplifier

(b) Thevenin's equivalent circuit of the amplifier.

Fig. 9.33.

This expression is a voltage-divider equation. It implies that the input voltage (v_{in}) drives two resistors R_s and $1/g_m$ with output voltage taken across resistor (R_s). It means that the output side of the amplifier appears as shown in figure 9.33 (*a*). It is obvious from this figure that the source resistor is driven by an a.c. source with an output resistance *i.e.*,

$$R_D = \frac{1}{g_m}$$

It may be noted that output resistance (R_o) is the value of resistance looking back into the source terminal of the JFET. Now, let us thevenize the output circuit. The resulting circuit will be as shown in figure 9.33 (b). It is evident from this figure that the resistance R_s is in parallel with $1/g_m$ and the output resistance of the amplifier stage will be

$$R'_o = R_s \| \frac{1}{g_m}$$

If the value of source resistance (R_s) is much larger than $1/g_m$, then the output resistance of the amplifier stage will be

$$R'_o = \frac{1}{g_m}$$

EXAMPLE 9.13 Figure 9.34 shows the circuit of a source follower. Determine the voltage gain of the amplifier.

Also determine the input and output resistance of the amplifier. Assume g_m = 800 µS, infinite input resistance and neglect FET output resistance.

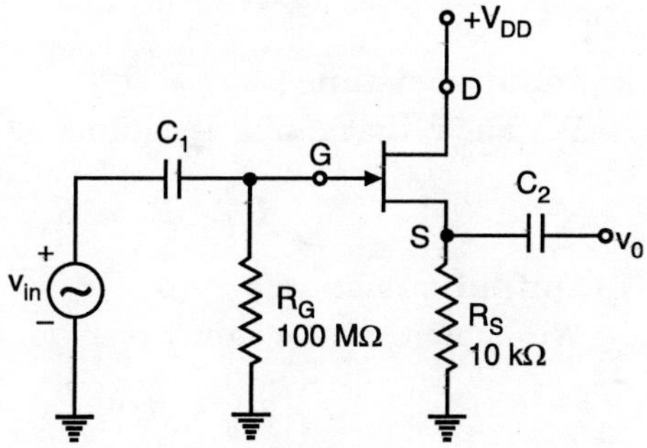

Fig. 9.34.

Solution: Given that $g_m = 8000\ \mu S = 8000 \times 10^{-6}$ S,

$R_s = 10\ k\Omega = 10 \times 10^3\ k\Omega$

and $R_G = 100\ M\Omega$.

Voltage gain

We know that the reciprocal of transconductance is given by

$$\frac{1}{g_m} = \frac{1}{8000 \times 10^{-6}} = 125\ \Omega$$

and the voltage gain will be

$$A_v = \frac{R_s}{R_s + \frac{1}{g_m}} = \frac{10 \times 10^3}{(10 \times 10^3) + 125} = 0.988 \quad \textbf{Ans.}$$

Input resistance

We know that the input resistance is expressed as

$$R'_i = R_G = 100\ M\Omega \quad \textbf{Ans.}$$

Output resistance

We also know that the output resistance is expressed as

$$R'_o = \frac{1}{g_m} = \frac{1}{8000 \times 10^{-6}}$$

$$= 125\ \Omega \quad \textbf{Ans.}$$

EXAMPLE 9.14 Find the voltage gain for the source follower shown in figure 9.35.

Also find the input and output resistance. If the input voltage is 2 mV, find the value of the output voltage. Assume g_m = 5500 µS.

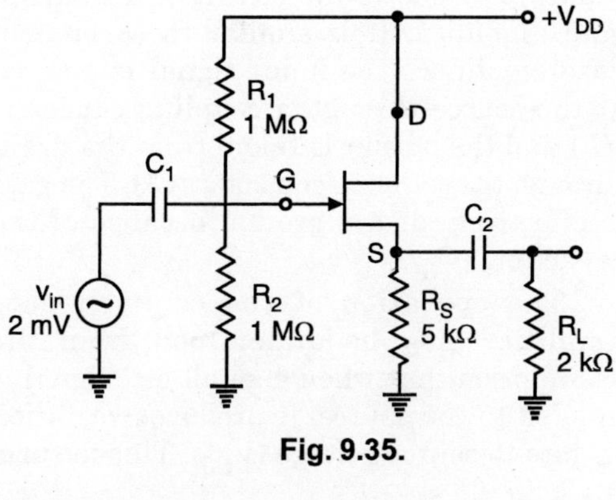

Fig. 9.35.

Solution: Given that v_{in} = 2 mA;

$g_m = 5500\ \mu S = 5500 \times 10^{-6}$ S

$R_1 = R_2 = 1\ M\Omega$

$$R_s = 5 \text{ k}\Omega = 5000 \ \Omega$$

and $$R_L = 2\text{k}\Omega = 2000 \ \Omega$$

Voltage gain

We know that the reciprocal of transconductance is given by

$$\frac{1}{g_m} = \frac{1}{5500 \times 10^{-6}} = 181.8 \ \Omega$$

and the voltage gain will be

$$A_v = \frac{R_s}{R_s + \frac{1}{g_m}} = \frac{5000}{5000 + 181.8} = 0.965 \quad \textbf{Ans.}$$

Input resistance

We know that input resistance is expressed as

$$R'_i = R_1 \,||\, R_2 = \frac{R_1 \cdot R_2}{R_1 + R_2} = \frac{1 \times 1}{1+1} \text{M}\Omega = 0.5 \text{ M}\Omega \quad \textbf{Ans.}$$

Output resistance

We know that the input resistance is given by

$$R'_o = R_s \,||\, \frac{1}{g_m} = 5000 \,||\, 181.8 = 175.4 \ \Omega \ \textbf{Ans}$$

Output voltage

Let v_o = Value of the output voltage.

Figure 9.36 shows the a.c. equivalent circuit of the output side of the source follower. It is evident from this figure, that an a.c. source of voltage $A_v \cdot v_{in}$ (equal to 0.965 × 2 or 1.93 V) is in series with an output resistance of 175.4 Ω.

We know that the a.c. voltage across the load resistor will be

$$v_o = \left(\frac{R_L}{R_L + R'_o}\right) \times A_v \cdot v_{in}$$

or $$v_0 = \left(\frac{2000}{2000 + 175.4}\right) \times 1.93 \text{ mV} = 1.77 \text{ mV} \quad \textbf{Ans.}$$

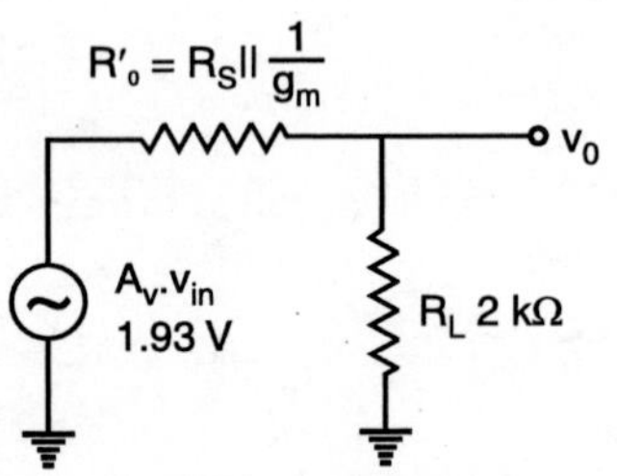

Fig. 9.36.

9.22. COMMON GATE AMPLIFIER

Figure 9.37 shows the circuit of a common gate amplifier. It is similar to a common base amplifier. The input signal is applied at the source through a coupling capacitor (C_1) and the output is taken from the drain through the coupling capacitor (C_2). The gate is effectively at a.c. ground because of the capacitor, (C_G).

The operation of the common gate amplifier may be understood from the assumption that when a small a.c. signal is applied to the source, it produces variation in gate-to-source voltage (V_{GS}). This in turn,

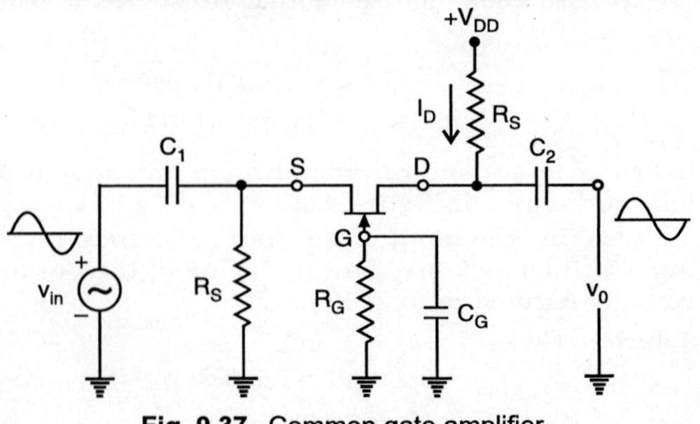

Fig. 9.37. Common gate amplifier.

produces the variations in drain current (I_D). As the gate-to-source voltage increases, the drain current also increases. As a result of this, the output voltage also increases. Thus, output voltage of a common gate amplifier is in phase with the input voltage.

A common gate amplifier has a low input resistance, high output resistance, high voltage gain and no phase reversal.

9.23. ANALYSIS OF A COMMON GATE AMPLIFIER

Figure 9.38 shows an a.c. equivalent circuit of a common gate amplifier. This circuit has been obtained from the amplifier circuit shown in figure 9.37 by short-circuiting the capacitors and the d.c. supply. The field-effect-transistor is also replaced by its low-frequency model.

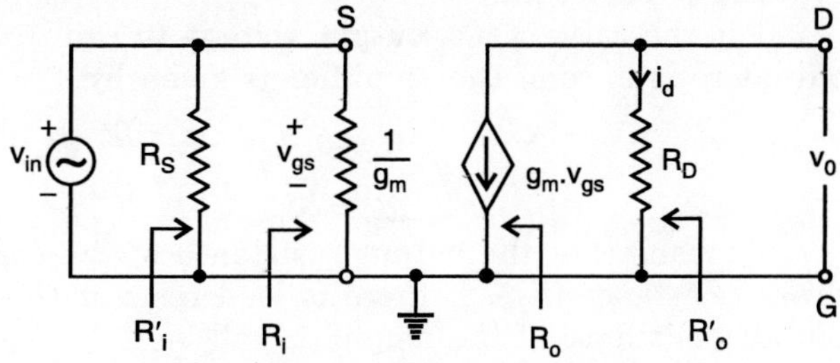

Fig. 9.38. Common gate amplifier.

Now we shall use this circuit to find the expressions for amplifier voltage gain, input resistance and output resistance.

1. Voltage gain

It is the ratio of a.c. output voltage (v_o) to the a.c. input voltage (v_{in}). Mathematically, the voltage gain,

$$A_v = \frac{v_o}{v_{in}}$$

We know that the a.c. input voltage is given by

$$v_{in} = v_{gs}$$

and the a.c. output voltage,

$$v_o = i_d \cdot R_D$$

∴ Voltage gain, $$A_v = \frac{i_d \cdot R_D}{v_{gs}}$$

Now, substituting the value of i_d (equal to $g_m \cdot v_{gs}$) in the above expression, we get

$$A_v = \frac{(g_m \cdot v_{gs}) \cdot R_D}{v_{gs}} = g_m \cdot R_D$$

It may be noted that the above expression for voltage gain is true only when there is no load resistor. If there is a load resistor (R_L) connected to the amplifier output, then the voltage gain will be given by

$$A_v = g_m\ (R_D \,||\, R_L)$$

2. Input resistance

It is the ratio of input voltage (v_{in}) to the input current (i_{in}). Mathematically, the input resistance,

$$R_i = \frac{v_{in}}{i_{in}}$$

For a common gate amplifier, the input voltage (v_{in}) is equal to the gate-to-source voltage (v_{gs}) and the input current (i_{in}) is equal to the drain current (i_d). Therefore, the input resistance will be given by

$$R_i = \frac{v_{gs}}{i_d} = \frac{v_{gs}}{g_m \cdot v_{gs}} \qquad \ldots (\because i_d = g_m \cdot v_{gs})$$

It may be noted that the resistance (R_i) is the value of input resistance looking directly into the source terminal of the JFET. The input resistance of the amplifier will be

$$R'_i = R_s \,||\, \frac{1}{g_m}$$

However, if R_s is much larger than the reciprocal of transconductance ($1/g_m$) then the input resistance of the amplifier

$$R'_i = R_s$$

It is evident from the above relation that a common gate amplifier has a low input resistance (equal to the value of external source resistance only). It is in contrast to the common source and common drain amplifier, which have extremely high output resistances.

3. Output resistance

It is the ratio of a.c. output voltage (v_o) to the a.c. output current (i_o). Mathematically, the output resistance of the amplifier is given by

$$R'_o = \frac{v_o}{i_o} = \frac{i_d \cdot R_D}{i_d} \qquad \ldots (\because v = i_d \cdot R_D \text{ and } i_o = i_d)$$

$$R'_o = R_D$$

It means that the output resistance of the common gate amplifier is equal to the external drain resistance (R_D). If there is load resistor (R_L) connected to the amplifier output, then the output resistance of the amplifier will be

$$R'_o = R_D \,||\, R_L$$

9.24. COMPARISON BETWEEN FET AND BJT

Table 9.4 shows some of the important points of comparison between the field effect transistor (FET) and bipolar junction transistor (BJT).

Table 9.4. Comparison Between Field-effect Transistor (FET) and Bipolar Junction Transistor (BJT) ***(Important)***

S.No.	Field-Effect Transistor (FET)	Bipolar junction transistor (BJT)
1.	It is an unipolar device *i.e.*, current in the device is carried either by electrons or holes.	It is bipolar device, *i.e.*, current in the device is carried by both electrons and holes.
2.	It is a voltage-controlled device, *i.e.*, voltage at the gate (or drain) terminal controls the amount of current flowing through the device.	It is a current-controlled device, *i.e.*, the base current controls the amount of collector current.
3.	Its input resistance is very high and is of the order of several megaohms.	Its input resistance is very low as compared to FET and is of the order of few kilo-ohms.
4.	It has a negative temperature coefficient at high current levels. It means that current decreases as the temperature increases. This characteristic prevents the FET from thermal breakdown.	It has a positive temperature coefficient at high current level. It means that collector current increases with the increase in temperature. This characteristic leads the BJT to thermal breakdown.
5.	It does not suffer from minority-carrier storage effects and therefore, has higher switching speeds and cut-off frequencies.	It suffers from minority carrier storage effects and therefore, has lower switching speed and cut-off frequencies than that of FET's.
6.	It is less noisy than a BJT or vacuum tube and is thus more suitable as an input amplifier for low-level signals. It is used extensively in high fidelity frequency modulated receivers.	It is comparatively more noisy than a field-effect transistor.
7.	It is much simpler to fabricate as an integrated circuit (IC) and occupies a less space on *IC* chip than that of BJT.	It is comparatively difficult to fabricate as an integrated circuit (IC) and occupies more space on IC chip than that of FET.

9.25. FET AS VOLTAGE VARIABLE RESISTANCE (VVR)

In almost all the linear applications of FET, it is operated in the constant current portion of its output characteristics (*i.e.* in the saturation region).

Now, let us observe the operation of JFET in the region just before pinch off. In this region, V_{DS} is small but I_D is large. This region is marked as ohmic region in the output characteristics of a JFET. In this region, the FET can be used as a voltage variable resistor (VVR) or voltage dependent resistor (VDR). The drain to source resistance r_{DS} in this region is dependent on the gate to source voltage. As V_{GS} becomes less negative, more drain current flows and r_{DS} decreases. For higher negative gate voltage, I_D decreases and r_{DS} will increase. Thus, the JFET acts as a voltage variable resistor (VVR).

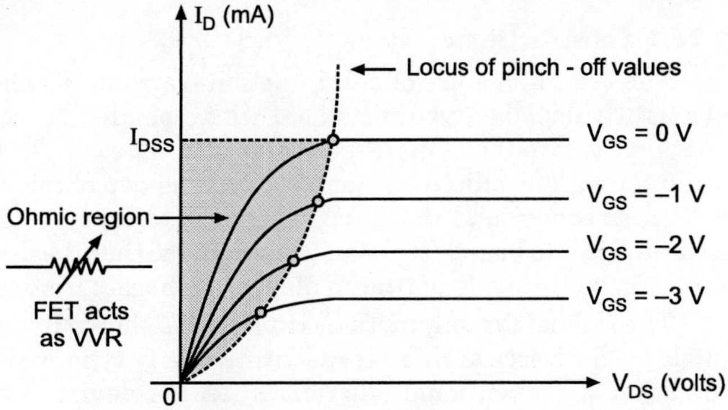

Fig. 9.39. Ohmic region in which FET acts as VVR

9.25.1. Application of VVR

(*i*) The FET used as VVR finds applications in many areas where this property can be used.

(*ii*) The VVR, for example, can be used to vary the voltage gain of a multistage amplifier A as the signal level is increased. This is called as **automatic gain control.**

(*iii*) A typical arrangement for controlling the gain of an amplifier automatically is shown in figure 9.40.

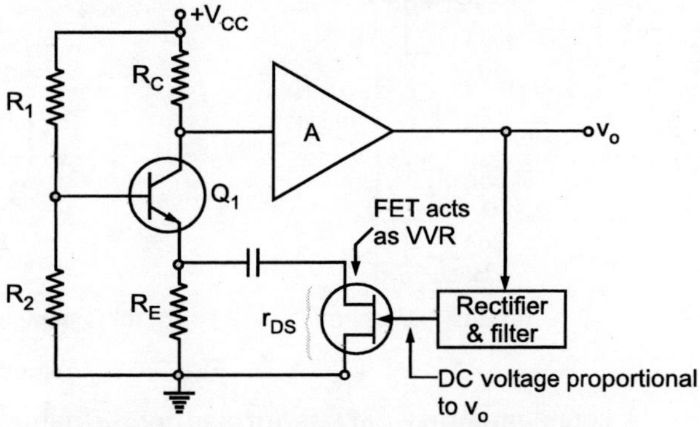

Fig. 9.40. AGC amplifier using FET as VVR.

Working Operation

The gain of the common emitter amplifier Q_1 is controlled. The output signal v_0 is rectified and filtered and the dc voltage proportional to v_0 is applied to the gate of FET operating in the ohmic region. Depending on the dc gate voltage, the resistance r_{DS} of FET will change, this will change the emitter resistance of Q_1 since R_E is in parallel with r_{DS}.

The gain of CE amplifier is approximately equal to $-\dfrac{R_C}{R_E}$. Hence, change in R_E will change the gain automatically. If v_0 increases, then dc voltage at the rectifier filter output will increase. Hence, V_{GS} becomes more negative. This decreases I_D or increases r_{DS} of FET. Hence, the parallel combination ($R_E \,||\, r_{DS}$) will increase and the gain of Q_1 will increase. Exactly opposite action takes place to increase gain of Q_1 if v_0 decreases. This is how the automatic gain control is achieved.

9.26 UNIJUNCTION TRANSISTOR (UJT)*

Unijunction transistor (abbreviated as UJT), also called the double-base diode is a 2-layer, 3-terminal

* Write short note on UJT characteristics and applications.

solid-state (silicon) switching device. The device has a unique characteristic that when it is triggered, its emitter current increases regeneratively (due to negative resistance characteristic) until it is restricted by emitter power supply. The low cost per unit, combined with its unique characteristic, have warranted its use in a wide variety of applications. A few include oscillators, pulse generators, saw-tooth generators, triggering circuits, phase control, timing circuits, and voltage-or current-regulated supplies. The device is in general, a low-power-absorbing device under normal operating conditions and provides tremendous aid in the continual effort to design relatively efficient systems.

9.26.1. Construction

The basic structure of a unijunction transistor is shown in figure 9.41(*a*). It essentially consists of a lightly-doped N-type silicon bar with a small piece of heavily doped p-type material alloyed to its one side to produce single P-N junction. The single *p-n* junction accounts for the terminology unijunction. The silicon bar, at its ends, has two ohmic contacts designated as base-1 (B_1) and base-2 (B_2), as shown and the p-type region is termed the emitter (E). The emitter junction is usually located closer to base 2 (B_2) than base-1 (B_1) so that the device is not symmetrical, because symmetrical unit does not provide optimum electrical characteristics for most of the applications.

The symbol for unijunction transistor is shown in figure 9.41 (*b*). The emitter leg is drawn at an angle to the vertical line representing the N-type material slab and the arrowhead points in the direction of conventional current when the device is forward biased, active or in the conducting state. The basic arrangement for the UJT is shown in figure 9.41 (*c*).

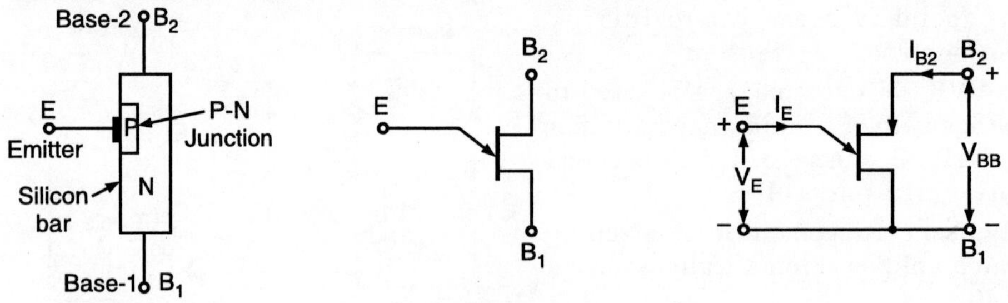

(a) Basic structure (b) Schematic symbol (c) Basic arrangement

Fig. 9.41. Unijunction transistor

A complementary UJT is formed by diffusing an N-type emitter terminal on a P-type base. Except for the polarities of voltage and current, the characteristics of a complementary UJT are exactly the same as those of a conventional UJT.

The worthnoting points about UJT are as under :

1. The device has only one junction, so it is called the unijunction device.
2. The device, because of one P-N junction, is quite similar to a diode but it differs from an ordinary diode that it has three terminals.
3. The structure of a UJT is quite similar to that of an N-channel JFET. The main difference is that *P*-type (gate) material surrounds the N-type (channel) material in case of JFET and the gate surface of the JFET is much larger than emitter junction of UJT.
4. In a unijunction transistor, the emitter is heavily doped while the N-region is lightly doped, so the resistance between the base terminals is relatively high, typically 4 to 10 kΩ when the emitter is open.
5. The N-type silicon bar has a high resistance and the resistance between emitter and base-1 is larger than that between emitter and base-2. It is because emitter is closer to base-2 than base-1.
6. UJT is operated with emitter junction forward-biased while the JFET is normally operated with the gate junction reverse-biased.

7. UJT does not have ability to amplify but it has the ability to control a large as power with a small signal.
8. It exhibits a negative resistance characteristic and so it can be employed as an oscillator.

9.26.2. Equivalent Circuit

The equivalent circuit of a UJT is shown in figure 9.42. The P-N junction is represented in the emitter by a diode.

Silicon bar being lightly doped has a high resistance and can be represented as two resistors connected in series- R_{B1} from base-1 to point A and R_{B2} *from* base-2 to point A, as shown in figure 5.62. The resistance R_{B1} *is* shown as a variable resistor since its magnitude varies with the emitter current I_E. In fact, for a typical UJT, R_{B1} may vary from 5 kΩ down to 50 Ω for a corresponding change of emitter current I_E from 0 to 50 µA. The total resistance of silicon bar from one end to the other end, *i.e.*, the resistance between B_2 and B_1 with emitter terminal *E* open is called the interbase resistance R_{BB}. From the equivalent circuit shown in figure 9.62 it is obvious that

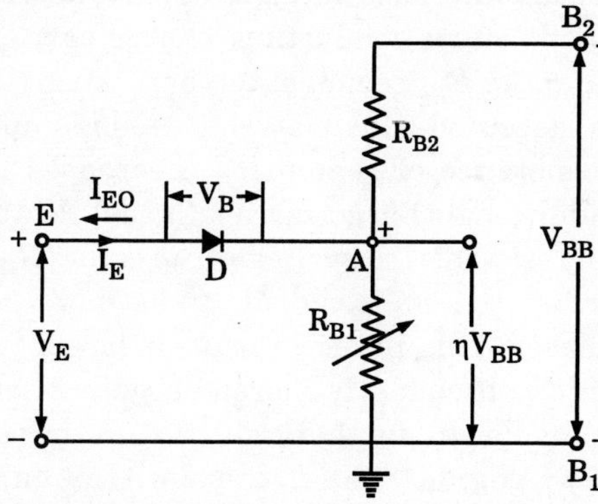

Fig. 9.42. Equivalent Circuit of a UJT

$$R_{BB} = R_{B_1} + R_{B_2} \text{ with emitter terminal } E \text{ open}$$

R_{BB} is typically within the range of 4 kΩ and 10 kW. Usually R_{B1} is greater than R_{B2} and R_{B1} = 60 percent of R_{BB}. With a voltage V_{BB} applied between B_2 and B_1, as shown in figure, the voltage at the junction of R_{B1} and R_{B2} (point A) is given by

$$V_A = V_{BB} \times \frac{R_{B_1}}{R_{B_1} + R_{B_2}} = \frac{R_{B_1}}{R_{BB}} V_{BB} \quad ...(9.80)$$

$$= \eta\, V_{BB} \quad ...(9.81)$$

The Greek letter η (eta) is called the intrinsic stand-off ratio of the device and is defined as

$$\eta = \frac{R_{B_1}}{R_{B_1} + R_{B_2}} \text{ with terminal } E \text{ open} = \frac{R_{B_1}}{R_{BB}} \quad ...(9.82)$$

The intrinsic stand-off ratio is the property of a UJT and is always less than unity. The typical range of η is from 0.5 to 0.8.

Voltage V_A is also the voltage at the cathode of the diode representing the P-N junction. With the emitter terminal open-circuited, the only current flowing is

$$I_{B2} = \frac{V_{BB}}{R_{BB}} \quad ...(9.83)$$

If the emitter terminal is grounded, the P-N junction is reverse-biased and a small emitter reverse current I_{E_0} flows.

9.26.3. Working Operation

Imagine that the emitter supply voltage is turned down to zero. Then the intrinsic stand-off voltage reverse-biases the emitter diode, as mentioned above. If V_B is the barrier voltage of the emitter diode, the total reverse bias voltage is $V_A + V_B + = \eta\, V_{BB} + V_B$. For silicon, $V_B = 0.7$ V.

Now, let the emitter supply voltage V_E be slowly increased. When V_E becomes equal to η V_{BB}, I_{EO} will be reduced to zero. With equal voltage levels on each side of the diode, neither reverse nor forward current will flow. When emitter supply voltage is further increased, the diode becomes forward-biased as soon as it exceeds the total reverse bias voltage (η $V_{BB} + V_B$). This value of emitter voltage V_E is called the total peak-point voltage and is denoted by V_P. When $V_E = V_P$, emitter current I_E starts to flow through R_{B1} to ground (*i.e.* B_1). This is the minimum current that is required to trigger the UJT. This is called the peak-point emitter current and denoted by I_P. I_P is inversely proportional to the interbase voltage, V_{BB}. Now, when the emitter diode starts conducting, charge carriers are injected into the R_{B_1} region of the bar. Since the resistance of a semiconductor material depends upon doping, the resistance of region R_{B_1} decreases rapidly due to additional charge carries (holds). With this decrease in resistance, the voltage drop across R_{B_1} also decrease, cause the emitter diode to be more heavily forward biased. This, in turn, results in larger forward current, and consequently more charge carriers are injected causing still further reduction in the resistance of the R_{B_1} region. Thus, the emitter current goes on increasing until it is limited by the emiltter power supply. Since V_A decreases with the increase in emitter current, the UJT is said to have **negative resistance characteristic.**

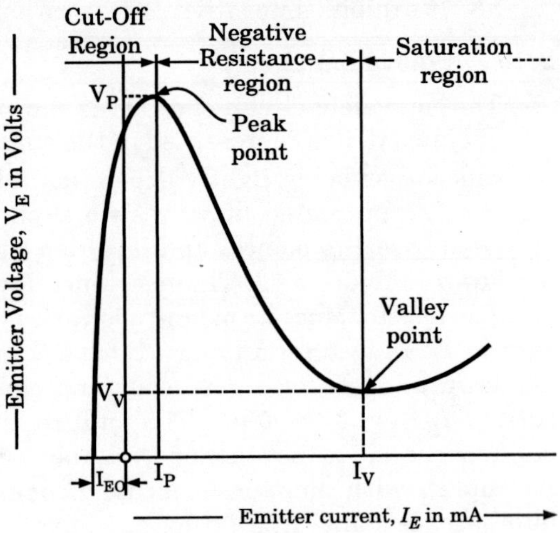

Fig. 9.43. Static emitter characteristic for a UJT

It is seen that the base-2 (B_2) is used only for applying external voltage V_{BB} across it. Terminals E and B_1 are the active terminals. UJT is usually triggered into conduction by applying a suitable positive pulse to the emitter. It can be turned off by applying a negative trigger pulse.

9.26.4. Characteristics*

The static emitter characteristic (a curve showing the relation between emitter voltage V_E and emitter current I_E) of a UJT at a given interbase voltage V_{BB} is shown in figure 9.43. It is noted that for emitter potentials to the left of peak point, emitter current I_E never exceeds I_{EO}. The current I_{EO} corresponds very closely to the reverse leakage current I_{CO} of the conventional BJT. This region, as shown in the figure, is called the cut-off region. Once conduction is established at $V_E = V_P$, the emitter potential V_E starts decreasing with the increase in emitter current I_E. This corresponds exactly with the decrease in resistance R_{B_1} for incresing current I_E. This device, therefore, has a negative resistance region which is stable enough to be used with a great deal of reliability in the areas of applications listed earlier. Eventually, the valley point reaches, and any further increase in emitter current I_E places the device in the saturation region, as shown in figure. Three other important parameters for the UJT are I_P, V_V and I_V (shown in figure 5.63) and are defined below :

Peak-point Emitter current I_P.

It is the emitter current at the peak point. It represents the minimum current that is required to trigger the device (UJT). It is inversely proportional to the interbase voltage V_{BB}.

Cut this Valley point voltage V_V.

The valley point is the emitter current at the valley point. The valley voltage increases with the increase in interbase voltage V_{BB}.

Valley point voltage V_V,

The valency point voltage is the emitter voltage at the valley point. The valley voltage increases with the increases in interbase voltage V_{BB}.

* Give the UJT characteristics and their application.

Valley point current I_V,

The valley point current is the emitter current at the valley point. It increases with the increase in inter-base voltage V_{BB}.

9.26.5. Special Features of UJT

The special features of a UJT are as under :

(*i*) A stable triggering voltage (V_P)– a fixed fraction of applied interbase voltage V_{BB}.
(*ii*) A very low value of triggering current.
(*iii*) A high pulse current capability.
(*iv*) A negative resistance characteristic.
(*v*) Low cost.

EXAMPLE 9.15. A silicon UJT has an interbase resistance of 8 kΩ and R_{B1} of 5 kΩ with I_E = 0. Determine (*i*) R_{B2} and (*ii*) intrinsic stand-off ratio. Determine the UJT current, stand-off voltage and peak-point voltage when a voltage of 15 V is applied between the bases. Take V_B = 0.7 V.

Solution : Interbase resistance, $R_{BB} = 8\text{ k}\Omega$

$$R_{B1} = 5\text{ k}\Omega$$

(*i*) $$R_{B2} = R_{BB} - R_{B1} = 8 - 5 = 3\text{ k}\Omega \textbf{ Ans.}$$

(*ii*) Intrinsic stand-off ratio, $\eta = \dfrac{R_{B_1}}{R_{BB}} = \dfrac{5}{8} = 0.625$ **Ans.**

Assuming V_E less than peak-point voltage, we have

$$\text{UJT current, } I_{B1} = I_{B2} = \frac{V_{BB}}{R_{BB}} = \frac{15}{8\text{ k}\Omega} = 1.875\text{ m A } \textbf{Ans.}$$

$$\text{Stand-off voltage} = \eta\, V_{BB} = 0.625 \times 15 = 9.375\text{ V} \quad \textbf{Ans.}$$

$$\text{Peak-point voltage, } V_P = \eta\, V_{BB} + V_B = 9.375 + 0.7 = 10.075\text{ V} \quad \textbf{Ans.}$$

9.27 UJT TRIGGERING OF AN SCR

One common application of the unijunction transistor is the triggering of the other devices such as the SCR, triac etc. The basic elements of such a triggering circuit are shown in figure 5.64. The resistor R_E is chosen so that the load line determined by R_E passes through the device characteristic in the negative resistance region, *i.e.*, to the right of the peak point but to the left of the valley point, as shown in figure 9.44. If the load line does not pass to the right of the peak point P, the device cannot turn on.

For ensuring turn -on of UJT, $R_E < \dfrac{V_{BB} - V_P}{I_P}$

This can be established as below:

Let us consider the peak point at which $I_{R_E} = I_P$ and $V_E = V_P$ (the equality $I_{R_E} = I_P$ is valid because the charging current of capacitor, at this instant is zero *i.e.*, the capacitor, at this particular instant, is changing from a charging state to a discharging state).

Then, we have $\quad V_E = V_{BB} - I_{R_E} R_E$

So, $$R_{E\,(\max)} = \frac{V_{BB} - V_E}{I_{R_E}}$$

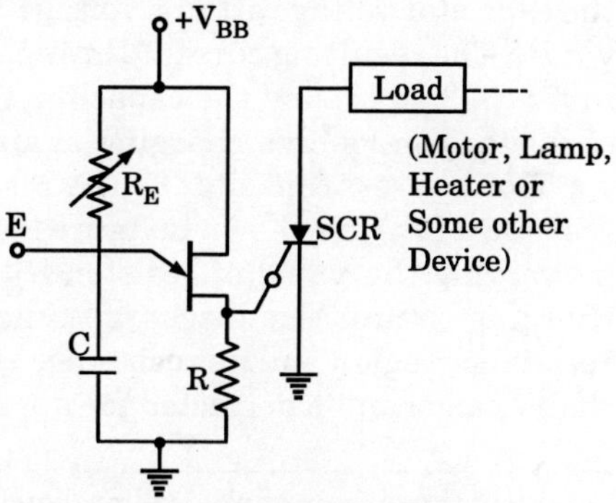

Fig. 9.44. UJT Triggering of an SCR

$$= \frac{V_{BB} - V_P}{I_P} \text{ at the peak point.}$$

or $$R_E < \frac{V_{BB} - V_P}{I_P}$$

At the valley point, V, we have

$$I_E = I_V \text{ and } V_E = V_V \text{, so that}$$

$$V_E = V_{BB} - I_{RE} R_E$$

So $$R_{E\,(min)} = \frac{V_{BB} - V_E}{I_{R_E}} = \frac{V_{BB} - V_V}{I_V}$$

or for ensuring turn off, we have

$$R_E > \frac{V_{BB} - V_V}{I_V}$$

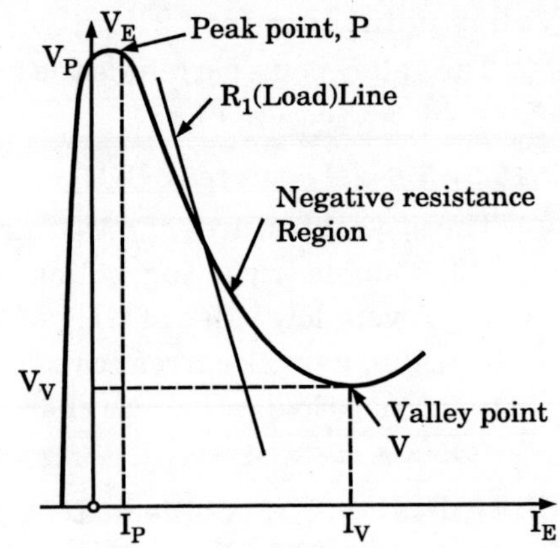

Fig. 9.45.

So, the range of resistor R_E is given as

$$\frac{V_{BB} - V_P}{I_P} > R_E > \frac{V_{BB} - V_V}{I_V}$$

The resistance R is chosen small enough so as to ensure that SCR is not turned on by voltage V_R when emitter terminal E is open or $I_E = 0$ (figure 9.45(a)).

$$\text{The voltage } V_R = \frac{R\,V_{BB}}{R + R_{BB}} \text{ for open-emitter terminal.}$$

The capacitor C determines the time interval between triggering pulses and the time duration of each pulse. By varying R_E, we can change the time constant $R_E\,C$ and alter the point at which the UJT fires. This allows us to control the conduction angle of the SCR, which means the control of load current.

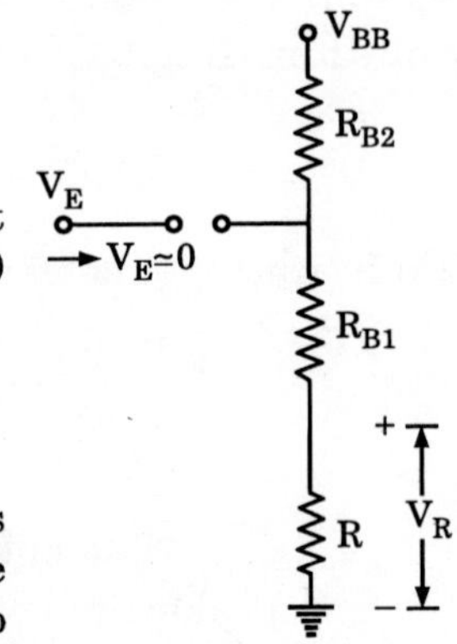

Fig. 9.45 (a).

9.28 UJT RELAXATION OSCILLATOR*

The relaxation oscillator shown in figure 9.46(a) consists of UJT and a capacitor C which is charged through resistor R_E when interbase voltage V_{BB} is switched on. During the charging period, the voltage across the capacitor increases exponentially until it attains the peak-point voltage V_P. When the capacitor voltage attains voltage V_P, the UJT switches on and the capacitor C rapidly discharges via B_1. The resulting current through the external resistor R develops a voltage spike, as illustrated in figure 9.46 (b) and the capacitor voltage drops to the value V_V. The device then cuts off and the capacitor commences charging again. The cycle is repeated continuously generating a saw-tooth waveform across capacitor C. The resulting waveforms of capacitor voltage V_C and the voltage across resistor R (V_B) are shown in figure 9.46(b). The frequency of the output saw-tooth wave can be varied by varying the value of resistor R_E as it controls the time constant ($T = R_E C$) of the capacitor charging circuit. The discharge time t_2 is difficult to calculate because the UJT is in its negative resistance region and its resistance is continually changing. However, t_2 is normally very much less than t_1 and can be neglected for approximation.

* Draw the circuit of an oscillator using an UJT. Discuss its working.

A kind heart is a fountain of gladness, making everything in its vicinity freshen into smiler.
— **Washington Irving**

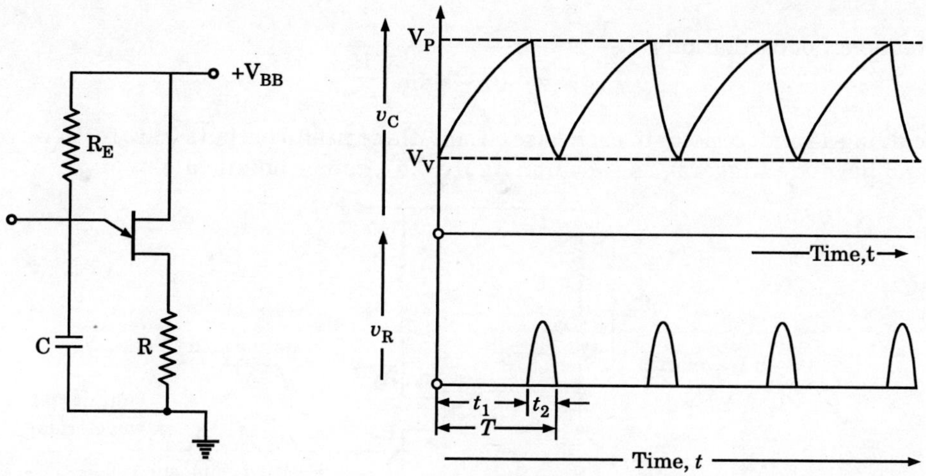

(a) Basic circuit (b) Voltage waveforms across C and R

Fig. 9.46. UJT relaxation oscillator

For satisfactory operation of the above oscillator, the following two conditions for the turn-on and turn-off of the UJT must be met, as derived earlier, *i.e.*,

$$R_E < \frac{V_{BB} - V_P}{I_P} \quad \text{and} \quad R_E > \frac{V_{BB} - V_V}{I_V}$$

i.e. the range of resistor R_E should be as given below

$$\frac{V_{BB} - V_P}{I_P} > R_E > \frac{V_{BB} - V_V}{I_V}$$

The time period and, therefore, frequency of oscillation can be derived as below.

During charging of capacitor, the voltage across the capacitor is given as

$$v_C = V_{BB}\,(1 - e^{-t/R_EC})$$

Where R_EC is the time constant of the capacitor charging circuit and t is the time from the commencement of the charging. The discharge of the capacitor commences at the end of charging period t_1, when the voltage across the capacitor v_C becomes equal to V_P *i.e.*, ($\eta\ V_{BB} + V_B$)

$$V_P = \eta\ V_{BB} + V_B = V_{BB}\,(1 - e^{-t_1/R_EC})$$

Neglecting V_B in comparison to $\eta\ V_{BB}$, we have

$$\eta\ V_{BB} = V_{BB}\,(1 - e^{-t_1/R_EC})$$

or $$e^{-t_1/R_EC} = 1 - \eta$$

or $$e^{t_1/R_EC} = \frac{1}{1-\eta}$$

or $$t_1 = R_EC \log_e \frac{1}{1-\eta}$$

So, charging time period, $t_1 = 2.3\ R_EC \log_{10}\left(\frac{1}{1-\eta}\right)$

Since discharging time duration t_2 is negligibly small as compared to charging time duration t_1, so taking time period of the wave, $T = t_1$

Time period of the saw-tooth wave, $T \cong 2.3\ R_EC\ \log_{10} \frac{1}{1-\eta}$

and frequency of oscillation $f = \frac{1}{T} \cong \frac{1}{2.3\, R_E C \log_{10} \frac{1}{1-\eta}}$ ...(5.84)

By including a small resistor in each base circuit, three useful outputs (saw-tooth waves, positive triggers, and negative triggers), as shown in figure 9.47, can be obtained.

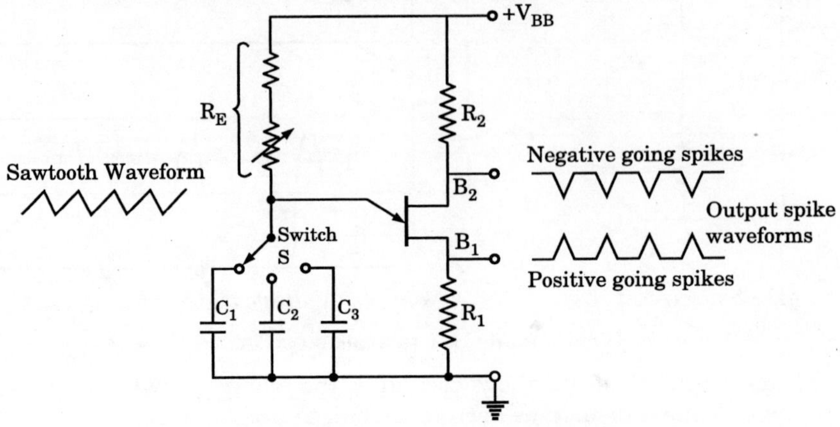

Fig. 9.47. Variable frequency UJT relaxation oscillator

When the UJT fires, the surge of current through B_1 causes a voltage drop across R_1 and produces the positive going spikes. Also at the UJT firing time, the fall of V_{EB1} causes I_{B2} to rise rapidly and generate the negative-going spikes across R_2, as shown in figure 5.67. R_1 and R_2 should be much smaller than R_{BB} to avoid altering the firing voltage of the UJT. A wide range of oscillation frequencies can be achieved by making R_E adjustable and including a switch to select different values of capacitance, as illustrated.

As already discussed, there is upper and lower limits to the signal source resistance R_E for the satisfactory operation of the UJT.

EXAMPLE 9.16. A UJT relaxation oscillator circuit shown in fig. 5.38(*a*) has R_E = 60 k Ω and C = 0.25 μ F. Determine the pulse repetition frequency. Take instrinsic stand-off ratio to be 0.65.

Solution: We know that Pulse repetition frequency is given by

$$f = \frac{1}{2.3 R_E C \log_{10} \frac{1}{1-\eta}}$$

$$f = \frac{1}{2.3 \times 60 \times 10^3 \times 0.25 \times 10^{-6} \log_{10} \frac{1}{1-0.65}} = 63.5 \text{ Hz} \quad \textbf{Ans.}$$

EXAMPLE 9.17 A transistor is connected in *CB* configuration. When the emitter voltage is changed by 200 mV, the emiter current changes by 5 mA. During this variation, the collector to base voltage is kept fixed. Calculate the dynamic input resistance of transistor.

Solution: The dynamic input resistance for *CB* configuration is given by

$$r_i = \left.\frac{\Delta v_{EB}}{\Delta i_E}\right|_{V_{CB} = \text{constant}}$$

Given that $\Delta v_{EB} = 200 \text{ mV}$

and $\Delta i_E = 5 \text{ mA}$

Therefore, $r_i = \dfrac{200 \text{ mV}}{5 \text{ mA}} = 40 \text{ Ohms.}$ **Ans.**

EXAMPLE 9.18 **Figure 9.48 shows the circuit of fixed bias using silicon transistor. Determine the following:**

(i) base current I_B
(ii) collector current I_C
(iii) V_{CE}
(iv) V_C and V_B
(v) V_{CB}

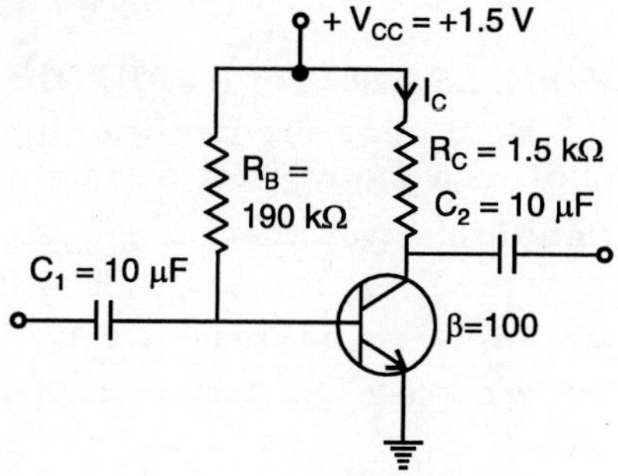

Fig. 9.48.

Solution: (i) We know that

$$V_{CC} = V_{BE} + I_B.R_B$$

Therefore, base current $I_B = \dfrac{V_{CC} - V_{BE}}{R_B}$

$$= \frac{15 - 0.7}{180 \times 10^3} = 79.44 \text{ μA} \quad \textbf{Ans.}$$

(ii) Collector Current , $I_C = \beta I_B$

$$I_C = 100 \times 79.44 \text{ mA} = 7.94 \text{ mA} \quad \textbf{Ans.}$$

(iii) Since $V_{CC} = V_{CE} + I_C R_C$, therefore

$$V_{CE} = V_{CC} - I_C.R_C$$
$$= 15 - 7.94 \times 10^{-3} \times 1.5 \times 10^3$$
$$V_{CE} = 15 - 11.9 = 3.1 \text{ Volt}$$

(iv)
$$V_C = V_{CE} = 3.1 \text{ Volt} \quad \textbf{Ans.}$$
$$V_B = V_{BE} = 0.7 \text{ Volt} \quad \textbf{Ans.}$$

(v)
$$V_{CB} = V_C - V_B = 3.1 - 0.7$$
$$V_{CB} = 2.4 \text{ Volt} \quad \textbf{Ans.}$$

FEW IMPORTANT SOLVED EXAMPLES

EXAMPLE 9.19 **Figure 9.49 shows the circuit of a common gate JEFT amplifier. The JFET has transconductance g_m = 2500 μS.**

Determine the amplifier voltage gain and input resistance.

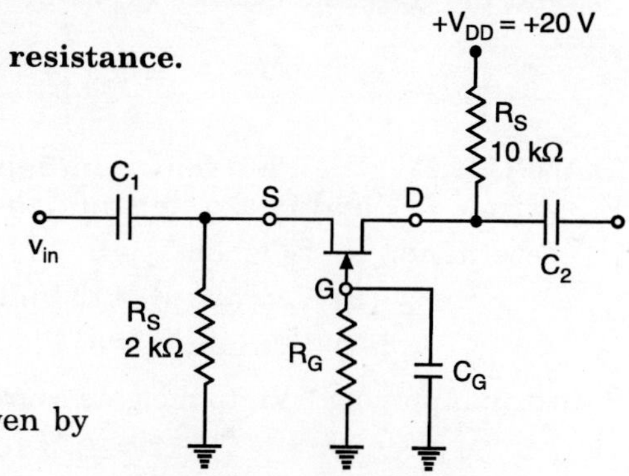

Fig. 9.49. Circuit for example 9.19.

Solution: Given that g_m = 2500 μS = 25000 × 10^{-6} S

R_s = 2 kΩ = 2000 Ω

and R_D = 10 kΩ = 10,000 Ω

Voltage gain

We know that voltage gain is given by

$A_v = g_m \cdot R_D$ = (2500 × 10^{-6}) × 10,000 = 25 **Ans.**

Input resistance

We know that reciprocal of transconductance is given by

$$\frac{1}{g_m} = \frac{1}{2500 \times 10^{-6}} = 400 \ \Omega$$

and the amplifier input resistance will be

$$R'_i = R_i \,||\, \frac{1}{g_m} = 2000 \,||\, 400\ \Omega$$

or $$R'_i = \frac{2000 \times 400}{2000 + 400} = 333\ \Omega \quad \textbf{Ans.}$$

EXAMPLE 9.20 **For a JFET, the transcondutance at zero gate-to-source voltage, g_{mo} = 5 mS. Determine the amplifier input resistance (R'_i) and a.c. voltage gain (v_o/v_{in}) for the circuit shown in figure 9.50. Assume the capacitors to be short circuit for a.c.**

Solution: Given that g_{mo} = 5 mS = 5 × 100^{-3} S; R_D = 1 kΩ = 1 × 10^3 Ω

R_s = 200 Ω and I_D = 5 mA = 5 × 10^{-3} A

Input resistance

We know that reciprocal of transconductance is given by

$$\frac{1}{g_{mo}} = \frac{1}{5 \times 10^{-3}} = 200\ \Omega$$

and the input resistance of the amplifier will be

$$R'_i = R_s \,||\, \frac{1}{g_{mo}}$$

$$= 200 \,||\, 200\ \Omega$$

$$= 100\ \Omega \quad \textbf{Ans.}$$

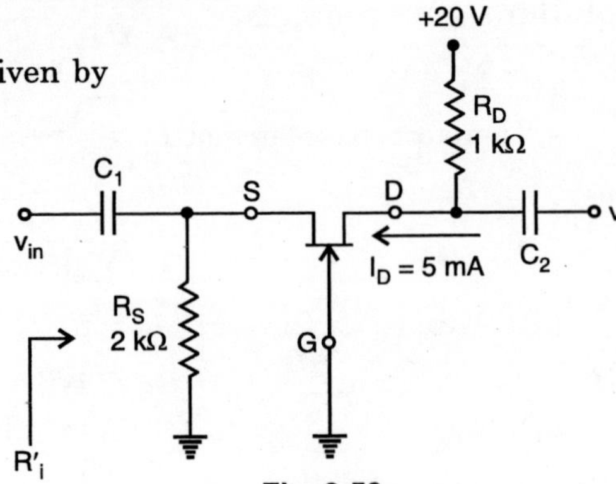

Fig. 9.50.

Voltage gain

We know that the d.c. sources voltage is given by

$$V_s = I_D \cdot R_s = (5 \times 10^{-3}) \times 200 = 1 \text{ Volt}$$

and the gate-to-source voltage will be

$$V_{GS} = V_s = 1\text{ V}$$

We also know that the maximum value of drain current will be given by

$$I_{DSS} = 2\, I_D = 2 \times (5 \times 10^{-3}) = 10 \times 10^{-3}\text{ A}$$

and gate-to-source cut-off voltage will be

$$V_{GS(\text{off})} = -\frac{2 I_{DSS}}{I_D} = -\frac{2 \times (10 \times 10^{-3})}{5 \times 10^{-3}} = -4\text{V}$$

and the transconductance (g_m) will be given by

$$= g_{mo}\left[1 - \frac{V_{GS}}{V_{GS(off)}}\right] = (5 \times 10^{-3})\left[1 - \frac{1}{4}\right] = 3.75 \times 10^{-3}\text{ S} \quad \textbf{Ans.}$$

EXAMPLE 9.21 **The FET shown in figure 9.51 has the following parameters: I_{DSS} = 5.6 mA, V_P = 4 Volt. If V and V_0 are constant, then**

determine: (*i*) V_0 when $V_i = 0$

(*ii*) V_0 when V_i = 10 Volts

(*iii*) V_i when $V_0 = 0$

Solution: Applying KVL to the gate source loop in figure 9.51, we have

$$12 = I_D \times 10 - V_{GS}$$

or $$V_{GS} = 10 \times I_D - 12$$

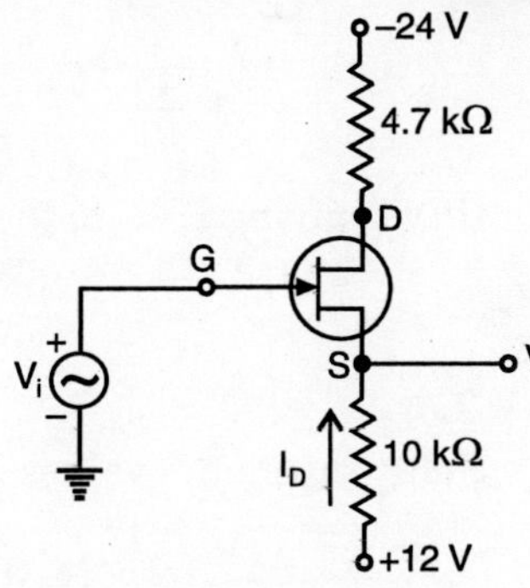

Fig. 9.51. Circuit for example 9.21.

or $$I_D = \frac{V_{GS}+12}{10} \quad ...(i)$$

Since, we do not know the value of I_D, therefore, let us find the value of I_D

$$I_D = I_{DSS}\left[1-\frac{V_{GS}}{V_P}\right]^2 = 5.6\left[1-\frac{V_{GS}}{4}\right]^2 \quad ...(ii)$$

Now substituting V_{GS} from equation (*i*) into equation (*ii*), we get

$$I_D = 5.6 \times 10^{-3}\left[1-\frac{10I_D-12}{4}\right]^2$$

or $$I_D = 5.6 \times 10^{-3} - \frac{56}{4}I_D + 3 \times 5.6 \times 10^{-3}$$

$\therefore$ $$15\ I_D = 22.4 \times 10^{-3}$$

or $$I_D = 1.49 \text{ mA} \quad ...(iii)$$

Substituting the value of I_D into equation (*i*) we get

$$V_{GS} = (10 \times 1.49) - 12 = 2.9 \text{ Volts}$$

The output voltage will be

$$V_0 = 12 - I_D R_S$$

Substituting values, we get

$$V_0 = 12 - (1.49 \times 10) = -29 \text{ Volts} \quad \textbf{Ans.}$$

EXAMPLE 9.22 **The JFET used in figure 9.52 has $V_P = -2$ V, $I_{DSS} = 1.65$ mA. It is desired to bias the circuit at $I_D = 0.8$ mA. If $V_{DD} = 24$ V and $r_d >> R_D$.**

Find: (*i*) V_{GS}, (*ii*) g_m, (*iii*) R_S.

Solution: (*i*) We know that

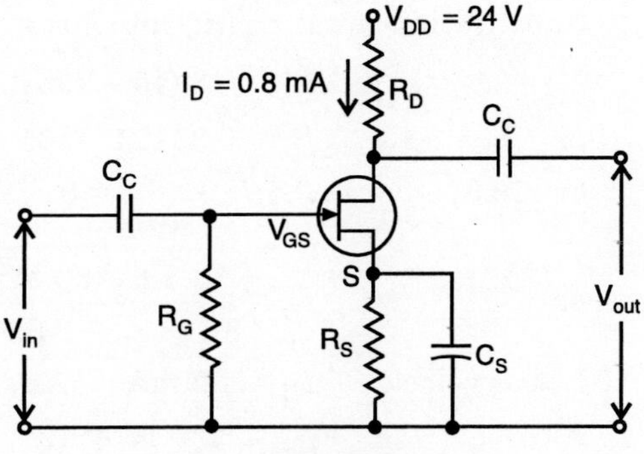

Fig. 9.52.

$$I_D = I_{DSS}\left[1-\frac{V_{GS}}{V_P}\right]^2 \quad ...(i)$$

or $$V_{GS} = V_P\left[1-\sqrt{\frac{I_D}{I_{DSS}}}\right]$$

Substituting the values, we get

$$V_{GS} = -2\left[1-\sqrt{\frac{0.8}{1.65}}\right] = -2\left[1-\frac{1}{\sqrt{2}}\right]$$

$$= +\ 0.6074 \text{ Volts} \quad \textbf{Ans.}$$

(*ii*) We know that $$g_m = g_{mo}\left[1-\frac{V_{GS}}{V_P}\right] \quad ...(ii)$$

But we do not know the value of g_{mo} so let us find it first as under:

$$g_{mo} = \frac{-2I_{DSS}}{V_P}$$

Substituting the values of I_{DSS} and V_P, we get

$$g_{mo} = \frac{-2 \times 1.65 \times 10^{-3}}{-2} = 1.65 \times 10^{-3} \quad ...(iii)$$

Substituting this value into equation (*iii*), we get

$$g_m = 1.65 \times 10^{-3}\left[1-\frac{0.6074}{2}\right]$$
$$= 1.15 \times 10^{-3} \text{ mho} \quad \textbf{Ans.}$$

(*iii*) Applying KVL to the gate to source circuit of figure 9.52, we get

$$0 = V_{GS} + I_D R_S$$

or
$$R_S = \frac{-V_{GS}}{I_D}$$

Substituting the values, we get

$$R_S = \frac{0.6074}{0.8 \times 10^{-3}} = 753.25\ \Omega \quad \textbf{Ans.}$$

EXAMPLE 9.23 The drain current in milliamperes of the enhancement type MOSFET shown in figure 9.53 is given by, $I_D = 0.3\ (V_{GS} - V_P)^2$ mA in the region $V_{DS} \geq V_{GS} - V_P$.

If $V_P = +4$ V determine I_{DSQ}, V_{GSQ} and V_{DSQ}.

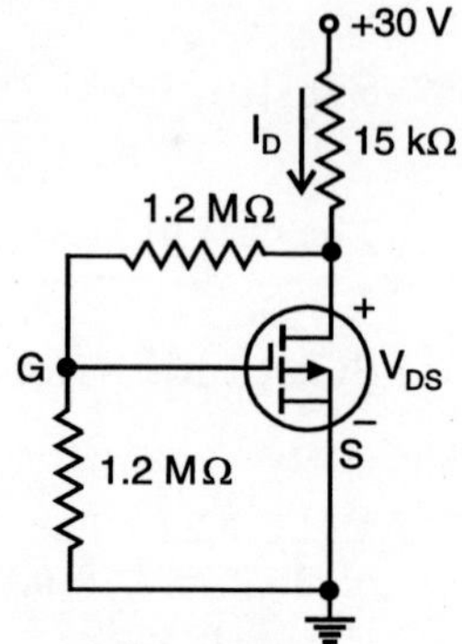

Fig. 9.53.

Solution: Since $I_G = 0$, therefore, we have

$$V_{GS} = \frac{1.2\ \text{M}\Omega}{(1.2\ \text{M}\Omega + 1.2\ \text{M}\Omega)} \times V_{DS} = 0.5\ V_{DS} \quad ...(i)$$

Applying KVL to the drain circuit, we get

$$V_{DS} = 30 - 15 \times I_D \quad ...(ii)$$

Substituting equation (*i*) into equation (*ii*), we get

$$V_{GS} = 0.5\ [30 - 15\ I_D] = 15 - 7.5 I_D \quad ...(iii)$$

Now given that $I_D = 0.3\ (V_{GS} - V_P)^2$, but $V_P = 4$

Substituting equation (*iii*) into last expression, we get

$$I_D = 0.3\ (15 - 7.5 I_D - 4)^2 = 0.3\ [11 - 7.5 I_D]^2$$

or
$$I_D = 0.3\ [121 - 165\ I_D + 56.25\ I^2_D] = 16.875\ I^2_D - 43.5\ I_D + 36.3$$

or $16.875\ I^2_D - 50.5\ I_D + 36.3 = 0$

or
$$I_D = \frac{50.5 \pm \sqrt{(50.5)^2 - 4(16.875 \times 36.3)}}{2 \times 16.875} = 4.46 \text{ mA} = 1.2 \text{ mA}$$

Let us select, $I_D = 1.2$ mA **Ans.**

Also, $V_{DS} = 30 - 15 \times 1.2 = 12$ Volt **Ans.**

$V_{GS} = 0.5 \times V_{DS} = 6$ Volt **Ans.**

EXAMPLE 9.24 Consider the following circuit (figure 9.54). Determine I_D. V_{GS} and V_{DS} for $|I_{DSS}| = 4$ mA, $V_P = 4$ V.

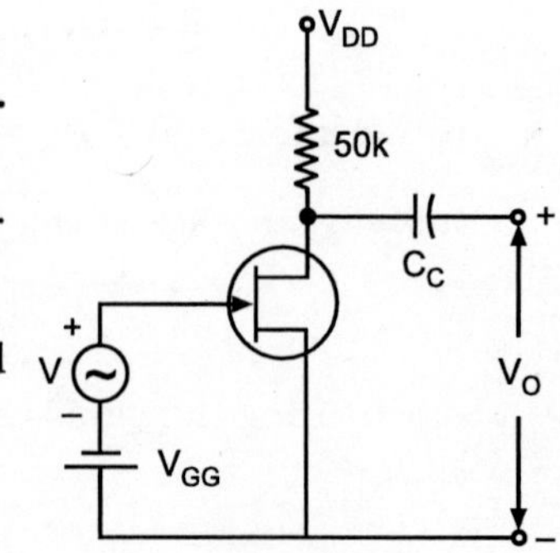

Fig. 9.54.

Solution: The given circuit is a CS amplifier with fixed bias circuit.

(*i*) Calculation of V_{GS}

For the DC analysis, the ac source acts as a short circuit and capacitors are equivalent to open circuit.

Now, applying KVL to the gate loop of figure 9.55, we get,

$$V_{GG} + V_{GS} = 0$$

or
$$V_{GS} = -V_{GG}$$

Let $V_{GG} = 2$ V

then, $V_{GS} = -2$ V

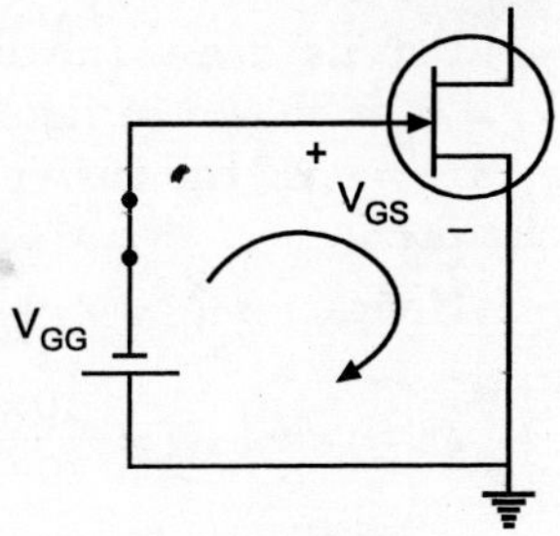

Fig. 9.55.

(ii) Calculation of I_D

We know that

$$I_D = I_{DSS}\left[I - \frac{V_{GS}}{V_P}\right]^2$$

or $$I_D = 4\left[1 - \frac{(-2)}{(-4)}\right]^2$$

or $I_D = 1$ mA. **Ans.**

***(iii)* Calculation of V_{DS}**

Applying KVL to the drain loop of figure 9.56, we obtain

$$V_{DD} = I_D R_D + V_{DS}$$

or $$V_{DS} = V_{DD} - I_D R_D$$

Let $V_{DD} = 10$ V

Therefore, $V_{DS} = 10 - (1 \times 5)$

$= 5$ V.

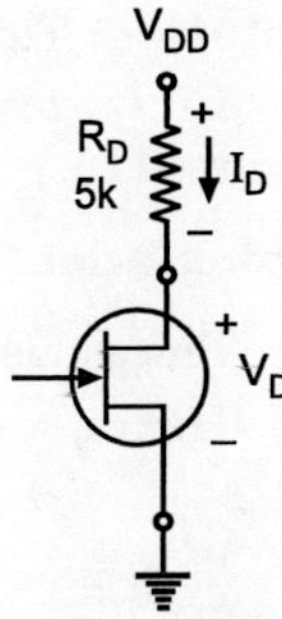

Fig. 9.56.

EXAMPLE 9.25 Evaluate the voltage gain and output resistance of the following circuit. Given that $g_m = 2$ mA/V and $r_d = 10$ k.

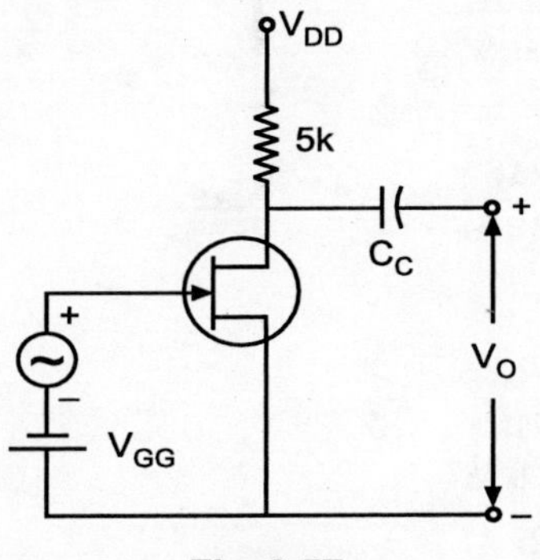

Fig. 9.57.

Solution: First, let us draw the small signal equivalent circuit as shown in figure 9.58.

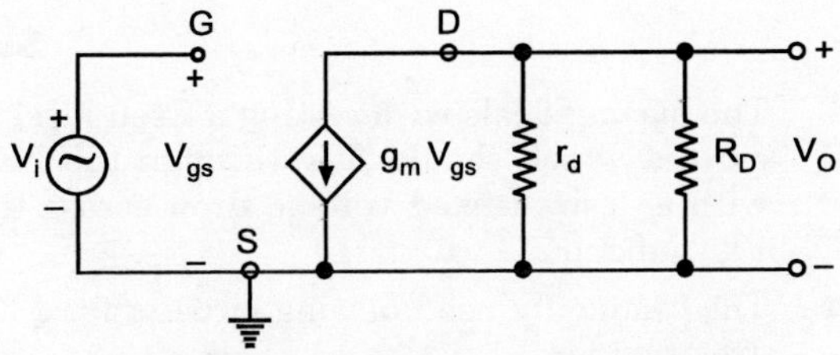

Fig. 9.58. Small signal equivalent circuit

***(i)* Voltage gain**

$$A_V = \frac{V_o}{V_i}$$

But $V_o = -g_m V_{gs}(r_d \parallel R_D)$

and $V_i = V_{gs}$

Therefore, $$A_V = \frac{-g_m V_{gs}(r_d \parallel R_D)}{V_{gs}} = -g_m V_{gs}(r_d \parallel R_D)$$

or $A_V = -2 \times 10^{-3}$ [10 k || 50 k) $= -2 \times 10^{-3} \times 8.33$ kΩ.

or $A_V = -16.67$ **Ans.**

***(ii)* Output Resistance**

Let us draw the equivalent circuit with $V_i = 0$ as shown in figure 9.59. As $V_i = V_{gs} = 0$, $g_m V_{gs} = 0$, so the current source acts as an open circuit.

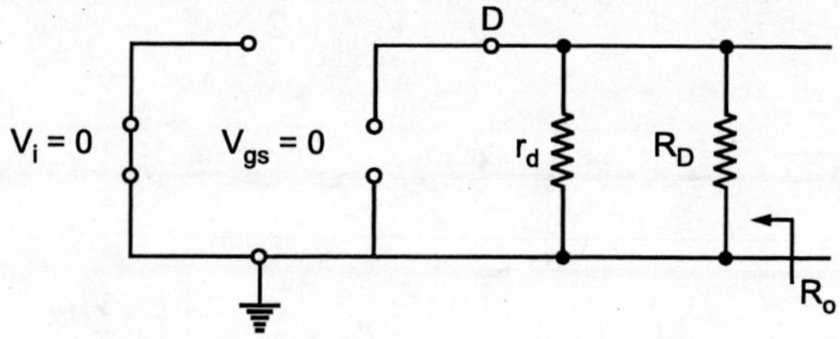

Fig. 9.59. Equivalent circuit to calculate R_o.

Hence $R_o = r_d \parallel R_D = 10 \text{ k} \parallel 50 \text{ k}$

or $R_o = \dfrac{10 \times 50}{10 + 50} = 8.33 \text{ k}\Omega$. **Ans.**

EXAMPLE 9.20 **Find A_v, R_i and R_o of the following circuit (figure 9.60)**

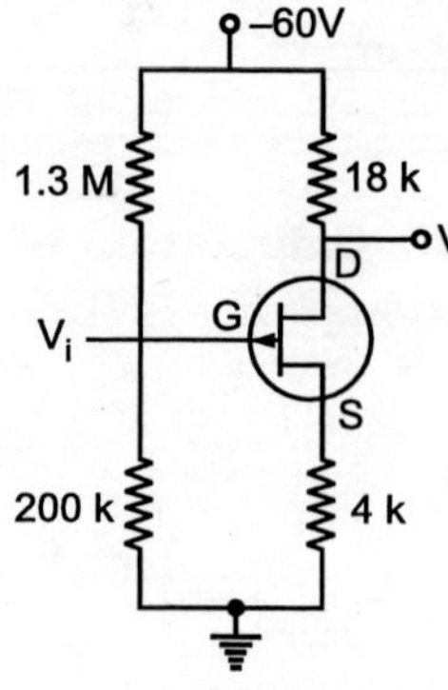

Fig. 9.60.

Solution: This is a CS amplifier with unbypassed R_s.

First, let us draw small signal equivalent circuit as shown in figure 9.61.

Calculation of A_v, R_i and R_o

Since r_d is not given, therefore, we shall use the approximate expressions

Hence, $R_i = R_1 \parallel R_2 = 1.3 \text{ M}\Omega \parallel 200 \text{ k}\Omega$

or $R_i = \dfrac{1300 \text{ k} \times 200 \text{ k}}{1500 \text{ k}}$

or $R_i = 173.33 \text{ k}\Omega$. **Ans.**

Also, $A_v = \dfrac{-g_m R_D}{1 + g_m R_s}$

Let $g_m = 2 \text{ ms}$

So, $A_v = \dfrac{-2 \times 18}{1 + (2 \times 4)} = -4$. **Ans.**

We also know that

$$R_o = \frac{R_D}{1 + g_m R_s}$$

$$R_o = \frac{18 \text{ k}}{1 + (2 \times 4)} = 2 \text{ k}\Omega. \textbf{ Ans.}$$

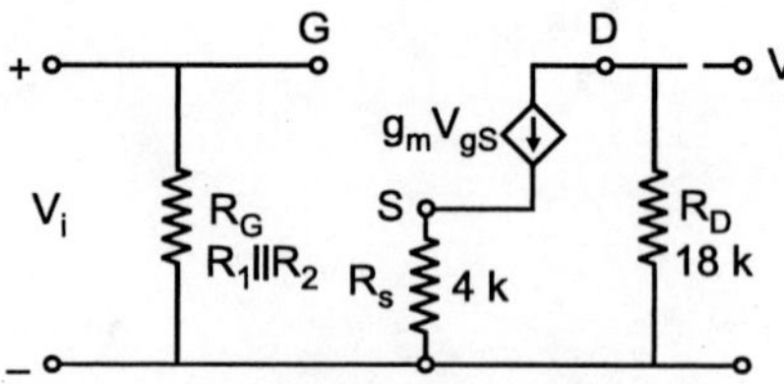

Fig. 9.61. Small signal equivalent circuit.

SUMMARY

- The biasing deals with setting a fixed level of current which should flow through the FET with a desired fixed voltage drop across the FET junctions.
- The commonly used biasing circuits for FET are as under:
  - *(i)* Fixed-bias circuit.
  - *(ii)* Self-bias circuit.
  - *(iii)* Voltage divider bias circuit.
- The Q-point (*i.e.*, quiescent or operating point or bias point) for a self-biased JFET is established by determining the value of drain current (I_D) for a desired value of gate-to-source voltage (V_{GS}) or vice-versa.
- In analytical method, we note down the values of the maximum drain current (I_{DSS}) and the gate-to-source cut-off voltage [$V_{GS(off)}$] from the data sheets of JFET supplied by the manufacturer.
- In graphical method, we draw a self-bias line such that it intersects the transfer characteristic curve near its mid-point.
- However, it has been observed that the JFET parameters like maximum drain current (I_{DSS})

and the gate-to-source cut-off [$V_{GS(off)}$] vary with the temperature.

- In order to provide compensation for the device parameter, variation the manufacturers provide two transistor characteristic curves.
- The first current corresponds to the maximum values of I_{DSS} and $V_{GS\ (off)}$.
- The self bias circuit is not suitable since it cannot provide the positive V_{GS} required for the operation of an N-channel depletion type MOSFET in the enhancement region.
- The transfer characteristics of the enhancement type MOSFET are completely different from that of a JFET.
- The other most popular biasing arrangement for the enhancement type MOSFET is voltage divider biasing circuit.
- The field-effect transistor (FET), is a voltage-controlled device.
- The output characteristics of FET are controlled by the applied input voltage and not by the input current.
- FET is an unipolar device.
- The current flowing through it is only due to one type of charge particles *i.e.*, holes or electrons.
- In a broad sense, there are two main types of field-effect transistors as under:
  (*i*) Junction field-effect transistors (JFET).
  (*ii*) Metal oxide semiconductor field-effect transistor (MOSFET) or insulated gate field-effect transistor (IGFET).
- The junction field-effect transistor (JFET) may be divided, depending on their structure, into the following two categories:
  (*i*) N-channel JFET.
  (*ii*) P-channel JFET.
- The space between the junction (*i.e.*, N-type regions) is known as channel.
- Whenever a voltage is applied across the drain and source terminals a current flows through the N-channel.
- The current carriers in JFET are the holes which flow through the P-type channel.
- This PN-junction is always reverse-bias in JFET operation.
- The reverse bias is applied by a voltage V_{GG} connected between the gate and source terminals.
- Whenever a *PN*-junction is reverse-biased, the electrons and holes diffuse across the junction and leave behind the positive ions on the N-side and negative ions on the *P*-side. The region containing these immobile ions is called depletion region.
- The reverse-bias across the gate-source junction of a JFET can also be achieved simply by applying a voltage across the drain and source terminals.
- In the presence of positive supply voltage V_{DD} (with gate open) the electrons flow from source to drain through the N-channel and constitutes a current known as drain current (I_D).
- The gate-to-source voltage (V_{GS}) at which the drain current is zero (or completely cut-off) is known as pinch off voltage.
- The value of pinch-off voltage V_P is negative for N-channel JFET. It depends upon
  (*i*)doping of the *N* and *P* regions of the device,
  (*ii*) width of the original channel structure.
- For a JFET there are two important characteristics as under:
  (*i*) *V-I* or drain characteristics
  (*ii*) Transfer Characteristics.
- The drain current in the pinch off region depends on the gate-to-source voltage and is given by the relation

$$I_D = I_{DSS}\left(1 - \frac{V_{GS}}{V_P}\right)^2$$

  This relation is known as Shockley's equation. The pinch-off region is the normal operating region of JFET, when used as an amplifier.
- The drain-to-source voltage corresponding to point *C* is known as the breakdown voltage.
- Mathematically, the d.c drain resistance is expressed as

$$R_{DS} = \frac{V_{DS}}{I_D}.$$

- The a.c drain resistance is expressed as

$$r_d = \frac{\Delta V_{DS}}{\Delta I_D}$$

  It is, sometimes, written as r_{ds}, which indicates that it is the resistance from drain-to-source terminals. Since r_d is the output resistance of JFET, it can also be expressed as an output admittance y_{DS} (*i.e.*, equal to $1/r_d$).
- The transconductance is expressed as

$$g_m = \frac{\Delta I_D}{\Delta V_{GS}}$$

The value of transconductance (g_m) is expressed in siemens (S) or mhos. We know that transfer characteristic of a JFET is a part of a parabola *i.e.,* it is non-linear. Therefore, the value of g_m varies depending upon the location on the curve as set by the gate-to-source voltage (V_{GS}). It has a greater value near the top of the curve than near the bottom.

- The amplification factor is expressed as

$$\mu = \frac{\Delta V_{DS}}{\Delta V_{GS}}.$$

- A field-effect transistor has a number of advantages over a bipolar junction transistor namely extremely high input resistance, less noisy, better thermal stability etc.
- In order to use the JFET as an amplifier, it is necessary to operate it in the "saturation region" of its drain characteristics. This region is very much similar to the active region of a BJT.

SHORT QUESTIONS WITH ANSWERS

Q. 1. What is junction field-effect transistor (JFET)? Explain.

Ans. The junction field-effect transistor (JFET) may be divided, depending on their structure, into the following two categories:

(*i*) N-channel JFET.
(*ii*) P-channel JFET.

Q. 2. What is depletion region in JFET? Expain.

Ans. Whenever a *PN*-junction is reverse-biased, the electrons and holes diffuse across the junction and leave behind the positive ions on the *N*-side and negative ions on the *P*-side. The region, containing these immobile ions, is called depletion region. As the reverse bias voltage across the junction is increased, the thickness of depletion region is also increased. If both the *P*-side and *N*-side of the junction are equally doped, the depletion region will extend equally in both the regions.

Q. 3. What is pinch-off region? Explain.

Ans. This region has been shown by the curve *BC*. It is also known as the saturation region or constant current region. In this region, the drain current remains constant at its maximum value (*i.e.,* I_{DSS}). The drain current in the pinch off region depends on the gate-to-source voltage and is given by the following relation:

$$I_D = I_{DSS}\left(1 - \frac{V_{GS}}{V_P}\right)^2$$

This relation is known as Shockley's equation. The pinch-off region is the normal operating region of JFET, when used as an amplifier.

Q. 4. Define breakdown region for JFET.

Ans. This region has been shown by the curve *CD*. In this region the drain current increases rapidly as the drain-to-source voltage is also increased. It happens because of the breakdown of gate-to-source junction due to the avalanche effect. The drain-to-source voltage corresponding to point *C* is known as the breakdown voltage.

Q. 5. How MOSFET works as a resistor?

Ans. The MOSFETs have an important property that they can be used as a resistor capacitor, amplifier and a switch. This makes the design of electronic circuits very simple, since the entire circuit consists of only MOSFETs and no other component. The examples of such circuits are microprocessors and memory circuits.

Q. 6. What is FET biasing? Explain.

Ans. We know the meaning and the need of biasing a device. We have already studied different biasing circuits for a bipolar junction transistor (BJT) in previous chapters. As a matter of fact, the biasing deals with setting a fixed level of current which should flow through the FET with a desired fixed voltage drop across the FET junctions.

Q. 7. What are the FET biasing circuits?

Ans. The commonly used biasing circuits for FET are as under:

(*i*) Fixed-bias circuit.
(*ii*) Self-bias circuit.
(*iii*) Voltage divider bias circuit.

Q. 8. What do you mean by field-effect transistor amplifier?

Ans. The field-effect transistor (FET) has a capability to amplify a.c. signals like a bipolar junction transistor (BJT). Depending upon the types of configuration, the FET amplifiers may be classified under the following three heads:

(*i*) Common source amplifier
(*ii*) Common drain amplifier, and
(*iii*) Common gate amplifier.

Q. 9. Why are field-effect transistors called unipolar transistors ?

Ans. In field-effect transistors, current conduction is by only one of type of majority carriers (either by electrons or by holes) and, therefore are called unipolar transistors.

Q. 10. Why the channel of a JFET is never completely closed at the drain end ?

Ans. If the channel is completely closed in JFET, then there will be no drain current, so there will be no voltage drop along the channel length and amount of reverse bias will become uniform and wedge-shaped depletion region will become rectangular one.

Q. 11. How is drain current controlled in a JFET ?

Ans. In a JFET, drain current is controlled by controlling the reverse bias given to its gate (i.e. V_{GS}).

Q. 12. What is meant by drain characteristics of FETs ?

Ans. The curve drawn between drain current and drain-source voltage with gate-to-source voltage as the parameter is called the drain characteristic.

Q. 13. What is meant by transfer characteristic of FETs ?

Ans. The curve drawn between drain current and gate-source voltage for a given value of drain-source voltage is called the transfer characteristic.

Q. 14. What is pinch-off voltage in a JFET ?

Ans. The value of drain-source voltage at which channel is pinched-off (*i.e.*, all the free charges from the channel get removed) is called the pinch-off voltage.

Q. 15. What is meant by saturation region ?

Ans. The region of drain characteristic of a FET in which drain current remains fairly constant is called the saturation or pinch-off region.

Q. 16. Why is input impedance of the FET very high ?

Ans. FET has very high input impedance because its input circuit (gate-to-source) is reverse biased and the input gate current is very small (of the order of few nano-amperes)

Q. 17. What is the significant difference between the construction of an enhancement type MOSFET and a depletion type MOSFET ?

Ans. In the depletion type MOSFET, a channel is physically constructed and a current between drain and source is due to voltage applied across the drain-source terminals while in enhancement type construction, no channel is formed during its construction. Voltage is applied to the gate in case of enhancement type MOSFET, to develop a channel of charge carriers so that a current results when a voltage is applied across the drain-source terminals.

Q. 18. Why E-MOSEFT is called sometimes normally-off MOSFET ?

Ans. E-MOSFET operates with large positive gate voltages only and does not conduct when gate-source voltage $V_{GS} = 0$, so it is called normally-off MOSFET.

Q. 19. Why MOSFETs are never connected or disconnected in the circuit when power is ON ?

Ans. If a MOSFET is connected or disconnected in circuit while the power is ON, transient voltages caused by inductive kickback and other effects may exceed $V_{GS(max)}$ and thus wipe out the MOSFET.

Q. 20. Name the factors which make the JFET superior to BJT?

Ans. The high input impedance, low output impedance and low noise level make JFET far superior to the BJT.

Q. 21. Explain FET is a voltage controlled device or current controlled device.

Ans. The voltage applied between gate and source (V_{GS}) controls the drain current I_D. Therefore, FET is a voltage controlled device. The transistor on the contrary is a current controlled device because the output current I_C is controlled by varying the base current I_B. The name field effect is derived from the fact that the current flow in FET is controlled by an electric field set by an external voltage.

Q.22. Explain FET is an unipolar device or a bipolar device.

Ans. FET is a unipolar device. This means that the current flowing through it is only due to one type of charge particles, holes or electrons. Transistor, on the other hand, is a bipolar device as holes and electrons both contribute to the flow of current. Just as there are NPN and PNP transistors, there are n-channel and p-channel field effect transistors. One of the most important characteristics of the FET is its very high input impedance. Typically, it is in the range of 1 to several megaohms, which is much higher than the input resistance of a BJT. FETs are more temperature stable as compared to the BJT and it requires less space than that for a BJT. Therefore, FETs are preferred in integrated circuits.

Q.23. Write the Advantages of JFET Over BJT

Ans. The FET enjoys several advantages over the conventional bipolar junction transistor (BJT). Some of them may be listed as under :

(*i*) It is a unipolar device so its operation depends on the flow of majority carriers only.

(*ii*) It is relatively immune to radiation.

(*iii*) FET has a very high input resistance (typically few megaohms)

(*iv*) FET is less noisy

(*v*) It does not exhibit any offset voltage at zero drain current hence it can be used as an excellent signal chopper.

(*vi*) FET has a better thermal stability.

Q.24. Write Disadvantage of JFET.

Ans. The main disadvantage of the JFET is that it has a relatively small gain-bandwidth product as compared with that of a conventional transistor (*i.e.* BJT).

REVIEW QUESTIONS

1. Illustrate with neat circuit diagrams the technique of biasing an enhancement MOSFET.
2. Explain giving necessary diagrams how JFET can be biased suitable to overcome the effect of unit to unit variation.
3. Explain briefly the technique of biasing a JFET for zero current drift with temperature.
4. Write notes on: Biasing techniques of MOSFET.
5. Explain any one method of biasing a single stage JFET amplifier.
6. With the help of transfer curve and bias line, explain the biasing techniques of N-channel FET.
7. With the help of neat diagrams explain the operation and drain characteristics of N-channel MOSFET.
8. Draw circuit diagram and describe the following applications of MOSFET:
   (*i*) Active load switching
   (*ii*) CMOS inverter.
9. Write a short note on: Depletion mode MOSFET.
10. Explain giving illustrative diagrams, how the pinch off condition occurs in a MOSFET.
11. Explain the difference between the operation of a JFET in ohmic and current saturation regions. Also explain the cut off condition.
12. What is thermal runway in BJT ? How can it be avoided? Is the thermal runway possible in JFET?
13. Draw the laboratory experimental setup for plotting output static characteristics of a JFET. Explain clearly, different operating regions in characteristics with proper reasoning.
14. Compare BJT and FET.
15. Define the parameters: Amplification factor, drain resistance, and mutual conductance for JFET and obtain the expression relating them.
16. Write a short note on MOSFET.
17. Compare JFET and MOSFET.
18. Give reason: FET behaves as a voltage controlled current source.
19. Sketch the drain characteristics of a JFET and explain its behaviour before and after pinch off.
20. Write a short note on: FET as *VVR*.
21. State true or false with reason.
    (*i*) The relationship between I_D and V_{GS} is non-linear.
    (*ii*) Gate current in FET is always neglected.
    (*ii*) In JFET, the P-N junction between gate and source is always kept in reverse biased condition.
    (*iv*) FET is a current controlled device.
22. What do you understand by "pinch off" voltage and cut-off voltage as applied to FET?
23. Draw the drain characteristics of depletion type MOSFET. Explain different operating regions in the characteristics with proper reasoning.
24. Draw the circuit diagram of JFET as an amplifier and explain its working.
25. Draw the a.c. equivalent circuit of JFET and mention important points.

26. Obtain the approximate equivalent circuit for JFET as an amplifier.

27. Draw the circuit diagram of JFET amplifier using fixed-bias configuration and explain the process.

28. What is the significant difference between the construction of an enhancement type MOSFET and a depletion type MOSFET ?

29. What is meant by gate-to-source threshold voltage in E-MOSFET ?

NUMERICAL PROBLEMS

1. An n-channel depletion type MOSFET is operated in the pinch off region. If I_{DSS} = 15 mA and $V_P = -4$ V, calculate I_D when: (*a*) $V_{GS} = -3$ V and (*b*) $V_{GS} = +2.5$ V.

   [Ans. I_D = 3.75 mA, 3961 mA]

2. A potential divider bias circuit of FET is shown in figure 9.62. Find the quiescent values of : (*i*) I_D, (*ii*) V_{GS}, (*iii*) V_{DS}, (*iv*) V_S.

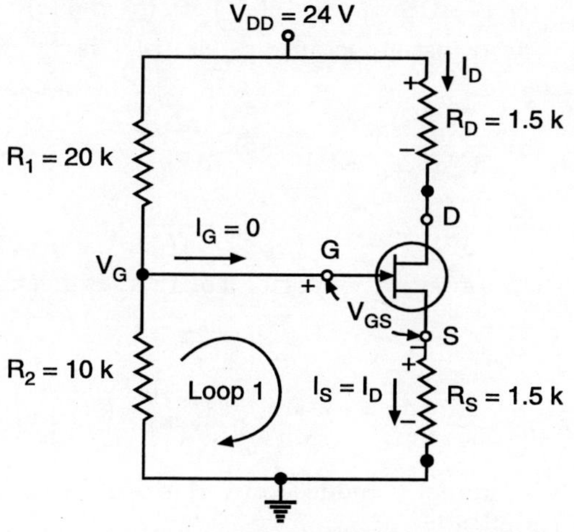

Fig. 9.62.

[Ans. (*i*) I_{DQ} = 3 mA, (*ii*) V_{GSQ} = –2 Volt (*iii*) V_{DS} = 1.5 Volt (*iv*) V_S = 6 Volt]

3. An N-channel JFET having a pinch off voltage of – 3 volts has a saturation current of 2.5 mA; when V_{GS} = – 1 V. What is its saturation current when,

   (*a*) V_{GS} = 0 Volts

   (*b*) V_{GS} = – 2 Volts **[Ans. (*i*) I_{DSS} = 5.625 mA at V_{GS} = 0 V, (*ii*) I_{DSS} = 0.625 mA at V_{GS} = –2 V]**

4. Given that for a certain JFET, I_{DSS} = 15 mA, V_P = – 5 V, and g_{mo} = 3 ms. *Calculate:* Transconductance and drain current I_D for V_{GS} = –3.5 Volts.

   [Ans. (*i*) $g_m = 0.9 \times 10^{-3}$ *AV*, (*ii*) I_D = 1.35 mA]

5. When a reverse gate voltage of 18 V is applied to a FET, the gate current is 0.9×10^{-3} mA. Determine the resistance between gate and source. **[Ans. 20,000 M]**

6. The following readings were obtained experimentally from a FET:

V_{GS}	0 V	0.V	– 0.3 V
V_{DS}	7 V	14.5 V	14.5 V
I_D	3.25 mA	10 mA	9.2 mA

   Determine r_d, g_m and μ.

   [Ans. 30 kΩ, 3500 μ*S*; 10^5]

7. Determine the value of transconductance of a FET, when the drain current changes from 1 mA to 1.5 mA with a change in gate voltage from – 2.125 V to – 2 V. **[Ans. 4,000 μ mhos]**

8. V_P = – 2 Volts I_{DSS} = 1.65 mA for the circuit shown in figure 9.63. It is desired to bias the circuit at I_D = 0.8 mA, V_{DD} = 24 Volts. Find V_{GS}. **[Ans. – 6073 V]**

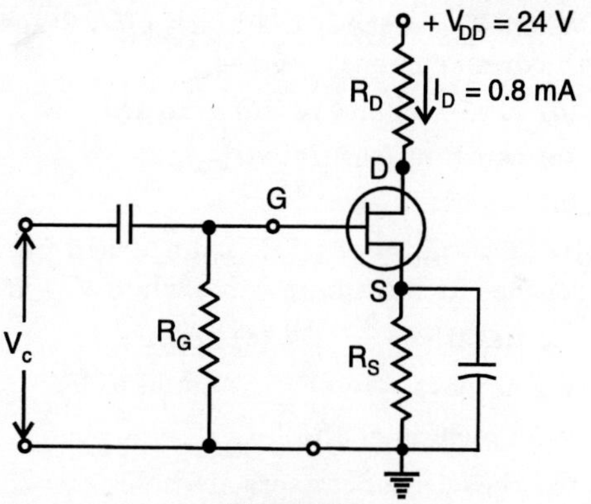

Fig. 9.63.

9. Find the pinch-off voltage of a silicon P-channel FET having half channel height of 2 microns and channel resistivity of 10 Ω-cm. Dielectric constant of silicon is 12 and mobility of holes is 500 cm²/V-sec. **[Ans. 3.77 V]**

10. (*a*) An N-channel JFET has I_{DSS} = 10 mA and V_P = – 4 V. Plot its transfer characteristics.

    (*b*) If a tangent is drain to the characteristics at V_G = 0, show that it cuts the X-axis at $-\frac{V_P}{2}$.

11. For an N-channel silicon FET with $a = 3 \times 10^{-4}$ cm and $N_D = 10^{15}$ electrons/cm^3, find (*a*) the pinch-off voltage and (*b*) the channel half-width for $V_{GS} = \frac{1}{2} V_P$ and $I_D = 0$.

[Ans. (*a*) 6.8 V (*b*) 0.87 × 10^{-4} cm]

12. A FET has a driven current of mA. If $I_{DSS} = 8$ mA and $V_{GS(off)} = -6$ V, find the values of V_{GS} and V_P. **[Ans. $V_{GS} = -1.76$ V, $V_P = 6$ V]**

OBJECTIVE TYPE QUESTIONS

Choose one correct answer in the following:

1. The JFET is
   (*a*) a bipolar device (*b*) unipolar device
   (*c*) current controlled device
   (*d*) voltage controlled device
   (*e*) *b* and *d* (*f*) *a* and *b*
2. The channel of a JFET exists between
   (*a*) gate and source (*b*) drain and source
   (*c*) gate and drain (*d*) input and output
3. For low values of V_{DS}, the JFET behaves like a
   (*a*) resistance
   (*b*) constant voltage device
   (*c*) constant current device
   (*d*) negative resistor
4. When V_{DS} reaches the pinch off voltage V_P, the drain current becomes
   (*a*) low (*b*) zero
   (*c*) constant (saturated)
   (*d*) reversed
5. In an *n*-channel JFET, the drain current reaches its maximum value when V_{GS} is
   (*a*) negative (*b*) zero
   (*c*) positive (*d*) equal to V_P
6. In an *n*-channel JFET
   (*a*) the current carriers are holes
   (*b*) the current carriers are electrons
   (*c*) V_{GS} is positive
   (*d*) the input resistance is very low
7. In a p-channel JFET
   (*a*) the current carriers are electrons
   (*b*) the current carriers are holes
   (*c*) V_{GS} is negative
   (*d*) the input resistance is very small
8. For an *n*-channel JFET
   (*a*) V_{GS} can vary between zero negatively to V_{GSO}
   (*b*) V_{GS} can vary between zero positively to V_{GSO}
   (*c*) pinch off occurs for positive V_{GS}
   (*d*) V_{DD} is negative
9. An FET cannot operate at $V_{GS} = 0$ V. The FET is
   (*a*) JFET (*b*) D-MOSFET
   (*c*) E-MOSFET (*d*) both (*a*) and (*b*)
10. The transconductance g_m of JFET is defined as
   (*a*) $\frac{\Delta I_D}{\Delta V_{GS}}$ (*b*) $\frac{\Delta I_D}{\Delta V_{DS}}$
   (*c*) $\frac{\Delta V_{GS}}{\Delta I_D}$ (*d*) $\frac{\Delta I_{DSS}}{\Delta I_D}$
11. The amplification factor μ of JFET is given by
   (*a*) $g_m \times r_d$ (*b*) $\frac{g_m}{r_d}$
   (*c*) $g_m \cdot I_{DSS}$ (*d*) $g_m\left(1 - \frac{V_{GS}}{V_P}\right)$
12. A channel is induced in a E-MOSFET by the application of a
   (*a*) $V_{GS} < (V_{GS})$th (*b*) $V_{GS} > V_P$
   (*c*) $V_{GS} > (V_{GS})$th (*d*) $V_{DS} > V_P$
13. The Shockley equation is given by
   (*a*) $I_D = I_{DSS}\left[1 - \frac{V_{GS}}{V_P}\right]^2$
   (*b*) $I_D = I_{DSS}\left[1 - \frac{V_{GS}}{V_P}\right]$
   (*c*) $g_m = g_{mo}\left[1 - \frac{V_{GS}}{V_P}\right]$
   (*d*) $I_D = C[V_{GS} - V_{GS}(\text{th})]^2$
14. Ideally, the equivalent circuit of an FET consists of
   (*a*) a resistance between drain and source
   (*b*) a current source between the gate and the source

(*c*) a current source between the drain and the source
(*d*) a current source between the gate and the drain

15. The magnitude of the current source in the ac equivalent circuit of an FET depends on
(*a*) the dc supply voltage
(*b*) V_{ds}
(*c*) external drain resistance
(*d*) transconductance and gate to source voltage

16. A CS amplifier has a drain load $R_D = 1\ k\Omega$ and an external load resistance of $R_L = 1\ k\Omega$. If $g_m = 10$ mS, the voltage gain will be
(*a*) 10 (*b*) 5
(*c*) 20 (*d*) 1

17. If the external load resistance is removed in problem 16, the voltage gain will be
(*a*) decrease (*b*) increase
(*c*) remain the same (*d*) become zero

18. In a *CD* amplifier, $R_S = 500\ \Omega$ and the transconductance is 9000 μS. The voltage gain will be
(*a*) 0.82 (*b*) 1
(*c*) 1.22 (*d*) 9

19. In a common gate amplifier, the transconductance is $g_m = 6000\ \mu S$, $R_G = 1\ M\Omega$. The total input resistance will be
(*a*) 1666 Ω (*b*) 166.6 Ω
(*c*) 1 MΩ (*d*) 100166.6 Ω

20. In a *CD* amplifier, $I_{DSS} = 5$ mA at $V_{GS} = 10$ V. If the resistor R_G is 20 *M*Ω, total input the resistance will be
(*a*) 20 MΩ (*b*) 200 MΩ
(*c*) 40 MΩ (*d*) 100 MΩ

21. If there is an internal open between the source and the drain in a CS amplifier, the drain voltage will be
(*a*) 0 V (*b*) V_{GS}
(*c*) V_{DD} (*d*) $\frac{V_{DD}}{2}$

Answers :

1. (*e*)	2. (*b*)	3. (*a*)	4. (*c*)	5. (*d*)	6. (*b*)
7. (*b*)	8. (*a*)	9. (*c*)	10. (*a*)	11. (*a*)	12. (*c*)
13. (*a*)	14. (*c*)	15. (*c*)	16. (*d*)	17. (*b*)	18. (*a*)
19. (*b*)	20. (*a*)	21. (*c*)			

UNIT—VIII

Special Purpose Electronic Devices

Principle of Operation and Characteristics of Tunnel Diode (with the help of Energy Band Diagram) and Varactor Diode, Principle of Operation of Schottky Barrier Diode, SCR, and Semiconductor Photo Diode

Above Syllabus has been covered in following Chapter:

CHAPTER 10 : Special Purpose Electronic Devices

Chapter 10

Special Purpose Electronic Devices

SYLLABUS

Special Purpose Electronic Devices : Principle of Operation and Characteristics of Tunnel Diode (with the help of Energy Band Diagram) and Varactor Diode, Principle of Operation of Schottky Barrier Diode, SCR, and Semiconductor Photo Diode.

Chapter Outline

- Introduction
- A Zener Diode
- Reverse Characteristic of a Zener
- Specifications of Zener Diode
- Zener Diode Equivalent Circuit
- Applications of Zener Diode
- The Tunnel Diode : Principle of Operation
- The Varactor (Varicap) Diode
- Specifications of Varactor Diode
- Applications of a Varactor Diode
- The Schottky Barrier Diode
- Advantages, Drawbacks And Applications
- The Step-recovery Diode
- A Pin Diode
- Applications of Pin Diode
- Light Emitting Diode (Led)
- Led Voltage Drop And Current
- Applications of Light Emitting Diode (Led)
- Multicolour Led's
- The Seven-segment Display
- Alpha Numeric Display
- Liquid Crystal Display (Lcd)
- Photo Diodes
- P-n Photodiode
- Introduction to Thyristors
- Classification of Thyristors
- Applications of Thyristors
- Silicon Controlled Rectifier (Scr)
- Basic Structure of Scr
- Working Operation of Scr
- Transistor Analogy: Two Transistor Equivalent Circuit
- Dynamic Characteristics of SCR

10.1. INTRODUCTION

We already know something about rectifier or general purpose diodes. Now, let us discuss several other types of diodes, which are used for specific applications. Such diodes are zener diode, varactor diode, tunnel diode, step recovery diode, photo diode and light emitting diodes. These diodes are also called *special purpose diodes* or simply special diodes. We shall discuss each of these diodes, one by one, in the sections to follow.

10.2. A ZENER DIODE

A zener diode is also called a voltage-reference, voltage regulator or breakdown diode. Like a rectifier diode, it is of utmost importance in several power applications. Figure 10.1(*a*) shows a schematic symbol for a zener diode. It will be interesting to know that the line at the end of the arrow looks like the letter *Z*.

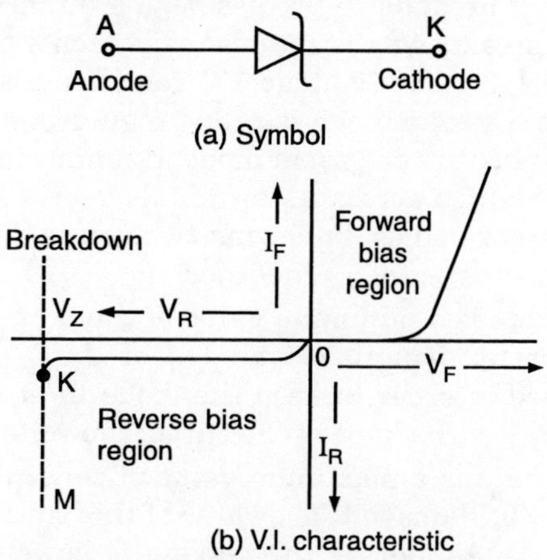

Fig. 10.1. Illustration of Zener diode.

The zener diode is a silicon PN junction device, which differs from a rectifier diode, in the sense, that it is operated in the reverse breakdown region. Also, the breakdown voltage of a zener diode is set by carefully controlling the *doping level* during manufacturing process. We know that when a reverse voltage across a diode is increased, a critical voltage called breakdown voltage is reached at which the reverse current increases sharply as depicted by the curve *KM* in figure 10.1(*b*).

> **DO YOU KNOW?**
>
> As with conventional diodes, the manufacturer places a band on the cathode end of the zener diode for terminal identification.

The reverse breakdown of a PN junction may occur either due to one of the two effects. These effects are called avalanche or zener effects or mechanisms. In PN-junctions, the avalanche breakdown occurs, when the accelerated free electrons acquire sufficient energy to ionize a lattice atom by bombardment. The additional free electrons, created in this way, are accelerated by the reverse field causing more and more ionization. The multiplication of the number of free carriers causes the reverse current to increase rapidly.

As a matter of fact, the zener breakdown occurs when the electric field across the junction, produced due to reverse voltage, is sufficiently high. This electric field exerts a force on the electrons in the outer most shell. This force is so high that the electrons are pulled away from their parent nuclei and become free carriers. This ionization, which occurs due to the electrostatic force of attraction, is known as **zener effect**. It causes an increase in the number of free carriers and hence an increase in the reverse current.

Important Point : *The zener diodes, with breakdown voltages of less than 6 V, operate predominantly in zener breakdown. Those with breakdown, voltages greater than 6 V, operate predominantly in avalanche breakdown. Strictly speaking, the first one should be called a zener diode and the second an avalanche diode. However, in actual practice, both the types are called zener diodes.*

10.3. REVERSE CHARACTERISTIC OF A ZENER DIODE

We established in the last article that a zener diode is operated only in the reverse-bias region. Therefore, we shall examine its characteristic in this region only. Figure 10.2 illustrates the reverse portion of the V-I characteristic of the zener diode. It may be noted from this curve that as the reverse voltage (generally written as V_R) is increased, the reverse current (usually called zener current, I_Z) remains negligibly small up to the 'knee' of the curve (point *K* in figure 10.2). At this point, the effect of breakdown process begins. From the bottom of the knee, the breakdown voltage (also called zener breakdown voltage this or simply zener voltage V_Z) remains essentially constant. Infact, ability of a diode is called regulating ability and is an important feature of a zener diode. It maintains an essentially a constant voltage across its terminals over a specified range of zener current values. Following two points are important from the characteristics of a zener diode :

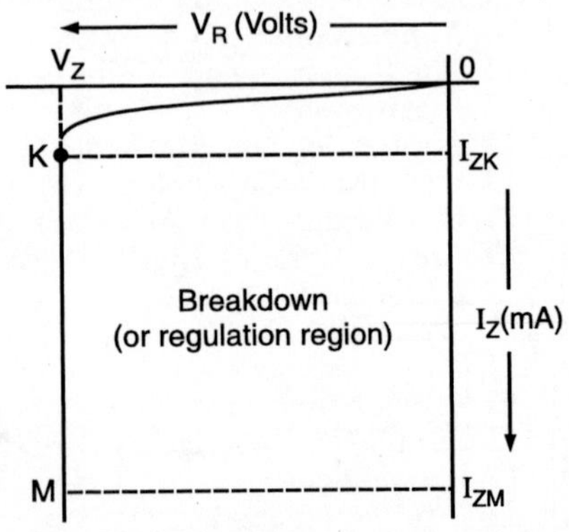

Fig. 10.2. Reverse characteristic of a zener diode.

(*i*) There is a minimum value of zener current called break over current designated as I_{ZK} or $I_{Z(min)}$ which must be maintained in order to keep the diode in breakdown (or regulation) region. When the current is reduced below the knee of the curve, the voltage changes drastically and the regulation is lost.

(*ii*) There is a maximum value of zener current designated as I_{ZM} or $I_{Z(max)}$ above which the diode may be damaged. The value of this current is given by the maximum power dissipation of the zener diode. As long as the maximum power dissipation is not exceeded, the diode would not be damaged. It will come out of the breakdown region, when the applied reverse voltage is reduced below the breakdown voltage.

10.4. SPECIFICATIONS OF ZENER DIODE

Generally, the zener diodes are specified in terms of four factors namely zener voltage (V_Z), maximum power dissipation (P_{ZM}, $P_{D\,(max)}$ or $P_{Z\,(max)}$), break over current (I_{ZK}) and zener resistance (r_Z). They are available with breakdown voltages ranging from 4.8V to 2000V. The power dissipation of a zener diode is the product of breakdown voltage (V_Z) and reverse current (I_Z). Mathematically the power dissipation,

DO YOU KNOW?

Zener diodes with breakdown voltages near 7 V have the smallest zener impedance.

$$P_Z = V_Z \times I_Z$$

The maximum value of power dissipation, which a zener can dissipate, without, failure, is called *power rating and is designated by* P_{ZM}. The zener diodes are available with power rating from 150 mW to 50 W. Generally, the data sheets of zener diodes include the maximum current, a zener can handle without exceeding its power rating. The value of this maximum current is given by the following relation:

$$I_{ZM} = \frac{P_{ZM}}{V_Z}$$

where P_{ZM} = Power rating of a zener diode, and

V_Z = Breakdown voltage.

10.5. ZENER DIODE EQUIVALENT CIRCUIT*

Figure 10.3(a) shows an ideal approximation of a zener diode in reverse breakdown condition. It shows that a zener diode is simply equivalent to a battery having a voltage equal to the zener voltage (V_Z) Figure 10.3 (b) shows a practical equivalent circuit of a zener diode. This circuit shows that a zener diode is equivalent to a battery with voltage, (V_Z) in series with a resistance (r_Z) called zener resistance. The zener resistance is also known as *dynamic resistance or a.c. resistance.*

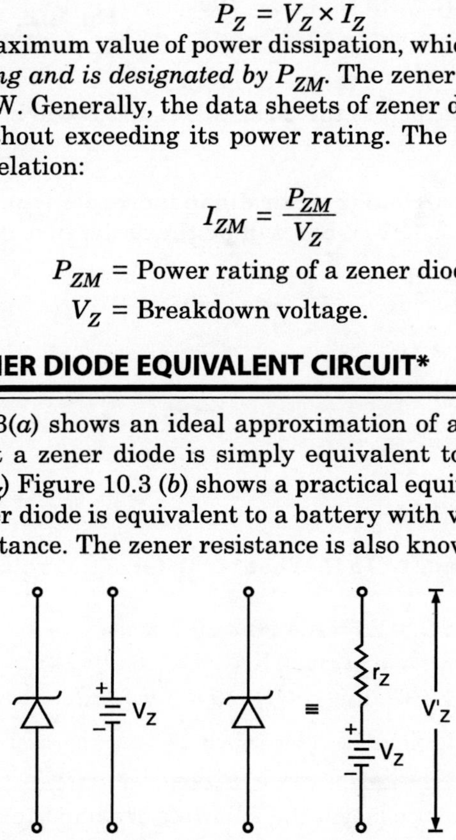

(a) (b)

Fig. 10.3. (a) Ideal zener equivalent diode circuit. (b) Practical zener diode equivalent circuit.

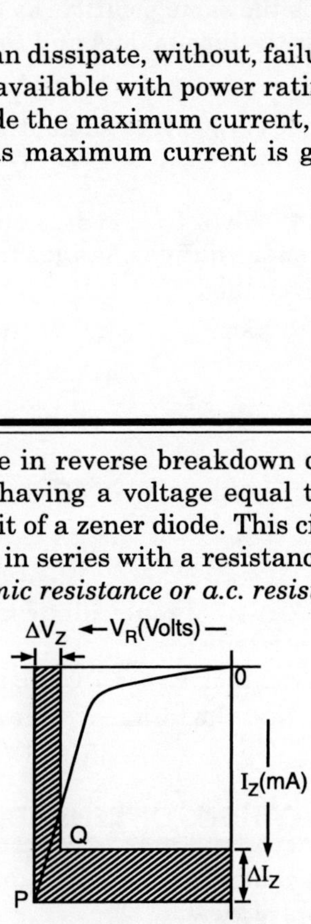

Fig. 10.4. Reverse V-I characteristic curve illustrating zener resistance.

The presence of resistance (r_Z) indicates that the reserve V-I characteristic of a zener diode is not ideally vertical but it is slightly tilted as shown in figure 10.4. The value of zener resistance may be obtained from the V-I characteristic as discussed below:

Let us consider any two point P and Q on the reverse V-I curve as shown in figure 10.4

Let ΔV_Z = Change in the values of zener voltage between the points P and Q, and

* Explain the V-I characteristics of a zener diode and give some of its applications.

ΔI_Z = Change in the values of zener current at the corresponding points.

We know that zener resistance is given by

$$r_Z = \frac{\Delta V_Z}{\Delta I_Z}$$

For a perfectly vertical breakdown curve, the value of zener resistance is zero. However, in actual practice, its value can vary from few ohms to several hundred ohms. The variation in its value depends upon the particular zener voltage and the value of operating current. It may be noted that if the zener current flows through the zener diode in the direction as shown in figure 10.5, then voltage drop (V_Z) across the zener resistance has the same polarity as that of a battery. In this case, the voltage across the terminals A and B, will be

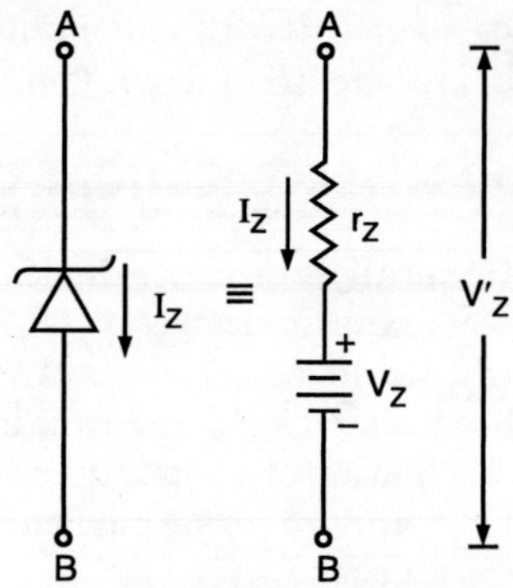

Fig. 10.5. Terminal voltage of a zener diode.

$$V_Z' = I_Z \cdot r_Z + V_Z$$

If the value of r_Z is very small, then the voltage across the terminals A and B will be given by

$$V_Z' = V_Z$$

EXAMPLE 10.1. When the reverse current in a particular zener diode increases from 20 mA to 30 mA, the zener voltage changes from 5.6 V to 5.65 V. What will be the resistance of device?

Solution. Given that $\Delta I_Z = 30 - 20$

$$\Delta I_Z = 10 \text{ mA} = 10 \times 10^{-3} \text{ A}$$

and $\Delta V_Z = 5.65 - 5.6 = 0.05$ V.

We know that resistance of the zener diode is given by

$$r_Z = \frac{\Delta V_Z}{\Delta I_Z}$$

$$r_Z = \frac{0.05}{10 \times 10^{-3}} = 5\ \Omega \textbf{ Ans.}$$

EXAMPLE 10.2 A 4.7 V zener diode has a resistance of 15 Ω. What is the terminal voltage, when the current is 20 mA?

Solution. Given that $V_Z = 4.7$ volts, $r_Z = 15\ \Omega$ and $I_Z = 20$ mA $= 20 \times 10^{-3}$ A.

We know that the terminal voltages of a zener diode is given by

$$V_Z' = (V_Z + I_Z \cdot r_Z) = 4.7 + (20 \times 10^{-3}) \times 15 = 5 \text{ V} \quad \textbf{Ans.}$$

10.6. APPLICATIONS OF ZENER DIODE

Although the zener diodes have a number of applications, yet the following are important from the subjects points of view:

(*i*) As a voltage regulator.
(*ii*) As a fixed reference voltage in transistor biasing circuits.
(*iii*) As peak clippers or limiters in waveshaping circuits.
(*iv*) For meter protection against damage from accidental applications.

10.7. THE TUNNEL DIODE: PRINCIPLE OF OPERATION

We know that the conventional PN junction diode has an impurity concentration of about 1 part in 10^8. This amount of doping results in the width of depletion region of the order of 5 microns (5×10^{-4} cm). The depleiton region results in the potential barrier at the junction. This potential barrier opposes the flow of carriers. If the impurity concentration (doping level) is increased to a great extent say 1 part in 10^3 then the device characteristics are completely changed. This new diode is called as **Tunnel Diode**. It was first invented in 1958 by Esaki. The operation of a tunnel diode is based on a special characteristics known as the **negative resistance**. The semiconductor materials used for constructing the tunnel diodes are Germanium or Gallium Arsenide. In fact, the construction of a tunnel diode is similar to that of a p-n junction diode but both *p* and *n* sides are very heavily doped than the conventional rectifier diodes. The heavy doping of both the sides of junction results in extremely narrow depletion region when the tunnel diode is reverse biased. This has been shown in the tunnel diode characteristics of figure 10.6 (*b*).

10.7.1. The Tunneling Phenomenon

The width of the depletion region is inversely proportional to the square root of impurity concentration (doping level). Therefore, with increase in the impurity concentration, the depletion region width will reduce from 5 microns to less than 100 Å (10^{-6} cm). Theoretically, the energy of a particle should be atleast equal to the height of the potential energy barrier if it is to move from one side of the junction to the other. But, the thickness of the depletion region of this diode is so small that the Schrodinger equation indicates that there is a large probability that an electron can penetrate through this barrier. This quantum mechanism behaviour is called as **tunneling**, and hence the name of the high impurity density PN junction is called as a **Tunnel Diode**.

10.7.2. Construction of Tunnel Diode

The material used for the construction of a tunnel diode is Germanium or Gallium Arsenide Figure 10.6 shows the basic construction of a tunnel diode.

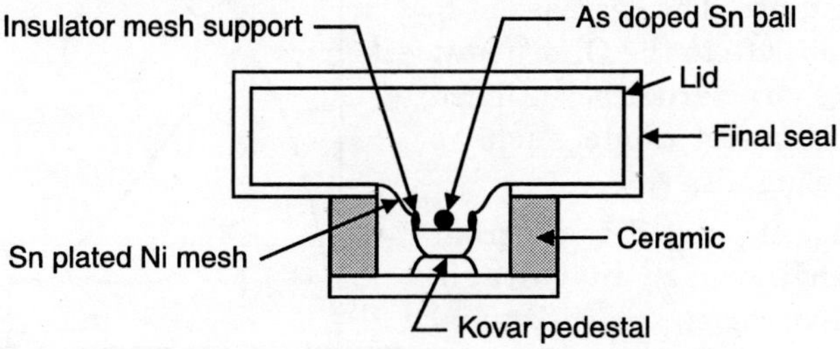

Fig. 10.6. Construction of a tunnel diode.

10.7.3. Characteristics of Tunnel Diode

The voltampere characteristics of a tunnel diode has been shown in figure 10.7 (*b*) and its symbols have been shown in figure 10.7 (*a*).

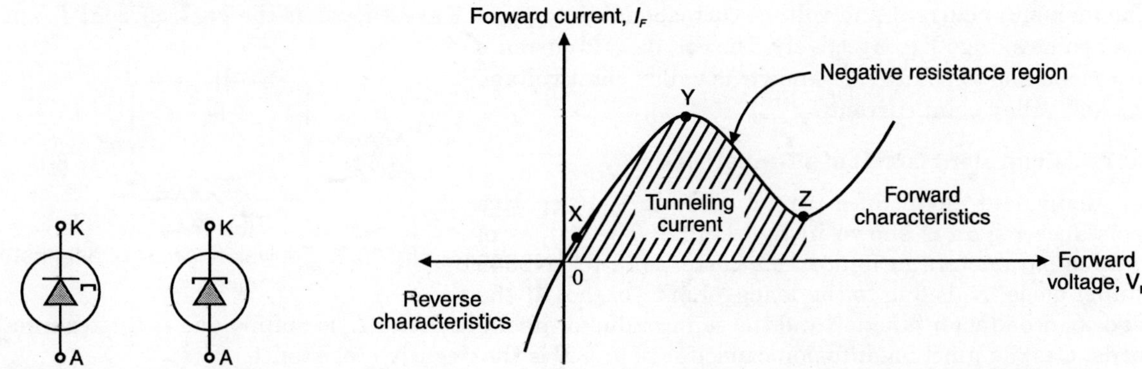

(a) Circuit symbols of tunnel diode

(b) Volt-ampere characteristics of a tunnel diode

Fig. 10.7.

(i) Reverse Characteristics

Due to heavy doping to *p* and *n* sides, the depletion region is extremely narrow when the tunnel diode is reverse biased. Therefore, the reverse blocking capacity of the junction is lost and reverse current will start flowing as soon as a very small reverse voltage is applied. Thus, the tunnel diode allows the conduction to take place for all the reverse voltages. There is no breakdown effect as observed in the conventional rectifier diode. Therefore, we can not use the tunnel diode as a rectifier.

DO YOU KNOW?

Tunnel diode exhibits a phenomenon known as negative resistance. This means that an increase in forward voltage produces a decrease in forward current.

(ii) Forward Characteristics

Forward characteristics can be divided into three regions namely *X* to *Y*, *Y* to *Z* and *Z* onwards.

Region *X* to *Y* : In this region the forward voltage V_F is extremely small. But heavy conductio will take place in this region because electrons "tunnel" through the *pn* junction. The "tunneling" takes place as a result of heavy doping.

Region *Y* to *Z* : In this region, at point *Y* the forward voltage begins to develop a barrier and the forward current starts decreasing inspite of continued increase in V_F. This region is called as the negative resistance region. The diode resistance in the region *Y* to *Z* is mathematically expressed as,

$$R_F = \frac{\Delta V_F}{\Delta I_F}$$

ΔV_F is positive but ΔI_F is negative. hence R_F is called as a negative resistance. The decrease in I_F with increase in V_F is opposite to the Ohm's law. The negative resistance characteristics is utilized in the applications of tunnel diode such as oscillator and microwave amplifier.

Region *Z* onwards : At point *Z*, the current starts increasing with increase in voltage. Therefore, this is a positive region and the tunnel diode acts as the conventional diode.

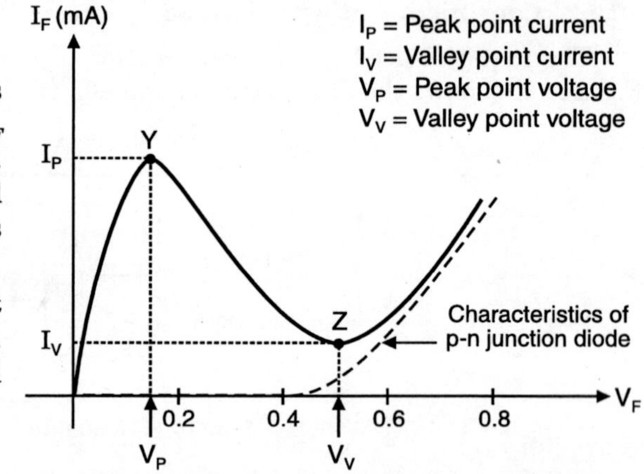

Fig. 10.8. Detailed characteristics of a tunnel diode.

10.7.4. Detailed Characteristics of a Tunnel Diode

Figure 10.8 shows the detailed characteristics of the tunnel diode. The forward characteristics of a conventional semiconductor diode has been shown by dotted lines for the purpose of comparison. The maximum current and voltage corresponding to point *Y* are known as the peak current I_P and peak point voltage V_P respectively. The voltage and currents corresponding to point *Z* are known as valley point voltage V_V and valley point current I_V.

10.7.5. Equivalent Circuit of a Tunnel Diode

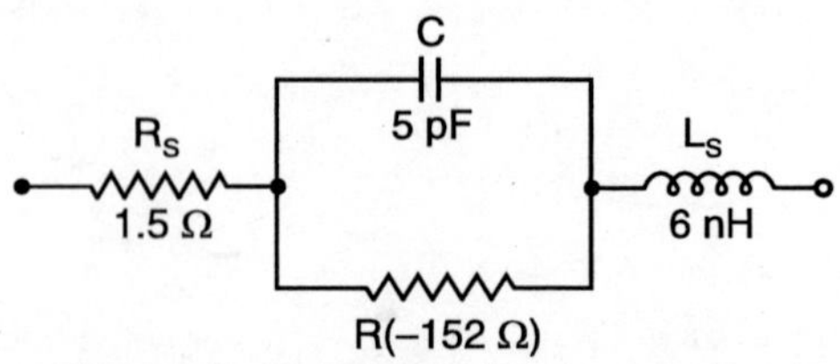

Fig. 10.9. Equivalent circuit of a tunnel diode.

Equivalent circuit of a tunnel diode in the negative resistance region is shown in figure 10.9. The values of various parameters in figure 10.9 correspond to 1N2939 tunnel diode. R_s is due to the leads, ohmic contact at the lead semiconductor junction and the semiconductor itself. Inductor L_s is mainly due to the terminal leads. *C* is the junction diffusion capacitance and *R* is the negative resistance.

10.7.6. Important Specifications of Tunnel Diode

Some of the important specifications of tunnel diode Ge 1N2939 are as given in Table 10.1.

Table 10.1. Specification of a Tunnel diode Ge 1N2939

S.No.	Specification	Value		
		Minimum	Typical	Maximum
1.	Forward current		5 mA	
2.	Reverse current		10 mA	
3.	Reverse voltage ($I_R = 1$ mA)			30 mV
4.	V_P	50	60	65 mV
5.	I_P	0.9	1.0	1.1 mA
6.	V_V		350 mV	
7.	I_V		0.1	0.14 mA

10.7.7. Salient Features of a Tunnel Diode

(*i*) Heavy doping of *p* and *n* sides.
(*ii*) Extremely narrow depletion region.
(*iii*) No reverse blocking capability, hence, it can not be used as rectifier.
(*iv*) Very low cut-in voltage. Conducts heavily at very small forward voltages.
(*v*) Exhibits a negative resistance characteristics when sufficiently forward biased.
(*vi*) Useful for very high speed applications.

10.7.8 Applications of Tunnel Diode as an Oscillator

Figure 10.10(*a*) shows a parallel resonant circuit which is also called as the tank circuit. If the switch SW is closed momentarily, then a current pulse is applied to the tank circuit. The capacitor charges and discharges sinusoidally to produce damped oscillations across the tank circuit as shown in figure 10.10 (*b*). The frequency of the damped oscillations is fixed and it is decided by the RLC components but the amplitude decreases as the time progresses.

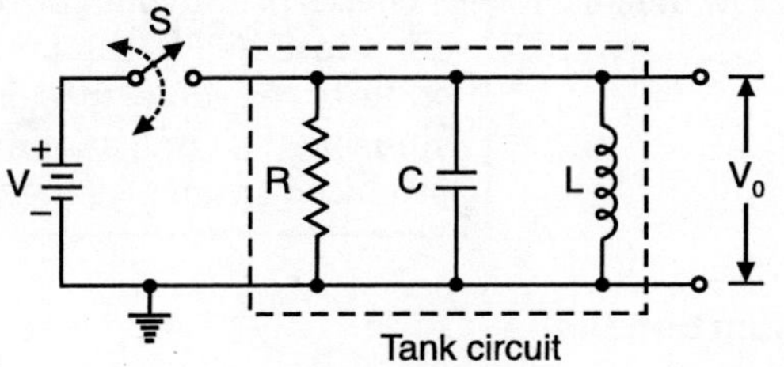

(a) Tank circuit

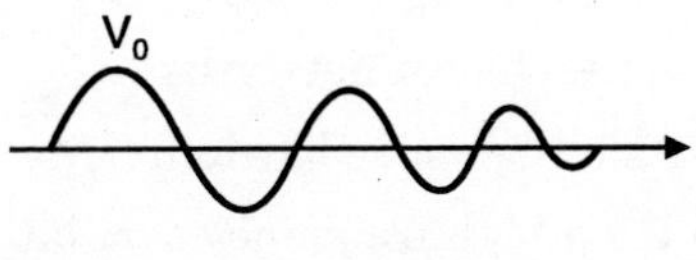

(b) Damped oscillations across the tank circuit

Fig. 10.10.

Basic Tunnel Diode Oscillator

Figure 10.11 (*a*) illustrates the basic tunnel diode oscillator. The tunnel is connected in series with the tank circuit. Resistors R_1 and R_2 are the biasing resistors. They forward bias the tunnel diode. The diode is biased at the center of the negative resistance region as shown in figure 10.11 (*b*).

The damped oscillations across the tank circuit occur due to the energy lost in the tank circuit resistance. The negative resistance characteristics of the tunnel diode counteracts the positive resistance characteristics of the tank circuit resistance. This results in sustained oscillations at the output of the oscillator as shown in figure 10.11 (*a*).

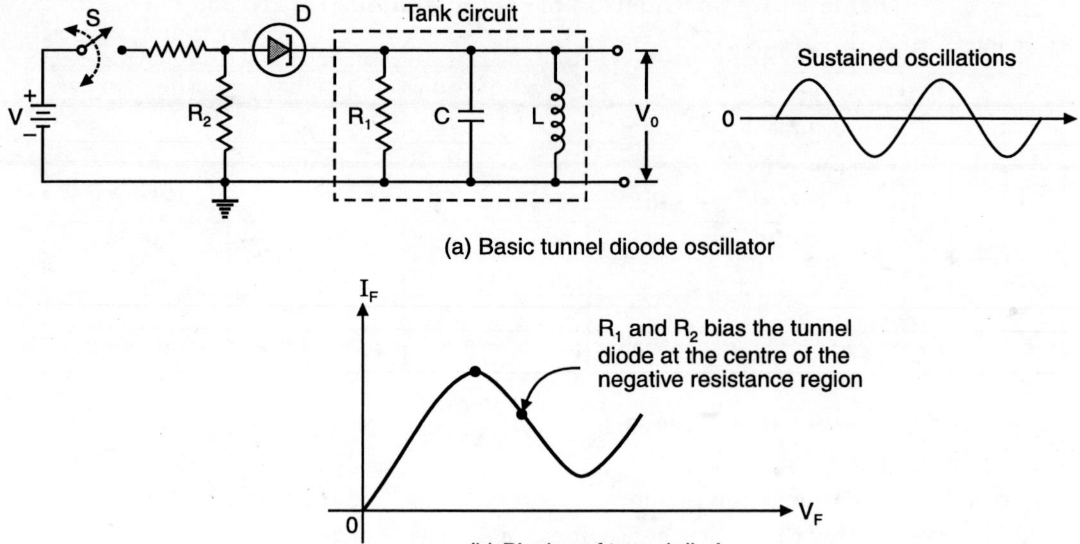

(a) Basic tunnel dioode oscillator

(b) Biasing of tunnel diode

Fig. 10.11.

10.7.9. Other Applications

(*i*) One of the important applications of a tunnel diode is in high speed computers where the switching times in the order of nanoseconds or picoseconds are desirable.

(*ii*) A tunnel diode can be used at such high speeds as a result of the electrons "punching through" at velocities that far exceed those in conventional diodes.

The punching through takes place due to narrow depletion region. Some other applications are :

(*i*) In the digital networks

(*ii*) As a high speed switch

(*iii*) As a high frequency oscillator.

> **DO YOU KNOW?**
>
> The negative resistance of tunnel diodes is useful in high-frequency circuits called oscillators. These circuits are able to generate a sinusoidal signal, similar to that produced by an ac generator.

10.7.10. Comparison of Tunnel Diode and p-n Junction Diode

Table 10.2. gives the comparison between tunnel diode and conventional diode.

Table 10.2. Comparison of tunnel diode and conventional diode

S.No.	Parameter	Conventional diode	Tunnel diode
1.	Cut-in-voltage	0.6 to 0.7 V for silicon	less than 0.1 volts
2.	Breakdown	Large reverse breakdown	Breakdown does not take place
3.	Reverse blocking capacity	Conventional diode can	Can not block the reverse voltage
4.	Doping levels	Moderate	Very high
5.	Width of depletion region	Wider than in tunnel diode	Narrower than that in the conventional.

6.	Characteristics of negative resistance	Not applicable	The region between V_P and V_V exhibits negative resistance
7.	Symbol		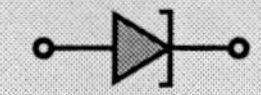
8.	Characteristics	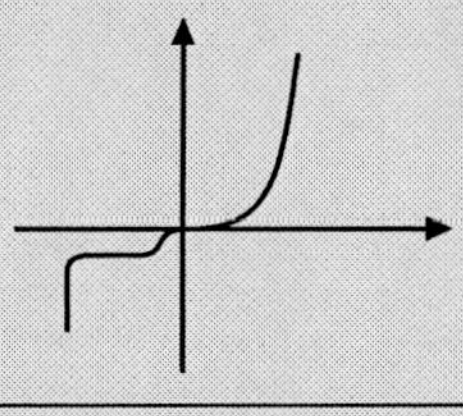	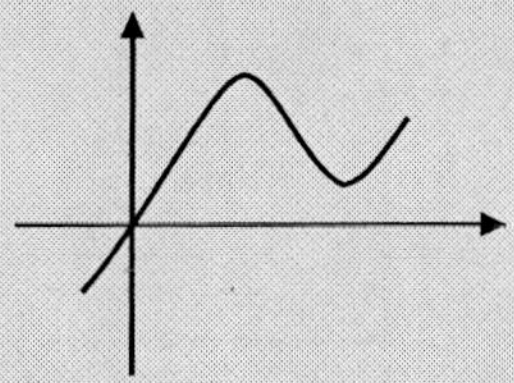
9.	Applications	Rectifiers, clipping and clamping circuits voltage multipliers.	Oscillator, high speed computers, microwave amplifiers.
10.	Speed of operation	Very slow, Suitable for operation at mains frequency.	Very fast due to the tunneling effect.
11.	Material	Silicon or Germanium	Gallium Arsenide.

10.7.11. Advantages of Tunnel Diode

(*i*) Low cost.

(*ii*) The peak point voltage and current (V_P and I_P) are not dependent on temperature.

(*iii*) Tunnel diode needs a small number of external components and a dc power supply for its operation.

(*iv*) It consumes less power.

(*v*) It produces small noise.

(*vi*) It is a high speed device and suitable for the high frequency applications.

10.7.12. Disadvantages

Some of the disadvantages may be listed as under :

(*i*) No isolation between the input and output.

(*ii*) Low output voltage swing. Hence, amplification is required region of the V-I characteristic

10.8. THE VARACTOR (VARICAP) DIODE

A varacter diode is, basically, a reverse biased PN junction, which utilizes the inherent capacitance of the depiction layer. It is also known as **varicap, voltcap or tunning diode**. It is used as a voltage-variable capacitor.

We already know about PN junction that the depletion layer created by the reverse bias acts as a capacitor dielectric, whereas the P-and N-regions act as the capacitor plates as shown in the figure 10.12 (*a*).

When the reverse bias voltage increases, the depiction layer widens. This increases the dielectric thickness, which in turn, reduces the capacitance. When the reverse bias voltage decreases, the depletion layer narrows down. This decreases the dielectric thickness, which in turn increases the capacitance. Figure 10.12 (*b*) shows the variation of capacitance with the reverse voltage. This indicates that the variation of capacitance is maximum when the reverse voltage is equal to zero. It reduces in a non-linear manner, as the value of reverse voltage is increased.

When individual interests become allied with public interests, then results are achieved.

— **J.L. Nehru**

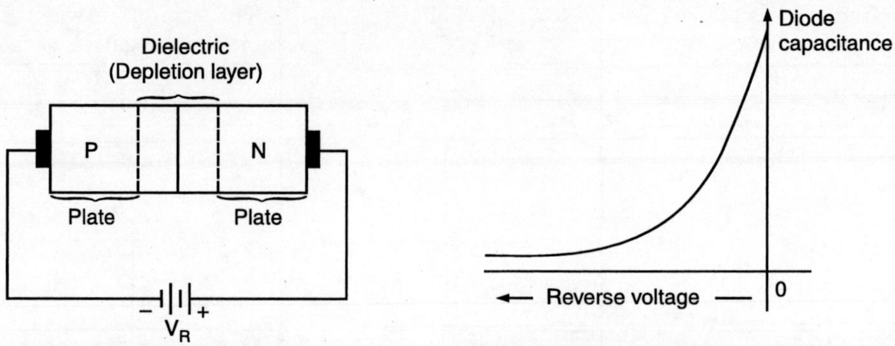

(a) Reverse biased PN junction (b) Capacitance versus reverse voltage

Fig. 10.12.

In a varactor diode, the capacitance parameters are controlled by the method of doping in the depletion layer, or the size and geometry of the diode construction. For example, figure 10.13 (*a*) shows the doping profile for an adrupt junction diode (*i.e.*, the normal PN junction). In this type, the doping is uniform on both sides of the junction. The range of capacitance variation (called tuning range) of an abrupt junction diode is 4 : 1. It means that if its maximum transition capacitance is 100 pF and the minimum is 25 pF, then its tuning range is 4 : 1.

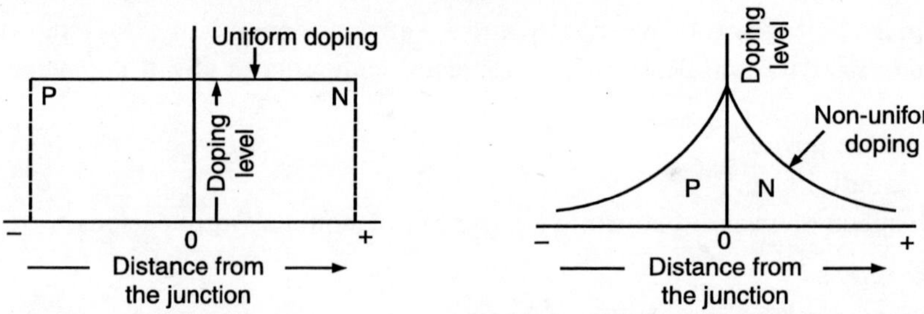

(a) Abrupt doping profile (b) Hyper abrupt profile

Fig. 10.13.

It may be noted that the higher tuning range is obtained for varactors, which have hyper abrupt junction. The doping profile of such a junction will be as shown in figure 10.13(*b*). In this type, the doping level increases as we approach the junction. The heavy doping at the junction results in a narrower depletion layer and hence a larger capacitance. Moreover, any change in reverse voltage produces a larger variation in capacitance than that of abrupt junction.

A hyper abrupt varactor diode has a tuning range of 10 : 1. This range is quite enough to tune a broadcast receiver, over its frequency range of medium wave band (*i.e.*, frequencies from 550 kHz to 1650 kHz).

Figure 10.13 (*a*) shows a sechematic symbol for a varactor diode and figure 10.13 (*b*) shows its equivalent circuit. Here, R_S is a reverse series resistance and C_T is a variable capacitance. The value of C_T ranges typically from 2 pF to 100 pF.

(a) Circuit symbol (b) Equivalent circuit

Fig. 10.14. Varactor diode.

10.9. SPECIFICATIONS OF VARACTOR DIODE

The specification sheet of any varactor diode will provide us with the information we need for any circuit analysis. Motorola MV 209 series varactor diodes are the most commonly used varactor diodes. Now, let us discuss birefly about various varactor diode parameter and characteristics.

> **DO YOU KNOW?**
>
> The tunning range of a varactor depends on the doping level. The tunning range of an abrupt-junction diode is between 3:1 and 4:1.

The maximum ratings for the varactor diode are the same as those that are used for the PN junction diode and the zener diode. The same holds true for the reverse breakdown voltage and reverse leakage current ratings which appear in the electrical characteristics section of the specification sheet. We already know about all these ratings for PN junction and zener diode. However, some of the additional ratings are given below:

(*i*) **The diode capacitance temperature coefficient (TC_C) rating**

This rating of a varactor diode provides the charge in the component's capacitance for each 1°C rise in temperature above 25°C. The TC_C rating for the MV 209 series varactor diodes is 300 ppm/°C. This means that varactor diode capacitance will increase by 300 parts per million for each 1°C rise in temperature above 25°C. For example, the normal capacitance rating of the MV 209 is typically 29 pF. Therefore, increase in capacitance for per 1°C rise in temperature will be,

$$\Delta C = (0.0003) \times 29 \text{ pF} = 0.0087 \text{ pF}$$

Hence, the MV 209 varactor diode will have to experience a temperature increase of nearly 115°C in order to experience an increase in capacitance of 1 pF. This leads us to conclude that the capacitance of a varacter diode is relatively independent of operating temperature.

(*ii*) **Diode capacitance rating (C_t)**

This rating for MV 209 series varactor diode is 26 to 32 pF when V_R is 3 V d.c. We know that the nominal value (29 pF) is the rated value of the component. In other words, MV 209 would be rated as 29 pF even though the actual value could fall anywhere within the specified range of V_R = 3 V dc.

(*iii*) **Capacitance ratio (C_R)**

This rating gives us the variation of the junction capacitance over the given range of voltages. For the MV 209 series, the C_R rating ranges from 5.0 to 6.5 for V_R = 3 to 25 V dc.

(*iv*) **Figure of merit (Q)**

It is the quality factor of a junction capacitance. We know that '*Q*' of a capacitor is the ratio of energy stored in the capacitor to the energy lost through leakage current. In other words, it is the ratio of power returned to the circuit by the capacitor to the power dissipated by the capacitor. For the MV 209, *Q* rating is 200 (minimum).

10.10. APPLICATIONS OF A VARACTOR DIODE *(Very Important)*

A major application of varactor diode is in tuning circuits. It is used in electronic tuners in radio, television and other commercial receivers. When a varactor is used in a tuning circuit, it allows a resonant frequency to be adjusted by a variable voltage level as shown in figure 10.20. In this circuit, two varactor diodes D_1 and D_2 provide the total variable capacitances in a parallel resonant circuit and V_C is a variable d.c. voltage, which controls the reverse bias and the capacitance of the diodes.

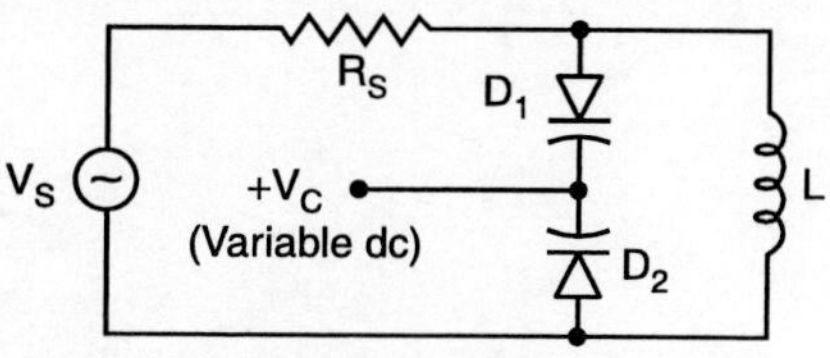

Fig. 10.15.

The resonant frequency of the tank circuit is given by the following relation,

$$f_0 = \frac{1}{2\pi\sqrt{LC}}$$

where L = value of the inductance,

C = value of total capacitance, which is given by the relation

$$\frac{C_1 \cdot C_2}{C_1 + C_2}$$

where C_1 and C_2 are the maximum and minimum values of the capacitances of diode.

The varactor diode is also used in automatic frequency control device, adjustable band-pass filter and parametric amplifier.

DO YOU KNOW?

How is a varactor used? It is connected in parallel with an inductor to form a parallel resonant circuit. This circuit has only one frequency at which maximum impedance occurs. This frequency is called the resonant frequency. If the dc reverse voltage to the varactor is changed, the resonant frequency is also changed. This is the principle behind electronic tunning of a radio station, a TV channel and so on.

EXAMPLE 10.3. The capacitance of certain varactor diode can be varied from 5 pF to 50 pF. The diode is used in a tuned circuit of a radio receiver as shown in figure 10.16.

Find the tuning rage for circuit, if the inductance is 10 mH.

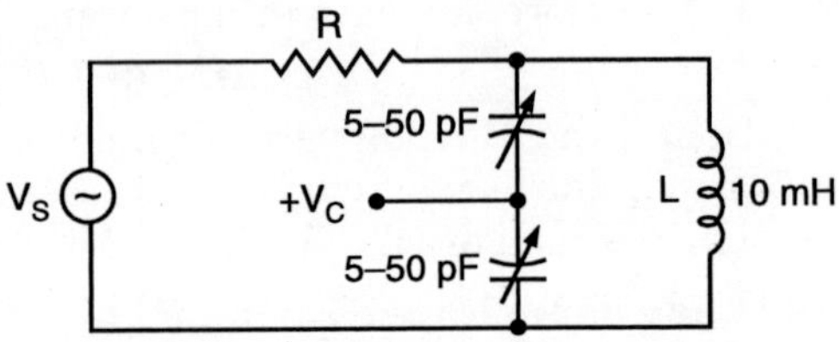

Fig. 10.16. Circuit for example 10.3

Solution: Given that C_1 (*i.e.*, C_{min}) = 5 pF, C_2 (*i.e.*, C_{max}) = 50 pF

and $\quad L = 10 \text{ mH} = 10 \times 10^{-3}$ H.

We know that the maximum resonant frequency is obtained, when the total capacitance of the varactor diode is minimum and its value is

$$C_{T\,(max)} = \frac{C_1 \cdot C_2}{C_1 + C_2} = \frac{5 \times 5}{5 + 5} = 2.5 \times 10^{-12} \text{ F} = 2.5 \text{ pF}$$

Hence, maximum resonant frequency will be

$$f_{o\,(max)} = \frac{1}{2\pi\sqrt{L \cdot C_{T(\text{min})}}} = \frac{1}{2\pi\sqrt{(10 \times 10^{-3}) \times (2.5 \times 10^{-12})}} \text{ Hz.}$$

or $\quad f_{o(max)} = 1 \times 10^6 \text{ Hz} = 1 \text{ MHz}$ **Ans.**

We also know that the minimum resonant frequency is obtained, when the capacitance of the varactor diode is maximum and its value is

$$C_{T\,(max)} = \frac{C_1 \cdot C_2}{C_1 + C_2} = \frac{50 \times 50}{50 + 50} = 25 \times 10^{-12} \text{ F} = 25 \text{ pF}$$

Hence, minimum resonant frequency, will be

$$f_{o\,(min)} = \frac{1}{2\pi\sqrt{L \cdot C_{T(\text{min})}}}$$

$$= \frac{1}{2\pi\sqrt{(10 \times 10^{-3}) \times (25 \times 10^{-12})}} \text{ Hz}$$

or $\quad f_{o(\text{min})} = 318 \times 10^3 \text{ Hz} = 318 \text{ kHz}$ **Ans.**

The very first step towards success in any occupation is to become interested in it.

— **Sir William Osler**

10.11. THE SCHOTTKY BARRIER DIODE

The schottky barrier diode is a two terminal device which is being increasingly used in applications such as ac-to-dc converters, radar systems, mixers and detectors in communication equipment, instrumentation and *A*-to-*D* converter.The schottky barrier diode is also called as **surface barrier diode** or **hot carrier diode**.

> **DO YOU KNOW?**
>
> Schottky diodes are relatively high-current devices, capable of switching quickly while providing forward currents in the neighbourhood of 50 A ! It is also worth noting that Schottky diodes normally have lower breakdown voltage ratings as compared to conventional pn junction rectifier diodes.

10.11.1 Construction of Schottky Diode

Figure 10.17 shows the construction of a schottky diode. The construction of a schottky diode is quite different from the conventional *p-n* junction diode. A metal semiconductor junction is formed between a metal and *n*-type semiconductor as shown in figure 10.17. Even though *n*-type semiconductor is normally used, sometimes a *p*-type semiconductor is also used.

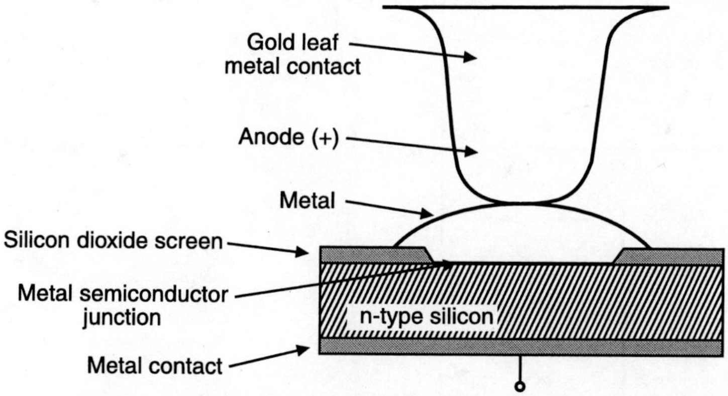

Fig. 10.17. Construction of schottky barrier diode.

The metals used are molybdenum, platinum, chrome or tungsten. Different construction techniques will result in a different set of characteristics for the device. The metal side acts as the anode and *n*-type semiconductor acts as cathode of the schottky diode.

> **DO YOU KNOW ?**
>
> The Schottky diode is sometimes called a hot-carrier diode. This name came about as follows. Forward bias increases the energy of the electrons on the *n* side to a higher level than that of the electrons on the metal side of the junction. This increase in energy inspired the name hot carrier for the *n*-side electrons. As soon as these high-energy electrons cross the junction, they fall into the metal, which has a lower-energy conduction band.

10.11.2. Principle of Operation of Schottky Diode

In the metal which acts as the anode and in the *n*-type semiconductor which acts as a cathode, **electrons** are the majority carriers. In the metal, the level of minority carriers (holes) is not significant. When the metal and semiconductor are joined to form the junction, the electrons in *n*-type material will immediately flow into the adjoining metal. This will establish a heavy flow of majority carriers. Since the injected carriers have a very high kinetic energy, compared to the electrons of the metal, they are commonly called as **hot carriers**. In the conventional *p-n* junction, the minority carriers get injected into the adjoining region. But here the electrons are injected into a region (metal) where electrons only are the majority carriers. Therefore the schottky diodes are different from the conventional diodes. In schottky diodes, the conduction entirely is due to the **majority carriers**.

The heavy flow of electrons into metal creates a region near the junction surface which is depleted of carriers in the silicon material. This is very similar to depletion region in the *p-n* junction diode. The additional carriers in the metal will establish a **negative wall** inside the metal at the boundary between the two materials. The result is a **surface barriers** between the two materials, preventing any further current. Now, when we apply a forward bias voltage, the strength of the negative barrier will reduce, because the external positive potential will attract electrons from the *p* side (metal). Due to this, a heavy flow of electrons across the junction will start. This forward current can be controlled by the level of applied bias potential.

10.11.3 V-I Characteristics of Schottky Diode

Figure 10.18 shows the V-I characteristics of a schottky diode. From this figure, we can compare the characteristics of schottky and conventional *p-n* junction diodes.

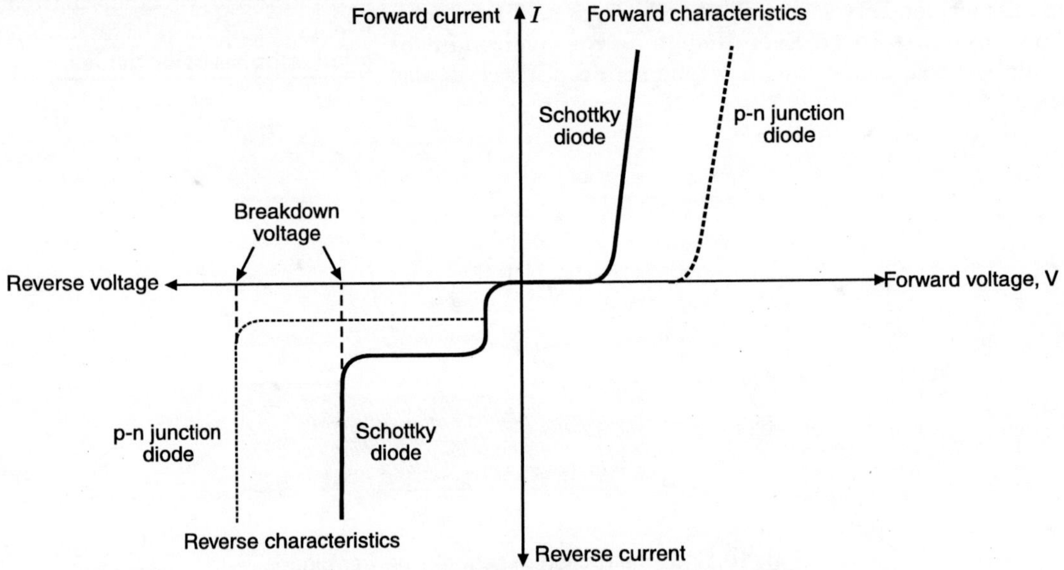

Fig. 10.18. V-I characteristics of a schottky diode.

The characteristics of a schottky diode is very similar to that of a *p-n* junction diode. But it has a very low cut-in-voltage (of the order of 0.2 Volts). The reverse breakdown voltage of a schottky diode is less than that of a *p-n* junction diode. Typically, it is about 50 V as compared to 150 V for a *p-n* junction diode. The leakage current in the reverse biased region is higher than that of a *p-n* junction diode. The reverse leakage current flows due to the electrons in metal passing into the semiconductor material.

10.11.4 Equivalent Circuit and Circuit Symbol of Schottky Diode

Figure 10.19(*a*) shows the approximate equivalent circuit of a schottky diode. Figure 10.19(*b*) shows the circuit symbol of a schottky diode. In the equivalent circuit, C_j is the junction capacitance which appears in parallel with an ideal diode.

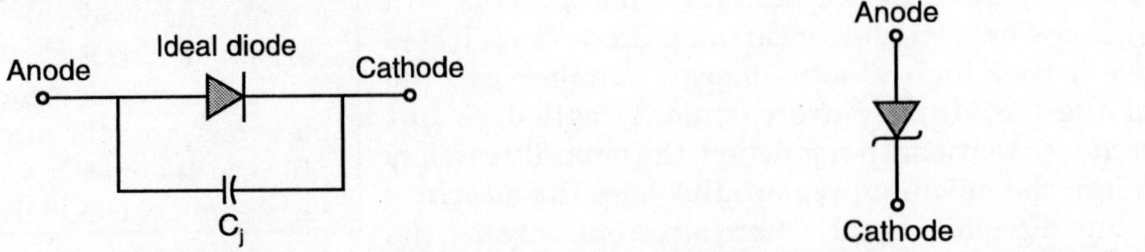

(a) Approximate equivalent circuit of a schottky diode

(b) Circuit symbol of a schottky diode

Fig. 10.19.

10.12. ADVANTAGES, DRAWBACKS AND APPLICATIONS

1. Advantages

(*i*) Low on state forward voltage drop.

(*ii*) Higher speed of switching. These diodes can be used upto frequencies in the gigahertz (GHz) range. [1 GHz = 10^9 Hz].

(*iii*) Total power dissipation taking place in the schottky diode is less than conventional diodes.

2. Drawbacks

(*i*) Low reverse breakdown voltage, therefore low PIV (peak inverse voltage) as compared to *p-n* junction diodes.

(*ii*) High leakage current.

3. Applications

(*i*) Switching power supplies (SMPS)

(*ii*) AC-to-DC converters.

(*iii*) Radar systems

(*iv*) Schottky TTL logic for computers.

(*v*) Mixers and detectors in communication equipments.

10.12.1 Comparison of Schottky Barrier Diode and *p-n* Junction Diode

S.No.	Parameter of comparison	p-n Junction Diode	Schottky Diode
1.	Circuit symbol	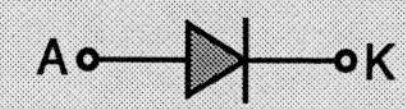	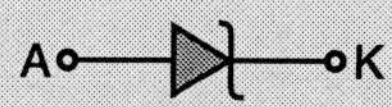
2.	Forward voltage drop	Typically 0.7 V for silicon diodes	Between 0 to 2.2 V
3.	Peak inverse voltage (PIV)	High (typically 150 V)	Low (typically 50 V)
4.	Leakage current	Low (in nano amperes)	High (in microamperes)
5.	Carriers	Conduction takes place due to majority and minority carriers.	Conduction takes place only due to majority carriers.
6.	Junction is formed between	*p* and *n* type semiconductors	A metal and *n*-type semiconductor.
7.	Swithcing speed	Very low. Therefore suitable only for low frequency operation.	Very high. Therefore, can operate upto GHz range.
8.	Applications	Rectifiers	SMPS, AC-to-DC converters, Schottky TTL logic etc.

10.13. THE STEP-RECOVERY DIODE *(Important)*

A step-recovery diode employs a graded doping, where the doping level of the semiconductor material is reduced as the junction is approached. Such a doping profile has been shown in figure 10.20(*a*). This profile indicates that the doping level (*i.e.*, the density) of charge carriers decreases as the junction is approached. Figure 10.20 (*b*) shows the schematic symbol of a step-recovery diode.

The unusual doping of the step-recovery diode affects the time required for the device to switch from OFF to ON and vice versa. The typical switching time for a step-recovery diode is in the low-picosecond range. This makes them ideal for ultra-high frequency switching applications.

When a step-recovery diode is connected to a sinusoidal a.c. input source, it produces an output as shown in figure 10.21. It may be noted that during the positive half cycle of the input signal, the diode conducts like a normal silicon diode. But, during the negative half cycle, reverse current exists for a short time and then suddenly drops to zero. The reverse current flows because of the stored charge in a diode, which takes some finite time for its removal. Since, the reverse recovery period is very sharp, therefore, the diode is called step-recovery diode. The step-recovery diode also be thought of as if the diode is suddenly snapped open. Therefore the step-recovery diode is also known as a **snap-diode**.

DO YOU KNOW?

The step-recovery diode has the unusual doping profile. This unusual doping profile indicates that the density of carriers decreases near the junction. This unusual distribution of carriers causes a phenomenon called reverse snap-off.

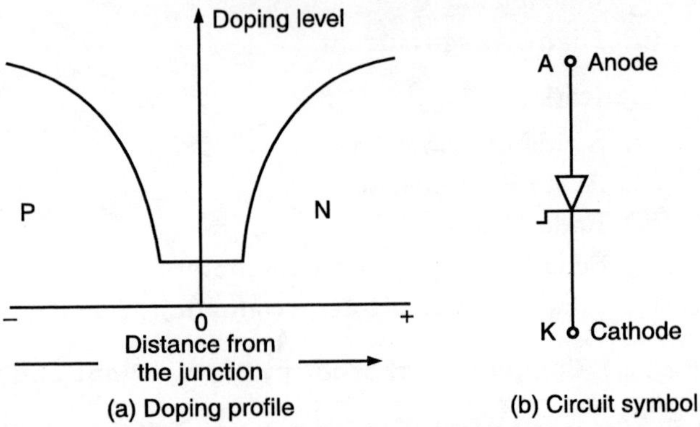

Fig. 10.20. Step-recovery diode.

A Fourier analysis of such a transition (*i.e.*, sharp reverse recovery) shows that the output consists of a number of frequency components called harmonics. These harmonics have frequencies of f, $2f$, $3f$, ...nf. The frequency f (10 MHz in the present case) is called fundamental frequency whereas $2f$, $3f$, ..., nf are called second, third..., harmonics. The output waveform can be filtered to produce a sine wave of higher frequency. Because of this fact, the step-recovery diodes are useful in frequency multipliers (*i.e.*, circuits whose output frequency is a multiple of the input frequency).

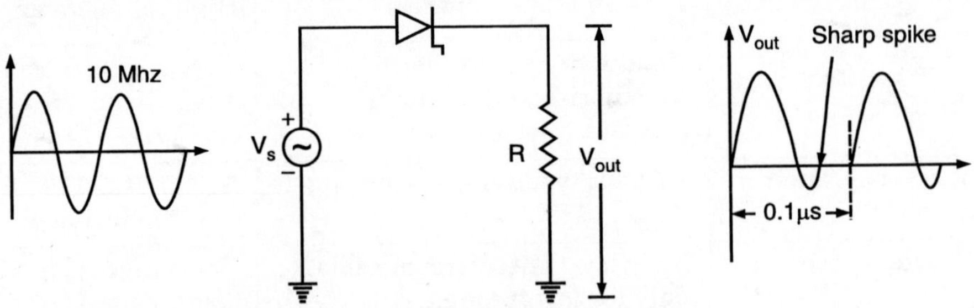

Fig. 10.21. Output showing a sharp spike.

10.14. A PIN DIODE

A PIN diode is made up of three semiconductor materials : two *heavily doped* P-and one N-type material separated by an *intrinsic** semiconductor (I) as shown in figure 10.22 (*a*). The intrinsic region offers high resistance to the current flowing through it.

* Undoped

The PIN diode has the following two advantages over the normal PN junction diode:

(*i*) The capacitance between the P and N region *decreases* because of the increased separation between the P and N regions. This advantage allows the PIN diode to have fast response time. Therefore, these diodes are useful at very-high-frequencies (*i.e.*, above 300 MHz).

(*ii*) There is a greater electron-hole pair generation because of the increased electric field between the P and N regions. This advantage allows the PIN diode to process even weak signals.

We know that when a PIN diode is unbiased (*i.e.*, no voltage is applied across the diode), there is a diffusion of electron and holes across the junction due to the different concentration of atoms in the P, I and N regions. The diffusion of electrons and holes produce a depletion layer across the PI and IN junctions as shown in figure 10.22(*b*). The depletion layers penetrate to a little distance in the P-type and N-type semiconductor regions but to a larger distance in the I-region. Under such condition, the device has a high value of resistance.

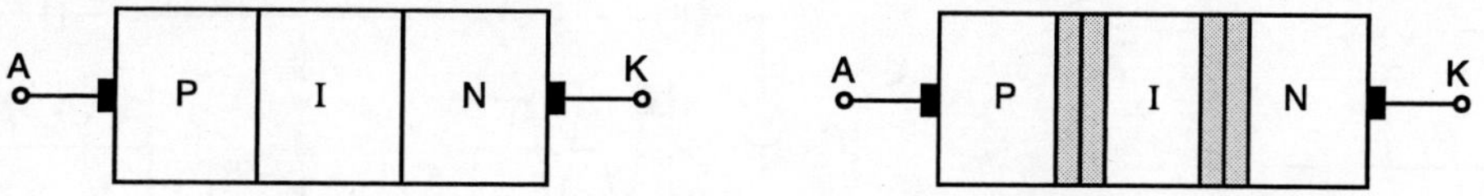

(a) Base scture (b) Formation of a depletion layer

Fig. 10.22.

When the PIN diode is forward biased, the width of depletion layers decreases. As a result of this, more carriers are injected into the I-region. This reduces the resistance of the I-region. If the depletion layers are not thick, then the I-region becomes flooded with the carriers at a stable bias. Thus, when a PIN diode is forward biased, it acts like a variable resistance as shown in figure 10.23 (*a*). The forward resistance of an intrinsic region decreases with the increasing current.

> **DO YOU KNOW?**
>
> When a PIN diode is forward-biased, it acts like a current-controlled resistance. When reverse-biased, the PIN diode acts like a fixed capacitor.

On the other hand when the PIN diode is reverse biased, the depletion layers become thicker. As the reverse bias is increased, the thickness of the depletion layer increases till the I-region becomes free of mobile carriers. The reverse bias, at which this happens, is called *swept out voltage*. At this stage, the PIN diode acts like an almost constant capacitance as shown in figure 10.23 (*b*).

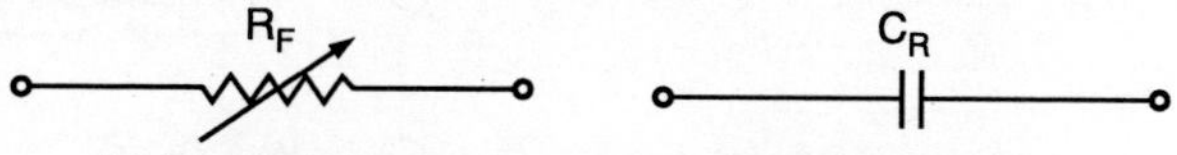

(a) Forward biased (b) Reverse bias

Fig. 10.23. Equivalent circuit of a PIN diode.

10.15. APPLICATIONS OF PIN DIODE

We know that when a PIN diode is forward biased, it acts like a variable resistance which decreases with the increasing current. Also when it is reverse biased, the PIN diode acts like an almost constant capacitance. These features make the PIN diode useful for a variety of applications as under :

(*i*) As a d.c. controlled microwave switch operated by rapid changes in bias or as a modulating device. This application takes advantage of the variable forward-resistance characteristic. Modulators are used to combine two signals of different frequencies into a single signal.

(*ii*) In attenuator applications because its resistance can be controlled by the current.

(*iii*) As an AM modulator of very high frequency signals.

10.15.1. High Frequency Switching Operation

The use of PIN diode in electronic high frequency switching has been illustrated in figure 10.24.

When the diode is reverse-biased, it looks like an open circuit as shown in figure 10.24 (*b*). The 500 MHz input signal voltage divides across the series-connected 100 Ω resistor and the diode in proportion to their resistances. Because the diode is open and hence has infinite resistance, the entire input signal appears across the open circuit. Therefore, the entire input signal passes out through coupling capacitance C_2 without any kind of loss or attenuation. Also, when the diode is forward-biased by a voltage source of +25 V dc, current I is given by

$$I = \frac{25}{1K} = 25 \text{ mA}$$

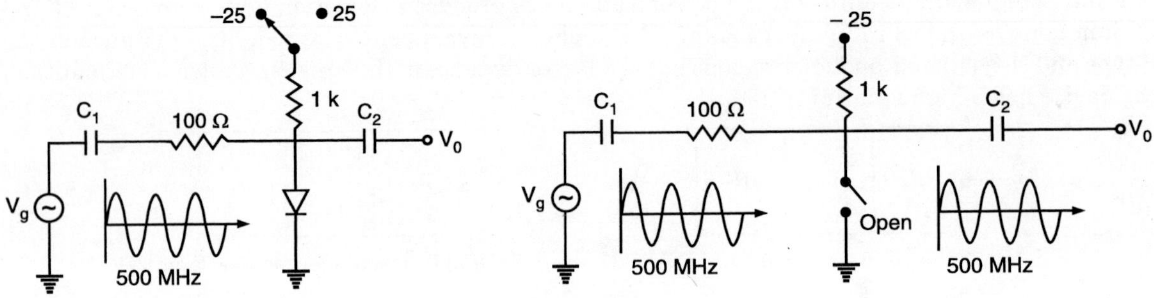

Fig. 10.24.

Because of this, the diode resistance becomes

$r_{ac} = \frac{50}{25} = 2\ \Omega$ as shown by its equivalent circuit shown in fiigure 10.24.

Now, in this case, almost all the input signal voltage drops across 100 Ω resistor and practically no voltage drops across the 2Ω resistor. Therefore, there is hardly any signal output.

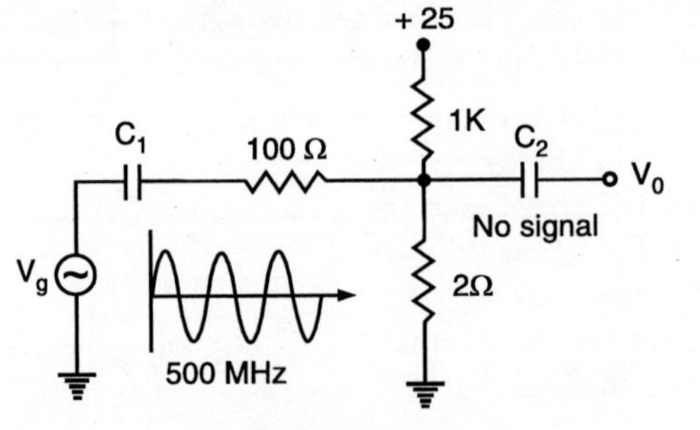

Fig. 10.25.

However, practically, instead of mechanically switching the diode-biasing supply voltage from – 25V to + 25V, a transistor is used to do this switching operation. Hence, in this manner, we can turn a very high frequency signal (MHz range) OFF and ON with the speed of a transistor switching circuit.

10.15.2. PIN Diode Operation as an AM Modulator

Figure 10.26 shows the method in which a high frequency 500 MHz signal is modulated at 1 kHz rate. Here, a 1 kHz signal is fed into a PNP transistor where it varies its dc output current at the same rate. This varying dc current is applied as biasing current to the PIN diode as depicted in figure 10.26. It varies the diode ac resistance as seen by the 500 MHz signal. Therefore, the signal is modulated at 1 kHz rate as shown.

A little knowledge that acts is worth infinitely more than much knowledge that is idle.

— **Khalil Gibran**

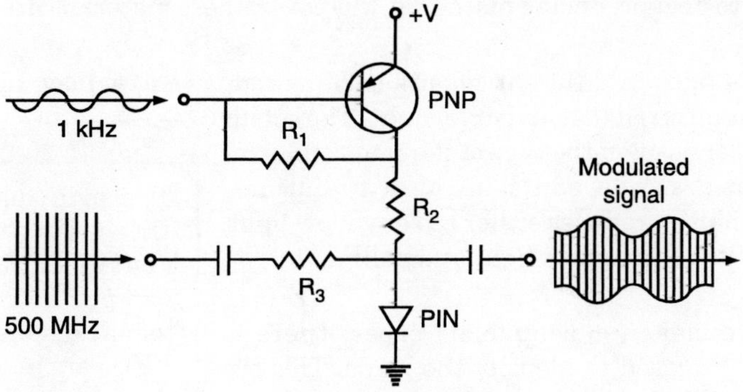

Fig. 10.26.

10.16. LIGHT EMITTING DIODE (LED) *(Very Important)*

A PN junction diode, which emits light when forward biased, is known as a light emitting diode (abbreviated as LED). The emitted light may be visible or invisible. The amount of light output is directly proportional to the forward current. Hence, higher the forward current, higher is the light output. The schematic symbol of a light emitting diode has been shown in figure 10.27(*a*). The arrows, pointing away from the diode symbol represent the light, which is being transmitted away from the junction.

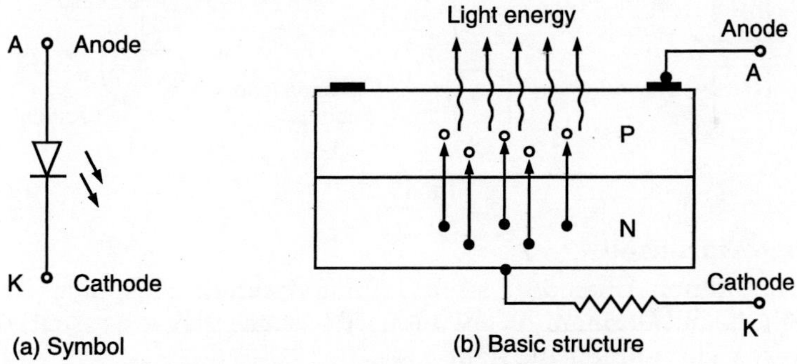

Fig. 10.27. Light emitting diode (LED).

Basic Structure

Figure 10.27(*b*) shows the basic structure of a light emitting diode. Here, an N-type layer is grown on a P-type substrate (not indicated in figure) by a diffusion process. Then, a thin P-type layer is grown on the N-type layer. The metal connections to both the layers make anode and cathode terminals as indicated. The light energy is released at the junction, when the recombination of electrons with holes take place. After passing through the P-region, the light is emitted through the window provided at the top of the surface.

It will be interesting to know that when the LED is forward biased, the electrons and holes move towards the junction and the recombination takes place. After recombination, the electrons, lying in the conduction bands of N-region, fall into the holes lying in the valence band of a P-region. The difference of energy between the conduction band and valence band is radiated in the form of light energy. In ordinary diodes, this energy is radiated in the form of heat.

The semiconducting materials used for manufacturing light emitting diodes are gallium arsenide, gallium arsenide phosphide. The silicon and germanium are not used to manufacture light emitting diodes because these are heat producing materials. Moreover, these materials are very poor in emitting light radiations.

The LED's radiate light in different colours such as red, green, yellow, blue, orange etc. Also, some of the LED's emit infrared (*i.e.* invisible) light. The colour of the emitted light depends upon the type of the semiconductor used. Hence, gallium arsenide emits infrared radiations, gallium arsenide phosphide produces either red or yellow light, gallium phosphide emits red or green light and gallium nitrite produces blue light.

Since LED's have clear (or semiclear) cases, there is normally no label on the cased to identify the leads. The two leads of a LED are identified using one of the several schemes as discussed below:

(*i*) The leads may have different lengths as shown in figure 10.28(*a*), when the scheme is used, the longer of the two leads is usually the anode.

(*ii*) One of the leads may be flattened as shown in figure 10.28(*b*). The flattened lead is usually the anode

(*iii*) One side of the case is flattened as shown in figure 10.28(*c*). The lead closest to the flattened side is usually the anode.

DO YOU KNOW?

The principal disadvantage of LEDs is that they draw considerable current in comparison to other types of visual displays. In many cases, LEDs are pulsed ON and OFF at a rapid rate, rather than being supplied with a steady drive current. LEDs appear to the eye to be on continuously, but they consume less power than if they ware ON continuously.

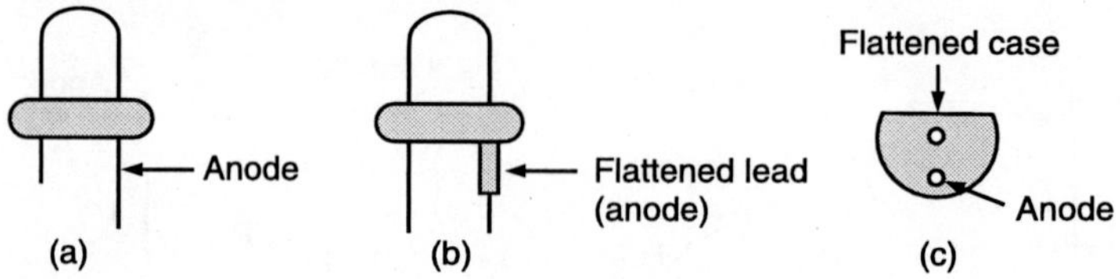

Fig. 10.28.

Semiconductor materials used

LEDs are made of Gallium Arsenide (GaAs), Gallium Arsenide phosphide (GaAsP) and Gallium phosphide (Ga P). Silicon and Germanium are not used because they are essentially heat producing materials and are very poor in producing light.

10.16.1. Principle of LED Operation

When the LED is forward biased, the electrons in the n-region will cross the junction and recombine with the holes in the p-type material. These free electrons reside in the conduction band and hence at a higher energy level than the holes in the valence band. When the recombination takes place, these electrons return back to the valence band which is at a lower energy level than the conduction band. While returning back, the recombining electrons give away the excess energy in the form of light. This is shown in figure 10.29. This process is called as **electro-luminescence**. In this way, an LED emits light. This is the principle of operation of LED.

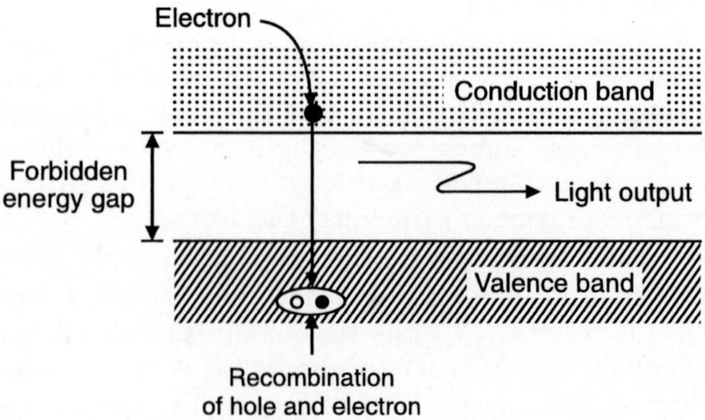

Fig. 10.29. Principle of operation of LED.

Colour of the emitted light

The colour of the emitted light is decided by its wavelength. The wavelength of emitted light (and hence its colour) depends on the value of forbidden gap (E_G). Forbidden gap value depends on the material used. Hence the colour of emitted light is dependent on the material used. The colours associated with different materials are as under.

Material used	Colour of the emitted light
1. Gallium Arsenide (Ga As)	Infrared (IR)
2. Ga As P	Red or Yellow
3. Gallium phosphide (Ga P)	Red or Green

Silicon and Germanium are not used for LEDs because their forbidden gaps do not allow the light emission in visible spectrum. The commonly available LEDs emit green, red, yellow, orange and blue colours. Some LEDs are capable of emiting infrared light which is not visible to our eyes. Some LEDs are capable of emitting infrared light which is not visible to our eyes. Various impurities are added during the doping process to get precisely the required colour of the emitted light.

10.17. LED VOLTAGE DROP AND CURRENT

The LED's have also a certain forward voltage drop, when they are used in the circuit like an ordinarily diode (which has a voltage drop of 0.7 V). This voltage drop depends upon the LED current, colour of the emitted light etc. Usually, the voltage drop has a typical value that varies from 1.5 to 2.5 V for currents between 10 and 50 mA. It will be interesting to know that LED's are not able to withstand reverse bias of even every small voltages. For this reason, it is necessary to assure that reverse bias is never applied to an LED.

Now, let us consider a LED circuit as shown in figure 10.30. In this circuit, the LED is connected to the supply voltage through a current limiting resistor (R). The current through the LED is given by the following relation :

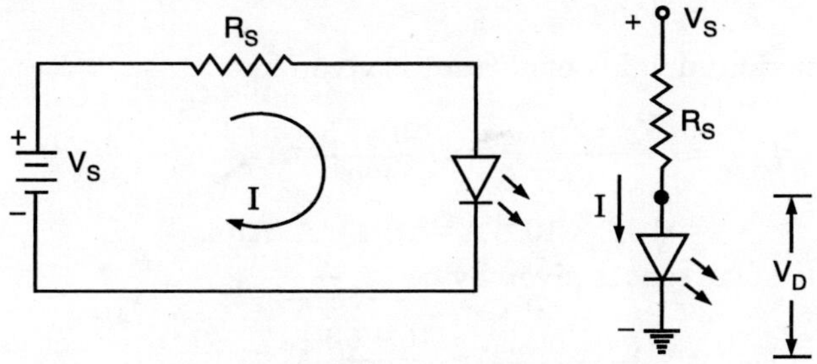

Fig. 10.30. LED circuit.

$$I = \frac{V_S - V_D}{R_S}$$

where V_S = Supply voltage,

V_D = Voltage drop across LED, and

R_S = Current limiting resistor.

The value of current limiting resistor is given by the following expression:

$$R_S = \frac{V_{out(peak)} - V_F}{I_F}$$

V_{out} *(peak)* = Peak output voltage

Opportunities are never lost. The other fellow takes those you miss. — **Anonymous**

V_F = LED voltage drop, and

I_F = LED current drop.

Because the voltage drop, across the diode, depends upon the value of current through it, therefore it controls the amount of light output (or brightness) of the LED. In order to avoid fluctuations in the brightness, the current through the LED must be kept constant. In actual practice, it is done by selecting a large supply voltage and a large value of current limiting resistor.

EXAMPLE 10.4. The driving circuit shown in figure 10.31 has a peak output voltage of 8 V. The LED has a rating of V_F = 1.8 to 2.0 V and $I_{F(max)}$ = 16 mA.

What will be the smallest standard resistor value that can be used as R_S.

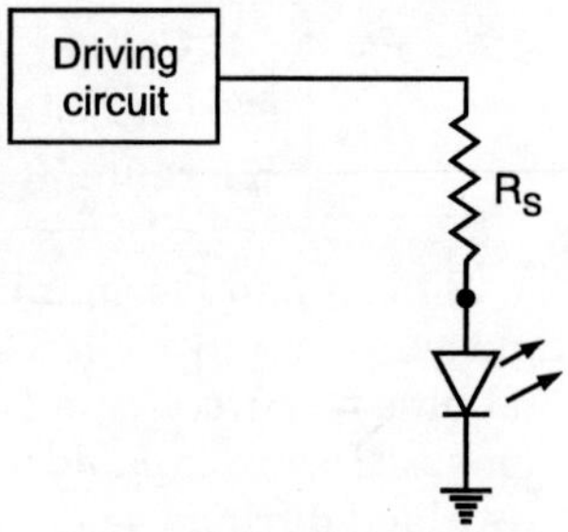

Fig. 10.31. Circuit for example 10.4

Solution: Given that $V_{out(peak)} = 8$ V; V_F = 1.8 to 2.0 V

and $I_{F\,(max)} = 16 \text{ mA} = 16 \times 10^{-3}$ A.

We know that current limiting resistor is given by

$$R_S = \frac{V_{out(pk)} - V_F}{I_F} = \frac{8-1.8}{16 \times 10^{-3}} = 387.5\ \Omega$$

and if $V_F = 2.0$ V then $R_S = 375\ \Omega$. In either case, the smallest standard-value resistor that has a value greater than 387.5 Ω or 375 Ω is the 390 Ω resistor. This is the component which we have to use in the circuit of figure 10.37.

EXAMPLE 10.5. Figure 10.31(*a*) shows an LED circuit. The LED has a minimum drop of 1.5 V and a maximum drop of 2.3 V. If the supply voltage is 10 V and R_1 = 470 Ω. What are the minimum and maximum values of LED current?

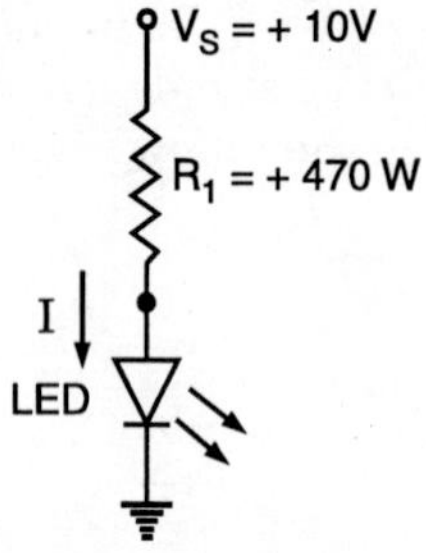

Fig. 10.31(a).

Solution: Given that $V_{D(min)} = 1.5$ volts;

$V_{D\,(max)} = 2.3$ volts;

$V_S = 10$ volts

and $R_1 = 470\ \Omega$.

We know that the maximum value of current is given by

$$I_{max} = \frac{V_S - V_{D(min)}}{R_1} = \frac{10-1.5}{470}$$

$$= 18.1 \times 10^{-3} \text{ A} = 18.1 \text{ mA} \quad \textbf{Ans.}$$

Also, minimum value of current is given by

$$I_{min} = \frac{V_S - V_{D(min)}}{R_1} = \frac{10-2.3}{470} \text{ A}$$

or $$I_{min} = 16.4 \times 10^{-3} \text{ A} = 16.4 \text{ mA} \quad \textbf{Ans.}$$

EXAMPLE 10.6. A light emitting diode with minimum and maximum voltage drops of 1.8V and 3 V respectively is connected alternately with 24 volt supply (with 820 Ω resistor), and 5 volt supply (with 120 Ω resistor). Indicate which supply voltage will keep the brightness of the diode as constant.

Solution: Given that $V_{D(min)} = 1.1$ volts

and $V_{D(max)} = 3$ volts

We know that minimum and maximum values of the current through the LED with 24 volt supply (V_S) and 820 Ω resistor (R_S) will be

$$I_{min} = \frac{V_S - V_{D(max)}}{R_S} = \frac{24-3}{820} = 25.6 \times 10^{-3} \text{ A} = 25.6 \text{ mA} \quad ...(i)$$

and $$I_{\max} = V_S - \frac{V_{D(\min)}}{V_S} = \frac{24-1.8}{820} = 27.1 \times 10^{-3}\text{ A} = 27.1\text{ mA} \qquad ...(ii)$$

Also, minimum and maximum values of the current through the LED with 5 volt supply (V_S) and 120 Ω resistor (R_S) will be

$$I_{min} = \frac{V_S - V_{D(\max)}}{R_S} = \frac{5-3}{120} = 16.7 \times 10^{-3}\text{ A} = 16.7\text{ mA} \qquad ...(iii)$$

and $$I_{max} = \frac{V_S - V_{D(\min)}}{R_S} = \frac{5-1.8}{120} = 26.7 \times 10^{-3}\text{ A} = 26.7\text{ mA} \qquad ...(iv)$$

It may be noted that the variation in current for the first case (27.1 – 25.6 = 1.5 mA) is small whereas in the second case it is (26.7 – 16.7 = 10.0 mA) large. Therefore, the brightness in the first case will remain constant whereas in the second case it will be changing. Therefore, in order to get an approximately constant brightness, we use as large a supply voltage as possible.

10.17.1. Spectral Response Curves of LED

The graph of output light (*Y*-axis) versus the wavelength is called as the spectral response curve of an LED. The shape and the wavelength at which maximum response is obtained varies depending on the material used as shown in figure 10.32.

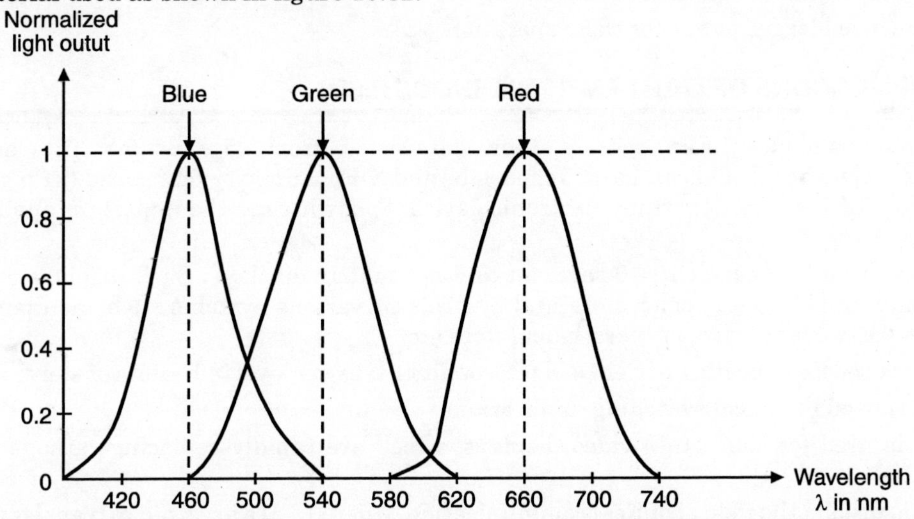

Fig. 10.32. Spectral response curves of LEDs operating in the visible portion of spectrum

10.17.2. Radiation Pattern of LED

As a matter fact, radiation pattern is a graph of light output plotted against angle as shown in figure 10.33. It gives us an idea about how directional the given light source is.

The LED is a highly directional source because it radiates only in a particular direction. Most of the radiated energy is confined within 20° of the direction of maximum light.

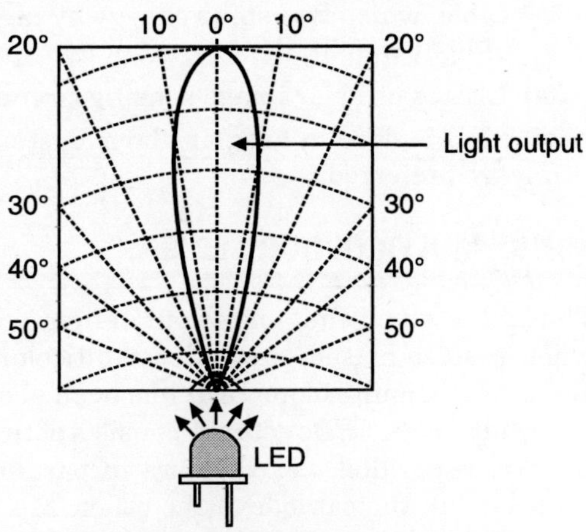

Fig. 10.33. Radiation pattern of LED.

10.17.3. Electrical and Optical Parameters of LED

Different electrical and optical parameters of LED are as follows:

- Reverse leakage current (I_R) = 50 nA.

- Reverse voltage (V_R) = 30 V
- Forward current (I_F) = 10 to 80 mA
- Forward voltage (V_F) = 1.2 V to 1.8 V
- Output power (P_0) = 200 to 500 μW.

10.17.4. Advantages of LED

(*i*) Small size and light weight.
(*ii*) Available in different spectral colours.
(*iii*) Longer life as compared to lamps.
(*iv*) Better linearity.
(*v*) Suitable for higher operating speeds.
(*vi*) Low cost.
(*vii*) Compatible with integrated circuits.

10.17.5. Disadvantages of LED

(*i*) Output power is affected by changes in temperature.
(*ii*) Overcurrent can damage it easily.
(*iii*) They need larger power for their operation.

10.18. APPLICATIONS OF LIGHT EMITTING DIODE (LED)

The light emitting diodes (LEDs) operate at low voltages *i.e.*, from 1.5 V to 2.5 V. They have a long life of about 10,000 hours and can be switched 'ON' and 'OFF' at a very fast speed (≅1 n sec). These features make LED's very important electronic device. Following are the important applications of the LED's :

(*i*) LED is used in 7-segment, 16-segment and dot matrix displays. Such displays are used to indicate alphanumeric characters and symbols in various systems such as digital clocks, micro-wave-ovens, stereo tuners, calculators etc.
(*ii*) LED is used for indicating power ON/OFF conditions, power level indicators or stereo amplifiers.
(*iii*) LED is used in optical switching applications.
(*iv*) LED is used for solid state video displays, which are rapidly replacing cathode ray tubes (CRT's).
(*v*) LED is used in the field of optical communication, where LED's are used to transfer (or couple) energy from one circuit to another. They are also used to send light energy to fibre optical cable, which transmits energy by means of total internal reflection. The fibre optical cable is of light weight, flexible, often transparent and as small a 0.043 mm in diameter.
(*vi*) LED is used for image sensing circuits in picturephon.
(*vii*) LED is used in burglar alarm systems. In such applications, LED's radiating infrared light are preferred.

10.19. MULTICOLOUR LED'S

The LED's which emit one colour of light when forward biased and another when reverse biased, are called multicolored LED's. One commonly used symbol for a multicolour LED has been shown in figure 10.34.

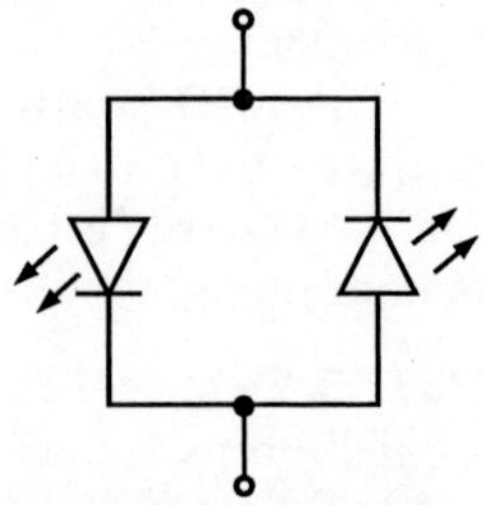

Fig. 10.34. Schematic circuit symbol for

Multicolour LEDs actually consists of two PN junction that are connected in reverse-parallel, *i.e.*, they are in parallel, with the anode of one being connected to the cathode of the other. Multicolour LEDs are typically red when biased in one direction and green when biased in the other direction.

Incidently, if a multicolour LED is switched fast, the LED will produce a third colour. For example a red/green LED will produce a yellow light when rapidly switched back and forth between biasing polarities.

10.20. THE SEVEN-SEGMENT DISPLAY *(Very Important)*

Figure 10.35(*a*) shows a seven segment display. It is used to display alphanumeric characters. It consists of 7 rectangular light emitting diodes designated by the letters *a, b, c, d, e, f* and *g*. Each LED is called a segment, because it forms a part of the character being displayed.

Figure 10.35(*b*) shows a schematic diagram of a seven-segment display. In this circuit, the anodes of all the diodes are connected together to the positive terminal of the dc voltage source. The cathodes are connected to the external resistors. The external resistor is essential to limit the current through the LED. By grounding the external resistors, we can form any decimal digit from 0 to 9. For example, by grounding *a, b, g, e* and *d*, we can form the digit '2' as shown in figure 10.35(*c*).

In a similar manner, by grounding *f, g,* and *c*, we can form the digit '4' and so on. The digit 0 can be formed by grounding all the terminals except *g*. A seven segment display can also display the capital letters, *A, C, E* and *F*. Besides this, it can also display the lower-case letters *b* and *d*.

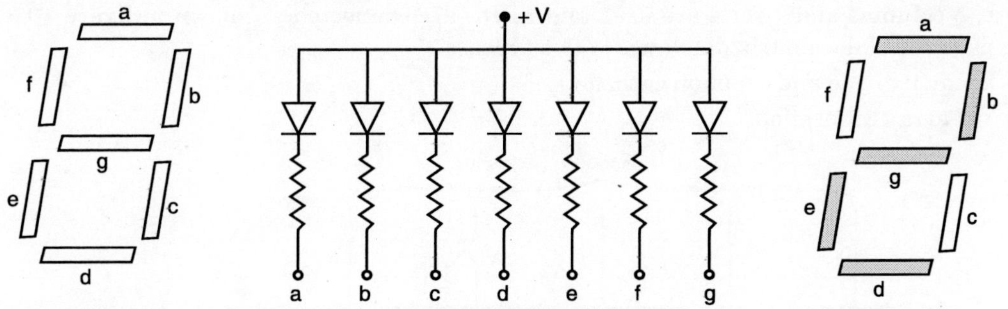

(a) 7-segments (b) Schematic diagram (c) Digit 2 in a 7-segment display

Fig. 10.35. Seven--segment display.

An average LED requires a current of about 20 mA for its operation. If we have a two-digit (*i.e.*, two seven-segment displays connected together) display in a circuit, then there are 14 LED's. If all the LEDs are glowing, the total current drawn from the supply voltage is 280 mA. With this much current, required to operate the display, alongwith the current, which is required to operate the remainder of the circuit, the battery life could be very short. In battery powered applications, it is a common practice to instal an electronic switch, which is used to activate the display for short periods of time. The battery life can be increased by turning OFF the display, when it is not needed. In an equipment, where power consumption is not a problem, the LED's make an excellent display. These days, the seven-segment displays are used in digital clocks, calculators, microwave ovens, stereo tuners, digital multimeters, microprocessor trainer kits etc.

The seven-segment displays are available in a single package as an integrated circuit (IC). In such a package, the emitted light is conducted through the pipes to make it more useful. Hence, TIL 312, TIL 313 and FND 500 seven-segment displays are available in a single package. Usually, these displays are referred to as common-anode or common cathode type of displays depending upon whether their anodes or cathodes are connected together. Both these types are useful in actual practice. In fact, the choice depends upon the type of application where it has to be used.

Knowledge is not something to be packed away in some corner of our brain.

— **S. Radhakrishnan**

10.21. ALPHA NUMERIC DISPLAY

As a matter of fact, the dot matrix displays are used to display the numeric and alphanumeric characteristics. These displays are made by using a number of LEDs. Dot matrix displays are available in 3 × 5, 5 × 7, 5 × 8 and 7 × 9 sizes. The basic font of dot matrix display is as shown in figure 10.36.

Here, each LED is connected for each dot. And according to the data to be displayed the LED's are switched ON.

A 3 × 5 dot matrix is used for display of numeric characters and 5 × 7 dot matrix is used to display alphanumeric characters.

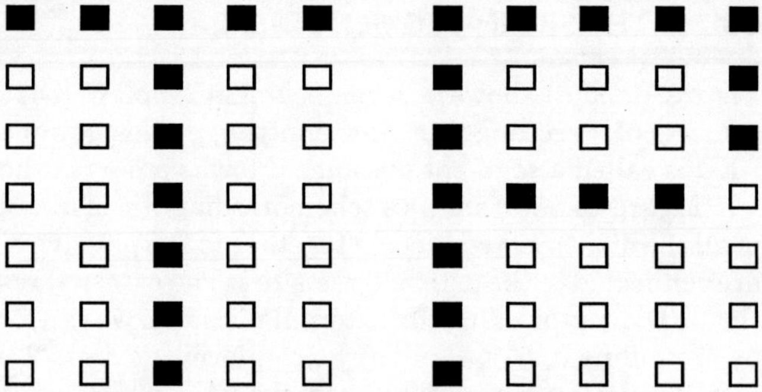

Fig. 10.36. Basic font of dot matrix display

10.21.1. 5 × 7 Dot Matrix Display

This display has been shown in figure 10.37.

Here, 5 columns and 7 rows are used and LED's are connected as shown in figure 10.37. The writing patterns of dot-matrix display is of two types.

(*i*) Common anode and common cathode.

(*ii*) *X*-*Y* array connection.

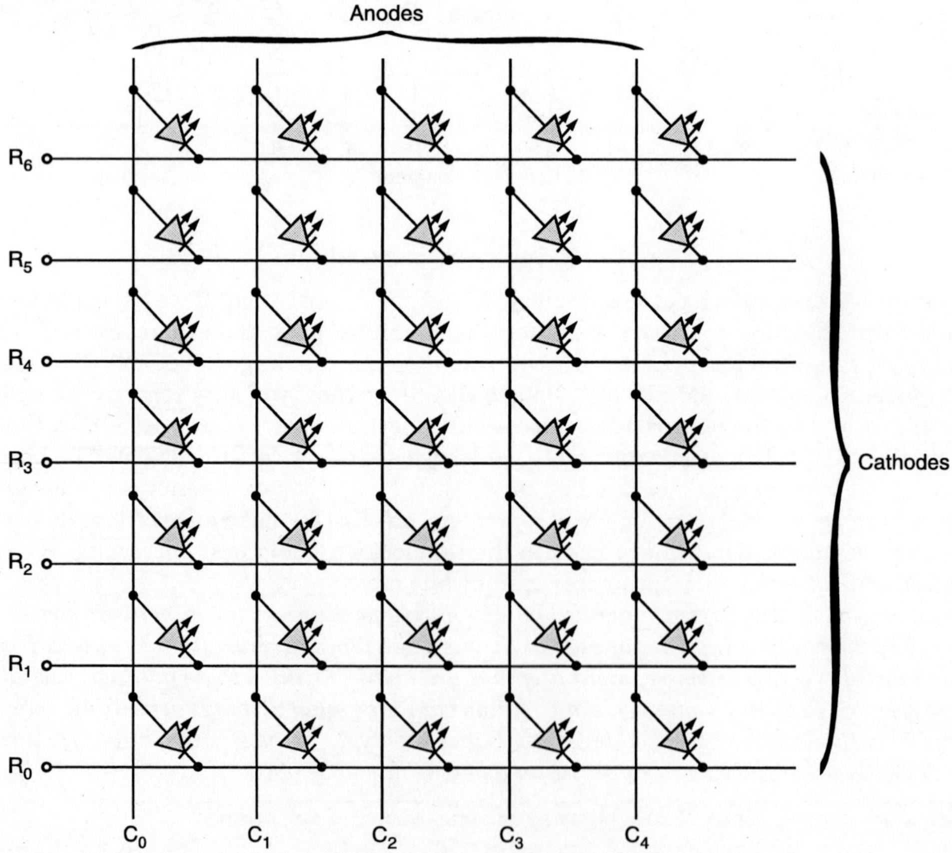

Fig. 10.37. A 5 × 7 dot matrix display

Due to a large number of LEDs used, it is not practical to use a separate driver for each LED. Therefore, multiplexing technique is used to lit the LEDs in a sequential manner. Techniques like dynamic display system can be used to improve the brightness of the display.

To drive the common anote seven segment display, connections are made as shown in figure 10.37.

10.22. LIQUID CRYSTAL DISPLAY (LCD)

The liquid crystal display (LCD) is a low power device as compared to LED. The power requirement is typically in the order of microwatts for the LCD as compared to the same order of milliwatts for LEDs. However, an LCD requires an external or internal light source, It is limited to a termperature range of about 0°C to 60°C and lifetime is an area of concern, because LCDs can chemically degrade with the passage of times.

10.22.1. Types

There are two major types of LCDs which are receiving great attention of the researchers :

(*i*) *Dynamic-scattering LCDs*, and

(*ii*) *Field-effect LCDs*.

A liquid crystal is a material (normally organic for LCDs) that will flow like a liquid but whose molecular structure has some properties normally associated with solids.

Field-effect LCDs are normally used in such applications where source of energy is a prime factor (*e.g.*, watches, portable instrumentation etc.).

They absorb considerably less power than the light-scattering type. However, the cost for field-effect units is typically higher, and their height is limited to 2 inches. On the other hand, light-scattering units are available up to 8 inches in height.

The turn-ON and turn-OFF time is an important consideration in all displays. LCDs are characteristically much slower then LEDs. The response time of LCDs is in the range of 100 to 300 ms, while for the LEDs, it is below 100 ns. However, there are numerous applications, such as in watch, where the difference between 100 ns and 100 ms is of little consequence. For such applications, the low power demand of LCDs is an important characteristic. The lifetime of LCDs is steadily increasing beyond 10,000 + hours limit. Since, the colour generated by LCD units is dependent on the source of illumination, there is a wide range of colour choice. These days, the LCDs are used in watches, pocket calculators, lap-top computers, data organizers, televisions, portable instrument displays.

10.22.2. Advantages of LCDs

(*i*) Less power per digit is required. So power saving takes place.

(*ii*) They have a good contrast ratio (typically 20 : 1).

(*iii*) They have low threshold voltage and require low operating voltage and current.

(*iv*) All the segments have uniform brightness.

(*v*) Low cost.

10.22.3. Disadvantages of LCDs

(*i*) Small operating temperature range.

(*ii*) Low switching speeds.

(*iii*) Backlighting is necessary.

(*iv*) Require larger area.

(*v*) Require an ac drive rather than dc.

10.22.4. Performance Comparison of LEDs and LCDs

S. No.	Parameter of comparison	LED display	LCD display
1.	Power required per digit	More (10 to 100 mW)	Less (about 100 μ W)
2.	Speed	Fast	Slow
3.	Life span	Longer life	Short life
4.	Colours	Red, green, yellow and orange	Depends on the illumination
5.	Contrast ratio	10 : 1	20 : 1
6.	Operating temperature range	Wide	Small
7.	Viewing angle	Wide	Narrow
8.	Operating voltage	Low	Low and moderate
9.	MOS compatibility	Yes for small displays only	Yes
10.	Ease of mounting	Very easy	Difficult
11.	Applications	7 segment display Indicating element	7 segment displays calculators, watches

10.22.5. Performance Comparison of p-n Junction Diode and LED

S.No.	Parameter of comparison	Conventional diode	LED
1.	Symbol	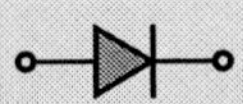	
2.	Material used	Silicon or Germanium	Gallium Arsenide
3.	Capacity to emit light	Cannot emit light	Can emit light when excited electrically
4.	On state voltage drop	0.7 V for silicon diode 0.3 V for germanium diode	Ranges between 1.2 to 2 volts
5.	Reverse breakdown voltage	High	Very low
6.	Applications	Rectifier clipper, clamper, etc.	As a light source in optical fibre applications. As indicator, in 7 segment displays.

10.23. PHOTO DIODES* *(JNTU Hyderabad, Sem. Exam; Dec. 2003)*

Basically, photodiodes are efficient light detectors. Photodiodes can be made in small sizes, have good linearity, have a high response speed, have simple biasing requirements and are relatively low-cost devices. As a matter of fact, the working function of a photodiode is just opposite to the working function of LEDs and laser diodes in the sense that photodiode converts light into an electrical signal. Photodiodes are faster than photoconductors since they have a *pn* junction to collect the carriers, but they do not have the internal gain which photo-conductors have since the charge carriers are counted only once.

Further, the photodetection mechanism in photodiodes is similar to that in photoconductors in the sense that an electron-hole pair is created by a photon for which $h\nu \geq E_g$. However, it differs in

* Write a short note on photo-diode. *(JNTU, Hyderabad, Sem. Exam; Dec. 2003)*

the sense that only the photocarriers which diffuse to the depletion layer are counted and only the minority carriers are collected.

10.23.1. Types

Usually, there are various types of photodiodes as under:

(*i*) *p-n* photodiodes

(*ii*) *p-i-n* photodiodes, and

(*iii*) Avalanche photodiodes (APD)

10.24. P-N PHOTODIODE

As a matter of fact, photodiodes are made of semiconductor *p-n* junctions under reverse-bias condition. Whenever, a photon having energy $E_p = h\nu = \dfrac{hc}{\lambda} \geq E_g$ strikes the material, the photon is absorbed and its energy is acquired by an electron. Therefore, the electron is excited at the conduction band and is now able to move. Hence, electron-hole pairs are generated within the depletion region. When a photon strikes a depletion region, its energy separates an electron from its hole as depicted in figure 10.38 (*a*). Now, the separated electron and hole are attracted by the positive and negative potentials of the depletion voltage respectively. Hence a flow of charge carriers generates the current. Further, under the influence of reverse bias, the flow of electrons and holes is enhanced. Figure 10.38 (*b*) shows the electrical circuit of a photodiode.

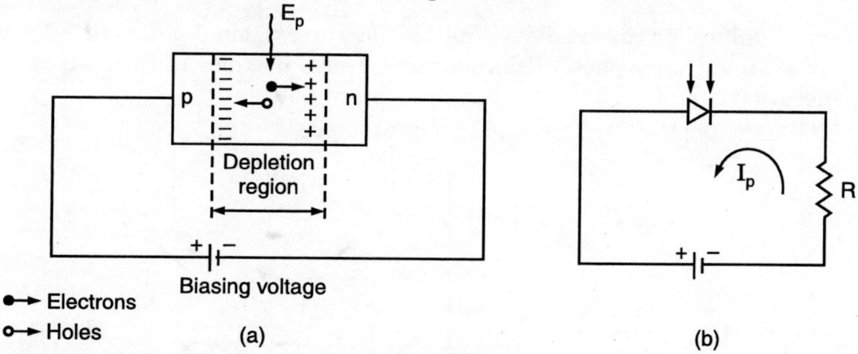

Fig. 10.38. Operating principle of a p-n photo diode (a) p-n junction under reverse biasing, (b) electrical circuit

10.24.1. Characteristics of a Photodetector

As a matter of fact, there are two important characteristics of a photodetector namely quantum efficiency and responsivity. In fact, these two characteristics or parameters depend upon the material band gap, the operating wavelength, the doping and the thickness of the p-n regions of the device.

(*i*) Quantum Efficiency

Quantum efficiency may be defined as the ratio of the number of electron-hole pairs (EHP) generated to the total number of incident photons and is expressed as

$$\eta = \frac{\text{Number of EHPs generated}}{\text{Total number of incident photons}} = \frac{I_p / q}{P_{op} / h\nu} = \frac{I_p h\nu}{q P_{op}}$$

where

I_p = the photogenerated current

P_{op} = the incident optical power

$h\nu$ = the photon energy

q = free carrier charge

Practically, 100 photons create 30 to 95 EHP, and hence giving η ranging from 30 to 95 percent. Also, to achieve a high quantum efficiency, the depletion layer must be thick enough to permit a large fraction of the incident light to be absorbed. Because of this, the photogenerated carriers have to drift a longer distance across the reverse-biased junction. Since, the carrier drift time determines the response speed of the photodiode, therefore, a compromise has to be made between quantum efficiency and response speed.

(*ii*) Responsivity

The responsivity may be defined as the ratio of photogenerated current to the incident optical power. It is expressed by the following expression:

$$R = \frac{I_p}{P_{op}}$$

where I_p = photogenerated current

P_{op} = the incident optical power

Further, the responsivity may be related to the quantum efficiency as under:

$$R = \frac{\eta q}{h\nu}$$

This parameter is very useful in the sense that it specifies the photocurrent generated per unit optical power.

Also, the typical values for responsivity R fall in the range from 0.5 A/W to 1.0 A/W. The value of responsivity R is provided by photodiode manufacturers. Because of this, we can evaluate the output photogeneration current with the given light-input power.

Figure 10.39 illustrates the responsivity of a photodiode.

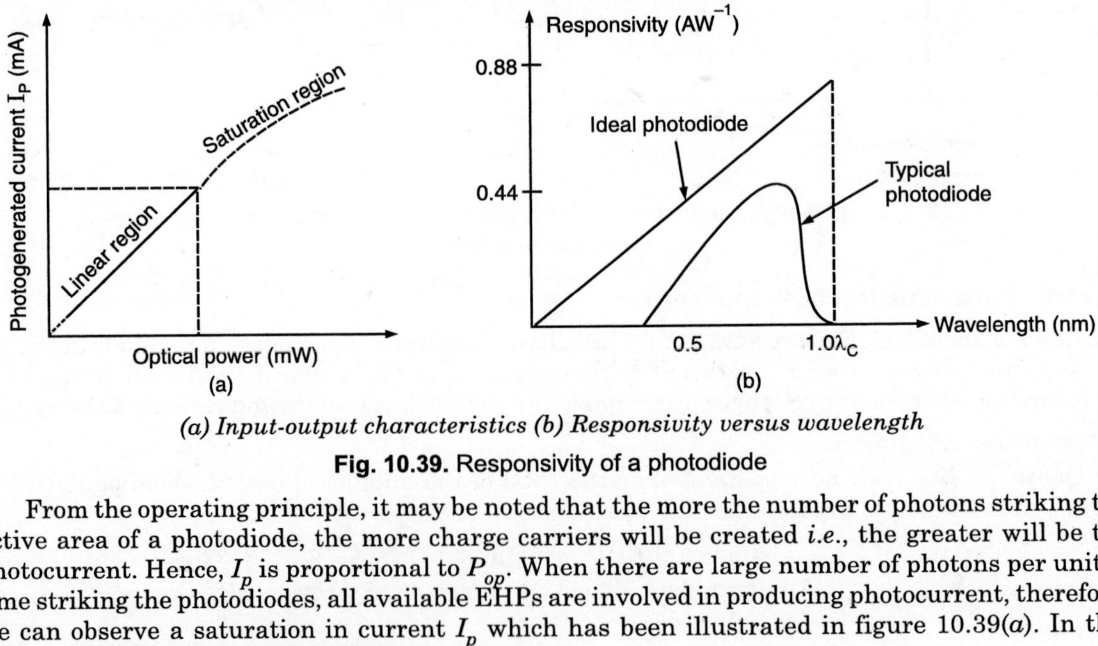

(a) Input-output characteristics (b) Responsivity versus wavelength

Fig. 10.39. Responsivity of a photodiode

From the operating principle, it may be noted that the more the number of photons striking the active area of a photodiode, the more charge carriers will be created *i.e.*, the greater will be the photocurrent. Hence, I_p is proportional to P_{op}. When there are large number of photons per unit of time striking the photodiodes, all available EHPs are involved in producing photocurrent, therefore, we can observe a saturation in current I_p which has been illustrated in figure 10.39(*a*). In this region, I_p is not proportional to P_{op}.

We know that the responsivity is expressed as

$$R = \frac{\eta q}{h\nu}$$

Above expression may be written as under:

$$R = \frac{\eta}{h\nu / q}$$

or $$R(A/W) = \left(\frac{\eta}{1248}\right)\lambda\,(nm)$$

Hence, according to above expression, the graph between R and λ must be a straight line as depicted in figure (10.39 (*b*)) by the dotted straight line. However, practically, the curve for responsivity R as a function of wavelength has been shown in figure (10.39(*b*)) as a solid line. In fact, this curve is very different from the expected straight line and also there exists a short and long cut-off waelengths.

DO YOU KNOW?

A photo diode has been optimized for its sensitivity to light. In this diode, a window lets light pass through the package to the junction. The incoming light produces free electrons and holes. The stronger the light, the greater the number of minority carriers and the larger the reverse current.

EXAMPLE 10.7. The responsivity of a photodiode is 0.85 *A/W* and the input power saturation is 1.5 mW. Calculate the photocurrent if the incident light power is

(*i*) 1 mW, and

(*ii*) 2 mW

Solution: We know that the Responsivity of a photodiode is given by

$$R = \frac{I_p}{P_{op}}$$

(*i*) Given that $R = 0.85\ A/W,\ P_{op} = 1\text{ mW}$

$$I_p = RP_{op} = 0.85\left(\frac{A}{W}\right) \times 1\text{ mW} = 0.85\text{ mA} \quad \textbf{Ans.}$$

(*ii*) $P_{op} = 2$ mW but given that input-power saturation is 1.5 mW so I_p is not proportional to P_{op}, hence we cannot find the value of photocurrent.

EXAMPLE 10.8. In a 100 ns pulse, 6 × 10⁶ photons at a wavelength of 1300 nm fall on an InGaAs phtodetector. On the average, 5.4 × 106 EHPs are generated. Calculate the quantum efficiency of photodetector.

Solution: We know that the quantum efficiency is expressed as

$$\eta = \frac{\text{Number of EHPs generated}}{\text{Number of incident photons}}$$

Substituting the given values, we get

$$\eta = \frac{5.4 \times 10^6}{6 \times 10^6} = 0.90$$

Hence, the quantum efficiency at 1300 nm is found to be 90%. **Ans.**

EXAMPLE 10.9. Determine the responsivity of an InGaAs photodiode if its quantum efficiency is equal to 70%. The energy gap of InGaAs is 0.75 eV.

Solution: We know that $$E_g = \frac{hc}{\lambda}$$

Substituting values, we get

$$\lambda = \frac{6.62 \times 10^{-34} \times 3 \times 10^8}{0.75 \times 1.6 \times 10^{-19}} = 1664\text{ nm}$$

Also, responsivity is given by

$$R = \left(\frac{\eta}{1248}\right)\lambda(\text{nm}) = \frac{0.70}{1248} \times 1664 = 0.933\text{ AW}^{-1}. \quad \textbf{Ans.}$$

10.24.2. Photodiode Bandwidth

As a matter of fact, the bandwidth of a photodiode is governed by following two factors:

(*i*) The transit time, τ_{tr}, which is the time needed or the electric field to sweep out the photoexcited carriers within the depletion region.

(*ii*) The RC time τ_{RC}, which is the time needed to discharge the junction capacitance (C_j) through a combination of internal and external resistances.

The transit time is expressed as

$$\tau_{tr} = \frac{W}{v_{sat}}$$

where v_{sat} = maximum drift velocity of charge carriers,
W = thickness of depletion region.

For silicon photodiodes, τ_{tr} may be estimated to a good approximation by using the expression

$$\tau_{tr} = \frac{\rho_n}{400} \text{ or } \frac{\rho_p}{1000} \text{ ns}$$

where ρ_n and ρ_p are the resistivities in ohm-cm of *n*-type or *p*-type materials.

The circuit defining the *RC* time constant of a silicon photodiode is shown in figure 10.40 in which R_S is the internal resistance, R_j is the junction resistance, R_L is the load resistance, C_j is the junction capacitance and *I* the photocurrent generated by the incident light.

R_S is comprised of three parts:

(*i*) the resistance of the undepleted region of the silicon chip,

(*ii*) the contact resistance,

(*iii*) collection resistance associated with the rersistivity of the front surface generated by diffusion.

Taking these factors into account, we get

$$\tau_{RC} = (R_S + R_L)\, C_j \cong R_L\, C_j$$

C_j can be estimated as $C_j = \frac{\varepsilon A}{W}$, where ε is the permittivity of a semiconductor and *A* is the active area (the photosensitive area) of the photodiode.

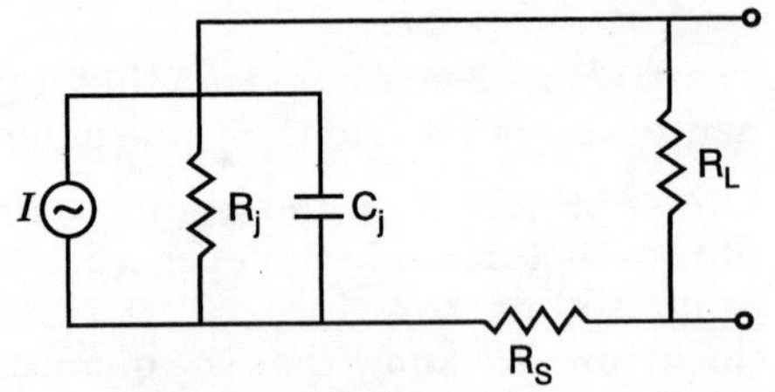

Fig. 10.40. Equivalent circuit of a photodiode.

Typical values of C_j are from 1 to 2 pF.

Hence, for $R_L = 50\ \Omega$, τ_{RC} is in the range of 50 to 100 ps.

Thus, the bandwidth of a photodiode is given by the following formula:

$$BW = \frac{1}{2\pi(\tau_{tr} + \tau_{RC})} = \frac{1}{2\pi\left[\left(\frac{W}{v_{sat}}\right) + R_L\left(\frac{\varepsilon A}{W}\right)\right]} \quad ...(i)$$

Hence, bandwidth of a photodiode is very much dependent upon the depletion region width (W). Becaue of this, *p-n* diode is an inefficient device for optical communication applications.

There is a third mechanism restricting the bandwidth of a *p-n* photodiode. As BW depends upon the depletion region width and since *p* and *n* regions are wider than the depletion region, many photons strike those regions, creating a diffusion photocurrent. But, diffusion flow is very slow compared with the drift flow of charge carriers. Therefore, the electric-output pulse of a *p-n* photodiode will be much wider than the optical input pulse and its tail will be determined by the diffusion photocurrent. This has been illustrated in figure 10.41.

Knowledge is not something to be packed away in some corner of our brain.

— **S. Radhakrishnan**

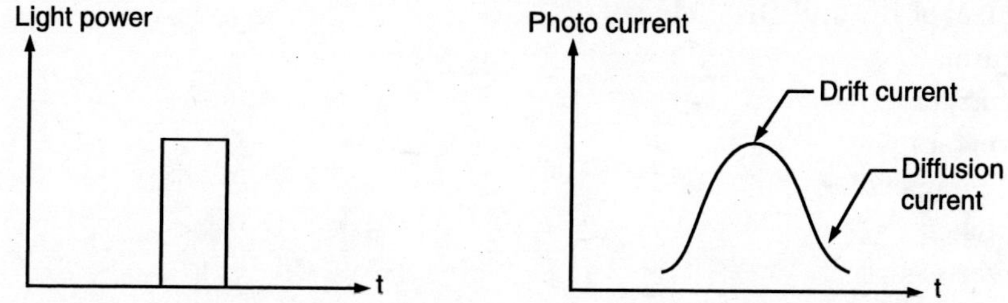

Fig. 10.41. Input and output pulses of a photodiode : (a) optical input, (b) electric-output pulse.

10.24.3. Performance Comparison of LED and Photodiode

S. No.	LED	Photodiode
1.	It is a light emitting device.	It is a light detecting device.
2.	Electrical energy is converted into light.	Electric current proportional to light intensity is produced.
3.	LED is always forward biased.	It is always reverse biased.
4.	GaAs or GaP or GaAsP are the materials used.	Silicon is used.
5.	Due to recombination of electrons and holes, light is emitted.	Due to generation of electron hole pairs, the photocurrent will flow.

10.25. INTRODUCTION TO THYRISTORS

As a matter of fact, thyristors is a general term which includes devices such as SCR, Triac, GTO, LASCR, RCT, SCS, ... and many more. Thus, thyristors is a family of devices. SCR *i.e.* silicon controlled rectifier is the most popular device of the thyristor family.

10.26. CLASSIFICATION OF THYRISTORS

Figure 10.42 illustrates the classification of thyristors.

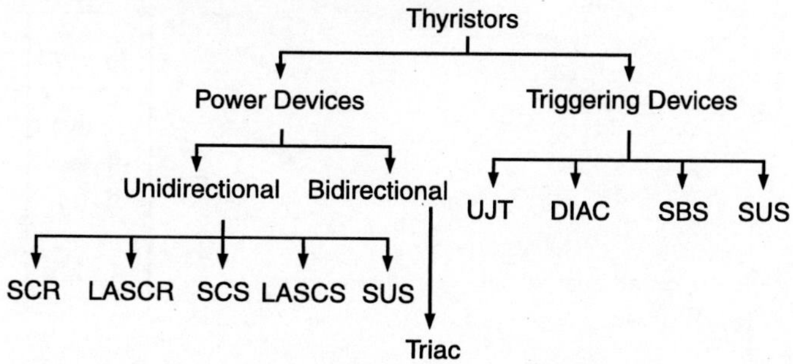

Fig. 10.42. Classification of thyristors.

10.27. APPLICATIONS OF THYRISTORS

Some of the typical applications of thyristors may be listed as under :

1. Temperature controllers
2. Level controllers
3. Intensity controllers

4. Speed control of AC and DC motors
5. Fan regulator
6. Battery charger
7. Power inverters
8. DC to DC choppers
9. Static switches
10. Cyclo converters
11. Controlled rectifiers (only SCRs)
12. HVDC transmission
13. AC voltage stabilizers.

10.28. SILICON CONTROLLED RECTIFIER (SCR)*

(JNTU, Hyderabad, Sem. Exam., June, 2004)

SCR is one of the oldest type of solid state power device. It was invented in 1957 by the General Electric Research Laboratories. SCRs have the highest power handling capacity of all the power semiconductor devices. They have a four layer construction with three user accessible terminals. SCR is a latching type device that can be turned on by the control terminal (gate) but once turned on, the gate loses control on it (*i.e.* it cannot be turned off by the gate).

10.28.1. Important Features of SCR

(*i*) It is a latching type device.
(*ii*) It can handle a very large power.
(*iii*) It is a current controlled device, because the gate current controls SCR.
(*iv*) It acts as an open or closed switch.
(*v*) The on state voltage drop is very low.
(*vi*) It can handle thousands of amperes of current.

10.29. BASIC STRUCTURE OF SCR

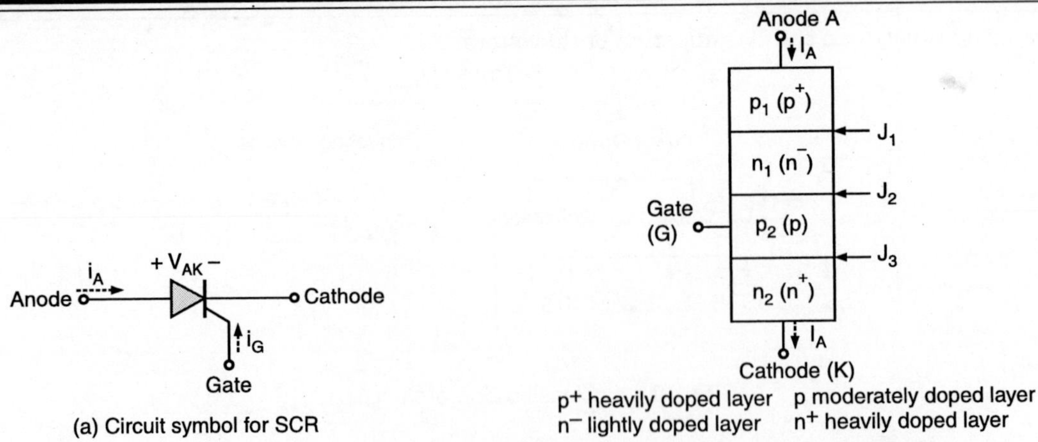

(a) Circuit symbol for SCR

(b) Basic structure of SCR

Fig. 10.43. Structural details of SCR.

The basic structure of SCR has been shown in figure 10.43(*b*). It is a four layer pnpn device, with three terminals brought out for the user, namely Anode, Cathode and Gate. The gate terminal is the

* Write a short note about construction of SCR. *(JNTU, Hyderabad, Sem. Exam., June, 2004)*

control terminal that can turn on the device whenever required. The symbol of SCR has been shown in figure 10.43 (*a*). It is essentially the symbol of a rectifying diode with a third control terminal, the gate added to it. The direction of forward anode current, voltage across the thyristor and direction of conventional gate current have been shown in figure 10.43(*a*).

Number of Junctions

As observed from figure 10.43(*b*), there are three junctions J_1, J_2 and J_3. In order to turn on SCR, the anode must be at a higher positive potential than cathode. This that means SCR must be forward biased.

Current Directions

The directions of the anode and gate current in figure 10.43(*a*) are conventional directions. It clearly indicates that SCR is a unidirectional device and that the gate current can be only positive. The gate current can flow only in one direction i.e. into the gate terminal.

10.29.1. Gate Cathode Layouts

Plan views of two different gate and cathode layouts are as shown in figure 10.44(*a*) and figure 10.44(*b*). The localised gate cathode structure is for smaller SCRs whereas the distributed gate structure is for larger diameter SCRs.

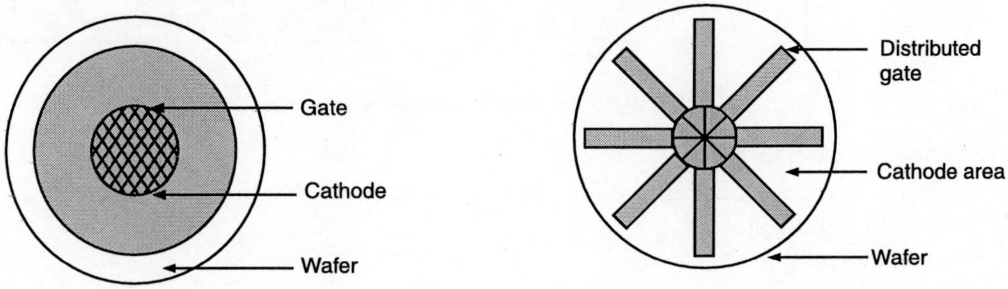

(a) Gate cathode layout for the localised gate electrode

(b) Gate cathode area layout for distributed gate electrode

Fig. 10.44. Structure details of an SCR.

10.30. I-V CHARACTERISTICS* *(JNTU, Hyderabad, Sem. Exam., June 2002, 2004; Dec., 2004)*

The I-V characteristics of SCR is a graph of anode current i_A on *y*-axis and anode to cathode voltage plotted on the *x*-axis as shown in figure 10.45. The I-V characteristics can be split into two parts namely the forward characteristics and reverse characteristics. The characteristics in the reverse direction (anode to cathode voltage negative) is similar to a reverse biased diode. For small reverse voltage, a small reverse leakage current flows until the avalanche breakdown takes place at reverse breakover voltage V_{BR}.

As soon as the reverse breakdown takes place due to avalanche breakdown, a large current flows through SCR whereas the voltage across the device remains constant. It is dangerous to operate SCR in the reverse breakdown state because it may get damaged due to overheating.

The region from 0 volts upto V_{BR} volts in which the SCR is reverse biased and non-conducting is called as *reverse blocking state.*

* (*i*) Explain the V-I characteristics of SCR. ***(JNTU, Hyderabad, Sem. Exam., June 2004)***

(*ii*) Draw the V-I characteristics of SCR and define all related terms. ***(JNTU, Hyderabad, Sem. Exam., June 2002)***

(*iii*) Explain V-I characteristics of an SCR. ***(JNTU, Hyderabad, Sem. Exam., Dec, 2004)***

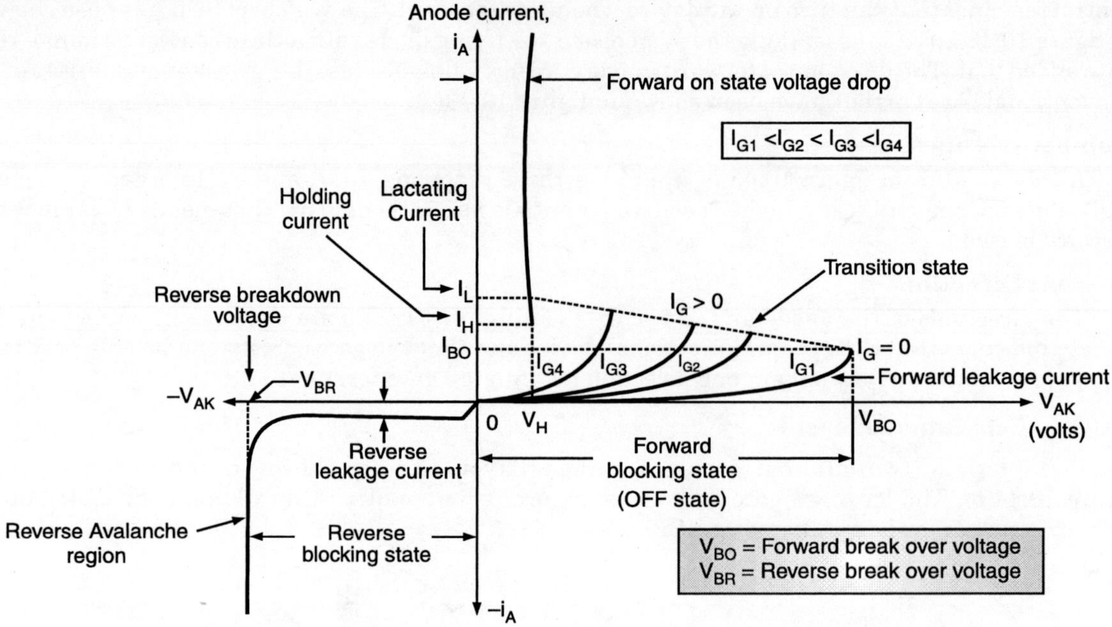

Fig. 10.45 I-V characteristics of SCR.

10.30.1. Forward Characteristics*

The forward characteristics may be divided into three regions of operation as under :

(*i*) Forward blocking state.

(*ii*) Transition state.

(*iii*) ON state.

(*i*) Forward Blocking State

This is the high voltage low current mode of operation in which SCR is in the OFF state. The current through it is *Forward Leakage Current*. This current flows due to thermally generated minority carriers.

(*ii*) Transition State

SCR remains in the blocking state as long as the forward anode to cathode voltage is less than the breakover voltage. As soon as V_{AK} becomes greater than the breakover value, the anode current i_A increases sharply to a high value and the voltage across SCR reduces sharply to a low value (on state voltage). The switching of SCR from off state to on state and vice versa takes place in a short time. This change over state is called as transition state. As this is an unstable state, it is shown by dotted line in figure 10.45.

(*iii*) The Low Voltage High Current Mode or the ON State

The voltage across SCR is low in this state. In the on state a high power SCR can conduct average currents as large as 2000 amp with on state voltage drop of a few volts.

10.30.2. Unstable State *(JNTU Hyderabad, Sem. Exam., June, 2004)*

These two stable starts or modes are connected together by an unstable mode (called as the transition mode) of operation that appears as a negative resistance on the I-V characteristics. (dotted line in figure 10.45).

---

* Sletch the V-I characteristics of an SCR and explain its working in detail.

(JNTU, Hyderabad, Sem. Exam., June, 2004)

The important voltage and current terms in the I-V characteristics may be listed as under :

(*i*) The forward breakover voltage (V_{BO})

This is the maximum forward voltage that can be applied between anode and cathode, without initiating forward conduction. This voltage is defined for a zero gate current. In short, this is the maximum forward voltage across SCR in its off state.

(*ii*) The forward leakage current

This is the small current flowing in the forward direction in the off state of the device. This current is generated due to the minority current carriers and therefore is dependent on the operating temperature.

(*iii*) The holding current (I_H)

It represents the minimum current that can flow through SCR and still *hold* it in the on state. The accompanying voltage is termed as holding voltage V_H. If the forward anode current is reduced below holding current, SCR will be turned off. The holding current is defined for zero gate current ($I_G = 0$).

(*iv*) ON state voltage

It may be noted that the voltage across SCR in its on state is very low as compared to the off state voltage.It will be proved later on that the on state voltage is equal to drop across one junction (*i.e.* of the order of 1 to 1.5 volts).

(*v*) Latching current (I_L)

It is minimum anode current that must flow through SCR to latch it into the on state. The latching current is higher than the holding current. The latching current is important when SCR is being turned on.

10.30.3. Difference Between Holding and Latching Currents

(*i*) The latching current is important only at the time of SCR turn on, whereas holding current is important only at the time of SCR turn off.

(*ii*) If the anode current goes below the holding current then the conducting SCR turns off. The holding current is defined for $I_G = 0$, whereas if anode current at the time of turn on is higher than latching current then only SCR will latch into its on state.

(*iii*) If $I_A < I_L$ then SCR does not latch. It will remain on as long as I_G is being supplied. As soon as I_G is removed, SCR will turn off.

(*iv*) Latching current is always higher than the holding current.

(*v*) The values of I_L and I_H are constant. They do not depend on the gate current magnitude.

10.30.4. Effect of Gate Current

Let us consider figure 10.45. It shows that as we increase the positive gate current from I_{G1} to I_{G4}, the breakover voltage of SCR reduces progressively. Thus, SCR will turn on at lower anode to cathode voltages as we increase the gate current.

10.30.5. How to Turn off SCR ?

The gate current can turn on SCR but once turned on, the gate loses control. Even if we reduce the gate current to zero, SCR continues to conduct. We can turn off a conducting SCR by :

(*i*) Reducing its anode current below the holding current or

(*ii*) Applying a reverse voltage across the SCR.

The experimental setup for plotting SCR characteristics has been shown in figure 10.46.

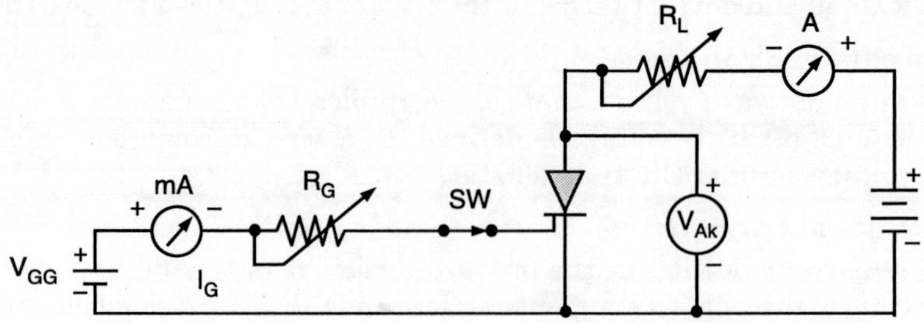

Fig. 10.46. Experimental setup to plot the I-V characteristics of SCR.

10.31. WORKING OPERATION OF SCR

Working operation of SCR can be explained under two different operating conditions. These two operating conditions are as under:

10.31.1. Working Operation Without Gate Current

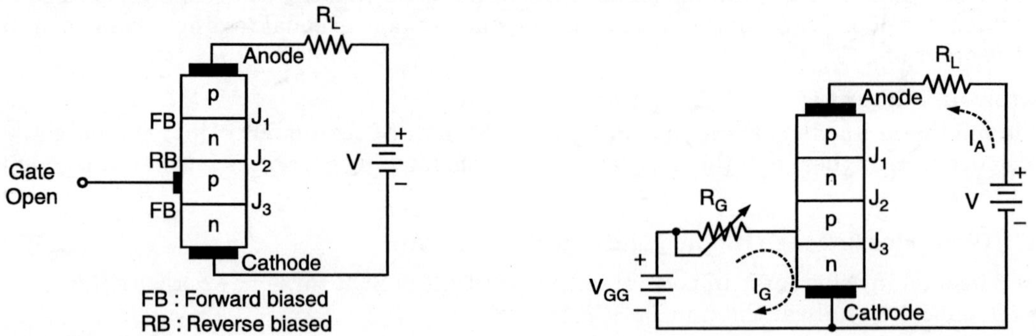

(a) Working operation of SCR without gate current

(b) Working operation of SCR with gate current

Fig. 10.47.

Let us consider figure 10.47(*a*). The gate terminal of SCR is left open so that $I_G = 0$. SCR is forward, biased by applying a positive voltage to anode with respect to cathode. Out of the three junctions, the junctions J_1 and J_3 are forward biased and junction J_2 is reverse biased. Therefore, current does not flow through the SCR. As the anode to cathode voltage is increased, the voltage across junction J_2 increases. At a certain voltage, this junction will breakdown and SCR will start conducting. This voltage is called as forward breakover voltage.

10.31.2. Operation With Gate Current

Let us consider figure 10.47(*b*). It illustrates that SCR is forward biased as before and the gate cathode junction also is forward biased using an external power source. The gate current I_G starts flowing which can be adjusted by the resistance R_G. As explained earlier, the value of gate current decides breakover voltage of SCR. As I_G increases, breakover voltage will decrease *i.e.,* SCR will turn on at lower voltages.

10.31.3. Reverse Blocking Capacity

In the reverse blocking state, anode is biased negative with respect to cathode. Therefore, the junctions J_1 and J_3 in figure 10.48 are reverse biased and junction J_2 is forward biased. Out of the two reverse biased junction J_1 and J_3, the lower junction J_3 has a low breakdown voltage due to the heavy doping on both sides of it. Hence, the reverse blocking capacity depends entirely on junction J_1. The reverse blocking capacity of junction J_1 is usually decided by the length of n^- (n_1) region.

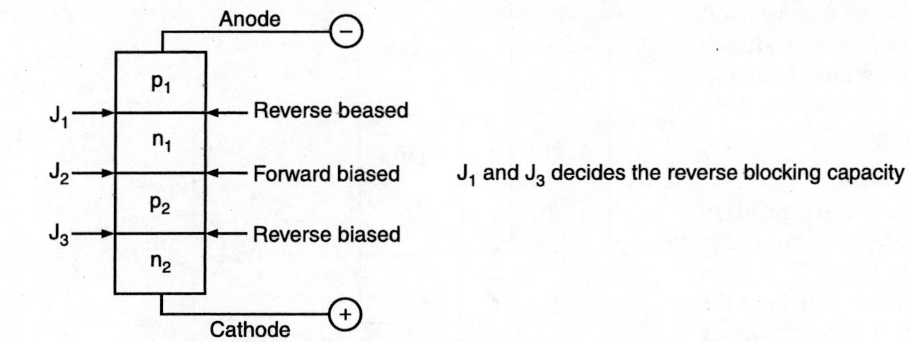

Fig. 10.48. Illustration of SCR in reverse blocking state.

10.31.4. Forward Blocking Capacity of SCR

In the forward blocking state, the thyristor has to block a forward voltage because anode is biased positive with respect to cathode. Therefore, the junctions J_1 and J_3 are forward biased, and junction J_2 is reverse biased. The junction J_2 has to block the entire anode to cathode voltage. In order to achieve this, the n^- layer (n_1) in figure 10.49 is lightly doped and made wide enough.

The depletion region of the reverse biased J_2 junction appears in this layer (n_1 layer). Thus, this region decides the forward blocking voltage.

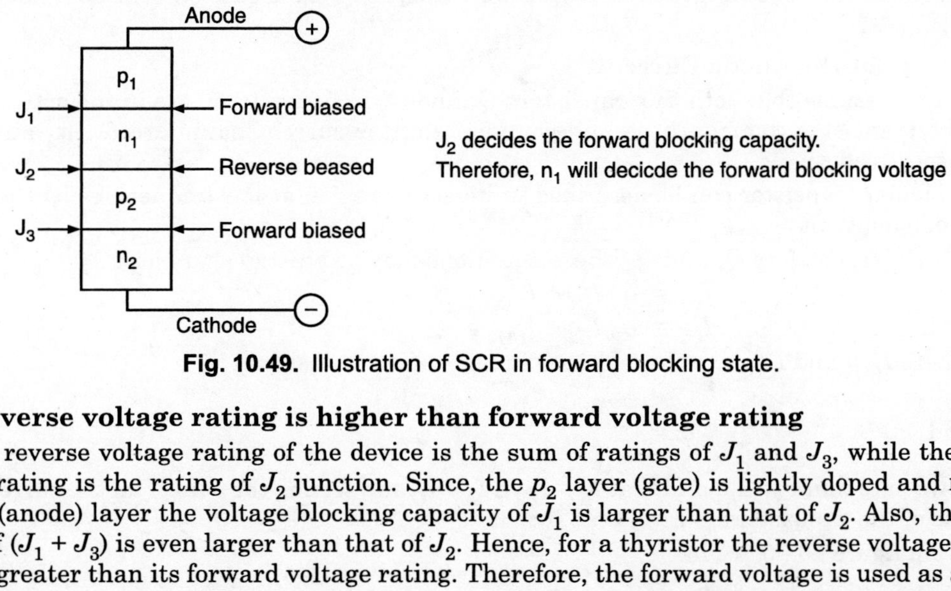

Fig. 10.49. Illustration of SCR in forward blocking state.

Why reverse voltage rating is higher than forward voltage rating

The reverse voltage rating of the device is the sum of ratings of J_1 and J_3, while the forward voltage rating is the rating of J_2 junction. Since, the p_2 layer (gate) is lightly doped and narrower than p_1 (anode) layer the voltage blocking capacity of J_1 is larger than that of J_2. Also, this means rating of ($J_1 + J_3$) is even larger than that of J_2. Hence, for a thyristor the reverse voltage rating is usually greater than its forward voltage rating. Therefore, the forward voltage is used as a criteria while designing thyristorised circuits. The forward and reverse blocking voltages are dependent on the junction temperature. They reduce with increase in junction temperature beyond 150°C. Therefore, most manufacturers specify the maximum junction temperature for their thyristors at 125°C.

10.32. TRANSISTOR ANALOGY: TWO TRANSISTOR EQUIVALENT CIRCUIT*

(JNTU, Hyderabad, Sem. Exam., June, 2004)

1. Equivalent Circuit

The working operation of SCR can be explained with the help of the one dimensional model as shown in figure 10.50. The device is considered to be an ideal device. A low frequency equivalent

* Briefly describe two transistor analogy. *(JNTU, Hyderabad, Sem. Exam., June, 2004)*

circuit made up of a *p-n-p* and *n-p-n* transistor has been shown in figure 10.50, which is easily derived from the one dimensional model.

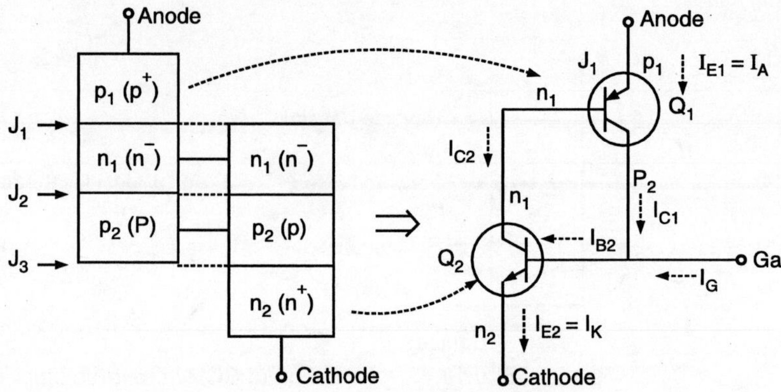

Fig. 10.50. Two transistor equivalent circuit of a thyristor.

2. SCR Operation

The anode is made positive with respect to cathode and gate current is supplied. The gate current acts as base current for Q_2 and it turns on. Its collector current $I_{C2} = \beta_2 I_G$ starts flowing. But, collector current of Q_2 acts as base current for Q_1 and Q_1 starts conducting. Its collector current $I_{C1} = \beta_1 I_{B1} = \beta_1 \beta_2 I_G$ now starts flowing. The collector current of Q_1 acts as base current of Q_2 and collector current of Q_2 in turn acts as base current of Q_1. Due to this type of connection, the current multiplication takes place which is called as cumulative current multiplication and both the transistors will saturate. The saturated transistors are equivalent to closed switches. Hence, the SCR will get latched into on state and acts as a closed switch. Even if the gate current is reduced to zero, the two internal transistors will supply base current to each other and hence will remain in saturation. Hence once located, removal of gate current does not turn off the SCR.

3. Expression for the Anode Current

(*i*) Let us assume that both the transistors Q_1 and Q_2 in figure 10.50 are in the active region. The transistor common base and common emitter current gains are α_1, α_2 and β_1, β_2 respectively.

(*ii*) A bipolar transistor can be described in its active region at low frequencies by the Ebers-Moll equations.

For the transistors Q_1 and Q_2, these equations may be written as under :

$$I_{C1} = \alpha_1 I_{E1} + I_{CO1} \quad ...(10.1)$$

and

$$I_{C2} = \alpha_2 I_{E2} + I_{CO2} \quad ...(10.2)$$

where I_{CO1} and I_{CO2} are the leakage currents of the transistors

and

$$\alpha = \frac{\beta}{(1+\beta)} \quad ...(10.3)$$

(*iii*) If we note that $I_A = I_{E1}$ and that $I_K = I_{E2} = I_A + I_G$ and setting the sum of all the currents into one of the transistors to zero, then, we have

From figure 10.50, we have

$$I_{E2} = I_{C2} + I_{B2} \quad ...(10.4)$$

(*iv*) But, $\quad I_{E2} = I_A + I_G$

therefore, $\quad I_A + I_G = I_{C2} + I_{B2} \quad ...(10.5)$

(*v*) But $\quad I_{E2} = I_{C1} + I_G$

therefore, $\quad I_A + I_G = I_{C2} + I_{C1} + I_G \quad ...(10.6)$

But $\quad I_{C2} = \alpha_2 I_{E2} + I_{CO2}$ and $I_{C1} = \alpha_1 I_{E1} + I_{CO1}$

therefore, $\quad I_A = \alpha_2 I_{E2} + I_{CO2} + \alpha_1 I_{E1} + I_{CO1} \quad ...(10.7)$

(*vi*) But $\quad I_{E2} = I_A + I_G$ and $I_{E1} = I_A$

Substituting all these values, equation (10.7) can be written as under :

$$I_A = \alpha_2 (I_A + I_G) + I_{CO2} + \alpha_1 I_A + I_{CO1} \quad ...(10.8)$$

or $\quad I_A - \alpha_2 I_A - \alpha_1 I_A = \alpha_2 I_G + I_{CO1} + I_{CO2}$

Therefore, we have
$$I_A = \left[\frac{\alpha_2 I_G + I_{CO1} + I_{CO2}}{1-(\alpha_1 + \alpha_2)}\right] \quad ...(10.9)$$

(*vii*) In the blocking state (off state), the sum of $(\alpha_1 + \alpha_2)$ is much less than unity, so anode current I_A is quite small, a few microamperes for low current devices and to a few milliamperes for the high current SCR.

4. How does SCR Turn On

In equation (10.9), as long as the values of α_1 and α_2 are small, the anode current I_A will remain low and SCR will be in the off state. If $(\alpha_1 + \alpha_2)$ approaches unity, the denominator of equation (10.9) will be very small and SCR will be at the breakdown point, ready to enter into the transition state. The current gains of the two transistors Q_1 and Q_2 would be high (greater than one) and the collector current of one transistor acts as base current for the other one and gets multiplied by its current gain β. This current then again acts as base current of the first transistor and gets multiplied by its current gain. This cumulative current multiplication action would soon increase the anode current to such a value that SCR will enter into its ON state. This is the turn on process of SCR.

5. How do the Values of α Increases

The key to the turn on process is to understand how the values of α_1 and α_2 of two transistors Q_1 and Q_2 increase from the small values in blocking state to the higher values required to turn SCR on. As the forward anode-cathode voltage is increased, the depletion region of junction J_2 grows into the n_1 (n^-) layer. This layer is the base for the upper *p-n-p* transistor Q_1. Due to the growth of depletion region of J_2 into n_1 layer the effective thickness of the base region of $p_1\, n_1\, p_2$ transistor (Q_1) reduces and hence α_1 *i.e.* α of Q_1 increases. Similarly, the extension of the depletion region layer J_2 into the P_2 layer (the base of the *n-p-n* transistor Q_2) will cause increase in the value of α_2.

10.32.1. SCR as a Switch

The SCR has only two states of operation *i.e.* it can be either fully ON or completely OFF. There is no *active* region of operation like transistor. Therefore, SCR is equivalent to a switch, as shown in figure 10.51. Since it cannot operate in the active region, it cannot be used as an amplifier.

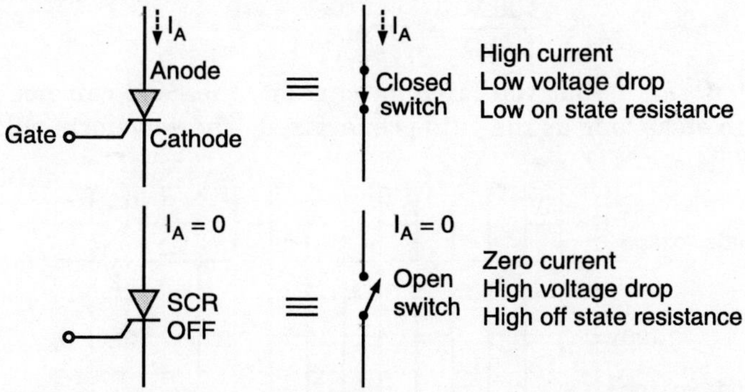

Fig. 10.51. Illustration of SCR as a switch.

EXAMPLE 10.10. An SCR, having average forward current rating of 25 Amp, holding and latching current of 20 mA is connected in series with a 10 kΩ resistor across a DC supply of V volts. At $t = 0$, a periodic pulse train at a frequency of 5 kHz with 50% duty cycle, varying from 0 to 5 Volts, capable of causing conduction in the device is applied to the SCR gate for 1 ms. Sketch the gate voltage, anode voltage and anode current waveforms by neglecting the turn on and turn off times of the SCR, for a time interval $t = 0$ to $t = 1.2$ ms, if (*i*) $V = 300$ Volts, (*ii*) $V = 100$ Volts.

Solution: The circuit diagram from the given description has been shown in figure 10.52.

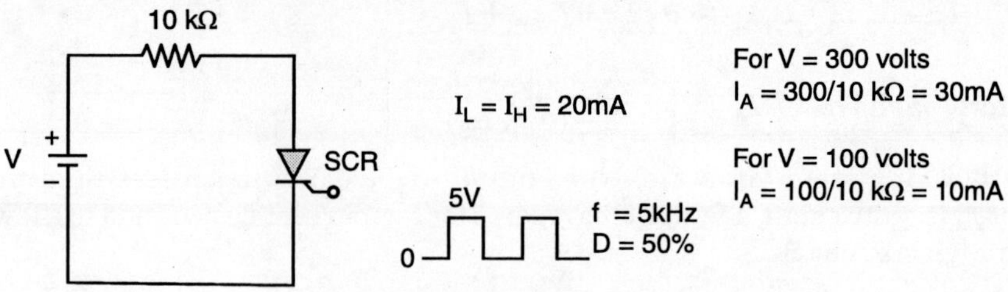

Fig. 10.52.

Case I : Waveforms with V = 300 Volts

If V = 300 volts, then the anode current is I_A = 300 V/10 kΩ = 30 mA. Since this current is higher than the latching current, the SCR is latched as soon as the first gate pulse is applied. The waveforms with V = 300 volts have been shown in figure 7.53.

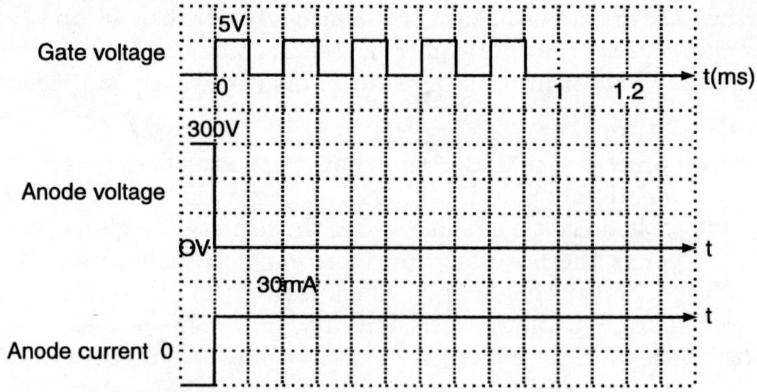

Fig. 10.53. Waveform with V = 300 volts.

Case II : Waveforms with V = 100 Volts

If V = 100 volts, then the anode current is given by

$$I_A = \frac{100\ V}{10\ k\Omega} = 10\ \text{mA}$$

Since this current is small that the latching current, the SCR can not latch into ON state. Therefore it will remain on as long as the gate is energized. The waveform with V_s = 100 volts have been shown in figure 7.54.

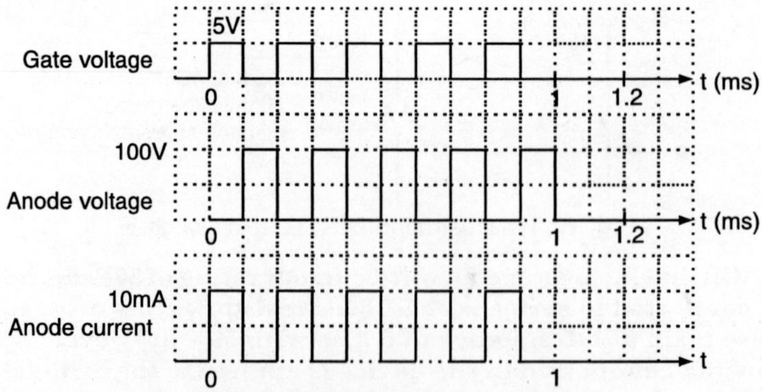

Fig. 10.54. Waveforms with V = 100 volts.

Little minds are termed and subdued by misfortune; but great minds rise about it.

— **Washington Irving.**

10.32.2. SCR Turn On Methods* *(JNTU Hyderabad, Sem. Exam., June 2003, 2004)*

SCR is turned on by increasing the anode current. This can be achieved with one of the following methods :

1. By raising the temperature
2. By focusing light
3. By applying high voltage
4. By applying voltage at high dv/dt
5. By increasing the gate current.

1. By raising the temperature

If the temperature of SCR increases, there is an increase in the number of electron-hole pairs. This will increase the leakage current. Therefore, the value of α_1 and α_2 will increase. If $(\alpha_1 + \alpha_2)$ tends to unity, then SCR may be turned on.

Note : SCR turn on due to increased temperature and it should be avoided because it may cause thermal runaway.

2. By focusing light

If light of adequate frequency and intensity is allowed to strike SCR junction, then the photons will strike the electrons and increase the number of electron hole pairs. SCR may turn on due to the increased leakage current. This principle is used to turn on the light activated SCR (LASCR).

3. By applying high voltage

SCR will turn on if the forward anode to cathode voltage is greater than the forward breakdown voltage V_{BO}, because then a sufficient leakage current will flow to turn on SCR. We should avoid this type of turn on because it may prove to be destructive.

4. By applying voltage at high *dv/dt*

We have already explained how a high dv/dt may result in the accidental turn on of a *SCR*. It may even damage SCR. Hence, SCR should be protected against high dv/dt. Care should be taken to limit the applied dv/dt below the value specified by the manufacturer.

5. By increasing the gate current

As discussed earlier, the breakdown voltage will decrease with increase in the gate current.

(i) The gate signal should be removed, once SCR is turned on. The gate signal should not be applied continuously in order to avoid gate power dissipation.

(ii) Gate current should not be applied when SCR is reverse biased, because then SCR may fail due to increased leakage current.

(iii) The width of the gate triggering pulse should be longer than the time taken by the anode current to the holding current I_h. This pulse width should be longer than the turn on time t_{on} of thyristor.

10.32.3. Salient Features of Gate Turn ON

The gate turn on has following salient features :

(i) Less power consumption.

(ii) It is possible to turn on the SCR precisely at the desired value of anode to cathode voltage.

(iii) If the supply voltage is AC, then it is possible to adjust the firing angle α to the desired value.

(iv) Triggering circuits are easy to design.

---

* Mention the various conditions to turn-on the thyristor and enumerate the requirements of ideal firing circuit.. *(JNTU Hyderabad, Sem. Exam., June 2003)*

(v) Turn on gain is very high. Turn on gain is the ratio of anode current to the gate current.
(vi) Pulsed triggering circuit provide greater flexibility.

10.33. DYNAMIC CHARACTERISTICS OF SCR

The I-V characteristics (static characteristics) of SCR, discussed earlier, can provide all the information about the behaviour of thyristor in the forward blocking, reverse blocking and on state. However, it cannot provide any information about the transition state. Therefore, to know about SCR behaviour at the time of turn on and the turn off, we must understand the dynamic characteristics of switching characteristics of SCR.

The dynamic characteristics can be further classified into two categories as under :

(i) Turn on characteristics
(ii) Turn off characteristics

The turn on characteristics is useful in knowing the behaviour of *SCR* at the time of turn on whereas the turn off characteristics is useful to understand its behaviour at the time of turn off.

10.33.1. Turn On Characteristics of SCR* *(JNTU Hyderabad, Sem. Exam., Dec. 2002, Dec. 2006)*

SCR is turned on by applying a current pulse of specified magnitude and duration to the gate of the device. The gate current is applied at $t = 0$ as shown in figure 7.14. The waveforms for the anode current and anode to cathode voltage at the time of turn on have been shown in figure 10.55. The forward anode current increases at a constant rate d_{iF}/dt. The total time taken by SCR to turn on completely can be subdivided into three times intervals. They are as under:

(i) The turn on delay time (t_d)
(ii) The rise time (t_r), and,
(iii) The spreading time (t_{ps})

Therefore, we have

$$t_{on} = t_d + t_r + t_{ps} \qquad ...(10.10)$$

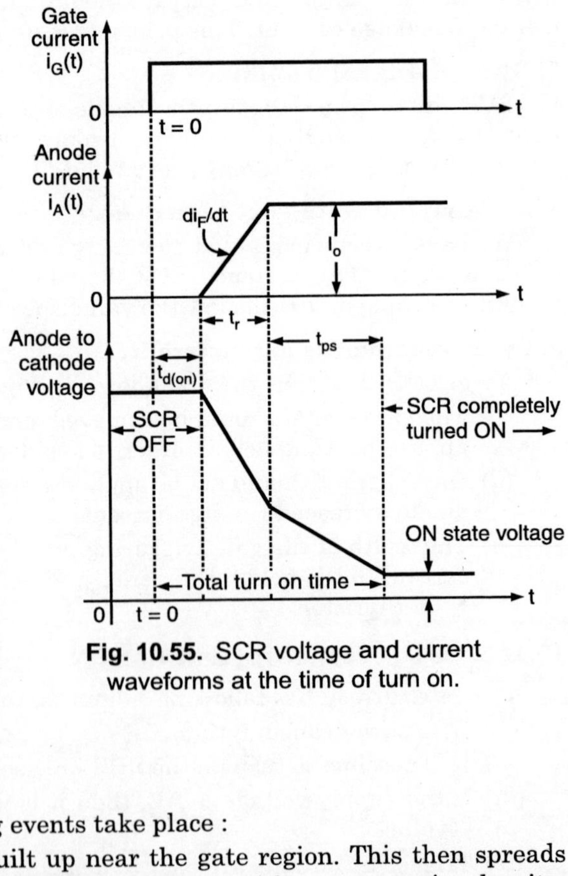

Fig. 10.55. SCR voltage and current waveforms at the time of turn on.

(i) Turn on delay time $t_{d\,(on)}$

During the turn on delay time $t_{d\,(on)}$, SCR appears to be in the blocking (off) state. The voltage across SCR is constant and high and current through it is zero as shown in figure 10.55. But as the gate current has already been applied, excess carriers are injected into the p_2 layer (the base of *n-p-n* transistor in the equivalent circuit). This increases the sum ($\alpha_1 + \alpha_2$). This sum finally reaches unity and SCR turns on, due to regeneration. The anode current begins to increase and this indicates the end of turn on delay time and start of rise time interval.

(ii) Rise time interval (t_r)

During the rise time interval (t_r), the following events take place :

A large excess carrier density or plasma is built up near the gate region. This then spreads laterally until the entire cross sectional area of the SCR is filled up with excess carrier density.

* Draw and explain the behaviour of switching characteristics of a thyristor during its turn-on and turn-off process.. *(JNTU Hyderabad, Sem. Exam., Dec. 2006)*

Simultaneously, the carrier injection begins from p_1 (anode) layer to n_1 layer and the number of carriers injected begins to grow. The anode current rises at the rate d_{iF}/dt. This rate is large enough so that the anode current reaches its steady state value (I_0) within a short time. The rise time interval comes (t_r) to an end as soon as the anode current reaches its steady state value (I_0). The anode to cathode voltage begins to drop rapidly during the rise time interval as shown in figure 10.55. Even after the rise time interval is over the plasma will continue to spread over the lateral area of SCR until SCR is completely shorted out by the large number of excess carriers.

***(iii)* Plasma spreading interval (t_{ps})**

The time required for the plasma to spread from the region around the gate to the entire device cross-section is the *Plasma Spreading Time* (t_{ps}). The rate of voltage drop during the plasma spreading time is slower than that in the rise time interval because t_{ps} is larger than t_r and because most of the voltage drop occurs during the rise time interval as shown in figure 10.56.

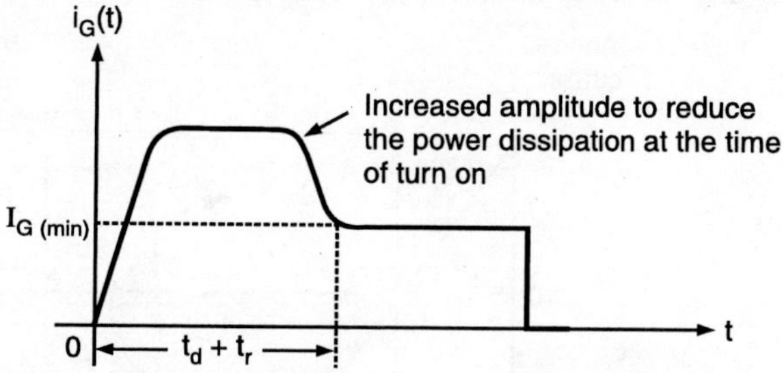

Fig. 10.56. Gate current with an initial large value.

10.33.2. Precautions to Avoid SCR Damage

(*i*) In order to avoid SCR damage or failure, the rate of anode current rise d_{iF}/dt should be kept less than the maximum value given in SCR specification sheet. The failure of the device may take place due to excessive power dissipation that takes place during the time interval t_r.

(*ii*) This power dissipation can be reduced by applying large values of gate current during the $t_{d\,(on)}$ and t_r intervals. For this reason, the gate current applied is often large at the start of the turn on interval and is gradually reduced as time proceeds, as shown in figure 10.56.

10.33.3. Turn Off Process

Once SCR is turned on, it gets latched into its on state and acts like a closed switch. The gate terminal loses control over SCR *i.e.* even though the gate pulses are removed SCR remains in the on state. A conducting SCR can be turned off with the help of an external circuit, by reducing the anode current below its holding current for a minimum specified time. During this time, the stored charge inside the device will be removed due to internal recombination and carrier sweep out. Due to this, the transistors Q_1 and Q_2 in the equivalent circuit of figure 10.50 come out of saturation and enter into active region. The values of α_1 and α_2 will decrease, the regeneration will decrease and the conducting SCR will turn off. The minimum time required for the recombination and carrier sweep out should be provided before the device is turned on again. This is essential for the successful turn off of SCR.

10.33.4. Turn Off Characteristics of SCR

The turn off characteristics using the forced commutation has been shown in figure 10.57.

The successful turn off a conducting SCR requires that it should be reverse biased by the external circuit for a minimum time period. The process to turn off SCR is also called as *commutation*. The turn on time of SCR is appreciably shorter than the turn off time. The turn off waveforms of SCR have been shown in figure 10.57.

(*i*) The current through the conducting SCR starts decreasing at $t = 0$ at a fixed rate d_{iR}/dt. This rate of change of current is decided by the external circuit. As the current decreases, the excess carrier in the four regions of SCR decrease from steady state values, due to the internal recombination and carrier sweep out.

(*ii*) Due to the two outer junctions J_1 and J_3, the turn off characteristics is similar to that of a diode. The peak reverse recovery current is denoted by I_{rr} and its magnitude is much higher than the normal reverse blocking current I_r.

***(iii)* Reverse recovery time t_{rr}**

The reverse recovery time t_{rr} is the time required for the minority carriers to recombine with the opposite charges and to be neutralized. It is the time from zero crossover point of SCR current to 25% of the peak reverse recovery current I_{rr}.

***(iv)* Recombination time t_{rc}**

This is the time required for the recombination of excess charges in the middle *p-n* junction J_2. A negative reverse voltage will reduce this time.

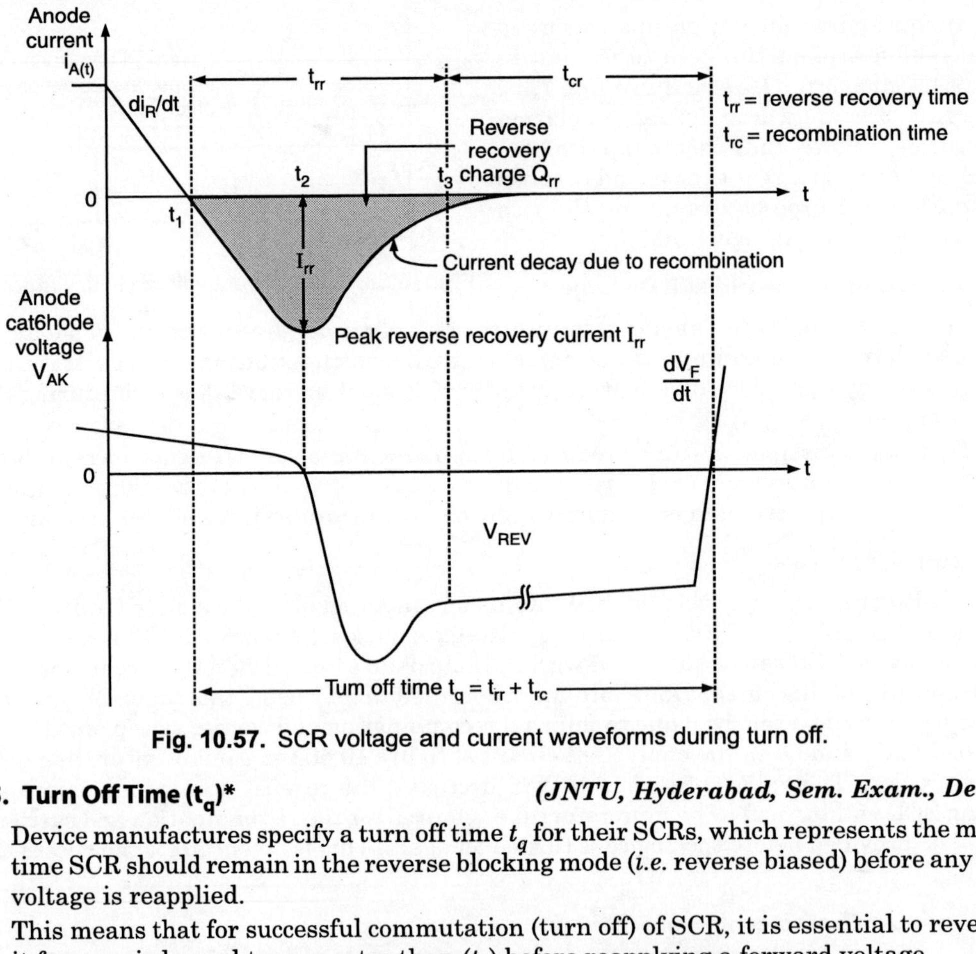

Fig. 10.57. SCR voltage and current waveforms during turn off.

10.33.5. Turn Off Time (t_q)*

(JNTU, Hyderabad, Sem. Exam., Dec. 2006)

(i) Device manufactures specify a turn off time t_q for their SCRs, which represents the minimum time SCR should remain in the reverse blocking mode (*i.e.* reverse biased) before any forward voltage is reapplied.

(ii) This means that for successful commutation (turn off) of SCR, it is essential to reverse bias it for a period equal to or greater than (t_q) before reapplying a forward voltage.

(iii) As shown in figure 10.57, the turn off time (t_q) of SCR is sum of reverse recovery period (t_{rr}) and the recombination time (t_{rc}) and is of the order of 40 μs to 200 μs. Thus, we have

$$t_q = t_{rr} + t_{rc} \quad \text{...(10.11)}$$

The turn off time of SCR is dependent upon following factors:

(i) The current through SCR (I_G)

(ii) di/dt at the time of commutation (turn off)

(iii) The junction temperature and

(iv) Magnitude of reverse voltage.

* Define the turn-on and turn-off time of an SCR. ***(JNTU, Hyderabad, Sem. Exam., Dec. 2006)***

(a) Increase in the values of factors 1, 2 and 3 will increase the turn off time, whereas with increase in the reverse voltage, the turn off time will decrease.

(b) The commutation circuit used to turn off SCR must have a turn off time t_{off} greater than t_q for successful commutation.

Hence, circuit turn-off time $t_{off} > t_q$ ...(10.12)

Difference between SCR turn off time and circuit turn off time

SCR turn off time is the time taken by the conducting SCR to completely turn off. It is denoted by t_q. The circuit turn off time t_{off} is the time for which SCR is reverse biased by an external commutation circuit, to ensure successful commutation.

Reverse recovery charge Q_{rr}

It may be defined as the amount of charge which has to be recovered during the turn off process. It is the area enclosed by the path of reverse recovery current (shaded portion of i_A (t) in figure 10.57). The value of reverse recovery charge Q_{rr} depends on the rate of fall of anode current and the peak value of on-state current before turn off. Reverse recovery charge decides the time required to turn off SCR. Larger Q_{rr}, more is the time required to turn off and slower is SCR.

> **Important Point :** *The turn on and turn off time of SCR are important because they decide the maximum switching frequency of SCR. These time periods also decide the power dissipation taking place in SCR at the time of turn on and turn off. Ideally the turn on and turn off time periods are zero and practically they should be as short as possible.*

EXAMPLE 10.11. Determine the peak reverse recovery current for SCR, if the reverse recovery time t_{rr} = 10 μ sec and reverse recovery charge Q_{rr} = 150 μC.

Solution: Given that $t_{rr} = 10\ \mu s$, $Q_{rr} = 150\ \mu c$

We know that peak reverse recovery current = $\dfrac{2 \times \text{Reverse recovery charge}}{\text{Reverse recovery time}}$

Therefore, $$I_{rr} = \frac{2 \times Q_{rr}}{t_{rr}} = \frac{2 \times 150 \times 10^{-6}}{10 \times 10^{-6}}$$

Solving, we get I_{rr} = 30 A **Ans.**

10.33.6. Turn Off Time and Effect of Reapplied dV_F / dt

Let us consider figure 10.57, which illustrates that even at instant t_3, substantial excess carriers remain in the n_1 and p_2 layers of a thyristor.

If a forward voltage is reapplied which increases at a rate dV_F/dt as shown in figure 10.58, a pulse of forward current (called as forward recovery current) will flow through the SCR. This current flows due to the remaining excess carriers. This forward recovery current may act as a pulse of gate current and therefore may turn on the thyristor even though the turn on is not desired. This is known as accidental turn on of thyristors. The accidental turn on of the thyristors is unwanted and can be avoided by taking the following measures :

(i) A thyristor must remain reverse biased for a minimum time equal to (t_q) specified by the device manufacturer. No forward voltage be reapplied during this time period.

(ii) The value of the rate of increase of the forward voltage dV_F/dt should be kept below the maximum value specified by the manufacturer. This is because larger the value of dV_F/dt is, larger will be the peak forward recovery current and hence more will be the possibility of accidental turn on.

(*iii*) The maximum value of reapplied dV_F/dt is decided by the value of space charge capacitance of junction J_2, because if the current through this capacitance due to dV_F/dt exceeds, the breakover current I_{BO}, accidental turn on will take place.

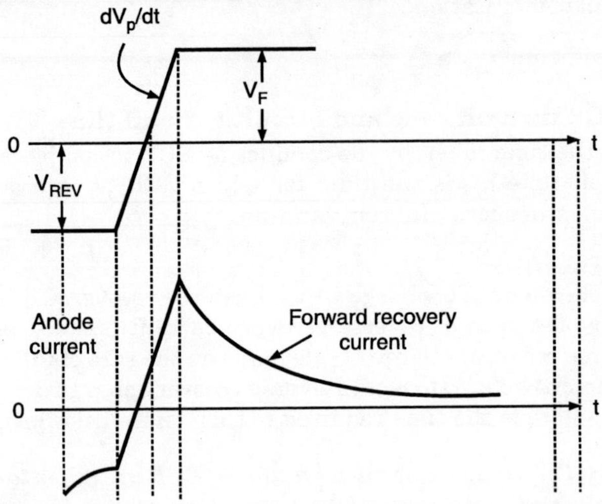

Fig. 10.58. Forward recovery current resulting from reapplication of a forward voltage across the thyristor before the end of the specified recovery time.

Types of SCRs

Depending on the turn off time of the SCR, they may be classified into following two categories:

(*i*) **Converter grade SCRs**

These SCRs have a large turn off time so they are suitable only for the low frequency applications like rectification.

(*ii*) **Inverter grade SCRs**

These SCRs have shorter turn off time. Hence, they can operate at higher frequencies, and hence are suitable for the high frequency applications like inverters.

10.33.7. Drawbacks of SCR

(*i*) It can conduct only in one direction. Hence, it can control power only during one half cycle of ac.

(*ii*) It can turn on accidentally due to high dv/dt of the source voltage.

(*iii*) It is not easy to turn off the conducting SCR. We have to use special circuits called commutation circuits to turn off SCR.

(*iv*) SCR can not be used at high frequencies. The maximum frequency of its operation is 400 Hz.

(*v*) Gate current can not be negative.

10.33.8. Advantages of SCR

(*i*) It can handle large voltages, currents and power.

(*ii*) The voltage drop across conducting SCR is small. This will reduce the power dissipation in the SCR.

(*iii*) Easy to turn on.

(*iv*) Triggering circuits are simple.

(*v*) It can be protected with the help of a fuse.

(*vi*) We can control the power delivered to the load.

10.33.9. Applications of SCR

(*i*) Controlled rectifiers
(*ii*) DC to DC converters or choppers
(*iii*) DC to AC converters or inverters
(*iv*) As static switch
(*v*) Battery charger
(*vi*) Speed control of DC and AC motors
(*vii*) Lamp dimmers
(*viii*) AC voltage stabilizers.

SUMMARY

- A zener diode is also called a voltage-reference, voltage regulator or breakdown diode. Like a rectifier diode, it is also important in many power applications.
- The zener diode is a silicon PN junction device, which differs from a rectifier diode, in the sense, that it is operated in the reverse breakdown region. Also, the breakdown voltage of a zener diode is set by carefully controlling the doping level during manufacturing process.
- As a matter of fact, the zener breakdown occurs when the electric field across the junction, produced due to reverse voltage, is sufficiently high. This electric field exerts a force on the electrons in the outer most shell. This force is so high that the electrons are pulled away from their parent nuclei and become free carriers.
- The zener diodes, with breakdown voltages of less than 6 V, operate predominantly in zener breakdown. Those with breakdown, voltages greater than 6 V, operate predominantly in avalanche breakdown.
- Although the zener diodes have a number of applications, yet the following are important from the subjects points of view:

  (*i*) As a voltage regulator.

  (*ii*) As a fixed reference voltage in transistor biasing circuits.

  (*iii*) As peak clippers or limiters in waveshaping circuits.

  (*iv*) For meter protection against damage from accidental applications.
- It has been observed that if the concentration of impurity atoms is greatly increased in a normal PN junction (by 1000 times or more), its characteristics are completely changed. This gives rise to a new type of diode known as *tunnel diode*. It was invented in 1958 by Dr. Leo Esaki. Because this, a tunnel diode is also known as Esaki diode.
- The phenomenon of penetrating the charge carriers, directly through the potential barrier, instead of climbing over it, is called tunneling. That is why, highly doped PN junction devices are called tunnel diodes. These diodes are usually made of germanium (Ge) or gallium-assenide (GaAs).
- If the tunnel diode is reversed biased (*i.e.*, P side of junction is made negative with respects to its N-side), it acts like an excellent conductor, *i.e.*, the reverse current increases with the increase in reverse voltage.
- The tunnel diodes exhibit a specific characteristic called negative resistance. They have extremely low values of inductance and capacitance. These features make them useful in a number of applications as under :

  (*i*) As an ultra high-speed switching device. It is possible due to the tunneling mechanism, which takes place at the speed of light. The switching time is of the order of nono-seconds (10^{-9} sec).

  (*ii*) As a logic memory storage device. It is possible due to the triple valued of the curve for current between I_P and I_V.

  (*iii*) As a microwave oscillator at frequencies in the order of 10 GHz. It is possible due to extremely low values of inductance and capacitance of the device.

  (*iv*) In relaxation oscillator. It is possible due to negative resistance of the device.
- A varacter diode is, basically, a reverse biased PN junction, which utilizes the inherent capacitance of the depiction layer. It is also known as varicap, voltcap or tunning diode. It is used as a voltage-variable capacitor.

- The maximum ratings for the varactor diode are the same as those that are used for the PN junction diode and the zener diode.
- A major application of varactor diode is in tuning circuits. It is used in electronic tuners in radio, television and other commercial receivers. When a varactor is used in a tuning circuit, it allows a resonant frequency to be adjusted by a variable voltage level.
- The resonant frequency of the tank circuit is given by the following relation,

$$f_0 = \frac{1}{2\pi\sqrt{LC}}$$

where L = value of the inductance,
C = value of total capacitance, which is given by the relation $\frac{C_1 \cdot C_2}{C_1 + C_2}$
where C_1 and C_2 are the maximum and minimum values of the capacitances of diode.
- The varactor diode is also used in automatic frequency control device, adjustable band-pass filter and parametric amplifier.
- The schottky diode is formed by joining a doped semiconductor region (usually N-type) with a metal such as gold, silver or platinum. Thus in a Schottky diode, there is a metal-to-semiconductor junction rather than a simple PN junction.
- At step-recovery diode employs a graded doping, where the doping level of the semiconductor material is reduced as the junction is approached.
- The unusual doping of the step-recovery diode affects the time required for the device to switch from off to on and vice versa. The typical switching time for a step-recovery diode is in the low-picosecond range. This makes them ideal for ultra-high frequency switching applications.
- A PIN diode is made up of three semiconductor materials : two *heavily doped* P-and one N-type material separated by an *intrinsic* (*i.e.*, undoped) semiconductor (I).
- A PN junction diode, which emits light when forward biased, is know as a light emitting diode (abbreviated as LED). The emitted light may be visible or invisible. The amount of light output is directly proportional to the forward current. Hence, higher the forward current, higher is the light output.
- The LED's which emit one colour of light when forward biased and another when reverse biased, are called multicolored LED's.
- The liquid crystal display (LCD) is a low power device as compared to LED. The power requirement is typically in the order of microwatts for the LCD as compared to the same order of milliwatts for LEDs.

SHORT QUESTIONS WITH ANSWERS

Q.1. What is a Zener Diode?

Ans. A zener diode is also called a voltage-reference, voltage regulator or breakdown diode. Like a rectifier diode, it is of utmost importance in several power applications.

Q.2. What do you mean by zener effect?

Ans. The zener breakdown occurs when the electric field across the junction, produced due to reverse voltage, is sufficiently high. This electric field exerts a force on the electrons in the outer most shell. This force is so high that the electrons are pulled away from their parent nuclei and become free carriers. This ionization, which occurs due to the electrostatic force of attraction, is known as zener effect. It causes an increase in the number of free carriers and hence an increase in the reverse current.

Q. 3. Write the applications of zener diode

Ans. Although the zener diodes have a number of applications, yet the following are important from the subjects points of view:

(*i*) As a voltage regulator.

(*ii*) As a fixed reference voltage in transistor biasing circuits.

(*iii*) As peak clippers or limiters in waveshaping circuits.

(*iv*) For meter protection against damage from accidental applications.

Q. 4. What is a the Tunnel Diode?

Ans. It has been observed that if the concentration of impurity atoms is greatly increased in a normal PN junction (by 1000 times or more), its characteristics are completely changed. This gives rise to a new type of diode known as *tunnel diode*. It was invented in 1958 by Dr. Leo Esaki. Because of this reason a tunnel diode is also known as Esaki diode.

Q. 5. What do you mean by tunneling?

Ans. The phenomenon of penetrating the charge carriers, directly through the potential barrier, instead of climbing over it, is called tunneling. That is why, highly doped PN junction devices are called tunnel diodes. These diodes are usually made of germanium (Ge) or gallium-assenide (Ga As).

Q.6. What is the significance of negative resistance of a tunnel diode?

Ans. It is the resistance of a tunnel diode which it offers when operated in a negative resistance region. The negative resistance is given by the following relation:

$$R_N = -\frac{\Delta V_F}{\Delta I_F}$$

where ΔV_F = change in forward voltage between any two points lying within the negative resistance region of the V-I characteristic, and

ΔI_F = corresponding change in forward current.

The negative resistance region may also be expressed in terms of slope of the characteristic in the negative resistance region. Its value depends upon the semiconductor material used for manufacturing tunnel diodes and the value may range from 10 Ω to 200 Ω.

Q. 7. Write the applications of a Tunnel diode?

Ans. The tunnel diodes exhibit a specific characteristic called negative resistance. They have extremely low values of inductance and capacitance. These features make them useful in a number of applications as under :

(*i*) Tunnel diode is used as an ultra high-speed switching device. It is possible due to the tunneling mechanism, which takes place at the speed of light. The switching time is of the order of nono-seconds (10^{-9} sec).

(*ii*) Tunnel diode is used as a logic memory storage device. It is possible due to the triple valued of the curve for current between I_P and I_V.

(*iii*) Tunnel diode is used as a microwave oscillator at frequencies in the order of 10 GHz. It is possible due to extremely low values of inductance and capacitance of the device.

(*iv*) Tunnel diode is used in relaxation oscillator. It is possible due to negative resistance of the device.

Q. 8. What do you mean by the Varactor Diode?

Ans. A varacter diode is, basically, a reverse biased PN junction, which utilizes the inherent capacitance of the depiction layer. It is also known as varicap, voltcap or tunning diode. It is used as a voltage-variable capacitor.

Q.9. What is the Schottky diode?

Ans. The schottky diode is formed by joining a doped semiconductor region (usually N-type) with a metal such as gold, silver or platinum. Thus in a Schottky diode, there is a metal-to-semiconductor junction rather than a simple PN junction

Q. 10. Write the applications of a Schottky Diode?

Ans. We know that Schottky diodes have two important features namely it operates only with majority carriers, and there is no charge storage when the device is switched from its 'ON' to 'OFF' state. These features make it useful for a variety of applications as under :

(*i*) To rectify very high frequency (above 300 MHz) signals.

(*ii*) As a switching device in digital computers. The speed of computers depends on how fast their diodes and transistors can turn 'ON' and 'OFF'.

(*iii*) In clipping and clamping circuits.

(*iv*) In low power Schottky transistor-transistor logic (Schottky TTL) circuits.

(*v*) In mixing and detecting circuits used in communication systems.

(*vi*) In low voltage power supply circuits. The Schottky diode has a threshold voltage of 0.25 V only. It is an advantage in low voltage rectifiers because we have to subtract only 0.25 V (instead of 0.7 V) for each diode from the low input voltage to get an output.

Q. 11. What is a PIN Diode?

Ans. A PIN diode is made up of three semiconductor materials : two *heavily doped* P-and one N-type material separated by an *intrinsic* (*i.e.*, undoped) semiconductor (I) as shown in figure 3.21 (*a*). The intrinsic region offers high resistance to the current flowing through it.

The PIN diode has the following two advantages over the normal PN junction diode:

(*i*) The capacitance between the P and N region *decreases* because of the increased separation between the P and N regions.

This advantage allows the PIN diode to have fast response time. Therefore, these diodes are useful at very-high-frequencies (*i.e.*, above 300 MHz).

(*ii*) There is a greater electron-hole pair generation because of the increased electric field between the *P* and *N* regions. This advantage allows the PIN diode to process even weak signals.

Q.12. Write the applications of PIN diode?

Ans. We know that when a PIN diode is forward biased, it acts like a variable resistance which decreases with the increasing current. Also when it is reverse biased, the PIN diode acts like an almost constant capacitance. These features make the PIN diode useful for a variety of applications as under:

(*i*) As a d.c. controlled microwave switch operated by rapid changes in bias or as a modulating device. This application takes advantage of the variable forward-resistance characteristic. Modulators are used to combine two signals of different frequencies into a single signal.

(*ii*) In attenuator applications because its resistance can be controlled by the current.

(*iii*) As an AM modulator of very high frequency signals.

Q. 13. What do you mean by liquid crystal display (LCD)?

Ans. The liquid crystal display (LCD) is a low power device as compared to LED. The power requirement is typically in the order of microwatts for the LCD as compared to the same order of milliwatts for LEDs. However, an LCD requires an external or internal light source, It is limited to a termperature range of about 0°C to 60°C and lifetime is an area of concern, because LCDs can chemically degrade with passage of fini.

Q. 14. Explain the basic operation of LCD. What are the relative difference in mode of operation between an LED and LCD?

Ans. In case of LED, the electric power is converted to light due to the process of recombination. But in case of an LCD it is not so. On application of electric power, the liquid crystal molecules orient in a different way to scatter light. This is the difference in their operation.

NUMERICAL PROBLEMS

1. Determine the power dissipating rating of Motorola's IN746 zener diode at 125°C. Assume that the dc power dissipation rating of the device is 500 mW and the derating factor is 3.33 mW/°C above 50°C. (**Ans.** 250.25 mW)
2. A 6.2 V zener diode has a resistance of 20Ω. What is the terminal voltage when the current in 25 mA? (**Ans.** 6.7 V)
3. A light emitting diode (LED) is connected across a voltage source + 10 volts through a series resistance of 820 Ω. Calculate the LED current. Assume the voltage drop across an LED of 1.5 volt. (**Ans.** 10.4 mA)
4. A step recovery diode is used in a frequency multiplier. The period of the rectified output signal is 0.05 sec. What is the frequency of the 5th harmonic? (**Ans.** 100 MHz)

OBJECTIVE TYPE QUESTIONS

1. The main advantage of a point contact diode is that
   (*a*) its reverse resistance is infinite
   (*b*) the ratio of forward current to reverse current is very high
   (*c*) its cut-in voltage is zero
   (*d*) none of the above
2. Avalance breakdown primarily depends on the phenomenon of
   (*a*) ionization (*b*) doping
   (*c*) collusion (*d*) recombination
3. Zener breakdown occurs
   (*a*) mostly in germanium junctions
   (*b*) due to rupture of covalent bonds
   (*c*) in lightly doped junctions
   (*d*) due to thermally generated minority carriers
4. For highly doped diode
   (*a*) Zener breakdown is likely to take place
   (*b*) Avalanche breakdown is likely to take place
   (*c*) either (*a*) or (*b*) will take place
   (*d*) neither (*a*) nor (*b*) will take place

5. The main cause, why electrons continual though a P-N junction is
   (*a*) their higher energy
   (*b*) very low barrier potential
   (c) low impurity level
   (*d*) extermely thin depletion layer
6. Schottky barrier diode can be used as
   (*a*) low noise amplifier
   (*b*) variable capacitance device
   (c) power supply rectifier
   (*d*) low level detector.
7. Which of the following diodes show the negative resistance region?
   (*a*) P-N junction diode
   (*b*) Zener diode
   (*c*) Tunnel diode
   (*d*) PIN
8. Which of the following is a unipolar device?
   (*a*) P-N junction diode
   (*b*) Zener diode
   (*c*) Tunnel diode
   (*d*) Schottky diode
9. Tunnel diode is made by following semiconductors
   (*a*) highly doped (*b*) sparsely doped
   (*c*) intrinsic
   (*d*) none of the above

Answers

1. (*b*) 2. (*c*) 3. (*b*) 4. (*a*) 5. (*d*) 6. (*d*) 7. (*c*) 8. (*d*)
9. (*a*).

❑❑❑

Objective Type Question Bank

OBJECTIVE TYPE QUESTION BANK

1. The resistivity of intrinsic semiconductor material is about
   (*a*) 10^5 ohm-m (*b*) 10^3 ohm-m
   (*c*) 100 ohm-m (*d*) 1 ohm-m
2. When a large number of atoms are brought together to form a crystal
   (*a*) the energy levels of inner shell electrons are affected appreciably by the presence of other neighbouring atoms.
   (*b*) the energy levels of outer shell electrons are affected appreciably by the presence of other neighbouring atoms.
   (*c*) the energy lends of both inner and outer shell electrons are affected appreciably by the presence of other neighbouring atoms.
   (*d*) none of above
3. In which material, do conduction and valence bands overlap
   (*a*) insulators (*b*) conductors
   (*c*) both conductor and semiconductor
   (*d*) semiconductors.
4. At 0 K, the forbidden energy gap in intrinsic semi conductor is about
   (*a*) 10 eV (*b*) 6 eV
   (*c*) 1 eV (*d*) 0.2 eV
5. In intrinsic semiconductor, the increase in conductivity per degree increase in temperature is about
   (*a*) 2% (*b*) 6%
   (*c*) 15% (*d*) 25%.
6. In intrinsic semiconductor, the fermi level
   (*a*) lies at the centre of forbidden energy gap
   (*b*) is near the conduction band
   (*c*) is near the valence band
   (*d*) may be anywhere in the forbidden energy gap.
7. The ratio of impurity atoms to intrinsic semiconductor atoms in an extrinsic semi conductor is about
   (*a*) 1 : 10 (*b*) 1 : 1000
   (*c*) 1 : 100000 (*d*) 1 : 10^8
8. The addition of *n* type impurity to intrinsic material creates allowable energy levels.
   (*a*) slightly below conduction band
   (*b*) slightly above conduction band
   (*c*) slightly below valence band
   (*d*) slightly above valence band.
9. The addition of *p* type impurity to intrinsic material creates allowable energy levels.
   (*a*) slightly below conduction band
   (*b*) slightly above conduction band
   (*c*) slightly below valence band
   (*d*) slightly above valence band.
10. The merging of a hole and an electron is called
    (*a*) recombination (*b*) covalent bonding
    (*c*) thermal union (*d*) none of the above
11. At room temperature, a semiconductor material is
    (*a*) perfect insulator
    (*b*) conductor
    (*c*) slightly conducting
    (*d*) any of above
12. The amount of time between the creation and disappearance of a hole in an intrinsic semiconductor material is called
    (*a*) life cycle
    (*b*) recombination time
    (*c*) life time (*d*) half life
13. When a voltage is applied to a semi-conductor crystal, the free electrons will flow
    (*a*) towards positive terminal
    (*b*) towards negative terminal
    (*c*) either towards positive terminal or negative terminal
    (*d*) towards positive terminal for 1 μs and towards negative terminal for next 1 μs.
14. As temperature increases, the number of tree electrons and holes in an intrinsic semiconductor
    (*a*) increases
    (*b*) decreases
    (*c*) remains the same
    (*d*) may increase or decrease
15. Holes act like
    (*a*) positive charges (*b*) neutral atoms
    (*c*) negative charges
    (*d*) crystals

16. In an intrinsic semiconductor, the intrinsic charge concentration at any absolute temperature *T* is proportional to
(*a*) T (*b*) T^2
(*c*) T^3 (*d*) T^4

17. Forbidden energy gap in germanium at 0 K is about
(*a*) 10 eV (*b*) 5 eV
(c) 2 eV (*d*) 0.78 eV

18. At absolute temperature
(*a*) the forbidden energy gap in germanium is higher than that in silicon
(*b*) the forbidden energy gap in germanium and silicon are equal
(*c*) the forbidden energy gap in silicon is higher than that in germanium.
(*d*) none of above

19. As temperature increases
(*a*) the forbidden energy gaps in silicon and germanium increase
(*b*) the forbidden energy gaps in silicon and germanium decrease.
(*c*) the forbidden energy gap in silicon decreases while that in germanium decreases
(*d*) the forbidden energy gap in silicon increases while that in germanium decreases

20. The intrinsic resistivity of silicon at 300 K is about
(*a*) 1 Ω-cm (*b*) 400 Ω-cm
(c) 10000 Ω-cm (*d*) 230000 Ω-cm

21. In intrinsic semiconductor at 300 K, the magnitude of free electron concentration in silicon is about
(*a*) 15×10^4 per cm^3
(*b*) 5×10^{12} percm^3
(*c*) 1.45×10^{10} per cm^3
(*d*) 1.45×10^6 per cm^3

22. In *n* type semiconductor, the free electron concentration.
(*a*) is nearly equal to density of donor atoms
(*b*) is much greater than density of donor atoms
(*c*) is much less than density of donor atoms
(*d*) may be equal or more or less than density of donor atoms.

23. In all metals
(*a*) conductivity decreases with increase in temperature
(*b*) current flow is by electrons as well as by holes
(*c*) resistivity decreases with increase in temperature
(*d*) The gap between valence and conduction bands is small

24. Which of the following has highest conductivity
(*a*) Silver (*b*) Aluminium
(*c*) Tungsten (*d*) Platinum

25. The product *kT* (where *k* is Boltzmann's constant and T is absolute temperature) at room temperature is about
(*a*) 0.051 eV (*b*) 0.026 eV
(*c*) 0.01 eV (*d*) 0.001 eV

26. The number *of p-n* junctions in a semiconductor diode are
(*a*) 0 (*b*) 1
(*c*) 2 (*d*) 1 or 2

27. At room temperature, the barrier potential in a silicon diode is
(*a*) 0.1 V (*b*) 0.3 V
(*c*) 0.7 V (*d*) 1 V

28. As the reverse voltage is increased, the depletion layer
(*a*) becomes narrow
(*b*) widens
(*c*) remains the same
(*d*) reduce to zero

29. A reverse voltage of 18 V is applied to a semiconductor diode. The voltage across the depletion layer is
(*a*) 0 (*b*) 0.7 V
(*c*) about 10 V (*d*) 18 V

30. The reverse saturation current in a semiconductor diode consists of
(*a*) avalanche current
(*b*) zener current
(*c*) minority carrier current
(*d*) minority carrier current and surface leakage current

31. A silicon diode is forward biased and total applied voltage is 5 V. The voltage across *p-n* junction is
(*a*) 5 V
(*b*) Slightly less than 5 V
(*c*) 0.7 V (*d*) 0

32. A potential of 7 V is applied to a silicon diode. A resistance of 1 k ohm is also in series with the diode. The current is

(*a*) 7 mA (*b*) 6.3 mA
(*c*) 0.7 mA (*d*) 0

33. A voltage of 9 V is applied in forward direction to a semiconductor diode in series with a load resistance of 1000 Ω. The voltage across the load resistance is zero. It indicates that
(*a*) diode is short circuited
(*b*) diode is open circuited
(*c*) resistor is open circuited
(*d*) diode is either o.c or s.c

34. The breakdown voltage in a zener diode
(*a*) is almost constant
(*b*) is very small
(*c*) may destroy the diode
(*d*) decreases with increase in current

35. The voltage across a zener diode
(*a*) is constant in forward direction
(*b*) is constant is reverse direction
(*c*) is constant in both forward and reverse directions
(*d*) none of above

36. Which of the following is basically a voltage controlled capacitance
(*a*) zener diode (*b*) diode
(*c*) varactor diode (*d*) LED

37. In a zener diode
(*a*) Ihe forward current is very high
(*b*) sharp breakdown occurs at a certain reverse voltage
(*c*) the ratio *v-i* can be negative
(*d*) there arc two *p-n* junctions.

38. If the reverse voltage across a *p-n* junction is increased three times, the junction capacitance
(*a*) will decrease (*b*) will increase
(*c*) will decrease by an approximate factor of about 2
(*d*) will increase by an approximate factor of about 2.

39. The depletion layer in a reverse biased *p-n* junction is due to the presence of
(*a*) electrons (*b*) holes
(*c*) both electrons and holes
(*d*) immobile ions

40. The cut in voltage
(*a*) is the same for silicon and germanium diodes
(*b*) is a forward voltage
(*c*) is a reverse voltage
(*d*) is a forward voltage below which the current is very small

41. When a *p-n* junction is reverse biased
(*a*) holes and electrons move away from the junction
(*b*) holes and electrons move towards the the junction
(*c*) holes more towards junction and electrons move away from junction
(*d*) holes move away from junction and electrons move towards junction

42. The saturation current in a semiconductor diode
(*a*) depends on forward voltage
(*b*) depends on reverse voltage
(*c*) is due to thermally generated minority carriers
(*d*) none of above

43. When a *p-n* junction is forward biased
(*a*) the width of depletion layer increases
(*b*) the width of depletion layer decreases
(*c*) the majority carriers move away from the junction
(*d*) the current is very small.

44. In a reverse biased *p-n* junction, the reverse bias is 4V. The junction capacitance is about
(*a*) 0.1 *F* (*b*) 4 μ*F*
(*c*) 10 *nF* (*d*) 20 *pF*

45. The dynamic resistance of a forward biased *p-n* diode
(*a*) varies inversely with current
(*b*) varies directly with current
(*c*) is constant
(*d*) is either constant or varies directly with current.

46. A *p-n* junction diode has
(*a*) low forward and high reverse resistance
(*b*) a non-linear *v-i* characteristics
(*c*) zero forward current till the forward voltage reaches cut in value
(*d*) all the above.

47. Typical value of reverse current in a semiconductor diode is
(*a*) 1 A (*b*) 0.1 A
(*c*) 1 mA (*d*) 1 μA

48. An increase in junction temperature of a semiconductor diode
(*a*) causes a small increase in reverse saturation current

(*b*) causes large increase in reverse saturation current
(*c*) does not affect reverse saturation current
(*d*) may cause an increase or decrease in reverse saturation current depending on rating of diode.

49. An extrinsic semiconductor sample has 6 billion silicon atoms and 3 million pentavalent impurity atoms. The number of electrons and holes is
(*a*) 3 million each
(*b*) 6 billion each
(*c*) 3 million free electrons and very small number of holes
(*d*) 3 million holes and very small number of free electrons.

50. At any point on the *v-i* characteristics of a semiconductor diode, the slope of *i-v* characteristics is called
(*a*) resistance of diode
(*b*) conductance of diode
(*c*) incremental resistance of diode
(*d*) incremental conductance of diode

51. In the equation $i = I_0\,(e^{V/\eta V_T} - 1)$, $V_T =$
(*a*) $\dfrac{11600}{T}$ (*b*) $\dfrac{T}{11600}$
(*c*) $T \times 11600$ (*d*) $\dfrac{11600}{T^2}$

52. A diode is operating in forward region and the forward voltage and current are $v = 3 + 0.3$ sin (or volts and $i = 5 + 0.2$ sin ωt mA. The average power dissipated is
(*a*) 15 mW (*b*) about 15 mW
(*c*) 1.5 mW (*d*) about 1.5 mW

53. A transistor has two *p-n* junctions. The batteries should be connected such that
(*a*) both junctions are forward biased
(*b*) both junction are reverse biased
(*c*) one junction is forward biased and the other is reverse biased
(*d*) none of above

54. The value of α in a transistor
(*a*) is always equal 1
(*b*) is less than 1 but more than 0.9
(*c*) is about 0.4 (*d*) is about 0.1

55. When a *p-n-p* transistor is properly biased to operate in active region the holes from emitter.
(*a*) diffuse through base into collector region
(*b*) recombinc with electrons in base
(*c*) recombine with electrons in emitter
(*d*) none of above

56. A zener diode is used in
(*a*) voltage regulator circuit
(*b*) amplifier circuits
(*c*) both voltage regulator and amplifier circuit
(*d*) none of above.

57. The number of doped regions in a bipolar transistor is
(*a*) 1 (*b*) 2
(*c*) 3 (*d*) 4

58. In an *n-p-n* transistor, the majority carriers in the base are
(*a*) electrons (*b*) holes
(*c*) both holes and electrons
(*d*) either holes or electrons

59. In a bipolar transistor, the barrier potential
(*a*) 0 (*b*) a total of 0.7 V
(*c*) 0.7 V across each depletion layer
(*d*) 0.35V

60. In a bipolar transistor, the emitter base junction has
(*a*) forward bias (*b*) reverse bias
(*c*) zero bias (*d*) zero or reverse bias

61. In a bipolar transistor, the base collector junction has
(*a*) forward bias (*b*) reverse bias
(*c*) zero bias
(*d*) zero or forward bias

62. The current gain of a transistor is the ratio of
(*a*) emitter current to base current
(*b*) emitter current to collector current
(*c*) collector current to base current
(*d*) collector current to emitter current

63. In a bipolar transistor which current is largest
(*a*) collector current
(*b*) base current
(*c*) emitter current
(*d*) base current or emitter current

64. In a bipolar transistor which current is smallest
(*a*) collector current
(*b*) base current
(*c*) emitter current
(*d*) any of the three currents

65. In a bipolar transistor
(*a*) the base is thick and highly doped
(*b*) the base is thin and highly doped
(*c*) the base is thick and lightly doped
(*d*) the base is thin and lightly doped

66. The power dissipation in a transistor is the product of
(*a*) emitter current and emitter to base voltage
(*b*) emitter current and collector to emitter voltage
(*c*) collector current and collector to emitter voltage
(*d*) none of above

67. In a bipolar transistor
(*a*) recombination in base regions of both *n-p-n* and *p-n-p* transistor is minimum.
(*b*) recombination in base regions of both *n-p-n* and *p-n-p* transistors is high
(*c*) recombination in base region of *n-p-n* transistor is minimum but that in *p-n-p* transistor is high
(*d*) recombination in base region of *p-n-p* transistor is minimum but that in *n-p-n* transistor is high

68. For bipolar transistor
(*a*) $\beta_{dc} = \frac{\alpha_{dc}}{1-\alpha_{dc}}$ (*b*) $\beta_{dc} = \frac{\alpha_{dc}}{1+\alpha_{dc}}$
(*c*) $\beta_{dc} = \frac{1-\alpha_{dc}}{\alpha_{dc}}$ (*d*) $\beta_{dc} = \frac{1+\alpha_{dc}}{\alpha_{dc}}$

69. The output characteristics of a bipolar transistor has three distinct regions. They are known as
(*a*) saturation region, active region and breakdown region
(*b*) inactive region, active region and breakdown region
(*c*) inactive region, saturation region and active region
(*d*) inactive region, saturation region and breakdown region

70. In which region of a *CE* bipolar transistor is collector current almost constant
(*a*) saturation region
(*b*) active region
(*c*) breakdown region
(*d*) both saturation and active regions

71. In a *CE* bipolar transistor operating in active region, collector current is independent of
(*a*) V_{CE} (*b*) I_B
(*c*) both V_{CE} and IB
(*d*) none of above

72. In the saturation region of *CE* output characteristics of *n-p-n* transistor, V_{CE} is about
(*a*) 0.5 V (*b*) 15 V
(*c*) –0.5 V (*d*) – 15V

73. *CE* saturation resistance of *n-p-n* transistor is
(*a*) $\frac{V_{CB}}{I_E}$ (*b*) $\frac{V_{CE}}{I_C}$
(*c*) $\frac{V_{CE}}{I_E}$ (*d*) $\frac{V_{CE}}{I_B}$

74. In forward active region of operation of a BJT
(*a*) both junctions are forward biased
(*b*) both junctions are reverse biased
(*c*) emitter base junction is forward biased and base collector junction is reverse biasd
(*d*) emitter base junction is reverse biased and base collector junction is reverse biased

75. In which mode of BJT operation are both junctions reverse biased
(*a*) active (*b*) saturation
(*c*) cut off (*d*) reverse active

76. In which mode of BJT operation are both junctions forward biased
(*a*) active (*b*) saturation
(*c*) cut off (*d*) reverse active

77. Between which regions does BJT act like switch
(*a*) cut off and saturation
(*b*) cut off and forward active
(*c*) forward active and cut off
(*d*) saturation and active.

78. In what condition does BJT act like an open switch
(*a*) cut off (*b*) saturation
(*c*) active (*d*) both (*b*) and (*c*)

79. The turn off time of a bipolar transistor is about
(*a*) 0.5 ns (*b*) 10 ns
(*c*) 70 ns (*d*) 150 ns

80. In *CE* connection, the leakage current of a transistor is about
(*a*) 10×10^{-9} A (*b*) 5×10^{-6} A
(*c*) 200×10^{-6} A (*d*) 5×10^{-3} A

81. In a bipolar junction transistor, the base region is made very thin so that
(*a*) recombination in base region is minimum
(*b*) electric field gradient in base is high
(*c*) base can be easily fabricated
(*d*) base can be easily biased.

82. In a *p-n-p* transistor, the main current carriers and the mechanism of flow respectively are
(*a*) electrons, drift (*b*) holes, drift
(*c*) holes, diffusion
(*d*) electrons, diffusion

83. In a bipolar junction transistor α_{dc} = 0.98, I_{CO}= 2 μA and I_B = 15 μA. The collector current I_C is
(*a*) 635 mA (*b*) 735 mA
(*c*) 835 mA (*d*) 935 mA

84. When a *p-n-p* transistor is operating in active region, the current in the *n* region is due to
(*a*) only holes (*b*) only electrons
(*c*) mainly holes (*d*) mainly electrons

85. In common base configuration, the input characteristics of bipolar junction transistor are drawn between
(*a*) V_{EB} and I_E (*b*) V_{CB} and I_C
(*c*) I_C and V_{CE} (*d*) I_E and V_{CE}

86. In common base connection, the output characteristics of a bipolar junction transistor is drawn between
(*a*) I_C and V_{CB} (*b*) I_C and V_{CE}
(*c*) I_E and V_{CB} (*d*) I_E and V_{CE}

87. In *CE* configuration, the output characteristics of a bipolar junction transistor is drawn between
(*a*) I_C and V_{CB} (*b*) I_E and V_{CB}
(*c*) I_C and V_{CE} (*d*) I_E and V_{CE}

88. In an *n-p-n* transistor biased for operation in forward active region
(*a*) emitter is positive with respect to base
(*b*) collector is positive with respect to base
(*c*) base is positive with respect to emitter and collector is positive with respect to base
(*d*) none of above

89. In a *p-n-p* transistor operating in forward active mode
(*a*) base is positive with respect to emitter and collector
(*b*) base is negative with respect to emitter and collector
(*c*) emitter is positive with respect to base and base is positive with respect to collector
(*d*) emitter is negative with respect to base and base is positive with respect to collector

90. In a transistor operating in forward active mode
(*a*) depletion layer between emitter and base is thin and that between base and collector is also thin
(*b*) both depiction regions are thick
(*c*) depletion layer between emitter and base is thin and that between base and collector is thick
(*d*) depletion layer between emitter and base is thick and that between base and collector is thin.

91. In the schematic representation of bipolar junction transistor, the direction of arrow shows the direction of flow of
(*a*) holes (*b*) electrons
(*c*) holes in *pnp* and electrons in *npn*
(*d*) electrons in *pnp* and holes in *npn*

92. When the gate terminal of MOSFET is positive, it is said to operate in
(*a*) depletion mode (*b*) conduction mode
(*c*) enhancement mode
(*d*) none of above

93. The channel of JFET consists of
(*a*) *p* type material only
(*b*) *n* type material only
(*c*) conducting material
(*d*) either *p* or *n* type material.

94. In *n* type MOSFET, the substrate
(*a*) is *p* type (*b*) is *n* type
(*c*) is metallic (*d*) may be *p* or *n* type

95. Which of the following exhibits negative resistance in a portion of its characteristics
(*a*) Tunnel diode (*b*) zener diode
(*c*) JFET
(*d*) MOSFET and tunnel diode

96. Which of the following is used for generating time-varying waveforms
(*a*) MOSFET (*b*) PIN diode
(*c*) tunnel diode (*d*) UJT

97. A JFET
(*a*) is current controlled device
(*b*) has low input resistance
(*c*) has high gate current
(*d*) is a voltage controlled device

98. In a JFET the width of channel is controlled by
(*a*) gate voltage (*b*) drain current
(*c*) source current (*d*) all the above

99. In *n* channel JFET, the gate voltage is made more negative
(*a*) the channel width will increase
(*b*) the channel width would decrease
(*c*) the channel width and drain current will decrease
(*d*) the channel width will decrease and drain current will increase

100. In a JFET
(*a*) the source gate junction is forward biased and gate drain junction is reverse biased
(*b*) both the junctions are forward biased
(*c*) both the junctions are reverse biased
(*d*) the source gate junction is reverse biased and gate drain junction is forward biased

101. The carriers is *n* channel JFET are
(*a*) free electrons and holes
(*b*) holes
(*c*) free electrons or holes
(*d*) free electrons

102. Compared to bipolar junction transistor, a JFET has
(*a*) lower input impedance
(*b*) high input impedance and high voltage gain
(*c*) higher voltage gain
(*d*) high input impedance and low voltage gain

103. In a JFET
(*a*) drain current is very nearly equal to source current
(*b*) drain current is much less than source current
(*c*) drain current may be equal to or less than source current
(*d*) drain current may be even more than source amount

104. The drain characteristics of JFET are drawn between.
(*a*) V_{GS} and V_{DS} for different values of drain current
(*b*) drain current and V_{GS} for different values of V_{DS}
(*c*) drain current and V_{DS} for different values of V_{GS}
(*d*) drain current and V_{GS} for one value of V_{DS}

105. In a JFET avalancha avalanchi breakdown occurs when V_{DS} = 22V and V_{GS} = 0. If V_{GS} = – 1 V, the avalanche breakdown will occur at
(*a*) V_{DS} = 22 V
(*b*) V_{DS} more than 22 V
(*c*) V_{DS} equal to or more than 22 V
(*d*) V_{DS} less than 22 V.

106. The normal operation of JFET is
(*a*) in constant voltage region
(*b*) constant current region
(*c*) both constant voltage and constant current regions
(*d*) either constant voltage or constant current region.

107. In a JFET, the drain current is maximum when
(*a*) V_{GS} = 0
(*b*) V_{GS} is not zero but is slightly negative
(*c*) V_{GS} is negative
(*d*) V_{GS} is negative and equal to V_{DS}

108. Which of the following is known as insulated gate FET
(*a*) JFET (*b*) MOSFET
(*c*) both JFET and MOSFET
(*d*) none of above

109. The units for transconductance are
(*a*) ohms (*b*) amperes
(*c*) volts (*d*) siemens.

110. In a JFET
(*a*) gate source cut off voltage is equal V_{DS}
(*b*) gate source cut off voltage is equal to pinch off voltage
(*c*) gate source cut off voltage is twice the pinch off voltage
(*d*) gate source cut off voltage is equal to negative of pinch off voltage.

111. *n* channel FETs are better as compared to *p*-channel FET because
(*a*) they are more efficient
(*b*) they have high switching time
(*c*) they have higher input impedance
(*d*) mobility of electrons is more than that of holes

112. In an insulated gate FET, the polarity of inversion layer is the same as that of
(*a*) minority carriers in source
(*b*) majority carriers in source
(*c*) charge on gate electrode
(*d*) minority carriers in drain.

113. The behaviour of a JFET is similar to that of
(*a*) NPN transistor (*b*) PNP transistor
(*c*) SCR (*d*) Vacuum triode

114. If the gate of JFET is revers biased, the width of depletion region
(a) becomes zero (*b*) is uniform
(*c*) is more near the source
(*d*) is more near the drain

115. The word enhancement mode is associated with
(*a*) Tunnel diode (*b*) MOSFET
(*c*) JFET (*d*) Photo diode

116. As compared to bipolar junction transistor, a FET
(*a*) is less noisy
(*b*) has better thermal stability
(*c*) has higher input resistance
(*d*) all the above

117. Which of the following are voltage controlled devices
(*a*) Vacum triode (*b*) FET
(*c*) SCR (*d*) both (*a*) and (*b*)

118. In an *n* channel JFET, the gate is
(*a*) *n* type (*b*) *p* type
(*c*) either *n* or *p* (*d*) party *n* and partly.

119. The depletion layer around *p-n* junction in JFET consists of
(*a*) hole (*b*) election
(*c*) immobile charges
(*d*) none of above

120. In an *n* channel JFET
(*a*) I_D, I_S and I_G are considered positive when flowing into the transistor
(*b*) I_D and I_S are considered positive when flowing into transistor and I_G is considered positive when flowing out of it
(*c*) I_D, I_S, I_G are considered positive when flowing out of transistor
(*d*) I_S and I_G are considered positive when flowing into transistor and I_D is considered positive when flowing out of it

121. Transconductance characteristics of JFET depict the relation between
(*a*) I_D and V_{GS} for different value of V_{DS}
(*b*) I_D and V_{DS} for different value of V_{GS}
(*c*) V_{GS} and V_{DS} for different values of I_D
(*d*) V_{DS} and I_D for different values of V_{GS}

122. In an *n* channel JFET, $V_{GS} = V_{GS(off)}$. Then
(*a*) I_D is zero
(*b*) I_D may be zero or positive
(*c*) I_D is positive
(*d*) I_D may be zero or negative

123. The drain characteristices of JFET in oerating region, are
(*a*) inclined upwards(*b*) almost flat
(*c*) inclined downwards
(*d*) Inclined upwards or downwards

124. In a JFET, V_{DS} exeeeds tghe rated value. Then it operates in
(*a*) active region (*b*) ohmic region
(*c*) cut off region
(*d*) either cut off or active region

125. A JFET operates in ohmic region when
(*a*) $V_{GS} = 0$
(*b*) V_{GS} is less than pinch off voltage
(*c*) V_{GS} = is positive
(*d*) $V_{GS} = V_{DS}$

126. A JFET behaves as a constant current source when
(*a*) $V_{GS} = 0$
(*b*) V_{GS} is less than pinch off voltage
(*c*) $V_{GS} = V_{DS}$
(*d*) V_{GS} is more than pinch off voltage

127. In a JFET, the drain current is maximum when
(*a*) $V_{GS} = V_{DS}$
(*b*) V_{GS} = pinch off voltage
(*c*) $V_{GS} = 0$ (*d*) V_{GS} is negative

128. In *n* channel JFET
(*a*) I_D and V_{DS} are positive but V_{GS} is negative
(*b*) I_D and V_{GS} are positive but V_{DS} is negative
(*c*) V_{DS} and V_{GS} are positive but I_D is negative
(*d*) I_D, V_{DS} and V_{GS} are all positive

129. Which of the following can be operated with positive as well as negative gate voltage
(*a*) JFET
(*b*) both JFET and MOSFET
(*c*) MOSFET
(*d*) neither JFET nor MOSFET

130. An *n* channel depletion type MOSFET has
(*a*) lightly doped *p* substrate and highly doped *n* source and drain
(*b*) highly doped *p* substrate and highly doped *n* source and drain

(c) highly doped p substrate and lightly doped n source and drain
(d) lightly doped n substrate and highly doped n source and drain.

131. Match List I and List II using the given codes

List I	List II
A NPN transistor	1. 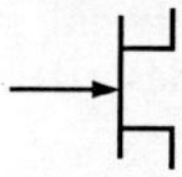
B PNP transistor	2. 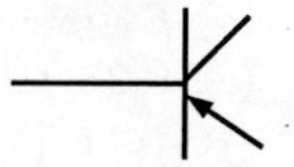
C n channel JFET	3. 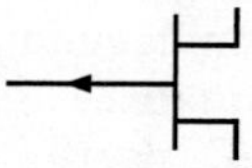
D p channel JFET	4. 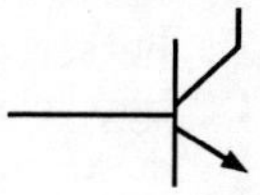

Codes :

	A	B	C	D
(a)	1	2	3	4
(b)	3	4	2	1
(c)	4	2	1	3
(d)	1	2	4	3

132. Match List I with II using the given code

List I	List II
A n Channel JFET	1.
B p Channel JFET	2.
C Depletion MOSFET	3.
D Enhancement MOSFET	4.

Codes :

	A	B	C	D
(a)	4	3	2	1
(b)	1	3	4	2
(c)	3	4	1	2
(d)	2	4	3	1

133. An enhancement mode MOSFET is off when the gate voltage is
(a) zero
(b) negative
(c) less than threshold value
(d) none of above

134. An enhancement mode MOSFET is on when the gate voltage is
(a) zero (b) positive
(c) high (d) more threshold value

135. Which of the following devices has a silicon dioxide layer
(a) NPN transistor (b) Tunnel diode
(c) JFET (d) MOSFET

136. In which n type device does p substrate extend upto silicion dioxide layer
(a) JFET
(b) Depletion type MOSFET
(c) Enhancement type MOSFET
(d) both (b) and (c)

137. Which of the following devices has substrate
(a) JFET
(b) Depletion Type MOSFET
(c) Enhancement type MOSFET
(d) both (b) and (c)

138. A digital voltmeter has a read out range from 0 to 999 counts. If the full scale reading is 9.999 V, the resolution is
(a) 1 V (b) 0.01 V
(c) 1 mV (d) 1 μV

139. The household energy meter is
(a) indicating instrument
(b) recording instrument
(c) integrating instrument
(d) none of the above

140. A meter has a resistance of 1Ω and full scale deflection current of 1 mA. To convert into an ammeter of 1 A range, the shunt resistance should be
(a) 1000 Ω (b) 1 Ω
(c) 0.01 Ω (d) 0.001 Ω

141. An ohmmeter is a
(a) moving coil instrument
(b) moving iron instrument
(c) dynamometer instrument
(d) induction instrument

142. In a CRO, the time base signal is applied to
(a) X plates (b) Y plates
(c) either X or Y plates
(d) alternately X and Y plates

143. The numbers of accelerating anodes in a CRO is
(a) 1 (b) 2
(c) 3 (d) 4

144. If E_b is the potential of accelerating anode, the energy of electrons as they emerge from electron gun is proportional to
(a) E_b (b) $\sqrt{E_b}$
(c) E_b^{15} (d) E_b^2

145. In a CRO the Y plates are oriented
(a) horizontally (b) vertically
(c) at an angle of 20° of horizontal
(d) at an angle of 70° the horizontal

146. If E_y is the voltage applied to Y plates, the vertical deflection of beam is proportional to

(a) E_y (b) $\sqrt{E_y}$

(c) $E_y^{1.5}$ (d) E_y^2

147. If two time varying signals of equal magnitude and zero phase displacement are applied to X and Y plates respectivley; of CRO, the display will be

(a) a staright line (b) a circle
(c) an ellipse (d) a square

148. If two time varying signals of equal magnitudes and 90° phase displacement are applied to X and Y plates of CRO, the display will be

(a) a straight line (b) a circle
(c) an ellipse (d) a rectangle

149. In electronic voltmeters the measuring device is a

(a) moving coil instrument
(b) moving iron instrument
(c) dynamometer instrument
(d) none of the above

150. When measuring phase angle between two waves using a CRO, the time base generator is connected to

(a) X plate (b) Y plats
(c) both X and Y plats
(d) neither X or Y plates

151. The accuracy of measurement using CRO is about

(a) ± 0.5% (b) ± 0.05 %
(c) ± 1% (d) ± 3%

152. In a CRO which of the following is not a part of electron gun

(a) cathode (b) grid
(c) accelerating anode(d) $X - Y$ plates

153. In a CRO, the frequency of output of time base generator

(a) is only one constant value
(b) is adjustable
(c) varies automatically
(d) (a) or (c)

154. A digital voltmeter can count from 0 to 9999. If full scale reading is 9.999 V, the resolution of full scale reading is

(a) 0.001 (b) 1000
(c) 2 digit (d) 1 mV

155. The Lissajous pattern observed on screen of CRO is a straight line inclined at 45° to x axis. If X-plate input is 2 sin ωt, the Y-plate input is

(a) 2 sin ωt (b) 2 sin (ωt + 45°)
(c) 2 sin (ωt – 45°)
(d) $2\sqrt{2}$ sine (ωt + 45°)

156. In a digital measuring device the input electrical signal is in the frequency range of dc to/Hz. It must be sampled at a rate of

(a) f times/sec (b) 0.5 f times/sec
(c) 0.25 f times/sec (d) 2 f times/sec

157. Digital instruments are preferred to analogue instruments due to the following qualites

A. narrow bandwidth
B. better accuracy
C. better resolution
D. cost

The correct arrangement of these qualities in order of their decreasing importance are

(a) A, B, C, D (b) B, C, D, A
(c) B, C, A, D (d) C, B, A, D

158. The time base of a high frequency CRO whose screen diameter is 10 cm is set at 10 μ sec/cm. The lowest frequency which can be fully displayed on the screen is

(a) more than 100 kHz
(b) equal to 100 kHz
(c) equal to 10 kHz
(d) less than 10 kHz

159. The Y plates of a CRO are excited by a voltage 2 sin 100 t and the X plates are not connected. The display would be

(a) a vertical line (b) a horizontal line
(c) sine wave (d) slanting line

160. Both X and Y plates of a CRO are excited by ac signals. The display is an ellipse which makes one rotation in 5 seconds. The means that

(a) the frequency of one input is 5 times that of other
(b) the difference between the frequencies of the two signals in one fifth of hertz
(c) the difference between the frequencies of the two signals is 5 Hz
(d) either (a) or (c)

161. A digital voltmeter has a 4½ digit display. The 1 V range can measure voltage up to

(a) 0.9999 V (b) 1.9999 V
(c) 0.199 V (d) 9.999 V

162. The deflection sensitivity of a CRO is generally expressed in
(*a*) meter/V (*b*) cm/V
(*c*) V/meter (*d*) V/cm

163. The load impedance Z_L of a *CE* amplifier has *R* and *L* in series. The phase difference between output and input will be
(*a*) 180° (*b*) 0
(*c*) more than 90° but less than 180°
(*d*) more than 180° but less than 270°

164. The maximum theoretical efficiency for class A amplifier is
(*a*) 12.5% (*b*) 25%
(*c*) 50% (*d*) 75%

165. If it is desired to have low output impedance in an amplifier circuit then we should use
(*a*) common collector circuit
(*b*) common base circuit
(*c*) common emitter circuit
(*d*) either (*b*) or (*c*)

166. In the graphical analysis of an amplifier circuit, the scope of dc load line depends on
(*a*) load resistance only
(*b*) collector supply voltage only
(*c*) both load resistance and collector supply voltage
(*d*) none of the above

167. Which amplifier has the higher power gain
(*a*) common emitter (*b*) common collector
(*c*) common base (*d*) both CC and CE

168. Which amplifier has a voltage gain of less than 1
(*a*) Common emitter
(*b*) Common collector
(*c*) Common base (*d*) both CB and CC

169. In the analysis of common emitter amplifier we may neglect
(*a*) h_{re} (*b*) h_{oe}
(*c*) h_{fe} (*d*) h_{ie}

170. In class C operation of an amplifier circuit, the collector current exists for
(*a*) 360° of input wave(*b*) 180° of input wave
(*c*) more than 180° of input wave
(*d*) less than 180° of input wave

171. Which of the following amplifier circuits has highest input impedance
(*a*) circuit using BJT
(*b*) circuit using JFET
(*c*) circuit using MOSFET
(*d*) either (*a*) or (*b*)

172. A transistor has two *p-n* junctions. The batteries should be connected such that
(*a*) both junctions are forward biased
(*b*) both junctions are reverse biased
(*c*) one junction is forward biased and the other is reverse biased
(*d*) either (*a*) or (*b*)

173. In a half wave diode rectifier circuit the current flows in the load circuit for
(*a*) 90° in each cycle
(*b*) 180° in each cycle
(*c*) 270° in each cycle
(*d*) 360° in each cycle

174. In a full wave diode rectifier circuit each diode carries current for
(*a*) 90° in each cycle
(*b*) 180° in each cycle
(*c*) 270° in each cycle
(*d*) 360° in each cycle

175. If input frequency is 50 Hz, the frequency of output wave in a half wave diode rectifier circuit is
(*a*) 25 Hz (*b*) 50 Hz
(*c*) 100 Hz (*d*) 200 Hz

176. If input frequency is 50 Hz, the frequency of output wave in a full wave diode rectifier circuit is
(*a*) 25 Hz (*b*) 50 Hz
(*c*) 100 Hz (*d*) 200 Hz

177. The dc load current in a bridge rectifier circuit is 10 mA. The dc load current through each diode is
(*a*) 40 mA (*b*) 20 mA
(*c*) 10 mA (*d*) 5 mA

178. In a bridge rectifier circuit the rms value of input ac voltage is 10 V. The PIV across each diode is
(*a*) 7.07 V (*b*) 14.14 V
(*c*) 10 V (*d*) 28.28 V

179. A bridge rectifier circuit has a dc load current of 10 mA and a filter capacitance of 1000 μF. The peak to peak ripple voltage is
(*a*) 0.01 V (*b*) 0.05 V
(*c*) 0.1 V (*d*) 0.5 V

180. In a rectifier circuit the filter capacitance is increased. Then the ripple will
(*a*) increase (*b*) decrease
(*c*) remain the same(*d*) either (*a*) or (*b*)

181. In an amplifier circuit the emitter is at ac ground potential in a

(a) CB circuit (b) CC circuit
(c) CE circuit
(d) both CB and CE circuit

182. In an amplifier the voltage gain is the ratio of
(a) output voltage and emitter resistance
(b) output voltage and input current
(c) output voltage and input voltage
(d) none of the above

183. In a self biase circuit for CE amplifier, the base voltage is
(a) equal to supply voltage
(b) more than supply voltage
(c) equal to or more than supply voltage
(d) less than supply voltage

184. The self bias provides
(a) stable Q point
(b) large voltage gain
(c) high input impedance
(d) high base current

185. In a self bias circuit for CE amplifier, an increase in emitter resistance R_E results in
(a) increase in emitter voltage
(b) decrease in emitter voltage
(c) increase in emitter current
(d) decrease in emitter current

186. In deriving ac equivalent circuit for an amplifier circuit we short circuit
(a) all resistors (b) all transistors
(c) all inductors (d) all capacitors

187. When the ac base voltage in a CE amplifier circuit is too high, the ac emitter current is
(a) zero (b) constant
(c) alternating (d) distorted

188. A CE amplifier has a high input signal. In the wave shape of emitter current
(a) positive and negative half cycles will be symmetrical
(b) positive half cycle will be elongated and negative half cycle will be compressed
(c) positive half cycle will be compressed and negative half cycle will be elongated
(d) either of the half cycles may be elongated or compressed

189. It is desired to reduce distortion in a CE amplifier circuit. We should
(a) reduce emitter current
(b) reduce collector current
(c) reduce ac base voltage
(d) reduce base emitter voltage

190. In a CE amplifier
(a) both ac and dc load lines have the same slope
(b) the ac load line has more slope than dc a load line
(c) the ac load line has less slope than dc load line
(d) either of the two load lines may have more slope than the other

191. In a CE amplifier circuit
(a) Q point must lie on both dc and ac load lines
(b) Q point lies on dc load line
(c) Q point lies on ac load line
(d) Q point lies may or may not lie on both dc and ac load lines

192. Which configuration has high input impedance and low output impedance
(a) CB (b) CE
(c) CC (d) CB and CE

193. If a CE amplifier drives a low load resistance directly the result will be
(a) overloading of amplifier
(b) underloading of amplifier
(c) overloading or underloading of amplifier depending on circuit
(d) none of the above

194. An amplifier has a large ac input signal. The clipping occurs on both the peaks. The output voltage will be nearly a
(a) sine wave (b) square wave
(c) tringular wave (d) (a) or (c)

195. A virtual ground is a ground for
(a) voltage (b) current
(c) both voltae and current
(d) voltage but not for current

196. In an inverting amplifier using op. amp. the current input to the inverting terminal
(a) is equal to the current through feedback resistance
(b) is less than the current through feedback resistance
(c) is more than the current through feedback resistance
(d) may be equal to, less or more than current through feedback resistance

197. An op-amp integrating circuit uses
(a) an inductor (b) a capacitor
(c) both inductor and capacitor
(d) none of the above

198. In op-amp integrating amplifier the input is feed to
(*a*) inverting input (*b*) non-inverting input
(*c*) both inverting and non-inverting input
(*d*) either inverting or non-inverting input

199. In the CE circuit shown in Fig. 1 the slope of ac load line will be

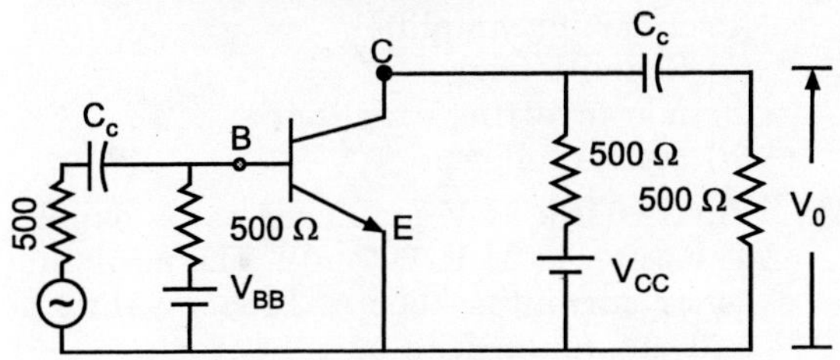

Fig. 1.

(*a*) $-\frac{1}{500}$ (*b*) $-\frac{1}{250}$
(*c*) $-\frac{1}{750}$ (*d*) $-\frac{1}{1000}$

200. In the CE circuit of Fig. 1, the slope of dc load line is
(*a*) $-\frac{1}{500}$ (*b*) $-\frac{1}{250}$
(*c*) $-\frac{1}{750}$ (*d*) $-\frac{1}{1000}$

201. Which of the following of parameters of CE circuit has the unit of ohms
(*a*) h_{ie} (*b*) h_{re}
(*c*) h_{fe} (*d*) h_{oe}

202. Which of the following h parameters is an admittance
(*a*) h_{ie} (*b*) h_{re}
(*c*) h_{fe} (*d*) h_{oe}

203. Match List I and List II and given the correct answer using the given codes.

List I (*h* parameter of BJT amplifier)	List II (Units)
A h_{ie}	1. siemens
B h_{oe}	2. no units
C h_{fe}	3. ohms

Codes :

	A	B	C
(*a*)	1	2	3
(*b*)	3	1	2
(*c*)	2	1	3
(*d*)	3	2	1

204. Match List I and List II and give correct answer using the given codes

List I (*h* parameter in CE configuration)	List II (Typical values)
A h_{ie}	1. 49
B h_{re}	2. 25×10^{-6} siemens
C h_{fe}	3. 2.4×10^{-4}
D h_{oe}	4. 1000 ohms

Codes :

	A	B	C	D
(*a*)	1	2	3	4
(*b*)	4	3	2	1
(*c*)	3	4	1	2
(*d*)	4	3	1	2

205. Match List I and List II and give correct answer using the given codes

List I (BJT amplifier)	List II (Characteristics)
A CE	1. good voltage gain, good current gain
B CC	2. voltage gain less than 1, high current gain
C CB	3. current gain less than 1, voltage gain high

Codes :

	A	B	C
(*a*)	1	2	3
(*b*)	1	2	3
(*c*)	3	1	2
(*d*)	3	2	1

206. In the circuit of Fig. 2, v_0 =

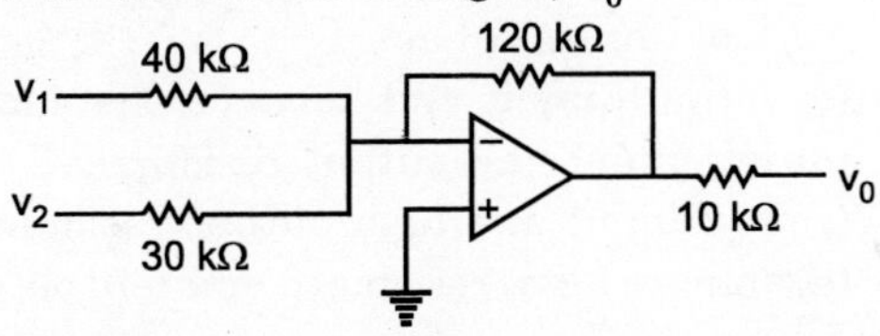

Fig. 2.

(*a*) $3\ v_1 + 4\ v_2$ (*b*) $-3\ v_1 - 4\ v_2$
(*c*) $4\ v_1 + 3\ v_2$ (*d*) $-4\ v_1 - 3\ v_2$

207. A difference amplifier using op-amp has closed loop gain = 50. If input is 2 V to each of inverting and non-inverting terminals, output is 5 mA. Then CMRR =
(*a*) 20000 (*b*) 2000
(*c*) 200 (*d*) 20

208. An op-amp has zero gain for common mode inputs. Then CMRR =
(*a*) zero (*b*) infinite
(*c*) high (*d*) low

209. The slew rate of an ideal op-amp is
(*a*) very slow (*b*) slow
(*c*) fast (*d*) infinitely fast

210. The output impedance of an ideal op-amp is

(*a*) zero (*b*) low
(*c*) high (*d*) infinite

211. Which of the following op-amp circuits has zero resistance in the feedback path

(*a*) inverting amplifier
(*b*) summing amplifier
(*c*) source follower
(*d*) integrating amplifier

212. In the circuit of Fig. 3, LED will be ON when v_1 is

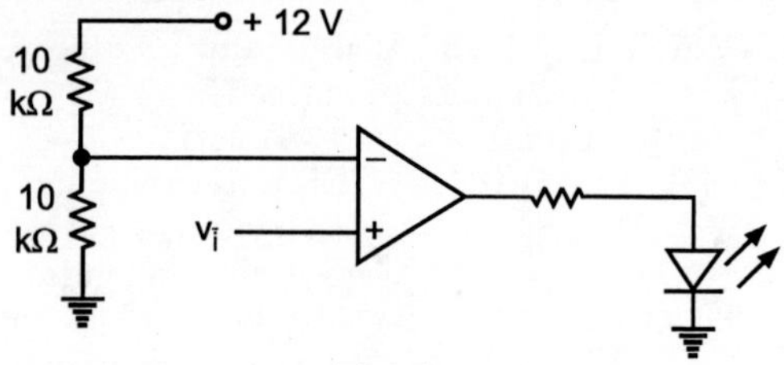

Fig. 3.

(*a*) less than 6 V (*b*) more than 6 V
(*c*) less than 12 V (*d*) more than 12 V

213. An ideal power supply has

(*a*) zero internal resistance
(*b*) very high internal resistance
(*c*) high output resistance
(*d*) both (*b*) and (*c*)

214. An ideal op-amp has

(*a*) infinite input and output resistance
(*b*) low input and output resistance
(*c*) low input and high output resistance
(*d*) infinite input resistance and zero output resistance

215. If the op-amp in Fig. 4 has a uity gain frequency of 1 MHz, the cutoff frequency is

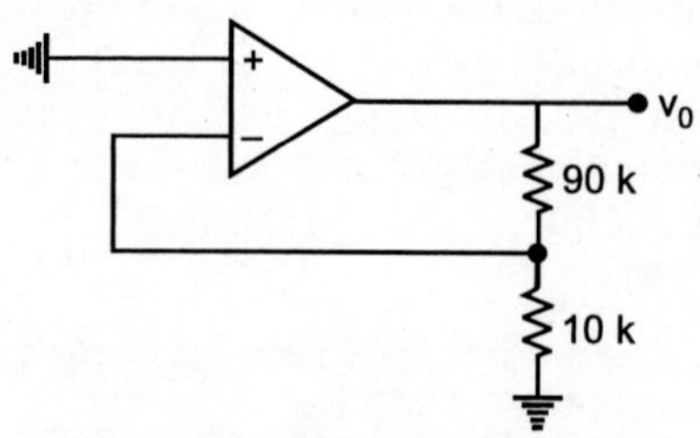

Fig. 4.

(*a*) 100 kHz (*b*) 1 MHz
(*c*) 10 MHz (*d*) 90 MHz

216. When $Z_1 << Z$, The circuit of Fig. 5 works as

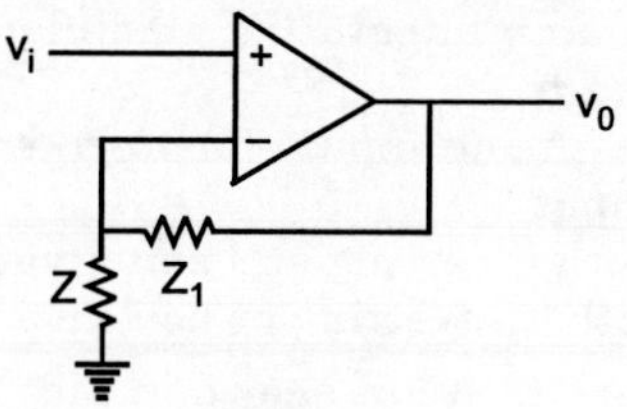

Fig. 5.

(*a*) inverting amplifier
(*b*) Schmitt trigger
(*c*) non-inverting amplifier
(*d*) source follower

217. Figure 6 is a 24 V_r stabilized power supply. The zener is 24 V, 600 mW. The minimum zener current is 10 mA. Proper values R and maximum load current are

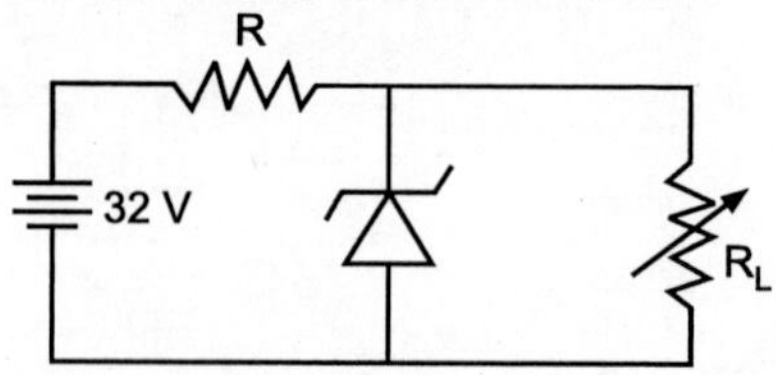

Fig. 6.

(*a*) 320 Ω, 10 mA (*b*) 400 Ω, 15 mA
(*c*) 400 Ω, 10 mA (*d*) 320 Ω, 15 mA

218. Figure 7 uses 10 V zener diode. The minimum and maximum current through series resistance are

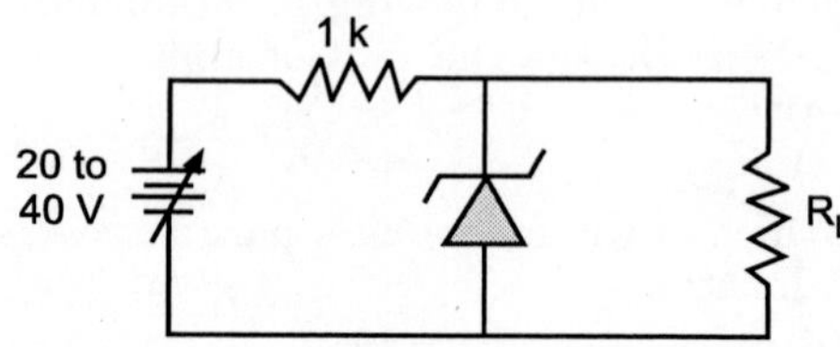

Fig. 7.

(*a*) 10 mA and 30 mA
(*b*) 20 mA and 40 mA
(*c*) 0 and 30 mA (*d*) 0 and 40 mA

219. In the op-amp circuit shown in Fig. 8 the current through load resistance R_L is

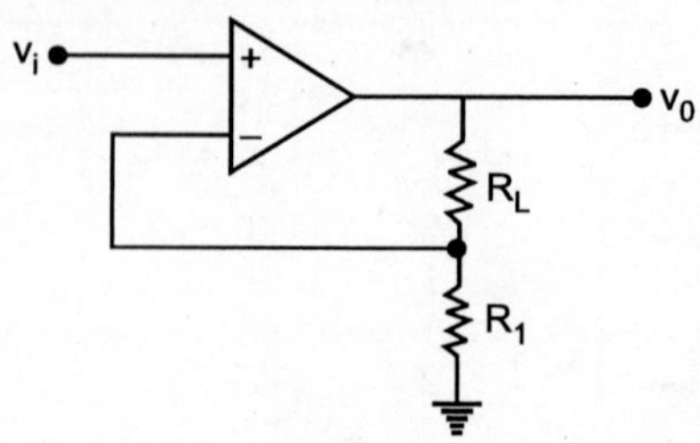

Fig. 8.

(a) $\frac{v_i}{R_1}$ (b) $\frac{v_i}{R_1 + R_L}$

(c) $\frac{(v_i) R_L}{R_1 + R_L}$ (d) $\frac{v_i (R_1 + R_L)}{R_1 R_L}$

220. In the op-amp circuit of Fig. 9, V_0 =

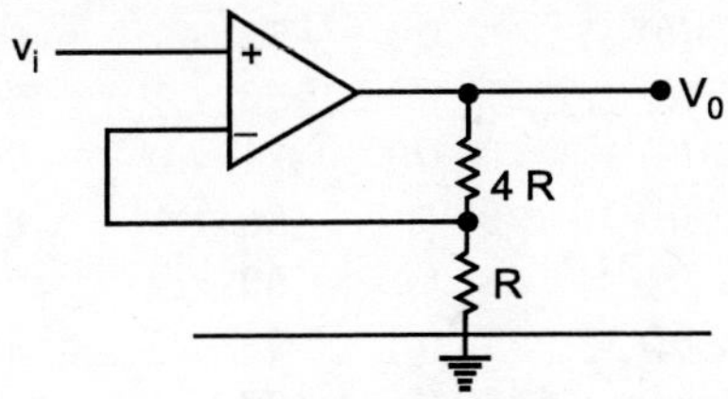

Fig. 9.

(a) V_i (b) 3 V_i
(c) 5 V_i (d) 0.5 V_i

221. In the op-amp circuit of Fig. 10, V_0 =

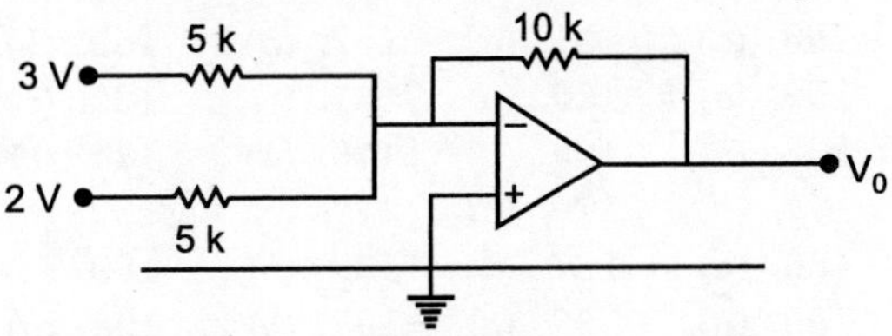

Fig. 10.

(a) 10 V (b) – 10 V
(c) 5 V (d) 5 V

222. In the op-amp circuit of Fig. 11, V_0 =

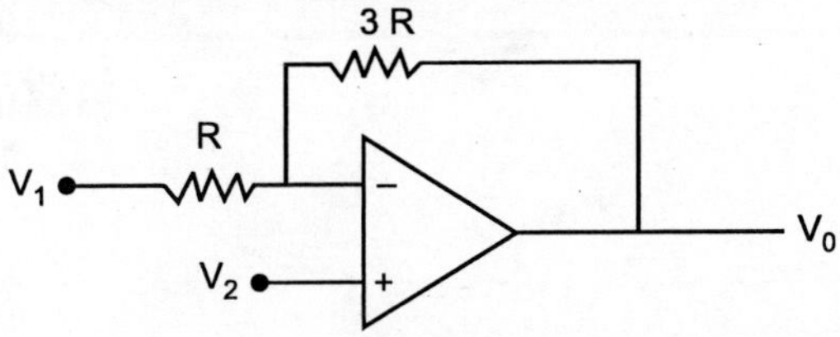

Fig. 11.

(a) 3 V_1 – 4 V_2 (b) – 3 V_1 + 4 V_2
(c) 3(V_1 – V_2) (d) – 3(V_1 – V_2)

223. A 120 V, 30 Hz source feeds a half wave rectifier circuit through a 4 : 1 step down transformer. the average output voltage is
(a) 30 V (b) 13.5 V
(c) 10 V (d) 7.07 V

224. In a full wave rectifier circuit using centre tapped transformer the peak voltage across half the secondary winding is 30 V. Then PIV is
(a) 30 V (b) 60 V
(c) 15 V (d) 10 V

225. In a full wave rectifier circuit using centre tapped transformer, the peak voltage across half the secondary winding is 40 V. If diodes are ideal, the average output voltage is
(a) 40 V (b) 25.5 V
(c) 20 V (d) 2.55 V

226. A full wave rectifier circuit using centre tapped transformer, the input frequency a 50 Hz. The frequency of output is
(a) 100 Hz (b) 50 Hz
(c) 25 Hz (d) none of the above

227. A full wave rectifier circuit using centre tapped transformer and a bridge rectifier are fed at 100 V, 50 Hz. The frequencies of outputs in these two rectifiers respectively are
(a) 100 Hz each (b) 50 Hz each
(c) 100 Hz and 50 Hz respectively
(d) 50 Hz and 100 Hz respectively

228. A half wave diode rectifier and full wave diode bridge rectifier and full wave diode bridge rectifier both have an input frequency of 50 Hz. The frequencies of outputs respectively are
(a) 100 Hz and 50 Hz
(b) 50 Hz and 100 Hz
(c) 100 Hz each (d) 50 Hz each

229. A half wave diode rectifier uses a diode having forward resistance of 50 ohms. The load resistance is also 50 ohms. Then the voltage regulation is
(a) 20% (b) 50%
(c) 100% (d) 200%

230. A full wave diode rectifier using centre tapped transformer has diodes with forward resistance of 50 ohms each. The load resistance of 50 ohms each. The load resistance is also 50 ohms. The voltage regulation is
(a) 20% (b) 50%
(c) 100% (d) 200%

231. A full wave bridge diode rectifier uses diodes having forward resistance of 50 ohms each. The load resistance is also 50 ohms. The voltage regulations is
(a) 20% (b) 50%
(c) 100% (d) 200%

232. Which of the following components is essential for a voltage multiplier circuit
(a) resistor (b) inductor
(c) capacitor (d) both (a) and (c)

ANSWERS

1. (*d*) **2.** (*b*) **3.** (*b*) **4.** (*c*) **5.** (*b*) **6.** (*a*) **7.** (*d*) **8.** (*a*) **9.** (*d*)
10. (*a*) **11.** (*c*) **12.** (*c*) **13.** (*a*) **14.** (*a*) **15.** (*a*) **16.** (*c*) **17.** (*d*) **18.** (*c*)
19. (*b*) **20.** (*d*) **21.** (*c*) **22.** (*a*) **23.** (*a*) **24.** (*a*) **25.** (*b*) **26.** (*b*) **27.** (*c*)
28. (*a*) **29.** (*d*) **30.** (*d*) **31.** (*c*) **32.** (*b*) **33.** (*b*) **34.** (*a*) **35.** (*b*) **36.** (*c*)
37. (*b*) **38.** (*c*) **39.** (*d*) **40.** (*d*) **41.** (*a*) **42.** (*c*) **43.** (*b*) **44.** (*d*) **45.** (*a*)
46. (*d*) **47.** (*d*) **48.** (*b*) **49.** (*c*) **50.** (*d*) **51.** (*b*) **52.** (*b*) **53.** (*c*) **54.** (*b*)
55. (*a*) **56.** (*a*) **57.** (*c*) **58.** (*a*) **59.** (*c*) **60.** (*a*) **61.** (*b*) **62.** (*c*) **63.** (*c*)
64. (*b*) **65.** (*d*) **66.** (*c*) **67.** (*a*) **68.** (*a*) **69.** (*a*) **70.** (*b*) **71.** (*a*) **72.** (*a*)
73. (*b*) **74.** (*c*) **75.** (*c*) **76.** (*b*) **77.** (*a*) **78.** (*a*) **79.** (*c*) **80.** (*c*) **81.** (*a*)
82. (*c*) **83.** (*b*) **84.** (*c*) **85.** (*a*) **86.** (*a*) **87.** (*c*) **88.** (*c*) **89.** (*c*) **90.** (*c*)
91. (*a*) **92.** (*c*) **93.** (*d*) **94.** (*a*) **95.** (*a*) **96.** (*d*) **97.** (*d*) **98.** (*a*) **99.** (*c*)
100. (*c*) **101.** (*d*) **102.** (*d*) **103.** (*a*) **104.** (*c*) **105.** (*d*) **106.** (*b*) **107.** (*a*) **108.** (*b*)
109. (*d*) **110.** (*d*) **111.** (*d*) **112.** (*b*) **113.** (*d*) **114.** (*d*) **115.** (*b*) **116.** (*d*) **117.** (*d*)
118. (*b*) **119.** (*c*) **120.** (*a*) **121.** (*a*) **122.** (*a*) **123.** (*b*) **124.** (*c*) **125.** (*b*) **126.** (*d*)
127. (*c*) **128.** (*a*) **129.** (*c*) **130.** (*a*) **131.** (*c*) **132.** (*c*) **133.** (*c*) **134.** (*d*) **135.** (*d*)
136. (*c*) **137.** (*d*) **138.** (*c*) **139.** (*c*) **140.** (*d*) **141.** (*a*) **142.** (*a*) **143.** (*b*) **144.** (*b*)
145. (*a*) **146.** (*a*) **147.** (*a*) **148.** (*b*) **149.** (*a*) **150.** (*d*) **151.** (*d*) **152.** (*d*) **153.** (*b*)
154. (*d*) **155.** (*a*) **156.** (*d*) **157.** (*c*) **158.** (*c*) **159.** (*a*) **160.** (*b*) **161.** (*c*) **162.** (*a*)
163. (*d*) **164.** (*b*) **165.** (*a*) **166.** (*a*) **167.** (*a*) **168.** (*b*) **169.** (*a*) **170.** (*d*) **171.** (*c*)
172. (*c*) **173.** (*b*) **174.** (*b*) **175.** (*b*) **176.** (*c*) **177.** (*d*) **178.** (*b*) **179.** (*c*) **180.** (*b*)
181. (*c*) **182.** (*c*) **183.** (*d*) **184.** (*a*) **185.** (*d*) **186.** (*d*) **187.** (*d*) **188.** (*b*) **189.** (*d*)
190. (*b*) **191.** (*a*) **192.** (*c*) **193.** (*a*) **194.** (*b*) **195.** (*d*) **196.** (*a*) **197.** (*b*) **198.** (*a*)
199. (*b*) **200.** (*a*) **201.** (*a*) **202.** (*d*) **203.** (*b*) **204.** (*d*) **205.** (*b*) **206.** (*b*) **207.** (*a*)
208. (*b*) **209.** (*d*) **210.** (*a*) **211.** (*c*) **212.** (*b*) **213.** (*a*) **214.** (*d*) **215.** (*a*) **216.** (*c*)
217. (*d*) **218.** (*a*) **219.** (*a*) **220.** (*c*) **221.** (*b*) **222.** (*b*) **223.** (*b*) **224.** (*b*) **225.** (*b*)
226. (*a*) **227.** (*a*) **228.** (*b*) **229.** (*c*) **230.** (*c*) **231.** (*d*) **232.** (*c*)

MODEL QUESTION PAPERS FOR PRACTICE

MODEL QUESTION PAPER-1

B. Tech.

ELECTRONIC DEVICES AND CIRCUITS

SECTION—A

1. **This question contains 20 objective type/Fill in the blanks/Matching type questions. Choose/fill/match correct answer** **[20 × 1 = 20]**

(*a*) To make silicon p-type, one can add the following dopant

(*i*) G_e (*ii*) S_i

(*iii*) Sb (*iv*) Ga

(*b*) Current flow in a semiconductor depends on the phenomenon of

(*i*) drift (*ii*) diffusion

(*iii*) recombination (*iv*) all of the above

(*c*) When a reverse bias is applied to a P-N junction, the width of the depletion layer

(*i*) remains the same (*ii*) is increased

(*iii*) is decreased (*iv*) may increase or decrease

(*d*) In a zener diode, large reverse current is due to

(*i*) collision (*ii*) presence of impurities

(*iii*) rupture of bonds

(*iv*) lower resistance in reverse biased region

(*e*) If V_m is the peak ac voltage of secondary of the transformer in a bridge rectifier, then PIV rating per diode is

(*i*) $\sqrt{2}\, V_m$ (*ii*) V_m

(*iii*) $2\, V_m$ (*iv*) $\frac{V_m}{\sqrt{2}}$

(*f*) $V_{BE\,sat}$ for Ge and Si P-N-P transistor is of the order of

(*i*) 0.1 V and 0.4 V (*ii*) 0.3 V and 0.7 V

(*iii*) – 0.3 V and – 0.7 V (*iv*) – 0.1 V and 0.2 V

(*g*) If $\alpha = 0.98$, $I_{CO} = 6\ \mu A$ and $I_B = 100\ \mu A$ for a transistor, then the value of I_C will be

(*i*) 2.3 mA (*ii*) 3.1 mA

(*iii*) 4.6 mA (*iv*) 5.2 mA

(*h*) Under saturated conditions, collector current is ______________ .

(*i*) The transistor configuration producing highest output resistance is

(*i*) CC (*ii*) CB

(*iii*) CE (*iv*) none of the above

(*j*) The drain-source voltage at which drain current becomes nearly constant is called

(*i*) barrier voltage (*ii*) breakdown voltage

(*iii*) pick-off voltage (*iv*) pinch-off voltage

SECTION—B

2. Attempt any Two parts of the following : **[15 × 2 = 30]**

(*a*) A 220 V, 50 Hz ac voltage is applied to the primary of 4 : 1 step-down transformer, which is used in a bridge rectifier, having a load resistance of 1 kΩ. Assuming the diodes to be an ideal, determine the following:

(*i*) DC output voltage (*ii*) DC power delivered to load

(*iii*) PIV of each diode (*iv*) output frequency

(*b*) Calculate I_B, I_C, V_E and V_{CE} for the following circuit if β = 130.

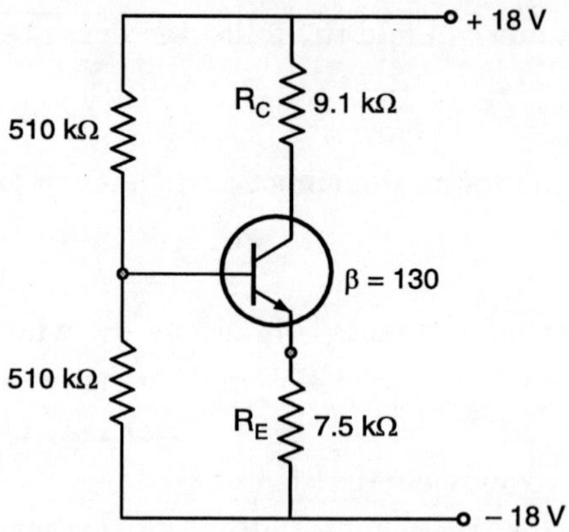

(*c*) What are the advantages of the FET over conventional bipolar junction transistor? Define pinch-off voltage, amplification factor and drain resistance of FET. Explain with the help of circuit diagram, how an FET is used as a voltage dependent resistor.

SECTION—C

3. Attempt any ONE part of the following : **[20 × 3 = 60]**

(*a*) Explain the behaviour of p-n junction at no bias, reverse bias and forward bias. Sketch the V-I characteristics of p-n junction diode.

(*b*) For the given zener diode network shown in figure, determine V_L, V_R, V_Z and I_R

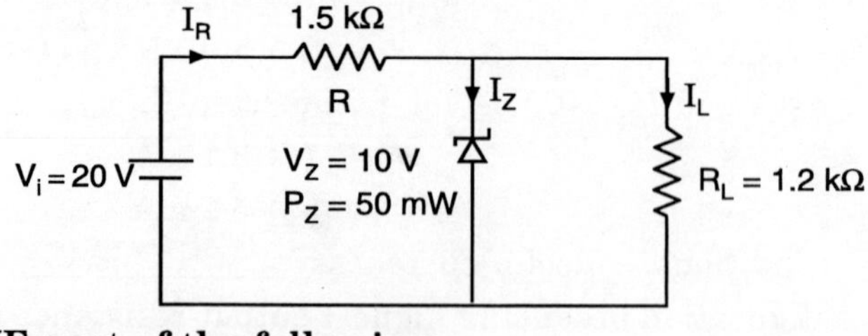

4. Attempt any ONE part of the following:

(*a*) Draw the circuit diagram of transistor in the CE configuration. Sketch the output characteristics and indicate the active, saturation and cut off region.

(*b*) Using h-parameter model, obtain the expression for a CE configuration circuit for

(*i*) A_i (*ii*) R_i

(*iii*) A_v and (*iv*) R_o

5. Attempt any ONE part of the following :

(*a*) What is the significant difference between the construction of an enhancement type MOSFET and depletion type MOSFET? Explain with suitable diagram.

(*b*) Define CMRR and slew rate. Draw the circuit diagram of differentiator circuit using op-amp and explain its working with the help of mathematical expressions.

SOLUTION

---

1. Answer to objective type Questions :

(*a*) (*iv*) (*b*) (*iv*) (*c*) (*ii*) (*d*) (*iii*) (*e*) (*ii*) (*f*) (*iii*) (*g*) (*iv*)

(*h*) (*i*) (*i*) (*ii*) (*j*) (*iv*)

■■■

MODEL QUESTION PAPER-2

B. Tech.

ELECTRONIC DEVICES AND CIRCUITS

SECTION—A

1. This question contains 20 objective type/Fill in the blanks/Matching type questions. Choose/fill/match correct answer **[10 × 1 = 10]**

(*a*) If an intrinsic semiconductor is doped with a very small amount of boron, then in the extrinsic semiconductor so formed, the number of electrons and holes will

(*i*) decrease (*ii*) increase and decrease

(*iii*) increase (*iv*) decrease and increase

(*b*) In a P-N junction without biasing, the p-side is

(*i*) negatively charged (*ii*) uncharged

(*iii*) None

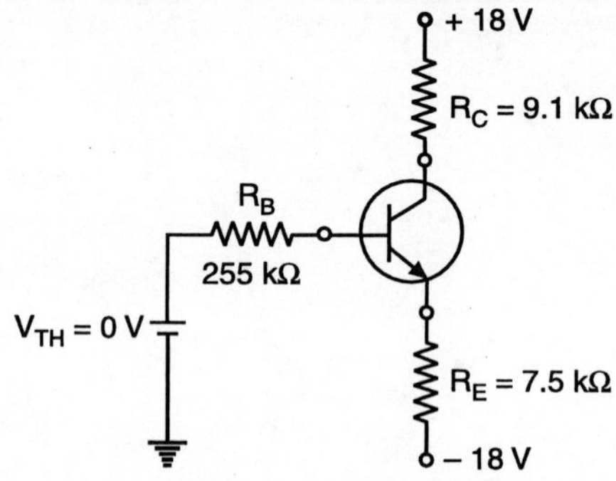

Fig. 1. Thevenin's equivalent circuit

(*c*) Maximum rectifying efficiency of half-wave rectifier is

(*i*) 0.406 (*ii*) 0.812

(*iii*) 1.2 (*iv*) 50

(*d*) Ripple factor of a full-wave rectifier without filter will be

(*i*) 0.2 (*ii*) 0.48

(*iii*) 0.24 (*iv*) 1.21

(*e*) I_{CBO} in a transistor can be reduced by reducing

(*i*) I_B (*ii*) V_{CC}

(*iii*) I_E (*iv*) temperature

(*f*) A common collector amplifier has the highest

(*i*) voltage gain (*ii*) current gain

(*iii*) power gain (*iv*) output impedance

(*g*) A source follower using a FET usually has a voltage gain which is

(*i*) greater than + 100 (*ii*) slightly less than unity but positive

(*iii*) exactly unity but negative (*iv*) about –10.

(*h*) The MOSFET switch in its on-state may be considered equivalent to

(*i*) resistor (*ii*) inductor

(*iii*) capacitor (*iv*) battery

(*i*) An NPN transistor operated at I_C = 2.5 mA has a collector load resistor of 1 kΩ and its emitter is grounded. The voltage gain of this stage is about

(*i*) 25 (*ii*) 100

(*iii*) 250 (*iv*) 1,000

(*j*) N-channel FETs are superior to p-channel FETs because of

(*i*) their high input impedance (*ii*) their high switching time

(*iii*) less power consumption

(*iv*) higher mobility of electrons in comparison to that of holes

SECTION-B

2. Attempt any Two parts of the following: **(15 × 2 = 30)**

(*a*) Calculate the following in the circuit given in figure 1:

I_S, I_L, I_Z and P_Z, when

(*i*) $R_L = 2$ kΩ (*ii*) $R_L = 10$ kΩ

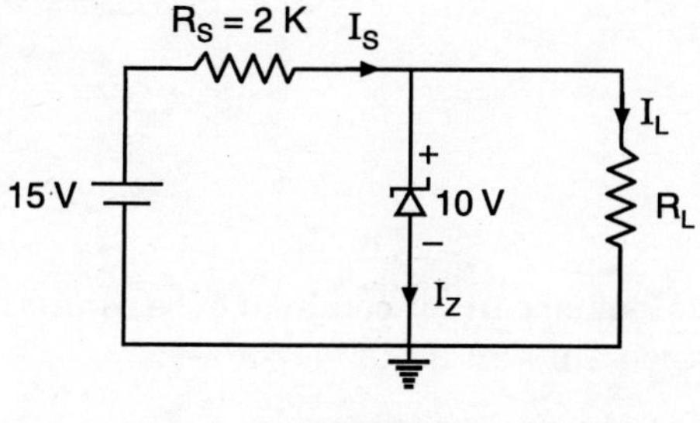

Fig. 1

(*b*) A transistor with β = 45 is used with collector to base resistor (*R*) biasing, with a quiscent value of 5 V for V_{CE}. If $V_{CC} = 24$ V, $R_L = 10$ kΩ and $R_E = 270$ Ω, find the value of (*i*) *R*, and (*ii*) stability factor.

(*c*) Evaluate the voltage gain and output resistance of the following circuit. Given that g_m = 2 mA/V and $r_d = 10$ k.

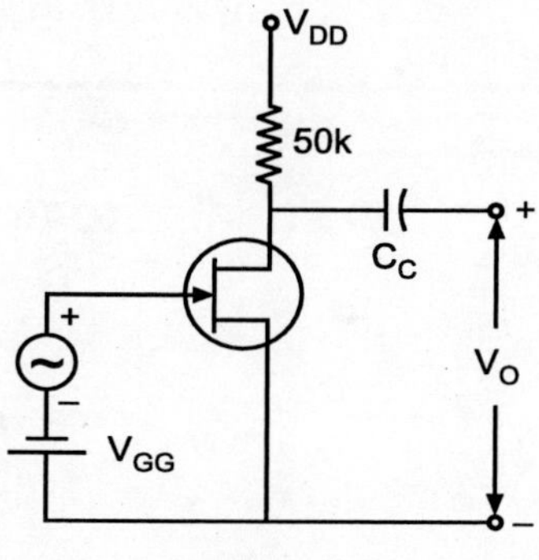

Fig. 2.

SECTION-C

3. Attempt any ONE part of the following: **(20 × 3 = 60)**

(*a*) What do you understand by average current, repetitive peak current, non-repetitive current, peak inverse voltage and reverse saturation current?

(*b*) Draw a circuit diagram which uses a breakdown diode to regulate the output voltage across a load and explain the working.

4. Attempt any ONE part of the following:

(*a*) Find I_C and V_{CE} for the following circuit. What will happen to V_{CE} if β increases due to temperature.

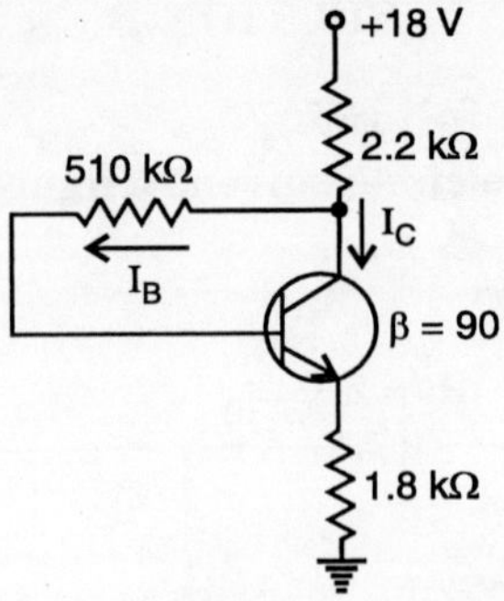

Fig. 3.

(*b*) Draw the hybrid equivalent circuit for common base configuration and find the expression for current gain, voltage gain and input impedance.

5. Attempt any ONE part of the following:

(*a*) What is the significant difference between the construction of an enhancement-type MOSFET and depletion type MOSFET? Explain with suitable diagram.

(*b*) With neat sketch, explain the working of:

(*i*) Varactor Diode

(*ii*) Tunnel Diode

SOLUTION

1. Answer to objective type Questions:

(*a*) (*iv*) (*b*) (*i*) (*c*) (*i*) (*d*) (*ii*) (*e*) (*iv*) (*f*) (*ii*) (*g*) (*ii*)

(*h*) (*iii*) (*i*) (*ii*) (*j*) (*iv*)

INDEX

A

B

E

F

H

J

L

M

N

O

P

Q

R

S

T

U

V

Z

OTHER DETAILS

(i) Who recommended you this book ? (Please tick (✓) in the box near the option relevant to you.

☐ Teacher ☐ Friends ☐ Bookseller

....

....

....

....

(ii) Name of the recommending teacher, his designation and address :

....

....

....

....

(iii) Name and address of the bookseller you purchased the book from :

....

....

....

....

(iv) Name and address of your institution (Please mention the University or Board as the case may be)

....

....

....

....

(v) Your name and complete postal address :

....

....

....

....

(vi) Course you are enrolled in :

....

....

....

....

The best assessment will be awarded each month: The award will be in the form of our publications as decided by the Editorial Board.

Please mail the filled up coupon at your earliest to :

S.K.KATARIA & SONS®
4885/109, Prakash Mahal, Dr. Subhash Bhargav Lane,
Opposite Delhi Medical Association, Daryaganj, New Delhi–110002 (INDIA)
Phone: +91-11-43551243; Mobile: +91-9871775858, 7678425209
e-mail: katariabook@gmail.com, katariabooks@yahoo.com
Website: www.skkatariaandsons.com